Fundamentals of Rock Mechanics

Fundamentals of Rock Mechanics

Fourth Edition

J. C. Jaeger, N. G. W. Cook, and R. W. Zimmerman

Blackwell
Publishing

BLACKWELL PUBLISHING
350 Main Street, Malden, MA 02148-5020, USA
9600 Garsington Road, Oxford OX4 2DQ, UK
550 Swanston Street, Carlton, Victoria 3053, Australia

First edition published 1969 by Methuen
Reprint published 1971 by Chapman and Hall
Second edition published 1976 by Chapman and Hall
Third edition published 1979
Fourth edition published 2007 by Blackwell Publishing Ltd

7 2013

Library of Congress Cataloging-in-Publication Data

Jaeger, J.C. (John Conrad), 1907-
Fundamentals of rock mechanics – J.C. Jaeger, N.G.W. Cook, and R.W. Zimmerman. – 4th ed.
p. cm.
Includes bibliographical references and index.
ISBN 978-0-632-05759-7 (pbk.: alk. paper)
1. Rock mechanics. I. Cook, Neville G.W.
II. Zimmerman, Robert W. III. Title.

TA706.J32 2007
624.1'513–dc22

2006036480

A catalogue record for this title is available from the British Library.

Set in 10.5 / 12.5pt Dante
by Newgen Imaging Systems (P) Ltd, Chennai, India

The publisher's policy is to use permanent paper from mills that operate a sustainable forestry policy, and which has been manufactured from pulp processed using acid-free and elementary chlorine-free practices. Furthermore, the publisher ensures that the text paper and cover board used have met acceptable environmental accreditation standards.

For further information on
Blackwell Publishing, visit our website:
www.blackwellpublishing.com

MIX
Paper from responsible sources
FSC® C013604

Contents

Preface to the Fourth Edition

When the first edition of this book appeared in 1969, rock mechanics had only recently begun to emerge as a distinct and identifiable scientific subject. It coalesced from several strands, including classical continuum mechanics, engineering and structural geology, and mining engineering. The two senior authors of *Fundamentals of Rock Mechanics* were perhaps uniquely qualified to play seminal roles in bringing about this emergence. John Jaeger had by that time already enjoyed a long and distinguished career as arguably the preeminent applied mathematician of the English-speaking world, and was the coauthor, with H. S. Carslaw, of one of the true classics of the scientific literature, *Conduction of Heat in Solids*. Neville Cook was at that time barely 30 years old, but was already the director of research at the South African Chamber of Mines, and well on his way to becoming acknowledged as the leading and most brilliant figure in this new field of rock mechanics.

The earlier editions of this book played a large role in establishing an identity for the field of rock mechanics and in defining what are now accepted to be the "fundamentals" of the field. These fundamentals consist firstly of the classical topics of solid mechanics – stress and strain, linear elasticity, plasticity, viscoelasticity, and elastic wave propagation. But rocks are much more complex than are most of the traditional engineering materials for which the classical mechanics theories were intended to apply. Hence, a book entitled *Fundamentals of Rock Mechanics* must also treat certain topics that are either unique to rocks, or at any rate which assume great importance for rocks, such as friction along rough surfaces, degradation and failure under compressive loads, coupling between mechanical deformation and fluid flow, the effect of cracks and pores on mechanical deformation, and, perhaps most importantly, the effect of fractures and joints on large-scale rock behavior.

Rock mechanics, thus defined, forms a cornerstone of several fields of science and engineering – from structural geology and tectonophysics, to mining, civil, and petroleum engineering. A search of citations in scientific journals shows that previous editions of this book have found an audience that encompasses not only these areas, but also includes material scientists and ceramicists, for example. It is hoped that this new edition will continue to be found useful by such a variety of researchers, students, and practitioners.

The extent to which the different chapters of this edition are new or expanded varies considerably, but aside from the brief, introductory Chapter 1, all have been revised and updated to one extent or another. The discussion of the basic theory of stress and strain in Chapter 2 has now been complemented by extensive use of vector and matrix notation, although all of the major results are also displayed in explicit component form. A discussion of rate-dependence has been added to Chapter 3 on friction. Chapter 4 on rock deformation has been updated, with more emphasis on true-triaxial failure criteria. Chapter 5 on linear elasticity now includes more discussion of anisotropic elasticity, as well as coverage of important general theorems related to strain energy. A detailed discussion of issues related to measurement of the strain-softening portion of the complete stress–strain curve has been added to Chapter 6 on laboratory measurements. Chapter 7 on poroelasticity is almost entirely new, and also includes a new section on thermoelasticity. Chapter 8 on stresses around cavities and inclusions, which is based heavily on the chapter in the 3rd edition that was entitled "Further Problems in Elasticity," has been simplified by moving some material to other more appropriate chapters, while at the same time adding material on three-dimensional problems. The chapters of the 3rd edition on ductile materials, granular materials, and time-dependent behavior have been combined to form Chapter 9 on inelastic behavior. Chapter 10, on micromechanical models, is a greatly enlarged and updated version of the old chapter on crack phenomena, with expanded treatment of effective medium theories. Chapter 11 on wave propagation has been doubled in size, with new material on reflection and refraction of waves across interfaces, the effects of pore fluids, and attenuation mechanisms. The important influence of rock fractures on the mechanical, hydraulic, and seismic behavior of rock masses is now widely recognized, and an entirely new chapter, Chapter 12, has been devoted to this topic. Chapter 13 on subsurface stresses collects material that had been scattered in various places in the previous editions. The final chapter, Chapter 14, briefly shows how the ideas and results of previous chapters can be used to shed light on some important geological and geophysical phenomena.

In keeping with the emphasis on fundamentals, this book contains no discussion of computational methods. Methods such as boundary elements, finite elements, and discrete elements are nowadays an indispensable tool for analyzing stresses and deformations around subsurface excavations, mines, boreholes, etc., and are also increasingly being used to study problems in structural geology and tectonophysics. But the strength of numerical methods has, at least until now, been in analyzing specific problems involving complex geometries and complicated constitutive behavior. Analytical solutions, although usually limited to simplified geometries, have the virtue of displaying the effect of the parameters of a problem, such as the elastic moduli or crack size, in a clear and transparent way. Consequently, many important analytical solutions are derived and/or presented in this book.

The heterogeneous nature of rock implies that most rock properties vary widely within a given rock type, and often within the same reservoir or quarry. Hence, rock data are presented in this work not to provide "handbook values" that could be used in specific applications, but mainly to illustrate trends, or to highlight the level of agreement with various models and theories. Nevertheless, this new edition contains slightly more actual rock data than did the previous edition, as measured by the number of graphs and tables that contain laboratory or field data. The reference list contains about 15% more items than in the 3rd edition, and more than half of the references are new. With only a few exceptions for some key references that originally appeared in conference proceedings or as institutional reports or theses, the vast majority of the references are to journal articles or monographs.

The ordering of the chapters remains substantially the same as in the 3rd edition. The guiding principle has been to minimize, as much as possible – in fact, almost entirely – the need to refer in one chapter to definitions, data or theoretical results that are not presented until a later chapter. In particular, then, the chapters are not structured so as to follow the workflow that would be used in a rock engineering project. For example, although knowledge of the *in situ* stresses would be required at the early stages of an engineering project, the chapter on subsurface stresses is placed near the end, because its presentation requires reference to analytical solutions that have been developed in several previous chapters.

The mathematical level of this edition is the same as in previous editions. The mathematical tools used are those that would typically be learned by undergraduates in engineering or the physical sciences. Thus, matrix methods are now extensively used in the discussion of stress and strain, as these have become a standard part of the undergraduate curriculum. Conversely, Cartesian tensor indicial notation, which is convenient for presenting the equations of stress, strain, and elasticity, has not been used, as it is not widely taught at undergraduate level. Perhaps the only exception to this rule is the use in Chapter 8 of functions of a complex variable for solving two-dimensional elasticity problems. But the small amount of complex variable theory that is required is presented as needed, and the integral theorems of complex analysis are avoided.

Rock mechanics is indeed a subfield of continuum mechanics, and my contribution to this book owes a heavy debt to the many excellent teachers of continuum mechanics and applied mathematics with whom I have been fortunate enough to study. These include Melvin Baron, Herbert Deresiewicz, and Morton Friedman at Columbia, and David Bogy, Michael Carroll, Werner Goldsmith, and Paul Naghdi at Berkeley. Although this book shows little obvious influence of Paul Naghdi's style of continuum mechanics, it was only after being inspired by his elegant and ruthlessly rigorous approach to this subject that I changed my academic major field to continuum mechanics, thus setting me on a path that led me to do my PhD in rock mechanics.

Finally, I offer my sincere thanks to John Hudson of Imperial College and Rock Engineering Consultants, and Laura Pyrak-Nolte of Purdue University for reading a draft of this book and providing many valuable suggestions.

R. W. Zimmerman
Stockholm, May 2006

Artwork from the book is available to instructors at:
www.blackwellpublishing.com/jaeger

To my wife, Jennifer, my partner in everything

Neville Cook,
Lafayette, Calif.
January 1998

1 Rock as a material

1.1 Introduction

Rock mechanics was defined by the Committee on Rock Mechanics of the Geological Society of America in the following terms: "Rock mechanics is the theoretical and applied science of the mechanical behavior of rock; it is that branch of mechanics concerned with the response of rock to the force fields of its physical environment" (Judd, 1964). For practical purposes, rock mechanics is mostly concerned with rock masses on the scale that appears in engineering and mining work, and so it might be regarded as the study of the properties and behavior of accessible rock masses due to changes in stresses or other conditions. Since these rocks may be weathered or fragmented, rock mechanics grades at one extreme into soil mechanics. On the other hand, at depths at which the rocks are no longer accessible to mining or drilling, it grades into the mechanical aspects of structural geology (Pollard and Fletcher, 2005).

Historically, rock mechanics has been very much influenced by these two subjects. For many years it was associated with soil mechanics at scientific conferences, and there is a similarity between much of the two theories and many of the problems. On the other hand, the demand from structural geologists for knowledge of the behavior of rocks under conditions that occur deep in the Earth's crust has stimulated much research at high pressures and temperatures, along with a great deal of study of the experimental deformation of both rocks and single crystals (Paterson and Wong, 2005).

An important feature of accessible rock masses is that they are broken up by joints and faults, and that pressurized fluid is frequently present both in open joints and in the pores of the rock itself. It also frequently happens that, because of the conditions controlling mining and the siting of structures in civil engineering, several lithological types may occur in any one investigation. Thus, from the outset, two distinct problems are always involved: (i) the study of the orientations and properties of the joints, and (ii) the study of the properties and fabric of the rock between the joints.

In any practical investigation in rock mechanics, the first stage is a geological and geophysical investigation to establish the lithologies and boundaries of the rock types involved. The second stage is to establish, by means of drilling or investigatory excavations, the detailed pattern of jointing, and to determine the mechanical and petrological properties of the rocks from samples. The third stage, in many cases, is to measure the *in situ* rock stresses that are present in the unexcavated rock. With this information, it should be possible to predict the response of the rock mass to excavation or loading.

This chapter presents a very brief introduction to the different rock types and the manner in which rock fabric and faulting influences the rock's engineering properties. A more thorough discussion of this topic can be found in Goodman (1993).

1.2 Joints and faults

Joints are by far the most common type of geological structure. They are defined as cracks or fractures in rock along which there has been little or no transverse displacement (Price, 1966). They usually occur in sets that are more or less parallel and regularly spaced. There are also usually several sets oriented in different directions, so that the rock mass is broken up into a blocky structure. This is a main reason for the importance of joints in rock mechanics: they divide a rock mass into different parts, and sliding can occur along the joint surfaces. These joints can also provide paths for fluids to flow through the rock mass.

Joints occur on all scales. Joints of the most important set, referred to as *major joints*, can usually be traced for tens or hundreds of meters, and are usually more or less planar and parallel to each other. The sets of joints that intersect major joints, known as *cross joints*, are usually of less importance, and are more likely to be curved and/or irregularly spaced. However, in some cases, the two sets of joints are of equal importance. The spacing between joints may vary from centimeters to decameters, although very closely spaced joints may be regarded as a property of the rock fabric itself.

Joints may be "filled" with various minerals, such as calcite, dolomite, quartz or clay minerals, or they may be "open," in which case they may be filled with fluids under pressure.

Jointing, as described above, is a phenomenon common to all rocks, sedimentary and igneous. A discussion of possible mechanisms by which jointing is produced is given by Price (1966) and Pollard and Aydin (1988). Joint systems are affected by lithological nature of the rock, and so the spacing and orientation of the joints may change with the change of rock type.

Another quite distinct type of jointing is *columnar jointing*, which is best developed in basalts and dolerites, but occasionally occurs in granites and some metamorphic rocks (Tomkeieff, 1940; Spry, 1961). This phenomenon is of some importance in rock mechanics, as igneous dykes and sheets are frequently encountered in mining and engineering practice. In rocks that have columnar jointing, the rock mass is divided into columns that are typically hexagonal, with side lengths on the order of a few tens of centimeters. The columns are intersected by cross joints that are less regular toward the interior of the body. The primary cause of columnar jointing appears to be tensile stresses that are created by thermal contraction during cooling. At an external surface, the columns run normal to the surface, and Jaeger (1961) and others have suggested that in the interior of the rock mass the columns run normal to the isotherms during cooling. The detailed mechanism of columnar jointing has been discussed by

Lachenbruch (1961); it has similarities to the cracks that form in soil and mud during drying, and to some extent to cracking in permafrost.

Faults are fracture surfaces on which a relative displacement has occurred transverse to the nominal plane of the fracture. They are usually unique structures, but a large number of them may be merged into a *fault zone*. They are usually approximately planar, and so they provide important planes on which sliding can take place. Joints and faults may have a common origin (de Sitter, 1956), and it is often observed underground that joints become more frequent as a fault is approached. Faults can be regarded as the equivalent, on a geological scale, of the laboratory shear fractures described in Chapter 4. The criteria for fracturing developed in Chapter 4 are applied to faults in §14.2.

From the point of view of rock mechanics, the importance of joints and faults is that they cause the existence of fairly regularly spaced, approximately plane surfaces, which separate blocks of "intact" rock that may slide on one another. In practice, the essential procedure is to measure the orientation of all joint planes and similar features, either in an exploratory tunnel or in a set of boreholes, and to plot the directions of their normal vectors on a stereological projection. Some typical examples are shown in the following figures taken from investigations of the Snowy Mountain Hydroelectric Authority in Australia.

Figure 1.1 is a stereographic projection plot of the normals to the fracture planes in the Headrace Channel for the Tumut 3 Project. The thick lines show the positions of the proposed slope cuts. In this case, 700 normal vectors were measured.

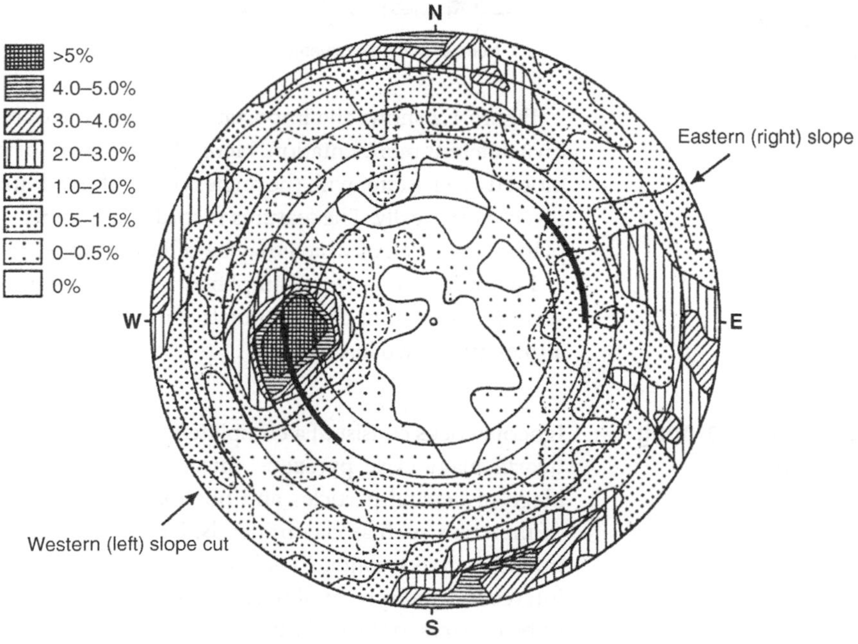

Fig. 1.1
Stereographic plot (lower hemisphere) of normals to fracture planes in Tumut 3 Headrace Channel. The contours enclose areas of equal density of poles.

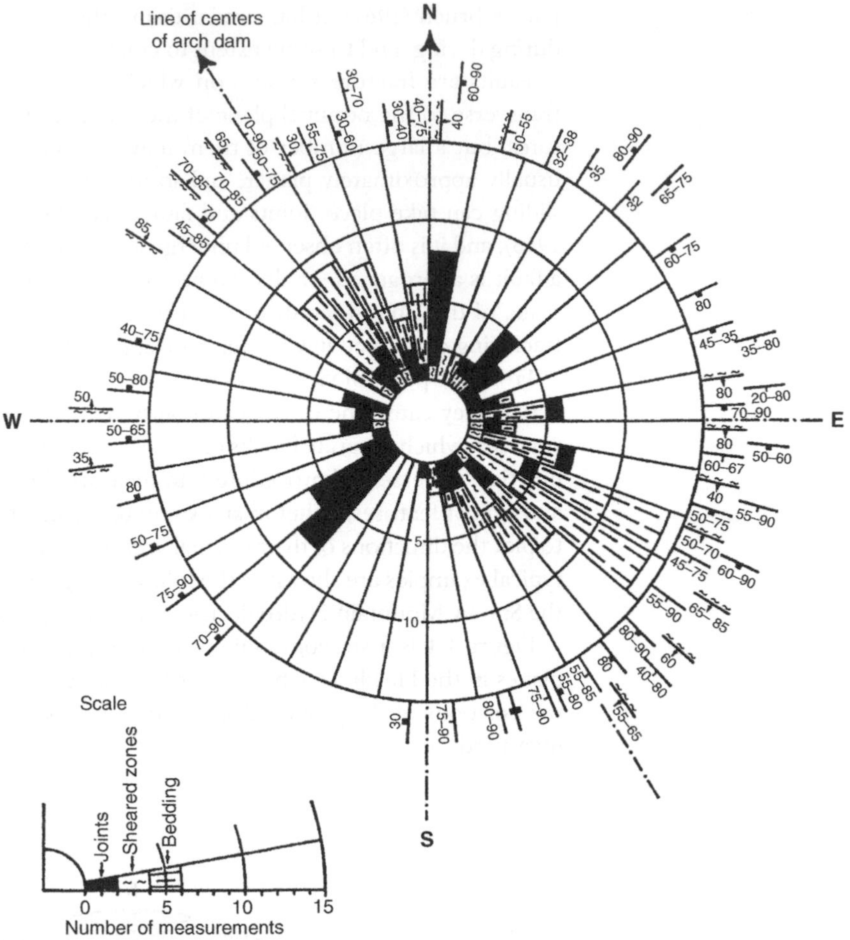

Fig. 1.2 Rosette diagram showing strikes of joints, sheared zones, and bedding planes at the Murray 2 dam site. The predominant dip for each strike is also shown.

Figure 1.2 shows the important geological features at the Murray 2 dam site on a different representation. Here, the directions of strike of various features are plotted as a rosette, with the angles of dip of the dominant features at each strike given numerically. The features recorded are joints, sheared zones, and bedding planes, any or all of which may be of importance.

Finally, Fig. 1.3 gives a simplified representation of the situation at the intersection of three important tunnels. There are three sets of joints whose dips and strikes are shown in Fig. 1.3.

1.3 **Rock-forming minerals**

Igneous rocks consist of a completely crystalline assemblage of minerals such as quartz, plagioclase, pyroxene, mica, etc. Sedimentary rocks consist of an assemblage of detrital particles and possibly pebbles from other rocks, in a matrix of materials such as clay minerals, calcite, quartz, etc. From their nature, sedimentary rocks contain voids or empty spaces, some of which may form an interconnected system of pores. Metamorphic rocks are produced by the action of heat, stress, or heated fluids on other rocks, sedimentary or igneous.

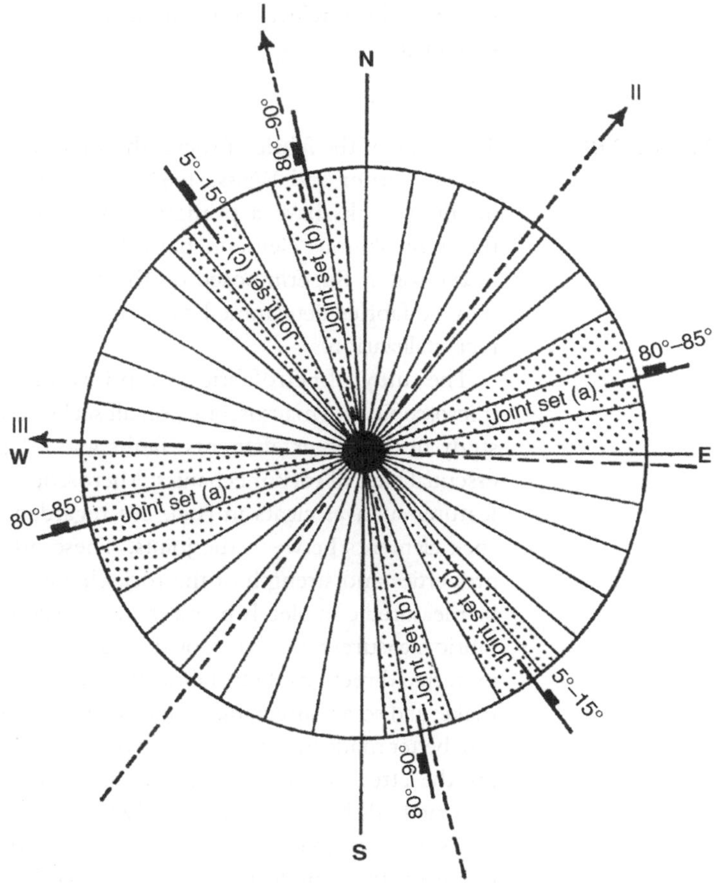

Fig. 1.3 Dips and strikes of three joint sets, (a) (b) and (c), at the intersection of three tunnels: I, Island Bend intake; II, Eucumbene-Snowy tunnel; III, Snowy-Geehi tunnel.

All of these minerals are anisotropic, and the elastic moduli of the more common ones, as defined in §5.10, are known numerically. If in a polycrystalline rock there are any preferred orientations of the crystals, this will lead to anisotropy of the rock itself. If the orientations of the crystals are random, the rock itself will be isotropic, and its elastic moduli may be estimated by the methods described in §10.2.

There are a number of general statistical correlations between the elasticity and strength of rocks and their petrography, and it is desirable to include a full petrographic description with all measurements. Grain size also has an effect on mechanical properties. In sedimentary rocks there are, as would be expected, some correlations between mechanical properties and porosity (Mavko et al., 1998).

A great amount of systematic research has been done on the mechanical properties of single crystals, both with regards to their elastic properties and their plastic deformation. Single crystals show preferred planes for slip and twinning, and these have been studied in great detail; for example, calcite (Turner et al., 1954) and dolomite (Handin and Fairbairn, 1955). Such measurements are an essential preliminary to the understanding of the fabric of deformed rocks, but they have little relevance to the macroscopic behavior of large polycrystalline specimens.

1.4 The fabric of rocks

The study of the fabric of rocks, the subject of *petrofabrics*, is described in many books (Turner and Weiss, 1963). All rocks have a fabric of some sort. Sedimentary rocks have a primary depositional fabric, of which the bedding is the most obvious element, but other elements may be produced by currents in the water. Superimposed on this primary fabric, and possibly obscuring it, may be fabrics determined by subsequent deformation, metamorphism, and recrystallization.

The study of petrofabrics comprises the study of all fabric elements, both microscopic and macroscopic, on all scales. From the present point of view, the study of the larger elements, faults and relatively widely spaced joints, is an essential part of rock mechanics. Microscopic elements and very closely spaced features such as cleats in coal, are regarded as determining the fabric of the rock elements between the joints. These produce an anisotropy in the elastic properties and strength of the rock elements. In principle, this anisotropy can be measured completely by mechanical experiments on rock samples, but petrofabric measurements can provide much useful information, in particular about preferred directions. Petrofabric measurements are also less time-consuming to make, and so are amenable to statistical analysis. Studies of rock fabric are therefore better made by a combination of mechanical and petrofabric measurements, but the latter cannot be used as a substitute for the former. Combination of the two methods has led to the use of what may be regarded as standard anisotropic rocks. For example, Yule marble, for which the calcite is known (Turner, 1949) to have a strong preferred orientation, has been used in a great many studies of rock deformation (Turner et al., 1956; Handin et al., 1960).

A second application of petrofabric measurements in rock mechanics arises from the fact that some easily measured fabric elements, such as twin lamellae in calcite and dolomite, quartz deformation lamellae, kink bands, and translation or twin gliding in some crystals, may be used to infer the directions of the principal stresses under which they were generated. These directions, of course, may not necessarily be the same as those presently existing, and so they form an interesting complement to underground stress measurements. Again, such measurements are relatively easy to make and to study statistically. The complete fabric study of joints and fractures on all scales is frequently used both to indicate the directions of the principal stresses and the large-scale fabric of the rock mass as a whole (Gresseth, 1964).

A great deal of experimental work has been concentrated on the study of the fabrics produced in rocks in the laboratory under conditions of high temperature and pressure. In some cases, rocks of known fabric are subjected to prescribed laboratory conditions, and the changes in the fabric are studied; for example, Turner et al. (1956) on Yule marble, and Friedman (1963) on sandstone.

Alternatively, specific attempts to produce certain types of fabrics have been made. Some examples are the work of Carter et al. (1964) on the deformation of quartz, Paterson and Weiss (1966) on kink bands, and Means and Paterson (1966) on the production of minerals with a preferred orientation.

Useful reviews of the application of petrofabrics to rock mechanics and engineering geology have been given by Friedman (1964) and Knopf (1957).

1.5 The mechanical nature of rock

The mechanical structure of rock presents several different appearances, depending upon the scale and the detail with which it is studied.

Most rocks comprise an aggregate of crystals and amorphous particles joined by varying amounts of cementing materials. The chemical composition of the crystals may be relatively homogeneous, as in some limestones, or very heterogeneous, as in a granite. Likewise, the size of the crystals may be uniform or variable, but they generally have dimensions of the order of centimeters or small fractions thereof. These crystals generally represent the smallest scale at which the mechanical properties are studied. On the one hand, the boundaries between crystals represent weaknesses in the structure of the rock, which can otherwise be regarded as continuous. On the other hand, the deformation of the crystals themselves provides interesting evidence concerning the deformation to which the rock has been subjected.

On a scale with dimensions ranging from a few meters to hundreds of meters, the structure of some rocks is continuous, but more often it is interrupted by cracks, joints, and bedding planes that separate different strata. It is this scale and these continuities which are of most concern in engineering, where structures founded upon or built within rock have similar dimensions.

The overall mechanical properties of rock depend upon each of its structural features. However, individual features have varying degrees of importance in different circumstances.

At some stage, it becomes necessary to attach numerical values to the mechanical properties of rock. These values are most readily obtained from laboratory measurements on rock specimens. These specimens usually have dimensions of centimeters, and contain a sufficient number of structural particles for them to be regarded as grossly homogeneous. Thus, although the properties of the individual particles in such a specimen may differ widely from one particle to another, and although the individual crystals themselves are often anisotropic, the crystals and the grain boundaries between them interact in a sufficiently random manner so as to imbue the specimen with average homogeneous properties. These average properties are not necessarily isotropic, because the processes of rock formation or alteration often align the structural particles so that their interaction is random with respect to size, composition and distribution, but not with respect to their anisotropy. Nevertheless, specimens of such rock have gross anisotropic properties that can be regarded as being homogeneous.

On a larger scale, the presence of cracks, joints, bedding and minor faulting raises an important question concerning the continuity of a rock mass. These disturbances may interrupt the continuity of the displacements in a rock mass if they are subjected to tension, fluid pressure, or shear stress that exceeds their frictional resistance to sliding. Where such disturbances are small in relation to the dimensions of a structure in a rock, their effect is to alter the mechanical properties of the rock mass, but this mass may in some cases still be treated as a continuum. Where these disturbances have significant dimensions, they must be treated as part of the structure or as a boundary.

The loads applied to a rock mass are generally due to gravity, and compressive stresses are encountered more often than not. Under these circumstances, the most important factor in connection with the properties and continuity of a rock mass is the friction between surfaces of cracks and joints of all sizes in the rock. If conditions are such that sliding is not possible on any surfaces, the system may be treated to a good approximation as a continuum of rock, with the properties of the average test specimen. If sliding is possible on any surface, the system must be treated as a system of discrete elements separated by these surfaces, with frictional boundary conditions over them.

2 Analysis of stress and strain

2.1 Introduction

In the study of the mechanics of particles, the fundamental kinematical variable that is used is the *position* of the body, and its two time derivatives, the *velocity* and the *acceleration*. The interaction of a given body with other bodies is quantified in terms of the *forces* that these other bodies exert on the first body. The effect that these forces have on the body is governed by Newton's law of motion, which states that the sum of the forces acting on a body is equal to the mass of the body times its acceleration. The condition for a body to be in equilibrium is that the sum of the external forces and moments acting on it must vanish.

These basic mechanical concepts such as position and force, as well as Newton's law of motion, also apply to extended, deformable bodies such as rock masses. However, these concepts must be altered somewhat, for various reasons. First, the fact that the force applied to a rock will, in general, vary from point to point, and will be distributed over the body must be taken into account. The idealization that forces act at localized points, which is typically used in the mechanics of particles, is not sufficiently general to apply to all problems encountered in rock mechanics. Hence, it is necessary to introduce the concept of *traction*, which is a force per unit area. As the traction generally varies with the orientation of the surface on which it acts, it is most conveniently represented with the aid of an entity known as the *stress tensor*.

Another fundamental difference between the mechanics of particles and deformable bodies such as rocks is that different parts of the rock may undergo different amounts of displacement. In general, it is the relative displacement of neighboring particles, rather than the absolute displacement of a particular particle, that can be equating in some way to the applied tractions. This can be seen from the fact that a rock sample can be moved *as a rigid body* from one location to another, after which the external forces acting on the rock can remain unaltered. Clearly, therefore, the displacement itself cannot be directly related to the applied loads. This relative displacement of nearby elements of the rock is quantified by an entity known as the *strain*.

The stress tensor is a symmetric second-order tensor, and many important properties of stress follow directly from those of second-order tensors. In the event that the relative displacements of all parts of the rock are small, the strain can also be represented by a second-order tensor called the *infinitesimal strain tensor*. A consequence of this fact is that much of the general theory of stresses applies also to the analysis of strains. The general theory of stress and strain is the topic of this chapter. Both of these theories can be developed without any reference to the specific properties of the material under consideration (i.e., the constitutive relationship between the stress and strain tensors). Hence, the discussion given in this chapter parallels, to a great extent, that which is given in many texts on elasticity, or solid mechanics in general. Among the many classic texts on elasticity that include detailed discussion of the material presented in this chapter are Love (1927), Sokolnikoff (1956), Filonenko-Borodich (1965), and Timoshenko and Goodier (1970). The chapter ends with a brief introduction to the theory of finite strains.

2.2 Definition of traction and stress

Consider a rock mass that is subject to some arbitrary set of loads. At any given point within this rock, we can imagine a plane slicing through the rock at some angle. Such a plane may in fact form an external boundary of the rock mass, or may represent a fictitious plane that is entirely internal to the rock. Figure 2.1a shows such a plane, along with a fixed (x, y) coordinate system. In particular, consider an element of that plane that has area A. Most aspects of the theory of stress (and strain) can be developed within a two-dimensional context, and extensions to three dimensions are in most cases straightforward. As most figures are easier to draw, and to interpret, in two dimensions than in three, much of the following discussion will be given first in two-dimensional form.

The plane shown in Fig. 2.1a can be uniquely identified by the unit vector that is perpendicular to its surface. The vector $\mathbf{n} = (n_x, n_y)$ is the *outward unit normal vector* to this plane: it has *unit* length, is *normal* to the plane, and points in the direction *away* from the rock mass. The components of this vector $\mathbf{n}$ are the direction cosines that the outward unit normal vector makes with the two coordinate axes. For example, a plane that is perpendicular to the x-axis would have $\mathbf{n} = (1, 0)$. As the length of any unit normal vector is unity, the Pythagorean theorem implies that $(n_x)^2 + (n_y)^2 = 1$. The unit normal vectors in the directions of the coordinate axes are often denoted by $\mathbf{e}_x = (1, 0)$ and $\mathbf{e}_y = (0, 1)$. The identification of a plane by its outward unit normal vector is employed frequently

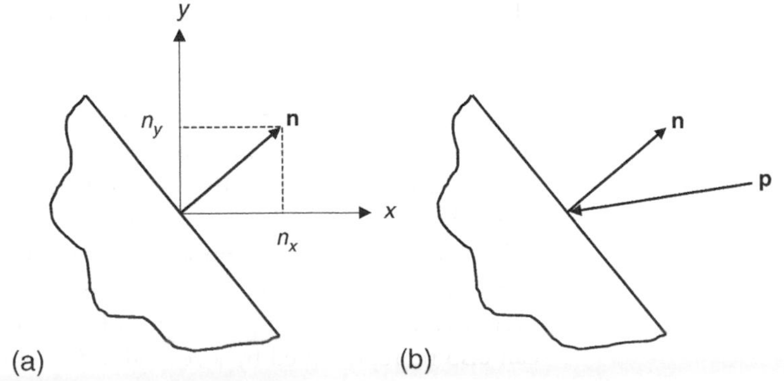

Fig. 2.1 Normal vector $\mathbf{n}$ and traction vector $\mathbf{p}$ acting on a surface.

in rock mechanics. It is important to remember that the vector **n** is perpendicular to the plane in question; it does not lie within that plane.

The action that the rock adjacent to the plane exerts on the rock that is "interior" to the plane can be represented by a resultant force **F**, which, like all forces, is a vector. The traction vector **p** is defined as the ratio of the resultant force **F** to the surface area A:

$$\mathbf{p}(\text{averaged over the area}) = \frac{1}{A}\mathbf{F}. \tag{2.1}$$

In order to define the traction that acts over a specific "point" in the rock, the area is now allowed to shrink down to a point, so that the magnitude A goes to zero. Following the convention often used in applied mathematics, the smallness of the area is indicated by the notation "dA," where the "d" denotes "differential," and likewise for the resultant force **F**. As the area shrinks down to a point, the traction at that point can then be defined by (Fig. 2.1b)

$$\mathbf{p}(\mathbf{x};\mathbf{n}) = \lim_{dA\to 0} \frac{1}{dA}d\mathbf{F}. \tag{2.2}$$

The notation $\mathbf{p}(\mathbf{x};\mathbf{n})$ denotes the traction vector at the point $\mathbf{x} \equiv (x, y, z)$, on a plane whose outward unit normal vector is **n**. In the following discussion, when the point **x** under consideration is either clear from the context, or immaterial to the particular discussion, the dependence of **p** on **x** will be suppressed in the notation.

At this point, it is necessary to introduce a sign convention that is inconsistent with the one used in most areas of mechanics, but which is nearly universal in the study of rocks and soils. The Cartesian component of the traction **p** in any given direction **r** is considered to be a positive number if the inner product (dot product) of **p** and a unit vector in the **r** direction is *negative*. One way to interpret this convention is that the traction is based on $-\mathbf{F}$, rather than **F**. The reason for utilizing this particular sign convention will become clear after the stresses are introduced.

It is apparent from the definition given in (2.2) that the traction is a vector, and therefore has two components in a two-dimensional system, and three components in a three-dimensional system. In general, this vector may vary from point to point, and is therefore a function of the location of the point in question. However, at any given point, the traction will also, in general, be different on different planes that pass through that point. In other words, the traction will also be a function of **n**, the outward unit normal vector. The fact that **p** is a function of two vectors, the position vector **x** and the outward unit normal vector **n**, is awkward. This difficulty is eliminated by appealing to the concept of *stress*, which was introduced in 1823 by the French civil engineer and mathematician Cauchy. The stress concept allows all possible traction vectors at a point to be represented by a single mathematical entity that does not explicitly depend on the unit normal of any particular plane. The price paid for this simplification, so to speak, is that the stress is not a vector, but rather a second-order tensor, which is a somewhat more complicated, and less familiar, mathematical object than is a vector.

Although there are an infinite number of different traction vectors at a point, corresponding to the infinity of possible planes passing through that point, all possible traction vectors can be found from knowledge of the traction vector on two mutually orthogonal planes (or three mutually orthogonal planes in three dimensions). To derive the relationship for the traction on an arbitrary plane, it is instructive to follow the arguments originally put forward by Cauchy. Consider a thin penny-shaped slab of rock having thickness h, and radius r (Fig. 2.2a). The outward unit normal vector on the right face of this slab is denoted by $\mathbf{n}$; the outward unit normal vector of the left face of the slab is therefore $-\mathbf{n}$. The total force acting on the face with outward unit normal vector $\mathbf{n}$ is equal to $\pi r^2\mathbf{p}(\mathbf{n})$, whereas the total force acting on the opposing face is $\pi r^2\mathbf{p}(-\mathbf{n})$. The total force acting on the outer rim of this penny-shaped slab will be given by an integral of the traction over the outer area, and will be proportional to $2\pi rh$, which is the surface area of the outer rim. Performing a force balance on this slab of rock yields

$$\pi r^2\mathbf{p}(\mathbf{n}) + \pi r^2\mathbf{p}(-\mathbf{n}) + 2\pi rh\mathbf{t} = 0, \tag{2.3}$$

where $\mathbf{t}$ is the mean traction over the outer rim. If the thickness h of this slab is allowed to vanish, this third term will become negligible, and the condition for equilibrium becomes

$$\mathbf{p}(-\mathbf{n}) = -\mathbf{p}(\mathbf{n}). \tag{2.4}$$

Equation (2.4), known as *Cauchy's first law*, essentially embodies a version of Newton's third law: if the material to the left of a given plane exerts a traction $\mathbf{p}$ on the material on the right, then the material on the right will exert a traction $-\mathbf{p}$ on the material to the left.

Now, consider a triangular slab of rock, as in Fig. 2.2b, with a uniform thickness w in the third (z) direction. Two faces of this slab have outward unit normal vectors that coincide with the negative x and y coordinate directions, respectively, whereas the third face has an outward unit normal vector of $\mathbf{n} = (n_x, n_y)$. The length of the face with outward unit normal vector $\mathbf{n}$ is taken to be h. The length of the face that has outward unit normal vector $\mathbf{n} = -\mathbf{e_x} = (-1, 0)$ is equal

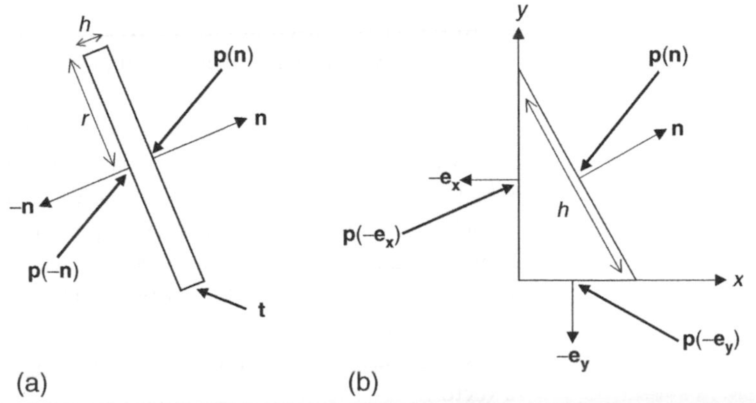

Fig. 2.2 (a) Thin slab used in derivation of Cauchy's first law; (b) triangular slab used in derivation of Cauchy's second law.

to hn_x, and so its area is hwn_x. The traction vector on this face is denoted by $\mathbf{p} = (-\mathbf{e_x})$, and so the total force acting on this face is $hwn_x\mathbf{p}(-\mathbf{e_x})$. Similar considerations show that the total force acting on the face with outward unit normal vector $-\mathbf{e_y}$ will be $hwn_y\mathbf{p}(-\mathbf{e_y})$. Hence, a force balance on this slab leads to

$$hwn_x\mathbf{p}(-\mathbf{e_x}) + hwn_y\mathbf{p}(-\mathbf{e_y}) + hw\mathbf{p}(\mathbf{n}) = 0. \tag{2.5}$$

Canceling out the common terms, and utilizing Cauchy's first law, (2.4), leads to *Cauchy's second law*:

$$\mathbf{p}(\mathbf{n}) = n_x\mathbf{p}(\mathbf{e_x}) + n_y\mathbf{p}(\mathbf{e_y}). \tag{2.6}$$

This result would remain unchanged if we consider the more general case in which a distributed *body force* acts on the tetrahedral-shaped element as in Fig. 2.2b. Whereas surface forces act over the outer surface of an element of rock, body forces act over the entire volume of the rock. The most obvious and common body force encountered in rock mechanics is that due to gravity, which has a magnitude of ρg (per unit volume), and is directed in the downward vertical direction. However, as will be shown in Chapter 7, gradients in temperature and pore fluid pressure also give rise to phenomena which have the same effect as distributed body forces. If there were a body-force density $\mathbf{b}$ per unit volume of rock, a total body force of $(1/2)h^2wn_xn_y\mathbf{b}$ would have to be added to the force balance in (2.5). If we divide through by h, and then let the size of the element shrink to zero (i.e., $h \rightarrow 0$), the body force term would drop out and $\mathbf{b}$ would not appear in the final result (2.6).

It is now convenient to recall that each traction is a vector, and therefore (in two dimensions) will have two components, one in each of the coordinate directions. The components of a traction vector such as $\mathbf{p}(\mathbf{e_x})$ are denoted using two indices – the first to indicate the direction of the outward unit normal vector and the second to indicate the component of the traction vector:

$$\mathbf{p}(\mathbf{e_x}) = \begin{bmatrix} \tau_{xx} \\ \tau_{xy} \end{bmatrix} = [\tau_{xx} \quad \tau_{xy}]^{\mathrm{T}}, \tag{2.7}$$

where we adhere to the algebraic convention that a vector is written as a column, and is therefore equivalent to the *transpose* of a row vector. Equation (2.6) can therefore be written in matrix form as

$$\mathbf{p}(\mathbf{n}) = n_x\begin{bmatrix} \tau_{xx} \\ \tau_{xy} \end{bmatrix} + n_y\begin{bmatrix} \tau_{yx} \\ \tau_{yy} \end{bmatrix} = \begin{bmatrix} \tau_{xx} & \tau_{yx} \\ \tau_{xy} & \tau_{yy} \end{bmatrix}\begin{bmatrix} n_x \\ n_y \end{bmatrix}. \tag{2.8}$$

In the first expression on the right in (2.8), n_x and n_y are treated as scalars that multiply the two traction vectors; in the second expression, the formalism of matrix multiplication is used. As the two components of the vector $p(\mathbf{n})$ are $p_x(\mathbf{n})$ and $p_y(\mathbf{n})$, (2.8) can be written in component form as

$$p_x(\mathbf{n}) = \tau_{xx}n_x + \tau_{yx}n_y, \tag{2.9}$$

$$p_y(\mathbf{n}) = \tau_{xy}n_x + \tau_{yy}n_y. \tag{2.10}$$

If we use the standard matrix algebra convention that the first subscript of a matrix component denotes the row, and the second subscript denotes the column, the matrix appearing in (2.8) is seen to actually be the *transpose* of the stress matrix, in which case we can rewrite (2.8) as

$$\mathbf{p}(\mathbf{n}) = \begin{bmatrix} \tau_{xx} & \tau_{xy} \\ \tau_{yx} & \tau_{yy} \end{bmatrix}^{\mathrm{T}} \begin{bmatrix} n_x \\ n_y \end{bmatrix}, \tag{2.11}$$

where the matrix that appears in (2.11), without the transpose operator, is the stress matrix, $\boldsymbol{\tau}$. Equation (2.11) can be written in direct matrix notation as

$$\mathbf{p} = \boldsymbol{\tau}^{\mathrm{T}}\mathbf{n}, \tag{2.12}$$

where $\mathbf{n}$ is a unit normal vector, $\mathbf{p}$ is the traction vector on the plane whose outward unit normal vector is $\mathbf{n}$, and $\boldsymbol{\tau}$ is the stress matrix, or *stress tensor*. In two dimensions the stress tensor has four components; in three dimensions it has nine. Equation (2.12) gives the traction on an arbitrarily oriented plane in terms of the stress matrix, relative to some fixed orthogonal coordinate system, and the direction cosines between the outward unit normal vector to the plane and the two coordinate axes. Note that a tensor can be written as a matrix, which is merely a rectangular array of numbers. However, a tensor has specific mathematical properties that are not necessarily shared by an arbitrary matrix-like collection of numbers. These properties relate to the manner in which the components of a tensor transform when the coordinate system is changed; these transformation laws are discussed in more detail in §2.3. The rows of the matrix that represents $\boldsymbol{\tau}$ are the traction vectors along faces whose outward unit normal vectors lie along the coordinate axes. In other words, the first row of $\boldsymbol{\tau}$ is $\mathbf{p}(\mathbf{e_x})$, the second row is $\mathbf{p}(\mathbf{e_y})$, etc.

The physical significance of the stress tensor is traditionally illustrated by the schematic diagram shown in Fig. 2.3a. Consider a two-dimensional square element of rock, whose faces are each perpendicular to one of the two coordinate axes. The traction vector that acts on the face whose outward unit normal vector is in the x direction has components (τ_{xx}, τ_{xy}). Each of these two components can be considered as a vector in its own right; they are indicated in Fig. 2.3a as arrows whose lines of action pass through the center of the face whose outward unit normal vector is $\mathbf{e_x}$. As the traction components are considered positive if they are oriented in the directions opposite to the outward unit normal vector, we see that the traction τ_{xx} is a positive number if it is *compressive*. Compressive stresses are much more common in rock mechanics than are tensile stresses. For example, the stresses in a rock mass that are due to the weight of the overlying rock are compressive. In most other areas of mechanics, tensile stresses are considered positive, and compressive stresses are reckoned to be negative. The opposite sign convention is traditionally used in rock (and soil) mechanics in order to avoid the frequent occurrence of negative signs in calculations involving stresses.

Many different notations have been used to denote the components of the stress tensor. We will mainly adhere to the notation introduced above, which has

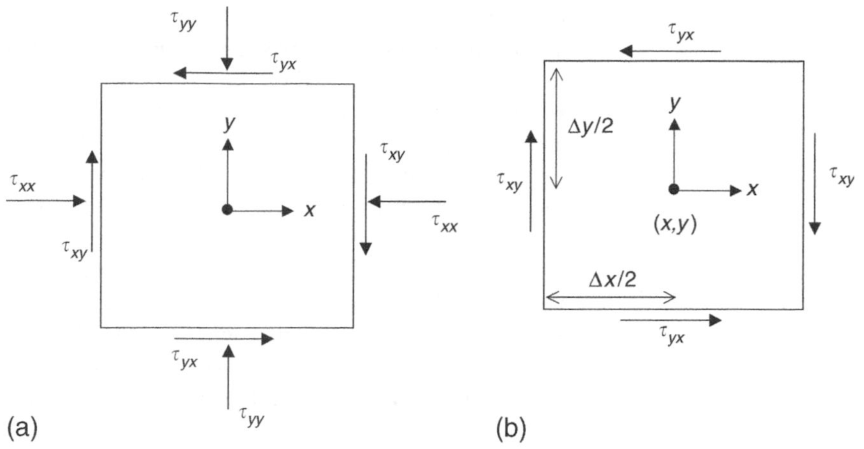

Fig. 2.3 (a) Stress components acting on a small square element. (b) Balance of angular momentum on this element shows that the stress tensor must be symmetric.

been used, for example, by Sokolnikoff (1956). Some authors use σ instead of τ as the basic symbol, but utilize the same subscripting convention. Many rock and soil mechanics treatments, including earlier editions of this book, denote shear stresses by τ_{xy}, etc., but denote normal stresses by, for example, σ_x rather than τ_{xx}. This notation, which has also been used by Timoshenko and Goodier (1970), has the advantage of clearly indicating the distinction between normal and shear components of the stress, which have very different physical effects, particularly when acting on fracture planes or other planes of weakness (Chapter 3). However, the $\{\sigma, \tau\}$ notation does not reflect the fact that the normal and shear components of the stress are both components of a single mathematical object known as the stress tensor. Furthermore, many of the equations in rock mechanics take on a simpler and more symmetric form if written in terms of a notation in which all stress components are written using the same symbol. However, a version of the Timoshenko and Goodier notation will occasionally be used in this book when discussing the traction acting on a specific plane. In such cases, for reasons of simplicity (so as to avoid the need for subscripts), it will be convenient to denote the normal stress by σ, and the shear stress by τ. Many classic texts on elasticity, such as Love (1927) and Filonenko-Borodich (1965), utilize the notation introduced by Kirchhoff in which τ_{xy} is denoted by X_y, etc. Green and Zerna (1954) use the notation suggested by Todhunter and Pearson (1886), in which τ_{xy} is denoted by $\widehat{xy}$, etc.

Equation (2.12) is usually written without the transpose sign, although strictly speaking the transpose is needed. The reason that it is allowable to write $\mathbf{p} = \boldsymbol{\tau}\mathbf{n}$ in place of $\mathbf{p} = \boldsymbol{\tau}^{\mathrm{T}}\mathbf{n}$ is that the stress matrix is in fact always *symmetric*, so that $\tau_{xy} = \tau_{yx}$, in which case $\boldsymbol{\tau} = \boldsymbol{\tau}^{\mathrm{T}}$. This property of the stress tensor is of great importance, if for no other reason than that it reduces the number of stress components that must be measured or calculated from four to three in two dimensions, and from nine to six in three dimensions. The symmetry of the stress tensor can be proven by appealing to the law of conservation of angular momentum. For simplicity, consider a rock subject to a state of stress that does not vary from point to point. If we draw a free-body diagram for a small rectangular element of rock, centered at point (x, y), the traction components acting on the four faces are shown in Fig. 2.3b. The length of the element is Δx in the x direction, Δy in the y direction, and Δz in the z direction (into the page). In order for this element of rock to be in equilibrium, the sum of all the moments about any point, such as (x, y), must be zero. Consider first the tractions that act on the right face of the element. The force vector represented by this traction is found by multiplying the traction by the area of that face, which is $\Delta y \Delta z$. The x-component of this force is therefore $\tau_{xx} \Delta y \Delta z$. However, the resultant of this force acts through the point (x, y), and therefore contributes no moment about that point. The y-component of this traction is τ_{xy}, and the net force associated with it is $\tau_{xy} \Delta y \Delta z$. The moment arm of this force is $\Delta x/2$, so that the total clockwise moment about the z-axis, through the point (x, y), is $\tau_{xy} \Delta x \Delta y \Delta z/2$. Adding up the four moments that are contributed by the four shear stresses yields

$$\tau_{xy} \Delta x \Delta y \Delta z/2 + \tau_{xy} \Delta y \Delta x \Delta z/2 - \tau_{yx} \Delta x \Delta y \Delta z/2 - \tau_{yx} \Delta y \Delta x \Delta z/2 = 0. \tag{2.13}$$

Canceling out the terms $\Delta x \Delta y \Delta z/2$ leads to the result

$$\tau_{xy} = \tau_{yx}. \tag{2.14}$$

In three dimensions, a similar analysis leads to the relations $\tau_{xz} = \tau_{zx}$ and $\tau_{yz} = \tau_{zy}$. This result should be interpreted as stating that at any specific point (x, y, z), the stress component $\tau_{xy}(x, y, z)$ is equal in magnitude and sign to the stress component $\tau_{yx}(x, y, z)$. There is in general no reason for the conjugate shear stresses *at different points* to be equal to each other.

Although the derivation presented above assumes that the stresses do not vary from point to point, and that the element of rock is in static equilibrium, the result is actually completely general. The reason for this is related to the fact that the result applies at each infinitesimal "point" in the rock. If we had accounted for the variations of the stress components with position, these terms would contribute moments that are of higher order in Δx and Δy. Dividing through the moment balance equation by $\Delta x \Delta y \Delta z$, and then taking the limit as Δx and Δy go to zero, would cause these additional terms to drop out, leading to (2.14). The same would occur if we considered the more general situation in which the element were not in static equilibrium, but rather was rotating. In this situation, the sum of the moments would be equal to the moment of inertia of the element about the z-axis through the point (x, y), which is $\rho \Delta x \Delta y \Delta z[(\Delta x)^2 + (\Delta y)^2]/12$, where ρ is the density of the rock, multiplied by the angular acceleration, $\dot{\omega}$. Hence, the generalization of (2.14) would be

$$\tau_{xy} \Delta x \Delta y \Delta z - \tau_{yx} \Delta x \Delta y \Delta z = \rho \Delta x \Delta y \Delta z[(\Delta x)^2 + (\Delta y)^2]\dot{\omega}/12. \tag{2.15}$$

Dividing through by $\Delta x \Delta y \Delta z$, and then taking the limit as the element shrinks down to the point (x, y), leads again to (2.14).

The symmetry of the stress tensor is therefore a general result. However, it is worth bearing in mind that although τ_{xy} and τ_{yx} are numerically equal, they are in fact physically distinct stress components, and act on different faces of an element of rock. Although the identification of τ_{xy} with τ_{yx} is eventually made when solving the elasticity equations, it is usually preferable to maintain a distinction between τ_{xy} and τ_{yx} when writing out equations, or drawing schematic diagrams such as Fig. 2.3a. This helps to preserve as much symmetry as possible in the structure of the equations.

The symmetry of the stress tensor followed from the principle of conservation of angular momentum. The principle of conservation of *linear* momentum leads to three further equations that must be satisfied by the stresses. These equations, which are known as the *equations of stress equilibrium* and are derived in §5.5, control the rate at which the stresses vary in space. However, much useful information about the stress tensor can be derived prior to considering the implications of the equations of stress equilibrium. Of particular importance are the laws that govern the manner in which the stress components vary as the coordinate system is rotated. These laws are derived and discussed in §2.3 and §2.5.

2.3 Analysis of stress in two dimensions

Discussions of stress are algebraically simpler in two dimensions than in three. In most instances, no generality is lost by considering the two-dimensional case, as the extension to three dimensions is usually straightforward. Furthermore, many problems in rock mechanics are essentially two dimensional, in the sense that the stresses do not vary along one Cartesian coordinate. The most common examples of such problems are stresses around boreholes, or around long tunnels. Many other problems are idealized as being two dimensional so as to take advantage of the relative ease of solving two-dimensional elasticity problems as compared to three-dimensional problems. Hence, it is worthwhile to study the properties of two-dimensional stress tensors. Various properties of two-dimensional stress tensors will be examined in this section; their three-dimensional analogues will be discussed in §2.5.

In order to derive the laws that govern the transformation of stress components under a rotation of the coordinate system, we again consider a small triangular element of rock, as in Fig. 2.4. The outward unit normal vector to

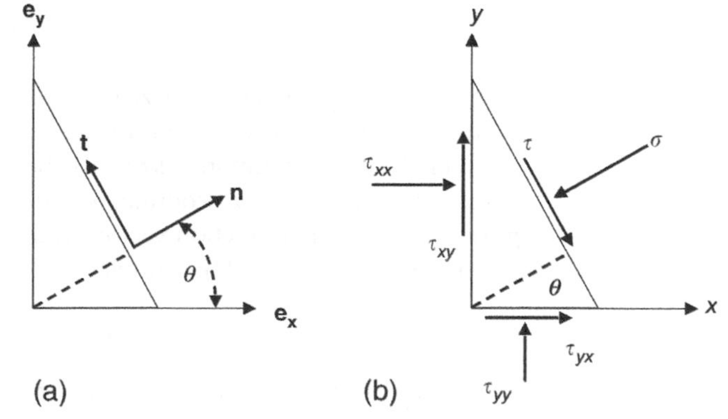

Fig. 2.4 Small triangular slab of rock used to derive the stress transformation equations.

the slanted face of the element is $\mathbf{n} = (n_x, n_y)$. We can construct another unit vector $\mathbf{t}$, perpendicular to $\mathbf{n}$, which lies along this face. Being of unit length, the components of $\mathbf{t}$ must satisfy the condition $\mathbf{t} \cdot \mathbf{t} = (t_x)^2 + (t_y)^2 = 1$. The orthogonality of $\mathbf{t}$ and $\mathbf{n}$ implies that $\mathbf{t} \cdot \mathbf{n} = t_x n_x + t_y n_y = 0$, which shows that $\mathbf{t} = \pm(n_y, -n_x)$. Finally, if we require that the pair of vectors $\{\mathbf{n}, \mathbf{t}\}$ have the same orientation relative to each other as do the pair $\{\mathbf{e_x}, \mathbf{e_y}\}$, the minus sign must be used, in which case $\mathbf{t} = (-n_y, n_x)$. This pair of vectors can be thought of as forming a new Cartesian coordinate system that is rotated from the original (x, y) system by a counterclockwise angle of $\theta = \text{arcos}(n_x)$. According to (2.9) and (2.10), the components of the traction vector $\mathbf{p(n)}$, expressed in terms of the (x, y) coordinate system, are given by

$$p_x = \tau_{xx} n_x + \tau_{yx} n_y, \tag{2.16}$$

$$p_y = \tau_{xy} n_x + \tau_{yy} n_y. \tag{2.17}$$

In order to find the components of $\mathbf{p}$ relative to the $\{\mathbf{n}, \mathbf{t}\}$ coordinate system, we take the inner products of $\mathbf{p}$ with respect to $\mathbf{n}$ and $\mathbf{t}$, in turn. For example,

$$p_n = \mathbf{p} \cdot \mathbf{n} = p_x n_x + p_y n_y = \tau_{xx} n_x^2 + \tau_{yx} n_y n_x + \tau_{xy} n_x n_y + \tau_{yy} n_y^2. \tag{2.18}$$

Utilization of the symmetry property $\tau_{yx} = \tau_{yx}$ allows this to be written as

$$p_n = \tau_{xx} n_x^2 + 2\tau_{xy} n_x n_y + \tau_{yy} n_y^2. \tag{2.19}$$

Similarly, the tangential component of the traction vector on this face, which is given by $p_t = \mathbf{p} \cdot \mathbf{t}$, can be expressed as

$$p_t = (\tau_{yy} - \tau_{xx}) n_x n_y + \tau_{xy} (n_x^2 - n_y^2). \tag{2.20}$$

The two unit vectors $\{\mathbf{n}, \mathbf{t}\}$ can be thought of as defining a new coordinate system that is rotated by a counterclockwise angle θ from the old coordinate system. This interpretation is facilitated by denoting these two new unit vectors by $\{\mathbf{e_{x'}}, \mathbf{e_{y'}}\}$. Equations (2.19) and (2.20) are therefore seen to give the components of the traction vector on the plane whose outward unit vector is $\mathbf{e_{x'}}$, that is,

$$p_{x'}(\mathbf{e_{x'}}) \equiv \tau_{x'x'} = \tau_{xx} n_x^2 + 2\tau_{xy} n_x n_y + \tau_{yy} n_y^2, \tag{2.21}$$

$$p_{y'}(\mathbf{e_{x'}}) \equiv \tau_{x'y'} = (\tau_{yy} - \tau_{xx}) n_x n_y + \tau_{xy} (n_x^2 - n_y^2), \tag{2.22}$$

where, for clarity, we reemphasize that these components pertain to the traction on the plane with outward unit normal vector $\mathbf{e_{x'}}$. According to the discussion given in §2.2, these components can also be interpreted as the components of the stress tensor in the (x', y') coordinate system. Specifically, $p_{x'}(\mathbf{e_{x'}}) = \tau_{x'x'}$, and $p_{y'}(\mathbf{e_{x'}}) = \tau_{x'y'}$. The traction vector on the plane whose outward unit normal vector is $\mathbf{e_{y'}}$ can be found by a similar analysis; the results are

$$p_{y'}(\mathbf{e_{y'}}) \equiv \tau_{y'y'} = \tau_{xx} n_y^2 - 2\tau_{xy} n_x n_y + \tau_{yy} n_x^2, \tag{2.23}$$

$$p_{x'}(\mathbf{e_{y'}}) \equiv \tau_{y'x'} = (\tau_{yy} - \tau_{xx}) n_x n_y + \tau_{xy} (n_x^2 - n_y^2). \tag{2.24}$$

Note that $p_{x'}(\mathbf{e}_{\mathbf{y}'}) = \tau_{y'x'} = p_{y'}(\mathbf{e}_{\mathbf{x}'}) = \tau_{x'y'}$, as must necessarily be the case, due to the general property of symmetry of the stress tensor.

Another common notation used for the stress transformation equations in two dimensions can be obtained by recalling that the primed coordinate system is derived from the unprimed system by rotation through a counterclockwise angle of $\theta = \arccos(n_x)$. Furthermore, the components (n_x, n_y) of the unit normal vector $\mathbf{n}$ can be written as $(\cos\theta, \sin\theta)$. In terms of the angle of rotation, the stresses in the rotated coordinate system are

$$\tau_{x'x'} = \tau_{xx}\cos^2\theta + 2\tau_{xy}\sin\theta\cos\theta + \tau_{yy}\sin^2\theta, \tag{2.25}$$

$$\tau_{y'y'} = \tau_{xx}\sin^2\theta - 2\tau_{xy}\sin\theta\cos\theta + \tau_{yy}\cos^2\theta, \tag{2.26}$$

$$\tau_{x'y'} = (\tau_{yy} - \tau_{xx})\sin\theta\cos\theta + \tau_{xy}(\cos^2\theta - \sin^2\theta). \tag{2.27}$$

This rotation operation can be represented by the rotation matrix $\mathbf{L}$, which has the defining properties that $\mathbf{L}^{\mathrm{T}}\mathbf{e}_{\mathbf{x}} = \mathbf{e}_{\mathbf{x}'}$, and $\mathbf{L}^{\mathrm{T}}\mathbf{e}_{\mathbf{y}} = \mathbf{e}_{\mathbf{y}'}$. In component form, relative to the (x, y) coordinate system, the two primed unit vectors are given by $\mathbf{e}_{\mathbf{x}'} = (\cos\theta, \sin\theta)$ and $\mathbf{e}_{\mathbf{y}'} = (-\sin\theta, \cos\theta)$. These two vectors therefore form the two columns of the matrix $\mathbf{L}^{\mathrm{T}}$ (Lang, 1971, p. 120), which is to say they form the *rows* of $\mathbf{L}$, that is,

$$\mathbf{L} = \begin{bmatrix} \cos\theta & \sin\theta \\ -\sin\theta & \cos\theta \end{bmatrix}. \tag{2.28}$$

Using this rotation matrix, the transformation equations (2.25)–(2.27) can be written in the following matrix form:

$$\begin{bmatrix} \tau_{x'x'} & \tau_{x'y'} \\ \tau_{y'x'} & \tau_{y'y'} \end{bmatrix} = \begin{bmatrix} \cos\theta & \sin\theta \\ -\sin\theta & \cos\theta \end{bmatrix} \begin{bmatrix} \tau_{xx} & \tau_{xy} \\ \tau_{yx} & \tau_{yy} \end{bmatrix} \begin{bmatrix} \cos\theta & -\sin\theta \\ \sin\theta & \cos\theta \end{bmatrix}, \tag{2.29}$$

which can also be expressed in direct matrix notation as

$$\boldsymbol{\tau}' = \mathbf{L}\boldsymbol{\tau}\mathbf{L}^{\mathrm{T}}. \tag{2.30}$$

The fact that the stresses transform according to (2.30) when the coordinate system is rotated is the defining property that makes the stress a *second-order tensor*. We note also that, using this direct matrix notation, the traction vector transforms according to $\mathbf{p}' = \mathbf{Lp}$. The appearance of *one* rotation matrix in this transformation law is the reason that vectors are referred to as *first-order tensors*.

The form of the stress transformation law given in (2.29) or (2.30) is convenient when considering a rotation of the coordinate system. However, from a more physically based viewpoint, it is pertinent to focus attention on the tractions that act on a given plane, such as the one shown in Fig. 2.4. The same equations are used in both situations, but their interpretation is slightly different. When focusing on a specific plane with unit normal vector $\mathbf{n}$, it is convenient to simplify the equations by utilizing the trigonometric identities $\cos^2\theta - \sin^2\theta = \cos 2\theta$, and $2\sin\theta\cos\theta = \sin 2\theta$. As long as attention is focused on a given plane, no confusion should arise if the normal stress acting on this plane is denoted by σ, and the shear stress is denoted by τ. After some algebraic manipulation, we arrive at the following equations for the normal and shear stresses acting on a plane whose outward unit normal vector is rotated counterclockwise from the x direction by an angle θ:

$$\sigma = \frac{1}{2}(\tau_{xx} + \tau_{yy}) + \frac{1}{2}(\tau_{xx} - \tau_{yy}) \cos 2\theta + \tau_{xy} \sin 2\theta, \tag{2.31}$$

$$\tau = \frac{1}{2}(\tau_{yy} - \tau_{xx}) \sin 2\theta + \tau_{xy} \cos 2\theta. \tag{2.32}$$

The variation of σ and τ with the angle of rotation is illustrated in Fig. 2.5, for the case where $\{\tau_{xx} = 4, \tau_{yy} = 2, \tau_{xy} = 1\}$.

An interesting question to pose is whether or not there are planes on which the shear stress vanishes, and where the stress therefore has purely a normal component. The answer follows directly from setting $\tau = 0$ in (2.32), and solving for

$$\tan 2\theta = \frac{2\tau_{xy}}{\tau_{xx} - \tau_{yy}}. \tag{2.33}$$

If $\tau_{xy} = 0$, then the plane with $\mathbf{n} = \mathbf{e_x}$ is already a shear-free plane, and (2.33) gives the result $\theta = 0$. In general, however, whatever the values of $\{\tau_{xy}, \tau_{xy}, \tau_{yy}\}$, there will always be two roots of (2.33) in the range $0 \le 2\theta < 2\pi$, and these roots will differ by π. Hence, there will be two values of θ that satisfy (2.33), differing by $\pi/2$, and lying in the range $0 \le \theta < \pi$; this situation will be discussed in more detail below. For now, note that if θ is defined by (2.33), it follows from elementary trigonometry that

$$\sin 2\theta = \pm[1 + \cos^2 2\theta]^{-1/2} = \pm\tau_{xy}[\tau_{xy}^2 + \frac{1}{4}(\tau_{xx} - \tau_{yy})^2]^{-1/2}, \tag{2.34}$$

$$\cos 2\theta = \pm[1 + \tan^2 2\theta]^{-1/2} = \pm\frac{1}{2}(\tau_{xx} - \tau_{yy})[\tau_{xy}^2 + \frac{1}{4}(\tau_{xx} - \tau_{yy})^2]^{-1/2}, \tag{2.35}$$

in which case the normal stress is found from (2.31) to be given by

$$\sigma = \frac{1}{2}(\tau_{xx} - \tau_{yy}) \pm [\tau_{xy}^2 + \frac{1}{4}(\tau_{xx} - \tau_{yy})^2]^{-1/2}. \tag{2.36}$$

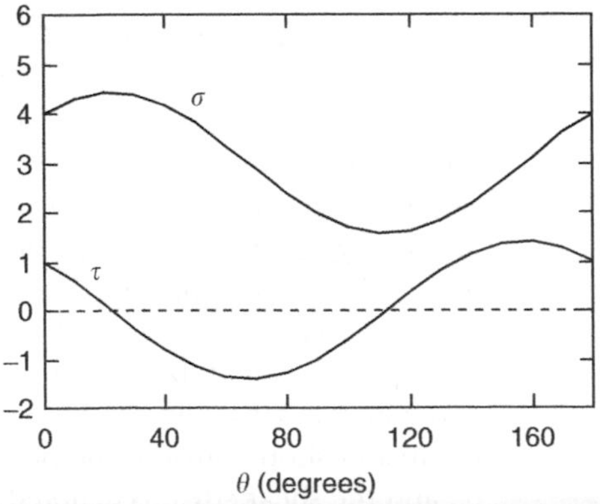

Fig. 2.5 Variation of normal and shear tractions with the angle θ (see Fig. 2.4a).

Equation (2.36) defines two normal stresses, σ_1 and σ_2, that are known as the *principal normal stresses*, or simply the *principal stresses*. These stresses act on planes whose orientations relative to the (x, y) coordinate system are given by (2.33). It is customary to set $\sigma_1 \geq \sigma_2$, in which case the + sign in (2.36) is associated with σ_1, that is,

$$\sigma_1 = \frac{1}{2}(\tau_{xx} + \tau_{yy}) + [\tau_{xy}^2 + \frac{1}{4}(\tau_{xx} - \tau_{yy})^2]^{1/2}, \tag{2.37}$$

$$\sigma_2 = \frac{1}{2}(\tau_{xx} + \tau_{yy}) - [\tau_{xy}^2 + \frac{1}{4}(\tau_{xx} - \tau_{yy})^2]^{1/2}. \tag{2.38}$$

These two principal normal stresses not only have the distinction of acting on planes on which there is no shear, but are also the minimum and maximum normal stresses that act on *any* planes through the point in question. This can be proven by noting that

$$\frac{d\sigma}{d\theta} = -(\tau_{xx} - \tau_{yy}) \sin 2\theta + 2\tau_{xy} \cos 2\theta = -2\tau, \tag{2.39}$$

so that any plane on which τ vanishes is also a plane on which σ takes on a locally extreme value. This is apparent from Fig. 2.5, which also shows, for example, that the shear traction τ will take on its maximum and minimum values on two orthogonal planes whose normal vectors bisect the two directions of principal normal stress.

Although it is clear from (2.37) and (2.38) which of the two principal stresses is largest, the direction in which the major principal stress acts is not so clear, due to the fact that (2.33) has two physically distinct solutions, that differ by $\pi/2$. The correct choice for σ_1 is the angle that makes the normal stress a local maximum, rather than a local minimum. To determine the correct value we examine the second derivatives of σ with respect to θ. From (2.39),

$$\frac{d^2\sigma}{d\theta^2} = -2(\tau_{xx} - \tau_{yy}) \cos 2\theta - 4\tau_{xy} \sin 2\theta. \tag{2.40}$$

Using (2.40), along with (2.33), eventually leads to the following results (Chou and Pagano, 1992, p. 10):

$$\tau_{xx} > \tau_{yy} \quad \text{and} \quad \tau_{xy} > 0 \quad \Rightarrow \quad 0 < \theta_1 < 45^\circ, \tag{2.41}$$

$$\tau_{xx} < \tau_{yy} \quad \text{and} \quad \tau_{xy} > 0 \quad \Rightarrow \quad 45^\circ < \theta_1 < 90^\circ, \tag{2.42}$$

$$\tau_{xx} < \tau_{yy} \quad \text{and} \quad \tau_{xy} < 0 \quad \Rightarrow \quad 90^\circ < \theta_1 < 135^\circ, \tag{2.43}$$

$$\tau_{xx} > \tau_{yy} \quad \text{and} \quad \tau_{xy} < 0 \quad \Rightarrow \quad 135^\circ < \theta_1 < 180^\circ. \tag{2.44}$$

The principal stresses and principal directions can also be found by a different method, which can more readily be generalized to three dimensions. We start again by asking whether or not there are planes on which the traction vector is purely normal, with no shear component. On such planes, the traction vector will be parallel to the outward unit normal vector, and can therefore be expressed as $\mathbf{p} = \sigma\mathbf{n}$, where σ is some (as yet unknown) scalar. From (2.27) it is known that $\mathbf{p} = \boldsymbol{\tau}^{\mathrm{T}}\mathbf{n}$, which, due to the symmetry of the stress tensor, can be written as $\mathbf{p} = \boldsymbol{\tau}\mathbf{n}$. Hence, any plane on which the traction is purely normal must satisfy the equation

$$\boldsymbol{\tau}\mathbf{n} = \sigma\mathbf{n}. \tag{2.45}$$

The left-hand side of (2.45) represents a matrix, $\boldsymbol{\tau}$, multiplying a vector, $\mathbf{n}$, whereas on the right-hand side the vector $\mathbf{n}$ is multiplied by a scalar, σ. If the 2×2 identity matrix is denoted by $\mathbf{I}$, then $\mathbf{n} = \mathbf{In}$, and (2.45) can be rewritten as

$$(\boldsymbol{\tau} - \sigma\mathbf{I})\mathbf{n} = 0. \tag{2.46}$$

Equation (2.46) can be recognized as a standard eigenvalue problem, in which σ is the *eigenvalue*, and $\mathbf{n}$ is the *eigenvector*. Much of the theory of stress follows immediately from the theory pertaining to eigenvectors and eigenvalues of a symmetric matrix. The main conclusions of this theory in an arbitrary number of dimensions N are (Lang, 1971) that there will always be N mutually orthogonal eigenvectors, each corresponding to a real eigenvalue σ, although the eigenvalues need not necessarily be distinct from each other. In the present case, the eigenvalues are the principal stresses, and the associated eigenvectors are the principal stress directions. These results, along with explicit expressions for the principal stresses and principal stress directions, can be derived from (2.46) without appealing to the general theory, however, as follows.

Equation (2.46) can be written in component form as

$$(\tau_{xx} - \sigma)n_x + \tau_{xy}n_y = 0, \tag{2.47}$$

$$\tau_{yx}n_x + (\tau_{yy} - \sigma)n_y = 0. \tag{2.48}$$

Using the standard procedure of Gaussian elimination, we multiply (2.47) by τ_{yx}, and multiply (2.48) by $(\tau_{xx} - \sigma)$, to arrive at

$$(\tau_{xx} - \sigma)\tau_{yx}n_x + \tau_{xy}\tau_{yx}n_y = 0, \tag{2.49}$$

$$(\tau_{xx} - \sigma)\tau_{yx}n_x + (\tau_{yy} - \sigma)(\tau_{xx} - \sigma)n_y = 0. \tag{2.50}$$

Subtraction of (2.49) from (2.50) yields

$$[\sigma^2 - (\tau_{xx} + \tau_{yy})\sigma + (\tau_{xx}\tau_{yy} - \tau_{xy}^2)]n_y = 0, \tag{2.51}$$

where use has been made of the relationship $\tau_{yx} = \tau_{xy}$. This equation will be satisfied if either the bracketed term vanishes, or if $n_y = 0$. In this latter case, we must have $n_x = 1$, since $\mathbf{n}$ is a unit vector. Equation (2.47) then shows that $\sigma = \tau_{xx}$, and (2.48) shows that $\tau_{xy} = 0$. This solution therefore corresponds to the special case in which the x direction is already a principal stress direction, and τ_{xx} is a principal stress. In general, this will not be the case, and we must proceed by setting the bracketed term to zero:

$$\sigma^2 - (\tau_{xx} + \tau_{yy})\sigma + (\tau_{xx}\tau_{yy} - \tau_{xy}^2) = 0. \tag{2.52}$$

The bracketed term in (2.51) is the *determinant* of the matrix $(\boldsymbol{\tau} - \sigma\mathbf{I})$, so (2.52) can be written symbolically as $\det(\boldsymbol{\tau} - \sigma\mathbf{I}) = 0$, which is the standard criterion for finding the eigenvalues of a matrix. This equation is a quadratic in σ, and will always have two roots, which will be functions of the two parenthesized coefficients that appear in (2.52). Before discussing these roots, we note that as the two principal stresses are scalars, their values should not depend on the coordinate system used. Therefore, the two coefficients $(\tau_{xx} + \tau_{yy})$ and $(\tau_{xx}\tau_{yy} - \tau_{xy}^2)$, must be independent of the coordinate system being used; this could also be shown more directly by adding (2.25) and (2.26). These two combinations of the stress components are known as *invariants*, and are discussed in more detail in a three-dimensional context in §2.8.

The quadratic formula shows that the two roots of (2.52) are given by the two values σ_1 and σ_2 from (2.37) and (2.38). If σ takes on one of these two values, (2.47) and (2.48) become linearly dependent. In this case, one of the two equations is redundant, and we can solve (2.47) to find

$$\tan\theta = \frac{n_y}{n_x} = \frac{2\tau_{xy}}{(\tau_{xx} - \tau_{yy}) \pm [4\tau_{xy}^2 + (\tau_{xx} - \tau_{yy})^2]^{1/2}}, \tag{2.53}$$

where the + sign corresponds to σ_1, and the − sign corresponds to σ_2. Using the trigonometric identity $\tan 2\theta = 2\tan\theta/(1 - \tan^2\theta)$, it can be shown that (2.53) is consistent with (2.33). These two directions, corresponding to the two orthogonal unit eigenvectors, will define a new coordinate system, rotated by an angle θ from the x direction, in which the shear stresses are zero. This coordinate system is often referred to as the *principal coordinate system*.

2.4 Graphical representations of stress in two dimensions

A simple graphical construction popularized by Mohr (1914) can be used to represent the state of stress at a point. Recall that (2.31) and (2.32) give expressions for the normal stress and shear stress acting on a plane whose unit normal direction is rotated from the x direction by a counterclockwise angle θ. Now imagine that we are using the principal coordinate system, in which the shear stresses are zero and the normal stresses are the two principal normal stresses. In this case we replace τ_{xx} with σ_1, replace τ_{yy} with σ_2, replace τ_{xy} with 0, and interpret θ as the angle of counterclockwise rotation from the direction of the maximum principal stress. We thereby arrive at the following equations that give the normal and shear stresses on a plane whose outward unit normal vector is rotated by θ from the first principal direction:

$$\sigma = \frac{(\sigma_1 + \sigma_2)}{2} + \frac{(\sigma_1 - \sigma_2)}{2}\cos 2\theta, \tag{2.54}$$

$$\tau = \frac{-(\sigma_1 - \sigma_2)}{2}\sin 2\theta. \tag{2.55}$$

These are the equations of a circle in the (σ, τ) plane, with its center at the point $\{\sigma = (\sigma_1 + \sigma_2)/2, \tau = 0\}$, and with radius $(\sigma_1 - \sigma_2)$. In contrast to the standard parameterization in which the angle θ is measured in the counterclockwise direction, this circle is parameterized in the clockwise direction, with angle 2θ. This becomes clear if we note that $\cos 2\theta$ can be replaced with $\cos(-2\theta)$ in (2.54),

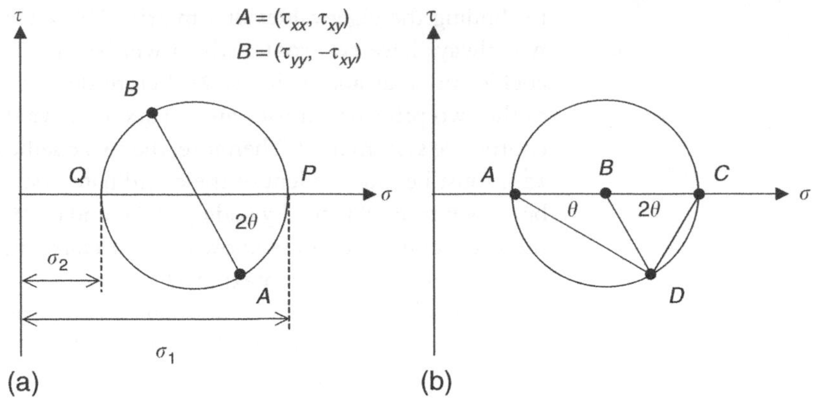

Fig. 2.6 Mohr's circle construction (see text for discussion).

and $-\sin 2\theta$ can be replaced with $\sin(-2\theta)$ in (2.55). A generic Mohr's circle is shown in Fig. 2.6a. A detailed discussion of the use of Mohr's circle in rock and soil mechanics is given by Parry (1995).

Many of the important properties of the two-dimensional stress tensor can be read directly off of the Mohr's circle. For example, at point P, when $\theta = 0$, there is no rotation from the σ_1 direction, and indeed Mohr's circle indicates that $(\sigma = \sigma_1, \tau = 0)$. Similarly, consider the plane for which $\theta = 90°$. This plane is rotated counterclockwise by 90° from the σ_1 direction, and therefore represents the σ_2 direction. This plane is represented on Mohr's circle by the point that is rotated clockwise by $2\theta = 180°$, which is point Q on Fig. 2.6a, where we find $(\sigma = \sigma_2, \tau = 0)$. This construction also clearly shows that the maximum shear stress has a magnitude equal to the radius of the Mohr's circle, and occurs on planes for which $2\theta = \pm 90°$, which is to say $\theta = \pm 45°$. These two planes bisect the two planes on which the principal normal stresses act, which are $\theta = 0°, 90°$.

Point A on the Mohr's circle in Fig. 2.6a shows the stresses acting on a generic plane whose unit normal vector is rotated by angle θ from the σ_1 direction. This direction can be denoted as the x direction, and these stresses can therefore be denoted by $(\sigma = \tau_{xx}, \tau = \tau_{xy})$. Now consider the plane that is rotated by an additional 90°. For this plane, the additional increment in 2θ is 180°, and the stresses are represented by the point B, which is located at the opposite end of a diameter of the circle from point A. This direction can be denoted as the y direction, in which case the x and y directions define an orthogonal coordinate system. However, the stresses at point B on Mohr's circle must be identified as $(\sigma = \tau_{yy}, \tau = -\tau_{yx})$. This is because it is implicit in (2.55) that the tangential direction is rotated 180° counterclockwise from the normal direction of the plane in question, which would then correspond to the $-x$ direction instead of the $+x$ direction.

It is also seen from Mohr's circle that the mean value of the two normal stresses, $(\tau_{xx} + \tau_{yy})/2$, is equal to the horizontal distance from the origin to the center of Mohr's circle, which is $(\sigma_1 + \sigma_2)/2$. This is another proof of the fact that the value of the mean normal stress is independent of the coordinate system used.

Mohr's circle can also be used to graphically determine the two principal stresses, and the orientations of the principal stress directions, given knowledge of the components of the stress tensor in some (x, y) coordinate system. We first plot the point (τ_{xx}, τ_{xy}) on the (σ, τ) plane, and note that these two stresses will be the normal and shear stresses on the plane whose outward unit normal vector is $\mathbf{e_x}$. This direction is rotated by some (as yet unknown) angle θ from the σ_1 direction. We next plot the stresses $(\tau_{yy}, -\tau_{yx})$ on the (σ, τ) plane, and note that these represent the stresses on the plane with outward unit normal vector $\mathbf{e_y}$. This direction is therefore rotated by an angle $\theta + 90°$ from the σ_1 direction. In accordance with the earlier discussion, the sign convention that is used for the shear stress on this second plane on a Mohr's diagram is opposite to that used when considering this as the second direction in an orthogonal coordinate system; hence, this second pair is plotted as $(\tau_{yy}, -\tau_{yx})$. As these two planes are rotated from one another by 90°, they will be separated by 180° on Mohr's circle; hence, the line joining these two points will be the diameter of Mohr's circle. Once this diameter is constructed, the circle can be drawn with a compass. The two points at which this circle intersects the σ-axis will be the two principal stresses, σ_1 and σ_2. The angle of rotation between the x direction and the σ_1 direction can also be read directly from this circle.

Mohr's circle can also be used to graphically find the orientation of the plane on which certain tractions act (Kuske and Robertson, 1974). Consider point D in Fig. 2.6b, at which the traction is given by (σ, τ). First note that $\angle DBA = \pi - \angle DBC$. Next, note that $\angle DAB$ and $\angle ADB$ are two equal angles of an isosceles triangle, the third angle of which is $\angle DBA$. It follows that $\angle DAB = \theta$. The chord AD therefore points in the direction of the outward unit normal vector to the plane in question. Since $\angle ADC$ is inscribed within a semicircle, we know that $\angle ADC = \pi/2$. Chords AD and DC are therefore perpendicular to each other, from which it follows that chord CD indicates the direction of the *plane* on which the tractions are (σ, τ). This construction is sometimes useful in aiding in the visualization of the tractions acting on various planes.

There are other geometrical constructions that have been devised to represent the state of stress at a point in a body. Most of these are less convenient than Mohr's circle, and to a great extent these graphical approaches, once very popular, have been superseded by algebraic methods. Nevertheless, we briefly mention Lamé's stress ellipsoid, which in two dimensions is a stress ellipse. To simplify the discussion, assume that we are using the principal coordinate system, in which case it follows from (2.9) and (2.10) that

$$p_1(\mathbf{n}) = \sigma_1 n_1 \quad \text{and} \quad p_2(\mathbf{n}) = \sigma_2 n_2, \tag{2.56}$$

where we have let $x \to 1, y \to 2$, and have noted that, by construction, $\tau_{12} = 0$. Since $\mathbf{n}$ is a unit vector, we see from (2.56) that

$$(p_1/\sigma_1)^2 + (p_2/\sigma_2)^2 = (n_1)^2 + (n_2)^2 = 1. \tag{2.57}$$

The point (p_1, p_2) therefore traces out an ellipse whose semimajor and semiminor axes are σ_1 and σ_2, respectively (Fig. 2.7). Each vector from the origin to a point

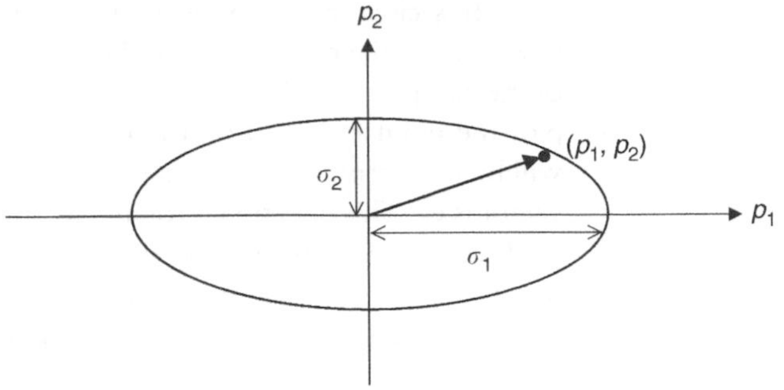

Fig. 2.7 Lamé's stress ellipse (see text for description).

on the ellipse represents a traction vector that acts on some plane passing through the point at which the principal stresses are σ_1 and σ_2. However, although the Lamé stress ellipse shows the various traction vectors that act on different planes, it does *not* indicate the plane on which the given traction acts. In general, only when the vector *OP* lies along one of the principal directions in Fig. 2.7 will the direction of the plane be apparent, since in these special cases the traction is known to be normal to the plane. In the more general case, the direction of the unit normal vector of the plane on which the traction is (p_1, p_2) can be found with the aid of the stress-director surface, which is defined by

$$(p_1^2/\sigma_1) + (p_2^2/\sigma_2) = \pm 1. \qquad (2.58)$$

For the case which is most common in rock mechanics, in which both principal stresses are positive, the + sign must be used in (2.58), and the surface is an ellipse with axes $\sqrt{\sigma_1}$ and $\sqrt{\sigma_2}$. The outward unit normal vector of the plane on which the traction is (p_1, p_2) is then given by the tangent to the stress-director ellipse at the point where it intersects the stress ellipsoid (Chou and Pagano, 1992, p. 200). Proof of this assertion, and more details of this construction, can be found in Timoshenko and Goodier (1970) and Durelli et al. (1958).

One interesting fact that is more apparent from the Lamé construction than from Mohr's circle is that not only does the magnitude of the normal component of the stress take on stationary values in the principal directions, but the magnitude of the *total* traction vector also takes on stationary values in these directions. In particular, the maximum value of $|\mathbf{p}|$ is seen to be equal to σ_1, and occurs in the direction of the major principal stress.

Most of the manipulations and transformations described above are concerned with the values of the stress and traction at a given "point" in the rock. In general, the state of stress will vary from point to point. The equations that govern these variations are described in §5.5. The state of stress in a rock mass can either be estimated based on a solution (either numerical or analytical) of these equations (Chapter 8), or from stress measurements (Chapter 13). In order to completely specify the state of stress in a two-dimensional rock mass, it is necessary either to know the values of τ_{xx}, τ_{yy}, and τ_{xy} at each point in the body, or, alternatively, to know at each point the values of the two principal stresses σ_1 and σ_2, along with the angle of inclination between the x direction and, say, the σ_1 direction. Although it is difficult to display all of this information graphically, there are a number of simple graphical representations that are useful in giving a partial picture of a stress field. Among these are:

1 *Isobars*, which are curves along which the principal stress is constant. There are two sets of isobars, one for σ_1 and one for σ_2. A set of isobars for one of the principal stresses, say σ_1, must by definition form a nonintersecting set of curves. However, an isobar of σ_1 may intersect an isobar of σ_2.
2 *Isochromatics*, which are curves along which the maximum shear stress $(\sigma_1 - \sigma_2)/2$, is constant. These curves can be directly found using the methods of photoelasticity, which is described by Frocht (1941) and Durelli et al. (1958).
3 *Isopachs*, which are curves along which the mean normal stress $(\sigma_1 + \sigma_1)/2$ is constant. It is shown in §5.5 that this quantity satisfies Laplace's equation, which is the same equation that governs, for example, steady-state temperature distributions, or steady-state distributions of the electric field, in isotropic conducting bodies. Hence, the isopachs can be found from analogue methods that utilize electrically conducting paper that is cut to the same shape as the rock mass under investigation. This procedure is discussed by Durelli et al. (1958).
4 *Isostatics*, or *stress trajectories*, are a system of curves which are at each point tangent to the principal axes of the stress. As the two principal axes are always orthogonal, the two sets of isostatic curves are mutually orthogonal. Since a free surface is always a principal plane (as it has no shear stress acting on it), an isostatic curve will intersect a free surface at a right angle to it.
5 *Isoclinics*, which are curves on which the principal axes make a constant angle with a given fixed reference direction. These curves can also be obtained by photoelastic methods.
6 *Slip lines*, which are curves on which the shear stress is a maximum. As the maximum shear stress at any point is always in a direction that bisects the two directions of principal normal stresses, these lines form an orthogonal grid.

2.5 Stresses in three dimensions

The theory of stresses in three dimensions is in general a straightforward extension of the two-dimensional theory. A generic plane in three dimensions will have a unit normal vector $\mathbf{n} = (n_x, n_y, n_z)$. The components of this vector satisfy the normalization condition $(n_x)^2 + (n_y)^2 + (n_z)^2 = 1$. A three-dimensional version of the argument accompanying Fig. 2.2b leads to the following generalization of (2.6):

$$\mathbf{p}(\mathbf{n}) = n_x\mathbf{p}(\mathbf{e_x}) + n_y\mathbf{p}(\mathbf{e_y}) + n_z\mathbf{p}(\mathbf{e_z}). \tag{2.59}$$

The components of the three traction vectors that act on planes whose outward unit normals are in the three coordinate directions are denoted by

$$\mathbf{p}(\mathbf{e_x}) = [\tau_{xx} \quad \tau_{xy} \quad \tau_{xz}]^T, \tag{2.60}$$

$$\mathbf{p}(\mathbf{e_y}) = [\tau_{yx} \quad \tau_{yy} \quad \tau_{yz}]^{\mathrm{T}}, \tag{2.61}$$

$$\mathbf{p}(\mathbf{e_z}) = [\tau_{zx} \quad \tau_{zy} \quad \tau_{zz}]^{\mathrm{T}}. \tag{2.62}$$

Substitution of (2.60)–(2.62) into (2.59) leads to

$$p_x(\mathbf{n}) = \tau_{xx}n_x + \tau_{yx}n_y + \tau_{zx}n_z, \tag{2.63}$$

$$p_y(\mathbf{n}) = \tau_{xy}n_x + \tau_{yy}n_y + \tau_{zy}n_z, \tag{2.64}$$

$$p_z(\mathbf{n}) = \tau_{xz}n_x + \tau_{yz}n_y + \tau_{zz}n_z, \tag{2.65}$$

which can be written in matrix form as $\mathbf{p}(\mathbf{n}) = \boldsymbol{\tau}^{\mathrm{T}}\mathbf{n}$, that is,

$$\begin{bmatrix} p_x(\mathbf{n}) \\ p_y(\mathbf{n}) \\ p_z(\mathbf{n}) \end{bmatrix} = \begin{bmatrix} \tau_{xx} & \tau_{yx} & \tau_{zx} \\ \tau_{xy} & \tau_{yy} & \tau_{zy} \\ \tau_{xz} & \tau_{yz} & \tau_{zz} \end{bmatrix} \begin{bmatrix} n_x \\ n_y \\ n_z \end{bmatrix}. \tag{2.66}$$

The three-dimensional analogue of the argument illustrated by Fig. 2.3b would show that the conjugate terms in the three-dimensional stress tensor are equal, that is,

$$\tau_{yx} = \tau_{xy}, \quad \tau_{yz} = \tau_{zy}, \quad \tau_{zx} = \tau_{xz}. \tag{2.67}$$

Hence, (2.66) can also be written as $\mathbf{p}(\mathbf{n}) = \boldsymbol{\tau}\mathbf{n}$.

The question can again be asked as to whether or not there are planes on which the shear stresses vanish. On such planes, the traction vector will be parallel to the outward unit normal vector, and therefore can be written as $\mathbf{p} = \sigma\mathbf{n}$, where σ is some scalar. But as $\mathbf{p}(\mathbf{n}) = \boldsymbol{\tau}\mathbf{n}$, we have $\boldsymbol{\tau}\mathbf{n} = \sigma\mathbf{n} = \sigma\mathbf{I}\mathbf{n}$, and therefore again arrive at the eigenvalue problem $(\boldsymbol{\tau} - \sigma\mathbf{I})\mathbf{n} = \mathbf{0}$, §2.3 (2.46), that is,

$$\begin{bmatrix} \tau_{xx} - \sigma & \tau_{xy} & \tau_{xz} \\ \tau_{yx} & \tau_{yy} - \sigma & \tau_{yz} \\ \tau_{zx} & \tau_{zy} & \tau_{zz} - \sigma \end{bmatrix} \begin{bmatrix} n_x \\ n_y \\ n_z \end{bmatrix} = \begin{bmatrix} 0 \\ 0 \\ 0 \end{bmatrix}. \tag{2.68}$$

From this point on, the development follows that for the two-dimensional theory. Although $(n_x, n_y, n_z) = (0, 0, 0)$ is obviously a solution to (2.68), it is inadmissible because it does not satisfy the condition that $\mathbf{n} \cdot \mathbf{n} = 1$. Admissible solutions can be found only if the determinant of the matrix $(\boldsymbol{\tau} - \sigma\mathbf{I})$ vanishes (Lang,1971). When the determinant is expanded out, it takes the form

$$\sigma^3 - I_1\sigma^2 - I_2\sigma - I_3 = 0, \tag{2.69}$$

where

$$I_1 = \tau_{xx} + \tau_{yy} + \tau_{zz}, \tag{2.70}$$

$$I_2 = \tau_{xy}^2 + \tau_{xz}^2 + \tau_{yz}^2 - \tau_{xx}\tau_{yy} - \tau_{xx}\tau_{zz} - \tau_{yy}\tau_{zz}, \tag{2.71}$$

$$I_3 = \tau_{xx}\tau_{yy}\tau_{zz} + 2\tau_{xy}\tau_{xz}\tau_{yz} - \tau_{xx}\tau_{yz}^2 - \tau_{yy}\tau_{xz}^2 - \tau_{zz}\tau_{xy}^2. \tag{2.72}$$

The fact that the stress tensor is symmetric ensures that (2.69) has three *real* roots. These roots are conventionally labeled such that $\sigma_1 \geq \sigma_2 \geq \sigma_3$. Each of these roots will correspond to an eigenvector that can be labeled as $\mathbf{n}^1 = (n_x^1, n_y^1, n_z^1)$, etc. Although, in general, eigenvectors are arbitrary to within a multiplicative constant, the condition that the vector must be of unit length leads to a unique determination of the vector components. The normalized eigenvector corresponding to σ_1 is the unit normal vector of the plane on which the normal stress is σ_1 and the shear stress is 0, and likewise for the other two principal stresses.

In two dimensions it was shown explicitly in (2.33) that the two principal stress directions are rotated by 90° from each other. The orthogonality of the principal directions continues to be the case in three dimensions; this can be proven as follows. First recall that the inner product of two vectors, $\mathbf{u} \cdot \mathbf{v}$, can also be expressed as $\mathbf{u}^{\mathrm{T}}\mathbf{v}$, since

$$\mathbf{u}^{\mathrm{T}}\mathbf{v} = \begin{bmatrix} u_x & u_y & u_z \end{bmatrix} \begin{bmatrix} v_x \\ v_y \\ v_z \end{bmatrix} = u_x v_x + u_y v_y + u_z v_z = \mathbf{u} \cdot \mathbf{v}. \quad (2.73)$$

Also, recall that the transpose operation has the properties that $(\mathbf{AB})^{\mathrm{T}} = \mathbf{B}^{\mathrm{T}}\mathbf{A}^{\mathrm{T}}$, and $(\mathbf{A}^{\mathrm{T}})^{\mathrm{T}} = \mathbf{A}$. Consider two principal stresses, $\sigma_1 \neq \sigma_2$, and the principal directions corresponding to these two stresses, that is, $\boldsymbol{\tau}\mathbf{n}^1 = \sigma_1\mathbf{n}^1$ and $\boldsymbol{\tau}\mathbf{n}^2 = \sigma_2\mathbf{n}^2$. Take the inner product of both sides of this first equation with $\mathbf{n}^2$:

$$(\mathbf{n}^2)^{\mathrm{T}}\boldsymbol{\tau}\mathbf{n}^1 = (\mathbf{n}^2)^{\mathrm{T}}\sigma_1\mathbf{n}^1 = \sigma_1(\mathbf{n}^2)^{\mathrm{T}}\mathbf{n}^1, \quad (2.74)$$

and the inner product of both sides of the second equation with $\mathbf{n}^1$:

$$(\mathbf{n}^1)^{\mathrm{T}}\boldsymbol{\tau}\mathbf{n}^2 = (\mathbf{n}^1)^{\mathrm{T}}\sigma_2\mathbf{n}^2 = \sigma_2(\mathbf{n}^1)^{\mathrm{T}}\mathbf{n}^2. \quad (2.75)$$

Now take the transpose of both sides of (2.75):

$$[(\mathbf{n}^1)^{\mathrm{T}}\boldsymbol{\tau}\mathbf{n}^2]^{\mathrm{T}} = (\mathbf{n}^2)^{\mathrm{T}}\boldsymbol{\tau}^{\mathrm{T}}\mathbf{n}^1 = [\sigma_2(\mathbf{n}^1)^{\mathrm{T}}\mathbf{n}^2]^{\mathrm{T}} = \sigma_2(\mathbf{n}^2)^{\mathrm{T}}\mathbf{n}^1. \quad (2.76)$$

But $\boldsymbol{\tau}^{\mathrm{T}} = \boldsymbol{\tau}$, so (2.76) shows that

$$(\mathbf{n}^2)^{\mathrm{T}}\boldsymbol{\tau}\mathbf{n}^1 = \sigma_2(\mathbf{n}^2)^{\mathrm{T}}\mathbf{n}^1. \quad (2.77)$$

Subtracting (2.77) from (2.74) yields

$$0 = (\sigma_1 - \sigma_2)(\mathbf{n}^2)^{\mathrm{T}}\mathbf{n}^1 = (\sigma_1 - \sigma_2)\mathbf{n}^2 \cdot \mathbf{n}^1. \quad (2.78)$$

As $\sigma_1 \neq \sigma_2$ by assumption, it follows that $\mathbf{n}^2 \cdot \mathbf{n}^1 = 0$, which is to say that $\mathbf{n}^1$ and $\mathbf{n}^2$ are orthogonal to each other. This argument applies to any pair of distinct principal stresses, proving that the three principal stresses act on three mutually orthogonal planes.

In certain cases, two of the roots of (2.69) may be equal, say $\sigma_1 > \sigma_2 = \sigma_3$. Assume that this is the case, and that we have found two vectors $\mathbf{n}^2$ and $\mathbf{n}^3$ that both satisfy $\boldsymbol{\tau}\mathbf{n} = \sigma_2\mathbf{n}$. Any vector $\mathbf{n}$ in the plane that is spanned by these two vectors can be written in the form $\mathbf{n} = c_2\mathbf{n}^2 + c_3\mathbf{n}^3$, so

$$\begin{aligned} \boldsymbol{\tau}\mathbf{n} &= \boldsymbol{\tau}(c_2\mathbf{n}^2 + c_3\mathbf{n}^3) = c_2\boldsymbol{\tau}\mathbf{n}^2 + c_3\boldsymbol{\tau}\mathbf{n}^3 = c_2\sigma_2\mathbf{n}^2 + c_3\sigma_2\mathbf{n}^3 \\ &= \sigma_2(c_2\mathbf{n}^2 + c_3\mathbf{n}^3) = \sigma_2\mathbf{n}, \end{aligned} \quad (2.79)$$

which proves that $\mathbf{n}$ is also an eigenvector of $\boldsymbol{\tau}$, with eigenvalue σ_2. Hence, all vectors in the plane spanned by $\mathbf{n}^2$ and $\mathbf{n}^3$ are principal stress directions associated with σ_2. No more than two of these vectors can be linearly independent, however, and it is conventional to pick two *orthogonal* directions in this plane as the principal stress directions associated with σ_2. In this way we maintain the orthogonality of the principal stress directions. In the special case in which all three principal stresses are equal, that is, $\sigma_1 = \sigma_2 = \sigma_3$, the same argument shows that *any* vector of unit length will be an eigenvector corresponding to σ_1, and will therefore define a principal stress direction. In other words, the traction on *all* planes will be a normal traction of magnitude σ_1.

In two dimensions, (2.39) showed that the principal stresses were also the maximum and minimum values of the normal stress that acted on any plane. This also follows directly from Mohr's circle, since the two locations at which the circle intersects the $\tau = 0$ axis are the extreme values of σ. In three dimensions, the three principal stresses also represent locally stationary values of σ, one of which is an absolute maximum, one an absolute minimum, and one a saddle point in the three-dimensional space of unit vectors $\mathbf{n}$. To prove this, we start with the fact that the normal component of the traction, p_n, can be found from

$$p_n = \mathbf{p} \cdot \mathbf{n} = \mathbf{n} \cdot \mathbf{p} = \mathbf{n}^{\mathrm{T}}\mathbf{p} = \mathbf{n}^{\mathrm{T}}\boldsymbol{\tau}\mathbf{n}. \tag{2.80}$$

By performing the indicated matrix multiplication, or by analogy with (2.19), we find

$$p_n = \tau_{xx}n_x^2 + \tau_{yy}n_y^2 + \tau_{zz}n_z^2 + 2\tau_{xy}n_xn_y + 2\tau_{xz}n_xn_z + 2\tau_{yz}n_yn_z. \tag{2.81}$$

It is now desired to find the extreme values of p_n, as a function of the three components of the unit normal vector $\mathbf{n}$, bearing in mind the constraint $n_x^2 + n_y^2 + n_z^2 = 1$. This constraint equation is of the form $f(n_x, n_y, n_z) = \text{constant}$. Geometrically, this is equivalent to considering the variation of p_n on the surface of the unit sphere in (n_x, n_y, n_z) space. According to the theory of Lagrange multipliers (Lang, 1973, pp. 140–4), the constrained maximum (or minimum) will occur at a point at which the gradient of p_n is parallel to the gradient of the constraint function f. The components of the gradient of p_n, considered as a function of (n_x, n_y, n_z), are calculated as follows:

$$\frac{\partial p_n}{\partial n_x} = 2\tau_{xx}n_x + 2\tau_{xy}n_y + 2\tau_{xz}n_z = 2(\tau_{xx}n_x + \tau_{xy}n_y + \tau_{xz}n_z) = 2p_x, \tag{2.82}$$

and similarly for the other two components. Hence, the gradient of p_n is

$$\text{grad}(p_n) = \left(\frac{\partial p_n}{\partial n_x}, \frac{\partial p_n}{\partial n_y}, \frac{\partial p_n}{\partial n_z}\right) = 2(p_x, p_y, p_z) = 2\mathbf{p}. \tag{2.83}$$

The gradient of the constraint function is

$$\text{grad}(f) = \text{grad}(n_x^2 + n_y^2 + n_z^2) = (2n_x, 2n_y, 2n_z) = 2\mathbf{n}. \tag{2.84}$$

If these two gradients are parallel, then $2\mathbf{p} = \sigma(2\mathbf{n})$ for some constant σ, which is to say $\mathbf{p} = \sigma\mathbf{n}$. But this is precisely the condition that defines the principal stresses and their associated principal directions. Hence, the principal directions also define those planes on which the normal traction takes on stationary values.

It is also of some interest to find the stationary values of the shear traction, and the planes on which these tractions act. For this purpose, it is convenient to work in the principal coordinate system. The traction vector on an arbitrary plane whose unit normal vector (in the principal coordinate system) is $\mathbf{n} = (n_1, n_2, n_3)$ is found from (2.66) to be

$$\mathbf{p} = \boldsymbol{\tau}\mathbf{n} = \begin{bmatrix} \sigma_1 & 0 & 0 \\ 0 & \sigma_2 & 0 \\ 0 & 0 & \sigma_3 \end{bmatrix} \begin{bmatrix} n_1 \\ n_2 \\ n_3 \end{bmatrix} = \begin{bmatrix} \sigma_1 n_1 \\ \sigma_2 n_2 \\ \sigma_3 n_3 \end{bmatrix}. \tag{2.85}$$

The normal traction on this plane, σ, is found by projecting $\mathbf{p}$ onto $\mathbf{n}$:

$$\sigma = \mathbf{p} \cdot \mathbf{n} = \mathbf{p}^{\mathrm{T}}\mathbf{n} = \begin{bmatrix} \sigma_1 n_1 & \sigma_2 n_2 & \sigma_3 n_3 \end{bmatrix} \begin{bmatrix} n_1 \\ n_2 \\ n_3 \end{bmatrix} = \sigma_1 n_1^2 + \sigma_2 n_2^2 + \sigma_3 n_3^2. \tag{2.86}$$

The magnitude (squared) of the total traction vector is given by

$$|\mathbf{p}|^2 = \mathbf{p} \cdot \mathbf{p} = \mathbf{p}^{\mathrm{T}}\mathbf{p} = \begin{bmatrix} \sigma_1 n_1 & \sigma_2 n_2 & \sigma_3 n_3 \end{bmatrix} \begin{bmatrix} \sigma_1 n_1 \\ \sigma_2 n_2 \\ \sigma_3 n_3 \end{bmatrix} = \sigma_1^2 n_1^2 + \sigma_2^2 n_2^2 + \sigma_3^2 n_3^2. \tag{2.87}$$

By the Pythagorean theorem,

$$\begin{aligned} \tau^2 &= |\mathbf{p}|^2 - \sigma^2 = \sigma_1^2 n_1^2 + \sigma_2^2 n_2^2 + \sigma_3^2 n_3^2 - (\sigma_1 n_1^2 + \sigma_2 n_2^2 + \sigma_3 n_3^2)^2 \\ &= (\sigma_1 - \sigma_2)^2 n_1^2 n_2^2 + (\sigma_2 - \sigma_3)^2 n_2^2 n_3^2 + (\sigma_3 - \sigma_1)^2 n_3^2 n_1^2. \end{aligned} \tag{2.88}$$

To find the local maximum and minimum values of the shear stress magnitude, we must optimize τ^2 subject to the constraint

$$f(n_1, n_2, n_3) = n_1^2 + n_2^2 + n_3^2 = 1. \tag{2.89}$$

By Lagrange's theorem, the stationary values of τ^2 occur where the gradient of τ^2 is parallel to the gradient of f; this leads to the following three scalar equations:

$$[(\sigma_1 - \sigma_2)^2 n_2^2 + (\sigma_1 - \sigma_3)^2 n_3^2] n_1 = c n_1, \tag{2.90}$$

$$[(\sigma_1 - \sigma_2)^2 n_1^2 + (\sigma_2 - \sigma_3)^2 n_3^2] n_2 = c n_2, \tag{2.91}$$

$$[(\sigma_2 - \sigma_3)^2 n_2^2 + (\sigma_1 - \sigma_3)^2 n_1^2] n_3 = c n_3, \tag{2.92}$$

where c is a Lagrange multiplier, the value of which is not needed in the present discussion.

Equation (2.90) can be satisfied by picking $n_1 = 0$, after which (2.91) and (2.92) show that $n_3^2 = n_2^2$. Invoking the normalization constraint (2.89) then shows that

$n_3^2 = n_2^2 = 1/2$. Hence, one family of planes on which τ will take on extreme values will correspond to

$$\mathbf{n} = (0, \pm 1/\sqrt{2}, \pm 1/\sqrt{2}). \tag{2.93}$$

On these planes, (2.88) shows that $|\tau| = (\sigma_2 - \sigma_3)/2$, and (2.86) shows that $\sigma = (\sigma_2 + \sigma_3)/2$. Another set of solutions to (2.90)–(2.92) is given by

$$\mathbf{n} = (\pm 1/\sqrt{2}, 0, \pm 1/\sqrt{2}), \tag{2.94}$$

on which planes $|\tau| = (\sigma_1 - \sigma_3)/2$ and $\sigma = (\sigma_1 + \sigma_3)/2$. Finally, a third set of solutions is

$$\mathbf{n} = (\pm 1/\sqrt{2}, \pm 1/\sqrt{2}, 0), \tag{2.95}$$

on which planes $|\tau| = (\sigma_1 - \sigma_2)/2$ and $\sigma = (\sigma_1 + \sigma_2)/2$. These extreme values of the shear stress are denoted by τ_1, τ_2, and τ_3, respectively. As the shear stresses are zero on the planes of the principal normal stresses, the extreme values found above are local maxima.

Thus far, we have found the magnitudes of the maximum shear stresses and the unit normals of the planes on which they act. To find the *directions within those planes* in which the shears act, consider one of these planes, and let **t** be the unit vector pointing in the direction of this shear traction. We decompose the total traction vector into a shear and a normal component, as follows:

$$\mathbf{p} = \sigma\mathbf{n} + \tau\mathbf{t}. \tag{2.96}$$

Solving (2.96) for **t**, and making use of (2.85), yields

$$\mathbf{t} = \frac{1}{\tau}(\mathbf{p} - \sigma\mathbf{n}) = [(\sigma_1 - \sigma)n_1/\tau \quad (\sigma_2 - \sigma)n_2/\tau \quad (\sigma_3 - \sigma)n_3/\tau]^{\mathrm{T}}, \tag{2.97}$$

in which τ, σ, and **n** are one set of the solutions to (2.90)–(2.92) described above. For example, on the plane whose unit normal vector is given by $\mathbf{n} = (0, 1/\sqrt{2}, 1/\sqrt{2})$, we have $\tau = (\sigma_2 - \sigma_3)/2$ and $\sigma = (\sigma_2 + \sigma_3)/2$, and (2.97) yields $\mathbf{t} = (0, 1/\sqrt{2}, -1/\sqrt{2})$.

2.6 Stress transformations in three dimensions

If the stress components are known in a given coordinate system, the stress components in a second coordinate system that is rotated from the first one can be found by matrix multiplication, as in (2.30). This was proven in a two-dimensional context in §2.3, and will now be generalized to three dimensions.

Consider a coordinate system (x, y, z), as in Fig. 2.8a, and another coordinate system (x', y', z') that is rotated by some angle in three-dimensional space. Let the direction cosines of the unit vector $\mathbf{e}_{x'}$, relative to the *unprimed* coordinate system, be (l_{11}, l_{12}, l_{13}), and similarly for the other two unit vectors $\mathbf{e}_{y'}$ and $\mathbf{e}_{z'}$, that is,

$$\mathbf{e}_{x'} = (l_{11}, l_{12}, l_{13}), \quad \mathbf{e}_{y'} = (l_{21}, l_{22}, l_{23}), \quad \mathbf{e}_{z'} = (l_{31}, l_{32}, l_{33}). \tag{2.98}$$

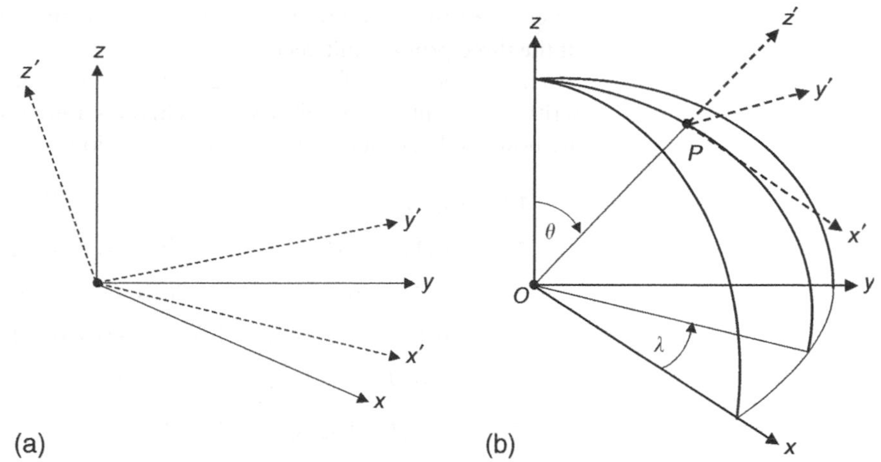

Fig. 2.8 (a) Original (unprimed) coordinate system, along with another rotated (primed) coordinate system, and (b) a system that utilizes the zenith and longitudinal angles.

Now consider the traction on the plane whose outward unit normal vector is $\mathbf{e}_{\mathbf{x}'}$. By (2.66), this traction is given by $\mathbf{p}(\mathbf{e}_{\mathbf{x}'}) = \boldsymbol{\tau}\mathbf{e}_{\mathbf{x}'}$, where $\boldsymbol{\tau}$ is the stress tensor in the unprimed coordinate system. The component of this traction in the $\mathbf{e}'_{\mathbf{x}}$ direction, which by definition is $\tau_{x'x'}$, is found, as in (2.6), by taking the inner product of $\mathbf{p}(\mathbf{e}_{\mathbf{x}'})$ with $\mathbf{e}_{\mathbf{x}'}$, that is,

$$\begin{aligned}\tau_{x'x'} &= \mathbf{e}_{\mathbf{x}'}\cdot\mathbf{p}(\mathbf{e}_{\mathbf{x}'}) = (\mathbf{e}_{\mathbf{x}'})^{\mathrm{T}}\mathbf{p}(\mathbf{e}_{\mathbf{x}'}) = (\mathbf{e}_{\mathbf{x}'})^{\mathrm{T}}\boldsymbol{\tau}\mathbf{e}_{\mathbf{x}'}\\ &= \begin{bmatrix} l_{11} & l_{12} & l_{13}\end{bmatrix}\begin{bmatrix}\tau_{xx} & \tau_{xy} & \tau_{xz}\\ \tau_{yx} & \tau_{yy} & \tau_{yz}\\ \tau_{zx} & \tau_{zy} & \tau_{zz}\end{bmatrix}\begin{bmatrix} l_{11}\\ l_{12}\\ l_{13}\end{bmatrix}\\ &= l_{11}^2\tau_{xx} + l_{11}l_{12}\tau_{xy} + l_{11}l_{13}\tau_{xz} + l_{12}l_{11}\tau_{yx} + l_{12}^2\tau_{yy}\\ &\quad + l_{12}l_{13}\tau_{yz} + l_{13}l_{11}\tau_{zx} + l_{13}l_{12}\tau_{zy} + l_{13}^2\tau_{zz}.\end{aligned} \tag{2.99}$$

Although use of the symmetry properties of the stress tensor would simplify (2.99) slightly, it would obscure the algebraic structure of the transformation equations, and is not introduced at this point. The other eight components of the stress tensor in the primed coordinate system can be found in the same manner; for example, $\tau_{x'y'} = \mathbf{e}_{\mathbf{y}'}\cdot\mathbf{p}(\mathbf{e}_{\mathbf{x}'})$, etc. The resulting equations, each of the form (2.99), can then be written in matrix form as

$$\begin{bmatrix}\tau_{x'x'} & \tau_{x'y'} & \tau_{x'z'}\\ \tau_{y'x'} & \tau_{y'y'} & \tau_{y'z'}\\ \tau_{z'x'} & \tau_{z'y'} & \tau_{z'z'}\end{bmatrix} = \begin{bmatrix} l_{11} & l_{12} & l_{13}\\ l_{21} & l_{22} & l_{23}\\ l_{31} & l_{32} & l_{33}\end{bmatrix}\begin{bmatrix}\tau_{xx} & \tau_{xy} & \tau_{xz}\\ \tau_{yx} & \tau_{yy} & \tau_{yz}\\ \tau_{zx} & \tau_{zy} & \tau_{zz}\end{bmatrix}\begin{bmatrix} l_{11} & l_{21} & l_{31}\\ l_{12} & l_{22} & l_{32}\\ l_{13} & l_{23} & l_{33}\end{bmatrix}. \tag{2.100}$$

This can be verified by carrying out the matrix multiplications indicated in (2.100), and comparing the results to (2.99). Equation (2.100) can be written symbolically as

$$\boldsymbol{\tau}' = \mathbf{L}\boldsymbol{\tau}\mathbf{L}^{\mathrm{T}}, \tag{2.101}$$

where the rows of the rotation matrix **L** are formed from the direction cosines of the three primed unit vectors.

The components of the stress tensor in the primed coordinate system can be written out explicitly as follows, wherein the symmetry properties of the stresses are now used in order to simplify the expressions:

$$\tau_{x'x'} = l_{11}^2\tau_{xx} + l_{12}^2\tau_{yy} + l_{13}^2\tau_{zz} + 2l_{11}l_{12}\tau_{xy} + 2l_{11}l_{13}\tau_{xz} + 2l_{12}l_{13}\tau_{yz}, \tag{2.102}$$

$$\tau_{y'y'} = l_{21}^2\tau_{xx} + l_{22}^2\tau_{yy} + l_{23}^2\tau_{zz} + 2l_{21}l_{22}\tau_{xy} + 2l_{21}l_{23}\tau_{xz} + 2l_{22}l_{23}\tau_{yz}, \tag{2.103}$$

$$\tau_{z'z'} = l_{31}^2\tau_{xx} + l_{32}^2\tau_{yy} + l_{33}^2\tau_{zz} + 2l_{31}l_{32}\tau_{xy} + 2l_{31}l_{33}\tau_{xz} + 2l_{32}l_{33}\tau_{yz}, \tag{2.104}$$

$$\tau_{x'y'} = l_{11}l_{21}\tau_{xx} + l_{12}l_{22}\tau_{yy} + l_{13}l_{23}\tau_{zz} + (l_{11}l_{22} + l_{12}l_{21})\tau_{xy} + (l_{12}l_{23} + l_{13}l_{22})\tau_{yz} + (l_{11}l_{23} + l_{13}l_{21})\tau_{xz} \tag{2.105}$$

$$\tau_{y'z'} = l_{21}l_{31}\tau_{xx} + l_{22}l_{32}\tau_{yy} + l_{23}l_{33}\tau_{zz} + (l_{21}l_{32} + l_{22}l_{31})\tau_{xy} + (l_{22}l_{33} + l_{23}l_{32})\tau_{yz} + (l_{21}l_{33} + l_{23}l_{31})\tau_{xz} \tag{2.106}$$

$$\tau_{x'z'} = l_{11}l_{31}\tau_{xx} + l_{12}l_{32}\tau_{yy} + l_{13}l_{33}\tau_{zz} + (l_{11}l_{32} + l_{12}l_{31})\tau_{xy} + (l_{12}l_{33} + l_{13}l_{32})\tau_{yz} + (l_{11}l_{33} + l_{13}l_{31})\tau_{xz} \tag{2.107}$$

It follows from these equations that the stress components in any particular plane transform according to the two-dimensional transformation laws of §2.3. For example, consider a rotation about the z-axis, which corresponds to a rotation matrix in which $l_{13} = l_{23} = l_{31} = l_{32} = 0$, and $l_{33} = 1$. Equations (2.102), (2.103), and (2.105) then essentially reduce to (2.107)–(2.109), aside from differences in notation. The maximum and minimum normal stresses in such a plane are often called the *subsidiary principal stresses*. Except in the special case in which z is a principal direction, these two subsidiary principal stresses will not correspond to two of the actual three-dimensional principal stresses, however.

If the unprimed coordinate system is the coordinate system of principal stresses, (2.102)–(2.107) take the form

$$\tau_{x'x'} = l_{11}^2\tau_{xx} + l_{12}^2\tau_{yy} + l_{13}^2\tau_{zz}, \tag{2.108}$$

$$\tau_{y'y'} = l_{21}^2\tau_{xx} + l_{22}^2\tau_{yy} + l_{23}^2\tau_{zz}, \tag{2.109}$$

$$\tau_{z'z'} = l_{31}^2\tau_{xx} + l_{32}^2\tau_{yy} + l_{33}^2\tau_{zz}, \tag{2.110}$$

$$\tau_{x'y'} = l_{11}l_{21}\tau_{xx} + l_{12}l_{22}\tau_{yy} + l_{13}l_{23}\tau_{zz}, \tag{2.111}$$

$$\tau_{y'z'} = l_{21}l_{31}\tau_{xx} + l_{22}l_{32}\tau_{yy} + l_{23}l_{33}\tau_{zz}, \tag{2.112}$$

$$\tau_{x'z'} = l_{11}l_{31}\tau_{xx} + l_{12}l_{32}\tau_{yy} + l_{13}l_{33}\tau_{zz}. \tag{2.113}$$

In some situations, it is more convenient to use the longitude angle, λ, and zenith angle, θ, to specify a plane, rather than the direction cosines of that plane relative to a particular Cartesian coordinate system (Fig. 2.8b). The axes associated with the primed coordinate system are Pz', which is in the radial direction; Px', which is in the plane OPz and is associated with the angle θ; and Py', which is chosen so as to complete the right-handed coordinate system, and which points in the direction of increasing λ. The components of these unit vectors, relative to the unprimed coordinate system, will be given by (2.98), which we now express in terms of λ and θ. It can be seen directly from Fig. 2.8b that the direction cosines of Pz' are given by

$$\mathbf{e}_{\mathbf{z}'} = (\sin\theta\cos\lambda, \sin\theta\sin\lambda, \cos\theta). \tag{2.114}$$

The components of the unit vector through O parallel to Px' can be found by replacing θ with $\theta + (\pi/2)$, which leads to

$$\mathbf{e}_{\mathbf{x}'} = (\cos\theta\cos\lambda, \cos\theta\sin\lambda, -\sin\theta). \tag{2.115}$$

Finally, a unit vector through O parallel to Py' is perpendicular to Oz, and makes an angle $\lambda + (\pi/2)$ with Ox, and an angle λ with Oy. Therefore,

$$\mathbf{e}_{\mathbf{y}'} = (-\sin\lambda,\ \ \cos\lambda,\ \ 0). \tag{2.116}$$

If the axes (x, y, z) are the principal axes, then the stress components on the plane $Px'y'$ are found from (2.112), (2.114), and (2.115) to be given by

$$\tau_{z'z'} = [\sigma_1\cos^2\lambda + \sigma_2\sin^2\lambda]\sin^2\theta + \sigma_3\cos^2\theta, \tag{2.117}$$

$$\tau_{y'z'} = -\frac{1}{2}(\sigma_1 - \sigma_2)\sin\theta\sin 2\lambda, \tag{2.118}$$

$$\tau_{x'z'} = \frac{1}{2}[\sigma_1\cos^2\lambda + \sigma_2\sin^2\lambda - \sigma_3]\sin 2\theta. \tag{2.119}$$

These relations have been used, for example, by Bott (1959) for defining tectonic regimes.

2.7 Mohr's representation of stress in three dimensions

Mohr's circle representation of two-dimensional stress states, described in §2.4, can be also be used in three dimensions. To simplify the notation, we denote the components of the unit normal vector of an arbitrary plane, relative to the principal coordinate system, by (l, m, n), rather than (n_1, n_2, n_3), and we denote the three principal directions by (x, y, z). By (2.86) and (2.88), the normal and shear tractions acting on this plane are

$$\sigma = l^2\sigma_1 + m^2\sigma_2 + n^2\sigma_3, \tag{2.120}$$

$$\tau^2 = l^2\sigma_1 + m^2\sigma_2 + n^2\sigma_3 - \sigma^2, \tag{2.121}$$

where

$$l^2 + m^2 + n^2 = 1. \tag{2.122}$$

Solving these equations for the direction cosines yields

$$l^2 = \frac{(\sigma_2 - \sigma)(\sigma_3 - \sigma) + \tau^2}{(\sigma_2 - \sigma_1)(\sigma_3 - \sigma_1)}, \tag{2.123}$$

$$m^2 = \frac{(\sigma_3 - \sigma)(\sigma_1 - \sigma) + \tau^2}{(\sigma_3 - \sigma_2)(\sigma_1 - \sigma_2)}, \tag{2.124}$$

$$n^2 = \frac{(\sigma_1 - \sigma)(\sigma_2 - \sigma) + \tau^2}{(\sigma_1 - \sigma_3)(\sigma_2 - \sigma_3)}. \tag{2.125}$$

Suppose that the direction cosine n is fixed, so that the normal vector to the plane makes a fixed angle $\theta = \arccos(n)$ with the z-axis, as in Fig. 2.9a. The intersection of the normal vector with the unit sphere will lie on the small circle $F'E'D'$. Equation (2.125) can be rearranged to yield

$$\tau^2 + \left[\sigma - \frac{1}{2}(\sigma_1 + \sigma_2)\right]^2 = \frac{1}{4}(\sigma_1 - \sigma_2) + n^2(\sigma_1 - \sigma_3)(\sigma_2 - \sigma_3), \tag{2.126}$$

which is the equation of a circle in the (σ, τ) plane. The center of this circle, A, is located at

$$\sigma = \frac{1}{2}(\sigma_1 + \sigma_2), \quad \tau = 0, \tag{2.127}$$

and the radius of this circle is

$$r = \left[\frac{1}{4}(\sigma_1 - \sigma_2) + n^2(\sigma_1 - \sigma_3)(\sigma_2 - \sigma_3)\right]^{1/2}. \tag{2.128}$$

As n varies from 0 to 1, the radius varies from $AQ = (\sigma_1 - \sigma_2)/2$ to $AR = [(\sigma_1 + \sigma_2)/2] - \sigma_3$. A typical circle for an intermediate value of n is shown in Fig. 2.9b as DEF.

In the same manner, holding l constant in (2.123) gives the family of circles

$$\tau^2 + \left[\sigma - \frac{1}{2}(\sigma_2 + \sigma_3)\right]^2 = \frac{1}{4}(\sigma_2 - \sigma_3) + l^2(\sigma_2 - \sigma_1)(\sigma_3 - \sigma_1). \tag{2.129}$$

The centers of these circles are at B, located at $\sigma = (\sigma_2 + \sigma_3)/2, \tau = 0$. The radii vary from $BQ = (\sigma_2 - \sigma_3)/2$ when $l = 0$, to $BP = \sigma_1 - [(\sigma_2 + \sigma_3)/2]$ when $l = 1$. A typical circle for intermediate values of l is GEH. In Fig. 2.9b, planes for which l = constant lie on a cone that makes an angle $\phi = \arccos(l)$ with the x-axis, and which intersects the unit sphere at $G'E'H'$.

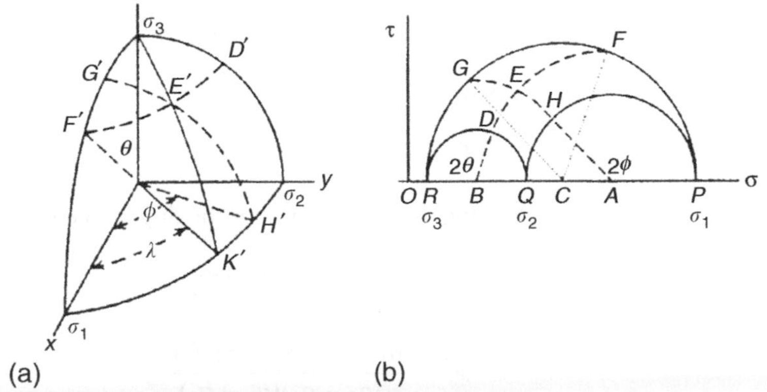

Fig. 2.9 Mohr's circle in three dimensions (see text for description).

Finally, holding m constant in (2.124) gives the family of circles

$$\tau^2 + \left[\sigma - \frac{1}{2}(\sigma_1 + \sigma_3)\right]^2 = \frac{1}{4}(\sigma_1 - \sigma_3) + m^2(\sigma_3 - \sigma_2)(\sigma_1 - \sigma_2). \qquad (2.130)$$

The centers of these circles are at C, located at $\sigma = (\sigma_1 + \sigma_3)/2$, $\tau = 0$. The radii vary from $CR = (\sigma_1 - \sigma_3)/2$, when $m = 0$, to $CQ = [(\sigma_1 + \sigma_3)/2] - \sigma_2$, when $m = 1$. These circles are not needed for the following analysis, and so are not shown in Fig. 2.9b.

Now consider a generic point on the unit sphere in Fig. 2.9a, such as E', which has direction cosines $l = \cos\phi$ and $n = \cos\theta$. The traction on this plane is indicated by the point E in Fig. 2.9b, which is at the intersection of the arc DEF defined by (2.126) with $n = \cos\theta$, and the arc GEH defined by (2.129) with $l = \cos\phi$. The points at which these two arcs intersect the two smaller circles in Fig. 2.9b are found as follows. The intersection point D corresponds to the point D' in Fig. 2.9a, for which $l = 0$, $m = \sin\theta$ and $n = \cos\theta$. Hence, by (2.120) and (2.121),

$$\sigma(D) = \sigma_2 \sin^2\theta + \sigma_3 \cos^2\theta = \frac{1}{2}(\sigma_2 + \sigma_3) - \frac{1}{2}(\sigma_2 - \sigma_3)\cos 2\theta, \qquad (2.131)$$

$$|\tau(D)| = [\sigma_2^2 \sin^2\theta + \sigma_3^2 \cos^2\theta - \sigma^2]^{1/2} = \frac{1}{2}(\sigma_2 - \sigma_3)\sin 2\theta. \qquad (2.132)$$

But (2.131) and (2.132) are essentially the equations of Mohr's circle in the (σ_1, σ_2) plane, and so it follows that $\angle RBD = 2\theta$. A similar argument shows that $\angle HAP = 2\phi$. Therefore, we can construct a diagram such as Fig. 2.10a, from which the normal and shear tractions acting on any plane can be found by locating the intersections of the proper circles.

The discussion given above dealt only with the magnitude of τ, and the plane on which it acts. The line of action of the shear traction can be found by the following construction (Zizicas, 1955). Consider the latitude λ of the point E' in Fig. 2.9a. From (2.77), on the great circle for which λ is constant,

$$l^2 = (1 - n^2)\cos^2\lambda. \qquad (2.133)$$

Using (2.123) and (2.125), it can be shown that (2.133) corresponds to the following locus in the (σ, τ) plane:

$$(\sigma_2 - \sigma_3)[(\sigma_2 - \sigma)(\sigma_3 - \sigma) + \tau^2] + (\sigma_2 - \sigma_1)[(\sigma_1 - \sigma_3)(\sigma_2 - \sigma_3) - (\sigma_1 - \sigma)(\sigma_2 - \sigma) - \tau^2]\cos^2\lambda = 0, \qquad (2.134)$$

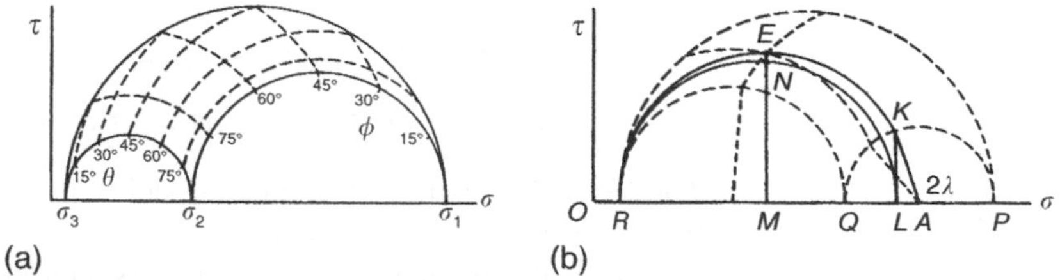

Fig. 2.10 Mohr's circle in three dimensions (see text for description).

which is a circle that passes through the point $R = (\sigma_3, 0)$. When $n = 0$, which corresponds to the point K' in Fig. 2.9b, we have $l = \cos\lambda$ and $m = \sin\lambda$. It then follows from (2.120) and (2.121), and from an argument along the lines that led to (2.131) and (2.132), that

$$\sigma(K) = \frac{1}{2}(\sigma_1 + \sigma_2) + \frac{1}{2}(\sigma_1 - \sigma_2)\cos 2\lambda, \tag{2.135}$$

$$|\tau(K)| = \frac{1}{2}(\sigma_1 - \sigma_2)\sin 2\lambda. \tag{2.136}$$

Therefore, the point K lies on the circle PQ of Fig. 2.10b, and $\angle KAP = 2\lambda$. The fact that this circle passes through the points R and K, and has its center on the σ-axis, is sufficient to unambiguously define it. It must pass through the point E, which is defined by the intersection of circles GEH and DEF of Fig. 2.9b. The lines of the construction of Fig. 2.9b are shown as dotted lines in Fig. 2.10b.

Consider now the stresses in the plane $OK'Z$ of Fig. 2.9a. The normal stresses in the directions OZ and OK' are given by the lengths of the line segments OR $(= \sigma_3)$ and OL in Fig. 2.10b. Therefore, the Mohr's circle for this plane has RL as its diameter. When the normal stress is OM, corresponding to the point E', the shear stress is NM. We have thus found the component of the shear stress in the plane $OK'Z$, and the total shear stress EM must make an angle $\cos^{-1}(NM/EM)$ with that plane. This analysis gives the line of action of the shear traction, but there is still an ambiguity of sign, since our calculation has only made use of the magnitude of τ. Determination of the actual signed orientation of the shear traction, using Mohr's circle, has been discussed by Almusallam and Taher (1995).

The calculations described above, along with Figs. 2.9a and 2.10b, have been carried out using latitude and longitude coordinates, with the σ_3-axis as the polar direction. A similar construction could be made using either of the other two principal directions as the poles.

There are numerous other applications of the Mohr's circle construction that are useful in rock and soil mechanics. Parry (1995) makes extensive use of Mohr's circle to discuss failure of geological materials. Werfel (1965) shows how to use Mohr's circle to determine the principal axes and the magnitudes of the principal stresses.

2.8 Stress invariants and stress deviation

In §2.5, the coefficients that appear in the characteristic polynomial that defines the three principal stresses were identified as the three stress invariants:

$$I_1 = \tau_{xx} + \tau_{yy} + \tau_{zz}, \tag{2.137}$$

$$I_2 = \tau_{xy}^2 + \tau_{xz}^2 + \tau_{yz}^2 - \tau_{xx}\tau_{yy} - \tau_{xx}\tau_{zz} - \tau_{yy}\tau_{zz}, \tag{2.138}$$

$$I_3 = \tau_{xx}\tau_{yy}\tau_{zz} + 2\tau_{xy}\tau_{xz}\tau_{yz} - \tau_{xx}\tau_{yz}^2 - \tau_{yy}\tau_{xz}^2 - \tau_{zz}\tau_{xy}^2. \tag{2.139}$$

The second invariant, I_2, is sometimes defined to be the *negative* of that defined by (2.138), in which case $+I_2$, rather than $-I_2$, would appear as the coefficient of σ in (2.69). It was argued in §2.5 that these coefficients must have the same value in all coordinate systems, since the three principal stresses, which are functions only of these three invariants, and which have a clear physical meaning that is independent of the choice of coordinate system, must themselves be invariant. A rigorous proof of this assertion is given by Spencer (1971). In fact, there are many combinations of stresses that form an invariant of the stress tensor, but all are dependent upon the set $\{I_1, I_2, I_3\}$. These three invariants can be defined in the terminology of matrix algebra by (Spencer, 1980)

$$I_1 = \mathrm{trace}(\boldsymbol{\tau}), \tag{2.140}$$

$$I_2 = \frac{1}{2}\{\mathrm{trace}(\boldsymbol{\tau}^2) - [\mathrm{trace}(\boldsymbol{\tau})]^2\}, \tag{2.141}$$

$$I_3 = \det(\boldsymbol{\tau}), \tag{2.142}$$

where the trace operator denotes the sum of the diagonal components, and the determinant is defined in the standard manner (Lang, 1971).

Since all shear stresses, which are the off-diagonal components of the stress matrix, vanish in the principal coordinate system, the invariants take on a simpler form when expressed in that coordinate system:

$$I_1 = \sigma_1 + \sigma_2 + \sigma_3, \tag{2.143}$$

$$I_2 = -(\sigma_1\sigma_2 + \sigma_2\sigma_3 + \sigma_1\sigma_3), \tag{2.144}$$

$$I_3 = \sigma_1\sigma_2\sigma_3. \tag{2.145}$$

Although I_1 has exactly the same algebraic form in all coordinate systems, it would be *incorrect* to apply (2.144) or (2.145) in a nonprincipal coordinate system, since these two equations only hold when all the shear stresses are zero. For example, it is not true in general that $I_3 = \tau_{xx}\tau_{yy}\tau_{zz}$. Among the many useful relations between the stress components that can be derived with the aid of these invariants is the following:

$$\tau_{xx}^2 + \tau_{yy}^2 + \tau_{zz}^2 + 2\tau_{xy}^2 + 2\tau_{xz}^2 + 2\tau_{yz}^2 = \sigma_1^2 + \sigma_2^2 + \sigma_3^2. \tag{2.146}$$

The stress invariants are useful in the construction of stress–strain laws and failure criteria, as will be seen in Chapters 5 and 9.

Another related set of quantities that appear frequently in the development of constitutive equations for rocks is the *octahedral stresses*. These are the normal and shear tractions that act on planes whose outward unit normal vectors, in the principal coordinate system, are given by

$$(n_1, n_2, n_3) = (\pm 1/\sqrt{3},\ \pm 1/\sqrt{3},\ \pm 1/\sqrt{3}). \tag{2.147}$$

These planes are equally inclined to each of the three principal directions. There are eight such planes, each associated with a different set of choices for the signs appearing in (2.147). These eight planes are each parallel to a side of an octahedron whose vertices are located on the principal axes (Nadai, 1950, p. 105). From (2.86), the normal traction acting on any octahedral plane is

$$\sigma_{\mathrm{oct}} = \sigma_1 n_1^2 + \sigma_2 n_2^2 + \sigma_3 n_3^2 = \frac{1}{3}(\sigma_1 + \sigma_2 + \sigma_3) = \frac{1}{3}I_1. \tag{2.148}$$

By (2.88), the magnitude of the shear traction acting on an octahedral plane is found to be given by

$$|\tau_{\text{oct}}| = \frac{1}{3}\{(\sigma_1 - \sigma_2)^2 + (\sigma_2 - \sigma_3)^2 + (\sigma_3 - \sigma_3)^2\}^{1/2} = \frac{\sqrt{2}}{3}\{I_1^2 + 3I_2\}^{1/2}. \quad (2.149)$$

The octahedral shear stress has the interesting physical interpretation of being equal to $\sqrt{(5/3)}$ times the root-mean-square shear stress, with all planes weighted equally.

Shear and normal stresses have different physical consequences, as the former act tangentially to a plane, and the latter act normal to it. The stress tensor, when written in a given coordinate system, explicitly contains only those shear stress components that act on planes whose normals are perpendicular to one of the three coordinate directions. Hence, when written in the principal coordinate system, the stress tensor seems to contain no shear stresses. However, in general it would be wrong to assume that no shear stresses are acting at that point, since the shears may be nonzero on oblique planes. Only in the special case in which all three principal stresses are *equal* is it true that the shear stresses on all planes are zero. This is most easily seen by referring to Mohr's circle, Fig. 2.9b, since in this case all three circles shrink down to a single point on the σ-axis.

It would be useful to have a way of representing the stress tensor that clearly showed whether or not there are *any* shear stresses acting at the point in question. To do this we decompose the stress tensor into an *isotropic* (or *hydrostatic*) part and a *deviatoric* part. The isotropic part of the stress is defined as

$$\tau^{\text{iso}} = \frac{1}{3} I_1 \mathbf{I} \equiv \tau_{\text{m}} \mathbf{I}, \quad (2.150)$$

where $\mathbf{I}$ is the 3×3 identity tensor, and τ_{m} is the mean value of the three principal stresses, or the *mean normal stress*. The mean normal stress is important in thermodynamic treatments of material deformation (McLellan, 1980), where it is analogous to the pressure that acts in a fluid. The deviatoric stress is obtained by subtracting the isotropic part of the stress tensor from the full stress tensor:

$$\mathbf{s} \equiv \boldsymbol{\tau}^{\text{dev}} = \boldsymbol{\tau} - \boldsymbol{\tau}^{\text{iso}} = \boldsymbol{\tau} - \tau_{\text{m}} \mathbf{I}. \quad (2.151)$$

These equations can be written out explicitly as

$$\mathbf{s} = \begin{bmatrix} \tau_{xx} - \tau_{\text{m}} & \tau_{xy} & \tau_{xz} \\ \tau_{yx} & \tau_{yy} - \tau_{\text{m}} & \tau_{yz} \\ \tau_{zx} & \tau_{zy} & \tau_{zz} - \tau_{\text{m}} \end{bmatrix}, \quad (2.152)$$

$$\tau_{\text{m}} = \frac{1}{3}(\tau_{xx} + \tau_{yy} + \tau_{zz}). \quad (2.153)$$

The principal stress deviations are found from solving the eigenvalue problem $\mathbf{sn} = s\mathbf{n}$, or $(\mathbf{s} - s\mathbf{I})\mathbf{n} = \mathbf{0}$, as in §2.5. But since $\mathbf{s} = \boldsymbol{\tau} - \tau_{\text{m}}\mathbf{I}$, the eigenvalue problem for the principal stress deviators is identical to that for the principal stresses themselves, except that the eigenvalue is now $s + \tau_{\text{m}}$ instead of σ. Hence, the principal directions of the deviatoric stress are the same as the principal stress directions, and the principal deviatoric stresses are

$$s_1 = \sigma_1 - \tau_m = (2\sigma_1 - \sigma_2 - \sigma_3)/3, \quad (2.154)$$

$$s_2 = \sigma_2 - \tau_m = (2\sigma_2 - \sigma_1 - \sigma_3)/3, \quad (2.155)$$

$$s_3 = \sigma_3 - \tau_m = (2\sigma_3 - \sigma_1 - \sigma_2)/3. \quad (2.156)$$

The usefulness of this decomposition arises from the fact, (5.7), that in the elastic range of deformation, the mean stress τ_m controls the volumetric change of a rock, whereas the deviatoric stress **s** controls the distortion. Moreover, many of the criteria for failure are concerned primarily with distortion, in which case these criteria are most conveniently expressed in terms of the invariants of the stress deviation. In an arbitrary coordinate system, these invariants take the form

$$J_1 = s_{xx} + s_{yy} + s_{zz} = 0, \quad (2.157)$$

$$J_2 = s_{xy}^2 + s_{xz}^2 + s_{yz}^2 - s_{xx}s_{yy} - s_{xx}s_{zz} - s_{yy}s_{zz}, \quad (2.158)$$

$$J_3 = s_{xx}s_{yy}s_{zz} + 2s_{xy}s_{xz}s_{yz} - s_{xx}s_{yz}^2 - s_{yy}s_{xz}^2 - s_{zz}s_{xy}^2. \quad (2.159)$$

The first invariant of the deviatoric stress is always identically zero. The other two can be written in terms of the principal deviatoric stresses as follows:

$$J_2 = -(s_1s_2 + s_1s_3 + s_2s_3), \quad (2.160)$$

$$J_3 = s_1s_2s_3. \quad (2.161)$$

Algebraic manipulation of the previous equations leads to the following alternative forms for J_2, which is the invariant that appears most often in failure criteria (Westergaard, 1952, p. 70):

$$J_2 = \frac{1}{2}(\tau_{xx}^2 + \tau_{yy}^2 + \tau_{zz}^2) + \tau_{xy}^2 + \tau_{yz}^2 + \tau_{xz}^2, \quad (2.162)$$

$$= \frac{1}{6}\left[(\tau_{xx} - \tau_{yy})^2 + (\tau_{xx} - \tau_{zz})^2 + (\tau_{yy} - \tau_{zz})^2\right] + \tau_{xy}^2 + \tau_{yz}^2 + \tau_{xz}^2, \quad (2.163)$$

$$= \frac{1}{6}\left[(\sigma_1 - \sigma_2)^2 + (\sigma_2 - \sigma_3)^2 + (\sigma_3 - \sigma_1)^2\right], \quad (2.164)$$

$$= \frac{1}{2}(s_1^2 + s_2^2 + s_3^2), \quad (2.165)$$

$$= 3(\tau_m)^2 + I_2, \quad (2.166)$$

$$= \frac{3}{2}(\tau_{oct})^2. \quad (2.167)$$

2.9 Displacement and strain

In rock mechanics, as in the mechanics of particles and rigid bodies, the fundamental kinematic variable is the *displacement*, which is the vector that quantifies the change in the position of a given particle of rock. In the typical rock mechanics problem, the position of each rock particle is labeled by its location, relative to some coordinate system, in some state that is taken to be the "initial" state of the rock. This position can be denoted by $\mathbf{x} = (x, y, z)$. Loads are then applied to the rock, causing the rock particle that was initially located at point $\mathbf{x}$ to be displaced to a new position, $\mathbf{x}^* = (x^*, y^*, z^*)$. The vector that connects the original position $\mathbf{x}$ and the final position $\mathbf{x}^*$ is known as the "displacement of the particle that was initially at point $\mathbf{x}$," or simply the "displacement at $\mathbf{x}$." This vector is denoted in vector notation by $\mathbf{u}$, and its components are (u, v, w). To be consistent with the sign convention used for tractions, in which a traction component is represented by a positive number if it points in the *negative* coordinate direction, the displacement vector must be defined according to

$$\mathbf{x}^* = \mathbf{x} - \mathbf{u}, \quad \text{that is,} \quad x^* = x - u, \quad y^* = y - v, \quad z^* = z - w. \tag{2.168}$$

The displacement $\mathbf{u}$ can be interpreted as a vector that points *from* the new position, $\mathbf{x}^*$, *toward* the original position, $\mathbf{x}$ (Fig. 2.11a). In general, the displacement will vary from point to point, so that each component (u, v, w) will vary with all three position coordinates, x, y, and z.

The objective in solving a rock mechanics problem is to calculate the displacement vector $\mathbf{u}$ at every point in the rock mass or rock specimen, based on knowledge of the applied surface tractions and body forces, and the boundary conditions. To do this, it is necessary to introduce a set of intermediate quantities known as the *strains*. The reason, as will be seen in §5.2, is that the stresses are more directly related to the strains than to the displacements themselves. The use of strain rather than displacement is somewhat analogous to the use of stress, rather than the more physically obvious variable, the traction. Consequently, much of the theory of solid mechanics, and in particular elasticity, deals with stresses and strains, rather than tractions (forces) and displacements. However, in contrast to many other areas of solid mechanics, in which the displacements are often not of much intrinsic interest, in rock mechanics the displacement itself is often extremely important. Examples of such situations include mine closure, wellbore breakouts, and surface subsidence above mines and reservoirs.

Strain is essentially a measure of the relative displacement of nearby particles, rather than a measure of their absolute displacement. The basic concept behind

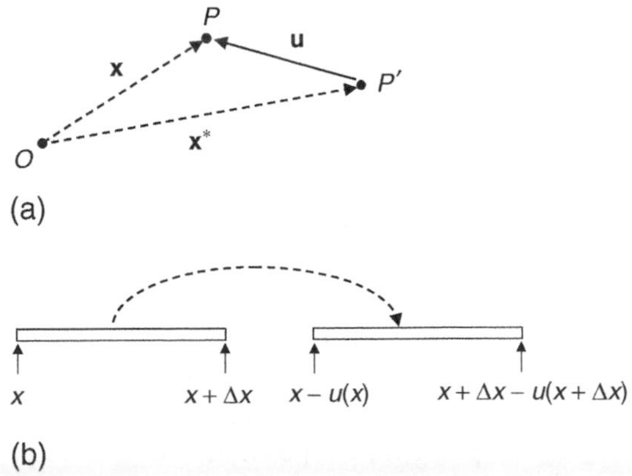

Fig. 2.11
(a) Displacement vector $\mathbf{u}$ of the piece of rock that is initially located at $\mathbf{x}$, (b) Generic displacement of a one-dimensional bar, used to define the normal strain.

the strain can be introduced in a one-dimensional context (Fig. 2.11b). Consider a short one-dimensional bar, initially of length L, whose left edge is initially located at point x, and whose right edge is located at point $x + \Delta x$. The initial length of this bar is given by $L = \Delta x$. This bar is now assumed to be deformed, such that the left edge of the bar moves to the location $x - u(x)$, and the right edge of the bar moves to the position $[x + \Delta x] - u(x + \Delta x)$. The new length of the bar is therefore equal to $L^* = \{[x + \Delta x] - u(x + \Delta x)\} - [x - u(x)]$. We now define the mean strain undergone by this bar as the *fractional decrease in the length of the bar*, that is,

$$\varepsilon = \frac{L-L^*}{L} = \frac{\Delta x - \{[x+\Delta x] - u(x+\Delta x) - [x - u(x)]\}}{\Delta x} = \frac{u(x+\Delta x) - u(x)}{\Delta x}. \tag{2.169}$$

According to this definition, the strain will be a positive number if the bar becomes shorter, and vice versa. Hence, positive strains represent contractions, and negative strains represent extensions. The strain *at the point* x is found by taking the limit of an infinitesimally short bar, which is mathematically equivalent to letting the initial length of the bar go to zero:

$$\varepsilon(x) = \lim_{L\to 0} \frac{L - L^*}{L} = \lim_{\Delta x \to 0} \frac{u(x + \Delta x) - u(x)}{\Delta x} \equiv \frac{du}{dx}. \tag{2.170}$$

The strain is therefore seen to be related to the spatial derivative of the displacement.

The type of strain described above, which is called a *normal strain*, can be generalized in an obvious way to two or three dimensions. However, in higher dimensions there are other types of strains, called the *shear strains*, which measure angular distortion, rather than stretching. The more general case of two- or three-dimensional strain is discussed in the following sections.

2.10 Infinitesimal strain in two dimensions

Consider a particle that is initially located at point $P = (x, y)$, as in Fig. 2.12a. Now consider a second particle that is initially located at $Q = (x + \Delta x, y)$, and a third particle located at $R = (x, y + \Delta x)$. The rock is then assumed to be deformed, such that these three particles move to positions P^*, Q^*, and R^*. The coordinates of these new locations are

$$\begin{aligned} P^* = P - \mathbf{u}(P) &= (x, y) - [u(x, y), v(x, y)] \\ &= [x - u(x, y), y - v(x, y)], \end{aligned} \tag{2.171}$$

$$\begin{aligned} Q^* = Q - \mathbf{u}(Q) &= (x + \Delta x, y) - [u(x + \Delta x, y), v(x + \Delta x, y)] \\ &= [x + \Delta x - u(x + \Delta x, y), y - v(x + \Delta x, y)], \end{aligned} \tag{2.172}$$

$$\begin{aligned} R^* = R - \mathbf{u}(R) &= (x, y + \Delta y) - [u(x, y + \Delta y), v(x, y + \Delta y)] \\ &= [x - u(x, y + \Delta y), y + \Delta y - v(x, y + \Delta y)]. \end{aligned} \tag{2.173}$$

Also shown in Fig. 2.12a are the points Q'', which is the projection of point Q^* onto the x-axis that passes through point P^*, and R'', which is the projection of point R^* onto the y-axis that passes through point P^*. For example, Q'' will have the same x-component as Q^* and the same y-component as P^* and likewise for R''.

We now express each of the displacements that appear in (2.171)–(2.173) as Taylor series taken about the point (x, y). As the increments Δx and Δy are infinitesimally small, all terms in the Taylor series higher than the first-order terms, which are linear in Δx and Δy, can be ignored. For example, $u(x, y + \Delta y) = u(x, y) + (\partial u/\partial y)\Delta y$, etc., where the partial derivative is understood to be evaluated at (x, y). The positions of the five points shown in Fig. 2.12a can therefore be expressed as

$$P^* = (x - u, y - v), \tag{2.174}$$

$$Q^* = (x + \Delta x - u - \frac{\partial u}{\partial x}\Delta x,\ y - v - \frac{\partial v}{\partial x}\Delta x), \tag{2.175}$$

$$R^* = (x - u - \frac{\partial u}{\partial y}\Delta y,\ y + \Delta y - v - \frac{\partial v}{\partial y}\Delta y) \tag{2.176}$$

$$Q'' = (x + \Delta x - u - \frac{\partial u}{\partial x}\Delta x,\ y - v), \tag{2.177}$$

$$R'' = (x - u,\ y + \Delta y - v - \frac{\partial v}{\partial y}\Delta y), \tag{2.178}$$

where it is understood that u, v, and their partial derivatives are evaluated at the point (x, y).

The normal strain in the x direction, which is denoted by ε_{xx}, is now defined, as in §2.9, as the fractional shortening of a line element that is initially oriented along the x-axis. In other words, the strain ε_{xx} at point (x, y) is equal to the fractional contraction undergone by the element PQ, in the limit as $\Delta x \to 0$. Initially, the length of element PQ, which we denote by $|PQ|$, is Δx. After deformation, the length $|P^*Q^*|$ is found using the Pythagorean theorem:

$$\begin{aligned} |P^*Q^*|^2 = |P^*Q''|^2 + |Q^*Q''|^2 &= \left(\Delta x - \frac{\partial u}{\partial x}\Delta x\right)^2 + \left(\frac{\partial v}{\partial x}\Delta x\right)^2 \\ &= (\Delta x)^2\left[1 - 2\frac{\partial u}{\partial x} + \left(\frac{\partial u}{\partial x}\right)^2 + \left(\frac{\partial v}{\partial x}\right)^2\right]. \end{aligned} \tag{2.179}$$

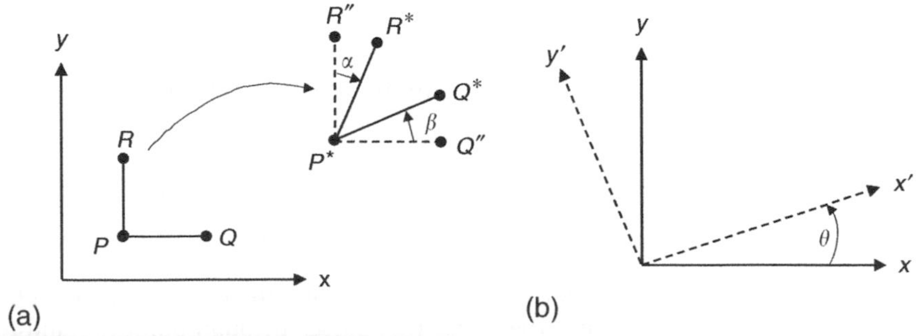

Fig. 2.12
(a) Displacement of two small line segments PQ and PR that are initially at right angles to each other. (b) New (primed) coordinate system rotated by angle θ from original (unprimed) coordinate system.

The crucial assumption is now made that all partial derivatives of the displacements are much smaller than unity, which is equivalent to restricting the displacements not to vary too abruptly from point to point. This assumption leads to the theory of *infinitesimal strain*, within which theory the strains are linear functions of the partial derivatives of the displacement components. This linearity is a tremendous aid in solving specific problems. Moreover, the strains that usually occur in situations of engineering interest, such as around boreholes, tunnels, mines, etc., as well as during processes such as seismic wave propagation, are in fact often very small. For these reasons, the theory of infinitesimal strain is almost universally used in rock mechanics analyses. The theory of finite strain, which is much more complicated, is nevertheless needed for some geological processes in which the small-strain assumption is not valid; this theory is briefly discussed in §2.15.

Under the assumption that the partial derivatives of the displacement are small, we neglect the squares of these derivatives in (2.179), and use the approximation that $(1-2\delta)^{1/2} \approx 1-\delta$ when $\delta \ll 1$, to find

$$\left|P^*Q^*\right| = \Delta x\left(1-\frac{\partial u}{\partial x}\right). \tag{2.180}$$

Hence, the normal strain in the x direction at the point (x, y) is given by

$$\varepsilon_{xx} = \lim_{\Delta x\to 0}\frac{|PQ|-|P^*Q^*|}{|PQ|} = \lim_{\Delta x\to 0}\frac{\Delta x-\Delta x(1-\partial u/\partial x)}{\Delta x} = \frac{\partial u}{\partial x}. \tag{2.181}$$

A similar analysis shows that the normal strain in the y direction, ε_{yy}, is given by

$$\varepsilon_{yy} = \lim_{\Delta x\to 0}\frac{|PR|-|P^*R^*|}{|PR|} = \lim_{\Delta x\to 0}\frac{\Delta y-\Delta y(1-\partial v/\partial y)}{\Delta y} = \frac{\partial v}{\partial y}. \tag{2.182}$$

There are other types of distortion, other than stretching or contraction, which can also be identified and quantified. Figure 2.12a shows that the three points $\{P, Q, R\}$ initially form a right angle, but, in the case shown, form an acute angle after deformation. The change in this angle is known as the *shear strain*. Specifically, the shear strain ε_{xy} is defined as one-half of the *increase* in the angle initially formed by two infinitesimal line segments that initially lie parallel to the x and y axes, that is,

$$\varepsilon_{xy} = \frac{1}{2}\lim_{\Delta x,\Delta y\to 0}\left(\angle R^*P^*Q^* - \angle RPQ\right). \tag{2.183}$$

From Fig. 2.12a we see that $\angle R^*P^*Q^* = 90° - \alpha - \beta$, whereas $\angle RPQ = 90°$, by construction. The angle α is calculated from

$$\tan\alpha = \frac{\left|R^*R''\right|}{|P^*R''|} = \frac{-(\partial u/\partial y)\Delta y}{\Delta y(1-(\partial v/\partial y))} \approx -\left(\frac{\partial u}{\partial y}\right), \tag{2.184}$$

where, due to the smallness of the partial derivatives, we have ignored the term $\partial v/\partial y$ in the denominator. But the smallness of the partial derivatives also allows us to approximate the angle α by $\tan\alpha$, so that $\alpha = -(\partial u/\partial y)$. Similarly, it can be shown that $\beta = -(\partial v/\partial x)$. Combining these results with (2.183) leads to

$$\varepsilon_{xy} = \frac{1}{2}\left(\frac{\partial u}{\partial y} + \frac{\partial v}{\partial x}\right). \tag{2.185}$$

The pattern embodied in (2.185) is that the shear strain ε_{xy} is equal to the mean of the partial derivative of the displacement in the x direction with respect to y and the partial derivative of the displacement in the y direction with respect to x. By this definition, we see that if a shear strain ε_{yx} is defined, it will necessarily be equal to ε_{xy}. These four strains can be thought of as the four components of the *strain matrix*, $\boldsymbol{\varepsilon}$:

$$\boldsymbol{\varepsilon} = \begin{bmatrix} \varepsilon_{xx} & \varepsilon_{xy} \\ \varepsilon_{yx} & \varepsilon_{yy} \end{bmatrix} = \begin{bmatrix} \dfrac{\partial u}{\partial x} & \dfrac{1}{2}\left(\dfrac{\partial u}{\partial y} + \dfrac{\partial v}{\partial x}\right) \\ \dfrac{1}{2}\left(\dfrac{\partial v}{\partial x} + \dfrac{\partial u}{\partial y}\right) & \dfrac{\partial v}{\partial y} \end{bmatrix}. \tag{2.186}$$

The shear strains are sometimes denoted by Γ_{xy} and Γ_{yx}. We will use the same basic symbol, ε, for all the strains, in order to emphasize that they are all components of the strain matrix. The shear strains will be distinguished by being the off-diagonal terms of this matrix, and therefore have mixed subscripts, whereas the normal strains are diagonal components, and have repeated subscripts. Another notation that had in the past been widely used, but is becoming obsolete, is to define "engineering" shear strains by $\gamma_{xy} = 2\varepsilon_{xy} = 2\Gamma_{xy}$. This definition seems to offer few advantages, and has the disadvantage that the matrix formed by the normal strains and the engineering shear strains does not constitute a tensor (see below).

The strain matrix is equal to the symmetric part of a matrix that is known as the *displacement gradient*, $\nabla\mathbf{u}$, whose components are the partial derivatives of the displacements with respect to the two coordinates, that is,

$$\boldsymbol{\varepsilon} = \mathrm{sym}(\nabla\mathbf{u}) \equiv \frac{1}{2}\left[\nabla\mathbf{u} + (\nabla\mathbf{u})^{\mathrm{T}}\right], \tag{2.187}$$

where

$$\nabla\mathbf{u} = \begin{bmatrix} \dfrac{\partial u}{\partial x} & \dfrac{\partial u}{\partial y} \\ \dfrac{\partial v}{\partial x} & \dfrac{\partial v}{\partial y} \end{bmatrix}. \tag{2.188}$$

The displacement gradient, and hence also the strain, is a second-order tensor, in the sense that if the coordinate system is rotated by an angle θ, the components of the strain in the new coordinate system are given by equations of the same form as (2.25)–(2.27). To prove this, consider a rotation of the coordinate system (Fig. 2.12b) such that the new x'-axis is rotated by counterclockwise angle θ from the original x-axis, that is,

$$x' = x\cos\theta + y\sin\theta, \quad y' = -x\sin\theta + y\cos\theta. \tag{2.189}$$

The inverse transformation, which is needed in the following calculations, is

$$x = x' \cos\theta - y' \sin\theta, \quad y = x' \sin\theta + y' \cos\theta. \tag{2.190}$$

As the displacement is also a vector, its components transform according to the same law:

$$u' = u\cos\theta + v\sin\theta, \quad v' = -u\sin\theta + v\cos\theta. \tag{2.191}$$

Now consider a component of the strain in the new coordinate system, such as, for example, $\varepsilon_{x'x'}$. Using the chain rule,

$$\begin{aligned}
\varepsilon_{x'x'} = \frac{\partial u'}{\partial x'} &= \frac{\partial u'}{\partial u}\frac{\partial u}{\partial x}\frac{\partial x}{\partial x'} + \frac{\partial u'}{\partial u}\frac{\partial u}{\partial y}\frac{\partial y}{\partial x'} + \frac{\partial u'}{\partial v}\frac{\partial v}{\partial x}\frac{\partial x}{\partial x'} + \frac{\partial u'}{\partial v}\frac{\partial v}{\partial y}\frac{\partial y}{\partial x'} \\
&= \frac{\partial u}{\partial x}\cos^2\theta + \frac{\partial u}{\partial y}\sin\theta\cos\theta + \frac{\partial v}{\partial x}\sin\theta\cos\theta + \frac{\partial v}{\partial y}\sin^2\theta. \\
&= \varepsilon_{xx}\cos^2\theta + 2\varepsilon_{xy}\sin\theta\cos\theta + \varepsilon_{yy}\sin^2\theta.
\end{aligned} \tag{2.192}$$

This is identical to the transformation law in two dimensions for τ_{xx}, (2.25). Similar analysis of the other components of the strain verifies that the strain in the primed coordinate system is related to that for the unprimed coordinate system by the same relationship that governs stress transformations, (2.30), that is,

$$\boldsymbol{\varepsilon}' = \mathbf{L}\boldsymbol{\varepsilon}\mathbf{L}^{\mathbf{T}}, \tag{2.193}$$

where the rotation matrix $\mathbf{L}$ is given by (2.28). This proves that the strain is a second-order tensor.

The strain is also *symmetric*, by (2.187). In consequence of the fact that strain is a symmetric second-order tensor, all theorems pertaining to principal stresses, principal stress directions, maximum shear stresses, etc., carry over directly to strains. For example, in two dimensions there will always be two mutually orthogonal directions of principal normal strain, such that the normal strain in one of these directions is the maximum normal strain and the other a minimum. In particular, the directions of principal normal strains are defined by

$$\tan 2\theta = \frac{2\varepsilon_{xy}}{\varepsilon_{xx} - \varepsilon_{yy}}, \tag{2.194}$$

and the two principal strains, ε_1 and ε_2, are given by

$$\varepsilon_1, \varepsilon_2 = \frac{1}{2}(\varepsilon_{xx} + \varepsilon_{yy}) \pm [(\varepsilon_{xy})^2 + \frac{1}{4}(\varepsilon_{xx} - \varepsilon_{yy})^2]. \tag{2.195}$$

Relative to the principal axes, the normal and shear strains in some direction that is rotated by an angle θ from the direction of the maximum normal strain, which in this context will be denoted by ε and Γ, are given by

$$\varepsilon = \varepsilon_1\cos^2\theta + \varepsilon_2\sin^2\theta, \tag{2.196}$$

$$\Gamma = -\frac{1}{2}(\varepsilon_1 - \varepsilon_2)\sin 2\theta. \tag{2.197}$$

The Mohr representation in the (ε, Γ) plane and the various constructions based on it, follow precisely as in §2.4.

The strain tensor quantifies the stretching and distortion undergone by the rock. There are also types of deformation that do not lead to any strain. One obvious type of such strain-free deformation is a *rigid-body translation*, in which all particles of the rock are displaced by the same amount, that is, $\mathbf{x}^* = \mathbf{x} - \mathbf{a}$, where the displacement vector $\mathbf{a}$ does not vary from point to point. In component form, a two-dimensional rigid-body translation is described by

$$x^* = x - a, \quad y^* = y - b, \tag{2.198}$$

where (a, b), the components of the vector $\mathbf{a}$, are constants. For this type of motion, $\nabla\mathbf{u} = \mathbf{0}$, and the strain is consequently zero.

It is clear from (2.187) that any deformation whose displacement gradient is antisymmetric will also lead to no infinitesimal strain. Such types of deformations are *infinitesimal rigid-body rotations*. To verify this, consider a rigid rotation of the rock in the counterclockwise direction, by some small angle φ. This is mathematically similar, but physically different, from the *rotation of the coordinate system* that was discussed above. This rigid-body rotation is described by

$$x^* = x\cos\varphi - y\sin\varphi, \quad y^* = x\sin\varphi + y\cos\varphi. \tag{2.199}$$

The coordinates (x^*, y^*) are the coordinates of the particle that was originally located at (x, y), *after* the rock has been rotated, but still referred to the *original* coordinate system. Since $\mathbf{x}^* = \mathbf{x} - \mathbf{u}$, the displacement components are given by

$$u = x - x^* = x(1 - \cos\varphi) + y\sin\varphi, \tag{2.200}$$

$$v = y - y^* = -x\sin\varphi + y(1 - \cos\varphi). \tag{2.201}$$

The displacement gradient can be calculated as

$$\nabla\mathbf{u} = \begin{bmatrix} \dfrac{\partial u}{\partial x} & \dfrac{\partial u}{\partial y} \\ \dfrac{\partial v}{\partial x} & \dfrac{\partial v}{\partial y} \end{bmatrix} = \begin{bmatrix} 1 - \cos\varphi & \sin\varphi \\ -\sin\varphi & 1 - \cos\varphi \end{bmatrix}. \tag{2.202}$$

As the angle of rotation is small, we can expand out the trigonometric terms in Taylor series, and ignore all terms that are higher than first-order in φ, yielding

$$\nabla\mathbf{u} = \begin{bmatrix} 0 & \varphi \\ -\varphi & 0 \end{bmatrix}. \tag{2.203}$$

The displacement gradient corresponding to an infinitesimal rigid-body rotation is therefore antisymmetric, in the sense that $(\nabla u)^{\mathrm{T}} = -(\nabla u)$. This type of rotation leads to no strain, since $2\boldsymbol{\varepsilon} = [(\nabla u) + (\nabla u)^{\mathrm{T}}] = \mathbf{0}$. In general, we can define the rotation tensor $\boldsymbol{\omega}$ as the antisymmetric part of the displacement gradient, that is,

$$\boldsymbol{\omega} = \mathrm{asym}(\nabla\mathbf{u}) \equiv \frac{1}{2}\left[\nabla\mathbf{u} - (\nabla\mathbf{u})^{\mathrm{T}}\right] = \begin{bmatrix} 0 & \dfrac{1}{2}\left(\dfrac{\partial u}{\partial y} - \dfrac{\partial v}{\partial x}\right) \\ -\dfrac{1}{2}\left(\dfrac{\partial u}{\partial y} - \dfrac{\partial v}{\partial x}\right) & 0 \end{bmatrix}, \tag{2.204}$$

in which case it follows from (2.187) and (2.204) that $\nabla\mathbf{u} = \boldsymbol{\varepsilon} + \boldsymbol{\omega}$.

The discussion just given shows that infinitesimal deformations can always be thought of as being composed of three additive components: (i) rigid-body translations, which give rise to no strain or rotation, (ii) stretching and/or distorting deformations, which give rise to a nonzero strain tensor, and (iii) infinitesimal rotations, which give rise to a nonzero rotation tensor. Because of the linear relationships between $\nabla\mathbf{u}$, $\boldsymbol{\varepsilon}$, and $\boldsymbol{\omega}$, the infinitesimal strain and infinitesimal rotation tensors are each additive in a manner that obeys the commutative law. In other words, the strain that arises as a result of two sequential displacements, $\mathbf{u}^1$ and $\mathbf{u}^2$, is equal to the sum of the two strains, $\boldsymbol{\varepsilon}^1$ and $\boldsymbol{\varepsilon}^2$, taken in either order. A similar superposition law also applies to the rotation tensor.

2.11 Infinitesimal strain in three dimensions

The theory of infinitesimal strain in three dimensions is a straightforward generalization of the two-dimensional theory. The initial location of a particle P will be denoted by $P = (x, y, z)$, relative to some Cartesian coordinate system. After the rock is deformed, the new location of that particle will be $P^* = (x^*, y^*, z^*)$, as in Fig. 2.12a. The vector that points *from* the new position *to* the old position is the displacement vector $\mathbf{u}$, that is,

$$\mathbf{u}(P) = P - P^*, \quad \text{that is,} \quad (u, v, w) = (x, y, z) - (x^*, y^*, z^*). \tag{2.205}$$

In most branches of mechanics, it is customary to define $\mathbf{u}(P) = P^* - P$. The change in sign used in rock mechanics is needed so as to be consistent with the "compression = positive" sign convention used for the stresses.

Following the definitions of the strain components given in §2.10, in three dimensions we define three normal strains as

$$\varepsilon_{xx} = \frac{\partial u}{\partial x}; \quad \varepsilon_{yy} = \frac{\partial v}{\partial y}; \quad \varepsilon_{zz} = \frac{\partial w}{\partial z}. \tag{2.206}$$

The normal strain in the x direction, ε_{xx}, has the same interpretation as in two dimensions: it is the fractional shortening of an infinitesimal line element that is initially oriented along the x-axis; likewise for the other two normal strains. Merely defining the three-dimensional strains in the same manner as in two dimensions does not of course guarantee that these strains will have the same physical interpretation. However, it is easy to see that the three-dimensional version of the calculation given in (2.179)–(2.181) would lead only to an additional term of $(\partial w/\partial z)^2$ in the bracketed part of (2.179). Under the assumption of infinitesimal strain, this term would be dropped, leading again to (2.181).

The three-dimensional shear strains are also defined in analogy with the two-dimensional case:

$$\varepsilon_{xy} = \varepsilon_{yx} = \frac{1}{2}\left(\frac{\partial u}{\partial y} + \frac{\partial v}{\partial x}\right), \quad \varepsilon_{xz} = \varepsilon_{zx} = \frac{1}{2}\left(\frac{\partial u}{\partial z} + \frac{\partial w}{\partial x}\right),$$
$$\varepsilon_{yz} = \varepsilon_{zy} = \frac{1}{2}\left(\frac{\partial v}{\partial z} + \frac{\partial w}{\partial y}\right). \tag{2.207}$$

These strains have the same physical interpretation as in two dimensions. For example, the shear strain ε_{xy} represents the increase in the angle formed by two infinitesimally small line segments that initially start at point (x, y, z), and lie along the x and y directions, respectively. However, the proof that this is indeed the correct interpretation of the definition of ε_{xy} given in (2.207) requires more than a simple extension of the two-dimensional version, since these two line segments will not necessarily continue to lie in the x–y plane after deformation. We start with a three-dimensional generalization of (2.171)–(2.173):

$$P^* = P - \mathbf{u}(P) = (x, y, z) - [u(x, y, x), v(x, y, z), w(x, y, z)], \tag{2.208}$$

$$Q^* = Q - \mathbf{u}(Q) = (x + \Delta x, y, z) - [u(x + \Delta x, y, x), v(x + \Delta x, y, z), w(x + \Delta x, y, z)], \tag{2.209}$$

$$R^* = R - \mathbf{u}(R) = (x, y + \Delta y, z) - [u(x, y + \Delta y, x), v(x, y + \Delta y, z), w(x, y + \Delta y, z)]. \tag{2.210}$$

Expanding each displacement component in a Taylor series around the point (x, y, z), and retaining only terms that are linear in the increments Δx and Δy, leads to

$$P^* = (x - u, y - v, z - w), \tag{2.211}$$

$$Q^* = (x + \Delta x - u - \frac{\partial u}{\partial x}\Delta x, y - v - \frac{\partial v}{\partial x}\Delta x, z - w - \frac{\partial w}{\partial x}\Delta x), \tag{2.212}$$

$$R^* = (x - u - \frac{\partial u}{\partial y}\Delta y, y + \Delta y - v - \frac{\partial v}{\partial y}\Delta y, z - w - \frac{\partial w}{\partial y}\Delta y), \tag{2.213}$$

where all of the partial derivatives are understood to be evaluated at the point $P = (x, y, z)$.

The angle formed by the line segments P^*Q^* and P^*R^* can be calculated from

$$(\overrightarrow{P^*Q^*}) \cdot (\overrightarrow{P^*R^*}) = |P^*Q^*|\;|P^*R^*| \cos(\angle R^*P^*Q^*), \tag{2.214}$$

where the superposed arrow indicates that the line segment must be treated as a *vector*. From (2.211) and (2.212), we have

$$P^*Q^* = \left(\Delta x - \frac{\partial u}{\partial x}\Delta x, -\frac{\partial v}{\partial x}\Delta x, -\frac{\partial w}{\partial x}\Delta x\right), \tag{2.215}$$

$$P^*R^* = \left(-\frac{\partial u}{\partial y}\Delta y, \Delta y - \frac{\partial v}{\partial y}\Delta y, -\frac{\partial w}{\partial y}\Delta y\right). \tag{2.216}$$

Again neglecting all products and higher powers of the displacement gradient components, we have

$$(\overrightarrow{P^*Q^*}) \cdot (\overrightarrow{P^*R^*}) = -\frac{\partial u}{\partial y}\Delta x\Delta y - \frac{\partial v}{\partial x}\Delta x\Delta y. \tag{2.217}$$

The lengths of the two segments P^*Q^* and P^*R^* are, from (2.180) and (2.182),

$$|P^*Q^*| = \Delta x\left(1 - \frac{\partial u}{\partial x}\right), \quad |P^*R^*| = \Delta y\left(1 - \frac{\partial v}{\partial y}\right). \tag{2.218}$$

Solving (2.214) for $\cos(\angle R^*P^*Q^*)$ yields, to first-order in the displacement gradient terms,

$$\cos(\angle R^*P^*Q^*) = \frac{-\left((\partial u)/(\partial x) + (\partial v)/(\partial y)\right)\Delta x \Delta y}{\Delta x\,(1 - (\partial u)/(\partial x))\,\Delta y\left(1 - (\partial v)/(\partial y)\right)} = -\left(\frac{\partial u}{\partial x} + \frac{\partial v}{\partial y}\right). \tag{2.219}$$

But as $\angle R^*P^*Q^*$ is very close to 90°, we can say that

$$\cos(\angle R^*P^*Q^*) = -\sin(\angle R^*P^*Q^*) \approx -(\angle R^*P^*Q^*). \tag{2.220}$$

Recalling the fundamental definition of shear strain, (2.183), and the fact that $\angle RPQ = 90°$, it follows from (2.219) and (2.220) that

$$\varepsilon_{xy} = \frac{1}{2}\lim_{\Delta x, \Delta y \to 0}(\angle R^*P^*Q^* - \angle RPQ) = \frac{1}{2}\left(\frac{\partial u}{\partial y} + \frac{\partial v}{\partial x}\right), \tag{2.221}$$

which verifies that the shear strains, as defined by (2.207), do indeed measure the angular distortions.

The nine strains defined in (2.206) and (2.207) are the components of the three-dimensional strain tensor, $\boldsymbol{\varepsilon}$:

$$\boldsymbol{\varepsilon} = \begin{bmatrix} \varepsilon_{xx} & \varepsilon_{xy} & \varepsilon_{xz} \\ \varepsilon_{yx} & \varepsilon_{yy} & \varepsilon_{yz} \\ \varepsilon_{zx} & \varepsilon_{zy} & \varepsilon_{zz} \end{bmatrix} = \begin{bmatrix} \frac{\partial u}{\partial x} & \frac{1}{2}\left(\frac{\partial u}{\partial y} + \frac{\partial v}{\partial x}\right) & \frac{1}{2}\left(\frac{\partial u}{\partial z} + \frac{\partial w}{\partial x}\right) \\ \frac{1}{2}\left(\frac{\partial u}{\partial y} + \frac{\partial v}{\partial x}\right) & \frac{\partial v}{\partial y} & \frac{1}{2}\left(\frac{\partial v}{\partial z} + \frac{\partial w}{\partial y}\right) \\ \frac{1}{2}\left(\frac{\partial u}{\partial z} + \frac{\partial w}{\partial x}\right) & \frac{1}{2}\left(\frac{\partial v}{\partial z} + \frac{\partial w}{\partial y}\right) & \frac{\partial w}{\partial z} \end{bmatrix}. \tag{2.222}$$

As in the two-dimensional case, the strain matrix is equal to the symmetric part of the *displacement gradient*, $\nabla\mathbf{u}$, whose components are the partial derivatives of the displacements with respect to the three coordinates, that is,

$$\boldsymbol{\varepsilon} = \mathrm{sym}(\nabla\mathbf{u}) \equiv \frac{1}{2}\left[\nabla\mathbf{u} + (\nabla\mathbf{u})^{\mathrm{T}}\right], \tag{2.223}$$

$$\nabla\mathbf{u} = \begin{bmatrix} \frac{\partial u}{\partial x} & \frac{\partial u}{\partial y} & \frac{\partial u}{\partial z} \\ \frac{\partial v}{\partial x} & \frac{\partial v}{\partial y} & \frac{\partial v}{\partial z} \\ \frac{\partial w}{\partial x} & \frac{\partial w}{\partial y} & \frac{\partial w}{\partial z} \end{bmatrix}. \tag{2.224}$$

The antisymmetric part of the displacement gradient is again defined to be the rotation tensor, $\boldsymbol{\omega}$, whose components are given by

$$\boldsymbol{\omega} = \text{asym}(\nabla\mathbf{u}) \equiv \frac{1}{2}\left[\nabla\mathbf{u} - (\nabla\mathbf{u})^{\mathrm{T}}\right] = \begin{bmatrix} \omega_{xx} & \omega_{xy} & \omega_{xz} \\ \omega_{yx} & \omega_{yy} & \omega_{yz} \\ \omega_{zx} & \omega_{zy} & \omega_{zz} \end{bmatrix}$$

$$= \begin{bmatrix} 0 & \frac{1}{2}\left(\frac{\partial u}{\partial y} - \frac{\partial v}{\partial x}\right) & \frac{1}{2}\left(\frac{\partial u}{\partial z} - \frac{\partial w}{\partial x}\right) \\ -\frac{1}{2}\left(\frac{\partial u}{\partial y} - \frac{\partial v}{\partial x}\right) & 0 & \frac{1}{2}\left(\frac{\partial v}{\partial z} - \frac{\partial w}{\partial y}\right) \\ -\frac{1}{2}\left(\frac{\partial u}{\partial z} - \frac{\partial w}{\partial x}\right) & -\frac{1}{2}\left(\frac{\partial v}{\partial z} - \frac{\partial w}{\partial y}\right) & 0 \end{bmatrix}. \tag{2.225}$$

Arguments exactly analogous to that given in (2.200)–(2.203) would show that, for example, ω_{xz} quantifies the infinitesimal rotation about the y-axis, etc. As there are only three independent components of the rotation tensor, these components are often identified as the elements of the *rotation vector*, $(\omega_x, \omega_y, \omega_z)$, where

$$\omega_x = \frac{1}{2}\left(\frac{\partial w}{\partial y} - \frac{\partial v}{\partial z}\right), \quad \omega_y = \frac{1}{2}\left(\frac{\partial u}{\partial z} - \frac{\partial w}{\partial x}\right), \quad \omega_z = \frac{1}{2}\left(\frac{\partial v}{\partial x} - \frac{\partial u}{\partial y}\right). \tag{2.226}$$

The particular choices of signs in these definitions are made such that, for example, ω_x will be positive if the rotation is counterclockwise when viewed *from* the positive x-axis, *toward* the origin (Chou and Pagano, 1992, p. 48). It is important to note that the strain and rotation components typically vary from point to point, so that the rotations we are speaking of actually are confined to an infinitesimal neighborhood of the point (x, y, z).

The three-dimensional demonstration that the strain is indeed a second-order tensor follows exactly the same lines as given in (2.192) for the two-dimensional case, except that there would be nine terms on the right-hand side, rather than four. The explicit expressions for the transformation of the six strain components are identical in form to (2.102)–(2.107). Hence, as the three-dimensional strain is a symmetric second-order tensor, all of the general theorems developed for three-dimensional stresses carry over to the strains. In particular, there will always be three mutually perpendicular directions in which the normal strain takes on locally stationary values, typically a maximum, a minimum, and a saddle point. These three principal normal strains are denoted by $\varepsilon_1 \geq \varepsilon_2 \geq \varepsilon_3$. In the *principal coordinate system*, in which the three principal directions serve as the coordinate axes, the shear strains are zero, and the strain tensor is purely diagonal. (The question of whether or not the principal directions of strain coincide with the principal directions of stress depends on the stress–strain law of the rock; see Chapter 5). Hence, at each point, there is always one coordinate system in which the deformation of the rock can be represented by stretching (or contraction) along three mutually perpendicular directions. Of course, there are also, in general, rigid-body translations and rotations occurring at each point. Therefore, the most general infinitesimal displacement at a point is a combination of stretching (by, in general, differing amounts) along three mutually perpendicular axes, followed by a rigid-body rotation and then a rigid-body translation. If the rotations are infinitesimal, it can be shown (Fung, 1965, p. 97) that subsequent rotations about the (x, y, z) axes by angles $(\omega_x, \omega_y, \omega_z)$ are equivalent to a single rotation about an axis that coincides with the rotation vector, the angle of which is given by $\omega = (\omega_x^2 + \omega_y^2 + \omega_z^2)^{1/2}$.

The values of the three principal strains are found by solving an equation analogous to (2.69):

$$\varepsilon^3 - I_1\varepsilon^2 - I_2\varepsilon - I_3 = 0, \tag{2.227}$$

in which the three *strain invariants* are defined by

$$I_1 = \text{trace}(\boldsymbol{\varepsilon}) = \varepsilon_{xx} + \varepsilon_{yy} + \varepsilon_{zz}, \tag{2.228}$$

$$I_2 = \frac{1}{2}\{\text{trace}(\boldsymbol{\varepsilon}^2) - [\text{trace}(\boldsymbol{\varepsilon})]^2\} = \varepsilon_{xy}^2 + \varepsilon_{xz}^2 + \varepsilon_{yz}^2 - \varepsilon_{xx}\varepsilon_{yy} - \varepsilon_{xx}\varepsilon_{zz} - \varepsilon_{yy}\varepsilon_{zz}, \tag{2.229}$$

$$I_3 = \det(\boldsymbol{\varepsilon}) = \varepsilon_{xx}\varepsilon_{yy}\varepsilon_{zz} + 2\varepsilon_{xy}\varepsilon_{xz}\varepsilon_{yz} - \varepsilon_{xx}\varepsilon_{yz}^2 - \varepsilon_{yy}\varepsilon_{xz}^2 - \varepsilon_{zz}\varepsilon_{xy}^2. \tag{2.230}$$

The first invariant of strain has the physical interpretation of representing the volumetric strain in the vicinity of the point (x, y, z). To prove this, consider a small cube of rock, one of whose corners is initially located at (x, y, z), and whose edges are aligned (before deformation) with the three principal directions of strain. Before the deformation, the edges of this cube each had length L, and the cube had volume V. After the deformation, the edges of this cube will have length $L(1 - \varepsilon_1)$, $L(1 - \varepsilon_2)$, and $L(1 - \varepsilon_3)$, and the new volume will be $V^* = L^3(1-\varepsilon_1)(1-\varepsilon_2)(1-\varepsilon_3)$. The volumetric strain, defined as the fractional *decrease* in the volume of the cube, is

$$\begin{aligned}
\varepsilon_v &= \lim_{L\to 0} \frac{V - V^*}{V} = \frac{L^3(1-\varepsilon_1)(1-\varepsilon_2)(1-\varepsilon_3) - L^3}{L^3} \\
&= (1-\varepsilon_1)(1-\varepsilon_2)(1-\varepsilon_3) - 1 \\
&= 1 - (\varepsilon_1 + \varepsilon_2 + \varepsilon_3) + (\varepsilon_1\varepsilon_2 + \varepsilon_1\varepsilon_3 + \varepsilon_2\varepsilon_3) - \varepsilon_1\varepsilon_2\varepsilon_3 - 1 \\
&\approx \varepsilon_1 + \varepsilon_2 + \varepsilon_3 = I_1,
\end{aligned} \tag{2.231}$$

where, in the last step, the smallness of the strains has been invoked. Another commonly used symbol for the volumetric strain is Δ. As I_1 is an invariant, the volumetric strain can be calculated from (2.228), in *any* coordinate system.

The strain tensor can be decomposed into an isotropic and a deviatoric part, in a manner exactly analogous to the decomposition of the stresses:

$$\boldsymbol{\varepsilon} = \boldsymbol{\varepsilon}^{\text{iso}} + \boldsymbol{\varepsilon}^{\text{dev}}, \tag{2.232}$$

$$\boldsymbol{\varepsilon}^{\text{iso}} = \frac{1}{3}I_1\,\mathbf{I} = \frac{1}{3}\varepsilon_v\mathbf{I}, \tag{2.233}$$

$$\mathbf{e} = \boldsymbol{\varepsilon}^{\text{dev}} = \boldsymbol{\varepsilon} - \boldsymbol{\varepsilon}^{\text{iso}} = \boldsymbol{\varepsilon} - \frac{1}{3}\varepsilon_v \mathbf{I}. \tag{2.234}$$

Equation (2.234) can be written out explicitly as

$$\mathbf{e} = \begin{bmatrix} \varepsilon_{xx} - \frac{1}{3}\varepsilon_v & \varepsilon_{xy} & \varepsilon_{xz} \\ \varepsilon_{yx} & \varepsilon_{yy} - \frac{1}{3}\varepsilon_v & \varepsilon_{yz} \\ \varepsilon_{zx} & \varepsilon_{zy} & \varepsilon_{zz} - \frac{1}{3}\varepsilon_v \end{bmatrix}. \tag{2.235}$$

The *mean normal strain* is defined by

$$\varepsilon_{\text{m}} = \frac{1}{3} I_1 = \frac{1}{3}\varepsilon_v. \tag{2.236}$$

The principal axes of the deviatoric strain coincide with the principal axes of strain, and the magnitudes of the principal deviatoric strains are

$$e_1 = \varepsilon_1 - \varepsilon_{\text{m}}, \quad e_2 = \varepsilon_2 - \varepsilon_{\text{m}}, \quad e_3 = \varepsilon_3 - \varepsilon_{\text{m}}, \tag{2.237}$$

so that $e_1 + e_2 + e_3 = 0$.

An important special type of deformation is a *homogeneous* deformation, in which the strains and rotations are the same at each point. In this case, it follows from (2.223)–(2.226) that the displacement gradient takes the form

$$\nabla \mathbf{u} = \begin{bmatrix} \frac{\partial u}{\partial x} & \frac{\partial u}{\partial y} & \frac{\partial u}{\partial z} \\ \frac{\partial v}{\partial x} & \frac{\partial v}{\partial y} & \frac{\partial v}{\partial z} \\ \frac{\partial w}{\partial x} & \frac{\partial w}{\partial y} & \frac{\partial w}{\partial z} \end{bmatrix} = \begin{bmatrix} \varepsilon_{xx}^0 & \varepsilon_{xy}^0 - \omega_z^0 & \varepsilon_{xz}^0 + \omega_y^0 \\ \varepsilon_{xy}^0 + \omega_z^0 & \varepsilon_{yy}^0 & \varepsilon_{yz}^0 - \omega_x^0 \\ \varepsilon_{xz}^0 - \omega_y^0 & \varepsilon_{yz}^0 + \omega_x^0 & \varepsilon_{zz}^0 \end{bmatrix}, \tag{2.238}$$

where the superscript 0 indicates that these components do not vary with position. The nine first-order partial differential equations represented by (2.238) can be integrated to yield

$$u = \varepsilon_{xx}^0 x + (\varepsilon_{xy}^0 - \omega_z^0)y + (\varepsilon_{xz}^0 + \omega_y^0)z + a, \tag{2.239}$$

$$v = (\varepsilon_{xy}^0 + \omega_z^0)x + \varepsilon_{yy}^0 y + (\varepsilon_{yz}^0 - \omega_x^0)z + b, \tag{2.240}$$

$$w = (\varepsilon_{xz}^0 - \omega_y^0)x + (\varepsilon_{yz}^0 + \omega_x^0)y + \varepsilon_{zz}^0 z + c, \tag{2.241}$$

where $\{a, b, c\}$ are constants. This deformation represents a stretching in the $\{x, y, z\}$ directions by fractional amounts $\{\varepsilon_{xx}^0, \varepsilon_{yy}^0, \varepsilon_{zz}^0\}$, followed by rotations about the $\{x, y, z\}$ axes by amounts $\{\omega_x^0, \omega_y^0, \omega_z^0\}$, finally followed by rigid-body translations in the $\{x, y, z\}$ directions by amounts $\{a, b, c\}$. It is easy to show that, under a homogeneous deformation: (i) a straight line remains straight, (ii) two parallel lines remain parallel, and (iii) spheres are transformed into ellipses. An interesting aspect of homogeneous deformations is that the above-listed properties hold *regardless* of whether or not the strain and rotation components are infinitesimal.

2.12 Determination of principal stresses or strains from measurements

Various methods exist to measure the state of stress underground. Many of these methods actually involve measurement of the strains, from which the stresses are estimated using known stress–strain relations for the given rock (see §5.2). These methods, which involve strain gauges or displacement meters, are capable of measuring normal strains in a given direction. The question then arises of using the measured normal strains to find the principal strains and the principal directions. As the problem is mathematically identical in the case of stresses or strains, we will discuss it in the context of finding the principal strains and their directions.

In two dimensions, there are three independent strain components, and so it seems plausible that three strain measurements will be needed to supply sufficient information to find these components. Another way of looking at this problem is that three pieces of data are needed to estimate the two principal strains and the angle of rotation of one of the principal directions with respect to some arbitrary direction. Assume that we have measured the normal strains in three directions, P, Q, and R, in some plane, as in Fig. 2.13. The angles of rotation between OP, OQ, and OR will be known, but the angle of rotation θ between, say, OP and the direction of the major principal stress will not be known a priori. In general, it will often not be the case that two principal stresses actually lie in the plane in question; in this case, we will actually be dealing with the *subsidiary principal stresses*, as defined in §2.6; this does not affect the following discussion.

From (2.54) we have

$$\varepsilon_{PP} = \frac{1}{2}(\varepsilon_1 + \varepsilon_2) + \frac{1}{2}(\varepsilon_1 - \varepsilon_2)\cos 2\theta, \tag{2.242}$$

$$\varepsilon_{QQ} = \frac{1}{2}(\varepsilon_1 + \varepsilon_2) + \frac{1}{2}(\varepsilon_1 - \varepsilon_2)\cos 2(\theta + \alpha), \tag{2.243}$$

$$\varepsilon_{PP} = \frac{1}{2}(\varepsilon_1 + \varepsilon_2) + \frac{1}{2}(\varepsilon_1 - \varepsilon_2)\cos 2(\theta + \alpha + \beta). \tag{2.244}$$

These equations can be solved for ε_1, ε_2, and θ. For the commonly used strain gauge rosette configuration in which $\alpha = \beta = 45°$, the solution is

$$\varepsilon_1 + \varepsilon_2 = \varepsilon_{PP} + \varepsilon_{RR}, \tag{2.245}$$

$$\varepsilon_1 - \varepsilon_2 = [(\varepsilon_{PP} - 2\varepsilon_{QQ} + \varepsilon_{RR})^2 + (\varepsilon_{PP} - \varepsilon_{RR})^2]^{1/2}, \tag{2.246}$$

$$\theta = \frac{1}{2}\arctan[(\varepsilon_{PP} - 2\varepsilon_{QQ} + \varepsilon_{RR})/(\varepsilon_{PP} - \varepsilon_{RR})]. \tag{2.247}$$

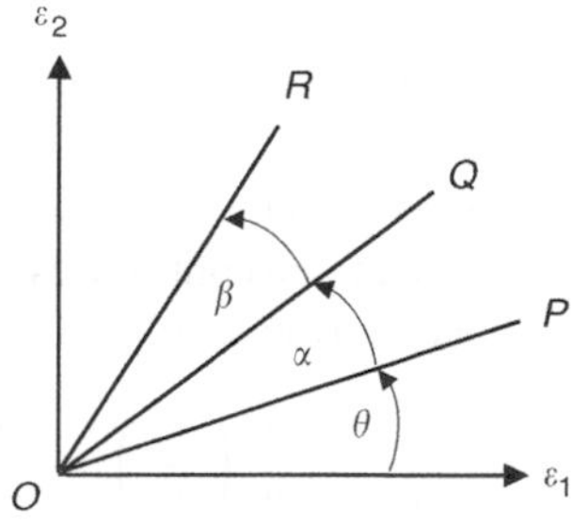

Fig. 2.13 Normal strains that are measured in three directions, OP, OQ, and OR can be used to find the principal normal strains and their orientation, as explained in the text.

Another common case is $\alpha = \beta = 60^\circ$, for which the solution is

$$\varepsilon_1 + \varepsilon_2 = \frac{2}{3}(\varepsilon_{PP} + \varepsilon_{QQ} + \varepsilon_{RR}), \tag{2.248}$$

$$\varepsilon_1 - \varepsilon_2 = \left[\frac{4}{3}(\varepsilon_{QQ} - \varepsilon_{RR})^2 + \frac{4}{9}(2\varepsilon_{PP} - \varepsilon_{QQ} - \varepsilon_{RR})^2\right]^{1/2}, \tag{2.249}$$

$$\theta = \frac{1}{2}\arctan[\sqrt{3}(\varepsilon_{QQ} - \varepsilon_{RR})/(\varepsilon_{QQ} + \varepsilon_{RR} - 2\varepsilon_{PP})]. \tag{2.250}$$

Solutions for other configurations are described by Hetényi (1950, p. 412). These equations could also be solved graphically, using Mohr's circle, as in §2.4.

2.13 Compatibility equations

In two dimensions, there are two displacement functions and four strain components. Two of these strains, the shear strains ε_{xy} and ε_{yx}, are equal to each other by definition. In three dimensions, there are essentially six strains and three displacements. If the displacement field $\mathbf{u}$ is known as a function of $\mathbf{x}$, the strains can be computed by simply taking partial derivatives. The converse problem, that of determining the displacements from the strains, requires a complicated integration procedure (Sokolnikoff, 1956, pp. 25–9). In those rock mechanics problems for which the displacements themselves are of interest, one would usually use a solution procedure that directly makes use of the displacements as the basic unknown variables. However, there are many situations in which only the stresses are desired but the displacements are not. For these problems, solution procedures exist that utilize the stresses as the basic unknowns but in which the displacements do not explicitly appear (see §5.5 and §5.7). Hence, in both types of situations the integration procedure itself is usually avoided. Nevertheless, when using a stress-based formulation of the elasticity equations, the number of equations will be less than the number of unknowns unless one considers an additional set of equations known as the *compatibility equations*. These equations are necessary conditions in order that the partial differential equations represented by (2.186) or (2.222) can be integrated to yield the displacements. These equations must be satisfied whether or not one attempts to perform the integration.

To understand the origin of the integrability condition, consider the simpler, but essentially equivalent, situation of a single function f of two variables, x and y. Assume that the two partial derivatives of f are given as specified functions, M and N, that is,

$$M(x,y) = \frac{\partial f}{\partial x}, \quad N(x,y) = \frac{\partial f}{\partial y}. \tag{2.251}$$

The two functions M and N cannot be specified arbitrarily, even if they are both assumed to be continuous, differentiable functions. This is because, as the mixed partial derivatives of f must be equal (Lang, 1973, p. 110), we must have

$$\frac{\partial M}{\partial y} = \frac{\partial}{\partial y}\left(\frac{\partial f}{\partial x}\right) = \frac{\partial^2 f}{\partial y \partial x} = \frac{\partial^2 f}{\partial x \partial y} = \frac{\partial}{\partial x}\left(\frac{\partial f}{\partial y}\right) = \frac{\partial N}{\partial x}. \tag{2.252}$$

Hence, M and N are not completely independent, but must satisfy the partial differential equation $\partial M/\partial y = \partial N/\partial x$.

Similar integrability conditions exist for the strains, in both two and three dimensions. The integrability condition is easy to derive in two dimensions, simply by taking two partial derivatives of each of the three strain components, as given by (2.181), (2.182), and (2.185), which leads to

$$2\frac{\partial^2 \varepsilon_{xy}}{\partial x \partial y} = \frac{\partial^2 \varepsilon_{xx}}{\partial y^2} + \frac{\partial^2 \varepsilon_{yy}}{\partial x^2}. \tag{2.253}$$

In three dimensions, six such relations can be derived among the six strain components. (The number of compatibility equations is *not* simply equal to the number of strains less the number of displacements, as is sometimes implied). These relations, first derived in 1860 by the great French elastician Saint Venant, are, in addition to (2.253),

$$2\frac{\partial^2 \varepsilon_{yz}}{\partial y \partial z} = \frac{\partial^2 \varepsilon_{yy}}{\partial z^2} + \frac{\partial^2 \varepsilon_{zz}}{\partial y^2}, \tag{2.254}$$

$$2\frac{\partial^2 \varepsilon_{zx}}{\partial z \partial x} = \frac{\partial^2 \varepsilon_{zz}}{\partial x^2} + \frac{\partial^2 \varepsilon_{xx}}{\partial z^2}, \tag{2.255}$$

$$\frac{\partial^2 \varepsilon_{xx}}{\partial y \partial z} = \frac{\partial}{\partial x}\left(-\frac{\partial \varepsilon_{yz}}{\partial x} + \frac{\partial \varepsilon_{zx}}{\partial y} + \frac{\partial \varepsilon_{xy}}{\partial z}\right), \tag{2.256}$$

$$\frac{\partial^2 \varepsilon_{yy}}{\partial z \partial x} = \frac{\partial}{\partial y}\left(\frac{\partial \varepsilon_{yz}}{\partial x} - \frac{\partial \varepsilon_{zx}}{\partial y} + \frac{\partial \varepsilon_{xy}}{\partial z}\right), \tag{2.257}$$

$$\frac{\partial^2 \varepsilon_{zz}}{\partial x \partial y} = \frac{\partial}{\partial z}\left(\frac{\partial \varepsilon_{yz}}{\partial x} + \frac{\partial \varepsilon_{zx}}{\partial y} - \frac{\partial \varepsilon_{xy}}{\partial z}\right). \tag{2.258}$$

These equations are necessary, and also sufficient, for the displacements to exist as continuous, single-valued functions (Fung, 1965, pp. 101–3). The importance of the compatibility equations in solving elasticity problems is illustrated in §5.7. Finally, it should be noted that the compatibility equations impose conditions only on the manner in which the strains vary from point to point. But there are no necessary conditions that must hold between the strain components at any one point; at any given point, the strain components ε_{xx}, ε_{yy}, and ε_{xy} can take on any three arbitrary values.

2.14 Stress and strain in polar and cylindrical coordinates

In many rock mechanics problems involving circular geometries, such as those involving tunnels, boreholes, or cylindrical rock cores used as test specimens, it is convenient to use polar coordinates (for two-dimensional problems) or cylindrical coordinates (for three-dimensional problems). We will first consider polar coordinates in a plane; cylindrical coordinates are created by simply adding a Cartesian axis that is perpendicular to the plane.

Consider a plane, with an (x, y) coordinate system attached to some origin O (Fig. 2.14a). Each point P, with coordinates (x, y) relative to the Cartesian coordinate system, can also be identified with two polar coordinates, r and θ.

The radial coordinate r measures the distance from P to the origin, whereas the angular coordinate θ measures the angle of counterclockwise rotation from the line segment OX to OP. These two sets of coordinates are related to each other by

$$x = r\cos\theta, \quad y = r\sin\theta, \tag{2.259}$$

$$r = (x^2 + y^2)^{1/2}, \quad \theta = \arctan(y/x). \tag{2.260}$$

At each point P, we can imagine a local pair of unit vectors, $\mathbf{e_r}$ and $\mathbf{e_\theta}$, pointing in the r and θ directions, respectively. An infinitesimal region of rock bounded by the arcs $r, r + \Delta r, \theta$, and $\theta + \Delta\theta$, is shown in Fig. 2.14b.

At each point P, the two unit vectors $\{\mathbf{e_r}, \mathbf{e_\theta}\}$ can be thought of as defining a Cartesian coordinate system that is rotated by an angle θ from the $\{\mathbf{e_x}, \mathbf{e_y}\}$ system. Hence, the stresses in the polar coordinate system can be found from those in the Cartesian system through equations (2.25)–(2.27):

$$\tau_{rr} = \tau_{xx}\cos^2\theta + 2\tau_{xy}\sin\theta\cos\theta + \tau_{yy}\sin^2\theta, \tag{2.261}$$

$$\tau_{\theta\theta} = \tau_{xx}\sin^2\theta - 2\tau_{xy}\sin\theta\cos\theta + \tau_{yy}\cos^2\theta, \tag{2.262}$$

$$\tau_{r\theta} = (\tau_{yy} - \tau_{xx})\sin\theta\cos\theta + \tau_{xy}(\cos^2\theta - \sin^2\theta). \tag{2.263}$$

At each point P, the stresses in polar coordinates constitute a symmetric second-order tensor.

The displacement vector in polar coordinates (Fig. 2.14b) will be denoted by (u', v'). Note that, according to the sign convention introduced in (2.171), the actual motion of the rock particle is in the direction of $(-u', -v')$. Because the (r, θ) coordinate system is (locally) merely a rotated Cartesian coordinate system, the displacement components in the (r, θ) system are related to those in the (x, y) system by (2.191):

$$u' = u\cos\theta + v\sin\theta, \quad v' = -u\sin\theta + v\cos\theta, \tag{2.264}$$

$$u = u'\cos\theta - v'\sin\theta, \quad v = u'\sin\theta + v'\cos\theta. \tag{2.265}$$

The situation is more complicated for relations that involve differentiation, such as the strain–displacement relations. For such relationships, it becomes necessary to account for the fact that, in contrast to the unit vectors $\{\mathbf{e_x}, \mathbf{e_y}\}$, the unit vectors $\{\mathbf{e_r}, \mathbf{e_\theta}\}$ point in different directions at different locations. An implication of this fact is that it is not proper to define the strains in a polar

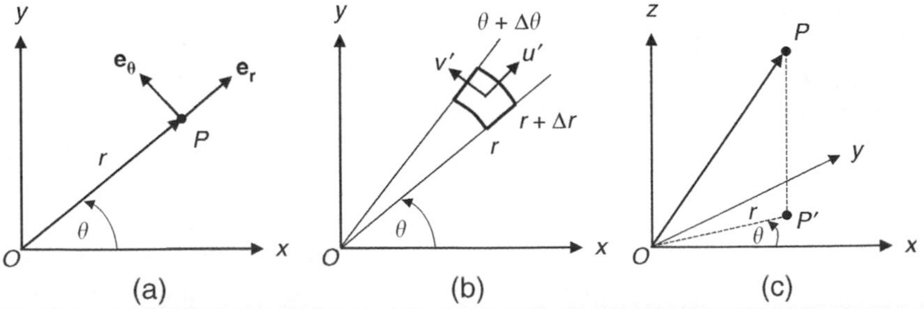

Fig. 2.14 (a) Polar coordinate system. (b) Infinitesimal element of rock in a polar coordinate system. (c) Cylindrical coordinate system.

coordinate system in the same manner as was done for Cartesian coordinates in §2.10. The proper definition of the strains in general, non-Cartesian coordinates is discussed by Green and Zerna (1954) and Amenzade (1979), among others. In polar coordinates, the proper definitions of strains can be arrived at by using the chain rule of partial differentiation, in conjunction with (2.259), (2.260), (2.264), and (2.265).

To carry out these calculations, we need the following partial derivatives:

$$\frac{\partial r}{\partial x} = \frac{\partial}{\partial x}(x^2 + y^2)^{1/2} = \frac{x}{(x^2 + y^2)^{1/2}} = \frac{x}{r} = \cos\theta, \tag{2.266}$$

$$\frac{\partial r}{\partial y} = \frac{\partial}{\partial y}(x^2 + y^2)^{1/2} = \frac{y}{(x^2 + y^2)^{1/2}} = \frac{y}{r} = \sin\theta, \tag{2.267}$$

$$\frac{\partial \theta}{\partial x} = \frac{\partial}{\partial x}[\arctan(y/x)] = \frac{-y/x^2}{[1 + (y/x)^2]} = \frac{-y}{r^2} = \frac{-\sin\theta}{r}, \tag{2.268}$$

$$\frac{\partial \theta}{\partial y} = \frac{\partial}{\partial y}[\arctan(y/x)] = \frac{1/x}{[1 + (y/x)^2]} = \frac{x}{r^2} = \frac{\cos\theta}{r}. \tag{2.269}$$

Now apply the chain rule to the strain component ε_{xx}:

$$\varepsilon_{xx} = \frac{\partial u}{\partial x} = \frac{\partial u}{\partial r}\frac{\partial r}{\partial x} + \frac{\partial u}{\partial \theta}\frac{\partial \theta}{\partial x} = \cos\theta\frac{\partial u}{\partial r} - \frac{\sin\theta}{r}\frac{\partial u}{\partial \theta}. \tag{2.270}$$

But, from (2.265),

$$\frac{\partial u}{\partial r} = \cos\theta\frac{\partial u'}{\partial r} - \sin\theta\frac{\partial v'}{\partial r}, \tag{2.271}$$

$$\frac{\partial u}{\partial \theta} = \cos\theta\frac{\partial u'}{\partial \theta} - u'\sin\theta - \sin\theta\frac{\partial v'}{\partial \theta} - v'\cos\theta. \tag{2.272}$$

Combining (2.270)–(2.272), and collecting terms, yields

$$\varepsilon_{xx} = \frac{\partial u'}{\partial r}\cos^2\theta - \left[\frac{\partial v'}{\partial r} + \frac{1}{r}\frac{\partial u'}{\partial \theta} - \frac{v'}{r}\right]\sin\theta\cos\theta + \left[\frac{1}{r}\frac{\partial v'}{\partial \theta} + \frac{u'}{r}\right]\sin^2\theta. \tag{2.273}$$

As the $\{\mathbf{e_x}, \mathbf{e_y}\}$ axes are obtained from the $\{\mathbf{e_r}, \mathbf{e_\theta}\}$ axes by a counterclockwise rotation of angle $-\theta$, the strain ε_{xx} can also be obtained from the strains in the polar coordinate system by the usual strain transformation equations, (2.192), using an angle of $-\theta$:

$$\varepsilon_{xx} = \varepsilon_{rr}\cos^2\theta - 2\varepsilon_{r\theta}\sin\theta\cos\theta + \varepsilon_{\theta\theta}\sin^2\theta. \tag{2.274}$$

Comparison of (2.273) and (2.274) indicates that

$$\varepsilon_{rr} = \frac{\partial u'}{\partial r}, \quad \varepsilon_{\theta\theta} = \frac{1}{r}\frac{\partial v'}{\partial \theta} + \frac{u'}{r}, \quad \varepsilon_{r\theta} = \frac{1}{2}\left(\frac{\partial v'}{\partial r} + \frac{1}{r}\frac{\partial u'}{\partial \theta} - \frac{v'}{r}\right). \tag{2.275}$$

These relations can also be derived by more sophisticated methods (Sokolnikoff, 1956, pp. 177–84).

The areal dilatation, which is the two-dimensional analogue of the volumetric strain, is given, as usual, by the sum of the diagonal components of the strain tensor:

$$\varepsilon_a = \varepsilon_{rr} + \varepsilon_{\theta\theta} = \frac{\partial u'}{\partial r} + \frac{u'}{r} + \frac{1}{r}\frac{\partial v'}{\partial \theta}, \tag{2.276}$$

The following compatibility criterion for the strains in polar coordinates can be derived by taking appropriate partial derivatives of the strains in (2.275):

$$2\frac{\partial^2(r\varepsilon_{r\theta})}{\partial r\partial\theta} = r\frac{\partial^2(r\varepsilon_{\theta\theta})}{\partial r^2} - r\frac{\partial\varepsilon_{rr}}{\partial r} + \frac{\partial^2\varepsilon_{rr}}{\partial\theta^2}. \tag{2.277}$$

In three-dimensional problems involving cylindrical geometry or axial symmetry, cylindrical polar coordinates are often used (Fig. 2.14c). The third unit vector $\mathbf{e_z}$ is perpendicular to the $\{\mathbf{e_r}, \mathbf{e_\theta}\}$ plane, and is oriented such that $\{\mathbf{e_r}, \mathbf{e_\theta}, \mathbf{e_z}\}$ form a right-handed coordinate system. The two coordinates r and θ are then taken to be the polar coordinates of the point P', which is the projection of point P onto the (x, y) plane, and the coordinate z is the perpendicular distance of P from the (x, y) plane. The displacement component in the z direction will be denoted by w', so that

$$\mathbf{u}' = u'\mathbf{e_r} + v'\mathbf{e_\theta} + w'\mathbf{e_z}. \tag{2.278}$$

The nine stress components are denoted by

$$\{\tau_{rr}, \tau_{r\theta}, \tau_{rz}, \tau_{\theta r}, \tau_{\theta\theta}, \tau_{\theta z}, \tau_{zr}, \tau_{z\theta}, \tau_{zz}\}, \tag{2.279}$$

where the subscripts play the usual role in indicating the direction of the traction and the plane on which it acts. The strain–displacement relations are given by (2.275), along with

$$\varepsilon_{zr} = \frac{1}{2}\left(\frac{\partial w'}{\partial r} + \frac{\partial u'}{\partial z}\right), \quad \varepsilon_{z\theta} = \frac{1}{2}\left(\frac{1}{r}\frac{\partial w'}{\partial\theta} + \frac{\partial v'}{\partial z}\right), \quad \varepsilon_{zz} = \frac{\partial w'}{\partial z}. \tag{2.280}$$

The volumetric strain is given by

$$\varepsilon_v = \varepsilon_{rr} + \varepsilon_{\theta\theta} + \varepsilon_{zz} = \frac{\partial u'}{\partial r} + \frac{u'}{r} + \frac{1}{r}\frac{\partial v'}{\partial\theta} + \frac{\partial w'}{\partial z}. \tag{2.281}$$

The five additional compatibility equations, which in the full three-dimensional case are needed in addition to (2.277), can be found in Rekach (1979, p. 27).

2.15 Finite strain

Most analyses in rock mechanics, including most of the commonly used computer codes, utilize the theory of infinitesimal strain. However, some problems, particularly those involving deformations that occur over geological time, necessarily involve strains that are not small compared to unity. To analyze these processes, the theory of finite strain must be used (Means, 1976; Oertel, 1996). This more general theory, which reduces to the theory of infinitesimal strain when the strain and rotation components are small, is more difficult to utilize due mainly to the nonlinear relationships that exist between the components of the displacement gradient and the components of the finite strain tensor. This nonlinearity is essentially different from that which arises when using nonlinear stress–strain relationships in conjunction with the infinitesimal strain assumption; this latter type of nonlinearity is quite common in rock mechanics analyses.

Recall that the original location of a rock particle is denoted by $\mathbf{x} = (x, y, z)$, and its position after the rock has been deformed is denoted by $\mathbf{x}^* = (x^*, y^*, z^*)$. For the time being, we will not utilize the concept of displacement, which is not fundamental to the theory of finite strain. Now consider a small line element in the undeformed configuration, represented by the vector $\mathbf{dx}$, whose tail is located at $\mathbf{x}$. After the deformation, this element will be deformed into a new line element $\mathbf{dx}^*$, located at point $\mathbf{x}^*$, whose components are given by

$$dx^* = \frac{\partial x^*}{\partial x}dx + \frac{\partial x^*}{\partial y}dy + \frac{\partial x^*}{\partial z}dz, \tag{2.282}$$

$$dy^* = \frac{\partial y^*}{\partial x}dx + \frac{\partial y^*}{\partial y}dy + \frac{\partial y^*}{\partial z}dz, \tag{2.283}$$

$$dz^* = \frac{\partial z^*}{\partial x}dx + \frac{\partial z^*}{\partial y}dy + \frac{\partial z^*}{\partial z}dz. \tag{2.284}$$

This relationship between the deformed and undeformed elements can be written in vector-matrix form as

$$\begin{bmatrix} dx^* \\ dy^* \\ dz^* \end{bmatrix} = \begin{bmatrix} \dfrac{\partial x^*}{\partial x} & \dfrac{\partial x^*}{\partial y} & \dfrac{\partial x^*}{\partial z} \\ \dfrac{\partial y^*}{\partial x} & \dfrac{\partial y^*}{\partial y} & \dfrac{\partial y^*}{\partial z} \\ \dfrac{\partial z^*}{\partial x} & \dfrac{\partial z^*}{\partial y} & \dfrac{\partial z^*}{\partial z} \end{bmatrix} \begin{bmatrix} dx \\ dy \\ dz \end{bmatrix}, \tag{2.285}$$

or, in a more condensed notation as

$$\mathbf{dx}^* = \mathbf{F}\mathbf{dx}, \tag{2.286}$$

where the matrix $\mathbf{F}$, whose elements are given in (2.285), is known as the *deformation gradient* (not to be confused with the displacement gradient, to which it is related, as shown below).

The deformation gradient $\mathbf{F}$ contains within it all of the relevant information concerning the stretching and rotation of the various elements of the deformed rock. It does not serve directly as a measure of strain, however, because a suitable measure of finite strain should have the property of vanishing if, for example, the rock undergoes a rigid-body rotation. To construct a suitable finite strain tensor, we examine the change in *length* undergone by the element $\mathbf{dx}$:

$$|\mathbf{dx}|^2 = \mathbf{dx} \cdot \mathbf{dx} = \mathbf{dx}^{\mathrm{T}}\mathbf{dx}, \tag{2.287}$$

$$|\mathbf{dx}^*|^2 = (\mathbf{dx}^*)^{\mathrm{T}}\mathbf{dx}^* = (\mathbf{F}\mathbf{dx})^{\mathrm{T}}(\mathbf{F}\mathbf{dx}) = \mathbf{dx}^{\mathrm{T}}(\mathbf{F}^{\mathrm{T}}\mathbf{F})\mathbf{dx}, \tag{2.288}$$

$$|\mathbf{dx}|^2 - |\mathbf{dx}^*|^2 = \mathbf{dx}^{\mathrm{T}}\mathbf{dx} - \mathbf{dx}^{\mathrm{T}}(\mathbf{F}^{\mathrm{T}}\mathbf{F})\mathbf{dx} = \mathbf{dx}^{\mathrm{T}}(\mathbf{I} - \mathbf{F}^{\mathrm{T}}\mathbf{F})\mathbf{dx} \equiv 2\mathbf{dx}^{\mathrm{T}}\mathbf{E}\mathbf{dx}, \tag{2.289}$$

where $\mathbf{E}$, as defined in (2.289), is the *Lagrangian finite strain tensor.* We have defined it here so as to be consistent with the "compression = positive" sign convention; it therefore differs from the usual definition given in continuum mechanics treatments, such as Fung (1965) or Malvern (1969). It can be seen from (2.289) that the elongation (or shortening, as the case may be) of a linear element can be computed from $\mathbf{E}$. Note that if the deformation is a rigid-body rotation, then it will necessarily be true that $\mathbf{F}^{\mathrm{T}} = \mathbf{F}^{-1}$, in which case $\mathbf{F}^{\mathrm{T}}\mathbf{F} = \mathbf{F}\,\mathbf{F}^{-1} = \mathbf{I}$, and the Lagrangian strain vanishes. This can be verified (in two dimensions) by considering a rigid-body rotation, such as given by (2.199), and explicitly calculating $\mathbf{F}^{\mathrm{T}}\mathbf{F}$. Although $\mathbf{F}$ will not necessarily be symmetric, it is easy to verify that $\mathbf{E}$ is symmetric.

Each normal component of $\mathbf{E}$ provides a measure of the strain of an element that initially lies along one of the coordinate axes. For example, consider an element that initially lies along the x-axis, that is, $\mathbf{dx} = (dx, 0, 0)$. From (2.287), we have

$$|\mathbf{dx}|^2 = \begin{bmatrix} dx & 0 & 0 \end{bmatrix} \begin{bmatrix} dx \\ 0 \\ 0 \end{bmatrix} = (dx)^2, \tag{2.290}$$

and from (2.289), we have

$$|\mathbf{dx}|^2 - |\mathbf{dx}^*|^2 = 2\begin{bmatrix} dx & 0 & 0 \end{bmatrix} \begin{bmatrix} E_{xx} & E_{xy} & E_{xz} \\ E_{yx} & E_{yy} & E_{yz} \\ E_{zx} & E_{zy} & E_{zz} \end{bmatrix} \begin{bmatrix} dx \\ 0 \\ 0 \end{bmatrix} = 2E_{xx}(dx)^2, \tag{2.291}$$

which is to say

$$|\mathbf{dx}^*| = (1 - 2E_{xx})^{1/2}|\mathbf{dx}|. \tag{2.292}$$

The fractional shortening, as defined in (2.169), is therefore given by

$$\frac{|\mathbf{dx}| - |\mathbf{dx}^*|}{|\mathbf{dx}|} = 1 - (1 - 2E_{xx})^{1/2}. \tag{2.293}$$

If E_{xx} is small, the right-hand side of (2.293) reduces to E_{xx}, and we recover (2.181).

Another commonly used measure of the elongation is the *stretch ratio*, λ, defined by

$$\lambda \equiv \frac{|\mathbf{dx}^*|}{|\mathbf{dx}|} = (1 - 2E_{xx})^{1/2}. \tag{2.294}$$

The stretch ratio of an element initially lying along the y direction would be $(1 - 2E_{yy})^{1/2}$, etc. As there may also, in general, be a rotational component to the deformation, a line segment that initially lies in the x direction will not necessarily continue to lie in that direction after the deformation.

Now consider two infinitesimal elements that are initially at right angles to each other, for example, $\mathbf{dx}_1 = (dx, 0, 0)$ and $\mathbf{dx}_2 = (0, dy, 0)$. These elements can be identified with *PQ* and *PR* of Fig. 2.12a. We now take the inner product of these two elements after deformation, as in the calculations given by (2.214)–(2.221):

$$\begin{aligned} |\mathbf{dx}_1^*||\mathbf{dx}_2^*| \cos \angle R^*P^*Q^* &= (\mathbf{dx}_1^*) \cdot (\mathbf{dx}_2^*) = (\mathbf{dx}_1^*)^{\mathrm{T}}(\mathbf{dx}_2^*) \\ &= (\mathbf{F}\mathbf{dx}_1)^{\mathrm{T}}(\mathbf{F}\mathbf{dx}_2) = (\mathbf{dx}_1)^{\mathrm{T}}\mathbf{F}^{\mathrm{T}}\mathbf{F}(\mathbf{dx}_2) = (\mathbf{dx}_1)^{\mathrm{T}}(\mathbf{I}-2\mathbf{E})(\mathbf{dx}_2) = [dx\ 0\ 0] \\ &\times \begin{bmatrix} 1-2E_{xx} & -2E_{xy} & -2E_{xz} \\ -2E_{yx} & 1-2E_{yy} & -2E_{yz} \\ -2E_{zx} & -2E_{zy} & 1-2E_{zz} \end{bmatrix} \begin{bmatrix} 0 \\ dy \\ 0 \end{bmatrix} = -2E_{xy}dxdy. \end{aligned} \tag{2.295}$$

The lengths of the deformed elements $|\mathbf{dx}_1^*|$ and $|\mathbf{dx}_2^*|$ are given by (2.292), so

$$\begin{aligned} \cos \angle R^*P^*Q^* &= \frac{-2E_{xy}dxdy}{(1-2E_{xx})^{1/2}dx(1-2E_{yy})^{1/2}dy} \\ &= \frac{-2E_{xy}}{(1-2E_{xx})^{1/2}(1-2E_{yy})^{1/2}}. \end{aligned} \tag{2.296}$$

Hence, the change in the angle formed by two initially perpendicular elements can be found in terms of the components of the finite strain tensor, although the relationship is not as simple as in the infinitesimal case. If all of the strain components are small, then, following along the lines of (2.220)–(2.221), it can be shown that the increase in the angle is given by $2E_{xy}$, and we recover (2.221).

The relationship between the finite strain tensor and the infinitesimal strain tensor can be investigated more thoroughly by recalling that, for example, $x^* = x - u$, in which case

$$\frac{\partial x^*}{\partial x} = 1 - \frac{\partial u}{\partial x}; \quad \frac{\partial x^*}{\partial y} = -\frac{\partial u}{\partial y}; \quad \frac{\partial x^*}{\partial z} = -\frac{\partial u}{\partial z}, \tag{2.297}$$

and similarly for y^* and z^*. Therefore, $\mathbf{F} = \mathbf{I} - \nabla\mathbf{u}$, and so

$$\begin{aligned} 2\mathbf{E} &= \mathbf{I} - \mathbf{F}^{\mathrm{T}}\mathbf{F} = \mathbf{I} - (\mathbf{I} - \nabla\mathbf{u})^{\mathrm{T}}(\mathbf{I} - \nabla\mathbf{u}) = \mathbf{I} - [\mathbf{I} - (\nabla\mathbf{u})^{\mathrm{T}}][\mathbf{I} - \nabla\mathbf{u}] \\ &= \mathbf{I} - [\mathbf{I} - (\nabla\mathbf{u})^{\mathrm{T}} - \nabla\mathbf{u} + (\nabla\mathbf{u})^{\mathrm{T}}\nabla\mathbf{u}] = (\nabla\mathbf{u})^{\mathrm{T}} + \nabla\mathbf{u} - (\nabla\mathbf{u})^{\mathrm{T}}\nabla\mathbf{u}, \\ &\text{that is,}\quad \mathbf{E} = \frac{1}{2}[\nabla\mathbf{u} + (\nabla\mathbf{u})^{\mathrm{T}} - (\nabla\mathbf{u})^{\mathrm{T}}\nabla\mathbf{u}]. \end{aligned} \tag{2.298}$$

Comparison of (2.298) with (2.223) shows that in addition to terms that are *linear* in the components of the displacement gradient, the finite strain tensor also contains terms that are *quadratic* in the components of the displacement gradient. If all of the components of $\nabla\mathbf{u}$ are small, the quadratic terms can be neglected, and $\mathbf{E}$ reduces to the infinitesimal strain tensor, $\boldsymbol{\varepsilon}$. The individual components of $\mathbf{E}$, as defined by (2.298), are

$$E_{xx} = \frac{\partial u}{\partial x} - \frac{1}{2}\left[\left(\frac{\partial u}{\partial x}\right)^2 + \left(\frac{\partial v}{\partial x}\right)^2 + \left(\frac{\partial w}{\partial x}\right)^2\right], \tag{2.299}$$

$$E_{yy} = \frac{\partial v}{\partial y} - \frac{1}{2}\left[\left(\frac{\partial u}{\partial y}\right)^2 + \left(\frac{\partial v}{\partial y}\right)^2 + \left(\frac{\partial w}{\partial y}\right)^2\right], \tag{2.300}$$

$$E_{zz} = \frac{\partial w}{\partial z} - \frac{1}{2}\left[\left(\frac{\partial u}{\partial z}\right)^2 + \left(\frac{\partial v}{\partial z}\right)^2 + \left(\frac{\partial w}{\partial z}\right)^2\right], \tag{2.301}$$

$$E_{xy} = E_{yx} = \frac{1}{2}\left[\frac{\partial u}{\partial y} + \frac{\partial v}{\partial x} - \left(\frac{\partial u}{\partial x}\frac{\partial u}{\partial y} + \frac{\partial v}{\partial x}\frac{\partial v}{\partial y} + \frac{\partial w}{\partial x}\frac{\partial w}{\partial y}\right)\right], \tag{2.302}$$

$$E_{xz} = E_{zx} = \frac{1}{2}\left[\frac{\partial u}{\partial z} + \frac{\partial w}{\partial x} - \left(\frac{\partial u}{\partial x}\frac{\partial u}{\partial z} + \frac{\partial v}{\partial x}\frac{\partial v}{\partial z} + \frac{\partial w}{\partial x}\frac{\partial w}{\partial z}\right)\right], \tag{2.303}$$

$$E_{yz} = E_{zy} = \frac{1}{2}\left[\frac{\partial v}{\partial z} + \frac{\partial w}{\partial y} - \left(\frac{\partial u}{\partial y}\frac{\partial u}{\partial z} + \frac{\partial v}{\partial y}\frac{\partial v}{\partial z} + \frac{\partial w}{\partial y}\frac{\partial w}{\partial z}\right)\right]. \tag{2.304}$$

3 Friction on rock surfaces

3.1 Introduction

Friction is the phenomenon by which a tangential shearing force is required in order to displace two contacting surfaces along a direction parallel to their nominal contact plane. The study of friction is of great importance in rock mechanics. Its effects arise on all scales: (i) on the microscopic scale, in which friction occurs between faces of minute "Griffith" cracks, (ii) on a somewhat larger scale in which it occurs between individual grains or pieces of aggregate, and (iii) on an even larger scale, on the order of many square meters, over which friction between fault surfaces occurs.

Modern knowledge of friction began with the work of Amontons in France at the end of the seventeenth century. Much experimental work on friction between solid surfaces, and theoretical attempts to explain friction in terms of the geometry and physical properties of the contacting surfaces, is summarized in the monograph of Bowden and Tabor (1985). More recently, technical advances such as atomic force microscopy have yielded new insights into the molecular basis of solid–solid friction (Krim, 1996). However, most of this understanding has been developed for metals and other engineering solids, and its applicability to rock surfaces remains uncertain. Hence, although an atomic-level explanation of rock friction remains elusive, many aspects of the frictional behavior of rock surfaces are indeed understood on a phenomenological level.

In §3.2, the classical theory of friction between solid (typically metallic) surfaces is discussed. Friction on rock surfaces is treated in §3.3. The interesting phenomenon of "stick–slip" between two sheared surfaces in contact is discussed in §3.4. The sliding of two adjacent pieces of rock along a contact surface or indeed any "plane of weakness" is analyzed in some detail in §3.5. Finally, §3.6 treats some aspects of rock friction, namely the effects of time and velocity, which are not accounted for in the simple classical theory.

Frictional forces act along faces of cracks, fractures, and faults in rock. These structures are discussed in more detail in later chapters. However, the basic ideas of rock friction present a simple context in which to utilize the Mohr's circle construction. They also serve as a starting point for the development of failure laws for intact rock. Finally, in terms of historical development, friction was perhaps the first aspect of rock behavior to be studied and partially understood.

For these reasons, it seems fitting to place the treatment of friction at this early point in the text.

3.2 Amontons' law

Suppose that two bodies in contact over an approximately planar surface are pressed together by a normal force N (Fig. 3.1a). The apparent (macroscopic, nominal) area of contact is A. If a tangential shearing force is applied parallel to the plane of contact, it is observed that this force must reach some critical value T in order for sliding to occur. The relationship between T and N may be written as

$$T = \mu N, \tag{3.1}$$

where μ is called the *coefficient of friction*. This coefficient will depend on the nature of the two surfaces. It might also be expected to depend on the normal load N, and/or on the nominal surface area A. However, to a good approximation, in many cases this coefficient is independent of both the normal load and the apparent area of contact – an empirical observation that is known as *Amontons' law*, first enunciated in 1699. In this case the shearing force needed to cause relative motion between the two surfaces is directly proportional to the normal force pressing the surfaces together. Dividing both sides of (3.1) by the nominal area A yields

$$\tau = \mu\sigma, \tag{3.2}$$

where τ is the shear component of the traction acting along the contact plane, and σ is the normal component. These are usually somewhat loosely referred to as the *shear stress* and *normal stress* acting on the contact plane.

Amontons had in fact found that μ was also independent of the materials and the surface finish of the contact region, with a value of about 1/3. More accurate experiments by other researchers later showed that μ may vary from values close to 0, for Teflon–Teflon contacts, to values as high as 1.5, for nickel–nickel contacts (Bowden and Tabor, 1985).

Equation (3.2) refers to the condition necessary to initiate sliding, starting from a state of rest. Hence, μ is more properly referred to as the coefficient of *static* friction. Once sliding has been initiated, however, it is found that the

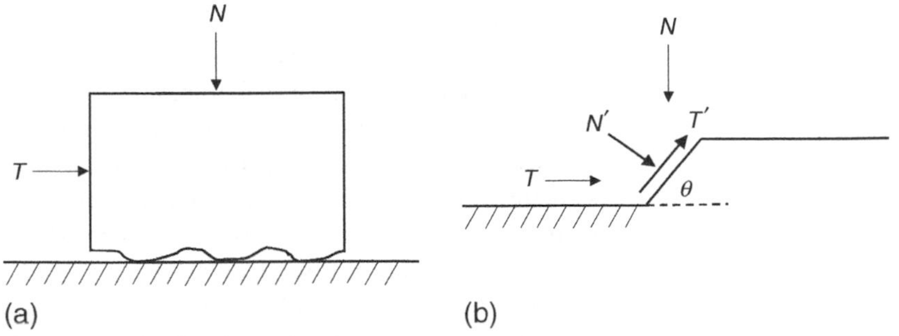

Fig. 3.1 Two bodies in contact along (a) a nominally planar surface, and (b) a surface having a local region inclined at an angle θ, acted on by a normal force N and a tangential force T.

shear stress required to maintain motion is somewhat less than $\mu\sigma$. In fact, the condition for maintaining relative motion at constant velocity is $\tau = \mu_d\sigma$, where $\mu_d < \mu$ is the coefficient of *dynamic* friction. For many engineering materials, μ_d is found to be nearly independent of the relative velocity of the two surfaces – an approximate "law" that was first noted by the French physicist Coulomb in 1785.

For irregular surfaces, it may happen that sliding takes place up a plane inclined by some small angle θ to the nominal plane of contact (Fig. 3.1b). The normal and tangential forces acting on this inclined surface are

$$N' = N\cos\theta + T\sin\theta, \quad T' = T\cos\theta - N\sin\theta. \tag{3.3}$$

If sliding along the incline is governed by $T' = \mu N'$, then the two force components acting along the nominal plane of contact are related by

$$T = N(\tan\theta + \mu)/(1 - \mu\tan\theta). \tag{3.4}$$

For small angles,

$$T = [\mu + (1 + \mu^2)\theta]N, \tag{3.5}$$

so that there is an increase in the *apparent* coefficient of friction, by an amount proportional to θ.

3.3 Friction on rock surfaces

Frictional effects are of importance in rock mechanics mainly in two connections. On a small scale, friction may occur between the facing surfaces of microcracks. On a larger scale, friction occurs between the surfaces of joints or fracture planes. The surfaces in question may be new surfaces, such as along a tension joint on which no sliding has yet occurred, or they may be old surfaces on which considerable relative motion has already taken place. The scale and condition of the surfaces on which laboratory measurements are made will usually lie somewhere between these two extremes.

Measurements of rock friction can be made using any of the configurations shown in Fig. 3.2. In configuration (a), essentially the same as shown in Fig. 3.1a, the two rock surfaces are pressed together by a normal force N and sheared by a tangential force T (Penman, 1953; Bowden, 1954; Horn and Deere, 1962). In practice, this configuration requires elaborate precautions to ensure that the normal load is applied uniformly across the surface. Hoskins et al. (1968) used

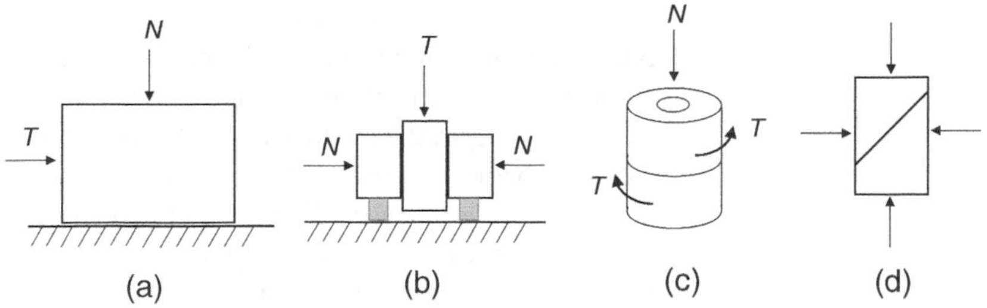

Fig. 3.2 Several configurations used in measuring friction along rock surfaces.

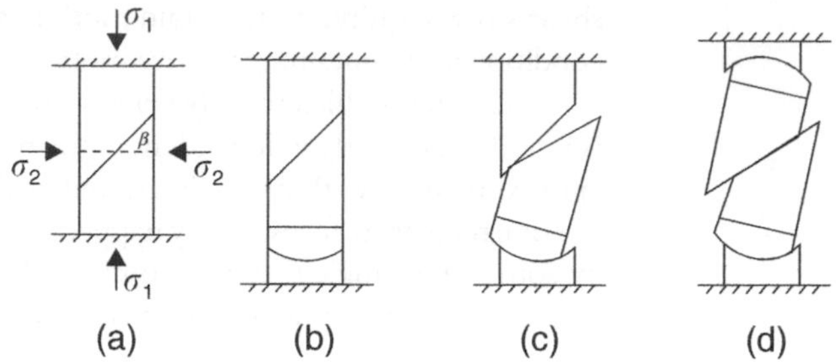

Fig. 3.3 Different configurations used for measuring friction in a triaxial testing machine.

configuration (b), with the normal forces N applied by flat pressure cells and the tangential force T applied by a standard rock-testing machine (§6.4). In this configuration, the amount of contact area will be constant, but the contact will not occur over precisely the same surfaces throughout the process. The rotary system (c) has the advantage that the same surfaces will be in contact, regardless of the amount of displacement that has occurred.

The most commonly used configuration is (d), in which a cylindrical core is sliced at an angle β to the diametral plane, and placed in a standard "triaxial" testing cell. The specimen is jacketed in rubber or copper, lateral pressure $\sigma_2 = \sigma_3$ is applied to its lateral curved surface by oil pressure, and an axial stress σ_1 is applied by the testing machine. This method allows high normal and tangential stresses to be applied to the sample, but has the disadvantage that a change in geometry, or stress, or both, occurs after any slippage between the two surfaces.

Three variations of this configuration may be used (Fig. 3.3). In configuration (a), the cylindrical specimen is situated between rigid platens whose faces are perpendicular to the maximum principal stress σ_1. After initial slip has occurred, a lateral stress is produced whose magnitude is not known. In (b), a single spherical seat is used, in which case the specimen rotates into configuration (c) as slip occurs, after which contact no longer exists over the original planar surfaces of elliptical cross section. If two spherical seats are used, as in (d), the specimen halves will rotate so as to maintain the angle of contact. However, the amount of contact area will decrease as slip proceeds, and lateral forces are again produced. Hence, in each configuration, either the geometry or the forces change as slippage occurs. Nevertheless, for small amounts of slip, consistent results can in fact be obtained using all three methods.

Jaeger (1959), Byerlee (1967) and others have used the method of a single spherical seat, (b). The experiments can be conducted using any angle β between about 30° and 65°. The inferred values of μ might be expected to vary with β, but in fact are found to be nearly independent of the angle, indicating that the changes in geometry shown in Fig. 3.3c do not greatly contaminate the results.

Consider again the configuration shown in Fig. 3.2b. According to the simple classical theory as expressed by Amontons' law, no sliding takes place when $T < \mu N$, although a small elastic deformation of the central block will occur. After T reaches the value μN, slippage will commence, starting at point P. If this slippage is constrained to occur at constant velocity, the tangential force will be

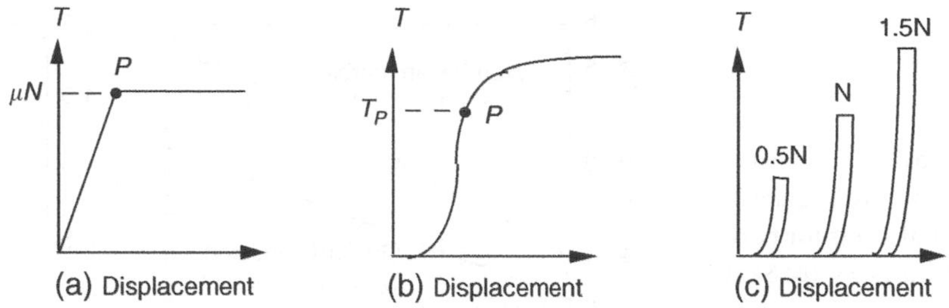

Fig. 3.4 (a) Force–displacement curve for idealized frictional behavior. (b) Schematized force–displacement curve obtained by Hoskins et al. (1968) for a moderately rough trachyte surface in an apparatus similar to that of Fig. 3.2b. (c) Subsequent force–displacement curves for the same surfaces, under normal loads that are 0.5, 1.0, and 1.5 times that used in (b).

$T = \mu_d N$, where μ_d is the *coefficient of dynamic friction*. This idealized behavior is shown schematically in Fig. 3.4a, where, for simplicity, we take $\mu_d = \mu$.

An example of actual data obtained by Hoskins et al. (1968) on a moderately rough trachyte surface is shown in Fig. 3.4b, although the results are typical of those found for other rocks. The initial elastic displacement is seen to be slightly nonlinear. Sliding begins at P, and if the coefficient of friction is defined based on the point of first slippage, we would find $\mu = T_P/N$. But the tangential force needed to produce continued sliding in fact increases, approaching an asymptotic value after a displacement of about 2 cm. This increase is attributable to the development of a new surface of sliding, which contains slickensides and detrital material. If sliding is reinitiated under a larger value of N, a constant value of T is attained more rapidly (Fig. 3.4c).

If the data obtained by Hoskins et al. (1968) are interpreted in accordance with (3.3), in which $\mu = T/N = \tau/\sigma$, it is found that the coefficient of friction decreases with increasing normal stress, stabilizing at high values of σ (Fig. 3.5b). However, if τ is plotted against σ, as in Fig. 3.5a, the data fall on nearly straight lines. This suggests using the following equation to describe rock friction:

$$\tau = S_0 + \mu\sigma, \qquad (3.6)$$

which was originally proposed by Coulomb in 1785. Borrowing the terminology used in soil mechanics, S_0 is called the *cohesion* of the surface, and μ the *coefficient of friction*. As this equation gives a different definition of μ than does (3.3), we henceforth use μ^* to refer to the ratio τ/σ. Thus, μ^* is the secant of the τ–σ curve, and μ is the (best-fit) tangent of that curve. If (3.6) is indeed obeyed, these two friction coefficients are related by $\mu^* = \mu + (S_0/\sigma)$.

Values of the friction coefficient μ of various rocks and minerals are shown in Table 3.1. The measured values vary with rock type, and also depend on surface finish and whether or not the surface is wet or dry. The presence of water increases μ in some cases, decreases it in others, and in some cases has little effect.

Byerlee (1978) found that the frictional behavior of a broad range of common rock types could be fit by the following bilinear empirical expression:

$$\begin{aligned} \tau &= 0.85\sigma \quad \text{for } \sigma < 200\text{ MPa}, \\ \tau &= 50\text{ MPa} + 0.6\sigma \quad \text{for } 200 < \sigma < 1700\text{ MPa}, \end{aligned} \qquad (3.7)$$

which is often referred to as *Byerlee's law*. This law is equivalent to taking a different pair of values for S_0 and μ in the two ranges $\{\sigma < 200\text{ MPa}\}$ and

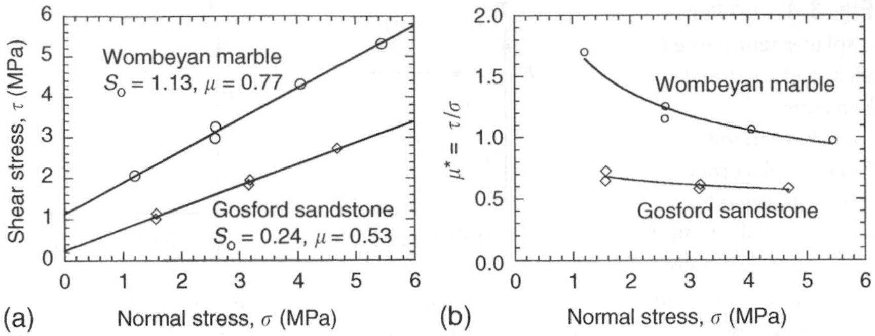

Fig. 3.5 (a) Shear stress plotted against normal stress for two rock surfaces. (b) Same data, with $\mu^* = \tau/\sigma$ plotted against σ (Hoskins et al., 1968).

Table 3.1 Coefficient of friction of some rocks and minerals.

Mineral	μ	Mineral	μ	μ (wet)
NaCl [B,s]	0.7	Quartz [HD]	0.11	0.42
PbS [B,s]	0.6	Quartz [P]	0.19	0.65
S [B,s]	0.5	Feldspar [HD]	0.11	0.46
Al_2O_3 [B,s]	0.4	Calcite [HD]	0.14	0.68
Ice [B,s]	0.5	Muscovite [HD]	0.43	0.23
Glass [B,s]	0.7	Biotite [HD]	0.31	0.13
Diamond [B,s]	0.1	Serpentine [HD]	0.62	0.29
Diamond [B,s,c]	0.3	Talc [HD]	0.36	0.16
Rock	**μ**	**Mineral**	**μ**	
Sandstone [R]	0.68	Trachyte [H,l,f]	0.63	
Sandstone [J,t,n]	0.52	Trachyte [H,l,c]	0.68	
Sandstone [H,l,r]	0.51	Trachyte [H,l,c,w]	0.56	
Sandstone [H,l,r,w]	0.61	Marble [H,l,f]	0.75	
Granite [By,t,n,c]	0.60	Marble [J,t,n]	0.62	
Granite [By,t,n,c,w]	0.60	Porphyry [J,t,n]	0.86	
Granite [H,l,g]	0.64	Dolomite [H,s,t,c]	0.40	
Gneiss [J,t,n]	0.71	Gabbro [H,l,f]	0.18	
Gneiss [J,t,n,w]	0.61	Gabbro [H,l,c]	0.66	

Notation: B = Bowden and Tabor (1985), By = Byerlee (1967), H = Hoskins et al., (1968), HS = Handin and Stearns (1964), HD = Horn and Deere (1962), J = Jaeger (1959), P = Penman (1953), R = Rae (1963); l = large surface, s = small surface, t = triaxial test, r = rough surface, c = coarsely ground surface, f = finely ground surface, n = natural shear surface, w = wet surface, d = dry surface.

$\{\sigma > 200\ \text{MPa}\}$. Notable exceptions to this law include clays and other sheet silicates (Lockner, 1995).

3.4 Stick–slip oscillations

If rock friction tests are conducted on finely ground or polished surfaces, a phenomenon known as "stick–slip" is sometimes found to occur (Fig. 3.6a). If a shear load T is applied as the normal load N is held constant, the displacement increases gradually as the shear force T increases, until at some point it jumps abruptly, after which the surfaces lock together again, and the force drops to a lower value. This process continues, with each sticking phase followed by a slip

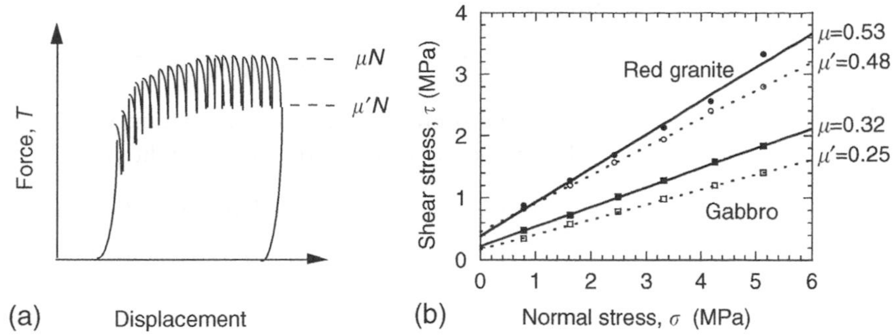

Fig. 3.6 (a) Force–displacement for a granite surface in the apparatus of Fig. 3.2b, under different normal loads. (b) Variation of maximum and minimum shear stress as a function of the normal load (Hoskins et al., 1968).

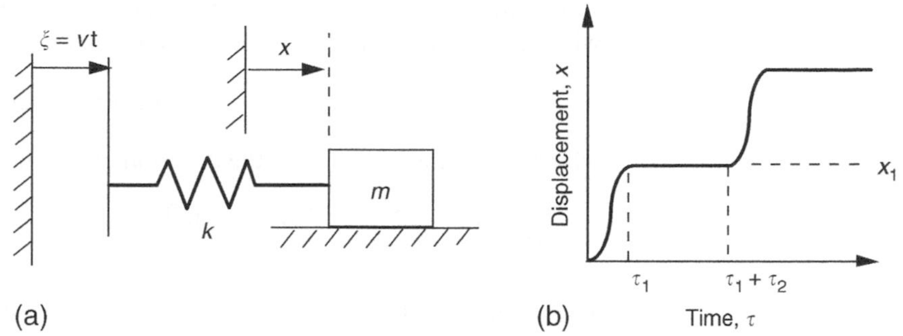

Fig. 3.7 (a) Model of a simple mechanical system that will exhibit stick–slip oscillations. (b) Motion of this system as a function of time.

phase of very short duration. After several cycles the maximum and minimum values of the shear load T each reach asymptotic values. If these asymptotic values of $T_{\max}$ and $T_{\min}$ are plotted against N, straight lines of the form

$$|\tau| = S_o + \mu\sigma, \quad \tau = S_o' + \mu'\sigma, \tag{3.8}$$

are obtained (Fig. 3.6b). As μ is obtained from the values of T that occur immediately before slip occurs, it represents the coefficient of *static* friction.

A simple model for a system exhibiting stick–slip oscillations can be formulated that requires only the assumption that $\mu_d < \mu$. Consider a body of mass m resting on a horizontal surface, attached on one side to a spring having stiffness k (Fig. 3.7a). The normal force exerted by the mass on the surface is mg. (An additional normal load N could be assumed to be pressing the mass against the surface, but this adds an additional parameter into the equations without giving qualitatively different behavior.) Let the free end of the spring be moved to the right at a constant velocity v, in which case the position of the free end is given by $\xi = vt$. If the position x of the mass is measured from its initial location, the spring is compressed by an amount $\xi - x = vt - x$, and the rightward force exerted by the spring on the mass will be $T = k(vt - x)$. If the mass is initially at rest, it will not move until the force T reaches μmg, which will occur at time $t^* = \mu mg/kv$. It is convenient to use the elapsed time as the independent variable since the initial movement of the mass, that is, $\tau = t - t^*$, so that $T = k(v\tau + vt^* - x)$.

At subsequent times, the mass will be in motion, and this motion will be resisted by a frictional force, $F = -\mu_d mg$. A horizontal force balance on the mass gives

$$m\ddot{x} = k(v\tau + vt^* - x) - \mu_d mg, \tag{3.9}$$

where the overdot denotes differentiation with respect to the elapsed time, τ. The solution to (3.9), for the initial conditions $x = \dot{x} = 0$ when $\tau = 0$, is

$$x = v\tau - (v/\omega)\sin\omega\tau - [(\mu - \mu_d)g/\omega^2]\cos\omega\tau + (\mu - \mu_d)g/\omega^2, \tag{3.10}$$

$$\dot{x} = v - v\cos\omega\tau + [(\mu - \mu_d)g/\omega]\sin\omega\tau. \tag{3.11}$$

where $\omega = \sqrt{k/m}$ is the undamped natural frequency of the system, and where use has been made of the fact that $kvt^* = \mu mg$.

The mass will move to the right in accordance with (3.10), coming to rest when $\dot{x} = 0$. From (3.11), the "slip" phase will therefore end when

$$-(\mu - \mu_d)g/\omega v = (1 - \cos\omega\tau_1)/\sin\omega\tau_1 \equiv \tan(\omega\tau_1/2), \tag{3.12}$$

which can be inverted to give the duration of the slip phase,

$$\tau_1 = \frac{2\pi}{\omega} - \frac{2}{\omega}\tan^{-1}\left[\frac{(\mu - \mu_d)g}{\omega v}\right]. \tag{3.13}$$

The displacement at this time is found from (3.10), (3.12), and (3.13) to be given by

$$x_1 \equiv x(\tau_1) = v\tau_1 + 2(\mu - \mu_d)g/\omega^2, \tag{3.14}$$

and the force exerted on the mass by the spring at this time will be

$$T_1 \equiv T(\tau_1) = k(v\tau_1 + vt^* - x_1) = \mu mg - 2k(\mu - \mu_d)g/\omega^2 = (2\mu_d - \mu)mg. \tag{3.15}$$

Noting that $\mu_d < \mu$, the spring force at time τ_1 will be less than μmg, and so the mass will be at rest. After this time, the spring will continue to compress, but the mass will be in its "sticking" phase. Hence, T_1 represents the *minimum* spring force that occurs at any time during the process, and so comparison with (3.8) shows that $2\mu_d - \mu = \mu'$, implying that the dynamic friction coefficient is equal to the mean value of $\mu_{max} = \mu$ and $\mu_{min} = \mu'$.

Slip will recommence when the spring force again reaches μmg. This will occur after an additional *elapsed* time of

$$\tau_2 = 2(\mu - \mu_d)mg/kv = 2(\mu - \mu_d)g/\omega v, \tag{3.16}$$

after which the cycle will repeat (Fig. 3.7b). The duration of the stick phase is τ_2, and so the stick–slip cycle will have a total period of $\Delta\tau = \tau_1 + \tau_2$. Comparison of (3.13) and (3.16) shows that

$$(\omega\tau_1/2) = \pi - \tan^{-1}(\omega\tau_2/2). \tag{3.17}$$

The usual situation is that the velocity v is "small," in the sense that $v \ll (\mu - \mu_d)g/\omega$, in which case (3.13) and (3.16), or (3.17), imply that $\tau_1 \ll \tau_2$. Hence, from (3.14), the displacement occurring over each cycle will be approximately given by

$$\Delta x \approx 2(\mu - \mu_d)g/\omega^2 = 2(\mu - \mu_d)mg/k. \tag{3.18}$$

But the maximum and minimum spring forces during each cycle are $T_{max} = \mu mg$ and $T_{min} = (2\mu_d - \mu)mg$, so

$$\Delta x(\text{cycle}) = (T_{max} - T_{min})/k. \tag{3.19}$$

Brace and Byerlee (1966) suggested that stick–slip between opposing faces of a fault may provide a mechanism for earthquakes. In this context, (3.19) would give a relationship between the amount of slip along a fault, the stress drop associated with that slip, and the "stiffness" of the surrounding rock mass. Burridge and Knopoff (1967) generalized the model by considering a chain of masses $\{m_1, m_2, \ldots, m_N\}$, connected in series by a set of springs with stiffnesses $\{k_1, k_2, \ldots, k_N\}$. This model predicts chaotic sequences consisting of many small slips and occasional large ones, which bear a resemblance to earthquake sequences. Further developments along these lines have been made by Mora and Place (1994) and others.

3.5 Sliding on a plane of weakness

In two dimensions, suppose that the rock has a preexisting plane of weakness whose outward unit normal vector makes an angle β with the direction of the maximum principal stress, σ_1 (Fig. 3.8a). The criterion for slippage to occur along this plane is assumed to be

$$|\tau| = S_0 + \mu\sigma, \tag{3.20}$$

as in (3.6), where σ is the normal traction component acting along this plane, and τ is the shear component. By (2.54) and (2.55), σ and τ are given by

$$\sigma = \frac{1}{2}(\sigma_1 + \sigma_2) + \frac{1}{2}(\sigma_1 - \sigma_2)\cos 2\beta, \tag{3.21}$$

$$\tau = -\frac{1}{2}(\sigma_1 - \sigma_2)\sin 2\beta. \tag{3.22}$$

These expressions may also be written as

$$\sigma = \sigma_m + \tau_m \cos 2\beta, \quad \tau = -\tau_m \sin 2\beta, \tag{3.23}$$

where $\sigma_m = (\sigma_1 + \sigma_2)/2$ is the mean normal stress, and $\tau_m = (\sigma_1 - \sigma_2)/2$ is the maximum shear stress.

If we define the *angle of internal friction*, ϕ, by the relation

$$\mu = \tan\phi, \tag{3.24}$$

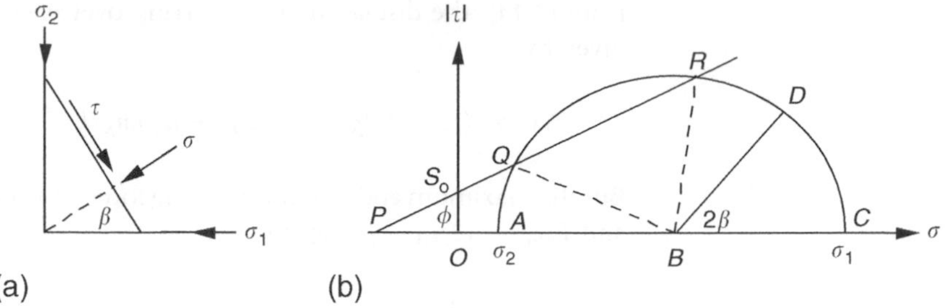

Fig. 3.8 (a) Plane of weakness with outward normal vector oriented at angle β to the direction of maximum principal stress. (b) Situation described on a Mohr diagram.

then condition (3.20) can be written in the following equivalent forms:

$$\tau_m(\sin 2\beta - \tan\phi\cos 2\beta) = S_o + \sigma_m \tan\phi, \tag{3.25}$$

$$\tau_m = (\sigma_m + S_o \cot\phi)\sin\phi \operatorname{cosec}(2\beta - \phi). \tag{3.26}$$

$$\sigma_1[\sin(2\beta - \phi) - \sin\phi] - \sigma_2[\sin(2\beta - \phi) + \sin\phi] = 2S_o\cos\phi, \tag{3.27}$$

$$\sigma_1 - \sigma_2 = \frac{2(S_o + \mu\sigma_2)}{(1 - \mu\cot\beta)\sin 2\beta}, \tag{3.28}$$

$$\sigma_1 = \frac{2S_o\cos\phi}{(1-k)\sin(2\beta - \phi) - (1+k)\sin\phi}, \tag{3.29}$$

where $k = \sigma_2/\sigma_1 \leq 1$. The stress difference that would be required to cause slippage can be found from (3.28), as a function of β, for a fixed value of the minor principal stress σ_2 (Fig. 3.9). If the major principal stress is aligned with the plane of weakness, then $\beta \to \pi/2$, and (3.28) shows that $\sigma_1 - \sigma_2 \to \infty$. The stress difference required to cause slippage also becomes infinite as $\beta \to \phi$. As the right-hand side of (3.28) must by definition be nonnegative, solutions can exist only for $\phi < \beta < \pi/2$. By differentiating (3.31), the minimum value of σ_1 needed to cause slippage is found to occur when

$$\tan 2\beta = -1/\mu, \tag{3.30}$$

and this minimum value is found to be

$$\sigma_1 = \sigma_2 + 2(S_o + \mu\sigma_2)[(1+\mu^2)^{1/2} + \mu]. \tag{3.31}$$

This problem can be analyzed with the aid of the Mohr diagram, Fig. 3.8b. The condition (3.20) for failure is represented by the straight line *PQR* that is oriented at angle β to the σ-axis, and intersects the $|\tau|$-axis at S_o and the σ-axis at $-S_o\cot\phi$. The normal and shear tractions along the plane of weakness are represented by the point *D* in the σ–τ plane. Slippage will not occur if the point *D* lies within either of the arcs *AQ* or *RC* of the Mohr circle, but will occur if *D* lies within arc *QR*. The limiting condition for slippage to occur can be found by imagining *D* to be located at either *Q* or *R*. Taking *D* to coincide with *R*, the angle *RBC* will equal 2β, and so the angle *PRB* will equal $2\beta - \phi$. Applying the law of sines to triangle *PRB* shows

$$\frac{BR}{\sin\phi} = \frac{PB}{\sin(2\beta - \phi)}. \tag{3.32}$$

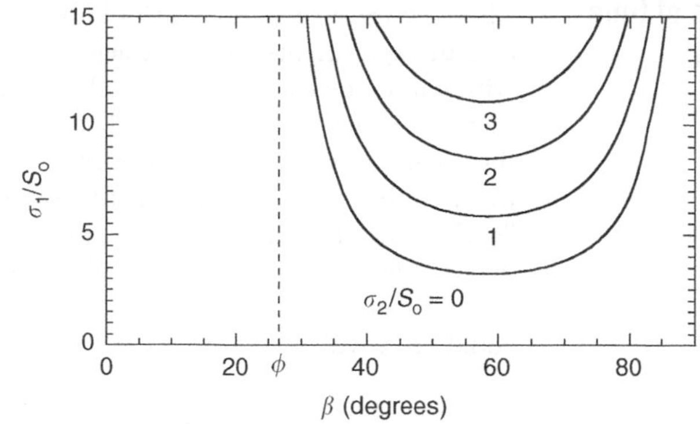

Fig. 3.9 Variation of σ_1 needed to cause sliding on a plane of weakness, for $\mu = 0.5$.

But $|BR| = \tau_m$, and $|PB| = |PO| + |OB| = S_o \cot\phi + \sigma_m$, so

$$\tau_m \sin(2\beta - \phi) = (\sigma_m + S_o \cot\phi)\sin\phi, \tag{3.33}$$

which is equivalent to (3.26). The same result would be obtained by imagining D to coincide with Q.

For a given state of stress and a given value of μ, there are two values of β that will satisfy (3.33), corresponding to whether D coincides with R or Q. These values are

$$2\beta_1 = \phi + \sin^{-1}\{[(\sigma_m + S_o \cot\phi)/\tau_m]\sin\phi\}, \tag{3.34}$$

$$2\beta_2 = \pi + \phi - \sin^{-1}\{[(\sigma_m + S_o \cot\phi)/\tau_m]\sin\phi\}. \tag{3.35}$$

Sliding will occur on any plane for which $\beta_1 \leq \beta \leq \beta_2$.

If the cohesion S_o were set to zero, the slippage line (3.20) would move to the right by the amount $|PO| = S_o \cot\phi = S_o/\mu$. If, at the same time, S_o/μ were added to each of the two principal stresses, the Mohr circle would translate to the right by this same amount, and the relative orientation of the Mohr circle and the slippage line would remain unchanged. Hence, results for the case of principal stresses σ_1 and σ_2, with a finite value of S_o, are identical to those for the case of principal stresses $\sigma_1 + (S_o/\mu)$ and $\sigma_2 + (S_o/\mu)$, with $S_o = 0$.

If the joint is filled with pore fluid under a pressure p, it is experimentally found that the theory will continue to apply if the amount p is first subtracted from each of the principal stresses (Byerlee, 1967). The values $\sigma_1 - p$, etc., are often referred to in this context as the *effective principal stresses*.

The basic theory outlined above has three fundamental applications of great importance. The most obvious is the study of sliding across an open joint or cut rock surface for which the criterion for slip has been experimentally found to be given by (3.6). If a joint is filled with a weaker material than that comprising the adjacent rock, the same theory will apply, with S_o being the cohesion and μ the coefficient of internal friction of the infilling material, in the sense of the Coulomb theory of failure described in §4.5. Finally, this theory will also describe the behavior of an anisotropic material possessing parallel planes of weakness.

3.6 Effects of time and velocity

Although most rocks obey a Coulomb-type friction law to a first approximation, careful experiments have revealed that the apparent friction coefficient varies with parameters such as time and velocity. Dieterich (1972) performed experiments in which τ and σ were held constant for an interval of time t, after which the shear stress was rapidly increased to the level required to produce slip. In this manner, the coefficient of static friction was measured as a function of the total "stick" time. Results for granite, greywacke, quartzite, and sandstone, under different levels of normal stress, all showed a small but measurable increase in μ as a function of the time of stick (Fig. 3.10a). The data from each experiment could be fit with an equation of the form

$$\tau = \mu\sigma \quad \mu = \mu_0 + a \ln t. \tag{3.36}$$

The parameter μ_0 was found to be in the range of 0.7–0.8, and decreased slightly as the normal stress was increased. The dimensionless parameter a, which is the rate of increase in μ with the natural logarithm of the stick time, was found to be insensitive to normal stress, and took on values of 0.0096, 0.0052, 0.0087, and 0.0069 for granite, greywacke, quartzite, and sandstone, respectively.

Equation (3.36) fits the data well, for stick times ranging from 1 to 10^5 s. Although stick times smaller than 1 s are probably not of physical interest, (3.36) nevertheless predicts unrealistic *negative* values of μ as t approaches zero. To remedy this, Dieterich (1978) proposed to represent the variation of μ with t by

$$\mu = \mu_0 + a \ln[1 + (t/t_0)], \tag{3.37}$$

which reduces to a form essentially equivalent to (3.36) for large t, but gives $\mu \approx \mu_0$ for $t \ll t_0$. Dieterich found that, for all of his data sets, $t_0 \approx 1$ s. Subsequent investigations by Scholz et al. (1972), Engelder et al. (1975) and others have shown that an increase in the coefficient of static friction with increasing stick time is a general characteristic of rock behavior under a variety of test conditions and surface properties.

The increase in the friction coefficient with time has been attributed to "indentation creep," which causes an increase in the actual area of asperity contact between the two contacting surfaces. Scholz and Engelder (1976) carried out microindentation tests on a natural quartz crystal and an olivine crystal, by

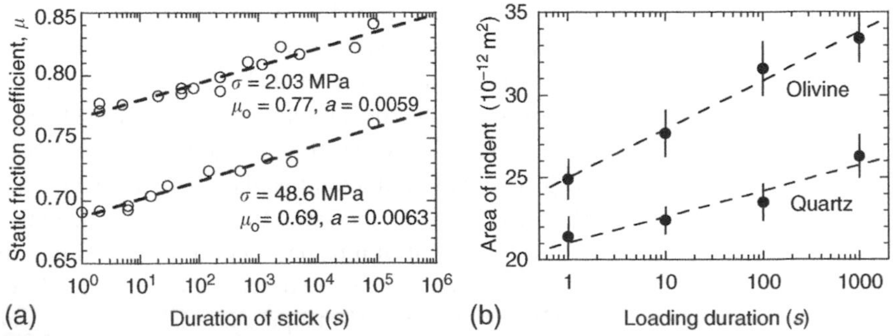

Fig. 3.10 (a) Dependence of the coefficient of static friction of quartz sandstone on the time of stick, for two different values of the normal stress (Dieterich, 1978). (b) Variation of the actual area of contact under a pyramidal indenter, as a function of loading time (Scholz and Engelder, 1976). Vertical lines indicate the range of values observed over twenty-five measurements.

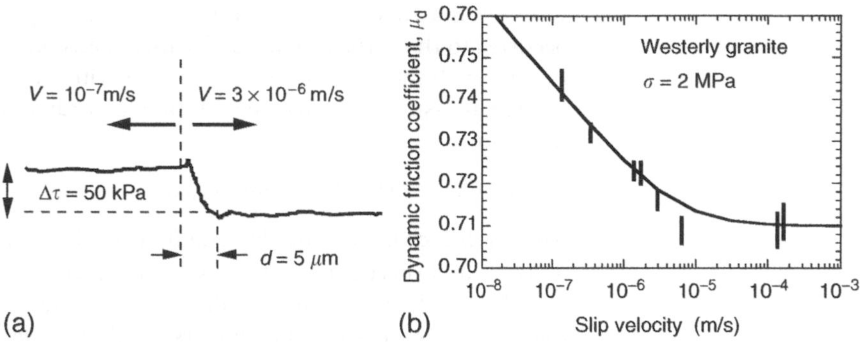

Fig. 3.11 (a) Variation of shear stress with displacement, before and after an abrupt change in slip velocity. (b) Variation of the coefficient of dynamic friction with slip velocity for Westerly granite at a normal stress of 20 MPa. Short vertical lines indicate the range of values observed at each given velocity; solid line is from (3.38), with $\mu_o = 0.71$, $t_o = 1$ s, $d = 5$ μm, and $a = 0.0087$ (Dieterich, 1978).

pressing a pyramidal "Vickers" diamond indenter into the surfaces under a constant normal load of 0.25 N. This load corresponds to a normal stress beneath the indenter of about 10–20 MPa. Twenty-five tests were conducted on each surface, for loading times of 1, 10, 100, and 1000 s. The actual area of the indentation was found to increase linearly with the logarithm of the indentation time (Fig. 3.11b). If the same process is assumed to occur at the asperities at which rough rock surfaces are in contact, and the friction coefficient is assumed to be proportional to the actual (as opposed to the nominal, macroscopic) area of contact, then this mechanism can qualitatively explain the increase in μ with time.

The dynamic friction coefficient acting between two rock surfaces typically exhibits a small velocity dependence (Scholz and Engelder, 1976; Dieterich, 1978), the origin of which can be explained as follows. As the two rock surfaces slide past each other, asperities that have been in contact will lose contact, and new asperities will come into contact. The mean lifetime of asperity contact is roughly given by $t \approx d/V$, where V is the relative velocity between the two surfaces, and d is a characteristic asperity diameter. If this asperity contact lifetime is identified with the contact time appearing in (3.37), a relationship is found between the dynamic friction coefficient and the sliding velocity:

$$\mu_d = \mu_o + a \ln[1 + (d/Vt_o)]. \quad (3.38)$$

The parameter d can also be identified with the slip distance required for μ to stabilize to a new value after an abrupt change in the velocity, and was found by Dieterich (1978) to equal about 5 μm for contact between two granite surfaces (Fig. 3.11a). Using this value of d, along with the values of μ_o and a obtained from the previously discussed static friction measurements, Dieterich (1978) found that (3.38) could accurately predict the measured variation of the coefficient of dynamic friction with velocity (Fig. 3.11b).

Ruina (1983), Tullis (1988) and others have proposed adding another term to (3.38) to account for the irreversible mechanical alteration of the rock surface that occurs during sliding. The resulting "rate/state" friction law can be expressed as

$$\tau = S_o + \mu_d \sigma, \quad \mu_d = \mu_o + a \ln(V/V_o) + b\psi, \quad (3.39)$$

where ψ is a state variable that in some sense represents the damage that has occurred to the surface, and a and b are dimensionless constants that reflect the rate at which the friction coefficient varies with changes in $\ln V$ and ψ. The state variable is assumed to change with time according to an evolution equation of the form

$$d\psi/dt = -(V/\lambda)[\psi + \ln(V/V_o)], \tag{3.40}$$

where λ is a characteristic slip length. The additional term in (3.39) allows the dynamic friction coefficient to vary as slip proceeds.

Although the variations in μ during sliding are usually small, it is mainly the *relative* magnitudes of the parameters a and b that determine whether or not the slip occurs in a stable or unstable (stick–slip) manner. The analysis given in §3.4 showed that, if the dynamic friction coefficient is independent of velocity and slip distance, the slip process will be unstable if $\mu_d < \mu$. Now consider an interface governed by a rate/state law of the form given by (3.39) and (3.40), with slip occurring at some constant velocity V_1. The state variable ψ will have already reached its steady-state value associated with velocity V_1. This steady-state value is found by setting $d\psi/dt = 0$ in (3.40), which gives $\psi_{ss1} = -\ln(V_1/V_o)$. Hence, the steady-state dynamic friction coefficient corresponding to velocity V_1 is given by (3.39) as

$$\mu_{d1} = \mu_o + (a - b)\ln(V_1/V_o). \tag{3.41}$$

Now imagine that the slip velocity changes abruptly to a new value $V_2 = V_1 + \Delta V$. Immediately after the change in velocity, ψ will be unchanged, so (3.39) shows that μ_d will at first increase or decrease depending on the sign of a. Experiments typically show that a is positive, which is consistent with the slight initial increase in τ that is seen in Fig. 3.11a. Eventually, the friction coefficient will again stabilize, at a value given by

$$\begin{aligned} \mu_{d2} &= \mu_o + (a - b)\ln[(V_1 + \Delta V)/V_o] \\ &= \mu_o + (a - b)\ln(V_1/V_o) + (a - b)\ln(\Delta V/V_1) \\ &= \mu_{d1} + (a - b)\ln(\Delta V/V_1). \end{aligned} \tag{3.42}$$

By analogy with the analysis of §3.4, slip would be expected to be stable if the friction coefficient increases following an increase in the velocity. This will occur if $b < a$, in which case the surface is said to be "velocity-strengthening." The more rigorous linear stability analysis given by Rice and Ruina (1983) leads to the same conclusion, for small increments in the slip velocity.

Unstable stick–slip oscillations may occur along a velocity-weakening surface, for which $b > a$, depending on the stiffness and mass of the system. For the model studied in §3.4, in which μ_d was independent of velocity, the motion was found to be unstable if $\mu_d < \mu$, regardless of the stiffness or mass. Gu et al. (1984) analyzed the system of Fig. 3.6a, with the frictional law given by (3.39) and (3.40), but with the normal load N not necessarily taken to be mg. They found that, for small driver velocities V (using the notation of §3.4), the motion will always be unstable if

$$k < (b - a)\sigma/d, \tag{3.43}$$

where σ is the normal stress acting between the mass and the surface. For stiffnesses greater than the critical value given by (3.43), the stiffness of the system will be sufficient to inhibit the onset of instability, if the driver velocity V is small. For larger driver velocities, dynamic instabilities may occur, the critical velocity increasing with the mass of the system (Gu and Wong, 1991).

4 Deformation and failure of rock

4.1 Introduction

The classical theories of continuum mechanics have for the most part been constructed so as to be in accord with experimental observations of the behavior of metals and other man-made engineering materials. These theories therefore describe the various types of behavior observed in metals, such as linear elastic, nonlinear elastic, plastic, brittle failure, etc. Although these concepts and models can be applied to rocks, in most cases the analogous behavior of rock is much more complex, undoubtedly because of its heterogeneous and porous nature. These classical constitutive models, and the modifications needed in order to apply them to rock, are discussed in detail in Chapters 5, 7, and 9.

In order to develop realistic constitutive models for rock deformation, it is necessary to begin with a discussion of the types of mechanical behavior that rocks may exhibit. Hence, in this chapter we first focus on describing, qualitatively, the type of stress–strain behavior observed when rocks are subjected to external loads. The stress–strain behavior is discussed phenomenologically in §4.2 and §4.3, and from a microscopic (albeit qualitative) viewpoint in §4.4. In §4.5, the widely used Coulomb failure criterion is introduced and discussed in detail. This criterion, which assumes that failure is controlled only by the maximum and minimum principal stresses, represents a simplification of actual rock behavior, but is nevertheless extremely useful for understanding the effects of stress state on rock failure. Mohr's generalization of Coulomb's law, in which failure is still assumed to be governed by the two extreme principal stresses, but possibly in a nonlinear manner, is discussed in §4.6. This section also contains some discussion of one particular nonlinear failure law, that of Hoek and Brown. The effects of pore pressure, and the concept of effective stress, are discussed in §4.7. Experimental data on rocks subjected to "polyaxial," or true-triaxial, stress states with $\sigma_1 > \sigma_2 > \sigma_3$ are presented in §4.8 and analyzed in light of the polyaxial failure criterion proposed by Mogi. Finally, the effects of material anisotropy are briefly treated in §4.9, in terms of the "single plane of weakness" theory.

4.2 The stress–strain curve

The most common method of studying the mechanical properties of rocks is by axial compression of a circular cylinder whose length is two to three times

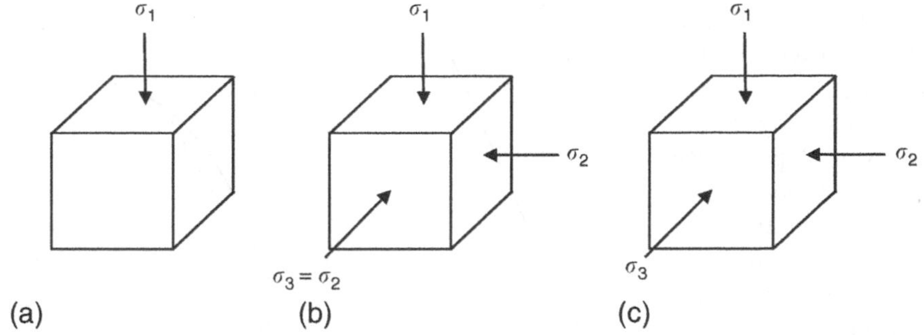

Fig. 4.1 Cubic specimen under (a) uniaxial stress, (b) traditional triaxial stress in which the two lateral confining stresses are equal, and (c) true-triaxial stress, in which all three principal stresses are possibly different.

its diameter. If the lateral surface of the rock is traction-free, the configuration is referred to as *uniaxial compression*, or *unconfined compression* (Fig. 4.1a). In this case, the resulting state of stress in the rock is $\{\sigma_1 > 0, \sigma_2 = \sigma_3 = 0\}$. If tractions are applied to the lateral surfaces, the experiment is referred to as one of *confined compression*. For tests done on a circular cylinder, the stresses applied in the two orthogonal directions perpendicular to the cylinder axis are necessarily equal (Fig. 4.1b), and the resulting state of stress in the rock is $\{\sigma_1 > \sigma_2 = \sigma_3 > 0\}$. This state is traditionally referred to as "triaxial," despite the fact that two of the principal stresses are equal. The more general state of stress, in which $\{\sigma_1 > \sigma_2 > \sigma_3 > 0\}$, can be achieved with cubical specimens, and is known either as "polyaxial" or "true triaxial" (Fig. 4.1c). The technical aspects of carrying out these experiments are discussed in Chapter 6.

We focus our attention for now on the so-called "triaxial" test carried out on a cylindrical specimen, in which the stresses are monitored, and the axial and lateral strains are measured by means of strain gauges attached to the specimens, or by deformation gauges attached to the end-caps (see §6.4). Typically, σ_2 and σ_3 are held constant, while σ_1 is increased. The results can be plotted in the form of a *stress–strain curve*, in which σ is plotted against ε. Strictly speaking, these variables are σ_1 and ε_1, but the subscripts will for now be dropped for simplicity of notation.

The simplest possible behavior is illustrated in Fig. 4.2a, in which the strain increases linearly with stress, ending in abrupt failure at some point F. This curve may be represented by the equation

$$\sigma = E\varepsilon, \tag{4.1}$$

where E, which has units of Pa, is called *Young's modulus* or the *elastic modulus*. Within the range of stress and strain prior to failure, this type of behavior is known as *linearly elastic*. The adjective "linear" refers to the mathematically linear relationship between stress and strain. In modern discussions of continuum mechanics, the term "elastic" means that the strain is a single-valued function of the stress, and does not depend on the stress history or stress path. For a linear elastic material, at any value of strain below point F in Fig. 4.2a, the slope of the curve will be given by $d\sigma/d\varepsilon = E$.

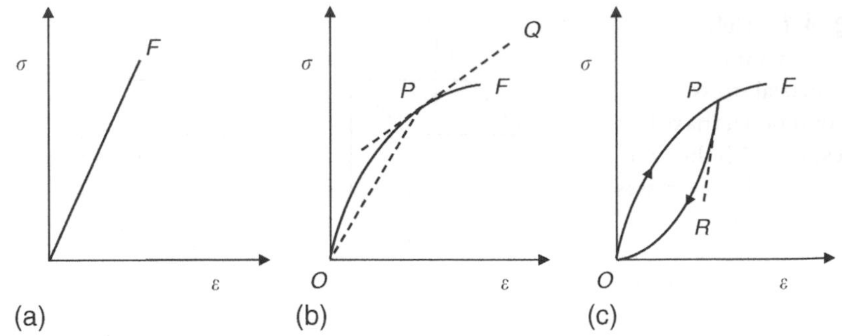

Fig. 4.2 (a) Linearly elastic behavior, with failure at F. (b) Nonlinearly elastic behavior: slope of OP is the secant modulus, slope of PQ is the tangent modulus. (c) Hysteretic material: unloading modulus at P is given by the slope of PR.

By this definition of elastic behavior, a material is elastic as long as the stress can be expressed as a single-valued function of the strain, which may or may not be linear, that is,

$$\sigma = f(\varepsilon). \tag{4.2}$$

Under this definition, the stress–strain behavior of an elastic material during "loading" is the same as during "unloading." The behavior illustrated in Fig. 4.2b is therefore referred to as *nonlinearly elastic*. For nonlinearly elastic materials, the slope of the stress–strain curve varies with the level of stress (or strain). Two types of "elastic moduli" can be defined for such materials, each of which will in general vary with both σ and ε. The *secant modulus* is defined to be the ratio of the total stress to the total strain, that is,

$$E_{sec} = \sigma/\varepsilon, \tag{4.3}$$

and is equal to the slope of line OP. The *tangent modulus*, on the other hand, is the local slope of the stress–strain curve,

$$E_{tan} = d\sigma/d\varepsilon, \tag{4.4}$$

and in Fig. 4.2b would be equal to the slope of line PQ. The secant and tangent moduli coincide for a linearly elastic material.

A rock is called *hysteretic* if it follows different stress–strain curves during loading and unloading, but returns to its original strain-free state when the stresses are removed. Figure 4.2c shows the stress–strain behavior of a hysteretic material, with the dashed line representing the unloading curve. Such a material exhibits a different tangent modulus during unloading than during loading; the loading modulus at P would be given by the tangent to curve OPF, whereas the unloading modulus at that same point P is given by the slope of the line PR. In §5.8, it is shown that the work done on the rock by the external loading agency during deformation, per unit volume of rock, is equal to the area under the stress–strain curve. Hence, for a rock exhibiting stress–strain behavior such as shown in Fig. 4.2c, the work done on the rock during loading would be greater than that done on it during unloading. The area between the loading and unloading portions of the stress–strain curve therefore represents energy that is

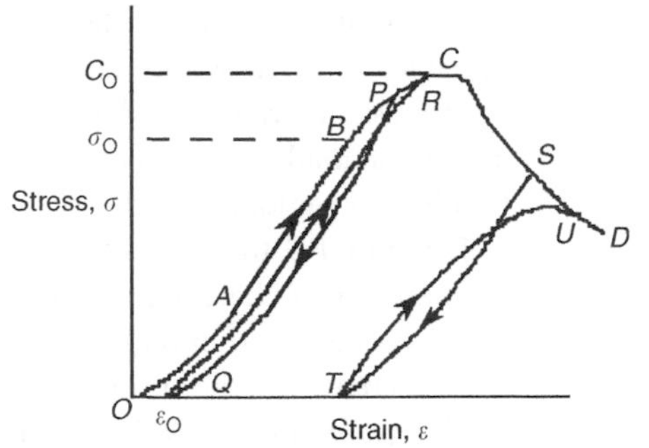

Fig. 4.3 Complete stress–strain curve for a rock under compression (see text for details).

dissipated (by friction along grain boundaries and crack faces, etc.). As the net amount of work done *on* the rock during a cycle of loading and unloading must be nonnegative, the unloading portion of the stress–strain curve cannot lie above the loading portion.

The idealized materials described in Fig. 4.2 each deform until *F*, at which point they fail abruptly if the applied stress is increased further. This type of abrupt failure is observed in materials under tension, but the behavior of a rock under the more commonly occurring compressive stress regime is more complicated (Fig. 4.3). The stress–strain curve for a rock under uniaxial compression can be divided conceptually into four regions. In region *OA*, the curvature, roughly indicated by the second derivative, is positive. In region *AB*, the curve is very nearly linear. The curve continues to rise in region *BC*, but the curvature is now negative. The strain reaches a maximum at *C*, after which it falls throughout region *CD*.

In the first two regions, *OA* and *AB*, the behavior is nearly elastic. Some slight hysteresis may be observed, but loading and unloading in this region will not produce irreversible changes in the structure or properties of the rock. In the third region, *BC*, which usually begins at a stress of about two-thirds the maximum stress at *C*, the slope of the stress–strain curve, that is, the tangent modulus, decreases steadily to zero as the stress increases. In this region, irreversible changes occur in the rock, and successive cycles of loading and unloading would trace out different curves. An unloading cycle such as *PQ* that starts in region *BC* would lead to a permanent strain ε_0 when the stress reaches zero. If the rock is reloaded, a curve such as *QR* would be traced out that lies below the original loading curve *OBC*, but which ultimately rejoins it, at a stress greater than the stress at *P*.

The fourth region, *CD*, begins at the point of maximum stress *C*, and has a negative slope. An unloading cycle such as *ST* that begins in this region would lead to a large permanent strain when the stress reaches zero. Subsequent reloading will trace out a curve in the $\{\sigma\text{–}\varepsilon\}$ plane that rejoins the initial loading curve at *U*, corresponding to a stress that is *lower* than that at the beginning of the cycle, point *S*. In this sense we can say that the ability of the rock to support a load has decreased. This region of the stress–strain curve cannot be observed in a testing machine in which the stress is the controlled variable, in which case violent failure of the specimen will occur near point *C*. But this region can be observed in a servo-controlled testing machine in which strain is the controlled variable, as discussed in detail in §6.5. This region is also of importance in a rock mass, in which the decreasing ability of one region of rock to support an additional load could be compensated for by some of the load being transferred to adjacent regions of rock.

In region *BC*, the rock is said to be in a *ductile state*, or simply to be *ductile*. Ductile behavior is characterized by the ability of the rock to support an increasing load as it deforms. In region *CD*, on the other hand, the load supported by the rock decreases as the strain increases. A rock exhibiting this behavior, which is qualitatively different from the ductile behavior described above, is said to be in a *brittle state*, or simply to be *brittle*. The range of stresses in which a rock exhibits either of these two types of behavior depends on the mineralogy, microstructure, and also on factors such as the temperature, as discussed in §4.3, and in more detail by Paterson (1978).

The value of the stress at point *B*, which marks the transition from elastic to ductile behavior, is known as the *yield stress* of the rock, and is usually denoted by σ_0. The value of the stress at point C, which marks the transition from ductile to brittle behavior, is known as the *uniaxial compressive strength* of the rock, and is usually denoted by C_0. The process of failure is regarded as a continuous process that occurs throughout the brittle region *CD*, in which the rock physically deteriorates, and its ability to support a load decreases. Failure therefore begins at C, and the *criteria for failure* for a rock subjected to uniaxial compression would simply consist of the condition that "failure occurs when $\sigma = C_0$." The failure criteria that are discussed in later sections of this chapter represent attempts to predict the onset of failure under stress states that are more general than uniaxial compression.

The foregoing discussion has shown that the axial strain that occurs under uniaxial compression can be quantified in terms of the Young's modulus, E. But a rock under a uniaxial compressive stress will not only deform in the direction of the load, it will also deform in each of the two directions perpendicular to the load. Figure 4.4 shows the strains measured on a cylindrical specimen by Hojem et al. (1975) during confined uniaxial compression of an argillaceous quartzite, with the lateral confining stress held constant at $\sigma_2 = \sigma_3 = 6.9$ MPa (1000 psi). As before, the convention used is that positive normal strains correspond to decreases in the linear dimensions of the specimen. The axial stress vs. axial strain curve exhibits most of the features described above, including a pronounced brittle regime. The strain in the two other directions (i.e., the radial strain, $\varepsilon_2 = \varepsilon_3$) is negative, which is to say that the specimen bulges outward as it is compressed. Within the elastic regime that corresponds to region *OB* of Fig. 4.3, the magnitude of the radial strain increases nearly in proportion to the axial strain. The negative of the ratio of the transverse strain to the axial strain, $-\varepsilon_2/\varepsilon_1$, is known as *Poisson's ratio*, and is denoted by ν. For a linear elastic material, this parameter is independent of stress, and is generally found to be in the range 0–0.5 (§5.2).

In the ductile regime, corresponding to region *BC* of Fig. 4.3, the transverse strains begin to grow (in magnitude) at a much faster rate than does the axial

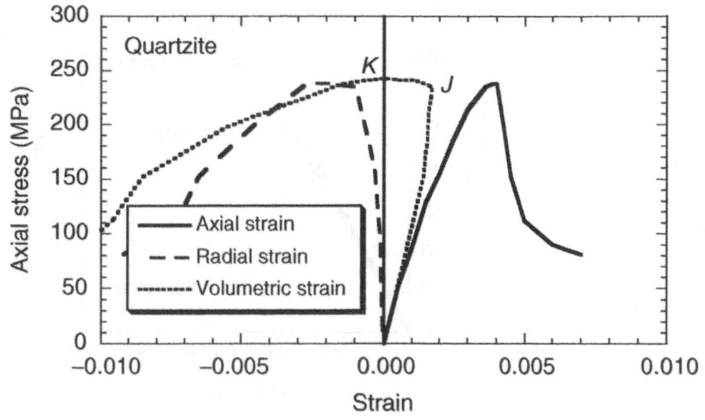

Fig. 4.4 Axial strain, radial strain, and volumetric strain as a function of axial stress, for a cylindrical sample of an argillaceous quartzite, tested under a confining stress of 6.9 MPa by Hojem et al. (1975).

strain. In terms of incremental strains, this behavior could be said to correspond to a value of Poisson's ratio that exceeds unity. As the volumetric strain, $\Delta V/V$, is equal to the sum of the strains in the axial and the two lateral directions, the volumetric strain begins to decrease with an increase in the axial stress. Bearing in mind our sign convention, this means that an incremental increase in the compressive axial stress causes an incremental *increase* in the volume. This first occurs at J in Fig. 4.4. Eventually, the lateral strains become sufficiently negative that the *total* volumetric strain becomes negative; this occurs at K in Fig. 4.4. The phenomenon by which the volume of the rock decreases under the action of an additional compressive stress is known as *dilatancy*. Dilatancy can be ascribed to the formation and extension of open microcracks whose axes are oriented parallel to the direction of the maximum principal stress (§4.4). By testing specimens in the form of thick-walled hollow tubes, Cook (1970) showed that dilatancy occurs pervasively throughout the entire volume of the rock, and is not a superficial phenomenon localized at the outer boundary. On the other hand, Spetzler and Martin (1974) and Hadley (1975) have shown that dilatancy is not uniformly distributed throughout the specimen, but rather becomes increasingly heterogeneous as failure is approached.

4.3 Effects of confining stress and temperature

It has been known since the end of the nineteenth century that if the confining stress applied to the sides of a cylindrical specimen during a triaxial compression test is increased, the axial stress required to cause failure will increase, and the rock will show a tendency toward greater ductility (Becker, 1893; Adams, 1912). In the classical experiments performed by von Kármán (1911) and Böker (1915), oil was used to apply a confining stress $\sigma_2 = \sigma_3$ to the sides of the specimen, while the axial stress σ_1 was slowly increased.

The effect that confining pressure has on the axial stress *vs.* axial strain curve is shown in Fig. 4.5a for a Rand quartzite. For each value of $\sigma_2 = \sigma_3$, the stress–strain curve initially exhibits a nearly linear elastic portion, with a slope (Young's modulus) that is nearly independent of the confining stress. But both the yield stress and the failure stress increase as the confining stress increases. Finally, there is a small descending portion of the curve, ending in brittle fracture.

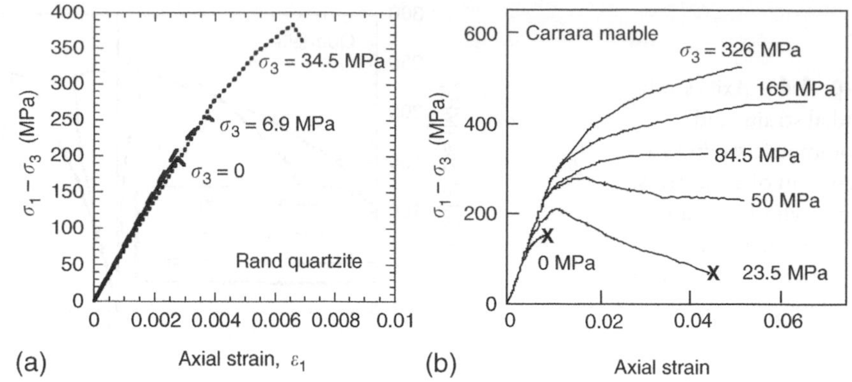

Fig. 4.5 Stress–strain curves for (a) Rand quartzite and (b) Carrara marble at various confining pressures. Crosses indicate abrupt brittle failure.

A different type of behavior is exhibited by other rocks, notably carbonates and some sediments. Figure 4.5b shows the data collected by von Kármán (1911) on a Carrara marble. For sufficiently low confining stresses, exemplified by the curve labeled $\sigma_3 = 0$ MPa, brittle fracture (denoted by X) occurs as for the quartzite described above. But at higher confining stresses, such as the curve labeled 50 MPa, the rock can undergo a strain as large as 7 percent, with no substantial loss in its ability to support a load (i.e., no decrease in the axial stress). In this case the rock is said to exhibit *ductile behavior*, which can be loosely defined as "the capacity for substantial change in shape without gross fracturing" (Paterson, 1978). The curve for $\sigma_3 = 23.5$ MPa can be said to show a transitional type of behavior, in that fairly substantial inelastic strain occurs, but the rock eventually fails by brittle fracture. Hence, there is a somewhat ill-defined value of the confining stress at which one can say there occurs a transition between brittle and ductile behavior. Heard (1960) proposed that this *brittle–ductile transition* be taken as that confining stress at which the strain at failure is, say, 3–5 percent. At still higher confining stresses, such as 165 MPa or above in Fig. 4.5b, the axial stress σ_1 continues to increase with strain after the yield point has been passed. Such behavior is known as *work hardening* in metallurgy, and more simply as *hardening* in rock mechanics. Following this nomenclature, the behavior shown in the descending portion of the stress–strain curve, such as that exhibited in Fig. 4.5b at a confining stress of 23.5 MPa, is often referred to as *softening*. Table 4.1 shows the measured brittle–ductile transition pressures for different rock types, at room temperature, as compiled by Paterson (1978) from various sources.

Higher temperatures generally have the effect, roughly speaking, of encouraging ductility. Figure 4.6a shows the stress–strain curves measured by Griggs et al. (1960) on a granite, at a confining stress of 500 MPa. At room temperature the rock is brittle, but at higher temperatures substantial amounts of permanent deformation may occur. By 800°C, the rock is almost fully ductile, in that the strain can continue to increase at a nearly constant load. Hence, for a fixed value of the constant confining stress, brittle behavior gives way to ductile behavior above a certain temperature. As both higher temperatures and higher confining stresses tend to favor ductility, the brittle–ductile transition temperature decreases as the confining stress increases. Heard (1960) developed a phase

Table 4.1 Brittle–ductile transition pressures under $\sigma_2 = \sigma_3$ compression, at room temperature (after Paterson, 1978).

Rock type	σ_2 (b → d) (MPa)	Source(s)
Limestones, marbles	30–100	Heard (1960), Rutter (1972a)
Dolomite	100–200+	Handin and Hager (1957), Mogi (1971)
Gypsum	40	Murrell and Ismail (1976)
Anhydrite	100	Handin and Hager (1957)
Rocksalt	<20	Handin (1953)
Talc	400	Edmond and Paterson (1972)
Serpentinite	300–500	Raleigh and Paterson (1965)
Chloritite	300	Murrell and Ismail (1976)
Argillaceous sandstone	200–300	Edmond and Paterson (1972), Schock et al. (1973)
Siltstones, shales	<100	Handin and Hager (1957)
Porous lavas	30–100	Mogi (1965)

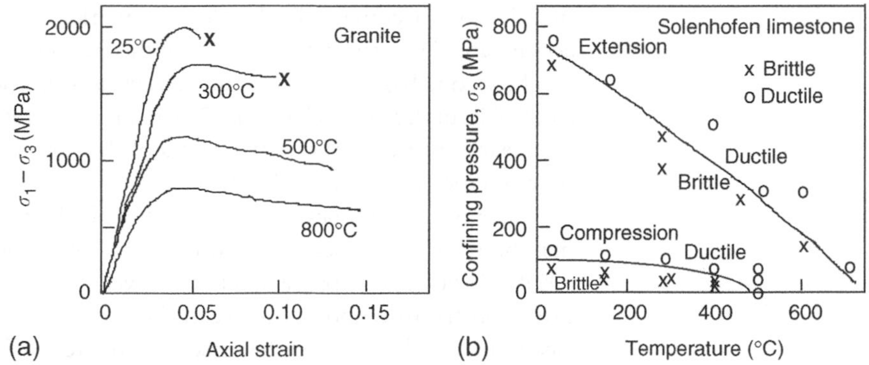

Fig. 4.6 (a) stress–strain curves for granite at various temperatures, at a confining pressure of 500 MPa, after Griggs et al. (1960). (b) Brittle–ductile phase diagram for Solenhofen limestone, after Heard (1960). "Extension" refers to loadings for which the axial stress is less compressive than the lateral stresses.

diagram for Solenhofen limestone in $\{T, \sigma_3\}$ space, which shows regions of ductile or brittle behavior, separated by a ductile–brittle transition curve (Fig. 4.6b). For this rock, ductility can be observed under zero confining stress if the temperature is above about 500°C. For most rocks, however, if there is no confining stress the behavior will be brittle up to the melting temperature (Murrell and Chakravarty, 1973).

4.4 Types of fracture

The different types of stress–strain behavior discussed in the previous two sections correspond to different physical processes occurring within the rock. Under unconfined compression, a rock tends to deform elastically, until failure occurs abruptly (Fig. 4.5). This failure is accompanied by somewhat irregular longitudinal splitting (Fig. 4.7a). With a moderate amount of confining pressure, longitudinal fracturing is suppressed, and failure occurs along a clearly defined plane of fracture (Fig. 4.7b). This plane is typically inclined at an angle less than 45° from

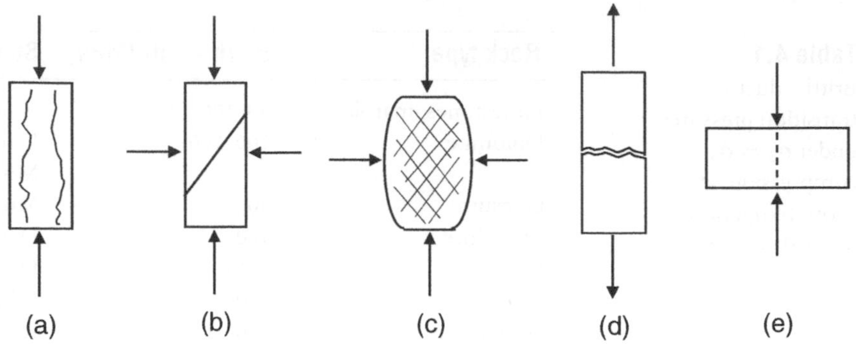

Fig. 4.7 (a) Longitudinal splitting under uniaxial tension, (b) shear fracture, (c) multiple shear fractures, (d) extension fracture, and (e) extension fracture produced by opposing line loads.

the direction of σ_1 (the axial direction, in this case). This plane is characterized by shearing displacement along its surface, and is referred to as a *shear fracture*. Under some circumstances, failure occurs along two conjugate shear planes located symmetrically with respect to the axial direction, but this seems to be an experimental artifact caused by the ends of the specimen being constrained against rotation (Paterson, 1978, p. 18). If the confining pressure is increased, so that the rock becomes fully ductile, a network of small shear fractures appears, accompanied by plastic deformation of the individual rock grains (Fig. 4.7c).

The second basic type of fracture, an *extension fracture*, typically appears when a rock fails under uniaxial tension. The main characteristic of this type of fracture is a clean separation of the two halves of the sample, with no tangential offset between the two surfaces (Fig. 4.7d).

Under more complicated systems of applied stress, fractures appear which may be regarded as belonging to one or another of these two basic types. If a slab of rock is compressed between two opposing line loads (Fig. 4.7e), an extension fracture appears between the loads. If these loads are caused by a jacket surrounding the core being squeezed into cracks in the rock, the resulting fracture has been described by Brace (1964) as an *intrusion fracture*. When the fracture surfaces are examined from a specimen that has undergone longitudinal splitting, as in Fig. 4.7a, parts of the surfaces will have the appearance of a shear fracture, and other parts will appear to be extension fractures.

In §4.2, attention was directed to the phenomenon of dilatancy, which occurs during the triaxial compression of rocks. Such tests are typically conducted under conditions of constant lateral confining stress. Under such conditions, the rock is relatively free to expand laterally. In a rock mass, however, such lateral expansion would be resisted, to some extent, by the adjacent rock. One would imagine that as a portion of rock expands laterally, the lateral compressive stress imposed by the adjacent rock would increase, thereby inhibiting the lateral expansion of the rock. Hence, the deformation of a specific portion of rock *in situ* would inevitably be coupled to the deformation of the adjacent mass of rock. This coupling does not typically occur in standard rock tests, for which the boundary conditions, be they ones of constant lateral stress or constant lateral strain, are imposed a priori.

In order to approximate more closely the situation that might occur *in situ*, Hallbauer et al. (1973) jacketed specimens of a fine-grained, argillaceous

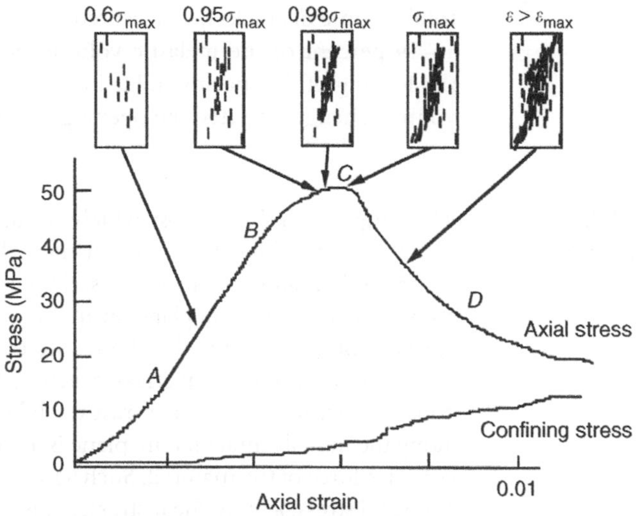

Fig. 4.8 Schematic representation of the axial stress and lateral confining stress measured by Hallbauer et al. (1973) on a set of argillaceous quartzite specimens, along with cartoons of the state of microcracking observed on specimens that were loaded to the indicated points along the stress–strain curve.

quartzite in a copper tube of 1 mm wall thickness, and tested them in a triaxial cell in which the confining stress was applied by a pressurized fluid. An initial confining stress of 100 bars was imposed on the specimens. Calibration of the system revealed that, due to the stiffness of the cell and the small volume of confining fluid, the confining pressure increased in proportion to the lateral strain of the specimen, that is, $\Delta\sigma_3 = c\,\Delta\varepsilon_3$, with the constant of proportionality found to be 1.122 GPa. Hence, the rock specimen can be thought of as being connected along its sides to linear springs of constant stiffness. The tests were conducted on a suite of specimens cut from a single block of quartzite, and were stopped at predetermined points along the stress–strain curve. Careful observation of longitudinal sections cut through the axes of the specimens allowed the growth of microcracks and fractures to be observed in relation to the stress–strain curve.

The results are illustrated in Fig. 4.8. In region *AB* of the stress–strain curve, the first visible structural damage appears as elongated microcracks having their axes oriented parallel (within $\pm 10°$) to the direction of maximum compressive stress (i.e., axially). The cracks were distributed throughout the sample, but were concentrated in the center. Toward the end of region *BC*, the number of microcracks increased drastically, and the cracks began to coalesce along a plane located in the central region of the specimen. At the point *C* of maximum axial stress, the microcracks begin to link up to form a macroscopic fracture "plane." Finally, in region *CD*, the fracture plane has extended through the entire specimen, and shear displacement begins to occur across the two faces of rock. In this region, the axial load carried by the specimen decreases as the rock continues to compress.

Measurements of the microcracks made after the specimens had been unloaded and sectioned showed them to be about 300 μm long and about 3 μm wide in their unloaded state. Their width under stress was presumably much greater than when unloaded. At each stage along the stress–strain curve, the total volume of cracks (as measured in their unloaded state) amounted to about 16–19 percent of the inelastic volumetric dilatancy that was observed during the loading of the specimen. Hallbauer et al. (1973) concluded that the dilatant volume change reflected the opening up of these microcracks.

4.5 Coulomb failure criterion

The simplest, and still most widely used, failure criterion is that of Coulomb (1773). Based on his extensive experimental investigations into friction, Coulomb assumed that failure in a rock or soil takes place along a plane due to the shear stress τ acting along that plane. In analogy with sliding along nonwelded surfaces (§3.3), motion is assumed to be resisted by a frictional-type force whose magnitude equals the normal stress σ acting along this plane, multiplied by some constant factor μ. But in contrast to sliding along nonwelded surfaces, motion along the initially intact failure plane is assumed also to be resisted by an internal cohesive force of the material. Such a force reflects the fact that, in the absence of a normal stress, a finite shear stress, S_0, is typically still needed in order to initiate failure. These considerations lead to the mathematical criterion that failure will occur along a plane if the following condition is satisfied:

$$|\tau| = S_0 + \mu\sigma. \tag{4.5}$$

(The sign of the shear stress only effects the direction of sliding after failure, so the absolute value of τ appears in the failure criterion, although it is often convenient to ignore the absolute value signs in mathematical manipulations.) Conversely, failure will not occur on any plane for which $|\tau| < S_0 + \mu\sigma$. The parameter S_0, also sometimes denoted by c, is known as the *cohesion*. The parameter μ is known as the *coefficient of internal friction*, as it applies along an imaginary surface that is internal to the rock before failure occurs. Although the term *coefficient of "internal" friction* derives from the mathematical analogy between (4.5) and (3.6), Savage et al. (1996) have argued that this effect is indeed due to sliding frictional forces acting along those microscale portions of the fracture surface that are not actually intact.

The form of criterion (4.5) suggests that the Mohr's circle construction will be useful in its analysis. Indeed, (4.5) defines a straight line on the $\{\sigma, -\tau\}$ plane that intercepts the τ-axis at $-S_0$, and has slope μ. The angle ϕ that this line makes with the σ-axis is given by $\phi = \tan^{-1}\mu$, and is known as the *angle of internal friction*. We temporarily assume a two-dimensional situation, but denote the minimum principal stress by σ_3 rather than σ_2, in preparation for generalizing the discussion to three dimensions. A stress state whose Mohr's circle lies *below* the line AL in Fig. 4.9b will not give rise to failure on any plane. If the principal stresses are such that the circle *touches* the failure line, the rock will fail in shear (Fig. 4.9b). Circles that extend *above* the failure line have no meaning in this context, since, if the stresses are assumed to increase slowly starting from some "safe" stress state that lies below the line, failure will occur as soon as the Mohr's circle first touches the line.

The point P at which the circle is tangent to the Coulomb line represents the stress state on the plane of failure. Hence, the angle by which the failure plane

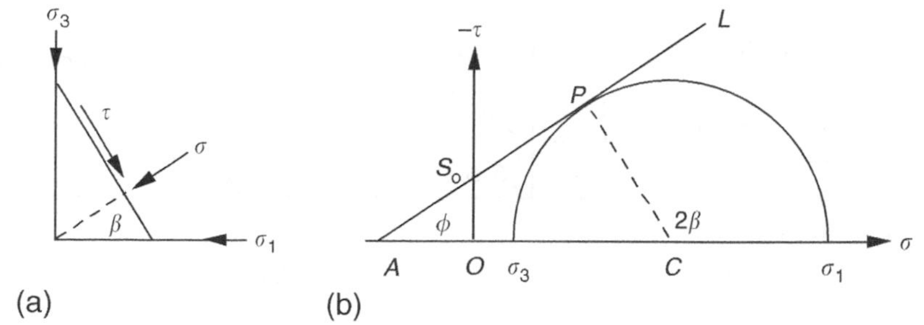

Fig. 4.9 (a) Normal and shear tractions on a plane whose outward normal is rotated from the σ_1 direction by an *arbitrary* angle β. (b) Mohr diagram, with failure curve (4.5) shown as line *AL*. Failure will occur on a *specific* plane whose angle β, demarcated by line *CP*, is given by (4.6).

is oriented to the σ_1 direction is given by one-half of the angle 2β that line *CP* makes with the horizontal axis. By considering the intersection of line *CP* with the horizontal axis, it is seen that $2\beta = 180° - \angle ACP$. From triangle *CPA*, it follows that $\angle ACP = 180° - \angle CPA - \angle PAC$. Hence, $2\beta = \angle CPA + \angle PAC$. Since *P* is the point of tangency of the circle and the failure line, line *CP* is perpendicular to the failure line, and so $\angle CPA = 90°$. Finally, $\angle PAC = \phi$, where ϕ is the angle of internal friction. Hence,

$$2\beta = 90° + \phi, \quad \text{or} \quad \beta = 45° + \frac{1}{2}\phi. \tag{4.6}$$

Recalling that the failure criterion (4.5) involves the absolute value $|\tau|$, it follows that a failure line rotated clockwise by angle ϕ from the horizontal could also be drawn in Fig. 4.9b, implying that the angle $-\beta$ must also represent a possible failure plane. As the angles on the Mohr diagram represent the normal vectors to the associated planes, we conclude that there are two possible planes of shear failure, each oriented at an acute angle of $\beta = 45° - (\phi/2)$ to the maximum principal stress. These two directions are referred to as the *conjugate directions* of shear failure.

The failure criterion (4.5) can also be written in many seemingly different but equivalent forms, each of which is convenient in certain circumstances. Figure 4.9b shows that $|CP| = (|AO| + |OC|)\sin\phi$, which can be written as

$$\frac{1}{2}(\sigma_1 - \sigma_3) = \left[S_o \cot\phi + \frac{1}{2}(\sigma_1 + \sigma_3)\sin\phi\right] = S_o \cos\phi + \frac{1}{2}(\sigma_1 + \sigma_3)\sin\phi. \tag{4.7}$$

In terms of the "two-dimensional mean stress" σ_m, and the maximum shear stress τ_m, which are given by

$$\sigma_m = \frac{1}{2}(\sigma_1 + \sigma_3), \quad \tau_m = \frac{1}{2}(\sigma_1 - \sigma_3), \tag{4.8}$$

the failure criterion (4.7) can be expressed as

$$\tau_m = S_o \cos\phi + \sigma_m \sin\phi. \tag{4.9}$$

In the $\{\sigma_m, \tau_m\}$ plane, this equation appears as a straight line that makes an angle $\tan^{-1}(\sin\phi)$ with the σ_m-axis, and intercepts that axis at $-S_o \cot\phi$ (Fig. 4.10a).

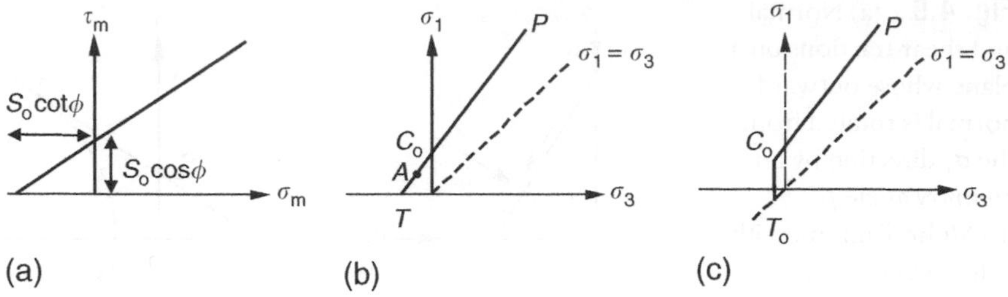

Fig. 4.10 (a) Failure curve in the $\{\sigma_m, \tau_m\}$ plane, from (4.9). (b) Failure curve in the $\{\sigma_3, \sigma_1\}$ plane, from (4.13), where region *TA* corresponds to *tensile* normal traction along the ostensible failure plane. (c) Same as (b), with the addition of a tension cutoff at T_o.

It would be convenient to express the Coulomb failure criterion directly in terms of the principal stresses $\{\sigma_1, \sigma_3\}$. First, (4.7) is rearranged as

$$\sigma_1 = 2S_0 \frac{\cos\phi}{1-\sin\phi} + \sigma_3 \frac{1+\sin\phi}{1-\sin\phi}. \tag{4.10}$$

Making use of (4.6), the coefficient of σ_3 can be transformed as follows:

$$\frac{1+\sin\phi}{1-\sin\phi} = \frac{1+\sin(2\beta-90°)}{1-\sin(2\beta-90°)} = \frac{1-\cos 2\beta}{1+\cos 2\beta} = \frac{2\sin^2\beta}{2\cos^2\beta} = \tan^2\beta. \tag{4.11}$$

Similarly, the coefficient of the term $2S_0$ can be transformed as follows:

$$\frac{\cos\phi}{1-\sin\phi} = \frac{\cos(2\beta-90°)}{1-\sin(2\beta-90°)} = \frac{\sin 2\beta}{1+\cos 2\beta} = \frac{2\sin\beta\cos\beta}{2\cos^2\beta} = \tan\beta. \tag{4.12}$$

Hence, the Coulomb failure criterion can also be expressed as

$$\sigma_1 = 2S_0 \tan\beta + \sigma_3 \tan^2\beta \equiv C_0 + \sigma_3 \tan^2\beta, \tag{4.13}$$

where $C_0 = 2S_0 \tan\beta$ is the uniaxial compressive strength. In the $\{\sigma_3, \sigma_1\}$ plane, the Coulomb failure criterion therefore appears as a straight line with slope $\tan^2\beta$, which intercepts the σ_3-axis at $-2S_0 \cot\beta$ (Fig. 4.10b).

The various forms of the Coulomb failure criterion can also be written in terms of the coefficient of internal friction, $\mu = \tan\phi$. Elementary trigonometry shows that

$$\cos\phi = 1/(1+\mu^2)^{1/2}, \quad \sin\phi = \mu/(1+\mu^2)^{1/2}, \tag{4.14}$$

which then allows (4.9) to be written as

$$(1+\mu^2)^{1/2}\tau_m = S_0 + \mu\sigma_m. \tag{4.15}$$

Similarly, starting from (4.10) or (4.13), and making use of (4.14), the failure criterion can also be expressed in the forms

$$[(1+\mu^2)^{1/2} - \mu]\sigma_1 = 2S_0 + [(1+\mu^2)^{1/2} + \mu]\sigma_3, \tag{4.16}$$

$$\sigma_1 = 2S_0[(1+\mu^2)^{1/2} + \mu] + [(1+\mu^2)^{1/2} + \mu]^2\sigma_3, \tag{4.17}$$

the latter of which shows that the uniaxial compressive strength is given by

$$C_o = 2S_o[(1 + \mu^2)^{1/2} + \mu]. \tag{4.18}$$

The similarity of the foregoing discussion with that given in §3.5 should be apparent. In that section, the rock mass was assumed to have a single, preexisting plane of weakness. Alternatively, if there are no particular planes of weakness, but rather all possible planes are equally weak due to random microcracks, grain boundaries and other small-scale imperfections, the rock mass will effectively choose its own plane of failure, according to the theory outlined above. Anisotropic rock masses, on the other hand, may possess certain directions in which fracture is more likely to occur than in others; this situation is discussed in §4.9.

To simplify the preceding discussion, a two-dimensional analysis was assumed. But appealing again to the Mohr's circle construction shows that, as long as failure is assumed to occur on a plane on which the normal and shear stresses satisfy condition (4.5), consideration of a fully three-dimensional stress state alters none of the conclusions reached above. Indeed, the first Mohr's circle to touch the failure line will necessarily be that corresponding to the largest and smallest principal stresses, σ_1 and σ_3. It therefore follows from Coulomb's assumption that the magnitude of the intermediate principal stress, σ_2, has no effect on failure. For many years, experiments could only readily be performed under the conditions $\sigma_2 = \sigma_3$, in which case any possible effect of the intermediate principal stress would not appear. More recent experiments under conditions of "true" triaxial conditions, $\sigma_1 > \sigma_2 > \sigma_3$, have revealed that this assumption is not correct. Criteria for failure under true-triaxial stress conditions are discussed in §4.8.

Coulomb's failure criterion (4.5) is essentially empirical, and thus perhaps it is not meaningful to discuss the conditions under which it is valid. Nevertheless, if it is considered that an implicit assumption contained within (4.5) is that the normal stress σ acting on the failure plane is positive (i.e., compressive), it would follow that certain portions of the failure curves shown in Fig. 4.10b should be ignored. The smallest value of σ_1 that satisfies the Coulomb failure condition and that corresponds to a nonnegative normal stress on the failure plane, is found by combining the expression for σ with relation (4.6), to yield

$$\sigma = \frac{1}{2}(\sigma_1+\sigma_3) + \frac{1}{2}(\sigma_1-\sigma_3)\cos 2\beta = \frac{1}{2}(\sigma_1+\sigma_3) - \frac{1}{2}(\sigma_1 - \sigma_3)\sin\phi > 0,$$

that is,

$$\sigma_1(1 + \sin\phi) > -\sigma_3(1 - \sin\phi). \tag{4.19}$$

Combining (4.19) and (4.10), and making use of (4.12), leads to the condition

$$\sigma_1 > S_o \tan\beta = \frac{1}{2}C_o. \tag{4.20}$$

It follows that the portion *TA* of failure line *TAP* in Fig. 4.10b does not correspond to a positive normal stress on the putative failure plane. Paul (1961) concluded that this portion of the curve could therefore not properly predict shear failure. Indeed, this critical value of σ_1 is positive, but (4.13) shows that it necessarily corresponds to a negative (tensile) value of σ_3. Experimentally, it is observed that for some negative values of σ_3, failure occurs by extensional fractures in planes perpendicular to this tensile stress. This mode of failure is entirely different from the shear failure that occurs under compression. The simplest modification of Coulomb's criterion that could account for this changeover in failure mode would be to truncate line *TAP* at point *A*, corresponding to $\sigma_1 = C_0/2$ and extend it as a vertical line until it meets the line $\sigma_1 = \sigma_3$. (As $\sigma_1 \geq \sigma_3$ by definition, only the region above and to the left of the line $\sigma_1 = \sigma_3$ is meaningful.) However, this would necessarily imply that the uniaxial tensile strength, T_0, is equal to one-half the uniaxial compressive strength, C_0. Experimental data typically shows that the ratio C_0/T_0 is much greater than 2. Consequently, Paul (1961) proposed using an experimentally measured value of a T_0 for the location of the vertical line in the $\{\sigma_3, \sigma_1\}$ plane (Fig. 4.10c). This is equivalent to replacing (4.13) with the bilinear failure criterion

$$\sigma_1 = C_0 + \sigma_3 \tan^2 \beta, \quad \text{for } \sigma_1 > C_0[1 - C_0 T_0/4S_0^2], \tag{4.21}$$

$$\sigma_3 = -T_0, \quad \text{for } \sigma_1 \leq C_0[1 - C_0 T_0/4S_0^2]. \tag{4.22}$$

4.6 Mohr's hypothesis

According to Coulomb's theory, failure will occur on a plane when the normal and shear stresses acting on that plane satisfy condition (4.5). In the $\{\sigma, |\tau|\}$ plane, this condition appears as a straight line with slope $\mu = \tan \phi$. The Mohr's circle corresponding to any state of stress that leads to failure will be tangent to this line. As discussed in §4.5, this theory ignores the effect of the intermediate principal stress. However, in principle, Coulomb's theory could be expected to apply to stress states in which $\sigma_2 = \sigma_3$. Leaving aside for now the issue of triaxiality, it is nevertheless the case that Coulomb's law is unrealistic in at least two respects.

Within the context of Coulomb's theory, the uniaxial tensile strength, T_0, can be found by setting $\sigma_1 = 0$ in (4.17), yielding

$$T_\sigma = -\sigma_3 = 2S_0 \cot \beta = \frac{2S_0}{(\mu^2 + 1)^{1/2} + \mu}. \tag{4.23}$$

According to (4.13), the unconfined compressive strength is given by $C_0 = 2S_0 \tan \beta$. Hence, the Coulomb theory predicts that the ratio of unconfined compressive strength to unconfined tensile strength will be

$$C_0/T_0 = \tan^2 \beta = [(\mu^2 + 1)^{1/2} + \mu]^2. \tag{4.24}$$

The Coulomb theory thereby predicts a relatively modest ratio of compressive to tensile strength. For example, small values of μ lead to ratios not much larger than unity, whereas a coefficient of internal friction as large as $\mu = 1$ leads to a strength ratio of only 5.83. Experimental values of this ratio, however, tend to be on the order of 10 or so. Roughly, this deficiency can be expressed by saying that the Coulomb failure line extends too far into the tensile region of the $\{\sigma, |\tau|\}$ plane. This empirical observation is entirely independent of the more

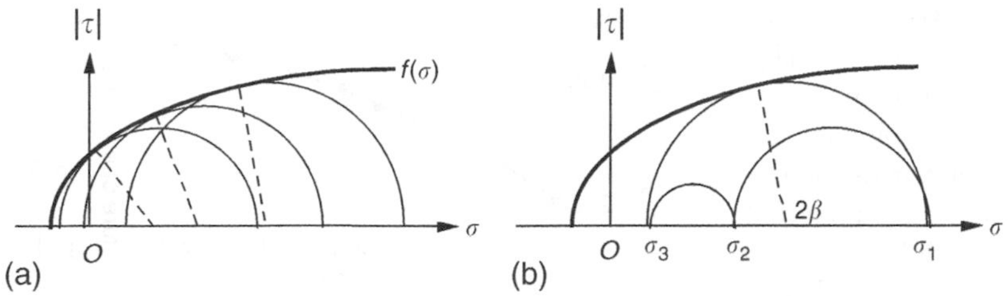

Fig. 4.11
(a) Nonlinear failure curve, defined as the envelope of all Mohr circles that cause failure. (b) Construction showing that, according to the Mohr hypothesis, the intermediate principal stress does not influence the onset of failure.

fundamental theoretical argument of Paul (1961), discussed in §4.5, regarding the supposed inapplicability of the Coulomb theory to planes on which the normal stress is tensile.

Coulomb's theory also predicts that the compressive stress required to cause failure, σ_1, will increase linearly with the confining stress, σ_3. Experiments typically show that σ_1 at failure increases at a less-than-linear rate with σ_3.

In order to correct these deficiencies in Coulomb's theory, Mohr (1900) suggested that Coulomb's equation, (4.5), be replaced by a more general, possibly nonlinear, relation of the form (Fig. 4.11a)

$$|\tau| = f(\sigma). \tag{4.25}$$

In principle, this curve can be determined experimentally as the envelope of all of the Mohr's circles that correspond to states of stress that cause failure. Aside from the fact that f may now be a nonlinear function, the basic ideas of Coulomb's model are retained. Specifically, failure is supposed to occur if one of the Mohr's circles touches the curve defined by (4.25). As shown in Fig. 4.11b, this will necessarily occur for the circle defined by σ_1 and σ_3, and so the value of the intermediate principal stress is not expected to affect the onset of failure.

Moreover, as the state of stress at the point of contact of the Mohr's circle and the failure curve represents the stresses acting on the failure plane, the generalized Mohr theory predicts, as did the Coulomb theory, that the failure *plane* passes through the direction of the intermediate principal stress, and its normal vector makes an angle β with the direction of maximum principal stress, where 2β is the angle by which line *PC* is rotated (counterclockwise) from the σ_1-axis (Fig. 4.11b). If the failure criterion (4.25) is concave downward, as is usually the case, the angle β of the failure plane will decrease with increasing confining stress, as indicated in Fig. 4.11a.

Experimentally, the failure criterion (4.25) can be determined by plotting the Mohr's circles for the stresses at failure, as found in a series of tests conducted under different confining stresses. The failure curve will then be given by the envelope of these circles (Fig. 4.11a). Alternatively, the stresses at failure could be plotted in the $\{\sigma_3, \sigma_1\}$ plane, thereby generating a nonlinear analogue to (4.13). Numerous mathematical formulae, each containing two or more adjustable parameters, have been proposed for the purposes of fitting such failure data. Many of these formulae are discussed by Andreev (1995) and Sheorey (1997).

Regardless of the specific form taken by the failure curve, an unambiguous prediction of the Mohr theory is that the orientation of the plane of failure

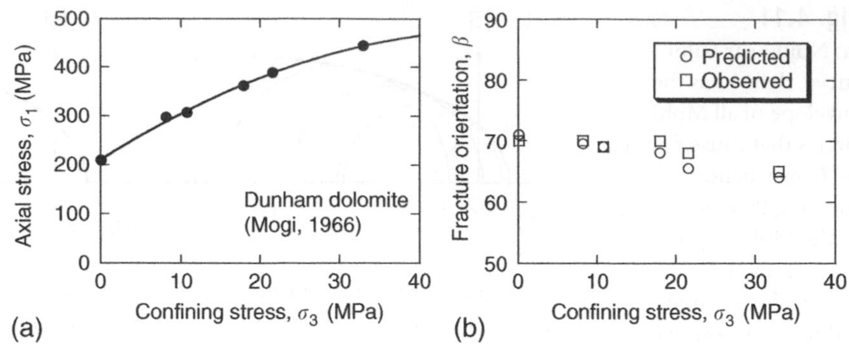

Fig. 4.12 (a) Stresses at failure, measured by Mogi (1966) on a Dunham dolomite. (b) Comparison of the failure angles predicted by the Mohr construction, and the observed angles (between the normal to the failure plane and the direction of maximum principal stress).

can be predicted by the construction shown in Fig. 4.11b. Data obtained under traditional $\sigma_2 = \sigma_3$ triaxial tests typically show reasonably good agreement with this aspect of the theory. For example, Mogi (1966) conducted a series of triaxial failure tests on a Dunham dolomite, determined the orientation angles β from the Mohr construction, and then compared these angles to the observed angles of the failure plane. As seen in Fig. 4.12, the failure curve in the $\{\sigma_3, \sigma_1\}$ plane was slightly nonlinear, but the observed failure angles were generally quite close to the predicted values.

All empirical failure criteria that follow Mohr's hypothesis will be expressible in some functional form $|\tau| = f(\sigma)$, or, alternatively, $\sigma_1 = g(\sigma_3)$. Either of these two functions, f or g, will suffice to determine the other. However, as will be clear from the analysis of the Coulomb theory given in §4.5, the relationship between these two functions will in general not be simple or obvious.

One such failure law that has become widely used, and which is capable of fitting data from many different rocks, is the Hoek–Brown criterion (Hoek and Brown, 1980). In terms of the two extreme principal stresses, this criterion takes the form (Fig. 4.13a)

$$\sigma_1 = \sigma_3 + (m\sigma_c\sigma_3 + \sigma_c^2)^{1/2}, \tag{4.26}$$

where m and σ_c are two fitting parameters. Setting $\sigma_3 = 0$ in (4.26) shows that σ_c is in fact equal to the uniaxial compressive strength, C_o. Setting $\sigma_1 = 0$ in (4.26) and solving the resultant quadratic equation for the uniaxial tensile strength $T_o = -\sigma_3$, gives

$$T_o = \frac{\sigma_c}{2}[(m^2 + 4)^{1/2} - m]. \tag{4.27}$$

The parameter m usually lies in the range $5 < m < 30$, in which case an expansion of (4.27) for "large" m leads to

$$T_o \approx \sigma_c/m, \quad \text{that is,} \quad C_o/T_o \approx m. \tag{4.28}$$

In practice, therefore, the Hoek–Brown model predicts a much larger ratio of compressive strength to tensile strength than does the Coulomb model, and so in that regard is in closer agreement with experimental data.

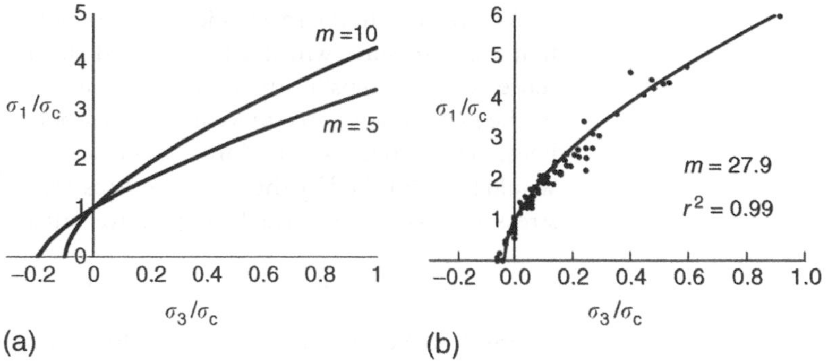

Fig. 4.13
(a) Hoek–Brown failure curves, for two different values of m.
(b) Experimental failure data for several different granites, fit with different values of σ_c, but one value of m (after Hoek and Brown, 1980).

Analysis of published strength data led Hoek and Brown (1980) to conclude that m takes on characteristic values for different rock types. They summarized these trends as follows:

1 $m \approx 7$ for carbonate rocks with well-developed crystal cleavage (dolomite, limestone, marble);
2 $m \approx 10$ for lithified argillaceous rocks (mudstone, siltstone, shale, slate);
3 $m \approx 15$ for arenaceous rocks with strong crystals and poorly developed crystal cleavage (sandstone, quartzite);
4 $m \approx 17$ for fine-grained polymineralllic igneous crystalline rocks (andesite, dolerite, diabase, rhyolite);
5 $m \approx 25$ for coarse-grained polymineralllic igneous and metamorphic rocks (amphibolite, gabbro, gneiss, granite, norite, quartz-diorite).

Figure 4.13b shows data on granites, collected from various sources by Hoek and Brown (1980) and plotted in normalized form. A single best-fit value of $m = 27.86$ was found for the combined data set, although specific values of σ_c were found for each different granite in the set. The resulting fit has a correlation coefficient of 0.99.

Although the Hoek–Brown model gives a nonlinear failure envelope on a Mohr diagram, in contrast to the linear relationship predicted by the Coulomb model, Hoek (1990) has presented equations that allow the Hoek–Brown failure envelope to be locally approximated by a Coulomb line. Various technical issues associated with accurately fitting failure data to different types of failure curves are discussed by Handy (1981), Sheorey (1997), and Pincus (2000).

4.7 Effects of pore fluids

The foregoing discussion of rock failure has ignored the fact that rocks are typically porous to some extent, and the pore space of a rock will *in situ* be filled with fluids under pressure. The pore fluid is usually water, but may be oil, gas, or rock melt. The pore fluid may affect the failure of the rock in two ways: due to the purely mechanical effect of pore pressure, or due to chemical interactions between the rock and the fluid.

With regards to the mechanical effect of pore fluid pressure, it seems plausible that pore pressure, which acts "outward" from the pore space, would in some sense act like a tensile stress. Moreover, in an isotropic rock, this effect should be the same in any three mutually orthogonal directions. Reasoning along these lines, the soil mechanician Karl Terzaghi (1936) proposed that the failure of a soil would be controlled by the "effective stresses," σ_i', which would be the principal stresses, reckoned positive if compressive, *minus* the pore pressure, that is,

$$\sigma_1' = \sigma_1 - P, \quad \sigma_2' = \sigma_2 - P, \quad \sigma_3' = \sigma_3 - P, \tag{4.29}$$

where P is the pore fluid pressure. In a more general formulation, P could be multiplied in (4.29) by some parameter α, which would be referred to as the *effective stress coefficient*. (It should be emphasized that the effective stress coefficient for failure processes has no particular connection to the effective stress coefficient that appears in the theory of linear poroelasticity, §7.4.)

Most experiments on rocks support the conclusion that the effective stress law (4.29) holds, which is to say that the effective stress coefficient for failure is unity. Despite many attempts to derive an effective stress law for failure, which have been contentious and inconclusive, this "law" is best viewed as an empirical observation. It is, however, consistent with the assumption that brittle failure is in some way controlled or initiated by the stress concentrations at the corners of thin microcracks (see §10.8). As shown in §8.10, these stress concentrations are indeed proportional to the difference between the far-field stress and the fluid pressure in the crack.

In the context of a Mohr diagram, replacing the stresses σ_i with the effective stresses σ_i' has the effect of *translating* all the stress circles to the *left* by the amount P. The Mohr's circle will therefore be shifted closer to the failure line. Hence, an *in situ* state of stress that is "safe" in the absence of a pore pressure may well cause the rock to fail if the pore pressure is increased by a sufficient amount (Fig. 4.14). This fact accounts for the increased occurrence of landslides in the aftermath of heavy rainfall.

A set of data that illustrates the effective stress principle of brittle failure is that of Murrell (1965), who conducted standard triaxial compression tests on a Darley Dale sandstone, at several different values of the pore pressure. The Darley Dale is a poorly graded feldspathic sandstone with 21 percent porosity. In each test, the pore pressure and the confining stress were held constant, while the axial stress was increased until failure occurred. When plotted in the $\{P, \sigma_1\}$ plane, the data fall on different curves, corresponding to the different values of σ_3 (Fig. 4.15a).

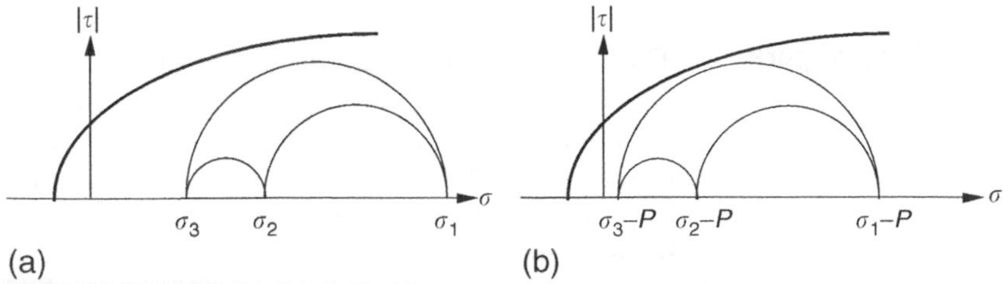

Fig. 4.14 (a) A stress state that lies below the failure curve. (b) Application of a pore pressure P causes the *effective* stress state to move closer to the failure curve.

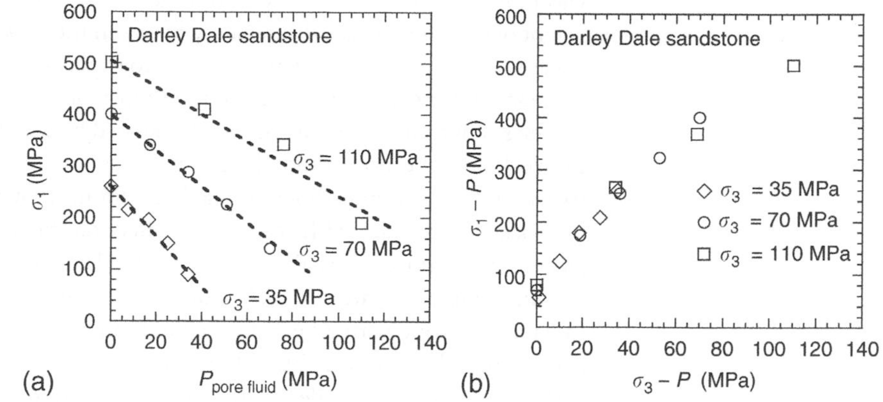

Fig. 4.15 (a) Stresses at failure in a Darley Dale sandstone (Murrell, 1965) as a function of pore pressure, for several different values of the confining stress. (b) Same data plotted in terms of the effective principal stresses, according to the effective stress law (4.29).

If plotted on the $\{\sigma_3, \sigma_1\}$ plane, the data would form three distinct vertical lines. However, when plotted in the $\{P, \sigma_1\}$ plane of maximum and minimum effective stresses, the failure data nearly form a single failure curve, which in this particular case is slightly concave downward (Fig. 4.15b).

It follows from (4.29) that the effective stress tensor $\boldsymbol{\sigma}'$ is related to the actual stress tensor $\boldsymbol{\sigma}$ by

$$\boldsymbol{\sigma}' = \boldsymbol{\sigma} - P\mathbf{I}, \tag{4.30}$$

where $\mathbf{I}$ is the identity tensor. The additional hydrostatic stress $-P\mathbf{I}$ gives rise to no shear stresses on any plane (§2.8), so the effective shear stresses are identical to the actual shear stresses. One implication of this fact is that the condition for sliding along a fault or other plane of weakness, (3.20), is replaced, in the presence of a pore fluid, by

$$|\tau| = S_0 + \mu(\sigma - P). \tag{4.31}$$

A simple physical interpretation of this condition is that, while the normal stress σ tends to strengthen the fault, by pushing the two opposing rock faces together, the fluid pressure acts to weaken it, by pushing the two opposing rock faces apart. Expression (4.31) has been verified experimentally by Byerlee (1967) on laboratory specimens of granite. Some geological implications of this concept have been discussed by Secor (1965) and Hubbert and Rubey (1959, 1960, 1961).

Extension failure in the presence of pore pressure has been discussed theoretically by Murrell (1964). The effective stress concept suggests that the criterion for failure should be

$$\sigma_3' = \sigma_3 - P = -T_0. \tag{4.32}$$

This result is consistent with the concept that tensile failure is caused by the stress concentrations at the edges of thin cracks oriented normal to the direction of the least compressive principal stress. Jaeger (1963) performed tensile failure tests on a fine-grained Tasmanian dolerite, and found that (4.32) held, although the data was fit best by multiplying the pore fluid pressure P by an effective stress coefficient of 0.95. The slight discrepancy between the experimental and theoretical effective stress coefficients was attributed to experimental error. Condition (4.32) for tensile failure is fundamental to the theory of hydraulic fracturing (§13.6).

The strength of rocks can also be influenced by *chemical* interactions between the rock and the pore fluid (Paterson, 1978, pp. 78–9). In many quartz-rich rocks, but also in limestones, it has been found that the strength decreases if the rock is in contact with water. This effect has been noted for sandstones by Jaeger (1943), for coal by Price (1960), for calcite rocks by Rutter (1972b), and for limestone by Parate (1973). Vukuturi (1974) measured the tensile strength of Indiana limestone with different pore fluids, including water, glycerine, benzene, and various alcohols and found that the strength decreased with an increase in surface tension. The tensile strength of the limestone when saturated with water was about 25 percent less than when saturated with ethyl alcohol and about 30 percent less than the value that could be extrapolated from the data for a hypothetical pore fluid with zero surface tension.

Pore fluids can also influence rock strength through the mechanism of *stress corrosion fracture* (Atkinson, 1979; Peck, 1983). In quartz-rich rocks, fracture proceeds through the rupture of Si–O bonds at the crack tip, and this rupture is accelerated when the bonds are strained by external stresses (Dove, 1995). The rate of crack growth can be modeled as a rapidly increasing function of the applied far-field stress, such as an exponential or power law function (Wiederhorn and Boltz, 1970; Lawn, 1993). Ojala et al. (2003) conducted standard triaxial compression tests on a Locharbriggs sandstone composed of 88 percent quartz and 6 percent K-feldspar, while flowing water through the sample at a rate of 30 ml/h. The concentration of silica in the effluent was found to correlate with the different stages of the deformation process: crack closure, linear elastic, and strain hardening. The yield stress increased with increasing strain rate, consistent with the idea that at higher strain rates the fluid has less time to react with the rock. The effects of rock/fluid interactions on rock failure have been reviewed by Atkinson and Meredith (1987).

4.8 Failure under true-triaxial conditions

Mohr's theory of failure is based on the assumption that failure is controlled by the minimum and maximum principal stresses, and is unaffected by the magnitude of the intermediate principal stress. As most "triaxial" rock testing is conducted under conditions of $\sigma_2 = \sigma_3$, such data do not allow a test of this aspect of Mohr's hypothesis. Nevertheless, there are situations in which $\sigma_2 = \sigma_3$ will be the exception in the subsurface rather than the rule. Hence, the question of whether or not Mohr's hypothesis is correct is pertinent.

Compression tests conducted under true-triaxial conditions (Mogi, 1971; Fig. 4.16a) and borehole breakout tests (Haimson and Song, 1995; Fig. 4.17a) in fact show that for many, although not all, rocks the intermediate principal stress has a pronounced influence on the value of σ_1 at failure. This suggests the need for failure criteria that depend on all three principal stresses. As it is known from traditional ($\sigma_2 = \sigma_3$) compression tests that a lateral confining stress of the form $\sigma_2 = \sigma_3$ has the effect of strengthening the rock, it is plausible that any increase of σ_2 above σ_3 may cause additional strengthening. This suggests replacing the failure criterion of the form $\sigma_1 = f(\sigma_2)$ by the more general form

$$\sigma_1 = f(\sigma_2, \sigma_3). \tag{4.33}$$

For an isotropic rock, such a failure criterion can also always be written in terms of the stress invariants (§2.8), and it is often convenient to do so.

In the context of metal plasticity, Nadai (1950) suggested that the "driving force" for failure will be J_2, the second invariant of the deviatoric stress. According to (2.164), this invariant is related to the three principal stresses by

$$J_2 = \frac{1}{6}[(\sigma_1 - \sigma_2)^2 + (\sigma_2 - \sigma_3)^2 + (\sigma_3 - \sigma_1)^2]. \tag{4.34}$$

According to (5.152), in a linearly elastic, isotropic material, J_2 is also a direct measure of the distortional strain energy. Nadai further suggested that failure was "opposed" by the mean stress in the material, which is consistent with the concept that confinement strengthens the rock. The mean stress can be represented by the first invariant of the stress, I_1, which according to (2.143) and (2.153) is related to the three principal stresses by

$$I_1 = \sigma_1 + \sigma_2 + \sigma_3 = 3\tau_{\rm m}, \tag{4.35}$$

where $\tau_{\rm m}$ is the mean normal stress. Hence, Nadai's assumption can be written in the form

$$J_2 = f(I_1), \tag{4.36}$$

where f is some *increasing* function of I_1. A factor of 2/3 is often included in front of J_2, in which case, according to (2.167), the failure criterion can be written in terms of the octahedral shear stress and the mean normal stress:

$$\tau_{\rm oct}^2 \equiv \frac{2}{3}J_2 = f(\tau_{\rm m}), \tag{4.37}$$

where f is some increasing function of the mean normal stress.

Many specific forms of a true-triaxial failure criterion have been proposed for rocks and soils. Drucker and Prager (1952) took the relationship between J_2 and I_1 at failure to be of the form

$$(J_2)^{1/2} = a + bI_1, \tag{4.38}$$

where a and b are material-dependent constants. Zhou (1994) extended this by adding a term that is quadratic in I_1:

$$(J_2)^{1/2} = a + bI_1 + cI_1^2. \tag{4.39}$$

Colmenares and Zoback (2002) referred to this as the "modified Wiebols and Cook" model in recognition of its similarities to the criterion proposed by Wiebols and Cook (1968) based on micromechanical analysis of sliding cracks (§10.6).

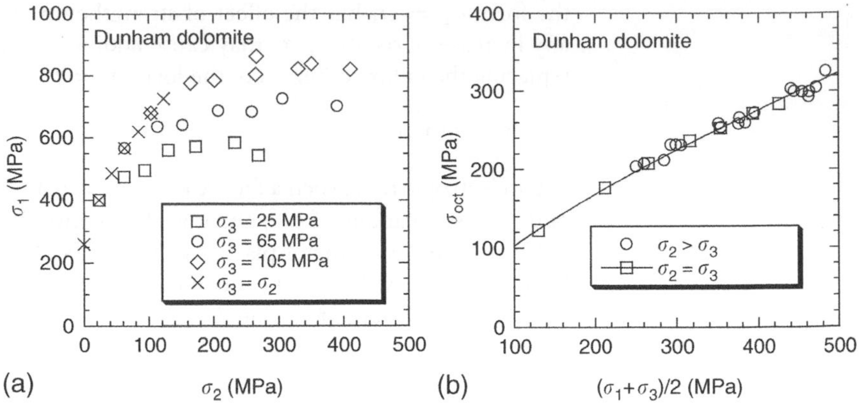

Fig. 4.16 (a) Stresses at failure in a Dunham dolomite (Mogi, 1971). If failure did not depend on σ_2, the values of σ_1 for fixed σ_3 would lie on horizontal lines. (b) Same data plotted in the $\{\tau_{m2}, \tau_{oct}\}$ plane, fit with a power law function.

Mogi (1971) modified the reasoning leading to (4.36) or (4.37) by the following argument. According to the Mohr hypothesis, the failure plane will lie parallel to the direction of the intermediate principal stress. Hence, it is plausible that fracture is resisted only by the mean normal stress in the plane normal to the failure plane, which is $(\sigma_1 + \sigma_3)/2$, rather than by the total mean normal stress. This suggests a brittle failure criterion of the general form

$$\tau_{oct} = f(\tau_{m2}), \quad \text{where } \tau_{m2} = (\sigma_1 + \sigma_3)/2. \tag{4.40}$$

To test this hypothesis, Mogi (1971) conducted true-triaxial compression tests on several rocks. The results showed that, for a fixed value of the minimum stress σ_3, the value of σ_1 at failure at first increases with an increase in σ_2, but eventually decreases slightly as σ_2 increases yet further (Fig. 4.16a). If the octahedral shear stress at failure is plotted against the mean stress on the plane parallel to σ_2, as suggested by (4.40), the results do indeed coalesce to a single line in the $\{\tau_{m2}, \tau_{oct}\}$ plane (Fig. 4.16b).

Haimson and Song (1995) compared the results of standard $\sigma_2 = \sigma_3$ confined compression tests with borehole breakout data measured on 10-cm cubical specimens containing central circular boreholes with a radius of 1 cm. The cubical specimens were subjected to various true-triaxial stress states, with the intermediate "far-field" principal stress always aligned parallel to the borehole. The local stresses at the borehole wall were calculated from the Kirsch solution, (8.113)–(8.115), for a circular hole in an infinite rock mass. The borehole wall was traction-free, so the minimum principal stress at the location of the borehole breakout, which in this case was the radial stress, was always zero, but σ_2 was always nonzero. The values of σ_1 at failure that were observed in the borehole breakout tests were 2–3 times greater than would be predicted by the Coulomb failure criterion that was derived from the standard compression tests (Fig. 4.17a). However, both the confined compression data and the borehole breakout data fell on a single line in the $\{\tau_{m2}, \tau_{oct}\}$ plane (Fig. 4.17b), in accordance with Mogi's model. For both Lac du Bonnet and Westerly granite, the failure data could be fit with curves of the form

$$\tau_{oct} = a + b\tau_{m2}. \tag{4.41}$$

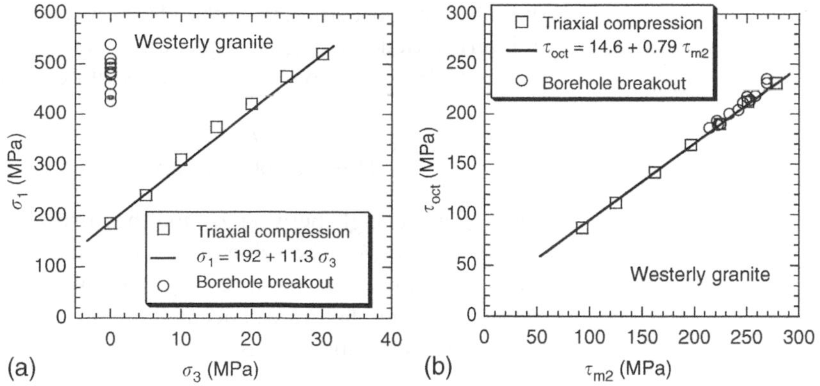

Fig. 4.17 (a) Stresses at failure in a Westerly granite, measured by Haimson and Song (1995) under standard "triaxial" confined compression and in borehole breakout tests, plotted in the $\{\sigma_3, \sigma_1\}$ plane. (b) Same data plotted in the $\{\tau_{m2}, \tau_{oct}\}$ plane.

The best-fitting values were found to be $\{a = 19.0\,\text{MPa}, b = 0.76\}$ for Lac du Bonnet, and $\{a = 14.6\,\text{MPa}, b = 0.79\}$ for Westerly.

Al-Ajmi and Zimmerman (2005) showed that a linear expression for the failure criterion in Mogi's $\{\tau_{m2}, \tau_{oct}\}$ space, such as is given by (4.41), reduces precisely to the Mohr–Coulomb criterion if any two of the principal stresses are equal. Thus, the linear Mogi law, which they referred to as the "Mogi–Coulomb" criterion, is in some sense a natural extension of the Mohr–Coulomb criterion into the polyaxial stress domain.

Colmenares and Zoback (2002) carried out a detailed analysis of several published true-triaxial data sets for brittle failure. Data were assembled on Solenhofen limestone, Dunham dolomite, Yuubari shale, Shirahama sandstone, and amphibolite. The data for each rock were fit to several triaxial failure models, including the Drucker–Prager criterion (4.38), Zhou's modified Wiebols and Cook criterion (4.39), and Mogi's criterion (4.40) in the specific form $f(\tau_{m2}) = c\tau_{m2}^n$, where c and n are fitting parameters. The linear Coulomb criterion (§4.5) and the nonlinear Hoek–Brown criteria (§4.6), which do not attempt to account for the effect of σ_2, were also used. In general, the modified Wiebols and Cook model was able to fit the data reasonably well, as was Mogi's model, whereas the Drucker–Prager model gave very poor fits. The Shirahama sandstone and Yuubari shale showed very weak dependence on the intermediate principal stress, and consequently these rocks could be fit by the Coulomb and Hoek–Brown criteria. These latter models did not provide as good fits for the other three rocks, which showed stronger dependence on σ_2. Although measured data could always be fit to Mogi's model, in some cases this model seemed to give multivalued predictions of the value of σ_1 at failure, at fixed values of $\{\sigma_2, \sigma_3\}$, which could be problematic when used as a predictive tool.

4.9 The effect of anisotropy on strength

Since most sedimentary and metamorphic rocks are anisotropic, the effect of anisotropy on strength is of great importance. The simplest situation, that of planar anisotropy in which a rock mass has a set of parallel planes of weakness, can now be addressed, by combining the results of §3.5 and §4.5. In §3.5, the problem of sliding along a preexisting plane of weakness was discussed. If S_w is the inherent shear strength of the planes of weakness, and μ_w is the coefficient of internal friction along those planes, then the condition for sliding along these planes, (3.28), can be written in the present notation as

$$\sigma_1 = \sigma_3 + \frac{2(S_w + \mu_w \sigma_3)}{(1 - \mu_w \cot\beta)\sin 2\beta}, \tag{4.42}$$

where β is the angle between σ_1 and the normal to the planes of weakness.

As shown in §3.5, the value of σ_1 required to cause failure, as given by (4.42), tends to infinity as $\beta \to \pi/2$ or $\beta \to \tan^{-1}\mu_w = \phi_w$. For angles between these two values, failure will occur at a finite value of σ_1 that varies with β. The minimum such value of σ_1 is

$$\sigma_1^{\min} = \sigma_3 + 2(S_w + \mu_w \sigma_3)[(1 + \mu_w^2)^{1/2} + \mu_w], \tag{4.43}$$

which occurs at a specific angle β_w that is given by

$$\tan 2\beta_w = -1/\mu_w. \tag{4.44}$$

If the plane of weakness is oriented from the direction of maximum principal stress by some angle other than β_w, failure can still occur, but only at a value of σ_1, as given by (4.42), that is greater than $\sigma_1^{\min}$. For values of $\beta < \phi_w$, failure along the plane of weakness is not possible, for any value of σ_1. The situation is illustrated in Fig. 3.9.

Thus far, only the possibility of failure along a plane of weakness has been considered. However, failure can occur on a plane other than the preexisting planes of weakness if the Coulomb failure criterion, (4.5), is satisfied, but with parameters that can be denoted as S_o and μ_o. As the planes of weakness are by definition weaker than the intact rock, it can be assumed that $S_w < S_o$ and $\mu_w < \mu_o$. According to (4.13), failure can occur on a plane other than a plane of weakness if σ_1 reaches the value

$$\sigma_1 = 2S_o \tan\beta_o + \sigma_3 \tan^2\beta_o, \tag{4.45}$$

where, according to (4.6), β_o is given by

$$\tan 2\beta_o = \tan(\phi_o + 90^\circ) = -1/\tan\phi_o = -1/\mu_o. \tag{4.46}$$

For a fixed value of σ_3, the value of σ_1 required to cause failure somewhere within the rock will then be equal to the *smaller* of two values given by (4.42) and (4.45). If, for a given orientation β of the normal to the planes of weakness relative to the direction of maximum principal stress, the value given by (4.42) is less than that given by (4.45), failure will occur along a plane of weakness. On the other hand, if the value given by (4.42) is greater than that given by (4.45), failure will occur along a plane within the intact rock whose orientation is defined by (4.46).

The value of σ_1 needed to cause failure is plotted in Fig. 4.18, for two values of σ_3. The concave-upward portions of the curves correspond to criterion (4.42), and represent failure along the plane of weakness. The horizontal portions of the

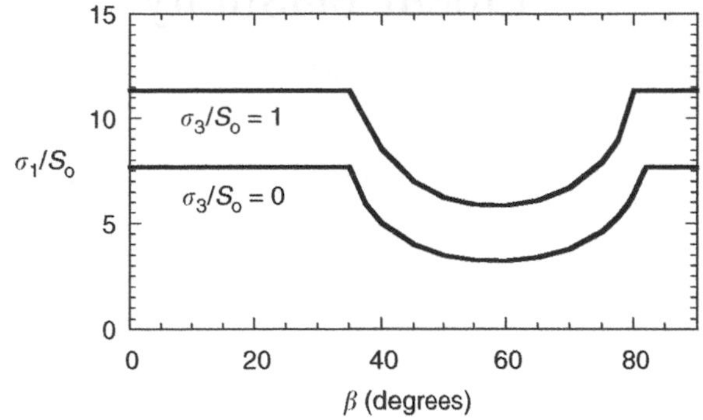

Fig. 4.18 Variation of the value of σ_1 needed to cause failure, as a function of the angle β between the normal to the plane of weakness and the maximum principal stress, for the case $\mu_w = 0.5$, $\mu_o = 0.7$, $S_o = 2S_w$.

curves are plotted from (4.45), and represent failure within the intact rock, along a plane defined by (4.46). Larger values of σ_3 would cause the curve to be shifted upward, as can be anticipated by examining Fig. 3.9 and (4.45). Experimental results (Donath, 1961; Hoek, 1964) show general agreement with that shown in Fig. 4.18.

5 Linear elasticity

5.1 Introduction

As demonstrated in Chapter 4, the stress–strain behavior of rock is quite complex, even in such seemingly simple situations as uniaxial compression of a cylindrical specimen. In order to generate analytical solutions to more complicated rock mechanics problems, it is usually necessary to idealize and simplify the stress–strain behavior. The most commonly used form for the stress–strain relationships for rocks is that of *linear elasticity*, in which the strain tensor is a linear function of the stress tensor. This assumption allows many important problems to be solved, such as stresses around boreholes and tunnels, stresses around faults and fractures, etc. Although no rocks are actually "linearly elastic" over a wide range of stresses, this approximation is often quite useful and accurate, since many rocks behave linearly for incremental changes in stress. The changes in stress brought about by any perturbation to the existing stress field, such as those caused by an excavation, are small throughout most of the region of rock, except in the immediate vicinity of the excavation itself. Linearity is therefore an excellent approximation throughout the bulk of the rock mass.

In order to solve a problem in rock mechanics, one needs a set of equations that are sufficient to ensure that the problem is "well posed" – that is, that it indeed has a mathematical solution. In particle mechanics, the most fundamental governing equation is Newton's law, which states that force is equal to mass times acceleration. The analogue of that law for a deformable body such as a rock is the law of *stress equilibrium*, which is derived in §5.5 and presented in cylindrical coordinates in §5.6. This law applies to all rocks, regardless of the relationship between stress and strain, and must be satisfied in all processes that occur in a rock specimen or rock mass. The stress equilibrium equations must be supplemented by a set of *constitutive equations* that describe the relationship between stress, strain, and possibly other parameters such as temperature, pore fluid pressure, etc. The constitutive equations of *isotropic* linear elasticity are presented and discussed in §5.2–§5.4. The final pieces of data that are needed to ensure the existence of a unique solution to a rock mechanics problem (assuming linear elastic behavior) are the *boundary conditions*, along with certain restrictions on the value of the elastic moduli. These restrictions are discussed in §5.9, where the uniqueness theorem of linear elasticity is proven. Finally, in §5.10, the equations of *anisotropic* linear elasticity are presented.

5.2 Stress–strain relations for an isotropic linear elastic solid

An isotropic solid can be loosely defined as one in which all directions are "equivalent." In other words, in an isotropic rock the relationship between vertical stress and vertical strain is the same as that between horizontal stress and horizontal strain, etc. An important consequence of the property of isotropy is that the principal axes of strain must *coincide* with the principal axes of stress. To see why this is true, note that along a principal axis of stress, the stresses are purely normal, and they vary symmetrically with any angular departure from this direction (see Fig. 2.7). In an isotropic rock, these symmetrical stresses must produce a symmetrical system of strains. But symmetrical strains exist (in general) only about a principal axis of strain; hence, the principal axes of stress and strain must coincide. An alternative proof of this result can be obtained using the methods presented in §5.10 for treating anisotropic materials.

The basic assumption underlying linear elasticity is that the components of stress are linear functions of the components of strain. It is also implicit in this assumption that the stress does not depend on the time rate of change of the strains, the past history of the strains, etc. In fact, given the obvious similarity between the stress transformation equations (2.101) and the strain transformation equations (2.227), the obvious first-order approximation to the stress–strain relationship is to assume a linear relation between the stress and strain tensors. Written in terms of the principal coordinate system, the stress–strain law of isotropic elasticity, often called "Hooke's law," takes the form

$$\sigma_1 = (\lambda + 2G)\varepsilon_1 + \lambda\varepsilon_2 + \lambda\varepsilon_3, \tag{5.1}$$

$$\sigma_2 = \lambda\varepsilon_1 + (\lambda + 2G)\varepsilon_2 + \lambda\varepsilon_3, \tag{5.2}$$

$$\sigma_3 = \lambda\varepsilon_1 + \lambda\varepsilon_2 + (\lambda + 2G)\varepsilon_3, \tag{5.3}$$

Any parameter that gives the ratio of one of the stress components to one of the strain components is generically called an "elastic modulus." The two elastic moduli appearing in (5.1)–(5.3), λ and G, are also known as the Lamé parameters. The parameter G is often denoted, particularly in mathematical elasticity treatments, by the symbol μ (λ and μ being the two Greek consonants in the surname of the French elastician who first developed the above equations, Gabriel Lamé). In order to avoid confusion with the coefficient of friction, however, we will use G. As will be shown below, G is the *shear modulus*, as it relates stresses to strains in a state of pure shear. If reference is made to the Lamé parameter (singular), this refers specifically to λ.

Recalling that the volumetric strain is the sum of the three principal normal strains, that is,

$$\varepsilon_v = \varepsilon_1 + \varepsilon_2 + \varepsilon_3. \tag{5.4}$$

the stress–strain equations (5.1)–(5.3) can be written as

$$\sigma_1 = \lambda\varepsilon_v + 2G\varepsilon_1, \quad \sigma_2 = \lambda\varepsilon_v + 2G\varepsilon_2, \quad \sigma_3 = \lambda\varepsilon_v + 2G\varepsilon_3. \tag{5.5}$$

Summing up the three principal normal stresses yields

$$3\tau_m = \sigma_1 + \sigma_2 + \sigma_3 = 3\lambda\varepsilon_v + 2G(\varepsilon_1 + \varepsilon_2 + \varepsilon_3) = (3\lambda + 2G)\varepsilon_v. \tag{5.6}$$

The volumetric strain is therefore related to the mean stress by

$$\tau_{\mathrm{m}} = \left(\lambda + \frac{2}{3}G\right)\varepsilon_{\mathrm{v}} \equiv K\varepsilon_{\mathrm{v}}, \tag{5.7}$$

where K is the *bulk modulus*. The multiplicative reciprocal of the bulk modulus, $1/K$, is known as the *bulk compressibility* and is usually denoted by β or C.

Hooke's law in the form (5.1)–(5.3) gives the stresses as functions of the strains. These equations can be inverted by using (5.7) to eliminate ε_{v} from (5.5), which gives

$$\varepsilon_1 = \frac{(\lambda + G)}{G(3\lambda + 2G)}\sigma_1 - \frac{\lambda}{2G(3\lambda + 2G)}\sigma_2 - \frac{\lambda}{2G(3\lambda + 2G)}\sigma_3, \tag{5.8}$$

$$\varepsilon_2 = -\frac{\lambda}{2G(3\lambda + 2G)}\sigma_1 + \frac{(\lambda + G)}{G(3\lambda + 2G)}\sigma_2 - \frac{\lambda}{2G(3\lambda + 2G)}\sigma_3, \tag{5.9}$$

$$\varepsilon_3 = -\frac{\lambda}{2G(3\lambda + 2G)}\sigma_1 - \frac{\lambda}{2G(3\lambda + 2G)}\sigma_2 + \frac{(\lambda + G)}{G(3\lambda + 2G)}\sigma_3. \tag{5.10}$$

Now consider a state of uniaxial stress in, say, the first principal direction. *Young's modulus, E*, also sometimes known as the *modulus of elasticity*, is defined as the ratio of this stress to the strain that results in the same direction, that is,

$$E \equiv \frac{\sigma_1}{\varepsilon_1} = \frac{G(3\lambda + 2G)}{(\lambda + G)}. \tag{5.11}$$

Poisson's ratio, ν, is defined as (the negative of) the ratio of the transverse strain to the longitudinal strain, under conditions of uniaxial stress, that is,

$$\nu \equiv \frac{-\varepsilon_2}{\varepsilon_1} = \frac{-\varepsilon_3}{\varepsilon_1} = \frac{\lambda}{2(\lambda + G)}. \tag{5.12}$$

Poisson's ratio is typically a positive number, in which case a longitudinal compression would be accompanied by transverse expansion and vice versa.

Equations (5.7), (5.11), and (5.12) give the elastic parameters E, K, and ν in terms of λ and G. Other useful relations between these parameters are as follows:

$$\lambda = \frac{E\nu}{(1 + \nu)(1 - 2\nu)}, \quad G = \frac{E}{2(1 + \nu)}, \quad K = \frac{E}{3(1 - 2\nu)}; \tag{5.13}$$

$$\lambda = \frac{2G\nu}{(1 - 2\nu)}, \quad E = 2G(1 + \nu), \quad K = \frac{2G(1 + \nu)}{3(1 - 2\nu)}; \tag{5.14}$$

$$\lambda = K - \frac{2}{3}G, \quad E = \frac{9KG}{3K + G}, \quad \nu = \frac{3K - 2G}{6K + 2G}. \tag{5.15}$$

Although numerous elastic parameters can be defined for an isotropic material, only two of them are independent; if any two are known, the others can be determined from equations such as those given above. A full listing of all thirty relations that can be obtained by expressing any three of the set $\{\lambda, K, G, E, \nu\}$ in terms of the other two is given by Davis and Selvadurai (1996).

Thus far, the discussion of Hooke's law for isotropic materials has been carried out only in the principal coordinate system. Hooke's law in an arbitrary orthogonal coordinate system can be derived by first writing (5.4) and (5.5) in matrix form as

$$\boldsymbol{\tau} = \lambda\,\mathrm{trace}(\boldsymbol{\varepsilon})\mathbf{I} + 2G\boldsymbol{\varepsilon}. \tag{5.16}$$

Now consider a second coordinate system, obtained from the principal coordinate system through a rotation matrix $\mathbf{L}$. According to the transformation laws for second-order tensors, (2.30) and (2.38), the stress and strain matrices in the new coordinate system are given by $\boldsymbol{\tau}' = \mathbf{L}\boldsymbol{\tau}\mathbf{L}^{\mathrm{T}}$ and $\boldsymbol{\varepsilon}' = \mathbf{L}\boldsymbol{\varepsilon}\mathbf{L}^{\mathrm{T}}$. Applying this transformation to (5.16) yields

$$\begin{aligned}\boldsymbol{\tau}' &= \mathbf{L}\boldsymbol{\tau}\mathbf{L}^{\mathrm{T}} = \mathbf{L}[\lambda\,\mathrm{trace}(\boldsymbol{\varepsilon})\mathbf{I} + 2G\boldsymbol{\varepsilon}]\mathbf{L}^{\mathrm{T}} = \lambda\,\mathrm{trace}(\boldsymbol{\varepsilon})\mathbf{L}\mathbf{I}\mathbf{L}^{\mathrm{T}} + 2G\mathbf{L}\boldsymbol{\varepsilon}\mathbf{L}^{\mathrm{T}} \\ &= \lambda\,\mathrm{trace}(\boldsymbol{\varepsilon}')\mathbf{L}\mathbf{L}^{\mathrm{T}} + 2G\boldsymbol{\varepsilon}' = \lambda\,\mathrm{trace}(\boldsymbol{\varepsilon}')\mathbf{I} + 2G\boldsymbol{\varepsilon}',\end{aligned} \tag{5.17}$$

where we have used the fact that $\mathrm{trace}(\boldsymbol{\varepsilon}) = \mathrm{trace}(\boldsymbol{\varepsilon}')$, because trace $(\boldsymbol{\varepsilon})$ is an invariant, along with the property that $\mathbf{L}\mathbf{L}^{\mathrm{T}} = \mathbf{I}$, since $\mathbf{L}$ is a rotation matrix. Comparison of the first and last terms in (5.17) shows that Hooke's law, as expressed in matrix form in (5.16), actually holds for an *arbitrary* orthogonal coordinate system. When written out term-by-term in the general case when the coordinate system is not aligned with the principal axes, (5.16) takes the form

$$\tau_{xx} = (\lambda + 2G)\varepsilon_{xx} + \lambda\varepsilon_{yy} + \lambda\varepsilon_{zz}, \tag{5.18}$$

$$\tau_{yy} = \lambda\varepsilon_{xx} + (\lambda + 2G)\varepsilon_{yy} + \lambda\varepsilon_{zz}, \tag{5.19}$$

$$\tau_{zz} = \lambda\varepsilon_{xx} + \lambda\varepsilon_{yy} + (\lambda + 2G)\varepsilon_{zz}, \tag{5.20}$$

$$\tau_{xy} = 2G\varepsilon_{xy}, \quad \tau_{xz} = 2G\varepsilon_{xz}, \quad \tau_{yz} = 2G\varepsilon_{yz}. \tag{5.21}$$

For a state of simple shear in which, say, τ_{xy} is the only nonzero stress, (5.18)–(5.21) show that $\gamma_{xy} = 2\varepsilon_{xy} = \tau_{xy}/G$ is the only nonzero strain. For this reason, G is known as the *shear modulus*.

Hooke's law has the same algebraic form in coordinate systems that are locally orthogonal, but not necessarily Cartesian, such as the cylindrical coordinate system defined in §2.15. In cylindrical coordinates,

$$\tau_{rr} = \lambda\varepsilon_{\mathrm{v}} + 2G\varepsilon_{rr}, \quad \tau_{\theta\theta} = \lambda\varepsilon_{\mathrm{v}} + 2G\varepsilon_{\theta\theta}, \quad \tau_{\phi\phi} = \lambda\varepsilon_{\mathrm{v}} + 2G\varepsilon_{\phi\phi}, \tag{5.22}$$

$$\tau_{r\theta} = 2G\varepsilon_{r\theta}, \quad \tau_{r\phi} = 2G\varepsilon_{r\phi}, \quad \tau_{\theta\phi} = 2G\varepsilon_{\theta\phi}, \tag{5.23}$$

where $\varepsilon_{\mathrm{v}} = \varepsilon_{rr} + \varepsilon_{\theta\theta} + \varepsilon_{\phi\phi}$, as in (2.281).

The inverse version of Hooke's law, in which the strains are expressed as functions of the stresses, is most naturally written as follows:

$$\varepsilon_{xx} = \frac{1}{E}\tau_{xx} - \frac{\nu}{E}\tau_{yy} - \frac{\nu}{E}\tau_{zz} = \frac{1}{E}[\tau_{xx} - \nu(\tau_{yy} + \tau_{zz})], \tag{5.24}$$

$$\varepsilon_{yy} = \frac{1}{E}\tau_{yy} - \frac{\nu}{E}\tau_{xx} - \frac{\nu}{E}\tau_{zz} = \frac{1}{E}[\tau_{yy} - \nu(\tau_{xx} + \tau_{zz})], \tag{5.25}$$

$$\varepsilon_{zz} = \frac{1}{E}\tau_{zz} - \frac{\nu}{E}\tau_{xx} - \frac{\nu}{E}\tau_{yy} = \frac{1}{E}[\tau_{zz} - \nu(\tau_{xx} + \tau_{yy})], \tag{5.26}$$

$$\tau_{xy} = 2G\varepsilon_{xy}, \quad \tau_{xz} = 2G\varepsilon_{xz}, \quad \tau_{yz} = 2G\varepsilon_{yz}. \tag{5.27}$$

In our subsequent presentation of the theory of linear elasticity, all five of the elastic parameters $\{\lambda, K, G, E, \nu\}$ will be used. Some formulae, such as (5.1)–(5.3), are most naturally written in terms of λ and G, whereas others, such as (5.24)–(5.26), take on a simpler form if written in terms of E and ν. Furthermore, different experimental configurations lead to the measurement of different parameters, as discussed in Chapter 6. For example, uniaxial compression or tension measures E and/or ν, whereas torsion gives G. It follows that relations amongst the elastic moduli, such as (5.7) and (5.11)–(5.15), are extremely useful and will be referred to frequently.

It seems plausible that a cylindrical rock specimen would shorten if subject to a compressive stress, in which case (5.24) indicates that $E > 0$. Similarly, if it is assumed that a hydrostatic compressive stress results in a decrease in volume, then (5.7) shows that $K > 0$. Finally, the condition that a positive shear stress leads to a positive shear strain, and vice versa, implies that $G > 0$. Each of these conditions is quite plausible, yet none are required by any known laws of mechanics or thermodynamics; they are merely conditions that guarantee that the material is *stable* (McLellan, 1980, pp. 247–50). In fact, these criteria *do not* hold in regimes such as the postfailure region of the uniaxial compression curve, such as shown in Fig. 4.3, where it is seen that, incrementally, an increase in strain is accompanied by a *decrease* in stress. Nevertheless, these conditions are usually assumed to hold when discussing the theory of linear elasticity. It will be seen in §5.8 that these criteria are closely related to the condition that any deformation of a rock results in energy being *stored* in the rock, rather than *released by* the rock. These criteria are also required in order that an elasticity problem will have a unique solution (see §5.9). Equation (5.13) shows that these criteria force the Poisson ratio to lie in the range $-1 < \nu < 0.5$. A negative Poisson ratio would imply that longitudinal extension is accompanied by transverse extension, rather than transverse contraction. Although it seems that no isotropic rocks have been found to have negative Poisson ratios, this peculiar behavior is not ruled out, even by stability arguments. Indeed, man-made foams have been produced which do have Poisson's ratios in the range $-1 < \nu < 0$ (Lakes, 1987).

In 1829, the French mathematical physicist S. D. Poisson developed a simplified model for atomic interactions in an elastic solid and concluded that $\lambda = \mu$. If this were the case, then we would also have

$$\lambda = G, \quad K = 5G/3, \quad E = 5G/2, \quad \nu = 1/4. \tag{5.28}$$

The condition $\lambda = \mu$, known as "Poisson's relation," is not a particularly accurate approximation for most rocks. In fact, Poisson's ratio takes on a range of values when various rock types are considered. Despite this fact, Poisson's relation is sometimes used, particularly in geophysics, to simplify the equations of elasticity. There is also a small body of literature on the problem of determining the solution to the elasticity equations for an arbitrary value of ν, using a solution which is known to hold for a particular value, such as 0.25 (Westergaard, 1952, pp. 137–9; Knops, 1958). Nevertheless, it must be said that the difficulties in solving elasticity problems are not caused by the appearance of two elastic parameters in the equations, rather than one, but by the structure of differential equations themselves. Hence, any advantages gained by putting $\lambda = \mu$ are outweighed by the resulting loss in generality.

Another particular type of idealized isotropic elastic material is the *incompressible solid*, which has $\beta = 0$, and hence $K = 1/\beta \to \infty$. For such materials, (5.15) shows that

$$K \to \infty, \quad \lambda \to \infty, \quad E \to 3G, \quad \nu \to 1/2, \tag{5.29}$$

whereas E and G can remain finite. A completely *rigid* material, on the other hand, is not only incompressible but also has infinite values of E and G. The limiting case of a *compressible fluid* is that in which the shear modulus vanishes, but the bulk modulus remains finite. In this case, (5.13)–(5.15) show that

$$G \to 0, \quad \nu \to 1/2, \quad E \to 0, \quad \lambda = K. \tag{5.30}$$

The elastic moduli are all ratios of stresses to strains. Since the strains are dimensionless, the moduli must have dimensions of stress. The Poisson ratio, which is an elastic parameter, although not quite an elastic modulus, is itself dimensionless. Many different units are used to quantify the moduli. In many engineering texts, as well as in the petroleum engineering industry, it is common to use pounds per square inch (psi). In geophysics, moduli are often quantified in dynes per square centimeter (dyne/cm^2), as well as in bars, where 1 bar = 10^6 dyne/cm^2. The official SI unit for stress is the Pascal, which is 1 Newton per square meter (Pa = N/m^2). As a Pascal is a much smaller value than usually occurs in rock mechanics, it is common to measure stresses in MegaPascals (1 MPa = 10^6 Pa) and moduli in GigaPascals (1 GPa = 10^9 Pa). The conversion factors between the various units are

$$1 \text{ bar} = 10^6 \text{ dyne/cm}^2 = 10^5 \text{ Pa} = 14.50 \text{ psi}. \tag{5.31}$$

5.3 Special cases

There are a number of special stress–strain states that are of sufficient practical importance to make it worthwhile to examine them explicitly. In the following discussions, it will be assumed that $0 < \nu < 1/2$, as is always the case in practice.

5.3.1 Hydrostatic stress, $\sigma_1 = \sigma_2 = \sigma_3 = P$

This is the state of stress that would occur if a rock specimen were surrounded by a fluid under a pressure of magnitude P. From (5.8)–(5.10), the strains are given by

$$\varepsilon_1 = \varepsilon_2 = \varepsilon_3 = \frac{P}{3K}. \tag{5.32}$$

It follows that the volumetric strain is

$$\varepsilon_v = \varepsilon_1 + \varepsilon_2 + \varepsilon_3 = \frac{P}{K}, \tag{5.33}$$

so that $1/K$ can be identified as the *compressibility* of the rock.

5.3.2 Uniaxial stress, $\sigma_1 \neq 0, \sigma_2 = \sigma_3 = 0$

This is the stress state that arises when a specimen is uniformly loaded in one direction, while its lateral boundaries are free from traction. As well as being commonly used in laboratory testing, this state will also be approximated in practical situations such as in a pillar in an underground mine, for example. The resulting strain state will be a contraction $\varepsilon_1 = \sigma_1/E$ in the direction of σ_1 and an expansion $\varepsilon_2 = \varepsilon_3 = -\nu\sigma_1/E$ in the two perpendicular directions. The fractional change in volume is found from (5.4) to be

$$\varepsilon_v = (1 - 2\nu)\sigma_1/E. \tag{5.34}$$

The volume decreases if the stress is compressive and increases if it is tensile.

5.3.3 Uniaxial strain, $\varepsilon_1 \neq 0, \varepsilon_2 = \varepsilon_3 = 0$

This state is often assumed to occur when, for example, fluid is withdrawn from a reservoir, in which the vertical strain is contractile, whereas lateral strain is inhibited by the rock that is adjacent to the reservoir. The stresses that accompany uniaxial strain are

$$\sigma_1 = (\lambda + 2G)\varepsilon_1, \quad \sigma_2 = \sigma_3 = \lambda\varepsilon_1 = [\nu/(1 - \nu)]\sigma_1. \tag{5.35}$$

In order for the lateral strains to be zero, nonzero lateral stresses must exist. The assumption of uniaxial strain is often used as a simple model for calculating *in situ* stresses below the Earth's surface (see §13.2).

5.3.4 The case $\sigma_1 \neq 0, \varepsilon_2 = 0, \sigma_3 = 0$

This state corresponds to an applied stress in one direction, with zero stress and zero strain in two mutually orthogonal directions that are each perpendicular to the direction of the applied load. Equations (5.8)–(5.10) can be manipulated in this case to yield

$$\varepsilon_1 = (1 - \nu^2)\sigma_1/E, \quad \sigma_2 = \nu\sigma_1, \quad \varepsilon_3 = -[\nu/(1 - \nu)]\varepsilon_1. \tag{5.36}$$

5.3.5 Biaxial stress or plane stress, $\sigma_1 \neq 0, \sigma_2 \neq 0, \sigma_3 = 0$

The strains in this case are found from (5.8)–(5.10) to be

$$\varepsilon_1 = \frac{1}{E}(\sigma_1 - \nu\sigma_2), \quad \varepsilon_2 = \frac{1}{E}(\sigma_2 - \nu\sigma_1), \quad \varepsilon_3 = \frac{-\nu}{E}(\sigma_1 + \sigma_2). \tag{5.37}$$

Plane stress occurs when a thin plate is loaded by forces acting in its own plane. It also occurs locally at any free surface, because the normal *and* shear stresses vanish on a free surface, and so the outward normal to a free surface is necessarily a direction of principal stress corresponding to $\sigma_3 = 0$. In plane stress, there is an expansion in the out-of-plane direction if $\sigma_1 + \sigma_2 > 0$ and a contraction if $\sigma_1 + \sigma_2 < 0$. When the biaxial stress state is one of pure shear, then $\sigma_1 + \sigma_2 = 0$ and the lateral strain is zero. The fractional volume change in plane stress is

$$\varepsilon_v = (1 - 2\nu)(\sigma_1 + \sigma_2)/E. \tag{5.38}$$

It follows from (5.1)–(5.3) that the stress–strain relations for plane stress conditions can be written as

$$\sigma_1 = \frac{4G(\lambda + G)}{(\lambda + 2G)}\varepsilon_1 + \frac{2G\lambda}{(\lambda + 2G)}\varepsilon_2, \tag{5.39}$$

$$\sigma_2 = \frac{2G\lambda}{(\lambda + 2G)}\varepsilon_1 + \frac{4G(\lambda + G)}{(\lambda + 2G)}\varepsilon_2. \tag{5.40}$$

It is important to note that this "two-dimensional" version of Hooke's law *cannot* be obtained from the three-dimensional form, (5.1)–(5.3), by simply ignoring the terms that contain the subscript "3."

5.3.6 Biaxial strain or plane strain, $\varepsilon_1 \neq 0$, $\varepsilon_2 \neq 0$, $\varepsilon_3 = 0$

The stresses in this case are found from (5.1)–(5.3) to be

$$\sigma_1 = (\lambda + 2G)\varepsilon_1 + \lambda\varepsilon_2 \quad \sigma_2 = (\lambda + 2G)\varepsilon_2 + \lambda\varepsilon_1, \tag{5.41}$$

$$\sigma_3 = \lambda(\varepsilon_1 + \varepsilon_2) = \frac{\lambda}{2(\lambda + G)}(\sigma_1 + \sigma_2) = \nu(\sigma_1 + \sigma_2). \tag{5.42}$$

In order for the out-of-plane strain to be zero, a nonzero out-of-plane stress whose magnitude is given by (5.42) is needed in order to counteract the Poisson effect due to the two in-plane stresses. The inverse form of Hooke's law for plane strain is

$$\varepsilon_1 = \frac{(1 - \nu^2)}{E}\sigma_1 - \frac{\nu(1 + \nu)}{E}\sigma_2, \tag{5.43}$$

$$\varepsilon_2 = \frac{(1 - \nu^2)}{E}\sigma_2 - \frac{\nu(1 + \nu)}{E}\sigma_1. \tag{5.44}$$

If the x- and y-axes are *not* principal axes, then (5.24)–(5.27) give

$$\varepsilon_{xx} = \frac{(1 - \nu^2)}{E}\tau_{xx} - \frac{\nu(1 + \nu)}{E}\tau_{yy}, \tag{5.45}$$

$$\varepsilon_{yy} = \frac{(1 - \nu^2)}{E}\tau_{yy} - \frac{\nu(1 + \nu)}{E}\tau_{xx}, \tag{5.46}$$

$$\varepsilon_{xy} = \frac{(1 + \nu)}{E}\tau_{xy} = \frac{1}{2G}\tau_{xy}. \tag{5.47}$$

The assumption of plane strain is very often invoked when analyzing the stresses around boreholes or elongated underground openings.

5.3.7 Combined formulae for plane stress and plane strain

If λ is replaced by $2G\lambda/(\lambda + 2G)$ in (5.41), then the resulting equations will be identical to (5.39) and (5.40). This suggests the possibility that the stress–strain relations for plane stress and plane strain could be written in a form that is applicable to both situations. Indeed, both sets of stress–strain relations can be written as

$$\varepsilon_1 = \frac{(\kappa + 1)}{8G}\sigma_1 + \frac{(\kappa - 3)}{8G}\sigma_2, \tag{5.48}$$

$$\varepsilon_2 = \frac{(\kappa - 3)}{8G}\sigma_1 + \frac{(\kappa + 1)}{8G}\sigma_2, \tag{5.49}$$

where "Muskhelishvili's coefficient," κ, is defined as

$$\kappa = 3 - 4\nu \quad \text{for plane strain,} \tag{5.50}$$

$$\kappa = \frac{3 - \nu}{1 + \nu} \quad \text{for plane stress.} \tag{5.51}$$

The correctness of (5.48) and (5.49) can be verified by substituting the appropriate value of κ and recalling that $E = 2G(1 + \nu)$. The general forms of (5.48) and (5.49) that are applicable in nonprincipal coordinate systems are

$$\varepsilon_{xx} = \frac{(\kappa + 1)}{8G}\tau_{xx} + \frac{(\kappa - 3)}{8G}\tau_{yy}, \tag{5.52}$$

$$\varepsilon_{yy} = \frac{(\kappa - 3)}{8G}\tau_{xx} + \frac{(\kappa + 1)}{8G}\tau_{yy}, \tag{5.53}$$

$$\varepsilon_{xy} = \frac{1 + \nu}{E}\tau_{xy} = \frac{1}{2G}\tau_{xy}. \tag{5.54}$$

It is useful to be able to convert solutions for plane stress into the corresponding solutions for plane strain, as many solutions in the literature are written out explicitly for one case or the other, but usually not for both cases. This is done most readily if the solutions are written in terms of G and ν, in which case a solution for plane strain maybe converted to the case of plane stress by replacing $3 - 4\nu$ with $(3 - \nu)/(1 + \nu)$,which is to say, by replacing ν with $\nu/(1 + \nu)$. Similarly, plane stress solutions may be converted to plane strain solutions by making the inverse substitution, which is to say, replacing ν with $\nu/(1 - \nu)$.

5.3.8 Constant strain along the z-axis

It is assumed here that w is independent of x and y, that both u and v are independent of z, and that

$$\varepsilon_{zz} = \frac{\partial w}{\partial z} = \varepsilon (= \text{constant}). \tag{5.55}$$

Under these circumstances, $\varepsilon_{xz} = \varepsilon_{yz} = 0$, and (5.55) and (5.24)–(5.25) give

$$E\varepsilon_{xx} = (1 - \nu^2)\tau_{xx} - \nu(1 + \nu)\tau_{yy} - E\nu\varepsilon, \tag{5.56}$$

$$E\varepsilon_{yy} = (1 - \nu^2)\tau_{yy} - \nu(1 + \nu)\tau_{xx} - E\nu\varepsilon, \tag{5.57}$$

$$2G\varepsilon_{xy} = \tau_{xy}. \tag{5.58}$$

These equations provide a simple generalization of the plain strain equations, (5.45)–(5.47) and are often used when considering the stresses around underground tunnels.

5.4 Hooke's law in terms of deviatoric stresses and strains

The stress–strain law for an isotropic, linear elastic solid takes on a particularly simple mathematical form when expressed in terms of the deviatoric stress and deviatoric strain. From (2.150) and (5.16), we have

$$\begin{aligned}\boldsymbol{\varepsilon}^{\text{iso}} &= \frac{1}{3}\text{trace}(\boldsymbol{\varepsilon})\mathbf{I} = \frac{1}{3}\text{trace}\left[\lambda\text{trace}(\boldsymbol{\varepsilon})\mathbf{I} + 2G\boldsymbol{\varepsilon}\right]\mathbf{I} \\ &= \frac{1}{3}\left[\lambda\,\text{trace}(\boldsymbol{\varepsilon})\text{trace}(\mathbf{I}) + 2G\,\text{trace}(\boldsymbol{\varepsilon})\right]\mathbf{I} \\ &= \frac{1}{3}\left[3\lambda\,\text{trace}(\boldsymbol{\varepsilon}) + 2G\,\text{trace}(\boldsymbol{\varepsilon})\right]\mathbf{I} = 3K\left[(1/3)\text{trace}(\boldsymbol{\varepsilon})\right]\mathbf{I} = 3K\boldsymbol{\varepsilon}^{\text{iso}},\end{aligned} \tag{5.59}$$

where we have made use of the fact that trace(**I**) $= 3$. Since $\boldsymbol{\tau}^{\text{iso}} = \tau_{\text{m}}\mathbf{I}$ and $\boldsymbol{\varepsilon}^{\text{iso}} = \varepsilon_{\text{m}}\mathbf{I}$, we also have

$$\tau_{\text{m}} = 3K\varepsilon_{\text{m}} = K\varepsilon_{\text{v}}. \tag{5.60}$$

Similarly, (2.151) and (5.16) give

$$\begin{aligned}\boldsymbol{\tau}^{\text{dev}} &= \boldsymbol{\tau} - \boldsymbol{\tau}^{\text{iso}} = \lambda\,\text{trace}(\boldsymbol{\varepsilon})\mathbf{I} + 2G\boldsymbol{\varepsilon} - K\,\text{trace}(\boldsymbol{\varepsilon})\mathbf{I} \\ &= \lambda\,\text{trace}(\boldsymbol{\varepsilon})\mathbf{I} + 2G\boldsymbol{\varepsilon} - \left(\lambda + \frac{2}{3}G\right)\text{trace}(\boldsymbol{\varepsilon})\mathbf{I} \\ &= 2G\boldsymbol{\varepsilon} - (2G/3)\text{trace}(\boldsymbol{\varepsilon})\mathbf{I} = 2G[\boldsymbol{\varepsilon} - (1/3)\text{trace}(\boldsymbol{\varepsilon})\mathbf{I}] \\ &= 2G[\boldsymbol{\varepsilon} - \boldsymbol{\varepsilon}^{\text{iso}}] = 2G\boldsymbol{\varepsilon}^{\text{dev}}.\end{aligned} \tag{5.61}$$

Hence, in terms of the deviatoric/isotropic decomposition, Hooke's law for an isotropic material can be written as

$$\boldsymbol{\tau}^{\text{iso}} = 3K\boldsymbol{\varepsilon}^{\text{iso}}, \quad \boldsymbol{\tau}^{\text{dev}} = 2G\boldsymbol{\varepsilon}^{\text{dev}}. \tag{5.62}$$

It follows from (5.62) that if the strain is either purely deviatoric or purely isotropic, then the stress tensor is essentially equal to the strain tensor, aside from a multiplicative scalar constant. In other words, deviatoric and isotropic strains are eigenvectors (in a nine-dimensional space) of the Hooke's law operator, with corresponding eigenvalues $2G$ and $3K$ (Gurtin, 1972). This seemingly abstract mathematical fact is actually crucial to many practically important calculations, such as the establishment of upper and lower bounds on the effective elastic moduli of heterogeneous or porous materials (Nemat-Nasser and Hori, 1993).

5.5 Equations of stress equilibrium

The basic problem of elasticity can be described as that of determining the stresses and displacements of a body of known shape that is subjected to prescribed tractions or displacements along its outer boundary and prescribed forces at its interior points. The forces that act at interior points may be localized point forces or forces that are distributed over the body, in which case they are called *body forces*. The most common body force is that due to gravity, although temperature and pore pressure gradients have the same effect as distributed body forces, as will be seen in Chapter 7. In order to find the state of stress and displacement that results from the application of certain loads or boundary displacements, it is necessary to solve a set of three coupled partial differential equations known as the *equations of stress equilibrium*. In more general situations for which the rock is not in static equilibrium, such as during seismic wave propagation, the governing equations are the *equation of motion*. These equations are found by applying the law of conservation of linear momentum, that is, Newton's second law, to the rock.

Newton's second law states that the total force applied to a body in a given direction is equal to the mass of the body multiplied by its acceleration in that direction. As there are three mutually orthogonal directions at any point in a rock mass, there will be three independent equations of motion. To derive the mathematical form of the laws of motion/equilibrium, consider an arbitrarily shaped body of finite size. We denote the region of three-dimensional space occupied by the body by B and the outer boundary of the body by ∂B. The total force acting on this body in, say, the x direction, consists of the sum of all the body forces that act over the internal portions of the body, plus the resultant force due to all of the surface tractions that act over the outer boundary of the body. If we let F_x be the body force, per unit mass, which acts at each element of the rock, then ρF_x is the body force per unit volume, and so the total body force is found by integrating the local body force over the entire body:

$$\text{total body-force component in } x \text{ direction} = \iiint_B \rho F_x dV. \tag{5.63}$$

To be consistent with the traditional rock mechanics sign conventions, the components of $\mathbf{F}$, namely (F_x, F_y, F_z), must be considered to be positive numbers if they act in the *negative* coordinate directions. Another common notation for the components of the body force vector is (X, Y, Z).

The total resultant of all of the surface tractions that are applied over the outer boundary of the body is found by integrating the surface tractions over the outer boundary:

$$\text{total } x\text{-component of force due to surface tractions} = \iint_{\partial B} p_x dA. \tag{5.64}$$

The x-component of the acceleration of each small element of rock is given by the second derivative with respect to time of u, the x-component of the displacement. The mass of each small element is given by ρdV, where dV is the incremental volume. The total x-component of the inertia term is therefore found by integrating the product of mass and acceleration over the entire body:

$$\text{total inertia component in } x \text{ direction} = \iiint_B \rho \frac{\partial^2 u}{\partial t^2} dV. \tag{5.65}$$

Equating the total force to the total inertia term yields

$$\iiint_B \rho F_x dV + \iint_{\partial B} p_x dA = \iiint_B \rho \frac{\partial^2 u}{\partial t^2} dV. \tag{5.66}$$

Equation (5.66) expresses Newton's law of motion in the x direction, in an integral form. To derive the more useful differential form of this equation, we first convert the surface integral into a volume integral over the entire body. To do this, we invoke the divergence theorem, also known as Green's theorem, which states that (Kellogg, 1970, pp. 37–9; Lang, 1973, pp. 327–33) for any vector **f**, with components (f_x, f_y, f_z),

$$\iint_{\partial B} (f_x n_x + f_y n_y + f_z n_z) dA = \iiint_B \left(\frac{\partial f_x}{\partial x} + \frac{\partial f_y}{\partial y} + \frac{\partial f_z}{\partial z} \right) dV, \tag{5.67}$$

where (n_x, n_y, n_z) are the components of the outward unit normal vector to the surface. The applicability of the divergence theorem requires certain assumptions about the differentiability of the functions in the integrand and the smoothness of the outer boundary ∂B; we will always assume that these conditions hold. Although the three functions (f_x, f_y, f_z) can be thought of as components of a vector, this is not necessary; they can in fact be any three differentiable functions.

To apply the divergence theorem to the surface integral in (5.66), we first use (2.63) to express the traction in terms of the stress components:

$$\iint_{\partial B} p_x dA = \iint_{\partial B} (\tau_{xx} n_x + \tau_{yx} n_y + \tau_{zx} n_z) dA. \tag{5.68}$$

Application of the divergence theorem to (5.68) now yields

$$\iint_{\partial B} p_x dA = \iint_{\partial B} (\tau_{xx} n_x + \tau_{yx} n_y + \tau_{zx} n_z) dA = \iiint_B \left(\frac{\partial \tau_{xx}}{\partial x} + \frac{\partial \tau_{yx}}{\partial y} + \frac{\partial \tau_{zx}}{\partial z} \right) dV. \tag{5.69}$$

Substitution of (5.69) into (5.66) gives

$$\iiint_B \rho F_x dV + \iiint_B \left(\frac{\partial \tau_{xx}}{\partial x} + \frac{\partial \tau_{yx}}{\partial y} + \frac{\partial \tau_{zx}}{\partial z} \right) dV = \iiint_B \rho \frac{\partial^2 u}{\partial t^2} dV, \tag{5.70}$$

which can be written as

$$\iiint_B \left[\frac{\partial \tau_{xx}}{\partial x} + \frac{\partial \tau_{yx}}{\partial y} + \frac{\partial \tau_{zx}}{\partial z} + \rho F_x - \rho \frac{\partial^2 u}{\partial t^2} \right] dV = 0. \tag{5.71}$$

Equation (5.71) must hold for any arbitrary subregion of the rock mass, because Newton's law of motion applies to any such subregion. In order for the integral of the bracketed term in (5.71) to vanish over any arbitrary region B, the integrand must be identically zero at all points of the rock mass. To prove this, assume that there is some point $\mathbf{x}$ at which the integrand is, say, positive rather than zero. By continuity, it would then also be positive in some small neighborhood of $\mathbf{x}$. We could then chose this small neighborhood as our region of integration, in which case the integral will be positive – in contradiction to (5.71). Hence, our assumption that the integrand is positive at point $\mathbf{x}$ must have been incorrect. We conclude that, at all points of the rock mass,

$$\frac{\partial \tau_{xx}}{\partial x} + \frac{\partial \tau_{yx}}{\partial y} + \frac{\partial \tau_{zx}}{\partial z} + \rho F_x = \rho \frac{\partial^2 u}{\partial t^2}. \tag{5.72}$$

Similar applications of Newton's second law in the y and z directions yield

$$\frac{\partial \tau_{xy}}{\partial x} + \frac{\partial \tau_{yy}}{\partial y} + \frac{\partial \tau_{zy}}{\partial z} + \rho F_y = \rho \frac{\partial^2 v}{\partial t^2}, \tag{5.73}$$

$$\frac{\partial \tau_{xz}}{\partial x} + \frac{\partial \tau_{yz}}{\partial y} + \frac{\partial \tau_{zz}}{\partial z} + \rho F_z = \rho \frac{\partial^2 w}{\partial t^2}. \tag{5.74}$$

These three equations of motion are coupled together, since each of the shear stresses appears in two of the equations. For example, $\tau_{xy} = \tau_{yx}$, so this stress component appears in both (5.72) and (5.73).

The equations of motion, (5.72)–(5.74), can be expressed in vector/matrix form by first recognizing that the gradient operator can be thought of as a 3×1 column vector, that is,

$$\text{gradient} = \nabla = \begin{bmatrix} \dfrac{\partial}{\partial x} \\ \dfrac{\partial}{\partial y} \\ \dfrac{\partial}{\partial z} \end{bmatrix}. \tag{5.75}$$

If we premultiply the stress matrix by the *transpose* of this gradient vector, the result is a 1×3 vector whose three components are given by the derivative terms in (5.72)–(5.74), as can easily be verified. Hence, (5.72)–(5.74) can also be written as

$$\nabla^{\mathrm{T}} \boldsymbol{\tau} + \rho \mathbf{F} = \rho \frac{\partial^2 \mathbf{u}}{\partial t^2}, \tag{5.76}$$

where we take the mathematical liberty of identifying $\mathbf{F}$ and $\mathbf{u}$ as 1×3 row vectors, rather than 3×1 column vectors. The term $\nabla^{\mathrm{T}} \boldsymbol{\tau}$ is often referred to as "div $\boldsymbol{\tau}$," although this terminology is only clear if we recall that div $\boldsymbol{\tau}$ must be calculated as $\nabla^{\mathrm{T}} \boldsymbol{\tau}$. If we let a superposed dot denote differentiation with respect to time, the equations of motion can be written succinctly as

$$\text{div}\, \boldsymbol{\tau} + \rho \mathbf{F} = \rho \ddot{\mathbf{u}}. \tag{5.77}$$

The set of three equations expressed by (5.72)–(5.74), (5.76), or (5.77), must be satisfied at all times, at all points in the rock mass, and are therefore the governing equations for rock deformation. However, six unknown stress components and three unknown displacements appear in these equations, so these three equations are in themselves not sufficient to enable the stresses and displacements to be found. In order to have a mathematically well-posed problem, there must be an equal number of equations and unknowns. The strain–displacement relations provide six additional equations, but also introduce six additional "unknowns," the six independent components of the strain tensor. An additional six equations are then supplied by the stress–strain law, which may be of any form: linearly elastic, nonlinearly elastic, plastic, viscoelastic, etc. The equations of motion, along with the strain–displacement relations and the stress–strain relations, constitute a set of equations in which the number of unknowns is equal to the number of equations. This is a necessary, but not sufficient, condition for a problem in rock mechanics to be mathematically well posed, in the sense of having a unique solution. Further conditions, involving the components of the elastic moduli and the boundary conditions, are discussed in §5.9.

In the frequently occurring case in which the rock is in static equilibrium, or in which the displacements are occurring very slowly, the right-hand sides of (5.72)–(5.74) can be neglected. In these cases, the equations of motion reduce to the *equations of equilibrium*:

$$\frac{\partial \tau_{xx}}{\partial x} + \frac{\partial \tau_{yx}}{\partial y} + \frac{\partial \tau_{zx}}{\partial z} + \rho F_x = 0, \tag{5.78}$$

$$\frac{\partial \tau_{xy}}{\partial x} + \frac{\partial \tau_{yy}}{\partial y} + \frac{\partial \tau_{zy}}{\partial z} + \rho F_y = 0, \tag{5.79}$$

$$\frac{\partial \tau_{xz}}{\partial x} + \frac{\partial \tau_{yz}}{\partial y} + \frac{\partial \tau_{zz}}{\partial z} + \rho F_z = 0. \tag{5.80}$$

The two sets of equations, (5.72)–(5.74) and (5.78)–(5.80), are completely general, in the sense that they contain no assumptions concerning the stress–strain behavior of the rock. If we assume that the rock is a linear elastic material, however, these equations can be expressed in terms of the displacements, as follows. For the remainder of this chapter, we consider only the equations of stress equilibrium; the full equations of motion will be reprised in Chapters 7 and 11. We first combine the strain–displacement relations, (2.222), and the stress–strain relations, (5.18)–(5.21), and then substitute the result into the equations of motion, (5.72)–(5.74), to find

$$\lambda\left(\frac{\partial^2 u}{\partial x^2} + \frac{\partial^2 v}{\partial x \partial y} + \frac{\partial^2 w}{\partial x \partial z}\right) + G\left(\frac{\partial^2 u}{\partial x^2} + \frac{\partial^2 v}{\partial x \partial y} + \frac{\partial^2 w}{\partial x \partial z} \right.$$
$$\left. + \frac{\partial^2 u}{\partial x^2} + \frac{\partial^2 u}{\partial y^2} + \frac{\partial^2 u}{\partial z^2}\right) + \rho F_x = 0. \tag{5.81}$$

$$\lambda\left(\frac{\partial^2 u}{\partial y\partial x}+\frac{\partial^2 v}{\partial y^2}+\frac{\partial^2 w}{\partial y\partial z}\right)+G\left(\frac{\partial^2 u}{\partial y\partial x}+\frac{\partial^2 v}{\partial y^2}+\frac{\partial^2 w}{\partial y\partial z}+\frac{\partial^2 v}{\partial x^2}\right.$$
$$\left.+\frac{\partial^2 v}{\partial y^2}+\frac{\partial^2 v}{\partial z^2}\right)+\rho F_y=0, \tag{5.82}$$

$$\lambda\left(\frac{\partial^2 u}{\partial z\partial x}+\frac{\partial^2 v}{\partial z\partial y}+\frac{\partial^2 w}{\partial z^2}\right)+G\left(\frac{\partial^2 u}{\partial z\partial x}+\frac{\partial^2 v}{\partial z\partial y}+\frac{\partial^2 w}{\partial z^2}+\frac{\partial^2 w}{\partial x^2}\right.$$
$$\left.+\frac{\partial^2 w}{\partial y^2}+\frac{\partial^2 w}{\partial z^2}\right)+\rho F_z=0. \tag{5.83}$$

Equations (5.81)–(5.83), known as the Navier equations, embody the equations of stress equilibrium, the stress–strain equations, and the strain–displacement identities. Assuming that the body forces are known, which is usually the case, these are equations for the three unknown displacements. As such, they form a set of differential equations that can be used to solve elasticity problems, as an alternative to the previously discussed set of stress-based equations. The relative advantages of the stress-based and displacement-based equations are discussed in §5.7.

The displacement form of the equilibrium equations can also be expressed in a more compact vector-matrix notation. First, note the following two identities:

$$\nabla^{\mathrm{T}}\mathbf{u}\equiv\nabla\cdot\mathbf{u}\equiv\operatorname{div}\mathbf{u}=\begin{bmatrix}\frac{\partial}{\partial x} & \frac{\partial}{\partial y} & \frac{\partial}{\partial z}\end{bmatrix}\begin{bmatrix}u\\ v\\ w\end{bmatrix}=\frac{\partial u}{\partial x}+\frac{\partial v}{\partial y}+\frac{\partial w}{\partial z}, \tag{5.84}$$

$$\nabla^{\mathrm{T}}\nabla\equiv\nabla^2=\begin{bmatrix}\frac{\partial}{\partial x} & \frac{\partial}{\partial y} & \frac{\partial}{\partial z}\end{bmatrix}\begin{bmatrix}\frac{\partial}{\partial x}\\ \frac{\partial}{\partial y}\\ \frac{\partial}{\partial z}\end{bmatrix}=\frac{\partial^2}{\partial x^2}+\frac{\partial^2}{\partial y^2}+\frac{\partial^2}{\partial z^2}. \tag{5.85}$$

The Laplacian operator ∇^2 is a scalar operator that can operate on, by pre-multiplication, a scalar or a vector. Using these identities, (5.81)–(5.83) can be written in the following hybrid form involving both scalar and vector notation:

$$(\lambda+G)\frac{\partial}{\partial x}(\nabla\cdot\mathbf{u})+G\nabla^2 u+\rho F_x=0, \tag{5.86}$$

$$(\lambda+G)\frac{\partial}{\partial y}(\nabla\cdot\mathbf{u})+G\nabla^2 v+\rho F_y=0, \tag{5.87}$$

$$(\lambda+G)\frac{\partial}{\partial z}(\nabla\cdot\mathbf{u})+G\nabla^2 w+\rho F_z=0. \tag{5.88}$$

From (5.75), we see that the three partial derivative operators appearing in these three equations form the components of the gradient row vector. Furthermore,

$$(\nabla^2 u,\nabla^2 v,\nabla^2 w)=\nabla^2(u,v,w)=\nabla^2\mathbf{u}. \tag{5.89}$$

Hence, (5.86)–(5.88) can be written as

$$(\lambda + G)\nabla(\nabla \cdot \mathbf{u}) + G\nabla^2\mathbf{u} + \rho\mathbf{F} = \mathbf{0}. \tag{5.90}$$

This form of the Navier equations not only has the advantage of compactness, but can also be generalized to any non-Cartesian coordinate system, such as cylindrical coordinates, in which the explicit representations of the gradient, divergence, and Laplacian operators are known (see Chou and Pagano, 1992).

Various important and useful general results can be found from the Navier equations. In situations in which there are no body forces, taking the partial derivative of (5.86) with respect to x, the partial derivative of (5.87) with respect to y, and the partial derivative of (5.88) with respect to z, adding the results, and interchanging the order of partial differentiation when appropriate, leads to

$$(\lambda + 2G)\nabla^2(\nabla \cdot \mathbf{u}) = 0. \tag{5.91}$$

But (2.231) shows that $\nabla \cdot \mathbf{u} = \varepsilon_{\mathrm{v}}$, the volumetric strain. Hence, in the absence of body forces, the volumetric strain satisfies Laplace's equation, that is,

$$\nabla^2\varepsilon_{\mathrm{v}} = 0. \tag{5.92}$$

Equations $\tau_{\mathrm{m}} = K\varepsilon_{\mathrm{v}}$ by (5.60), in which case (5.92) implies that the mean normal stress also satisfies Laplace's equation:

$$\nabla^2\tau_{\mathrm{m}} = 0. \tag{5.93}$$

Differentiation of (5.86) with respect to x, again with the body force taken to be zero, gives, after interchanging the orders of some of the derivatives,

$$(\lambda + G)\frac{\partial^2\varepsilon_{\mathrm{v}}}{\partial x^2} + G\nabla^2\varepsilon_{xx} = 0. \tag{5.94}$$

Invoking (5.92), (5.93), and (5.18) leads to

$$\nabla^2\tau_{xx} = -\frac{6(\lambda + G)}{(3\lambda + 2G)}\frac{\partial^2\tau_{\mathrm{m}}}{\partial x^2} = \frac{-3}{(1 + \nu)}\frac{\partial^2\tau_{\mathrm{m}}}{\partial x^2}, \tag{5.95}$$

with similar equations holding for the other two normal stresses. Likewise, it can be shown that

$$\nabla^2\tau_{xy} + \frac{3}{(1 + \nu)}\frac{\partial^2\tau_{\mathrm{m}}}{\partial x\partial y} = 0, \tag{5.96}$$

as well as two similar equations for the other two shear stresses. These six equations, known as the Beltrami–Michell equations, are equivalent to the strain-compatibility equations of §2.13, although they are expressed in terms of the stresses. If a purely stress-based formulation is desired for the elasticity equations, which is useful in cases where the boundary conditions are known in terms of stresses rather than displacements, the Beltrami–Michell equations give a complete set of six equations for the six stresses. If nonconstant body forces act on the rock, additional terms (see Chou and Pagano, 1992, p. 78) involving the derivatives of the body-force vector **F** appear on the right-hand sides of (5.95) and (5.96).

In situations in which the displacement components u and v depend only on x and y, and the strain in the z direction is constant with z, the stress equilibrium equations (5.78)–(5.80) reduce to following pair of equations:

$$\frac{\partial \tau_{xx}}{\partial x} + \frac{\partial \tau_{yx}}{\partial y} + \rho F_x = 0, \tag{5.97}$$

$$\frac{\partial \tau_{xy}}{\partial x} + \frac{\partial \tau_{yy}}{\partial y} + \rho F_y = 0. \tag{5.98}$$

The two-dimensional form of the displacement-based Navier equations can be found from (5.81)–(5.83) by ignoring all terms that involve either w or partial differentiation with respect to z.

5.6 Equations of stress equilibrium in cylindrical and spherical coordinates

In rock mechanics problems involving cylindrical boundaries, such as tunnels or boreholes, it is convenient to use cylindrical coordinates rather than Cartesian coordinates. In this coordinate system, the governing equations of elasticity take on a very different form than that given in §5.5. The equilibrium equations could be transformed into cylindrical coordinates by starting with a vector formulation such as (5.77) or (5.90), and using the cylindrical coordinate version of the divergence, gradient, and Laplacian operators (Chou and Pagano, 1992). This procedure is complicated by the need to account for the fact that the directions of the unit vectors $\{\mathbf{e_r}, \mathbf{e_\theta}, \mathbf{e_z}\}$ vary with position. The equations can also be derived in the following more physically intuitive manner by starting with a differential form of Newton's second law, rather than an integral form.

Consider (Fig. 5.1a) a small element of rock that is bounded by three sets of surfaces, on each of which one of the three cylindrical coordinates is constant. The front and rear surfaces are the cylindrical surfaces corresponding to the radii r and $r + \delta r$; the two lateral surfaces are planes defined by two values of the angular variable θ and $\theta + \delta\theta$; and the upper and lower surfaces are the planes that correspond to z and $z + \delta z$. The cylindrical coordinates of point A are $\{r, \theta, z\}$. Now consider the sum of, say, all of the forces that act in the z direction. As each of the increments $\{\delta r, \delta\theta, \delta z\}$ is small, the mean traction acting over each face can be approximated by the traction that acts at the centroid of that face. The component in the z direction of the traction that acts over the outer cylindrical surface $A'B'C'D'$ is therefore

$$-\tau_{rz}\left(r + \delta r, \theta + \frac{1}{2}\delta\theta, z + \frac{1}{2}\delta z\right), \tag{5.99}$$

and the area of the surface over which this traction acts is $(r + \delta r)\delta\theta\delta z$. Likewise, the z-component of the traction acting on the inner cylindrical surface $ABCD$ is

$$\tau_{rz}\left(r, \theta + \frac{1}{2}\delta\theta, z + \frac{1}{2}\delta z\right), \tag{5.100}$$

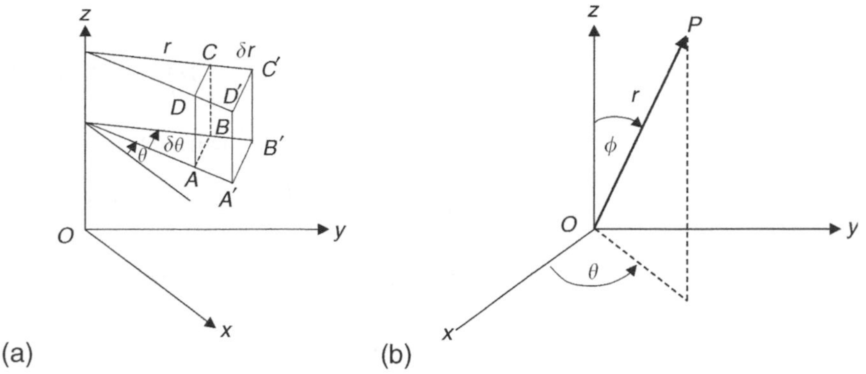

Fig. 5.1 (a) Elementary volume of rock in a cylindrical coordinate system. (b) Spherical coordinate system.

and the area of the surface over which this traction acts is $r\delta r\delta\theta\delta z$. The traction component in the z direction acting on the top surface, $CC'D'D$, is

$$-\tau_{zz}\left(r+\frac{1}{2}\delta r,\theta+\frac{1}{2}\delta\theta,z+\delta z\right),\tag{5.101}$$

and the area of this surface is $r\delta r\delta\theta$. The traction component in the z direction acting on the bottom surface, $AA'B'B$, is

$$\tau_{zz}\left(r+\frac{1}{2}\delta r,\theta+\frac{1}{2}\delta\theta,z\right),\tag{5.102}$$

and the area of this surface is also $r\delta r\delta\theta$. The traction component in the z direction acting on the rectangular surface $CC'B'B$ is

$$-\tau_{\theta z}\left(r+\frac{1}{2}\delta r,\theta+\delta\theta,z+\delta z\right),\tag{5.103}$$

and the area of this surface is $\delta r\delta z$. Lastly, the traction component in the z direction acting on the surface $AA'D'D$ is

$$\tau_{\theta z}\left(r+\frac{1}{2}\delta r,\theta,z+\frac{1}{2}\delta z\right),\tag{5.104}$$

and the area of this surface is also $\delta r\delta z$.

The body force in the z direction, per unit mass, is denoted by $-F_z$ (recalling the sign convention in which the force component is positive if it points in the *negative* coordinate direction), so the body force component per unit volume is $-\rho F_z$. The volume of the element is $r\delta r\delta\theta\delta z$, so the total body force acting on this element in the z direction is $-\rho rF_z\delta r\delta\theta\delta z$. For this element to be in static equilibrium, the sum of all of the forces acting on it in the z direction must vanish, that is,

$$\begin{aligned}
&-\tau_{rz}\left(r+\delta r,\theta+\frac{1}{2}\delta\theta,z+\frac{1}{2}\delta z\right)[(r+\delta r)\delta\theta\delta z]\\
&+\tau_{rz}\left(r,\theta+\frac{1}{2}\delta\theta,z+\frac{1}{2}\delta z\right)[r\delta\theta\delta z]\\
&-\tau_{zz}\left(r+\frac{1}{2}\delta r,\theta+\frac{1}{2}\delta\theta,z+\delta z\right)[r\delta r\delta\theta]\\
&+\tau_{zz}\left(r+\frac{1}{2}\delta r,\theta+\frac{1}{2}\delta\theta,z\right)[r\delta r\delta\theta]\\
&-\tau_{\theta z}\left(r+\frac{1}{2}\delta r,\theta+\delta\theta,z+\frac{1}{2}\delta z\right)[\delta r\delta z]\\
&+\tau_{\theta z}\left(r+\frac{1}{2}\delta r,\theta,z+\frac{1}{2}\delta z\right)[\delta r\delta z]-\rho F_z[r\delta r\delta\theta\delta z]=0.
\end{aligned}\tag{5.105}$$

Dividing the entire equation by $r\delta r\delta\theta\delta z$ yields

$$\begin{aligned}
&\frac{\tau_{rz}(r+\delta r,\theta+(1/2)\delta\theta,z+(1/2)\delta z)-\tau_{rz}(r,\theta+(1/2)\delta\theta,z+(1/2)\delta z)}{\delta r}\\
&+\frac{\tau_{rz}(r+\delta r,\theta+(1/2)\delta\theta,z+(1/2)\delta z)}{r}\\
&+\frac{\tau_{zz}(r+(1/2)\delta r,\theta+(1/2)\delta\theta,z+\delta z)-\tau_{zz}(r+(1/2)\delta r,\theta+(1/2)\delta\theta,z)}{\delta z}\\
&+\frac{\tau_{\theta z}(r+(1/2)\delta r,\theta+\delta\theta,z+(1/2)\delta z)-\tau_{\theta z}(r+(1/2)\delta r,\theta,z+(1/2)\delta z)}{r\delta\theta}\\
&-\rho F_z=0.
\end{aligned}\tag{5.106}$$

Taking the limit as all three of the coordinate increments become infinitesimally small, we arrive at the partial differential equation that embodies Newton's second law for the z direction:

$$\frac{\partial\tau_{rz}}{\partial r}+\frac{1}{r}\frac{\partial\tau_{\theta z}}{\partial\theta}+\frac{\partial\tau_{zz}}{\partial z}+\frac{\tau_{rz}}{r}+\rho F_z=0.\tag{5.107}$$

Similar applications of Newton's second law in the r and θ directions yield

$$\frac{\partial\tau_{rr}}{\partial r}+\frac{1}{r}\frac{\partial\tau_{\theta r}}{\partial\theta}+\frac{\partial\tau_{zr}}{\partial z}+\frac{\tau_{rr}-\tau_{\theta\theta}}{r}+\rho F_r=0,\tag{5.108}$$

$$\frac{\partial\tau_{r\theta}}{\partial r}+\frac{1}{r}\frac{\partial\tau_{\theta\theta}}{\partial\theta}+\frac{\partial\tau_{z\theta}}{\partial z}+\frac{2\tau_{r\theta}}{r}+\rho F_\theta=0.\tag{5.109}$$

In two dimensions, with all stress components containing a subscript "z" equal to zero, the equations of stress equilibrium reduce to

$$\frac{\partial\tau_{rr}}{\partial r}+\frac{1}{r}\frac{\partial\tau_{\theta r}}{\partial\theta}+\frac{\tau_{rr}-\tau_{\theta\theta}}{r}+\rho F_r=0,\tag{5.110}$$

$$\frac{\partial\tau_{r\theta}}{\partial r}+\frac{1}{r}\frac{\partial\tau_{\theta\theta}}{\partial\theta}+\frac{2\tau_{r\theta}}{r}+\rho F_\theta=0.\tag{5.111}$$

The Navier equations can be expressed in cylindrical coordinates by combining (5.107)–(5.109) with the stress–strain relations, (5.22)–(5.23), and the strain–displacement relations, (2.275) and (2.280):

$$(\lambda+G)\frac{\partial \varepsilon_v}{\partial r}+G\left(\frac{\partial^2 u}{\partial r^2}+\frac{1}{r}\frac{\partial u}{\partial r}-\frac{u}{r^2}+\frac{1}{r^2}\frac{\partial^2 u}{\partial \theta^2}-\frac{2}{r^2}\frac{\partial v}{\partial \theta}+\frac{\partial^2 u}{\partial z^2}\right)+\rho F_r=0, \tag{5.112}$$

$$(\lambda+G)\frac{1}{r}\frac{\partial \varepsilon_v}{\partial \theta}+G\left(\frac{\partial^2 v}{\partial r^2}+\frac{1}{r}\frac{\partial v}{\partial r}-\frac{v}{r^2}+\frac{1}{r^2}\frac{\partial^2 v}{\partial \theta^2}-\frac{2}{r^2}\frac{\partial v}{\partial \theta}+\frac{\partial^2 v}{\partial z^2}\right)+\rho F_\theta=0, \tag{5.113}$$

$$(\lambda+G)\frac{\partial \varepsilon_v}{\partial z}+G\left(\frac{\partial^2 w}{\partial r^2}+\frac{1}{r}\frac{\partial w}{\partial r}+\frac{1}{r^2}\frac{\partial^2 w}{\partial \theta^2}+\frac{\partial^2 w}{\partial z^2}\right)+\rho F_z=0, \tag{5.114}$$

where (u, v, w) are the displacements in the $\{r, \theta, z\}$ directions, and the volumetric strain is given by

$$\varepsilon_v=\frac{\partial u}{\partial r}+\frac{u}{r}+\frac{1}{r}\frac{\partial v}{\partial \theta}+\frac{\partial w}{\partial z}. \tag{5.115}$$

In problems involving spherical cavities or inclusions, it is convenient to use a *spherical* coordinate system (Fig. 5.1b), defined by

$$x=r\sin\phi\cos\theta, \quad y=r\sin\phi\sin\theta, \quad z=r\cos\phi. \tag{5.116}$$

The general form of the elasticity equations in spherical coordinates can be found in Timoshenko and Goodier (1970) or Sokolnikoff (1956). The most common and important special case is that of radial symmetry, in which case the only nonzero displacement component will be the radial component, $u(r)$. The only nonzero strains will be the three normal strains, which are related to the radial displacement by

$$\varepsilon_{rr}=\frac{\partial u}{\partial r}, \quad \varepsilon_{\theta\theta}=\varepsilon_{\phi\phi}=\frac{u}{r}. \tag{5.117}$$

The stress–strain relations in this case are, from (5.16),

$$\tau_{rr}=(\lambda+2G)\varepsilon_{rr}+\lambda(\varepsilon_{\theta\theta}+\varepsilon_{\phi\phi}), \tag{5.118}$$

$$\tau_{\theta\theta}=(\lambda+2G)\varepsilon_{\theta\theta}+\lambda(\varepsilon_{rr}+\varepsilon_{\phi\phi}), \tag{5.119}$$

$$\tau_{\phi\phi}=(\lambda+2G)\varepsilon_{\phi\phi}+\lambda(\varepsilon_{rr}+\varepsilon_{\theta\theta}), \tag{5.120}$$

where $\tau_{\theta\theta}$ and $\tau_{\phi\phi}$ are the normal stresses in the two directions perpendicular to r. Combining (5.117) and (5.118)–(5.120) yields the stress–displacement equations for spherically symmetric deformations:

$$\tau_{rr}=(\lambda+2G)\frac{\partial u}{\partial r}+2\lambda\frac{u}{r}, \quad \tau_{\theta\theta}=\tau_{\phi\phi}=\lambda\frac{\partial u}{\partial r}+2(\lambda+G)\frac{u}{r}. \tag{5.121}$$

The only nontrivial equation of stress equilibrium is

$$\frac{\partial \tau_{rr}}{\partial r}+\frac{2(\tau_{rr}-\tau_{\theta\theta})}{r}=0. \tag{5.122}$$

Substituting (5.121) into (5.122) yields

$$\frac{\partial^2 u}{\partial r^2} + \frac{2}{r}\frac{\partial u}{\partial r} - \frac{2u}{r^2} = 0, \tag{5.123}$$

which is the Navier equation for spherically symmetric problems with no body forces.

5.7 Airy stress functions

Elasticity problems can be approached by working with the three Navier equations, (5.81)–(5.83), in which the dependent variables are the three displacements. Alternatively, the six Beltrami–Michell equations, (5.95)–(5.96), in which the stresses are the dependent variables, also form a complete set of partial differential equations for solving elasticity problems. The Navier equations would appear to be simpler to use, as they are three, rather than six, coupled equations. In situations in which the displacements are specified on the outer boundaries of the rock mass, the Navier equations are convenient to use. However, it is more commonly the case that tractions, rather than displacements, are the known boundary data. In this case, the Navier equations suffer from the disadvantage that traction boundary conditions will inevitably take on a complicated form when expressed in terms of the unknown displacements. For these problems, it is often convenient to use a stress-based formulation of the elasticity equations.

A stress-based method of solving the elasticity equations becomes particularly simple for two-dimensional problems involving plane stress or plane strain, in which case the governing equations reduce to a *single* partial differential equation. It is not easy to see that the six Beltrami–Michell equations reduce to a single equation in this case; rather, it is easier to start with the single strain-compatibility equation for two-dimensional problems, (2.253):

$$2\frac{\partial^2 \varepsilon_{xy}}{\partial x \partial y} = \frac{\partial^2 \varepsilon_{xx}}{\partial y^2} + \frac{\partial^2 \varepsilon_{yy}}{\partial x^2}. \tag{5.124}$$

Substitution of the two-dimensional stress–strain relations, (5.52)–(5.54), into (5.124) yields

$$(\chi + 1)\left[\frac{\partial^2 \tau_{xx}}{\partial y^2} + \frac{\partial^2 \tau_{yy}}{\partial x^2}\right] + (\chi - 3)\left[\frac{\partial^2 \tau_{xx}}{\partial x^2} + \frac{\partial^2 \tau_{yy}}{\partial y^2}\right] = 8\frac{\partial^2 \tau_{xy}}{\partial x \partial y}, \tag{5.125}$$

where $\chi = 3 - 4\nu$ for plane strain and $\chi = (3 - \nu)/(1 + \nu)$ for plane stress. Next, differentiation of the two equilibrium equations (5.97)–(5.98) yields

$$\frac{\partial^2 \tau_{xy}}{\partial x \partial y} = -\frac{\partial^2 \tau_{xx}}{\partial x^2} - \rho\frac{\partial F_x}{\partial x} = -\frac{\partial^2 \tau_{yy}}{\partial y^2} - \rho\frac{\partial F_y}{\partial y}, \tag{5.126}$$

which is to say

$$-8\frac{\partial^2 \tau_{xy}}{\partial x \partial y} = 4\left[\frac{\partial^2 \tau_{xx}}{\partial x^2} + \frac{\partial^2 \tau_{yy}}{\partial y^2}\right] + 4\rho\left[\frac{\partial F_x}{\partial x} + \frac{\partial F_y}{\partial y}\right]. \tag{5.127}$$

Substitution of (5.127) into (5.126) yields

$$\left(\frac{\partial^2}{\partial y^2}+\frac{\partial^2}{\partial x^2}\right)[\tau_{xx}+\tau_{yy}]=\frac{-4\rho}{(\chi+1)}\left(\frac{\partial F_x}{\partial x}+\frac{\partial F_y}{\partial y}\right), \tag{5.128}$$

which can also be written in vector form as

$$\nabla^2\tau_{\mathrm{m}}=\frac{-2\rho}{(\chi+1)}\nabla\cdot\mathbf{F}, \tag{5.129}$$

where τ_{m} refers here to the two-dimensional mean normal stress, $(\tau_{xx}+\tau_{yy})/2$.

It is often the case that the body force can be expressed as the gradient of a potential function, V, that satisfies Laplace's equation, $\nabla^2 V=\nabla\cdot(\nabla V)=0$; a common example is the body force due to gravity, for which $V=-gz$. In these situations, we have $\mathbf{F}=-\nabla V$, that is,

$$F_x=-\frac{\partial V}{\partial x},\quad F_y=-\frac{\partial V}{\partial y}. \tag{5.130}$$

But then the right-hand side of (5.129) vanishes, because $\nabla\cdot\mathbf{F}=-\nabla\cdot(\nabla V)=0$, that is,

$$\frac{\partial F_x}{\partial x}+\frac{\partial F_y}{\partial y}=-\frac{\partial}{\partial x}\left(\frac{\partial V}{\partial x}\right)-\frac{\partial}{\partial y}\left(\frac{\partial V}{\partial y}\right)=-\left(\frac{\partial^2 V}{\partial x^2}+\frac{\partial^2 V}{\partial y^2}\right)=0. \tag{5.131}$$

Equation (5.129) then reduces to the requirement that the two-dimensional mean normal stress must satisfy Laplace's equation,

$$\nabla^2\tau_{\mathrm{m}}=0. \tag{5.132}$$

Although it may appear that (5.132) accounts for the equations of stress equilibrium, in fact only the derivatives of these equations have been used in deriving (5.132); the stress equilibrium equations must yet be satisfied. To do this, we note that the stress equilibrium equations will *automatically* be satisfied if we define the three independent stress components in terms of some function U, as follows:

$$\tau_{xx}=\frac{\partial^2 U}{\partial y^2}+\rho V,\quad \tau_{yy}=\frac{\partial^2 U}{\partial x^2}+\rho V,\quad \tau_{xy}=-\frac{\partial^2 U}{\partial x\partial y}. \tag{5.133}$$

It can be verified by inserting (5.133) into (5.97)–(5.98) that, if the stresses are defined as in (5.133), the equilibrium equations will be identically satisfied. Finally, insertion of (5.133) into (5.132) gives

$$0=\nabla^2\tau_{\mathrm{m}}=\frac{1}{2}\nabla^2\left[\frac{\partial^2 U}{\partial x^2}+\frac{\partial^2 U}{\partial y^2}+2\rho V\right]=\frac{1}{2}\nabla^2\left(\nabla^2 U\right)+\rho\nabla^2 V. \tag{5.134}$$

But $\nabla^2 V=0$, so we see that the function U must satisfy the *biharmonic equation*,

$$\nabla^2(\nabla^2 U)\equiv\nabla^4 U=0. \tag{5.135}$$

In Cartesian coordinates, this equation takes the form

$$\frac{\partial^4 U}{\partial x^4} + 2\frac{\partial^4 U}{\partial x^2 \partial y^2} + \frac{\partial^4 U}{\partial y^4} = 0. \tag{5.136}$$

In summary, if a function U satisfies the biharmonic equation, (5.135), then the stresses that are derived from U via (5.133) will automatically satisfy the two-dimensional elasticity equations (including Hooke's law, strain compatibility, and stress equilibrium). Hence, the mathematical process of solving the elasticity equations has been reduced to the solution of a single fourth-order partial differential equation. This approach is convenient if the boundary conditions are all given in terms of tractions, rather than displacements; otherwise, a displacement-based formulation must be used. The biharmonic function U is known as an *Airy stress function*, after G. B. Airy, the British astronomer who first suggested this approach (Airy, 1863; Maxwell, 1870).

In polar coordinates (and assuming no body forces), the stress components can be derived from the Airy stress function $U(r,\theta)$ by

$$\tau_{rr} = \frac{1}{r}\frac{\partial U}{\partial r} + \frac{1}{r^2}\frac{\partial^2 U}{\partial \theta^2}, \quad \tau_{\theta\theta} = \frac{\partial^2 U}{\partial r^2}, \quad \tau_{r\theta} = -\frac{\partial}{\partial r}\left(\frac{1}{r}\frac{\partial U}{\partial \theta}\right), \tag{5.137}$$

and the biharmonic equation for U takes the form

$$\left(\frac{\partial^2}{\partial r^2} + \frac{1}{r}\frac{\partial}{\partial r} + \frac{1}{r^2}\frac{\partial^2}{\partial \theta^2}\right)\left(\frac{\partial^2 U}{\partial r^2} + \frac{1}{r}\frac{\partial U}{\partial r} + \frac{1}{r^2}\frac{\partial^2 U}{\partial \theta^2}\right) = 0. \tag{5.138}$$

The stress function approach can in principle also be used for three-dimensional problems. Two different types of three-dimensional stress functions have been developed by Maxwell (1870) and by Morera (1892); see also Bradley (1990) and Michelitsch and Wunderlin (1996). However, in three dimensions this approach loses much of its simplicity, as in general three different stress functions are needed to solve a given elasticity problem.

5.8 Elastic strain energy and related principles

When an elastic body is deformed, the forces (body or surface) that cause the deformation do work on the body as they deform it. In accordance with the principle of conservation of energy, this work is stored in the deformed body in the form of *elastic strain energy*. Elastic strain energy plays an important role in rock mechanics. Physically, strain energy is important because it is available to potentially cause adverse phenomena such as rock bursts, borehole collapses, etc. (Cook et al., 1966; Salamon, 1984). Mathematically, strain energy is important because many widely used solution methods are based on energy-related principles. In §5.9, it is shown that strain energy considerations are crucial to the establishment of the uniqueness theorem for elastic boundary-value problems.

The strain energy stored in a deformed rock is essentially equivalent to the elastic energy stored in a deformed spring, for example. To calculate the stored strain energy, we make use of the principle of conservation of energy, which states that the strain energy will be equal to the work done on the body by the external forces that cause the deformation. This work can be calculated from the elementary definition that work is equal to force multiplied by the displacement of the point at which that force acts; in this calculation, only the component of the force that acts in the direction of the displacement must be used. Consider first a cube of rock, of length a on each side, that is subjected to a uniaxial stress σ_1 in one direction. Imagine that this stress is applied slowly, so that it increases from 0 to σ_1 in a quasi-static manner. This can be represented mathematically by taking the stress to be $k\sigma_1$, where k is a scalar parameter that increases from 0 to 1. If the rock is *linearly* elastic, the strain will increase at a rate proportional to the stress, from 0 to $k\varepsilon_1$. The stress acts over a face of area a^2, so its associated force is $k\sigma_1 a^2$. The final displacement of this face in the direction of the applied stress will be $a\varepsilon_1$, so that at any intermediate time during the deformation, the displacement is $ak\varepsilon_1$. The incremental work done by this force as the parameter k increases from k to $k + dk$ is therefore given by

$$dW = \text{force} \times \text{displacement} = (k\sigma_1 a^2)(a\varepsilon_1 dk) = \sigma_1\varepsilon_1 a^3 k dk. \tag{5.139}$$

The total work done by the force through the entire process of deformation is

$$W = \int dW = \sigma_1\varepsilon_1 a^3 \int_0^1 k dk = \frac{1}{2}\sigma_1\varepsilon_1 a^3. \tag{5.140}$$

This work is now stored in the body as elastic strain energy, $\mathcal{E}$, so that $\mathcal{E}_{\text{TOT}} = \sigma_1\varepsilon_1 a^3/2$. The elastic strain energy per unit volume of rock, which will be referred to as the elastic strain energy *density*, is $\mathcal{E} = \mathcal{E}_{\text{TOT}}/a^3 = \sigma_1\varepsilon_1/2$.

If normal tractions are also applied over the other two pairs of faces, their contributions to the total work will be additive. No cross products appear in the final expression for W, since, for example, the force due to the stress that acts on the second face, σ_2, has no component in the direction of the displacement of the first face, etc. Hence, the elastic strain energy stored in this cube of rock, per unit volume, will be given by

$$\mathcal{E} = \frac{1}{2}(\sigma_1\varepsilon_1 + \sigma_2\varepsilon_2 + \sigma_3\varepsilon_3). \tag{5.141}$$

In this thought experiment, no shear tractions were applied to the cube of rock, so the stresses and strains appearing in (5.141) are indeed principal stresses and strains. The bracketed term in (5.141) can be shown to be equal to the trace of the product of the stress and the strain matrices, when expressed in terms of the principal coordinate system. But the trace of a matrix is an invariant and therefore has the same value when calculated in any coordinate system. It is therefore true, in general, that

$$\begin{aligned}
\mathcal{E} &= \frac{1}{2}\text{trace}(\tau\varepsilon) \\
&= \frac{1}{2}(\tau_{xx}\varepsilon_{xx} + \tau_{yy}\varepsilon_{yy} + \tau_{zz}\varepsilon_{zz} + \tau_{xy}\varepsilon_{xy} + \tau_{xz}\varepsilon_{xz} + \tau_{yz}\varepsilon_{yz} + \tau_{yx}\varepsilon_{yx} \\
&\quad + \tau_{zx}\varepsilon_{zx} + \tau_{zy}\varepsilon_{zy}).
\end{aligned} \tag{5.142}$$

Although this derivation of (5.142) seems to depend on the tacit assumption that the principal directions of stress and strain are equal, which is not necessarily the case for anisotropic rocks, it can be shown that (5.142) is in fact true in all cases.

The six shear terms in (5.142) could be grouped together into three terms, using the symmetry of the stresses and strains. In some situations, this is convenient, such as when discussing Hooke's law for anisotropic materials, as in §5.10. However, some issues are obscured by this simplification, such as the fact (shown below) that the stresses and strains are conjugate thermodynamic variables. In any event, $\mathcal{E}$ can either be considered to be a function of nine stress components, as in (5.142), or can be considered to be a function of six stress components, if use is made of the relations $\tau_{xy} = \tau_{yx}$, etc. This latter approach will be taken in §5.10.

For isotropic rocks, the stored strain energy density can be written in many different forms. Substitution of (5.24)–(5.27) into (5.142) yields

$$\mathcal{E} = \frac{1}{2E}[(\tau_{xx}^2 + \tau_{yy}^2 + \tau_{zz}^2) - 2\nu(\tau_{xx}\tau_{yy} + \tau_{xx}\tau_{zz} + \tau_{yy}\tau_{zz}) + (1+\nu)(\tau_{xy}^2 + \tau_{xz}^2 + \tau_{yz}^2 + \tau_{yx}^2 + \tau_{zx}^2 + \tau_{zy}^2)]. \tag{5.143}$$

In terms of the principal stresses, (5.143) takes the form

$$\mathcal{E} = \frac{1}{2E}[(\sigma_1^2 + \sigma_2^2 + \sigma_3^2) - 2\nu(\sigma_1\sigma_2 + \sigma_2\sigma_3 + \sigma_3\sigma_1)]. \tag{5.144}$$

Alternatively, $\mathcal{E}$ can be written in terms of the strains as

$$\mathcal{E} = \frac{1}{2}[\lambda(\varepsilon_{xx} + \varepsilon_{yy} + \varepsilon_{zz})^2 + 2G(\varepsilon_{xx}^2 + \varepsilon_{yy}^2 + \varepsilon_{zz}^2) + 2G(\varepsilon_{xy}^2 + \varepsilon_{xz}^2 + \varepsilon_{yz}^2 + \varepsilon_{yx}^2 + \varepsilon_{zx}^2 + \varepsilon_{zy}^2)], \tag{5.145}$$

which, in terms of principal strains, takes the form

$$\mathcal{E} = \frac{1}{2}[\lambda(\varepsilon_1 + \varepsilon_2 + \varepsilon_3)^2 + 2G(\varepsilon_1^2 + \varepsilon_2^2 + \varepsilon_3^2)]. \tag{5.146}$$

Recalling (2.154)–(2.156) and (2.237), the elastic strain energy as given by (5.142), (5.144), and (5.146) can also be written in terms of the deviatoric stresses and deviatoric strains as (Mal and Singh, 1991, p. 154)

$$\mathcal{E} = \frac{1}{2}(s_1e_1 + s_2e_2 + s_3e_3) + \frac{3}{2}\tau_{\mathrm{m}}\varepsilon_{\mathrm{m}}, \tag{5.147}$$

$$= \frac{1}{4G}(s_1^2 + s_2^2 + s_3^2) + \frac{1}{2K}\tau_{\mathrm{m}}^2, \tag{5.148}$$

$$= G(e_1^2 + e_2^2 + e_3^2) + \frac{9}{2}Ke_{\mathrm{m}}^2, \tag{5.149}$$

where $\{s_1, s_2, s_3\}$ are the principal deviatoric stresses, $\{e_1, e_2, e_3\}$ are the principal deviatoric strains, τ_{m} is the mean normal stress, and ε_{m} is the mean normal strain (which equals one-third of the volumetric strain, ε_{v}).

Equations (5.148) and (5.149) show that the elastic strain energy can be split into two parts: a term that depends on the volume change,

$$\mathcal{E}_{\mathrm{v}} = \frac{1}{2K}\tau_{\mathrm{m}}^2 = \frac{9K}{2}\varepsilon_{\mathrm{m}}^2, \tag{5.150}$$

and a term that depends on the deviatoric stresses (or strains),

$$\mathcal{E}_{\mathrm{d}} = \frac{1}{4G}(s_1^2 + s_2^2 + s_3^2) = G(e_1^2 + e_2^2 + e_3^2). \tag{5.151}$$

This latter term is known as the *distortional strain energy*. It follows from (2.165) and (2.167) that

$$\mathcal{E}_{\mathrm{d}} = \frac{1}{2G}J_2 = \frac{3}{4G}(\tau_{\mathrm{oct}})^2. \tag{5.152}$$

The assumption is often made that the elastic strain energy should be zero if all of the strain components are zero, but should be positive whenever at least one strain component is nonzero. Mathematically, this condition is equivalent to saying that the strain energy function is a "positive-definite" function of the strains (or stresses). The motivation behind this assumption is that it seems reasonable to assume that work must be done *on* the body in order to deform it, rather than for the body to do work on its surroundings during a deformation that begins from the unstrained state. Assuming this to be the case, consider a hydrostatic stress state, in which case the deviatoric stresses are zero. Equation (5.148) shows that in order for the strain energy to be positive, the bulk modulus K must satisfy $K > 0$. Similarly, if the stress is purely deviatoric, positive-definiteness of the strain energy implies that $G > 0$. This reasoning could not have been applied to (5.145), since in that equation the terms involving G and those involving λ are *not* independent of each other. The fact that the strain energy can be written as a sum of two *uncoupled* terms involving K and G is related to the fact that these two moduli are the eigenvalues of the Hooke's law operator for an isotropic material.

If $\mathcal{E}$ is thought of as a function of the nine stresses, differentiation of (5.143) yields

$$\frac{\partial \mathcal{E}}{\partial \tau_{xx}} = \frac{1}{E}[\tau_{xx} - \nu(\tau_{yy} + \tau_{zz})] = \varepsilon_{xx}, \tag{5.153}$$

$$\frac{\partial \mathcal{E}}{\partial \tau_{xy}} = \frac{(1+\nu)}{E}\tau_{xy} = \frac{1}{2G}\tau_{xy} = \varepsilon_{xy}, \tag{5.154}$$

and likewise for the other seven stress components. Similarly, differentiation of (5.145) leads to

$$\frac{\partial \mathcal{E}}{\partial \varepsilon_{xx}} = \tau_{xx}, \quad \frac{\partial \mathcal{E}}{\partial \varepsilon_{xy}} = \tau_{xy}, \quad \text{etc.} \tag{5.155}$$

Equations (5.153)–(5.155) show that the stresses and strains are conjugate thermodynamic variables (McLellan, 1980), in the same sense that pressure and volume are conjugate variables for a fluid, surface tension and interface area are conjugate variables for an interface between two fluids, etc. Although the derivation given above assumed that the material obeys the isotropic version of Hooke's law, (5.153)–(5.155) are in fact true in general for all elastic materials, whether isotropic or not. Taking cross partial derivatives of equations of the form (5.155) leads to a series of reciprocity equations such as

$$\frac{\partial \tau_{xx}}{\partial \varepsilon_{yy}} = \frac{\partial^2 \mathcal{E}}{\partial \varepsilon_{yy}\partial \varepsilon_{xx}} = \frac{\partial^2 \mathcal{E}}{\partial \varepsilon_{xx}\partial \varepsilon_{yy}} = \frac{\partial \tau_{yy}}{\partial \varepsilon_{xx}}, \quad \text{etc.} \tag{5.156}$$

These relations are trivially satisfied in the isotropic case, as (5.18) and (5.19) show that both the first and the last term in (5.156) are equal to 2G. However, relations such as (5.156) are very useful in supplying nonobvious constraints on the elastic moduli when attempting to formulate *anisotropic* versions of Hooke's law (§5.10).

The total elastic strain energy stored in a body can be found by integrating the elastic strain energy density over the entire body:

$$\mathcal{E}_{\text{TOT}} = \iiint\limits_{B} \mathcal{E} dV = \frac{1}{2}\iiint\limits_{B} \text{trace}(\boldsymbol{\tau\varepsilon})dV. \tag{5.157}$$

In practice, calculation of this integral is often difficult, due to the fact that it is a three-dimensional integral whose integrand is a complicated quadratic function of the stresses (or strains). Calculation of the total stored elastic strain energy can be simplified by appealing to the equivalence of the stored strain energy and the work that was done on the body in bringing about the deformation. The total work done by the applied body forces and surface tractions can be calculated by generalizing the argument used in conjunction with (5.140), to yield

$$W = \frac{1}{2}\iint\limits_{\partial B} \mathbf{p}\cdot\mathbf{u}dA + \frac{1}{2}\iiint\limits_{B} \rho\mathbf{F}\cdot\mathbf{u}dV, \tag{5.158}$$

where the first integral represents the work done by the surface tractions applied to the outer surface of the body, and the second integral represents the work done by the body forces. In (5.158), the forces, tractions, and displacements are those that exist at equilibrium, that is, after the loads have been applied and the body comes to rest; they do *not* refer to intermediate values that occur during the loading process.

Equation (5.158) was written down in analogy with (5.140). To prove that this expression for the total work done on the body is indeed equal to expression (5.142) for the total stored elastic strain energy, we proceed as follows. We first transform the surface integral in (5.158) into a volume integral, by utilizing

(2.63)–(2.65) and (5.67):

$$
\begin{aligned}
\iint_{\partial B} \mathbf{p}\cdot\mathbf{u}\mathrm{d}A &= \iint_{B} (p_x u_x + p_y u_y + p_z u_z)\mathrm{d}A \\
&= \iint_{\partial B} (\tau_{xx}n_x+\tau_{yx}n_y+\tau_{zx}n_z)u_x+(\tau_{xy}n_x+\tau_{yy}n_y+\tau_{zy}n_z)u_y \\
&\quad + (\tau_{xz}n_x + \tau_{yz}n_y + \tau_{zz}n_z)u_z\mathrm{d}A \\
&= \iint_{\partial B} (\tau_{xx}u_x+\tau_{xy}u_y+\tau_{xz}u_z)n_x+(\tau_{yx}u_x+\tau_{yy}u_y+\tau_{yz}u_z)n_y \\
&\quad + (\tau_{zx}u_x + \tau_{zy}u_y + \tau_{zz}u_z)n_z\mathrm{d}A \\
&= \iiint_{B} \frac{\partial}{\partial x}(\tau_{xx}u_x+\tau_{xy}u_y+\tau_{xz}u_z)+\frac{\partial}{\partial y}(\tau_{yx}u_x+\tau_{yy}u_y+\tau_{yz}u_z) \\
&\quad + \frac{\partial}{\partial z}(\tau_{zx}u_x + \tau_{zy}u_y + \tau_{zz}u_z)\mathrm{d}V \\
&= \iiint_{B} u_x\left(\frac{\partial\tau_{xx}}{\partial x}+\frac{\partial\tau_{yx}}{\partial y}+\frac{\partial\tau_{zx}}{\partial z}\right)+u_y\left(\frac{\partial\tau_{xy}}{\partial x}+\frac{\partial\tau_{yy}}{\partial y}+\frac{\partial\tau_{zy}}{\partial z}\right) \\
&\quad + u_z\left(\frac{\partial\tau_{xz}}{\partial x}+\frac{\partial\tau_{yz}}{\partial y}+\frac{\partial\tau_{zz}}{\partial z}\right)\mathrm{d}V \\
&\quad + \iiint_{B}\left[\tau_{xx}\frac{\partial u_x}{\partial x}+\tau_{xy}\frac{\partial u_y}{\partial x}+\tau_{xz}\frac{\partial u_z}{\partial x}+\tau_{yx}\frac{\partial u_x}{\partial y}+\tau_{yy}\frac{\partial u_y}{\partial y}\right. \\
&\quad \left. + \tau_{yz}\frac{\partial u_z}{\partial y}+\tau_{zx}\frac{\partial u_x}{\partial z}+\tau_{zy}\frac{\partial u_y}{\partial z}+\tau_{zz}\frac{\partial u_z}{\partial z}\right]\mathrm{d}V \\
&= \iiint_{B} u_x(-\rho F_x) + u_y(-\rho F_y) + u_z(-\rho F_z)\mathrm{d}V \\
&\quad + \iiint_{B} (\tau_{xx}\varepsilon_{xx}+\tau_{xy}\varepsilon_{yx}+\tau_{xz}\varepsilon_{zx}+\tau_{yx}\varepsilon_{xy}+\tau_{yy}\varepsilon_{yy}+\tau_{yz}\varepsilon_{zy} \\
&\quad + \tau_{zx}\varepsilon_{xz} + \tau_{zy}\varepsilon_{yz} + \tau_{zz}\varepsilon_{zz})\mathrm{d}V \\
&= \iiint_{B} \left[-\rho(\mathbf{F}\cdot\mathbf{u}) + \mathrm{trace}(\boldsymbol{\tau\varepsilon})\right]\mathrm{d}V. \qquad (5.159)
\end{aligned}
$$

In the next-to-last step taken above, we have used the fact that $\partial u_x/\partial y = \varepsilon_{xy} + \omega_{xy}$, after which all terms involving the rotations cancel out, due to the fact that $\omega_{xy} = -\omega_{yx}$. Combining (5.158) and (5.159), and comparing the result with (5.157), yields

$$
W = \frac{1}{2}\iint_{\partial B} \mathbf{p}\cdot\mathbf{u}\mathrm{d}A + \frac{1}{2}\iiint_{B} \rho\mathbf{F}\cdot\mathbf{u}\mathrm{d}V = \frac{1}{2}\iiint_{B} \mathrm{trace}(\boldsymbol{\tau\varepsilon})\mathrm{d}V = \mathcal{E}_{\mathrm{TOT}}. \qquad (5.160)
$$

The work expression in (5.158) is sometimes written *without* the factor of 1/2, in which case it does *not* represent the actual work that would be done by loads that are applied quasi-statically, starting from the unloaded state, but rather represents a hypothetical work term that corresponds to the work that would be done if the "final" equilibrium forces acted throughout the total displacement process. If this definition were used, (5.160) would be written as $\mathcal{E}_{\mathrm{TOT}} = 2W$, in which form it is known as *Clapeyron's theorem*. Regardless of the definition used for W, the integrals that appear on the right-hand sides of (5.157) and (5.158) are nevertheless equal, as the terms in their integrands always refer to the final equilibrium values of the stresses, displacements, etc.

Another important theorem, the Maxwell–Betti reciprocal theorem, can be proven in a similar manner. Consider two sets of forces, consisting of surface tractions and body forces, which may be applied to a given body. Let these sets of forces be denoted by $\{\mathbf{F}^1, \mathbf{p}^1\}$ and $\{\mathbf{F}^2, \mathbf{p}^2\}$, and let the stresses, strains, and displacements due to these two sets of forces be denoted by $\{\boldsymbol{\tau}^1, \boldsymbol{\varepsilon}^1, \mathbf{u}^1\}$ and $\{\boldsymbol{\tau}^2, \boldsymbol{\varepsilon}^2, \mathbf{u}^2\}$. The reciprocal theorem states that *the work W_{12} that would be done by the first set of forces if they acted through the displacements that are due to the second set of forces is equal to the work W_{21}that would be done by the second set of forces if they acted through the displacements that are due to the first set of forces*. Both of these work terms are hypothetical work terms, computed by assuming that the final equilibrium loads act throughout the total displacement process.

To prove the reciprocal theorem, we start with, from the verbal definition given above,

$$W_{12} = \frac{1}{2}\iint_{\partial B} \mathbf{p}^1 \cdot \mathbf{u}^2 dA + \frac{1}{2}\iiint_{B} \rho \mathbf{F}^1 \cdot \mathbf{u}^2 dV. \tag{5.161}$$

By precisely the same steps that led from (5.158) to (5.160), this expression can be transformed into

$$W_{12} = \iiint_{B} \mathrm{trace}(\boldsymbol{\tau}^1 \boldsymbol{\varepsilon}^2) dV. \tag{5.162}$$

Utilization of the matrix form of Hooke's law, (5.16), yields

$$\begin{aligned} W_{12} &= \iiint_{B} \mathrm{trace}\{[\lambda \mathrm{trace}(\boldsymbol{\varepsilon}^1)\mathbf{I} + 2G\boldsymbol{\varepsilon}^1]\boldsymbol{\varepsilon}^2\} dV \\ &= \iiint_{B} \mathrm{trace}\{[\lambda \mathrm{trace}(\boldsymbol{\varepsilon}^1)\boldsymbol{\varepsilon}^2 + 2G\boldsymbol{\varepsilon}^1\boldsymbol{\varepsilon}^2]\} dV \\ &= \iiint_{B} [\lambda \mathrm{trace}(\boldsymbol{\varepsilon}^1)\mathrm{trace}(\boldsymbol{\varepsilon}^2) + 2G\, \mathrm{trace}(\boldsymbol{\varepsilon}^1\boldsymbol{\varepsilon}^2)] dV, \end{aligned} \tag{5.163}$$

where use has been made of the linearity of the trace operation. Furthermore, since $\mathrm{trace}(\mathbf{AB}) = \mathrm{trace}(\mathbf{BA})$, the last expression on the right-hand side of (5.163) is symmetric with respect to the superscripts 1 and 2. Hence, (5.163) shows that $W_{12} = W_{21}$.

The reciprocal theorem is useful in numerous areas of rock mechanics. For example, Geertsma (1957a) used it to develop general relations between the pore and bulk compressibilities of porous rocks (see §7.2). Selvadurai (1982) used it to find the displacements caused by a point load applied to a rigid foundation on an elastic half-space. It also plays a crucial role in the development of the boundary-element method, which is one of the more widely used computational methods for solving elasticity problems (Brady, 1979).

5.9 Uniqueness theorem for elasticity problems

It was seen in §5.5 that all of the relevant differential equations of linear elasticity for an isotropic rock, that is, stress equilibrium, Hooke's law, and strain compatibility, can be distilled into either the three Navier equations, (5.81)–(5.83), or the six Beltrami–Michell equations, (5.95)–(5.96). The unknown functions in the three Navier equations are the three displacement components, whereas the unknown functions in the six Beltrami–Michell equations are the six independent stress components. In either case, the number of unknowns is equal to the number of equations, which is a necessary requirement in order that a given elasticity problem be solvable. However, there will in general be an *infinite* number of mathematical functions that satisfy either of these two sets of equations. In order to find the actual stresses and displacements in a given situation, some knowledge of the stress and/or displacements along the outer boundary of the rock mass is required. It is important to know a priori the amount of such boundary information that is needed. This question is answered by the *uniqueness theorem of linear elasticity*, which roughly states that if either the tractions or the displacements are known at each point of the outer boundary of the rock mass, there will be a unique solution to the governing elasticity equations. The practical importance of this theorem is that the conditions of the theorem *specify the amount of boundary information that must be known* before attempting to solve the elasticity equations in a given situation. The validity of the uniqueness theorem depends crucially on the assumption of infinitesimal strain and on the assumption that the stored elastic strain energy is a positive-definite function of the strains. There is no uniqueness theorem for finite strain elasticity problems, for example; indeed, multiple finite strain equilibrium states *can* exist for an elastic body under a given set of body and surface tractions.

The uniqueness theorem for linear elasticity was proven in 1850 by the German physicist Gustav Kirchhoff, essentially as follows (Kirchhoff, 1850). Consider a piece or rock that occupies a region B in space, with its outer boundary ∂B subjected to a known body-force distribution, $\mathbf{F}(\mathbf{x})$. The elastic moduli of the rock, K and G, are both assumed to be positive, so that the stored elastic strain energy function is a positive-definite function of the stresses or strains. Assume that the displacements are known at each point on the outer boundary of the rock, that is,

$$\mathbf{u}(\mathbf{x}) = \mathbf{u}^{\mathrm{o}}(\mathbf{x}) \quad \text{for all } \mathbf{x} \in \partial B, \tag{5.164}$$

where $\mathbf{u}^{o}(\mathbf{x})$ is a known vector-valued function of position. Any solution to this elasticity problem must satisfy boundary condition (5.164), as well as the Navier equations (5.90). Now assume that there are *two different solutions* to this problem, $\mathbf{u}^1$ and $\mathbf{u}^2$, and let $\mathbf{u}^{\delta}(\mathbf{x})$ be the hypothetical displacement field that is the difference between these two displacement fields, that is, $\mathbf{u}^{\delta}(\mathbf{x}) = \mathbf{u}^1(\mathbf{x}) - \mathbf{u}^2(\mathbf{x})$. As both $\mathbf{u}^1$ and $\mathbf{u}^2$ satisfy (5.90), we have

$$
\begin{aligned}
(\lambda + G)\nabla\nabla^{T}\mathbf{u}^{\delta} + G\nabla^2\mathbf{u}^{\delta} &= (\lambda + G)\nabla\nabla^{T}(\mathbf{u}^1 - \mathbf{u}^2) + G\nabla^2(\mathbf{u}^1 - \mathbf{u}^2) \\
&= [(\lambda + G)\nabla\nabla^{T}\mathbf{u}^1 + G\nabla^2\mathbf{u}^1] \\
&\quad - [(\lambda + G)\nabla\nabla^{T}\mathbf{u}^2 + G\nabla^2\mathbf{u}^2] \\
&= -\rho\mathbf{F} - (-\rho\mathbf{F}) = \mathbf{0}. \qquad (5.165)
\end{aligned}
$$

Hence, the displacement field $\mathbf{u}^{\delta}$ satisfies the Navier equations in the region B, for the case where the *body forces are zero*. As both $\mathbf{u}^1$ and $\mathbf{u}^2$ satisfy the boundary condition (5.164), we also have

$$\mathbf{u}^{\delta}(\mathbf{x}) = \mathbf{u}^1(\mathbf{x}) - \mathbf{u}^2(\mathbf{x}) = \mathbf{u}^{o}(\mathbf{x}) - \mathbf{u}^{o}(\mathbf{x}) = \mathbf{0} \quad \text{for all } \mathbf{x} \in \partial B, \qquad (5.166)$$

which is to say that this hypothetical displacement field vanishes along the outer boundary of the rock.

From (5.158), the stored strain energy that is associated with the displacement field $\mathbf{u}^{\delta}$ can be calculated as

$$
\begin{aligned}
\mathcal{E}^{\delta} &= \frac{1}{2}\iint_{\partial B} \mathbf{p}^{\delta} \cdot \mathbf{u}^{\delta} dA + \frac{1}{2}\iiint_{B} \rho\mathbf{F}^{\delta} \cdot \mathbf{u}^{\delta} dV = \frac{1}{2}\iint_{\partial B} \mathbf{p}^{\delta} \cdot \mathbf{0} dA \\
&\quad + \frac{1}{2}\iiint_{B} \rho\mathbf{0} \cdot \mathbf{u}^{\delta} dV = 0, \qquad (5.167)
\end{aligned}
$$

since the boundary displacements vanish by (5.166), and the body-force vector vanishes by (5.165). The stored elastic strain energy associated with the displacement field $\mathbf{u}^{\delta}$ is therefore zero. But this energy can also be calculated from (5.149) and (5.157) as

$$\mathcal{E}^{\delta} = \frac{1}{2}\iiint_{B} [2G(e_1^2 + e_2^2 + e_3^2) + 9Ke_{\mathrm{m}}^2] dV. \qquad (5.168)$$

The integrand in (5.168) is a positive-definite function of the strains, so in order for the integral to be zero, all of the strain components must vanish at each point of the body. Hence, the strain fields corresponding to the two solutions 1 and 2 must be equal throughout the body, that is,

$$\mathbf{0} = \boldsymbol{\varepsilon}^{\delta} = \boldsymbol{\varepsilon}^1 - \boldsymbol{\varepsilon}^2, \quad \text{so} \quad \boldsymbol{\varepsilon}^1 = \boldsymbol{\varepsilon}^2 \quad \text{for all } \mathbf{x} \in B. \qquad (5.169)$$

If the strains $\boldsymbol{\varepsilon}^1$ and $\boldsymbol{\varepsilon}^2$ are equal, then the displacements $\mathbf{u}^1$ and $\mathbf{u}^2$ can differ by at most a rigid-body displacement. But these two displacement fields coincide on the boundary, so this rigid-body displacement must vanish. Therefore, the two displacement fields $\mathbf{u}^1$ and $\mathbf{u}^2$ are in fact equal at all points of the body, as are their associated stresses and strains. The assumption that there were two different solutions to this problem has thus been shown to be false – the two assumed solutions are in fact identical. This completes the proof of the uniqueness theorem in the case where displacements are specified over the entire outer boundary of the rock.

The key step in the proof given above was (5.167), in which the stored strain energy associated with the δ solution was shown to vanish. Any other boundary conditions that cause the two integrands in (5.167) to vanish would also be sufficient to prove uniqueness. For example, if the tractions are specified along the outer boundary of the body, reasoning analogous to (5.165) shows that the δ solution satisfies the Beltrami–Michell equations with zero body force. An argument analogous to (5.166) then shows that the boundary tractions associated with the δ solution vanish, after which an argument analogous to (5.167) shows that the strain energy of the δ solution vanishes. The stress-based version of (5.168), which utilizes (5.133), shows that the stresses of the δ solution vanish at all points of the body, in which case Hooke's law shows that the strains associated with solutions 1 and 2 are equal throughout the body. The displacement fields $\mathbf{u}^1$ and $\mathbf{u}^2$, and their associated stresses and strains, are again seen to be equal, except for a possible (stress-free and strain-free) rigid-body displacement.

In the most general case, uniqueness requires that sufficient data pertaining to the boundary displacements and boundary tractions are known such that the term $\mathbf{p}^\delta \cdot \mathbf{u}^\delta$ appearing in the integrand of the surface integral in (5.167) vanishes at each point of the boundary; the integrand of the volume integral always vanishes, because the body force vector associated with the δ solution is zero as long as the solutions 1 and 2 each satisfy the elasticity equations. For example, $\mathbf{p}^\delta \cdot \mathbf{u}^\delta$ could be made to vanish at a given point on the boundary if the normal component of the displacement and the two tangential components of the traction are specified. This would be the case, for example, in a contact problem in which a rigid indenter is pushed into a lubricated surface. In this situation, the normal component of the displacement vector would take on some known value, and the two tangential traction components would be zero.

5.10 Stress–strain relations for anisotropic materials

Most rocks are anisotropic to one extent or another. For example, if cylindrical cores are cut from a rock in the horizontal and a vertical direction, and Young's moduli are then measured under uniaxial compression, the two values thus measured will in general differ from one another. Common cases of anisotropic rocks include sedimentary rocks that have different elastic properties in and perpendicular to the bedding planes, or metamorphic rocks such as slates that have a well-defined plane of cleavage. In contrast to the situation for an isotropic rock, the generalized Hooke's law for an anisotropic rock will have more than two independent elastic coefficients. It is therefore more difficult to experimentally characterize the elastic properties of an anisotropic rock and more difficult to solve boundary-value problems for such materials. Consequently, most rock mechanics analyses have been conducted under the assumption of isotropy, despite the fact that this assumption strictly holds in very few instances. Increasingly, however, anisotropic versions of Hooke's law are being used in rock mechanics, particularly in the analysis of *in situ* subsurface stress measurements (Amadei, 1996) and seismic wave propagation (Helbig, 1994; Schoenberg and Sayers, 1995).

As both the stress and strain are second-order tensors, with nine components each, the most general linear relationship between the stresses and strains could be expressed via a fourth-order tensor that has $9 \times 9 = 81$ components. This relationship is often written as $\boldsymbol{\tau} = \mathbf{C}\boldsymbol{\varepsilon}$, where $\mathbf{C}$ is a fourth-order tensor whose 81 components are known as the *elastic stiffnesses*. Although this notation is often used in theoretical studies, it is awkward for two reasons. As there are actually only six independent stress components and six independent strain components, there cannot be more than 36 independent coefficients in the most general version of Hooke's law. It is therefore notationally wasteful to use for this purpose a mathematical entity that has eighty-one components. Moreover, there is no straightforward way to write out the eighty-one components of a fourth-order tensor in the form of a matrix, so use of the tensor notation causes the convenience of matrix multiplication to be sacrificed.

A more concise approach is that due to Voigt (1928), in which the stress and strain tensors are each converted into 1×6 column vectors, and the elastic stiffnesses are then represented by 36 stiffness coefficients that can be written as a 6×6 matrix. Hooke's law can be written in the Voigt notation as

$$\begin{bmatrix} \tau_{xx} \\ \tau_{yy} \\ \tau_{zz} \\ \tau_{yz} \\ \tau_{xz} \\ \tau_{xy} \end{bmatrix} = \begin{bmatrix} c_{11} & c_{12} & c_{13} & c_{14} & c_{15} & c_{16} \\ c_{21} & c_{22} & c_{23} & c_{24} & c_{25} & c_{26} \\ c_{31} & c_{32} & c_{33} & c_{34} & c_{35} & c_{36} \\ c_{41} & c_{42} & c_{43} & c_{44} & c_{45} & c_{46} \\ c_{51} & c_{52} & c_{53} & c_{54} & c_{55} & c_{56} \\ c_{61} & c_{62} & c_{63} & c_{64} & c_{65} & c_{66} \end{bmatrix} \begin{bmatrix} \varepsilon_{xx} \\ \varepsilon_{yy} \\ \varepsilon_{zz} \\ 2\varepsilon_{yz} \\ 2\varepsilon_{xz} \\ 2\varepsilon_{xy} \end{bmatrix}, \qquad (5.170)$$

where the factors of "2" arise because Voigt originally worked in terms of the engineering shear strains, γ, rather than the tensor shear strains. Equation (5.170) can also be written symbolically as $\boldsymbol{\tau} = \mathbf{C}\boldsymbol{\varepsilon}$, although in this instance $\boldsymbol{\tau}$ and $\boldsymbol{\varepsilon}$ must be interpreted as 6×1 row vectors rather than 3×3 matrices and $\mathbf{C}$ as a 6×6 matrix. The inverse version of Hooke's law, in which the strains are expressed as linear functions of the stresses, can be symbolically written as $\boldsymbol{\varepsilon} = \mathbf{S}\boldsymbol{\tau}$, where $\mathbf{S} = \mathbf{C}^{-1}$. The components of the matrix $\mathbf{S}$ are referred to as the *elastic compliances*.

The Voigt notation has the advantage of allowing the use of matrix methods, but is slightly inelegant in that the stiffness matrix $\mathbf{C}$ as defined in (5.170) is no longer a tensor, in the sense that its components do not obey a tensor-like transformation law such as (2.30) when the coordinate system is rotated. Mehrabadi and Cowin (1990) showed that if (5.170) is written in the form

$$\begin{bmatrix} \tau_{xx} \\ \tau_{yy} \\ \tau_{zz} \\ \sqrt{2}\tau_{yz} \\ \sqrt{2}\tau_{xz} \\ \sqrt{2}\tau_{xy} \end{bmatrix} = \begin{bmatrix} c_{11} & c_{12} & c_{13} & \sqrt{2}c_{14} & \sqrt{2}c_{15} & \sqrt{2}c_{16} \\ c_{21} & c_{22} & c_{23} & \sqrt{2}c_{24} & \sqrt{2}c_{25} & \sqrt{2}c_{26} \\ c_{31} & c_{32} & c_{33} & \sqrt{2}c_{34} & \sqrt{2}c_{35} & \sqrt{2}c_{36} \\ \sqrt{2}c_{41} & \sqrt{2}c_{42} & \sqrt{2}c_{43} & 2c_{44} & 2c_{45} & 2c_{46} \\ \sqrt{2}c_{51} & \sqrt{2}c_{52} & \sqrt{2}c_{53} & 2c_{54} & 2c_{55} & 2c_{56} \\ \sqrt{2}c_{61} & \sqrt{2}c_{62} & \sqrt{2}c_{63} & 2c_{64} & 2c_{65} & 2c_{66} \end{bmatrix} \begin{bmatrix} \varepsilon_{xx} \\ \varepsilon_{yy} \\ \varepsilon_{zz} \\ \sqrt{2}\varepsilon_{yz} \\ \sqrt{2}\varepsilon_{xz} \\ \sqrt{2}\varepsilon_{xy} \end{bmatrix}, \tag{5.171}$$

then the stiffness matrix that appears in (5.171) does in fact obey the tensor transformation law for a second-order tensor in a six-dimensional space. Nevertheless, the Voigt version of Hooke's law is typically written in form (5.170) rather than (5.171).

The matrices that appear in (5.170) and (5.171) are always symmetric, so in fact at most only twenty-one of the stiffness coefficients can be independent. This is a consequence of the relations (5.156), which reflect the fact that two of the cross partial derivatives of the strain energy function with respect to two different strains must be equal. For example,

$$c_{12} = \frac{\partial \tau_{xx}}{\partial \varepsilon_{yy}} = \frac{\partial^2 \mathcal{E}}{\partial \varepsilon_{yy}\partial \varepsilon_{xx}} = \frac{\partial^2 \mathcal{E}}{\partial \varepsilon_{xx}\partial \varepsilon_{yy}} = \frac{\partial \tau_{yy}}{\partial \varepsilon_{xx}} = c_{21}. \tag{5.172}$$

Fourteen similar relations exist, one for each of the terms above the diagonal of the **C** matrix; hence, there are only twenty-one independent stiffnesses, and Hooke's law can be written in the Voigt notation as

$$\begin{bmatrix} \tau_{xx} \\ \tau_{yy} \\ \tau_{zz} \\ \tau_{yz} \\ \tau_{xz} \\ \tau_{xy} \end{bmatrix} = \begin{bmatrix} c_{11} & c_{12} & c_{13} & c_{14} & c_{15} & c_{16} \\ c_{12} & c_{22} & c_{23} & c_{24} & c_{25} & c_{26} \\ c_{13} & c_{23} & c_{33} & c_{34} & c_{35} & c_{36} \\ c_{14} & c_{24} & c_{34} & c_{44} & c_{45} & c_{46} \\ c_{15} & c_{25} & c_{35} & c_{45} & c_{55} & c_{56} \\ c_{16} & c_{26} & c_{36} & c_{46} & c_{56} & c_{66} \end{bmatrix} \begin{bmatrix} \varepsilon_{xx} \\ \varepsilon_{yy} \\ \varepsilon_{zz} \\ 2\varepsilon_{yz} \\ 2\varepsilon_{xz} \\ 2\varepsilon_{xy} \end{bmatrix}. \tag{5.173}$$

The greatest possible number of independent stiffnesses is 21, and some materials (triclinic crystals) do fall into this category. However, if a material exhibits any physical symmetry, the number of independent stiffnesses can be reduced further. Fumi (1952a,b) devised a systematic method for deducing the number of independent components of the stiffness matrix from the symmetry elements of the material, using group theory. However, these results can also be found in the following manner, by considering each symmetry element of the material in turn. For example, consider a rock mass that has a series of evenly spaced parallel fractures, in which case any plane parallel to these fractures will be a plane of symmetry with respect to reflection across that plane. (It is implicitly assumed here that the effective elastic moduli of the rock mass are defined for length scales that are much larger than the fracture spacing.) A Cartesian coordinate system is chosen such that the x-axis is perpendicular to the plane of the fractures. Now consider a second "primed" coordinate system whose unit vectors are $\{-\mathbf{e_x}, \mathbf{e_y}, \mathbf{e_z}\}$, which is to say it differs from the first system only in that its x-axis points in the opposite direction.

The stress components in the new coordinate system can be found by the transformation (2.30), $\boldsymbol{\tau}' = \mathbf{L}\boldsymbol{\tau}\mathbf{L}^{\mathrm{T}}$, where, as usual, the rows of the matrix $\mathbf{L}$ are composed of the components of the new unit vectors in terms of the old unit vectors. (Although this coordinate transformation is a reflection rather than a rotation a (2.30) holds as long as $\mathbf{L}$ is "unitary" in the sense of satisfying $\mathbf{L}^{\mathrm{T}}\mathbf{L} = \mathbf{I}$ and $\mathbf{L}^{\mathrm{T}}\mathbf{L} = \mathbf{I}$.) Hence, the three rows of $\mathbf{L}$ are given by $\mathbf{e'_x} = -\mathbf{e_x} = (-1, 0, 0)$, $\mathbf{e'_y} = \mathbf{e_y} = (0, 1, 0)$, and $\mathbf{e'_z} = \mathbf{e_z} = (0, 0, 1)$, and so the stress matrix in the new coordinate system is given by

$$\begin{bmatrix} \tau'_{xx} & \tau'_{xy} & \tau'_{xz} \\ \tau'_{yx} & \tau'_{yy} & \tau'_{yz} \\ \tau'_{zx} & \tau'_{zy} & \tau'_{zz} \end{bmatrix} = \begin{bmatrix} -1 & 0 & 0 \\ 0 & 1 & 0 \\ 0 & 0 & 1 \end{bmatrix} \begin{bmatrix} \tau_{xx} & \tau_{xy} & \tau_{xz} \\ \tau_{yx} & \tau_{yy} & \tau_{yz} \\ \tau_{zx} & \tau_{zy} & \tau_{zz} \end{bmatrix} \begin{bmatrix} -1 & 0 & 0 \\ 0 & 1 & 0 \\ 0 & 0 & 1 \end{bmatrix}$$
$$= \begin{bmatrix} \tau_{xx} & -\tau_{xy} & -\tau_{xz} \\ -\tau_{yx} & \tau_{yy} & \tau_{yz} \\ -\tau_{zx} & \tau_{zy} & \tau_{zz} \end{bmatrix}, \qquad (5.174)$$

and similarly for the strains. The "16" component of the Voigt stiffness matrix in the new coordinate system is related to the "16" component in the old coordinate system by

$$c'_{16} = \frac{\partial \tau'_{xy}}{\partial \varepsilon'_{xx}} = \frac{-\partial \tau_{xy}}{\partial \varepsilon_{xx}} = -c_{16}. \qquad (5.175)$$

But since the x-axis is a plane of symmetry of this rock, the "16" coefficient in Hooke's law *must be the same in the two coordinate systems*, and so (5.175) implies that $c_{16} = 0$. Another way to reach this conclusion is to note from (5.173) and (5.142) that the stored strain energy function in the old coordinate system will contain the term $c_{16}(\varepsilon_{xy})^2$, whereas the stored strain energy function in the new coordinate system will contain the term $c'_{16}(\varepsilon'_{xy})^2 = -c_{16}(-\varepsilon_{xy})^2 = -c_{16}(\varepsilon_{xy})^2$. But the strain energy, being a scalar, cannot depend on the choice of coordinate system, so it again follows that $c_{16} = 0$. Similar arguments show that $c_{15} = c_{25} = c_{26} = c_{36} = c_{36} = c_{45} = c_{46} = 0$, so we see that a single plane of symmetry will reduce the number of independent stiffnesses to $21 - 8 = 13$ (Amadei, 1983, p. 11).

If a rock mass contains three mutually perpendicular sets of fractures, the preceding argument can be used with respect to a reflection across the x–y or x–z planes. This leads to the conclusion that $c_{14} = c_{24} = c_{34} = c_{56} = 0$, leaving only nine nonzero stiffnesses. Such a rock mass is called an *orthotropic* (or *orthorhombic*) *material*, and its stress–strain law can be written as

$$\begin{bmatrix} \tau_{xx} \\ \tau_{yy} \\ \tau_{zz} \\ \tau_{yz} \\ \tau_{xz} \\ \tau_{xy} \end{bmatrix} = \begin{bmatrix} c_{11} & c_{12} & c_{13} & 0 & 0 & 0 \\ c_{12} & c_{22} & c_{23} & 0 & 0 & 0 \\ c_{13} & c_{23} & c_{33} & 0 & 0 & 0 \\ 0 & 0 & 0 & c_{44} & 0 & 0 \\ 0 & 0 & 0 & 0 & c_{55} & 0 \\ 0 & 0 & 0 & 0 & 0 & c_{66} \end{bmatrix} \begin{bmatrix} \varepsilon_{xx} \\ \varepsilon_{yy} \\ \varepsilon_{zz} \\ 2\varepsilon_{yz} \\ 2\varepsilon_{xz} \\ 2\varepsilon_{xy} \end{bmatrix}; \qquad (5.176)$$

or, in nonmatrix form:

$$\tau_{xx} = c_{11}\varepsilon_{xx} + c_{12}\varepsilon_{yy} + c_{13}\varepsilon_{zz}, \tag{5.177}$$

$$\tau_{yy} = c_{12}\varepsilon_{xx} + c_{22}\varepsilon_{yy} + c_{23}\varepsilon_{zz}, \tag{5.178}$$

$$\tau_{zz} = c_{13}\varepsilon_{xx} + c_{23}\varepsilon_{yy} + c_{33}\varepsilon_{zz}, \tag{5.179}$$

$$\tau_{yz} = 2c_{44}\varepsilon_{yz}, \quad \tau_{xz} = 2c_{55}\varepsilon_{xz}, \quad \tau_{xy} = 2c_{66}\varepsilon_{xy}. \tag{5.180}$$

Hooke's law will take this form for an orthotropic rock *only* if the coordinate system is aligned with the planes of symmetry of the rock. This is in contrast to the situation for an isotropic rock, for which Hooke's law takes the same form for any locally orthogonal coordinate system, regardless of its orientation. In some situations, such as when a borehole is drilled at an oblique angle to the axes of symmetry of the rock, it may be convenient to align the coordinate system with the borehole, rather than the symmetry axes of the rock mass. In this event, the stress–strain equations in the new coordinate system would take on a much more complicated form, which could be found by transforming both the stresses and strains into the new coordinate system using the transformation laws $\boldsymbol{\tau}' = \mathbf{L}\boldsymbol{\tau}\mathbf{L}^{\mathrm{T}}$ and $\boldsymbol{\varepsilon}' = \mathbf{L}\boldsymbol{\varepsilon}\mathbf{L}^{\mathrm{T}}$. In general, there will be many more nonzero elements in the stiffness matrix in an arbitrary coordinate system, as opposed to the "natural coordinate system" that is aligned with the symmetry axes; nevertheless, only nine of these components will be independent.

The compliance matrix $\mathbf{S}$ of an orthotropic rock can be found by inverting the stiffness matrix $\mathbf{C}$. As $\mathbf{C}$ is block-diagonal, its inverse is readily found to be

$$s_{11} = (c_{22}c_{33} - c_{23}^2)/D, \quad s_{22} = (c_{11}c_{33} - c_{13}^2)/D, \quad s_{33} = (c_{11}c_{22} - c_{12}^2)/D, \tag{5.181}$$

$$s_{12} = (c_{12}c_{23} - c_{12}c_{33})/D, \quad s_{13} = (c_{12}c_{23} - c_{22}c_{13})/D, \quad s_{23} = (c_{12}c_{13} - c_{11}c_{23})/D, \tag{5.182}$$

where D is the determinant of the upper-left-corner block of $\mathbf{C}$, that is,

$$D = \det \begin{bmatrix} c_{11} & c_{12} & c_{13} \\ c_{12} & c_{22} & c_{23} \\ c_{13} & c_{23} & c_{33} \end{bmatrix}, \tag{5.183}$$

and all other components of $\mathbf{S}$ are zero. The inverse form of Hooke's law for an orthotropic material is therefore

$$\begin{bmatrix} \varepsilon_{xx} \\ \varepsilon_{yy} \\ \varepsilon_{zz} \\ 2\varepsilon_{yz} \\ 2\varepsilon_{xz} \\ 2\varepsilon_{xy} \end{bmatrix} = \begin{bmatrix} s_{11} & s_{12} & s_{13} & 0 & 0 & 0 \\ s_{12} & s_{22} & s_{23} & 0 & 0 & 0 \\ s_{13} & s_{23} & s_{33} & 0 & 0 & 0 \\ 0 & 0 & 0 & s_{44} & 0 & 0 \\ 0 & 0 & 0 & 0 & s_{55} & 0 \\ 0 & 0 & 0 & 0 & 0 & s_{66} \end{bmatrix} \begin{bmatrix} \tau_{xx} \\ \tau_{yy} \\ \tau_{zz} \\ \tau_{yz} \\ \tau_{xz} \\ \tau_{xy} \end{bmatrix}, \tag{5.184}$$

where the s coefficients are given by (5.181)–(5.183).

The compliance coefficient s_{11} is the ratio of ε_{xx} to τ_{xx} in a uniaxial stress test, so it can be identified as $1/E_x$, where E_x is the "Young's modulus of the rock in the x direction"; similarly for s_{22} and s_{33}. Likewise, s_{44} can be identified as $1/G_{yz}$, where G_{yz} is the "shear modulus in the y–z plane," etc. Finally, making the obvious identifications of the off-diagonal components of **S** with the "Poisson effect," we can write Hooke's law for an orthotropic material as (Amadei, 1983, p. 11)

$$\varepsilon_{xx} = \frac{1}{E_x}\tau_{xx} - \frac{\nu_{yx}}{E_y}\tau_{yy} - \frac{\nu_{zx}}{E_z}\tau_{zz}, \tag{5.185}$$

$$\varepsilon_{yy} = -\frac{\nu_{yx}}{E_y}\tau_{xx} + \frac{1}{E_y}\tau_{yy} - \frac{\nu_{zy}}{E_z}\tau_{zz}, \tag{5.186}$$

$$\varepsilon_{zz} = -\frac{\nu_{zx}}{E_z}\tau_{xx} - \frac{\nu_{zy}}{E_z}\tau_{yy} + \frac{1}{E_z}\tau_{zz}, \tag{5.187}$$

$$\varepsilon_{yz} = \frac{1}{2G_{yz}}\tau_{yz}, \quad \varepsilon_{xz} = \frac{1}{2G_{xz}}\tau_{xz}, \quad \varepsilon_{xy} = \frac{1}{2G_{xy}}\tau_{xy}. \tag{5.188}$$

In the special case in which the three directions (x, y, z) are all elastically equivalent, an orthotropic material does *not* reduce to an isotropic material. For although it would be true that $E_x = E_y = E_z$, $G_{xz} = G_{yz} = G_{xy}$, and $\nu_{xz} = \nu_{yz} = \nu_{xy}$, it will in general not be the case that $E = 2G(1 + \nu)$. This type of material, which is said to possess *cubic symmetry*, will therefore have three independent elastic moduli, rather than two. A physical explanation for this fact is that although, for example, the Young's modulus of a cubic material will be the same in the x and y directions, there is no reason for E to have the same value in a direction that is oriented at an arbitrary angle to one of the three coordinate axes.

Although the nine stiffnesses of an orthotropic rock are said to be independent, the requirement that the stored strain energy function be positive-definite imposes some constraints on their numerical values. In terms of the engineering moduli E, G, and ν, these constraints are (Amadei et al., 1987)

$$E_x, E_y, E_z, G_{yz}, G_{xz}, G_{xy} > 0, \tag{5.189}$$

$$\nu_{xy}\nu_{yx} < 1, \quad \nu_{yz}\nu_{zy} < 1, \quad \nu_{xz}\nu_{zx} < 1, \tag{5.190}$$

$$\nu_{yx}\nu_{xy} + \nu_{yz}\nu_{zy} + \nu_{yz}\nu_{zy} + \nu_{yx}\nu_{xz}\nu_{zy} + \nu_{zx}\nu_{xy}\nu_{yz} < 1, \tag{5.191}$$

where the Poisson ratios that appear in (5.189)–(5.191) but not in (5.185)–(5.188) are defined by $\nu_{xy}/E_x = \nu_{yx}/E_y$, etc.

Another common form of anisotropy observed in rocks is the case when one of the three axes is an axis of rotational symmetry, in the sense that all directions perpendicular to this axis are elastically equivalent. In this case, the rock is isotropic within any plane normal to this rotational symmetry axis. A rock possessing this type of symmetry is known as "transversely isotropic" – a somewhat misleading term, as a *transversely isotropic* rock is actually *anisotropic*. If we identify the axis of rotational symmetry as the z-axis, the two Young's moduli E_x and E_y are obviously equal to each other. Likewise, it must also be true that $G_{xz} = G_{yz}$ and $\nu_{xz} = \nu_{yz}$. Finally, the requirement that the stiffness coefficients be invariant with respect to an arbitrary rotation of the coordinate system about the z-axis can be shown to lead to the relation $E_x = 2G_{xy}(1 + \nu_{xy})$. Hence, there are only five independent coefficients in Hooke's law for a transversely isotropic rock:

$$\varepsilon_{xx} = \frac{1}{E_x}\tau_{xx} - \frac{\nu_{yx}}{E_x}\tau_{yy} - \frac{\nu_{zx}}{E_z}\tau_{zz}, \tag{5.192}$$

$$\varepsilon_{yy} = -\frac{\nu_{yx}}{E_x}\tau_{xx} + \frac{1}{E_x}\tau_{yy} - \frac{\nu_{zx}}{E_z}\tau_{zz}, \tag{5.193}$$

$$\varepsilon_{zz} = -\frac{\nu_{zx}}{E_z}\tau_{xx} - \frac{\nu_{zx}}{E_z}\tau_{yy} + \frac{1}{E_z}\tau_{zz}, \tag{5.194}$$

$$\varepsilon_{yz} = \frac{1}{2G_{xz}}\tau_{yz}, \quad \varepsilon_{xz} = \frac{1}{2G_{xz}}\tau_{xz}, \quad \varepsilon_{xy} = \frac{(1+\nu_{yx})}{E_{xy}}\tau_{xy}. \tag{5.195}$$

A commonly used notation for transversely isotropic materials is for the elastic coefficients pertaining to the (x, y) plane of isotropy to be referred to as $\{E, G, \nu\}$, with $E = 2G(1+\nu)$, and those involving the z direction denoted by $\{E', G', \nu'\}$. This allows the stress–strain relations to be written as (Amadei, 1996)

$$\varepsilon_{xx} = \frac{1}{E}\tau_{xx} - \frac{\nu}{E}\tau_{yy} - \frac{\nu'}{E'}\tau_{zz}, \tag{5.196}$$

$$\varepsilon_{yy} = -\frac{\nu}{E}\tau_{xx} + \frac{1}{E}\tau_{yy} - \frac{\nu'}{E'}\tau_{zz}, \tag{5.197}$$

$$\varepsilon_{zz} = -\frac{\nu'}{E'}\tau_{xx} - \frac{\nu'}{E'}\tau_{yy} + \frac{1}{E'}\tau_{zz}, \tag{5.198}$$

$$\varepsilon_{yz} = \frac{1}{2G'}\tau_{yz}, \quad \varepsilon_{xz} = \frac{1}{2G'}\tau_{xz}, \quad \varepsilon_{xy} = \frac{(1+\nu)}{E}\tau_{xy}. \tag{5.199}$$

The condition that the stored strain energy function be a positive-definite function of the stresses or strains leads to the following constraints on the values of the stiffnesses (Pickering, 1970):

$$E > 0, \quad E' > 0, \quad G' > 0, \quad -1 < \nu < 1, \tag{5.200}$$

$$E'(1-\nu) - 2E(\nu')^2 > 0, \tag{5.201}$$

Intact transversely isotropic rocks are typically stiffer in the plane of isotropy than in the direction of the axis of symmetry. Amadei (1996) analyzed 98 sets of data on the elastic moduli of anisotropic rocks collected by various researchers and found that in most cases $1 < E/E' < 4$; in no case did this ratio fall below 0.7. The ratio of the in-plane shear modulus to the out-of-plane shear modulus, G/G', was in all cases observed to be in the range of 1–3. On the other hand, anisotropic rock masses whose anisotropy is due to a set (or sets) of fractures may exhibit much greater elastic anisotropy ratios.

As many rock mechanics problems are idealized as being two dimensional, it is worthwhile to discuss the model of a *two-dimensional orthotropic* material. Although the stress–strain equations in this case could be found by specializing (5.185)–(5.188) to plane stress or plane strain, it is easier to derive Hooke's law for this case ab initio, using symmetry arguments similar to those presented above for the three-dimensional case. The results can be written as

$$\varepsilon_{xx} = \frac{1}{E_x}\tau_{xx} - \frac{\nu_{xy}}{E_x}\tau_{yy}, \tag{5.202}$$

$$\varepsilon_{yy} = -\frac{\nu_{yx}}{E_y}\tau_{xx} + \frac{1}{E_y}\tau_{yy}, \tag{5.203}$$

$$\varepsilon_{xy} = \frac{1}{2G}\tau_{xy}. \tag{5.204}$$

The reciprocity relations for the elastic moduli imply that $\nu_{xy}/E_x = \nu_{yx}/E_y$; however, there is in general no relationship between G and the Young's moduli and Poisson's ratios. The equations for the stresses in terms of the strains are

$$\tau_{xx} = \frac{E_x}{1 - \nu_{xy}\nu_{yx}}(\varepsilon_{xx} + \nu_{yx}\varepsilon_{yy}), \tag{5.205}$$

$$\tau_{yy} = \frac{E_y}{1 - \nu_{xy}\nu_{yx}}(\varepsilon_{yy} + \nu_{xy}\varepsilon_{xx}), \tag{5.206}$$

$$\tau_{xy} = 2G\varepsilon_{xy}. \tag{5.207}$$

6 Laboratory testing of rocks

6.1 Introduction

The mechanical properties of a piece of rock depend on its mineral composition, the arrangement of the mineral grains, and any cracks that may have been introduced into it during its long geological history by diagenesis or tectonic forces. Consequently, the mechanical properties of rock vary not only between different rock types but also between different specimens of the nominally same rock. Hence, unlike "reproducible" engineering materials such as steels, for which property values can be measured on standard specimens and listed in handbooks, only very rough approximate values of the mechanical properties of a given rock can be estimated from tabulated handbook data. For this reason, laboratory testing necessarily plays a large role in rock mechanics.

In this chapter we describe the basic types of laboratory measurements that are routinely conducted to measure the mechanical properties of rocks. Each particular experimental apparatus and/or procedure subjects the rock specimen to a certain state of stress. The chapter is structured in such a way that successively more complex stress states are considered in each subsequent section. We start in §6.2 with a discussion of hydrostatic tests that can be performed on porous rocks. Uniaxial compression tests are discussed in §6.3. Traditional triaxial compression tests, in which the two lateral stresses are equal to each other and less than the axial stress, are discussed in §6.4. The effect of the mechanical stiffness of the testing machine is examined in §6.5. True-triaxial, or polyaxial tests, in which three different stresses may be applied to the sample, are discussed in §6.6. In all of the aforementioned tests, the stress state induced in the sample is nominally homogeneous.

There are several other important test configurations in which an inhomogeneous state of stress is induced in the rock. The so-called "Brazilian test," which is used to create a tensile stress within a rock, is described in §6.7. Torsion of a cylindrical specimen is discussed in §6.8, along with the mathematical solution for the stresses and displacements. Bending of a beam-like specimen is treated in §6.9, again along with a brief mathematical derivation of the stresses and displacements. Finally, compression tests on hollow cylinders are discussed in §6.10.

In each case, the discussion will be quite general, focusing on the salient features of the experimental apparatus, the state of stress involved in the tests, and the interpretation of the test results for the purpose of extracting numerical values of the relevant rock properties. More specific details of the design of the apparatuses, and other technical issues that arise during these tests, can be found in several major review articles, such as Tullis and Tullis (1986). Specifications of standard testing procedures and practices, such as specimen size, suggested strain rates, etc., can be found in the various "ISRM Suggested Methods," which are prepared under the authority of the International Society for Rock Mechanics and published in the *International Journal of Rock Mechanics and Mining Sciences*.

Although the focus will be on describing the measurement systems and procedures, some representative data will be discussed. Many data sets on rock deformation were originally measured using British Imperial units, with stresses measured in pounds per square inch (psi). Such units are still used in some countries and within the petroleum industry. Geophysicists, on the other hand, usually quantify stresses in units of bars. Modern scientific convention, as codified in the Système Internationale (SI), requires stresses to be measured in Pascals, defined by $1\ \text{Pa} = 1\ \text{N/m}^2$. Conversion between these units can be achieved through the relations $1\ \text{psi} = 6895\ \text{Pa}$, $1\ \text{bar} = 10^5\ \text{Pa}$.

6.2 Hydrostatic tests

The simplest type of boundary traction that can be applied to the outer boundary of a piece of rock is a uniform normal traction, such as would be exerted if the boundary of the rock were in contact with a fluid. In a homogeneous solid, such boundary conditions would give rise to a state of uniform hydrostatic stress throughout the body. The ratio of the magnitude of the stress to the volumetric strain of the sample would then, according to (5.7), give the bulk modulus K of the rock.

This type of test can be conducted in a pressure vessel filled with a pressurized fluid (Fig. 6.1). The pressurizing fluid is connected to a pump or piston located outside the pressure cell. The pressure of the fluid, also referred to as the *confining pressure*, is measured by a manual pressure gauge or electronic pressure transducer. As there will be no pressure gradient in the fluid, aside from a negligible gravitational gradient, the pressure gauge or transducer can be located outside the vessel.

The rock sample is usually machined into a cylindrical shape. In order to prevent the pressurizing fluid from entering the pore space of the rock, the specimen

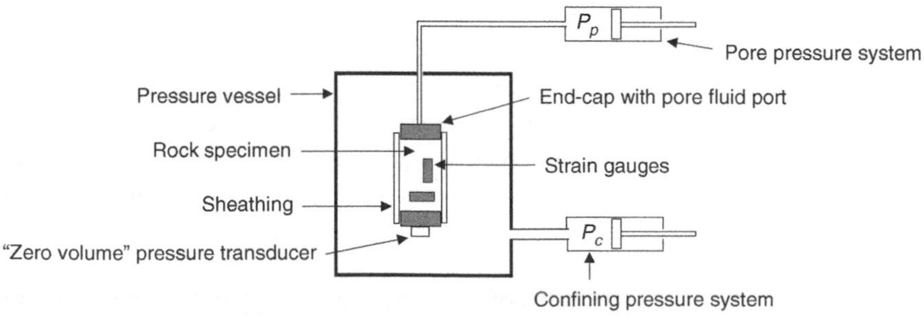

Fig. 6.1 Schematic diagram of typical experimental system used to measure compression of a porous rock subjected to hydrostatic confining pressure and pore pressure (after Hart and Wang, 2001).

must be covered with a tight-fitting, impermeable sheathing, such as heat-shrink tubing. The volumetric strain of the specimen can be measured by strain gauges glued onto the sides. If the rock is isotropic, the radial, circumferential, and axial strains should all be equal. Nevertheless, it is advisable to measure the strains in more than one direction.

If the rock is porous and permeable, a variable pore pressure can also be applied to the specimen. Furthermore, other poroelastic parameters, described in §7.2, can be measured under these hydrostatic conditions. In a typical configuration (Fig. 6.1), the two flat faces of the cylindrical specimen are fitted with metal end-caps, which have small holes drilled into them, through which the pore fluid can flow. The pore fluid is collected outside the pressure vessel in a piston-like device that allows the pore fluid pressure, and the extruded volume of the pore fluid, to be controlled and monitored. Such experimental configurations are described in more detail by Zimmerman et al. (1986), Hart and Wang (2001), and Lockner and Stanchits (2002).

As discussed in §7.2, a poroelastic rock has four fundamental compressibilities, which relate changes in the hydrostatic confining stress and pore pressure to the resulting pore or bulk strains. The bulk compressibility C_{bc} can be found by measuring the bulk strain that occurs in response to a change in confining pressure, with the pore pressure held constant. The other bulk compressibility, C_{bp}, is found from the bulk strain that occurs when the pore pressure is changed and the confining pressure is held constant. These two measurements pose no major difficulties.

The pore compressibility C_{pc} quantifies the pore strain that results from changing the confining pressure, with the pore pressure held constant. If the pore pressure is constant, then the volume of pore fluid in the system is constant, and so the change in the pore volume of the rock is exactly equal to the volume of pore fluid that enters or leaves the pore pressure piston device. Hence, measurement of C_{pc} poses no fundamental difficulty.

Measurement of the other pore compressibility, C_{pp}, which quantifies the change in pore volume caused by a change in pore pressure, with the confining pressure held constant, is not so straightforward. As the pore fluid is varied, the total volume of pore fluid will change, through the relation $\Delta V_f = -C_f V_f \Delta P_p$, where C_f is the compressibility of the pore fluid. Some of this volume change will occur in the pore space of the rock, some will occur in the tubing leading from the specimen to the pore pressure piston, and some will occur within the piston itself. Specifically,

$$\Delta V_{\text{fluid}} = \Delta V_{\text{pore}} + \Delta V_{\text{tubing}} + \Delta V_{\text{piston}}. \tag{6.1}$$

The term ΔV_{piston} is measured directly, whereas the desired quantity is ΔV_{pore}. Estimation of the actual pore volume change therefore requires knowledge of the other two terms in (6.1). This can in principle be achieved by first performing calibration tests, for example using an effectively rigid specimen such as one made of steel, to determine the compliance of the tubing and the total storativity, $C_f V_f$, of the pore fluid. However, the two unwanted terms in (6.1) are generally at least as large as the change in pore volume, rendering the estimation of the pore volume change quite problematic (Hart and Wang, 1995).

Another related "hydrostatic" elastic parameter of a porous rock is Skempton's B coefficient (Skempton, 1954). This parameter is defined, in (7.28), as the ratio of the pore pressure increment to the confining pressure increment, when the confining pressure is varied under "undrained" conditions, in which no fluid is permitted to leave the specimen. However, in a configuration such as that of Fig. 6.1, some pore fluid must indeed leave the specimen in order to enter the piston device. If the piston is replaced by a pressure transducer, it is nevertheless true that the pressure response of the pore fluid is influenced by both the compliance of the pore space of the specimen and by the compliance of the tubing and transducer. These effects can again in principle be accounted for by proper calibration, but in practice, this is quite difficult to achieve accurately, as the system compliance may be of the same magnitude as that of the pore space. Accurate measurements of B can presumably be obtained by placing a "zero volume" pressure transducer in immediate contact with the specimen, inside the pressure vessel (Hart and Wang, 2001), thus eliminating the effects of system compliance.

6.3 Uniaxial compression

The uniaxial compression test, in which a right circular cylinder or prism of rock is compressed between two parallel rigid plates (Fig. 6.1), is the oldest and simplest mechanical rock test and continues to be widely used. This test is used to determine the Young's modulus, E, and also the unconfined compressive strength, C_o.

In the simplest version of this test, Fig. 6.2a, a cylindrical core is compressed between two parallel metal platens. Hydraulic fluid pressure is typically used to apply the load. The intention of this test is to induce a state of *uniaxial stress* in the specimen, that is,

$$\tau_{zz} = \sigma, \quad \tau_{xx} = \tau_{yy} = \tau_{xy} = \tau_{yz} = \tau_{xz} = 0. \tag{6.2}$$

The axial stress σ is the controlled, independent variable, and the axial strain is the dependent variable. The longitudinal strain can be measured by a strain gauge glued to the lateral surface of the rock. Alternatively, the total shortening of the core in the direction of loading can be measured by an extensiometer that

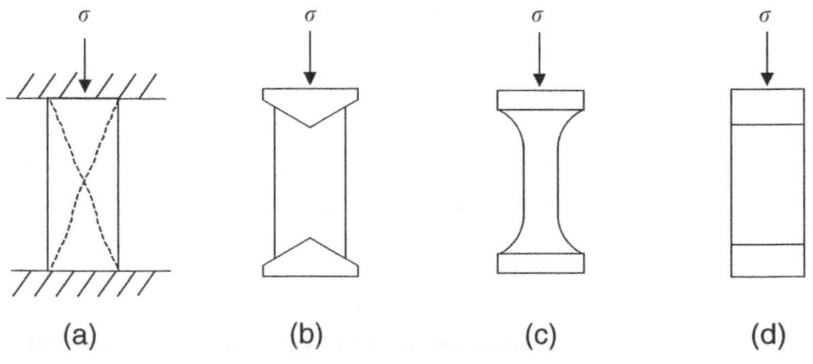

Fig. 6.2 Unconfined uniaxial compression of a rock: (a) standard configuration, with failure initiating at the corners, (b) conical end-pieces to eliminate frictional restraint, (c) tapered specimen, (d) matched end-pieces.

monitors the change in the vertical distance between the platens. In this case, the longitudinal strain is calculated from the relative shortening of the core, that is, $\varepsilon = -\Delta L/L$. If the stress state were indeed uniaxial, then the Young's modulus of the rock could be estimated from $E = \sigma/\varepsilon$. The stress can be increased slowly until failure occurs, as discussed in §4.2. The stress at which the rock fails is known as the *unconfined*, or *uniaxial*, compressive strength of the rock.

Unfortunately, the actual state of stress in the rock core, in a configuration such as that of Fig. 6.2a, will not be a homogeneous state of uniaxial compression. This is due primarily to the constraining influence of the frictional forces acting along the interface between the core and the platens. A true uniaxial stress state would lead to lateral expansion associated with the Poisson effect, (5.12). But this lateral expansion is hindered at the platens due to friction. A more realistic boundary condition to assume for the rock core is that of uniform vertical displacement and no lateral displacement (Filon, 1902; Pickett, 1944; Edelman, 1949). Hence, in a testing configuration such as shown in Fig, 6.2a, the rock core would bulge outward away from the end platens, but would be constrained against such bulging at the platens, thereby taking on a barrel-shape.

This lack of homogeneity in the stress state has implications both for the measurement of the elastic modulus and the compressive strength. Although the stress state indeed approaches that of uniaxial stress away from the platens, in the middle of the core, it is much more complex and inhomogeneous at the ends. Hence, it is not obvious that $-\sigma L/\Delta L$ will yield a correct estimate of E. Chau (1997) presented an accurate approximate solution to this problem, for the case in which the friction between the end platens and the rock is sufficiently large that no lateral motion of the rock can occur at the two boundaries. He expressed his results in terms of a parameter λ, defined as the ratio of the "true" Young's modulus to the "apparent" value estimated from $E = -\sigma L/\Delta L$. As would be expected, this factor approaches unity as the Poisson ratio goes to zero, since in this case the tendency for lateral expansion does not arise. For cores in which the length is at least as large as the diameter, and for which the Poisson ratio is less than 0.3 (which will usually be the case), the factor λ was found to lie in the range 0.97–1.0. Greenberg and Truell (1948) carried out a similar analysis for a rectangular prism compressed in plane strain conditions, with a Poisson ratio of 0.33, and found $\lambda = 0.96$. Hence, as far as the calculation of E is concerned, the issue of friction along the rock/platen interface is probably not of engineering significance.

Nevertheless, this frictional restraint leads to a stress concentration at the corners of the rock core, at the points where it meets the platen. This causes a shear fracture to initiate at that point, as shown in Fig. 6.2a, at an applied (nominal) stress σ that is actually less than the "true" uniaxial compressive strength. Several methods have been proposed to avoid this problem. One suggested approach is to machine the specimens to have hollow conical ends and then to compress them between conical end-pieces, the surfaces of which are inclined to the diameter of the specimen at the angle of friction (Fig. 6.2b).

Barnard (1964), Murrell (1965) and others have used shaped specimens that have a smaller diameter in the necked midregion than near the ends (Fig. 6.2c).

The shape of the specimen is carefully chosen, based on photoelasticity studies or finite element analysis, so that the stress distribution is uniform across the section in the neck. This permits the true Young's modulus to be calculated from the longitudinal strains measured in the neck with strain gauges, and also avoids the problem of shear fractures initiating at the point of the stress concentration near the platens. These shaped specimens are difficult to prepare, however, and tend by necessity to have short necked regions.

Another approach to mitigating the problems of stress concentrations at the platens is to compress the rock core between metal end-pieces that have the same diameter as the core and are made of a metal that has the same ratio of ν/E as does the rock (Fig. 6.2d). In this case, the lateral expansion of the rock at its ends should match that of the platens, eliminating the unwanted stress concentrations. This approach has been used by Cook (1962) and others.

Labuz and Bridell (1993) carried out compression tests on granite cores, with various lubricants applied between the core and the platens. Radial strains were measured near the ends and in the central portion of the core, to investigate the barreling effect. In the absence of lubrication, the radial hoop strains were as much as 50 percent higher in the central region of the core than near the ends. By testing various lubricants, including graphite and molybdenum disulfide, they found that this stress inhomogeneity could essentially be eliminated by the application of a mixture of stearic acid and petroleum jelly to the rock-platen interface.

6.4 Triaxial tests

One of the most widely used and versatile rock mechanics tests is the traditional "triaxial" compression test. Indeed, much of the current understanding of rock behavior has come from such tests. Despite the name, which would seem to imply a state of three independent principal stresses, in a triaxial test, a rock specimen is subjected to a homogeneous state of stress in which two of the principal stresses are of equal magnitude. Typically, all three stresses are compressive, with the unequal stress more compressive than the two equal stresses, so that $\sigma_1 > \sigma_2 = \sigma_3 > 0$.

The restriction of traditional triaxial tests to stress states in which two of these stresses are necessarily equal in magnitude is imposed by experimental limitations. Consequently, despite the ubiquitous nature of triaxial tests and triaxial compression data on rocks, it should not be erroneously concluded that stress states in which two principal stresses are of equal magnitude are particularly common in the subsurface. Indeed, there is no particular reason for σ_2 and σ_3 to be equal, either in undisturbed rock or in the vicinity of an excavation.

A triaxial stress state can be achieved by subjecting a cylindrical rock specimen to uniaxial compression by a piston, as described in §6.3, in the presence of hydrostatic compression applied by a pressurized fluid, as described in §6.2. Depending on the experimental configuration, the hydrostatic pressure may act in all three directions or only over the two lateral surfaces of the rock. In either case, the value of the two equal lateral stresses, $\sigma_2 = \sigma_3$, is known in this context as the *confining stress* and the other principal stress is referred to as the *axial stress*. The difference between the axial stress and the confining stress, $\sigma_1 - \sigma_3$, is referred to as the *differential stress.*

The classic triaxial compression tests on a rock were those performed by von Kármán (1911) on specimens of Carrara marble, using an apparatus that can be said to have served since as the prototype for triaxial testing machines. His machine and procedure, along with subsequent improvements, are described in detail by Paterson (1978, Chapter 2), upon which some of the following discussions are based.

Triaxial tests are usually conducted with cylindrical specimens having a length-to-diameter ratio of between 2:1 and 3:1. It is imperative that the flat surfaces of the specimen be as nearly parallel as possible, to avoid bending of the specimen under the axial stress. The core is jacketed in rubber or thin copper tubing so that the confining fluid does not penetrate into the pore space (Fig. 6.3a). If the effect of pore pressure is to be investigated, pore fluid would be introduced into the rock through a small hole in one of the end-pieces, as described in §6.2.

A simple triaxial apparatus is the one developed at the US Bureau of Reclamation (Fig. 6.3b). A spherical seat is used on one end-piece to correct for the possibility that the platens are not parallel. However, this apparatus has two disadvantages. Firstly, the confining pressure acts against the loading piston, so that the applied axial force must be large enough to overcome this force, in addition to creating the axial stress. Secondly, as the specimen compresses, the volume of the confining fluid in the cell decreases, making it difficult to control the confining pressure. This effect can be greatly diminished by having the pistons and end-pieces be of the same diameter as the specimen and by minimizing the volume available to the confining fluid (Donath, 1966).

Griggs et al. (1960) and Paterson (1964) avoid interaction between the axial displacement of the end-pieces and the confining pressure completely, by using two pistons connected by a yoke, only one of which applies load to the specimen. This arrangement allows the volume of confining fluid to remain constant as the axial load is increased. It also allows the confining fluid pressure to act in the axial direction, not only in the lateral directions, so that the loading piston needs only

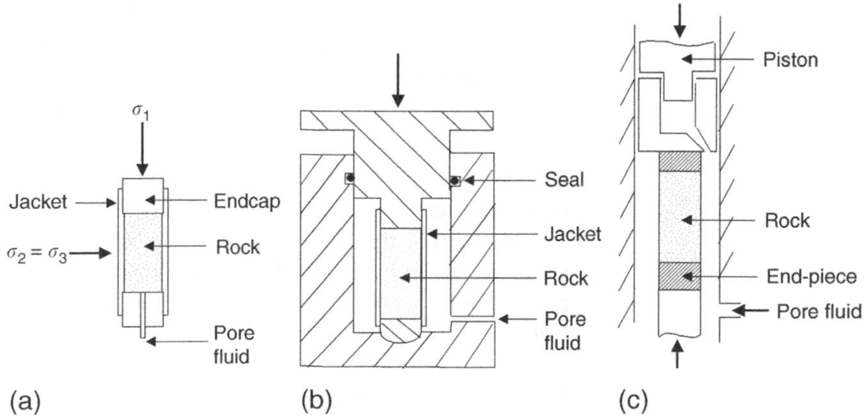

Fig. 6.3 Triaxial testing apparatus: (a) jacketed cylindrical rock specimen with end-pieces and provision for pore fluid, (b) US Bureau of Reclamation cell, (c) central portion of a constant-volume triaxial cell.

to supply enough force to create the differential load and to overcome friction, but does not need to overcome the confining pressure (Fig. 6.3c).

The variables that must be measured in a triaxial test include the confining stress, the axial stress (or the differential stress), the axial strain, and the lateral strain. The confining stress acting on the rock is easily measured by measuring the pressure of the confining fluid with a pressure gauge or an electronic pressure transducer, which may for convenience be placed outside the cell. The axial stress can be calculated from the pressure of the oil in the loading jack, after correcting for the area ratio, although such a calculation ignores friction along the sides of the piston. Alternatively, the axial load can be measured by placing in series with the rock specimen a *load cell*, which is essentially a metal element of known elastic modulus to which strain gauges are attached (Davis and Gordon, 1968). The axial (ε_{zz}) and lateral ($\varepsilon_{\theta\theta}$) normal strains of the rock are most accurately measured by strain gauges glued to the outer face of the cylindrical rock core. To avoid end effects, these gauges are usually placed midway between the two end-pieces.

If the axial stress is smaller in magnitude than the two lateral stresses, but nevertheless still compressive, the resulting state of stress, $\sigma_1 = \sigma_2 > \sigma_3 > 0$, is referred to as *triaxial extension* (Heard, 1960). Such tests are useful in testing Mohr's assumption that failure is not influenced by the magnitude of the intermediate principal stress. Triaxial extension tests are readily conducted in a triaxial testing apparatus, provided that the piston can be suitably attached to the end-pieces.

6.5 Stability and stiff testing machines

As discussed in §4.2, many rocks exhibit a postpeak, strain-softening regime in which the tangent modulus, $E_{\tan} = d\sigma/d\varepsilon$, is *negative*. This does not conform to one of the basic assumptions of the theory of elasticity, which is that the elastic modulus should be positive in order for the stored strain energy function to be positive-definite; see §5.8. Although positivity of E is not required by any thermodynamic law, a negative tangent modulus can, under certain situations, give rise to unstable behavior. This has important implications during laboratory compression tests.

To understand the inherent instability of a rock having a negative tangent modulus, consider a cylindrical rock specimen of length L and cross-sectional area A, compressed under a weight, W, as in Fig. 6.4a. The stress–strain behavior of the rock will be represented in the idealized form of Fig. 6.4b, in which a linear elastic regime with modulus E is followed by a strain-softening regime with modulus $E_{\tan} = -|E_{ss}|$, where this notation is used to underscore the fact that the tangent modulus is negative in this regime.

Imagine that the load W is precisely large enough so that the rock is loaded to its elastic limit, denoted by point B in Fig. 6.4b. At this point, the stress in the rock is $\sigma = W/A$, the strain is $\varepsilon = W/AE$, and the stored strain energy is $\mathcal{E} = W^2L/2AE$. Now imagine that the rock somehow compresses by an additional amount Δz, without the introduction of any additional energy into the system, so that it moves to point C on the stress–strain curve. The change in

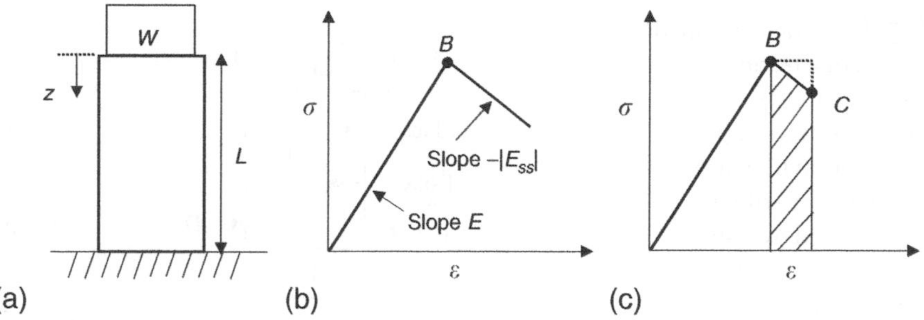

Fig. 6.4 (a) Rock cylinder loaded by a weight, W; (b) idealized stress–strain curve of a rock exhibiting strain-softening behavior; (c) work done on rock by additional compression from B to C.

the total energy of the system must be zero, that is,

$$\Delta\mathcal{E} = \Delta\mathcal{E}_{\text{elastic}} + \Delta\mathcal{E}_{\text{gravitational}} + \Delta\mathcal{E}_{\text{other}} = 0, \tag{6.3}$$

where $\mathcal{E}_{\text{other}}$ represents energy that may be available to cause additional damage to the rock in the form of microcracking, etc. The gravitational potential energy of the load decreases by $W\Delta z$, so $\Delta\mathcal{E}_{\text{grav}} = -WL\Delta\varepsilon$. The strain energy stored in the rock, per unit volume, *increases* by an amount equal to the shaded area in Fig. 6.4c. This area is equal to the area of the rectangle of height σ and width $\Delta\varepsilon$, *minus* the area of the small triangle having base $\Delta\varepsilon$ and height $|E_{ss}|\Delta\varepsilon$. After multiplying by the volume of the rock,

$$\Delta\mathcal{E}_{\text{elastic}} = AL[\sigma\Delta\varepsilon - \frac{1}{2}|E_{ss}|\Delta\varepsilon^2] = WL\,\Delta\varepsilon - \frac{AL}{2}|E_{ss}|\Delta\varepsilon^2. \tag{6.4}$$

From (6.3), the change in the "other" energy that is available to further degrade the rock will be

$$\begin{aligned}\Delta\mathcal{E}_{\text{other}} &= -\Delta\mathcal{E}_{\text{elastic}} - \Delta\mathcal{E}_{\text{grav}} \\ &= -WL\Delta\varepsilon + \frac{AL}{2}|E_{ss}|\Delta\varepsilon^2 - (-WL\Delta\varepsilon) = \frac{AL}{2}|E_{ss}|\Delta\varepsilon^2 > 0. \end{aligned} \tag{6.5}$$

Hence, a small additional compression of the rock will liberate a positive amount of energy, which will be available to cause further microscale degradation of the rock, thereby causing further softening of the tangent modulus, etc. It is clear that this is an unstable process that will inevitably lead to complete disintegration of the specimen.

If the tangent modulus were positive at point B, the quadratic term in (6.4) would be positive, and (6.5) would show the energy available for cracking the rock to be negative, which by definition is not possible. In this case, the additional compression of the rock would *not* spontaneously occur without the addition of external energy to the system (such as by increasing the load, W). Hence, a negative tangent modulus is necessary for this type of instability to occur.

Having established that a negative value of the tangent modulus may lead to instabilities under certain experimental conditions, we now consider a more realistic model for a traditional rock-testing machine, as used in uniaxial compression tests, by considering the effect of the compliance of the machine (Salamon, 1970; Hudson et al., 1972; Hudson and Harrison, 1997, pp. 89–92). Consider a

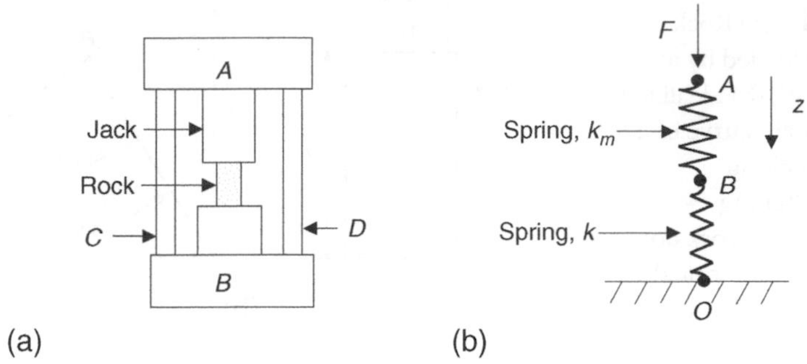

Fig. 6.5 (a) Simplified model of a testing machine used to compress a rock. (b) Idealization in which the machine stiffness is represented by a spring k_m and the rock is represented by a (nonlinear) spring, k.

rock-testing apparatus such as shown in Fig. 6.5a. As the rock specimen is compressed by the hydraulic jack, the jack exerts a downward force F on the rock, and, according to Newton's third law, the rock exerts an upward force F on the jack. As the load F increases, additional elastic energy is of course stored in the rock, but it is also stored in the hydraulic system, the platens, the vertical bars C and D, and other parts of the apparatus. For conceptual simplicity, the entire loading system can be represented by an elastic spring of stiffness k_m, defined such that if the load is F, the energy stored in the loading system is $F^2/2k_m$. The rock specimen can also be thought of as a spring, with stiffness $k = EA/L$. At equilibrium, the two springs are subjected to the same force, F, so they can be assumed to be in series (Fig. 6.5b).

Assume again that the system is in equilibrium, with the rock compressed to point B in Fig. 6.4c. In this state, the compressive force in both springs is F. Now imagine that the rock spontaneously undergoes an additional small compression, moving to point C on its stress–strain curve. With respect to the spring model in Fig. 6.5b, point B is displaced downward by an amount Δz_B. If this occurs without the addition of any energy to the system, this is equivalent to specifying that no displacement can occur at point A in Fig. 6.5b. By definition, no displacement occurs at point O.

Following the same argument as was used to derive (6.4), but replacing the load W with F, and noting that $\Delta\varepsilon = \Delta z/L$, the change in the amount of strain energy stored in the rock specimen is

$$\Delta\mathcal{E}_{\text{specimen}} = F\Delta z_B - \frac{1}{2}\frac{A|E_{ss}|}{L}(\Delta z_B)^2. \tag{6.6}$$

For a small displacement Δz_B, this term is clearly positive, reflecting the fact that the rock continues to absorb energy, even as it deforms into its strain-softening regime. A similar argument for the elastic energy stored in the testing machine gives

$$\Delta\mathcal{E}_{\text{machine}} = -F\Delta z_B + \tfrac{1}{2}k_m(\Delta z_B)^2. \tag{6.7}$$

The energy stored in the testing machine decreases, as the machine is undergoing *unloading* in this regime. This can be seen from Fig. 6.5b, where a downward displacement of point B will *decrease* the amount of compression in the spring representing the testing machine. Hence, the amount of additional energy that is liberated, and is therefore available to cause further degradation of the rock, is

$$\Delta\mathcal{E}_{\text{other}} = -\Delta\mathcal{E}_{\text{specimen}} - \Delta\mathcal{E}_{\text{machine}} = \frac{1}{2}\left[\frac{A|E_{ss}|}{L} - k_m\right](\Delta z_B)^2. \qquad (6.8)$$

If this term is positive, the system will be unstable, as this energy will cause further microcracking of the rock, etc. Hence, the condition for *stability* is that the stiffness of the testing machine, k_m, be *greater* than $A|E_{ss}|/L$, where A is the cross-sectional area of the specimen, L is its length, and E_{ss} is the slope of the stress–strain curve of the rock in the strain-softening regime. A testing machine is categorized as being *stiff* or *soft*, with respect to a given rock specimen, depending on whether or not it satisfies this criterion.

The condition for the onset of instability of this system can also be derived by the following simple argument (Salamon, 1970). At equilibrium, the compressive forces in the two springs in Fig. 6.5b are equal, so $k_m(z_A - z_B) = k(z_B - z_O) = kz_B$, where the displacements of each of the three points (A, B, O) are measured starting from their values at $F = 0$. Let the force F be increased slightly by ΔF. If we require that the system move to a new equilibrium state, then the displacements must satisfy the constraint of force equilibrium, and so $k_m(\Delta z_A - \Delta z_B) = k\Delta z_B$. In general, $k = E_{\tan}A/L$, and so, since $\Delta z_B = L\Delta\varepsilon$, this relation can be solved to give

$$\Delta\varepsilon = \frac{k_m}{Lk_m + AE_{\tan}}\Delta z_A. \qquad (6.9)$$

In the elastic regime of the rock's behavior, $E_{\tan}$ will be positive, and (6.9) can be solved to uniquely determine the additional incremental strain in the rock. However, if the rock softens sufficiently that $E_{\tan} = -Lk_m/A$, there will be no finite solution to (6.9), implying that the rock cannot deform to a new equilibrium state – it will fail catastrophically. As was the case for the derivation based on energy considerations, the condition for stability is $k_m > A|E_{\tan}|/L$.

In reality, the transition from a positive tangent modulus to a negative tangent modulus occurs gradually, not abruptly as in Fig. 6.4b. If a rock is compressed in a "soft" machine, unstable disintegration of the rock will commence when the slope of the stress–strain curve first becomes sufficiently negative that $|E_{\tan}|$ equals Lk_m/A. This will typically occur at a point very near the peak of the stress–strain curve; the rock will fail abruptly and explosively, and it will not be possible to observe and measure the strain-softening portion of the stress–strain curve. An understanding of the role of machine compliance in obscuring the softening portion of the stress–strain curve was first developed by Whitney (1943) and others in the context of concrete testing, but was not fully appreciated in the field of rock mechanics until the 1960s (Hudson et al., 1972).

There are many sources of elastic compliance in a testing machine, such as the hydraulic system, the vertical columns (C and D in Fig. 6.5a), the crossheads (A and B), etc. As each of these are subject to the same load, the compliances are additive. The individual stiffnesses k_i are therefore combined by adding their reciprocals, so that $k_m = [\Sigma(1/k_i)]^{-1}$. One approach to solve this instability problem is to minimize the individual sources of elastic compliance in the system.

The largest contributions to the compliance typically come from the hydraulic system and the columns (Cook and Hojem, 1966). Cook and Hojem constructed a machine with a stiff frame and minimized the compliance of the hydraulic system by using for the hydraulic system a short column of mercury with a large cross-sectional area. This apparatus was used by Crouch (1970,1972) to study the compressional behavior of quartzite and norite.

A quite different approach to stiffening a testing machine is to add a stiff element in *parallel* with the rock specimen, so that the stiffener and the specimen undergo the same displacement. In this case the overall stiffness will be the sum of that of the machine and that of the stiffener. Cook (1965) stiffened a conventional testing machine by loading a steel ring in parallel with the rock specimen and was able to significantly reduce the explosive nature of the failure of a specimen of Tennessee marble. Bieniawski et al. (1969) used a similar apparatus to study the compression of sandstone and norite.

A third approach is to use the thermal expansion/contraction of the columns in the testing machine to supply the force needed to compress the rock. Cook and Hojem (1966) constructed a testing machine in which a hydraulic jack was used to prestress the specimen and the remaining displacement was induced by thermal contraction of the vertical columns.

Each of the proposed solutions to the machine stiffness problem has serious drawbacks, however. There are practical limits to the extent to which one can eliminate sources of compliance within a testing machine. Adding a stiffening element in parallel has the unwanted effect of decreasing the effective load capacity of the system, as much of this capacity will be used to compress the stiffening member. Finally, it is very difficult to control the rate at which the load is applied when thermal contraction of the columns is used to compress the rock.

The unstable collapse and disintegration of the specimen are caused by the rapid flow into the specimen of some of the energy that had been stored in the machine. Much of this energy is stored in the hydraulic system. If, for example, fluid could be drawn out of the hydraulic system rapidly and in a controlled manner, this problem could be avoided. This can indeed be achieved with servo-controlled testing machines (Bernhard, 1940; Rummel and Fairhurst, 1970). A main idea behind the performance of these machines is that, to trace out the full stress–strain curve beyond the point of peak strength, the strain in the rock specimen must be the controlled variable. With regards to the loading platens, this implies that the displacement, rather than the load, is the variable that must be controlled. In a servo-controlled testing machine, the deformation of the rock is monitored and then compared to the desired strain. Any difference between the desired and current strain is used to create a "correction signal" that adjusts the hydraulic pressure so as to bring the actual strain closer to the desired value; see Hudson et al. (1972) and Hudson and Harrison (1997) for details. The response time of such systems is in the order of a few milliseconds, which is sufficiently rapid to be able to arrest the unstable disintegration of the rock (Rummel and Fairhurst, 1970). Thus, the full stress–strain curve can be obtained, provided that the strain increases monotonically.

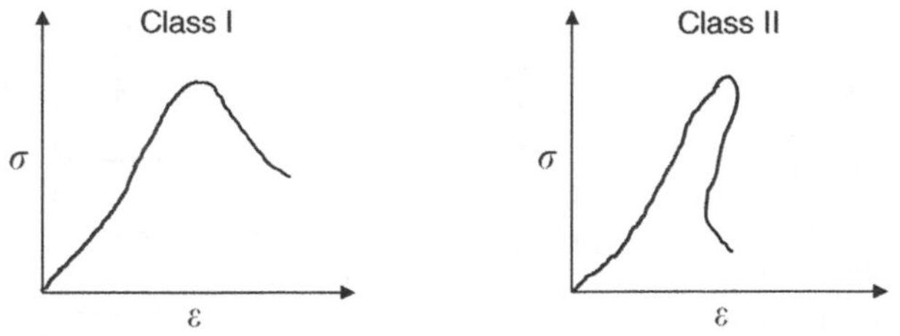

Fig. 6.6 Class I and Class II stress–strain curves.

Some rocks, however, exhibit a complete stress–strain curve in which neither the stress nor the strain increases monotonically. For such rocks, denoted by Wawersik and Fairhurst (1970) as "Class II," the stress and strain each decreases as the rock begins to fail (Fig. 6.6). This occurs as the rock continues to deteriorate on a microscopic scale and is fundamentally different from elastic unloading with hysteresis. Okubo and Nishimatsu (1985) showed that by using a linear combination of stress and strain as the feedback signal in a servo-controlled testing machine, the complete stress–strain curves of both Class I and Class II rocks can be obtained.

6.6 True-triaxial tests

Traditional "triaxial" compression tests, such as described in §6.4, involve states of stress in which $\sigma_1 > \sigma_2 = \sigma_3 > 0$. Such tests are incapable of probing the effects of the intermediate principal stress. In order to investigate rock behavior over the full range of stresses that may occur in the subsurface, it would be desirable to conduct tests in which all three principal stresses may have different (compressive) values. Such tests have sometimes been referred to as "polyaxial," although this name has the disadvantage of not being self-explanatory. More recently, the term "true-triaxial," which is inelegant but less open to misinterpretation, has gained acceptance.

Several researchers have constructed testing cells that attempt to produce states of homogeneous stress in which the three principal stresses, $\sigma_1 \geq \sigma_2 \geq \sigma_3 \geq 0$, are independently controllable (Fig. 6.7a). Although the designs differ in various ways, in each case a "rectangular" (i.e., parallelepiped-shaped) specimen is used, in contrast to the cylindrical specimens used in traditional triaxial tests. Hojem and Cook (1968) constructed a cell in which the two lateral stresses σ_2 and σ_3 were applied to the specimen by two pairs of thin copper flat jacks and the axial load was applied by a traditional loading piston. However, it was difficult to apply high lateral stresses with this apparatus, thus limiting its range of usefulness.

Mogi (1971) built an apparatus in which the minimum stress, σ_3, was applied by a pressurized fluid, and the two other stresses were applied by opposing sets of flat jacks (Fig. 6.7b,c). The choice of having the minimum stress applied by fluid pressure was made so that this stress could be measured with the greatest accuracy. The specimen was in the form of a rectangular prism, 1.5 cm × 1.5 cm in cross section and 3.0 cm long in the σ_1 direction. The steel end-pieces over which σ_1 was applied were connected to the specimen by epoxy, whereas the end-pieces over which σ_2 was applied were coupled to the specimen through thin rubber lubricating sheets. The sides of the specimen were jacketed with

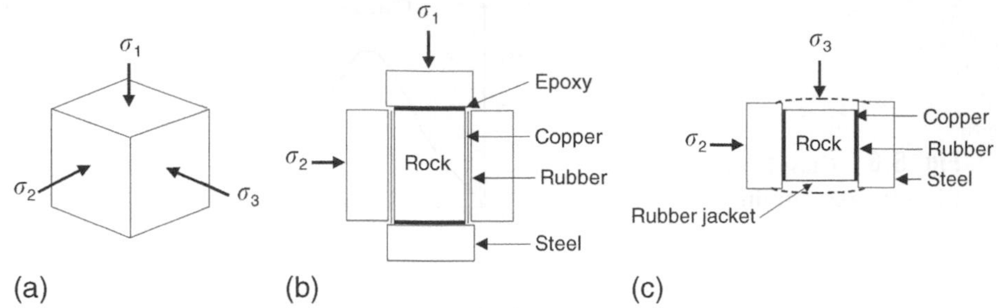

Fig. 6.7 (a) True-triaxial state of stress applied to a cubical specimen; (b) view along the σ_3 direction of the apparatus used by Mogi (1971); (c) view along the σ_1 direction.

thin copper sheets to prevent the rubber from intruding into the rock, and a silicon rubber jacket was used to prevent the pressurizing fluid from entering the pores of the rock. Mogi used this apparatus to investigate the influence of the intermediate principal stress on the yield and fracture of several rock types.

Haimson and Chang (2000) built a compact and portable true-triaxial cell based on Mogi's design. Their apparatus can subject a specimen to values of σ_1 and σ_2 as high as 1600 MPa and σ_3 as high as 400 MPa. Normal strains in the direction of maximum and intermediate stress were measured with strain gauges glued to the respective faces of the specimen, whereas the third strain was measured with a beryllium-copper beam fitted with a strain gauge. The center of the beam makes contact with a pin affixed to the face of the specimen, and as the rock expands in the σ_3 direction, the beam bends outward, and its deflection is measured by the strain gauge. This apparatus was used to investigate the influence of the intermediate stress on the failure of Westerly granite, which was found to be significant. No such effect of σ_2 was found for a hornfels and a metapelite from the Long Valley caldera in California (Chang and Haimson, 2005).

Hunsche and Albrecht (1990) describe an apparatus that uses three pairs of double-acting pistons to apply three independently variable normal stresses to the faces of a cubical specimen. Heaters placed between the specimen and the pistons allowed the rock to be heated to 400°C. The forces applied by each pair of pistons were calculated from pressure gauges in the hydraulic lines. Deformation of the specimen in the three directions was measured with linear variable displacement transducers (LVDTs), which essentially measure the change in the distance between the opposing platens. Paraffin wax (at room temperature) and graphite (at elevated temperatures) were used as lubricants between the platens and rock. This apparatus was used to study the deformation of rock salt, and it was found that the observed strength of the rock, defined as the maximum value of the octahedral shear stress, depended sensitively on the ratio of specimen size to platen size. Sayers et al. (1990) describe a similar apparatus that is fitted with ultrasonic transducers in each end-piece, so as to be able to measure shear and compressional wavespeeds under states of true-triaxial stress.

6.7 Diametral compression of cylinders

The difficulties associated with performing a direct uniaxial tension test on rock have led to the development of a number of "indirect" methods for assessing the tensile strength. Such methods are called indirect because they do not involve

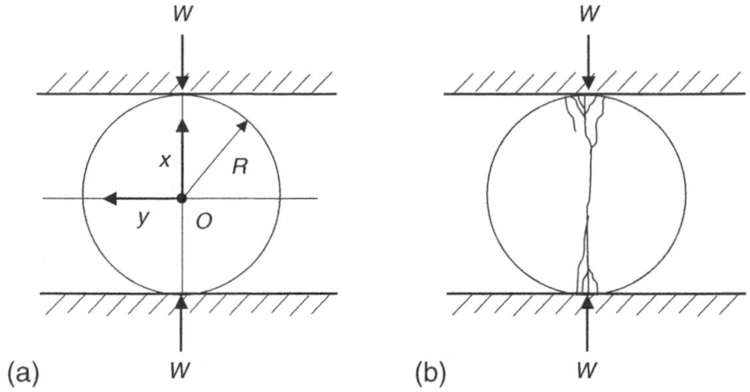

Fig. 6.8 (a) Cylinder compressed between parallel surfaces by a line load W (per unit length into page); (b) typical fracture pattern resulting from this loading.

the creation of a homogeneous state of tensile stress in rock, but rather involve experimental configurations that lead to inhomogeneous stresses that are tensile in some regions of the specimen. The precise value of the tensile stress at the location where failure initiates must be found by solving the equations of elasticity.

The most popular of these tests is the so-called *Brazilian test*, developed by the Brazilian engineer Fernando Carneiro in 1943 for use in testing concrete. A thin circular disk of rock is compressed between two parallel platens, so that the load is directed along the diameter of the disk (Fig. 6.8a). As the platens are relatively rigid compared to the rock, they can be assumed to apply a point load W (per axial length of the cylinder) to the two opposing loading points. With the coordinate system taken as in Fig. 6.8a, the stresses along this vertical diameter are, by (8.165),

$$\tau_{xx} = \frac{-W}{\pi R}, \quad \tau_{yy} = \frac{W(3R^2 + x^2)}{\pi R(R^2 - x^2)}, \tag{6.10}$$

whereas along the y-axis, perpendicular to the load, the stresses are

$$\tau_{xx} = \frac{-W(R^2 - y^2)^2}{\pi R(R^2 + y^2)^2}, \quad \tau_{yy} = \frac{W(R^2 - y^2)(3R^2 + y^2)}{\pi R(R^2 + y^2)^2}. \tag{6.11}$$

By symmetry, these stresses are principal stresses. The largest and smallest principal stresses occur on the vertical axis, through which the load passes. The minimum principal stress, τ_{xx} in (6.10), is uniform and *tensile* along this entire axis. The maximum principal stress, τ_{yy} in (6.10), is compressive and becomes unbounded near the platens, but varies only weakly near the center of the disk. At the center of the disk, the two principal stresses are, by setting $x = 0$ in (6.10),

$$\tau_{xx} = \frac{-W}{\pi R}, \quad \tau_{yy} = \frac{3W}{\pi R}. \tag{6.12}$$

As the disk is in a state of plane stress, the third principal stress, normal to the plane of the disk, is zero, and consequently is the intermediate principal stress.

When a cylindrical rock specimen is compressed in this way, failure typically occurs by an extension fracture in, or close to, the loaded diametral plane, at some value of the applied load W, as in Fig. 6.8b. It is generally assumed that the failure is the result of the tensile stress $\tau_{xx} = \tau_{\theta\theta} = -W/\pi R$, and so the tensile strength is given by the value of $W/\pi R$ at failure. Tensile strengths measured in this way are very reproducible and are in reasonable agreement with values obtained in uniaxial tension. The Brazilian indirect tension test has been used to determine the tensile strength of coal by Berenbaum and Brodie (1959), and of various sandstones and siltstones by Hobbs (1964).

If the applied load is actually a uniform normal stress of magnitude σ, distributed over a small arc of angle 2α, then the state of stress near the points of contact will be a uniform compression of magnitude σ. This will decrease the likelihood of failure by shear fracture at the contact points, but has virtually no effect on the stresses near the center of the disk. Hence, Brazilian tests conducted with loads distributed over a narrow arc, such as 15°, yield values of the tensile strength that are little different from those obtained using line loads and give rise to similar diametral extension fractures.

If jacketed cylinders are subjected to confining pressure p applied by a pressurized fluid, as well as to diametral compression, then at the center of the disk a hydrostatic stress p would be added to the three principal stresses discussed above, leading to

$$\sigma_1 = (3W/\pi R) + p, \quad \sigma_2 = p, \quad \sigma_3 = p - (W/\pi R). \tag{6.13}$$

This configuration gives a means of studying failure in situations where all three stresses are compressive, but σ_3 is small, as is often the case near an underground excavation. The three principal stresses will be connected by the relation

$$\sigma_1 - 4\sigma_2 + 3\sigma_3 = 0, \tag{6.14}$$

so this test will determine a curve defined by the intersection of the failure surface with the surface defined by (6.14). Jaeger and Hoskins (1966a) found that the values of σ_1 and σ_3 obtained from these tests, using (6.13), agreed reasonably well with those obtained in standard triaxial compression tests, although the values of σ_1 tended to be consistently higher than those measured in the triaxial tests for the same value of σ_3. They attributed this to the strengthening effect of the intermediate principal stress, as discussed in §4.8.

The analysis presented above assumes that the rock is isotropic, which may not be the case. Chen et al. (1998) developed a mathematical solution for the diametral compression of a thin disk of rock that is transversely isotropic in the plane of the disk. In this case, the analysis of the results is complicated by the fact that the two principal stresses at the center of the disk depend, in a complicated and implicit manner, on the values of the elastic moduli. Claesson and Bohloli (2002) analyzed this solution further and derived accurate approximate expressions for these stresses. Lavrov and Vervoort (2002) presented a solution that accounts for the influence of transverse tractions applied at the rock–platen interface, caused by friction, and showed that such tractions would have little effect on the stresses at the center of the disk and hence little effect on the interpretation of tests in which failure initiated at or near the center.

6.8 Torsion of circular cylinders

As discussed in §6.3, Young's modulus, E, can be measured by subjecting the rock to a state of homogeneous uniaxial compression. The shear modulus, G, could in principle be measured by inducing a state of homogeneous *shear* stress in a rock specimen. The shear modulus would be found from the ratio of the shear stress to the shear strain. However, it is not easy to induce a homogeneous state of shear in a piece of rock. But an inhomogeneous state of shear can be induced in a circular cylinder by subjecting it to torsion. Analytical solution of the elasticity equations for this configuration yields a simple relationship between the applied torque, the angle of twist, and the shear modulus. Hence, measurement of the applied torque and the resulting angle of twist will permit G to be calculated.

In a torsion experiment, loads are applied to the two ends of a cylindrical specimen of radius a and length L, so as to create a torque M about the longitudinal axis (Fig. 6.9a). It is convenient to imagine that the $z = 0$ face is fixed and that the $z = L$ face rotates within its plane by an angle α. A reasonable assumption for the displacement field within the cylinder is that each plane normal to the axis of the cylinder also rotates, by an angle that increases linearly from 0 at $z = 0$ to α at $z = L$. In cylindrical coordinates, this displacement field is

$$u = 0, \quad v = \alpha rz/L, \quad w = 0. \tag{6.15}$$

From (2.275) and (2.280), the stresses and strains associated with these displacements are

$$\varepsilon_{z\theta} = \varepsilon_{\theta z} = \alpha r/2L, \quad \tau_{z\theta} = \tau_{\theta z} = G\alpha r/L, \tag{6.16}$$

with all other stress and strain components vanishing. It is easy to verify that these stresses satisfy the equations of stress equilibrium, (5.107)–(5.109), and also give zero tractions along the outer surface of the cylinder, $r = a$. The total moment applied to the end of the cylinder, about the z-axis, is found from

$$M = \int_0^a \int_0^{2\pi} \tau_{z\theta} r^2 \mathrm{d}r \mathrm{d}\theta = \frac{2\pi G\alpha}{L} \int_0^a r^3 \mathrm{d}r = \frac{\pi G a^4 \alpha}{2L}. \tag{6.17}$$

Hence, measurement of M and α will allow G to be found.

Elimination of G between (6.16) and (6.17) yields $\tau_{z\theta} = 2Mr/\pi a^4$, which shows that the shear stresses within the cylinder vary from 0 at the center of

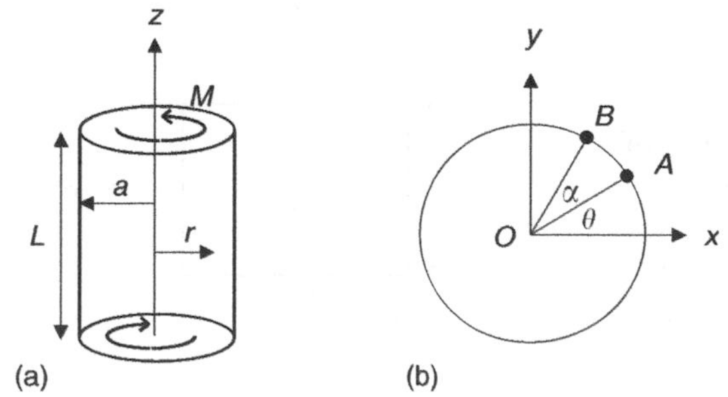

Fig. 6.9 (a) Torsion of a circular cylinder by moments applied over the opposing faces; (b) view of the $z = L$ face, showing rotation of point A to point B by angle α.

the cylinder, to a maximum of $\tau_{z\theta} = 2M/\pi a^3$ at the outer surface. At the outer surface, the principal stresses are, from (2.37)–(2.38), seen to be $\pm 2M/\pi a^3$.

More complex, but predictable, stress fields can be obtained using hollow cylinders. Talesnick and Ringel (1999) developed an apparatus that can apply torsion to a hollow cylinder, superposed on a traditional triaxial stress state, and used it to determine the five independent elastic moduli of several transversely isotropic rocks: Loveland sandstone, Indiana limestone, Lac du Bonnet granite, and Marasha chalk. Paterson and Olgaard (2000) developed an apparatus that is capable of combining traditional triaxial stresses with large-angle torsion and used it to study the rheological properties of Carrara marble under large shear strains.

6.9 Bending tests

Bending is used in rock testing, for measurement of E and for tensile strength (Pomeroy and Morgans, 1956; Berenbaum and Brodie, 1959; Evans, 1961; Coviello et al., 2005). It is also a very sensitive method for studying creep and transient behavior (Phillips, 1931; Price, 1964). This type of loading produces regions of tensile stress and compressive stress in the rock. The stress and displacement distributions can be found from elementary beam theory, as outlined below.

Consider first a rectangular beam of width b, height h, and length L, as in Fig. 6.10. A moment of magnitude M is applied to the beam about the x-axis. According to the classical Euler–Bernoulli theory, each planar section in the x–y plane remains planar, but rotates about the x-axis, as shown in Fig. 6.10b. Lines of constant-y in the y–z plane, which were initially horizontal, now form circular arcs with C at their center. The upper fibers of the beam, $y > 0$, are in compression, and the lower fibers, $y < 0$, are in tension. The so-called *neutral axis*, $y = 0$, is neither in tension nor compression, so the deformed length of OO' is L and the radius of curvature of the neutral axis is $R = L/\theta$. The deformed length of the upper face of the beam, BB', is $(R - h)\theta$, and its original length was $L = R\theta$, so the longitudinal compressive strain in the upper fibers is $\varepsilon_{zz} = h/R$. Similarly, the strain of the lowermost fibers is $\varepsilon_{zz} = -h/R$. The same analysis for an arbitrary value of y shows that, in general, $\varepsilon_{zz} = y/R$, and so the longitudinal stress is $\tau_{zz} = yE/R$.

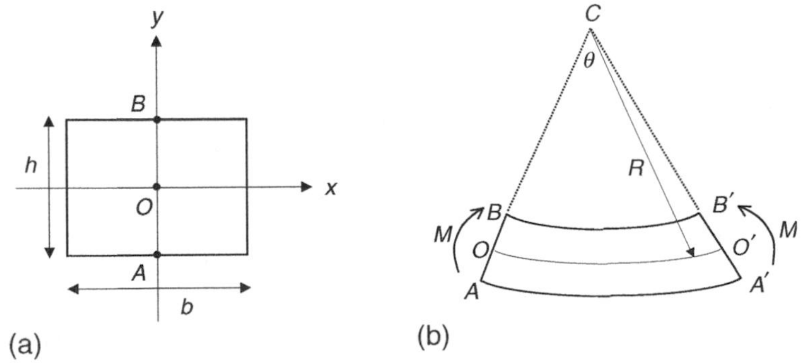

Fig. 6.10 Bending of a prismatic beam by moments applied at its ends: (a) cross section normal to the z-axis; (b) side view, normal to the x-axis.

The total moment about the x-axis is given by

$$M = \int_{-h/2}^{h/2}\int_{-b/2}^{b/2} \tau_{zz} y \mathrm{d}x \mathrm{d}y = \frac{bE}{R}\int_{-h/2}^{h/2} y^2 \mathrm{d}y = \frac{bh^3 E}{12R} \equiv \frac{EI}{R}, \tag{6.18}$$

where $I = bh^3/12$ is the moment of inertia of the cross section, about the x-axis. Measurement of the applied moment and the radius of curvature of the deformed beam therefore provides a value for E. The greatest tensile stress, which occurs at the lower face of the beam, is equal to

$$\tau_{zz}(\text{tensile max}) = -hE/2R = -Mh/2I. \tag{6.19}$$

If the moment is increased until failure occurs, this relation can be used to give the tensile strength.

In practice, the applied loading is somewhat different from the case of pure bending by end-couples. To treat the loading configurations actually used in the laboratory, the following generalization of (6.18) is needed. For small deformations, the radius of curvature can be approximated as $1/R = \mathrm{d}^2y'/\mathrm{d}z^2$, where y' is the deformed position of the neutral axis. The initial position of this axis was $y = 0$, so $y' = v$, where v is the y-component of the displacement of the neutral axis, in which case (6.18) can be written as

$$M = EI\frac{\mathrm{d}^2 v}{\mathrm{d}z^2}. \tag{6.20}$$

This form of the equation can be used for cases in which the moment M varies along the z-axis. The following two cases are of importance in rock mechanics testing.

6.9.1 Three-point loading

Consider a beam that is simply supported at its two ends and loaded by a point load F at its center (Fig. 6.11a). It will be convenient to place the origin at the midpoint of the beam and denote the total length by $2L$. By symmetry, the reaction forces at the two ends will each have magnitude $F/2$. By performing a moment balance on a segment of the beam located between some generic point $0 < z < L$ and the right edge of the beam, it follows that the internal moment acting along the face of the beam, normal to the z-axis, must be

$$M(z) = \tfrac{1}{2}F(L - z). \tag{6.21}$$

The maximum tensile stress occurs again at the lower face of the beam, where it is given by (6.19). According to (6.21), this will occur at the midpoint of the beam, where $z = 0$ and $M_{\max} = FL/2$. Hence, the greatest tensile stress will be

$$\tau_{zz} = -\frac{hFL}{4I}. \tag{6.22}$$

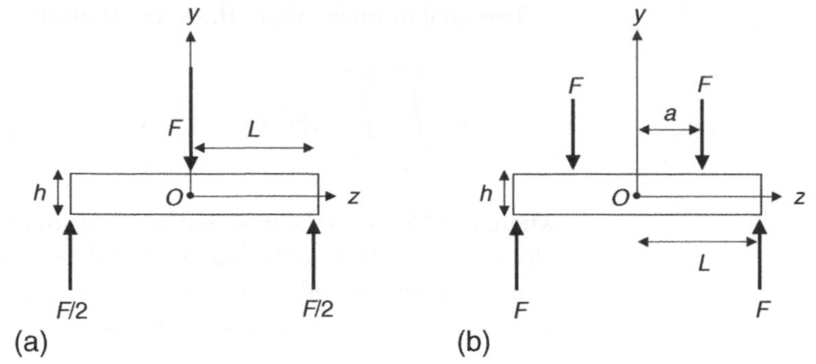

Fig. 6.11
(a) Three-point loading and (b) four-point loading of a beam.

The differential equation (6.20) for the deflection of the beam becomes

$$EI\frac{d^2v}{dz^2} = \frac{1}{2}F(L - z). \tag{6.23}$$

Integration of (6.23), using the boundary conditions $v = 0$ and (from symmetry) $dv/dz = 0$ when $z = 0$, gives

$$v = \frac{F}{12EI}(3Lz^2 - z^3). \tag{6.24}$$

The displacement at $z = L$ represents the deflection of the midpoint of the beam in the direction of the applied force F:

$$v_{max} = \frac{FL^3}{6EI}. \tag{6.25}$$

This relation provides a means to estimate E from the deflection of the beam.

6.9.2 *Four-point loading*

An objection to the use of the three-point loading configuration to estimate tensile strength arises from the fact that the maximum stress (6.22) occurs immediately beneath the point of application of one of the loads, and it is not reasonable to expect elementary beam theory to be very accurate at such locations. This problem can be avoided by using four-point loading (Fig. 6.11b), in which two loads of magnitude F are applied at $z = \pm a$, for some value $0 < a < L$. By symmetry, the reaction forces at the two ends will each have magnitude F. In this case, taking a moment balance for a segment of the beam to the right of the midpoint yields

$$M = F(L - a) \quad \text{for } 0 < z < a, \tag{6.26}$$

$$M = F(L - z) \quad \text{for } a < z < L, \tag{6.27}$$

with similar expressions for the region $z < 0$. Hence, the moment is uniform and equal to $F(L - a)$ throughout the entire region $-a < z < a$. So, the magnitude of the maximum tensile stress is

$$|\tau_{zz}|_{max} = \frac{Mh}{2I} = \frac{Fh(L - a)}{2I}. \tag{6.28}$$

In particular, this stress will occur at the midpoint of the beam, $z = 0$, which is not located immediately under any of the concentrated loads.

The maximum deflection of beam can be calculated in this case to be

$$v_{\max} = \frac{F}{6EI}(2L^3 - 3a^2L + a^3). \tag{6.29}$$

The theory described above assumes linear elastic behavior of the rock. Exadaktylos et al. (2001a,b) gave an analysis of bending that accounted for nonlinearity in the stress–strain behavior and also for the possibility that the elastic modulus E may be different in tension than in compression. The model was applied to three-point bending tests conducted on a Dionysos marble, for which $E_c = 0.8E_t$. They found that tensile failure occurred at the lower edge of the beam, at a (local) stress that was consistent with the tensile strength measured under direct uniaxial tension.

6.10 Hollow cylinders

A hollow cylinder subjected to an axial load and an external or internal fluid pressure along its curved surfaces provides a ready method for studying the strength and fracture of rock under a variety of principal stresses. Among the earliest tests on hollow cylinders of rock were those of Adams (1912), who observed failure by spalling at the inner surface of the cylinder. His results, along with some of their geological implications, were discussed by King (1912). Robertson (1955) used cylinders of rock with different ratios of their inner and outer diameter, stressed by fluid pressure applied to their ends and outer surfaces. He discussed his results, in which failure started at the inner surface, in terms of elastic–plastic theory. Hollow cylinders subjected to axial load and external fluid pressure have been used since then on a variety of rock types (Hobbs, 1962; Obert and Stephenson, 1965; Santarelli and Brown, 1989; Lee et al., 1999).

The solutions for the stresses in a pressurized hollow cylinder are given in §8.4. Consider first the case of a hollow cylinder of inner radius, a, and outer radius, b, subjected to a compressive axial stress, σ, and an external fluid pressure, p_o. At the inner surface, the three principal stresses will be

$$\tau_{zz} = \sigma, \quad \tau_{\theta\theta} = 2p_o/(1 - \rho^2), \quad \tau_{rr} = 0, \tag{6.30}$$

and at the outer surface, the principal stresses will be

$$\tau_{zz} = \sigma, \quad \tau_{\theta\theta} = p_o(1 + \rho^2)/(1 - \rho^2), \quad \tau_{rr} = p_o, \tag{6.31}$$

where $\rho = a/b$. The axial stress is the same at both surfaces, and as $\rho < 1$ by definition, it follows that the maximum and minimum principal stresses will always occur at the inner surface, where the minimum principal stress is zero. Depending on the numerical values of σ and p_o, the maximum principal stress may be either τ_{zz} or $\tau_{\theta\theta}$. According to the common failure theories discussed in Chapters 4 and 10, this specimen would be expected to fail at its inner surface, at a value of the maximum principal stress that differs from the uniaxial strength of the rock by the strengthening influence, if any, of the intermediate principal stress.

If $\sigma > 2p_o/(1-\rho^2)$, then the principal stresses at the inner surface of the cylinder are

$$\sigma_1 = \sigma, \quad \sigma_2 = 2p_o/(1-\rho^2), \quad \sigma_3 = 0. \tag{6.32}$$

For relatively small values of the outer confining pressure p_o and the inner radius a, failure will occur much as it does for a solid cylinder under triaxial compression, forming a single shear fracture across the entire cylinder at some small angle to the longitudinal axis (Fig. 6.12a). For larger values of p_o and a, failure will occur in the form of a conical fracture whose axis lies along that of the cylinder (Fig. 6.12b). The conical fracture surface will be tangential to the direction of the intermediate principal stress (i.e., the θ direction).

If $\sigma < 2p_o/(1-\rho^2)$, then the principal stresses at the inner surface of the cylinder are

$$\sigma_1 = 2p_o/(1-\rho^2), \quad \sigma_2 = \sigma, \quad \sigma_3 = 0. \tag{6.33}$$

In this case, failure occurs by spiral fractures that are parallel to the axis of the cylinder and consequently parallel to the direction of the intermediate principal stress (Fig. 6.12c).

Consider now a hollow cylinder subjected to an internal pressure p_i along its inner surface and an axial stress σ. If $\sigma < p_i$, the principal stresses at the inner surface are

$$\sigma_1 = \tau_{rr} = p_i, \quad \sigma_2 = \tau_{zz} = \sigma, \quad \sigma_3 = \tau_{\theta\theta} = -p_i(1+\rho^2)/(1-\rho^2), \tag{6.34}$$

and failure usually occurs as a planar, diametral extension fracture. If $\sigma > p_i$, the principal stresses at the inner surface are

$$\sigma_1 = \tau_{zz} = \sigma, \quad \sigma_2 = \tau_{rr} = p_i, \quad \sigma_3 = \tau_{\theta\theta} = -p_i(1+\rho^2)/(1-\rho^2). \tag{6.35}$$

In this case the intermediate principal stress is radial and helicoidal fractures are observed.

In the general case, both internal and external pressures can be applied to the cylinder, along with an axial stress. By using various combinations of σ, p_i,

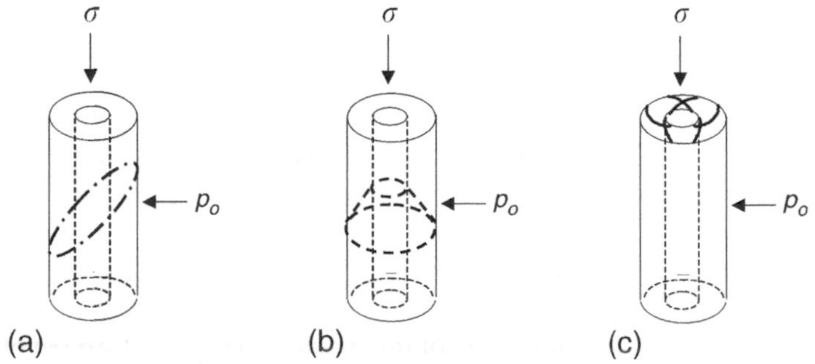

Fig. 6.12 Different systems of fracture in a hollow cylinder subjected to axial stress and external pressure (see text for details).

and p_o, large regions of the failure surface $\sigma_1 = f(\sigma_2, \sigma_3)$ can be probed. Alsayed (2002) modified a traditional Hoek triaxial cell (Hoek and Franklin, 1968) so as to accept hollow cylinders and used it to study the behavior of Springwell sandstone under a variety of stress conditions. Hollow cylinder tests such as those described above are of particular value in the analysis of borehole stability problems (Ewy et al., 2001).

7 Poroelasticity and thermoelasticity

7.1 Introduction

Subsurface rocks are, by their nature, filled with cracks and pores that are saturated with one or more fluid phases (water, air, oil, etc.). These pore fluids will have a major influence on the mechanical behavior of a rock mass. In §4.7, it was seen that if a rock were under compression, pore fluid pressure would cause the state of stress to move closer to the failure surface. Aside from this influence, pore fluid pressures also give rise to macroscopic elastic deformation of the rock. The mechanical deformation of a rock is therefore coupled to the pore fluid pressure. Pore fluids flow through the rock in response to gradients in the pore pressure, but can also flow due to changes in the macroscopic stresses due to natural causes such as tectonic forces, and man-made causes such as the drilling of boreholes, etc. Hence, the mechanical and hydrological behavior of rocks is fully coupled.

Most analyses of rock mechanics problems, and subsurface flow problems, ignore this coupling. In particular, the majority of work on subsurface flow problems in hydrology, petroleum engineering, or geophysics is conducted under the assumption that the rock mass is porous but completely rigid. Similarly, most rock mechanics analyses either ignore pore fluid effects or assume that the pore pressures can be found independently of the mechanical deformation. Although such assumptions are often acceptable, there are many situations in which the coupling between deformation and pore fluid pressure and fluid flow must be accounted for. For example, pore pressure effects play an important role in the deformation around a borehole (Detournay and Cheng, 1988), hydraulic fracturing of boreholes (Detournay et al., 1989), and slip along active faults (Rudnicki and Hsu, 1988).

The general theory that accounts for this coupled hydromechanical behavior is *poroelasticity*. This theory was put forth by Biot (1941), and developed further by, among others, Verruijt (1969), Rice and Cleary (1976), and Detournay and Cheng (1993). A theory of poroelasticity for hydrostatic loading is presented in §7.2 and §7.3. Although this theory is restricted in that it does not account for deviatoric loading, the hydrostatic case can be developed in a fully nonlinear form. Although several nonlinear theories of nonhydrostatic poroelasticity have been developed, most poroelastic analyses utilize the linearized "Biot" theory. This theory, which accounts for deviatoric stresses as well as hydrostatic stresses and pore fluid pressures, is presented in §7.4 and §7.5. Some applications of the theory of poroelasticity are discussed in §7.6 and §7.7.

Rock mass deformation is also influenced by thermal effects. In fact, the theory of thermoelasticity can be developed along lines that are entirely analogous to poroelasticity, with the temperature playing the role of the pore pressure, etc. (Norris, 1992). One difference between the two theories stems from the fact that in most situations thermomechanical coupling is unilateral, in the sense that the temperature field has an effect on the mechanical deformation, but the stresses and strains have a negligible effect on the temperature (Boley and Weiner, 1960). On the other hand, the coupling in poroelasticity between mechanical deformation and pore pressures generally cannot be ignored (Fahrenthold and Cheatham, 1986; Zimmerman, 2000). The theory of thermoelasticity is discussed in §7.8.

7.2 Hydrostatic poroelasticity

Theories for the poroelastic behavior of rocks can be developed on the basis of two conceptual models of porous rock: a solid material permeated with an interconnected collection of voids (Geertsma, 1957a; Zimmerman, 1991) or an aggregation of grains in partial contact with each other at various points (Gassmann, 1951a; Brandt, 1955; Digby, 1981). The latter model is more appropriate for soils, whereas the former has proven to be more fruitful for studying rocks. There is no clear demarcation between these two types of geological media (Anagnostopoulos, 1993), and the behavior of poorly consolidated sedimentary rocks is in many ways similar to that of some soils. However, a theory of poroelasticity that applies to most rock-like materials can be constructed by starting with the idealization of a rock as a connected mineral phase permeated with voids. These voids may be interconnected, or may exist as isolated vugs; the latter do not contribute to the fluid flow processes, and for the present purposes can be ignored.

The theory of poroelasticity has been developed in great detail for the special case of hydrostatic loading (Geertsma, 1957a; Zimmerman, 1991). This restricted theory is applicable to many problems in rock mechanics and petroleum engineering, and also provides a simple context in which to introduce concepts that appear in the more general theory. To develop this theory, consider a porous rock, as shown in Fig. 7.1a. The macroscopic "bulk" volume of the rock is V_b, the volume occupied by the pore space is V_p, and the volume occupied by the solid mineral component is V_m, where

$$V_b = V_m + V_p. \tag{7.1}$$

The mineral phase of the rock is often referred to in poroelasticity as the "matrix," although this usage should not be confused with the use of that word by petroleum geologists (Pettijohn et al., 1987, p. 140) to refer to certain specific types of intergranular material in sandstones. The relative amounts of void space and solid component can be quantified either by the porosity, ϕ, which is defined by

$$\phi = V_p/V_b, \tag{7.2}$$

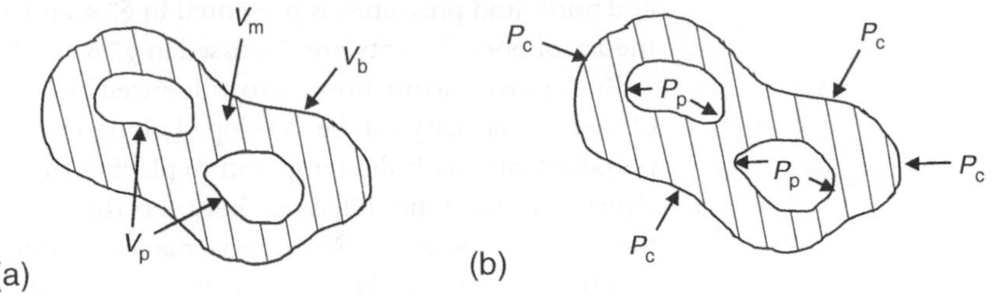

Fig. 7.1 Generic porous rock, showing (a) the bulk volume, pore volume, mineral/matrix volume (shaded), and (b) the pore pressure and confining pressure.

or by the void ratio, e, defined by

$$e = V_p/V_m = \phi/(1-\phi). \tag{7.3}$$

The porosity is restricted, by definition, to the range $0 \leq \phi < 1$, whereas the void ratio can take on any positive value.

Imagine that this piece of porous rock is subjected externally to a purely normal traction of magnitude P_c, where the subscript c denotes "confining" pressure, and the internal pore walls are subjected to a pore pressure of magnitude P_p, exerted by the pore fluid (Fig. 7.1b). As a pore fluid cannot sustain a shear stress under static conditions, no shear traction can be transmitted to the pore walls. There are two independent pressures that may act on the rock and two independent volumes (taken here to be V_b and V_p), so four compressibilities can be defined (Zimmerman et al., 1986):

$$C_{bc} = \frac{-1}{V_b^i}\left(\frac{\partial V_b}{\partial P_c}\right)_{P_p}, \quad C_{bp} = \frac{1}{V_b^i}\left(\frac{\partial V_b}{\partial P_p}\right)_{P_c}; \tag{7.4}$$

$$C_{pc} = \frac{-1}{V_p^i}\left(\frac{\partial V_p}{\partial P_c}\right)_{P_p}, \quad C_{pp} = \frac{1}{V_p^i}\left(\frac{\partial V_p}{\partial P_p}\right)_{P_c}. \tag{7.5}$$

where the superscript "i" denotes the initial, unstressed state. The bulk and pore strain increments can be expressed in terms of the porous rock compressibilities as follows:

$$d\varepsilon_b = \frac{-dV_b}{V_b^i} = C_{bc}dP_c - C_{bp}dP_p, \tag{7.6}$$

$$d\varepsilon_p = \frac{-dV_p}{V_p^i} = C_{pc}dP_c - C_{pp}dP_p, \tag{7.7}$$

where, as usual, a decrease in the volume is considered to be a positive strain. The bulk strain $d\varepsilon_b$ is equivalent to the macroscopic volumetric strain $d\varepsilon_v$ defined by (2.231); the new notation is needed in poroelasticity so as to distinguish between the pore and bulk volumes. Alternative definitions and notations for porous rock compressibilities have been reviewed by Chilingarian and Wolf (1975) and Raghavan and Miller (1975).

These four porous rock compressibilities are typically stress-dependent: decreasing with increasing stress and leveling off to constant values at confining pressures that are on the order of 50 MPa (1 MPa = 145 psi). Figure 7.2a

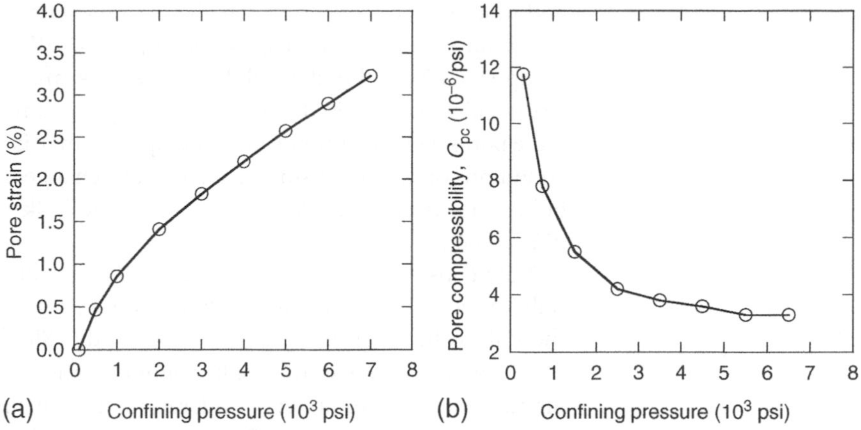

Fig. 7.2 Pore strain (a) and pore compressibility (b) of a Frio sandstone from East Texas, measured at zero pore pressure (after Carpenter and Spencer, 1940).

shows the pore strain vs. confining stress for a Frio sandstone from East Texas (after Carpenter and Spencer, 1940), measured with the pore pressure held constant at 0 MPa. At low pressures, the curve is quite nonlinear, but becomes linear at pressures above 40 MPa. The compressibility C_{pc} is shown in Fig. 7.2b. The behavior of the other porous rock compressibilities would be qualitatively similar. Some rocks, on the other hand, such as limestones (Hart and Wang, 1995), have compressibilities that are nearly independent of stress.

Relationships between the four porous rock compressibilities can be derived under the assumption that the matrix can be treated as if it were composed of an isotropic, homogeneous elastic material. This assumption can often be justified by the fact that although many rocks consist of more than one type of mineral, the elastic moduli of most rock-forming minerals do not differ by large amounts (see Clark, 1966; Simmons and Wang, 1971), so that the "effective" elastic moduli of the matrix are usually more well-constrained than the variegated mineralogical composition would at first imply. In fact, if the mineralogical composition of a given rock is known, fairly accurate effective elastic moduli can be estimated using the Voigt-Reuss-Hill average (§10.2). And although individual mineral grains are usually intrinsically anisotropic, in many cases they are arranged with random orientations so that the bulk behavior is nearly isotropic.

In the following derivation, all applied pressures and their resulting strains will be *incremental* changes superimposed on an already-existing state of stress and strain. The loading state consisting of a uniform hydrostatic stress (i.e., normal traction) of magnitude dP_c applied over the entire outer surface of the porous rock and uniform hydrostatic pressure of magnitude dP_p applied over the entire interior pore surface, will be denoted by $\{dP_c, dP_p\}$. If a stress increment $\{dP, dP\}$ is applied to the surface of the body (that is to say, $dP_c = dP_p = dP$), the resulting incremental stress state in the rock is that of uniform hydrostatic stress of magnitude dP throughout the matrix (Geertsma, 1957a). This can be verified by noting that uniform hydrostatic stress of magnitude dP satisfies the stress equilibrium equations, because the divergence of a spatially uniform tensor is identically zero, and also satisfies the boundary conditions on both the outer and inner surfaces of the matrix. This stress state leads to a uniform isotropic dilatation of magnitude $d\varepsilon_m = dP/K_m = C_m dP$, where C_m and K_m are the compressibility and bulk modulus of the rock matrix material. But this state of stress and strain within the matrix is exactly the same as that which would occur if the pores were hypothetically filled up with matrix material, and the boundary conditions on the outer surface were left unchanged. In this latter case, the total bulk strain is equal to $d\varepsilon_b = C_m dP$, so the bulk volume change is given by $dV_b = -C_m V_b^i dP$.

Now consider the stress increment $\{dP, 0\}$, which corresponds to a change only in the confining pressure. By definition, this will give rise to a change in the bulk volume given by $dV_b = -C_{bc} V_b^i dP$. Similarly, a stress increment of $\{0, dP\}$ would give rise to a bulk volume change of $dV_b = C_{bp} V_b^i dP$. For the infinitesimal changes under consideration, the principle of superposition is valid, so the stress increment $\{0, dP\}$ can be separated into the difference of the two increments $\{dP, dP\}$ and $\{dP, 0\}$, as illustrated in Fig. 7.3. The strains resulting from the stress increment $\{0, dP\}$ will therefore be equal to the difference between the strains that result from the stress increments $\{dP, dP\}$ and $\{dP, 0\}$. Using the notation $dV_b(dP_c, dP_p)$ to refer to the bulk volume change resulting from the stress increment $\{dP_c, dP_p\}$, we have

$$dV_b(0, dP) = dV_b(dP, dP) - dV_b(dP, 0), \tag{7.8}$$

$$\text{so,} \quad C_{bp} V_b^i dP = -C_m V_b^i dP + C_{bc} V_b^i dP, \tag{7.9}$$

$$\text{hence,} \quad C_{bp} = C_{bc} - C_m. \tag{7.10}$$

The following analogous relation between the two pore compressibilities can be derived in a similar manner:

$$C_{pp} = C_{pc} - C_m. \tag{7.11}$$

The pore compressibilities and the bulk compressibilities can be related to each other by applying the Maxwell–Betti reciprocal theorem (§5.8) to the two sets of loads $\{dP, 0\}$ and $\{0, dP\}$. The work that would be done by the first set of loads acting through the displacements due to the second set is given by

$$W^{12} = -dP[dV_b(0, dP)] = -dP[C_{bp} V_b^i dP] = -C_{bp} V_b^i (dP)^2, \tag{7.12}$$

where the minus sign accounts for the fact that the confining pressure acts in the $-\mathbf{n}$ direction, where $\mathbf{n}$ is the outward unit normal vector to the external surface,

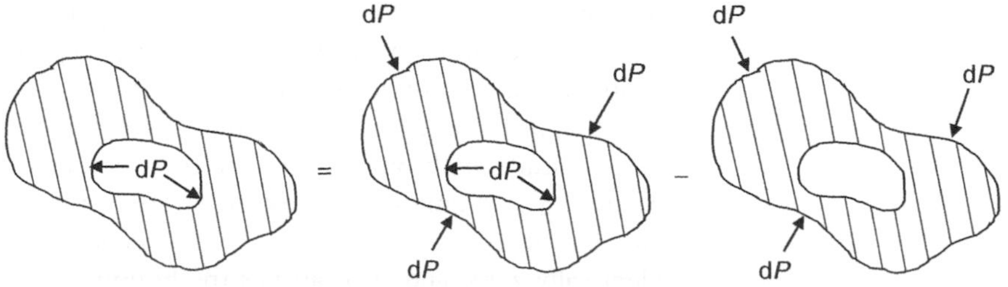

Fig. 7.3 Illustration of the superposition concept used in deriving relationships between the various porous rock compressibilities.

whereas the bulk volume would increase if the displacement were in the $+\mathbf{n}$ direction. Similarly,

$$W^{21} = \mathrm{d}P[\mathrm{d}V_{\mathrm{p}}(\mathrm{d}P, 0)] = \mathrm{d}P[-C_{\mathrm{pc}}V_{\mathrm{p}}^{i}\mathrm{d}P] = -C_{\mathrm{pc}}V_{\mathrm{p}}^{i}(\mathrm{d}P)^{2}, \tag{7.13}$$

No minus sign is needed in the defining equation for W^{21}, because the pore pressure acts in the same direction as that of increasing pore volume.

The reciprocal theorem implies that $W^{21} = W^{12}$, and so comparison of (7.12) and (7.13) reveals that $C_{\mathrm{bp}}V_{\mathrm{b}}^{i} = C_{\mathrm{pc}}V_{\mathrm{p}}^{i}$. But $V_{\mathrm{p}}^{i} = \phi^{i}V_{\mathrm{b}}^{i}$, and so

$$C_{\mathrm{bp}} = \phi^{i}C_{\mathrm{pc}}. \tag{7.14}$$

This relation requires for its validity only the assumption of elastic behavior; it does not require that the matrix be either homogeneous or isotropic.

Equations (7.10), (7.11), and (7.14) provide three relations between the four compressibilities $\{C_{\mathrm{pc}}, C_{\mathrm{pp}}, C_{\mathrm{bc}}, C_{\mathrm{bp}}\}$, with ϕ^{i} and C_{m} as the only other parameters explicitly involved. In terms of $\{C_{\mathrm{bc}}, C_{\mathrm{m}}, \phi^{i}\}$, the other three compressibilities are

$$C_{\mathrm{bp}} = C_{\mathrm{bc}} - C_{\mathrm{m}}, \tag{7.15}$$

$$C_{\mathrm{pc}} = (C_{\mathrm{bc}} - C_{\mathrm{m}})/\phi^{i}, \tag{7.16}$$

$$C_{\mathrm{pp}} = [C_{\mathrm{bc}} - (1+\phi^{i})C_{\mathrm{m}}]/\phi^{i}. \tag{7.17}$$

The approximate validity of (7.15)–(7.17) has been demonstrated for several consolidated sandstones (Zimmerman et al., 1986), as well as several limestones (Laurent et al., 1993). Note that it follows from (7.15)–(7.17) that the bulk compressibility cannot be written as a volumetrically weighted average of the pore and matrix compressibilities, as has occasionally been supposed (Dobrynin, 1962).

Although the porous rock compressibilities vary with P_{c} and P_{p}, they actually depend only on the single parameter $P_{\mathrm{d}} = P_{\mathrm{c}} - P_{\mathrm{p}}$, known as the *differential pressure* (Nur and Byerlee, 1971). The physical reason for this is that the stress-dependence of the compressibilities is due to the closure of cracks, seating of grain contacts, or other processes that involve the closure of thin, crack-like voids, and pore pressure has essentially an equal-but-opposite effect on the deformation of a thin void as does the confining pressure (see §8.9). The dependence of the compressibilities on the differential stress is also a necessary condition in order for the volumetric strain to be independent of the path taken in stress space. Indeed, applying the Euler condition for exactness of a differential to the bulk strain increment gives

$$\frac{\partial^2 \varepsilon_{\mathrm{b}}}{\partial P_{\mathrm{p}}\partial P_{\mathrm{c}}} = \frac{\partial}{\partial P_{\mathrm{p}}}\left(\frac{\partial \varepsilon_{\mathrm{b}}}{\partial P_{\mathrm{c}}}\right) = \frac{\partial C_{\mathrm{bc}}}{\partial P_{\mathrm{p}}}, \tag{7.18}$$

$$\frac{\partial^2 \varepsilon_{\mathrm{b}}}{\partial P_{\mathrm{c}}\partial P_{\mathrm{p}}} = \frac{\partial}{\partial P_{\mathrm{c}}}\left(\frac{\partial \varepsilon_{\mathrm{b}}}{\partial P_{\mathrm{p}}}\right) = \frac{-\partial C_{\mathrm{bp}}}{\partial P_{\mathrm{c}}} = \frac{-\partial (C_{\mathrm{bc}} - C_{\mathrm{m}})}{\partial P_{\mathrm{c}}} = \frac{-\partial C_{\mathrm{bc}}}{\partial P_{\mathrm{c}}}, \tag{7.19}$$

where, since the bulk compressibilities of rock-forming minerals do not change by more than a few percent over the range of pressures up to about 100 MPa

(Anderson et al., 1968), C_m can be taken to be a constant. Equating the two mixed partial derivatives shows that C_{bc} satisfies the partial differential equation $\partial C_{bc}/\partial P_c = -\partial C_{bc}/\partial P_p$, the general solution to which is $C_{bc}(P_c, P_p) = f(P_c - P_p)$, where f is any function that depends on the two pressures only through the combination $P_c - P_p$. Equations (7.15)–(7.17) then show that the other three porous rock compressibilities also depend only on the differential pressure.

In the development of poroelasticity given by Biot (1941), no assumption was made concerning the isotropy or homogeneity of the mineral phase. Biot's analysis reproduces (7.10) and (7.14), with C_m identified as an effective solid-phase compressibility, but not (7.11). In Biot's formulation, $C_{pc} - C_{pp} = C_\phi$, where the additional parameter C_ϕ is identified with the fractional change in pore volume resulting from equal increments of the pore and confining pressures (Brown and Korringa, 1975). If the matrix were homogeneous, C_ϕ should equal C_m. Hart and Wang (1995) reported data on Berea sandstone and Indiana limestone for which C_ϕ exceeded C_m by factors of about six and nine, respectively, although no explanation was given for this large discrepancy.

There have been other treatments of hydrostatic poroelasticity that include additional restrictive assumptions or simplifications, as opposed to greater generality. Domenico (1977) assumed that the volumetric strain in the matrix is negligible compared to the pore strain and arrived at equations that can be derived from (7.15)–(7.17) by setting C_m equal to zero. The two pore compressibilities are then equal to each other, as are the two bulk compressibilities. This assumption is acceptable for soils and unconsolidated sands (see Newman, 1973) but not for most consolidated sandstones or hard rocks.

The assumption that the mineral phase effectively behaves like an isotropic elastic medium can be tested experimentally through so-called "unjacketed tests," in which the rock is pressurized by a fluid which is allowed to seep into its pores, resulting in equal pore pressure and confining pressure. If $dP_p = dP_c$, then (7.6) and (7.10) give

$$d\varepsilon_b = C_{bc}dP_c - C_{bp}dP_c = (C_{bc} - C_{bp})dP_c = C_m dP_c. \tag{7.20}$$

In an unjacketed test, the stress-dependence disappears, and the rock deforms as a linear elastic medium with compressibility C_m (Fabre and Gustkiewicz, 1997). Zimmerman (1991) used the Voigt-Reuss-Hill average (§10.2) to compute an effective matrix compressibility of Berea sandstone of 2.54×10^{-5}/MPa, which is very close to the unjacketed compressibility of 2.58×10^{-5}/MPa measured by Andersen and Jones (1985).

Instead of using V_b and V_p as the kinematic variables, Carroll and Katsube (1983) developed a theory of hydrostatic poroelasticity in terms of ϕ and V_m. Using the identities $V_m = V_b - V_p$ and $\phi = V_p/V_b$, along with (7.15)–(7.17), it follows that the porosity and matrix volume are governed by

$$d\phi = -[(1 - \phi^i)C_{bc} - C_m]d(P_c - P_p), \tag{7.21}$$

$$d\varepsilon_m = \frac{C_m}{(1 - \phi^i)}d(P_c - \phi^i P_p). \tag{7.22}$$

One interpretation of (7.22) is that the average stress in the mineral phase is given by

$$\langle P_{\mathrm{m}} \rangle = \frac{P_{\mathrm{c}} - \phi^{i} P_{\mathrm{p}}}{1 - \phi^{i}}. \tag{7.23}$$

This result is actually true in general and does not require that the matrix be isotropic, homogeneous, or elastic (Dewers and Ortoleva, 1989; Zimmerman et al., 1994).

7.3 Undrained compression

The equations developed and discussed in the previous section are appropriate for processes in which the pore pressure and confining pressure vary independently. An example is the reduction in pore pressure due to the withdrawal of fluid from a petroleum or groundwater reservoir. Since the confining stresses acting on the reservoir rock are in general due to the lithostatic gradient, along with tectonic forces, they will not change while fluid is being withdrawn. There are other situations, however, in which the pore and confining pressures are not independent. For example, during drilling, the *in situ* stresses will be altered in the vicinity of the borehole. This will induce changes in the pore volume within a certain region surrounding the borehole, which will initially lead to a change in the pore pressure. If the rock is highly permeable, pore fluid will quickly flow in such a manner as to reestablish pore pressure equilibrium with the adjacent regions of rock. For a rock that is relatively impermeable, such as a shale, the time required for pore pressure equilibrium to be reestablished may be very large, so for a certain period of time, the pore fluid must be considered as being "trapped" inside the stressed region near the borehole. This phenomenon is discussed by Black et al. (1985), who relate it to the poor drillability properties of shales (see also Warren and Smith, 1985; Peltier and Atkinson, 1987). This type of "undrained" compression is also relevant to seismic wave propagation processes, in which the viscosity of the pore fluid will not permit it to travel between the pores within the time frame (i.e., one period) of the stress oscillations (Mavko and Nur, 1975; Cleary, 1978).

The deformation of a fluid-saturated porous rock at first takes place in an undrained manner and eventually, after the pore pressure has had sufficient time to reequilibrate itself, in a "drained" manner. The transition between these two cases, and the time needed for this transition to occur, can be studied within the context of the more general theory of poroelasticity developed in §7.4. However, an analysis of the limiting case of completely undrained compression can be conducted by starting with the expressions for the bulk and pore strains given by (7.6)–(7.7). If the pore space is completely saturated with a fluid that is "trapped," then the pore strain is also equal to the strain undergone by the pore fluid, that is,

$$d\varepsilon_{\mathrm{p}} = d\varepsilon_{\mathrm{f}} = -C_{\mathrm{f}} dP_{\mathrm{p}}. \tag{7.24}$$

If the confining pressure is changed, then (7.24) and (7.6)–(7.7) form a set of three equations for the three unknowns $\{d\varepsilon_{\mathrm{b}}, d\varepsilon_{\mathrm{p}}, dP_{\mathrm{p}}\}$. Solution of these equations yields the following expression for the *undrained bulk compressibility*:

$$C_{bu} = \left(\frac{\partial \varepsilon_b}{dP_c}\right)_{undrained} = C_{bc} - \frac{C_{bp}C_{pc}}{C_{pp} + C_f}. \tag{7.25}$$

The undrained bulk compressibility is less than the drained bulk compressibility, due to the additional stiffness that the trapped pore fluid imparts to the rock/fluid system.

The expression for C_{bu} given by (7.25) is independent of any assumption concerning the microstructure and stress–strain behavior of the matrix. If the matrix is assumed to be elastic, then the use of (7.14) allows (7.25) to be written as (Brown and Korringa, 1975)

$$C_{bu} = \left(\frac{\partial \varepsilon_b}{dP_c}\right)_{undrained} = \frac{\phi^i C_{bc}(C_f - C_\phi) + C_m(C_{bc} - C_m)}{\phi^i(C_f - C_\phi) + (C_{bc} - C_m)}. \tag{7.26}$$

If the rock matrix is also homogeneous, then $C_\phi = C_m$, and the undrained compressibility would be given by (Gassmann, 1951b)

$$C_{bu} = \left(\frac{\partial \varepsilon_b}{dP_c}\right)_{undrained} = \frac{\phi^i C_{bc}(C_f - C_m) + C_m(C_{bc} - C_m)}{\phi^i(C_f - C_m) + (C_{bc} - C_m)}. \tag{7.27}$$

The undrained bulk compressibility is an increasing function of the fluid compressibility, all other parameters being constant. The variation of C_{bu} with C_f, for the Fort Union sandstone described by Murphy (1984), is shown in Fig. 7.4. This sandstone had a porosity of 0.085, a matrix compressibility of $C_m = 0.286 \times 10^{-4}$/MPa, a drained bulk compressibility of $C_{bc} = 1.31 \times 10^{-4}$/MPa, and a pore compressibility of $C_{pp} = 11.8 \times 10^{-4}$/MPa. If the pores were filled with air at atmospheric pressure, which has a compressibility of $C_f = 9.87$/MPa, then the undrained bulk compressibility would equal (to three significant figures) the drained value, 1.31×10^{-4}/MPa. If the rock was saturated with water, which has a compressibility of $C_f = 5.0 \times 10^{-4}$/MPa, the undrained bulk compressibility would be 0.573×10^{-4}/MPa. A hypothetical "incompressible" pore fluid, on

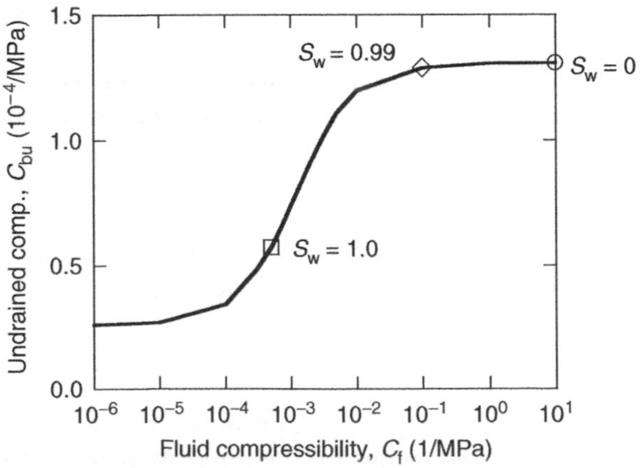

Fig. 7.4 Undrained bulk compressibility of Fort Union sandstone, as a function of the pore fluid compressibility (Murphy, 1984; Zimmerman, 1985a). If the pore fluid is a mixture of water and air at atmospheric pressure, the symbols show the three cases of 0%, 99%, and 100% water saturation.

the other hand, would lead to an undrained compressibility of 0.261×10^{-4}/MPa. Hence, the assumption that water is incompressible, which is acceptable in many engineering situations, would yield a grossly incorrect value for the undrained bulk compressibility (Zimmerman, 1985a).

Under quasi-static loading, the effective compressibility of the pore fluid is the volumetrically weighted average of the compressibilities of the different fluid phases. Hence, a small amount of air greatly increases the effective value of C_f and will thereby cause the undrained bulk compressibility to increase.

If a fluid-saturated porous rock undergoes undrained compression, the confining pressure causes the pores to contract, thereby pressurizing the trapped pore fluid. The magnitude of this induced pore pressure increment can be found from (7.24) and (7.6)–(7.7) to be given by

$$\left(\frac{\partial P_p}{dP_c}\right)_{\text{undrained}} \equiv B = \frac{C_{pc}}{C_{pp} + C_f} = \frac{C_{pp} + C_\phi}{C_{pp} + C_f}, \tag{7.28}$$

where B is the *Skempton coefficient* (Skempton, 1954; Mesri et al., 1976). If the matrix is assumed to be microhomogeneous, then C_ϕ can be replaced by C_m. In practice, the pore fluid is more compressible than the rock matrix, so the Skempton coefficient B will lie between 0 and 1. Furthermore, since it is also usually the case that $C_{pp} \gg C_m$,

$$B = \frac{C_{pp} + C_m}{C_{pp} + C_f} \approx \frac{C_{pp}}{C_{pp} + C_f} = \frac{1}{1 + (C_f/C_{pp})}. \tag{7.29}$$

If the pore fluid is a gas, then $C_f \gg C_{pp}$, and B will approach zero; induced pore pressures will therefore be negligible. For stress-sensitive rocks, C_{pp} decreases as a function of the differential pressure, so B will also decrease with P_d.

The induced pore pressure and Skempton coefficient for a Tunnel City sandstone are shown in Fig. 7.5 (after Green and Wang, 1986). The Skempton coefficient decreases from 0.95 at low differential pressures, to 0.6 at a differential pressure of 15 MPa. Green and Wang also measured B for a Berea sandstone and

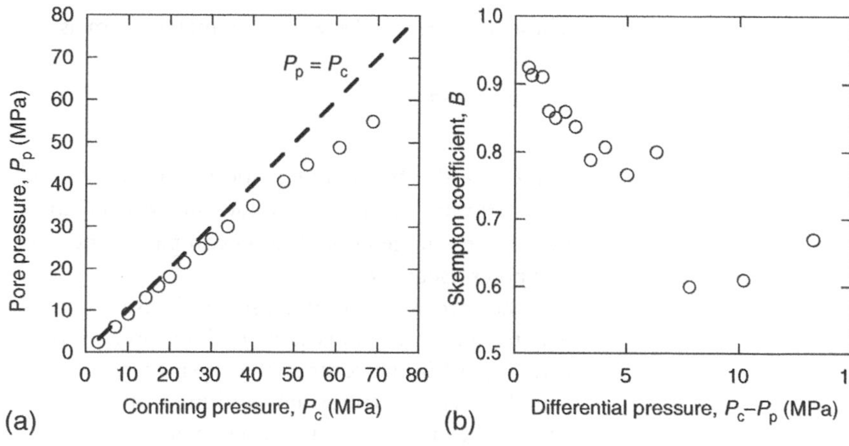

Fig. 7.5 (a) Induced pore pressure, and (b) Skempton coefficient, in a hydrostatically stressed sample of Tunnel sandstone (Green and Wang, 1986).

compared the values to those predicted from the equation

$$B = \frac{(C_{bc} - C_m)}{\phi^i (C_f - C_m) + (C_{bc} - C_m)}, \tag{7.30}$$

which is the form taken by (7.28) if the matrix is assumed to be homogeneous. The measured value of B decreased from 0.99 to 0.87 as P_d varied from 0 to 2 MPa, whereas the value predicted by (7.30) decreased from 1.0 to 0.85.

An undrained pore compressibility can also be found from (7.24) and (7.6)–(7.7) in the form

$$C_{pu} = \left(\frac{\partial \varepsilon_p}{dP_c}\right)_{\text{undrained}} = \frac{C_{pc} C_f}{C_{pp} + C_f} = \frac{C_{pc}}{1 + (C_{pp}/C_f)}. \tag{7.31}$$

The presence of trapped pore fluid therefore causes the compressibility of the pore space with respect to the confining pressure to decrease. The undrained pore compressibility appears in the analysis of the effect of pore fluids on seismic wave propagation (O'Connell and Budiansky, 1974).

7.4 Constitutive equations of poroelasticity

To develop a linearized, nonhydrostatic theory of poroelasticity, first recall the stress–strain relations of a nonporous material, (5.24)–(5.27), expressed in terms of the shear modulus and Poisson ratio:

$$\varepsilon_{xx} = \frac{1}{2G}\left[\tau_{xx} - \frac{\nu}{(1+\nu)}(\tau_{xx} + \tau_{yy} + \tau_{zz})\right], \tag{7.32}$$

$$\varepsilon_{yy} = \frac{1}{2G}\left[\tau_{yy} - \frac{\nu}{(1+\nu)}(\tau_{xx} + \tau_{yy} + \tau_{zz})\right], \tag{7.33}$$

$$\varepsilon_{zz} = \frac{1}{2G}\left[\tau_{zz} - \frac{\nu}{(1+\nu)}(\tau_{xx} + \tau_{yy} + \tau_{zz})\right], \tag{7.34}$$

$$\varepsilon_{xy} = \tau_{xy}/2G, \quad \varepsilon_{xz} = \tau_{xz}/2G, \quad \varepsilon_{yz} = \tau_{yz}/2G. \tag{7.35}$$

These relations can be written in matrix form as

$$\boldsymbol{\varepsilon} = \frac{1}{2G}\boldsymbol{\tau} - \frac{\nu}{2G(1+\nu)}\text{trace}(\boldsymbol{\tau})\mathbf{I}. \tag{7.36}$$

In these, and subsequent equations, the stresses and strains at each point should be interpreted as average values taken over an "infinitesimal" region that is nevertheless large enough to encompass some suitably large number of grains and pores (Bear, 1988).

If the porous rock is macroscopically isotropic, a pore pressure increment will lead to equal extensions along each of three mutually orthogonal directions, but cannot (within the context of a linear theory) cause any shear strains. Since the total bulk volumetric strain resulting from an applied pore pressure is $-C_{bp}P_p$, the coefficient that relates each macroscopic longitudinal strain to the pore pressure must be $-C_{bp}/3$. Hence, a term $-C_{bp}P_p/3$ must be added to each of the longitudinal strains, yielding

$$\boldsymbol{\varepsilon} = \frac{1}{2G}\boldsymbol{\tau} - \frac{\nu}{2G(1+\nu)}\text{trace}(\boldsymbol{\varepsilon})\mathbf{I} - \frac{C_{bp}}{3}P_p\mathbf{I}. \tag{7.37}$$

Recalling that $C_{bp} = C_{bc} - C_m$, where C_m is the effective compressibility of the matrix, and noting that $C_{bc} = 1/K_{bc} = 1/K$, where K is the macroscopic bulk modulus, (7.37) can be written as

$$\boldsymbol{\varepsilon} = \frac{1}{2G}\boldsymbol{\tau} - \frac{\nu}{2G(1+\nu)}\text{trace}(\boldsymbol{\varepsilon})\mathbf{I} - \frac{\alpha}{3K}P_p\mathbf{I}, \tag{7.38}$$

where the Biot coefficient α is defined by (Biot and Willis, 1957; Nur and Byerlee, 1971)

$$\alpha = 1 - \frac{C_m}{C_{bc}} = 1 - \frac{K_{bc}}{K_m} = 1 - \frac{K}{K_m}. \tag{7.39}$$

Taking the trace of both sides of (7.38) yields

$$\text{trace}(\boldsymbol{\varepsilon}) = \frac{(1-2\nu)}{2G(1+\nu)}\text{trace}(\boldsymbol{\tau}) - \frac{\alpha}{K}P_p. \tag{7.40}$$

But $\text{trace}(\boldsymbol{\varepsilon}) = \varepsilon_b$, $\text{trace}(\boldsymbol{\tau}) = 3\tau_m \equiv 3P_c$ (in the present poroelastic terminology), and $2G(1+\nu) = 3K(1-2\nu)$ from (5.14), so (7.40) is equivalent to

$$\varepsilon_b = \frac{1}{K}(P_c - \alpha P_p) = C_{bc}(P_c - \alpha P_p). \tag{7.41}$$

As the Biot coefficient always satisfies the inequality $\alpha < 1$, by definition (7.39), the pore pressure is not totally effective in counteracting the effect of the confining pressure in changing the bulk volume. For this reason, the Biot coefficient is also known as the *effective stress coefficient*. Different effective stress coefficients apply to different processes (Robin, 1973, Berryman, 1992), so it is sometimes convenient to denote α by n_b, where the subscript indicates that n_b is the effective stress coefficient for *bulk* volumetric deformation. Using this notation, for example, (7.21) and (7.22) show that $n_\phi = 1$, and $n_m = \phi^i$.

The equations for the stresses in terms of the strains are found by inverting (7.38) with the aid of (7.40):

$$\boldsymbol{\tau} - \alpha P_p\mathbf{I} = 2G\left[\boldsymbol{\varepsilon} + \frac{\nu}{(1-2\nu)}\text{trace}(\boldsymbol{\varepsilon})\mathbf{I}\right] = 2G\boldsymbol{\varepsilon} + \lambda\text{trace}(\boldsymbol{\varepsilon})\mathbf{I}, \tag{7.42}$$

which can be written explicitly as

$$\tau_{xx} - \alpha P_p = 2G\varepsilon_{xx} + \lambda(\varepsilon_{xx} + \varepsilon_{yy} + \varepsilon_{zz}), \tag{7.43}$$

$$\tau_{yy} - \alpha P_p = 2G\varepsilon_{yy} + \lambda(\varepsilon_{xx} + \varepsilon_{yy} + \varepsilon_{zz}), \tag{7.44}$$

$$\tau_{zz} - \alpha P_p = 2G\varepsilon_{zz} + \lambda(\varepsilon_{xx} + \varepsilon_{yy} + \varepsilon_{zz}), \tag{7.45}$$

$$\tau_{xy} = 2G\varepsilon_{xy}, \quad \tau_{xz} = 2G\varepsilon_{xz}, \quad \tau_{yz} = 2G\varepsilon_{yz}. \tag{7.46}$$

The stress–strain equations are identical to those for a nonporous isotropic elastic rock, except that αP_{p} is subtracted from each of the normal stresses. The terms on the left-hand sides of (7.43)–(7.45) are known as the *effective stresses*. Constitutive equations and effective stress coefficients for *anisotropic* poroelastic rocks have been studied by Carroll (1979), Thompson and Willis (1991), and Cheng (1997).

The stress–strain relations for isotropic poroelasticity have a particularly simple form when expressed in terms of the deviatoric stresses and deviatoric strains. From (7.41), the mean normal stress is related to the mean normal strain by

$$\varepsilon_{\text{m}} = \frac{1}{3K}(\tau_{\text{m}} - \alpha P_{\text{p}}). \tag{7.47}$$

Using (2.152), (2.235), and (7.42) and (7.47), the deviatoric stresses can be expressed in terms of the deviatoric strains as

$$\begin{aligned} s_{xx} &= 2Ge_{xx}, \quad s_{yy} = 2Ge_{yy}, \quad s_{zz} = 2Ge_{zz}, \quad s_{xy} = 2Ge_{xy}, \\ s_{xz} &= 2Ge_{xz}, \quad s_{yz} = 2Ge_{yz}. \end{aligned} \tag{7.48}$$

In matrix notation, these relations take the form

$$\boldsymbol{\varepsilon}^{\text{iso}} = \frac{1}{3K}\boldsymbol{\tau}^{\text{iso}} - \frac{\alpha}{3K}P_{\text{p}}\mathbf{I}, \quad \boldsymbol{\varepsilon}^{\text{dev}} = \frac{1}{2G}\boldsymbol{\tau}^{\text{dev}}. \tag{7.49}$$

Consider now a region of rock with bulk volume V_{b}, whose pore space contains an amount of fluid of mass m. If the density of the pore fluid is ρ_{f}, the volume occupied by this fluid is m/ρ_{f}. If the pores are fully saturated with a single fluid component, this volume is also equal to the pore volume, V_{p}. The incremental change in the pore volume is therefore given by

$$\text{d}V_{\text{p}} = \text{d}(\text{m}/\rho_{\text{f}}) = \frac{\text{dm}}{\rho_{\text{f}}} - \frac{\text{m}\text{d}\rho}{\rho_{\text{f}}^2} = \frac{\text{dm}}{\rho_{\text{f}}} - \frac{\text{m}}{\rho_{\text{f}}}\frac{\text{d}\rho_{\text{f}}}{\rho_{\text{f}}}. \tag{7.50}$$

But $\text{d}\rho_{\text{f}}/\rho_{\text{f}} = C_{\text{f}}\text{d}P_{\text{p}}$, so (7.50) can be written as

$$\text{d}V_{\text{p}} = \frac{\text{dm}}{\rho_{\text{f}}} - V_{\text{p}}C_{\text{f}}\text{d}P_{\text{p}}. \tag{7.51}$$

Dividing all terms by the bulk volume gives

$$\frac{\text{d}V_{\text{p}}}{V_{\text{b}}} = \frac{1}{V_{\text{b}}}\frac{\text{dm}}{\rho_{\text{f}}} - \phi C_{\text{f}}\text{d}P_{\text{p}}. \tag{7.52}$$

The change in the volumetric fluid content of a certain region of rock can therefore be broken up into two parts – one due to additional fluid moving into the region of rock (the first term on the right), and another due to compression or expansion of the fluid that is already in that region (the second term on the right). This first term is denoted by $\text{d}\zeta$, and is defined as that portion of the change in volumetric fluid content that is due solely to mass transfer. From (7.52),

$$\text{d}\zeta \equiv \frac{1}{V_{\text{b}}}\frac{\text{dm}}{\rho} = \frac{\text{d}V_{\text{p}}}{V_{\text{b}}} + \phi C_{\text{f}}\text{d}P_{\text{p}}. \tag{7.53}$$

Recalling (7.7) for the pore volume change, the increment in ζ can be expressed as

$$d\zeta = -\phi[C_{pc}dP_c - (C_{pp} + C_f)dP_p]. \tag{7.54}$$

This equation allows a simple rederivation of Skempton's induced pore pressure coefficient. In an undrained process, the fluid content increment must be zero, so setting $d\zeta = 0$ yields

$$B = \left(\frac{dP_p}{dP_c}\right)_{d\zeta=0} = \frac{C_{pc}}{C_{pp} + C_f}. \tag{7.55}$$

Using the various relations between the porous rock compressibilities, (7.54) can also be written as

$$d\zeta = -\alpha C_{bc}(dP_c - \frac{1}{B}dP_p) = -\frac{\alpha}{K}(dP_c - \frac{1}{B}dP_p), \tag{7.56}$$

which shows that $1/B$ is the effective stress coefficient for ζ.

In the linearized theory of poroelasticity, the constitutive parameters are assumed to be independent of stress, so (7.56) can be integrated, making use of the fact that, by definition, $\zeta = 0$ in the unstressed state, to yield

$$\zeta = -\frac{\alpha}{K}\left(P_c - \frac{1}{B}P_p\right). \tag{7.57}$$

Inverting (7.57) by using (7.41) to eliminate P_c yields

$$P_p = \frac{BK}{\alpha(1 - \alpha B)}(\zeta + \alpha\varepsilon_b) \equiv M(\zeta + \alpha\varepsilon_b), \tag{7.58}$$

which relates the pore pressure to the bulk volumetric strain and the excess fluid content. The parameter M, also sometimes denoted by Q, is known as the *Biot modulus*.

Combining (7.41) and (7.58) yields

$$\varepsilon_b = \frac{(1 - \alpha B)}{K}P_c - B\zeta. \tag{7.59}$$

This equation has several important implications. It shows that the "strength" of the coupling between the mechanical deformation and pore pressure is quantified by the Skempton coefficient B, in the sense that the limiting case $B \to 0$ leads back to the nonporous version of the relationship between bulk strain and mean normal stress. It also directly implies that the ***undrained bulk modulus***, defined by $K_u = 1/C_{bu}$, can be expressed as (Wang, 1993)

$$K_u = \frac{K}{(1 - \alpha B)}. \tag{7.60}$$

The constitutive equations of poroelasticity for an isotropic elastic rock contain numerous coefficients. Aside from those introduced in the discussion above, other poroelastic parameters have been defined and used by Biot (1941), Biot and Willis (1957), and Rice and Cleary (1976). The relations between these parameters have been examined by Kümpel (1991), Detournay and Cheng (1993), Hickey et al. (1995), and Wang (2000). However, only four of these parameters are independent (Hart and Wang, 1995). This independent set can be taken to consist of any two of the standard elastic moduli $\{\lambda, G, K, E, \nu\}$, plus the two moduli K_m and K_ϕ. The Biot effective stress coefficient can then be found from (7.39), and the Skempton induced pore pressure coefficient is given by (7.55). This latter coefficient also involves the fluid compressibility, so, strictly speaking, it is not a property of the rock, but rather of the rock/fluid system. Under the assumption of microscopic matrix homogeneity, $K_\phi = K_m$, and the number of independent poroelastic moduli reduces to three.

The physical significance of the constitutive parameters can be examined by considering a few simple types of loading or deformation.

7.4.1 Unjacketed hydrostatic compression; $\tau_{xx} = \tau_{yy} = \tau_{zz} = P_p = P$

In this case, (7.39) and (7.41) yield

$$\varepsilon_b = \frac{(1-\alpha)P}{K} = \frac{P}{K_m}. \tag{7.61}$$

7.4.2 Drained uniaxial compression, with no lateral strain; $\tau_{zz} > 0$, $\varepsilon_{xx} = \varepsilon_{yy} = P_p = 0$

This is a very common state applied in soil mechanics tests. From (7.43), the strain in the loading direction is given by

$$\varepsilon_{zz} = \frac{1}{\lambda + 2G}\tau_{zz}. \tag{7.62}$$

7.4.3 Undrained uniaxial compression, with no lateral strain; $\tau_{zz} > 0$, $\varepsilon_{xx} = \varepsilon_{yy} = \zeta = 0$

This situation is relevant to the problem of consolidation, which is discussed in §7.5. Since $\zeta = 0$ and $\varepsilon_b = \varepsilon_{zz}$ in this case, (7.58) gives

$$P_p = \alpha M \varepsilon_{zz} = \frac{BK}{1-\alpha B}\varepsilon_{zz}, \tag{7.63}$$

after which (7.43) gives

$$\tau_{zz} = (\lambda + 2G + \alpha^2 M)\varepsilon_{zz}. \tag{7.64}$$

The pore pressure induced by the axial stress is therefore given by

$$P_p = \frac{\alpha M}{(\lambda + 2G + \alpha^2 M)}\tau_{zz}. \tag{7.65}$$

7.5 Equations of stress equilibrium and fluid flow

In §7.4, a few simple homogeneous loading states were examined, using only the stress–strain relations. To find the stress and strain distributions in more complicated, nonhomogeneous problems, a governing partial differential equation for the displacement vector is needed. In poroelasticity, the strains are related to the displacement vector in the usual manner, with the understanding that the displacement represents the mean displacement of the solid matrix material located at each infinitesimal point. To derive the poroelastic version of the Navier equations for the displacements, we combine the stress–strain law (7.42) with the stress equilibrium equations (5.76), and the strain–displacement equations (2.222), to arrive at

$$G\nabla^2\mathbf{u} + (\lambda + G)\nabla(\nabla\cdot\mathbf{u}) = -\mathbf{f} - \alpha\nabla P_p, \tag{7.66}$$

where $\mathbf{f}$ is the body-force vector (due, say, to gravity) per unit volume. The *gradient* of the pore pressure, multiplied by Biot's effective stress coefficient, therefore acts as an additional body force.

In a laboratory experiment, the pore pressure can be controlled, and therefore can be considered known. Equation (7.66) then gives three scalar equations for the three unknown displacement components. In the field, however, the pressure usually varies from point to point and may vary with time. In order to find the pore pressure distribution, a governing differential equation for P_p is also needed. This equation is found by considering conservation of mass for the pore fluid, along with a constitutive equation that relates the fluid flux to the pore pressure gradient.

Consider a small planar surface of rock, with area dA and outward unit normal vector $\mathbf{n}$ (Fig. 7.6a). The fluid flux vector, $\mathbf{q}$, which has dimensions of [m/s], is defined such that the total volume of fluid that passes through this surface, per unit time, is $(\mathbf{q}\cdot\mathbf{n})dA$; this quantity has dimensions of [m^3/s]. Only the normal component of $\mathbf{q}$ represents fluid that crosses the surface. The components of $\mathbf{q}$ are defined to be positive if they point in the direction of the coordinate axes; unlike stresses, they are not defined with respect to the direction of $\mathbf{n}$. To prove that $\mathbf{q}$ is indeed a vector, consider the two-dimensional prismatic element shown in Fig. 7.6b, whose three faces have outward unit normal vectors $-\mathbf{e}_x$, $-\mathbf{e}_y$, and $\mathbf{n}$, where the n direction is rotated by a counterclockwise angle θ from the x direction. The length of the slanted edge of this prism is h, and the thickness

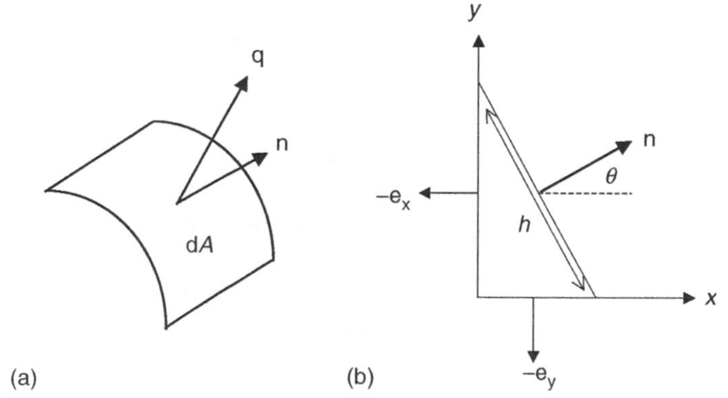

Fig. 7.6 (a) Small planar surface of area dA, with outward unit normal vector $\mathbf{n}$, and volumetric fluid flow vector $\mathbf{q}$. (b) Elementary prism used in derivation of (7.69).

in the z direction is t. Assuming the flow field to be independent of time, the total volumetric flux out of this prismatic element, per unit time, must be zero. For the orientation shown in the figure, the outward unit normal vector of the bottom face is $-\mathbf{e_y}$, and so the flux out of the prism through this face is

$$Q(\text{out of bottom face}) = (\mathbf{q} \cdot \mathbf{n})dA = [\mathbf{q} \cdot (-\mathbf{e_y})]ht \sin\theta = -q_y ht \sin\theta. \quad (7.67)$$

Following the same steps for the other two faces, we find that the total flux out of the element, which must equal zero, is given by

$$Q(\text{total}) = -q_y ht \sin\theta - q_x ht \cos\theta + q_n ht = 0. \quad (7.68)$$

Solving (7.68) for q_n gives

$$q_n = q_x \cos\theta + q_y \sin\theta, \quad (7.69)$$

which shows that the components of $\mathbf{q}$ do indeed obey the transformation law of a vector (i.e., a first-order tensor). This result would also hold if the flow field were time-dependent, since the flux terms are proportional to h, whereas the rate of accumulation of fluid within the prism is proportional to h^2, so in the limit as the size of the prism vanishes, the storage term becomes negligible, and we again arrive at (7.69).

Now consider a piece of rock that occupies a region of space R, with outer boundary ∂R. The total volumetric flux of fluid leaving this region, per unit time, is given by the integral of $(\mathbf{q} \cdot \mathbf{n})dA$ over the outer surface. The total increment of fluid volume stored within that region, due to mass transfer across the outer boundary, is found by integrating ζ over the region. The time rate of change of this integral must equal the total flux *into* the region, so

$$\iiint_R \frac{\partial \zeta}{\partial t} dV = -\iint_{\partial R} (\mathbf{q} \cdot \mathbf{n}) dA = -\iint_{\partial R} (q_x n_x + q_y n_y + q_z n_z) dA. \quad (7.70)$$

Using the divergence theorem on the surface integral and collecting terms gives

$$\iiint_R \left(\frac{\partial \zeta}{\partial t} + \frac{\partial q_x}{\partial x} + \frac{\partial q_y}{\partial y} + \frac{\partial q_z}{\partial z} \right) dV = 0. \quad (7.71)$$

For (7.71) to be true for all regions R, the bracketed integrand must vanish, that is,

$$\frac{\partial \zeta}{\partial t} + \nabla \cdot \mathbf{q} = 0. \quad (7.72)$$

Equation (7.72) represents conservation of *mass* for the pore fluid, despite the fact that its derivation seemed to involve only fluid *volumes*. This is because ζ is defined to be that portion of the additional fluid volume content that is due solely to mass transfer; the compressibility effect is accounted for separately, as shown by (7.53).

It now remains to find a relationship between the flux vector **q** and the pore pressure. For the low flow rates that are usually encountered in the subsurface, this relation is given by Darcy's law, which in the most general anisotropic case takes the form (Bear, 1988)

$$\mathbf{q} = -\frac{\mathbf{k}}{\mu}\nabla(P_\mathrm{p} - \rho_\mathrm{f}\mathbf{g}\cdot\mathbf{x}), \tag{7.73}$$

where **k** is the permeability tensor, with units of [m^2], μ is the fluid viscosity, with units of [Pa s], P_p is the pore fluid pressure, ρ_f is the fluid density, and **g** is the gravitational acceleration vector. The permeability tensor **k** is a second-order tensor, and as such it obeys the usual transformation rules when the coordinate system is rotated.

All other factors being equal, the magnitude of the fluid flux vector **q** is inversely proportional to the viscosity of the fluid. In general, the viscosities of liquids are much greater than those of gases. For example, the viscosity of air at 20°C is 1.81×10^{-5} Pa s, whereas that of water is 0.001 Pa s. The viscosities of most liquid hydrocarbons are usually higher than that of water, by a factor of about 1–100 (Matthews and Russell, 1967, p. 155).

If the z direction is taken to point downward, then $\mathbf{g} = g\mathbf{e}_z$, where $g = 9.81\ \mathrm{m/s^2}$. In this case, (7.73) takes the form

$$\mathbf{q} = -\frac{\mathbf{k}}{\mu}\nabla(P_\mathrm{p} - \rho_\mathrm{f} g z), \tag{7.74}$$

which can be written explicitly as

$$\begin{bmatrix} q_x \\ q_y \\ q_z \end{bmatrix} = -\frac{1}{\mu}\begin{bmatrix} k_{xx} & k_{xy} & k_{xz} \\ k_{yx} & k_{yy} & k_{yz} \\ k_{zx} & k_{zy} & k_{zz} \end{bmatrix}\begin{bmatrix} (\partial P_\mathrm{p}/\partial x) \\ (\partial P_\mathrm{p}/\partial y) \\ (\partial P_\mathrm{p}/\partial z) - \rho_\mathrm{f} g \end{bmatrix}, \tag{7.75}$$

For simplicity of notation, the gravitational term will be ignored in the remainder of this chapter.

For many types of anisotropy, it can be proven that **k** must be symmetric. It is also often argued that the symmetry of **k** follows as a consequence of the Onsager reciprocal relations of irreversible thermodynamics (Prigogine, 1961). However, the issue of whether or not a second-order tensor that governs an irreversible transport process such as fluid flow through porous media must necessarily be symmetric remains a matter of some controversy (Nye, 1957). Nevertheless, it seems to be an empirical fact that the permeability tensor is symmetric, in which case $k_{yx} = k_{xy}$, etc.

The symmetry of **k** implies the existence of a principal coordinate system in which the permeability tensor assumes a diagonal form. In this coordinate system Darcy's law (7.75) takes the form

$$\begin{bmatrix} q_{x'} \\ q_{y'} \\ q_{z'} \end{bmatrix} = -\frac{1}{\mu}\begin{bmatrix} k_{x'x'} & 0 & 0 \\ 0 & k_{y'y'} & 0 \\ 0 & 0 & k_{z'z'} \end{bmatrix}\begin{bmatrix} (\partial P_\mathrm{p}/\partial x') \\ (\partial P_\mathrm{p}/\partial y') \\ (\partial P_\mathrm{p}/\partial z') \end{bmatrix}. \tag{7.76}$$

In this coordinate system, the flux in one of the principal directions depends only on the pressure gradient in that direction; the coupling between the fluxes and pressure gradients in orthogonal directions disappears. In scalar form,

$$q_{x'} = -\frac{k_{x'x'}}{\mu}\frac{\partial P_p}{\partial x'}, \tag{7.77}$$

$$q_{y'} = -\frac{k_{y'y'}}{\mu}\frac{\partial P_p}{\partial y'}, \tag{7.78}$$

$$q_{z'} = -\frac{k_{z'z'}}{\mu}\frac{\partial P_p}{\partial z'}. \tag{7.79}$$

In most sedimentary rocks, two of the principal directions of permeability lie in the bedding plane, and the other is perpendicular to the bedding (de Marsily, 1986, p. 64). The permeability in the direction normal to the bedding plane is typically larger than in the other two principal directions, usually by a factor of 1–100. In fractured rock masses, the principal directions of permeability are controlled by, but are not necessarily coincident with, the directions of the major fracture sets. However, if there are three mutually orthogonal fracture sets, then the principal directions of the permeability tensor will lie in these three directions. The relationship between the permeability of fractured rock masses and the orientation, spacing, and interconnectivity of the fractures is discussed by, among others, Long et al. (1982) and Lee and Farmer (1993).

Although rock masses are rarely hydrologically isotropic, most analyses of subsurface flow processes are conducted under the assumption of isotropy. If the permeability tensor is isotropic, then $\mathbf{k} = k\mathbf{I}$, where the scalar coefficient k is referred to simply as "the permeability." The numerical value of the permeability depends strongly on the size of the pores or fractures, as well as on the interconnectivity of the pore space. For porous but unfractured rocks, the permeability varies as the fourth power of the mean pore size (Dullien, 1992). A consequence of this strong pore-size dependence is that permeabilities vary by many orders of magnitude from one rock to another. Table 7.1, compiled from de Marsily (1986) and Guéguen and Palciauskas (1994), shows the range of values that may be expected for various rock types. The table shows that permeability

Table 7.1 Expected ranges of permeabilities and hydraulic conductivities of various rock types.

a – de Marsily (1986); b – Guéguen and Palciauskas (1994).

Rock Type	k (m²)	k (Darcies)	K_h (m/s)	Reference
Coarse gravels	10^{-9}–10^{-8}	10^{3}–10^{4}	10^{-2}–10^{-1}	a
Sands, gravels	10^{-12}–10^{-9}	10^{0}–10^{3}	10^{-5}–10^{-2}	a
Fine sands, silts	10^{-16}–10^{-12}	10^{-4}–10^{0}	10^{-9}–10^{-5}	a
Clays, shales	10^{-23}–10^{-16}	10^{-11}–10^{-4}	10^{-16}–10^{-9}	a,b
Dolomites	10^{-12}–10^{-10}	10^{0}–10^{2}	10^{-5}–10^{-3}	a
Limestones	10^{-22}–10^{-12}	10^{-10}–10^{0}	10^{-15}–10^{-5}	a,b
Sandstones	10^{-17}–10^{-11}	10^{-5}–10^{1}	10^{-10}–10^{-4}	a,b
Granites, gneiss	10^{-20}–10^{-16}	10^{-8}–10^{-4}	10^{-13}–10^{-9}	a,b
Basalts	10^{-19}–10^{-13}	10^{-7}–10^{-1}	10^{-12}–10^{-6}	b

can vary widely, even within the same rock type. Hence, it is rarely possible to use tabulated handbook values for k; rock-specific values must usually be measured.

In SI units, permeability is expressed in square meters. Permeabilities are also frequently expressed in Darcies, where 1 Darcy $= 0.987 \times 10^{-12}\,\text{m}^2$, or in milliDarcies (mD). A related parameter that is often used by groundwater hydrologists and civil engineers is the "hydraulic conductivity," which is defined by $K_h = \rho_f g k/\mu$, and has the dimensions of velocity. When numerical values of the hydraulic conductivity are reported, it is always implicitly assumed that the fluid is water at 20°C, in which case $\rho_f = 998\,\text{kg/m}^3$ and $\mu = 0.001$ Pa s. For example, since $g = 9.81\,\text{m/s}^2$, a rock having a permeability of 1 Darcy will have an hydraulic conductivity of $9.66 \times 10^{-6}\,\text{m/s}$.

The values given in Table 7.1 are for unfractured rocks. These are the values that would be measured in the laboratory on an intact core. In fractured rock masses, the macroscopic permeability is determined by the aperture and interconnectedness of the fracture network, and is usually much larger than the permeability of the matrix rock. According to de Marsily (1986), the permeability of a fractured limestone may be in the range 10^2–10^4 Darcies, whereas that of a fractured crystalline rock mass will be in the range 10^{-3}–10^1 Darcies.

For isotropic rock masses, insertion of (7.74) into (7.72) yields

$$\frac{\partial \zeta}{\partial t} = \frac{k}{\mu}\nabla^2 P_p. \tag{7.80}$$

If the assumption is made that the rock is rigid and undeformable, as is often done in hydrology and petroleum engineering, then (7.54) reduces to $d\zeta = \phi C_f dP_p$, and (7.80) takes the form

$$\frac{\partial P_p}{\partial t} = \frac{k}{\phi \mu C_f}\nabla^2 P_p. \tag{7.81}$$

Equation (7.81) is a diffusion equation, with the hydraulic diffusivity given by $D = k/\phi\mu C_f$. In the case of a rigid porous medium, the fluid content increment ζ also satisfies (7.81), as can be proven by substituting (7.54) into (7.81).

In the general case of a nonrigid poroelastic medium, combining (7.58) with (7.80) gives

$$\frac{\partial P_p}{\partial t} = \frac{kM}{\mu}\nabla^2 P_p + \alpha M \frac{\partial \varepsilon_b}{\partial t}. \tag{7.82}$$

If the three displacement components and the pore pressure are taken as the four basic field variables of poroelasticity, then (7.66) and (7.82) supply the four governing equations needed to provide a well-posed mathematical problem. The diffusion equation for the pore pressure can also be expressed entirely in terms of pressures/stresses by using (7.41) to eliminate ε_b from (7.82):

$$\frac{\partial P_p}{\partial t} = \frac{kBK}{\alpha \mu}\nabla^2 P_p + B\frac{\partial \tau_m}{\partial t}. \tag{7.83}$$

The pore pressure therefore satisfies a diffusion-type equation containing an additional coupling term that relates the pore pressure to the isotropic part of the stress (or strain) tensor. This term represents a transient Skempton-type effect, in which the pore pressure rises as a result of bulk compression of the rock. It is clear from (7.82) and (7.83) that the form of the diffusivity coefficient for the pressure depends on the type of process that is occurring. If the confining stresses are constant, (7.83) shows that the diffusivity is given by $kBK/\alpha\mu$, where $K \equiv 1/C_{bc}$ is the drained bulk modulus, k is the permeability, B is the Skempton coefficient, and α is the Biot effective stress coefficient. If the strains are constant, (7.82) shows that the diffusivity is kM/μ.

A diffusion equation can also be derived from (7.80) that contains the excess fluid content ζ as the only dependent variable. To do this, we first find a relationship between the Laplacian of P_p and the Laplacian of ζ as follows. Taking the divergence of both sides of (7.66), in the case where the body-force vector **f** is zero, gives

$$G\nabla \cdot (\nabla^2 \mathbf{u}) + (\lambda + G)\nabla \cdot \nabla(\nabla \cdot \mathbf{u}) = -\alpha\nabla \cdot \nabla P_p. \tag{7.84}$$

But $\nabla \cdot (\nabla^2 \mathbf{u}) = \nabla^2(\nabla \cdot \mathbf{u}) = \nabla^2 \varepsilon_b$, and $\nabla \cdot \nabla = \nabla^2$, so (Geertsma, 1957b)

$$(\lambda + 2G)\nabla^2 \varepsilon_b = -\alpha\nabla^2 P_p. \tag{7.85}$$

Equation (7.85) is interesting, and perhaps unanticipated, as it relates the spatial variation in bulk strain to the spatial variation in pore pressure, without any reference to the mean normal stress. But it follows directly from (7.58) that

$$\nabla^2 \varepsilon_b = \frac{1}{\alpha M}\nabla^2 P_p - \frac{1}{\alpha}\nabla^2 \zeta. \tag{7.86}$$

Eliminating the bulk strain between (7.85) and (7.86) yields

$$\nabla^2 P_p = \frac{(\lambda + 2G)M}{\lambda + 2G + \alpha^2 M}\nabla^2 \zeta, \tag{7.87}$$

which, combined with (7.80), yields (Green and Wang, 1990)

$$\frac{\partial \zeta}{\partial t} = \frac{k}{\mu S}\nabla^2 \zeta, \tag{7.88}$$

where the storage coefficient S is given by

$$S = \frac{1}{M} + \frac{\alpha^2}{\lambda + 2G} \equiv \frac{1}{M} + \frac{\alpha^2}{K + 4G/3}. \tag{7.89}$$

The combination $\lambda + 2G$ is the elastic modulus that governs uniaxial strain, as shown in (5.60). For this reason, it is occasionally stated that the storage coefficient S is defined for, or is relevant to, *only* those processes in which the macroscopic strains are zero in two directions, such as is usually assumed to occur during depletion of a reservoir. On the contrary, the derivation of (7.88) shows that the diffusivity coefficient for the fluid content does not depend on the specific process that is occurring, and is always given by $k/\mu S$, where S is defined by (7.89). The storage coefficient can be expressed in numerous other equivalent forms, such as

$$S = \left[\frac{1}{K_f} - \frac{1}{K_m}\right]\phi + \left[1 - \frac{2(1-2\nu)}{3(1-\nu)}\alpha\right]\frac{\alpha}{K}, \tag{7.90}$$

where ν is the Poisson ratio of the rock under drained conditions. In the limiting case of a rigid medium, the storage coefficient reduces to ϕC_f, and the diffusivity becomes $k/\phi\mu C_f$, as in (7.81). Equations (7.66) and (7.88) represent four equations for the four kinematic field variables $\{u, v, w, \zeta\}$.

Finally, mention should be made of the undrained Poisson ratio. Bearing in mind that the drained shear modulus is identical to the undrained shear modulus, the drained and undrained Poisson's ratios are

$$\nu = \frac{3K - 2G}{6K + 2G}, \quad \nu_u = \frac{3K_u - 2G}{6K_u + 2G}. \tag{7.91}$$

Since $K_u > K$, and ν is an increasing function of K when G is held constant (Zimmerman, 1992), it follows from (7.91) that $\nu_u > \nu$. It also follows from (7.91) that $\nu_u < 1/2$, as is the case for the drained Poisson ratio. The Skempton coefficient B, Biot modulus M, and storativity coefficient S can be expressed in terms of the drained and undrained Poisson ratios as follows:

$$B = \frac{3(\nu_u - \nu)}{\alpha(1-2\nu)(1+\nu_u)}, \tag{7.92}$$

$$M = \frac{3K(\nu_u - \nu)}{\alpha^2(1-2\nu_u)(1+\nu)}, \tag{7.93}$$

$$S = \frac{(1-\nu_u)(1-2\nu)(1+\nu)\alpha^2}{3(1-\nu)(\nu_u - \nu)K}. \tag{7.94}$$

Table 7.2 lists the poroelastic parameters of several rocks, as compiled by Detournay and Cheng (1993) from Rice and Cleary (1976), Fatt (1958), Yew and Jogi (1978), and Yew et al. (1979).

7.6 One-dimensional consolidation

The problem of consolidation, in which a porous layer of rock or soil is subjected to an instantaneously applied normal load at its upper surface, is one of the most important problems in geotechnical engineering. This problem was originally formulated and solved in an ad hoc manner by Terzaghi in 1923 (see Terzaghi et al., 1996), under certain restrictive assumptions. Solution of this problem within the context of poroelasticity clarifies the meaning of the parameters appearing in Terzaghi's solution, and also serves as an illustrative example of the procedure needed to solve boundary-value problems in poroelasticity.

Consider a fluid-filled poroelastic layer extending from the surface $z = 0$ down to a depth $z = h$. In the context of the linearized theory developed in §7.5, the initial state of stress and pore fluid pressure can be taken to be zero. At time $t = 0$, a normal traction of magnitude P is applied at the upper surface. Initially, the layer deforms as an elastic layer with *undrained* elastic moduli, and an excess

Table 7.2 Poroelastic parameters for various rocks. Moduli are expressed in units of GPa; permeabilities in units of milliDarcies. The pore fluid in each case is water at 20°C.

Rock	G	K	ν	K_u	ν_u	K_m	α	B	M	ϕ	k
Ruhr sandstone	13	13	0.12	30	0.31	36	0.65	0.88	41	0.02	0.2
Tennessee marble	24	40	0.25	44	0.27	50	0.19	0.51	81	0.02	0.0001
Charcoal granite	19	35	0.27	41	0.30	45	0.27	0.55	84	0.02	0.0001
Berea sandstone	6.0	8.0	0.20	16	0.33	36	0.79	0.62	12	0.19	190
Westerly granite	15	25	0.25	42	0.34	45	0.47	0.85	75	0.01	0.0004
Weber sandstone	12	13	0.15	25	0.29	36	0.64	0.73	28	0.06	1.0
Ohio sandstone	6.8	8.4	0.18	13	0.28	31	0.74	0.50	9.0	0.19	5.6
Pecos sandstone	5.9	6.7	0.16	14	0.31	39	0.83	0.61	10	0.20	0.8
Boise sandstone	4.2	4.6	0.15	8.3	0.31	42	0.85	0.50	4.7	0.26	800

pore pressure is induced in the layer as a result of the Skempton effect. Gradually, the pore fluid drains out at, say, the upper surface, and the pore pressure relaxes back to its initial value. As this occurs, the layer continues to deform vertically downward. Eventually, the state of stress and strain in the layer is that of an elastic layer having the elastic moduli of the *drained* medium.

To solve this problem in the context of poroelasticity theory, it is necessary to first determine the proper initial conditions to be imposed an infinitesimally small time after $t = 0$. These conditions are those that are due to the undrained compression of the layer, which, in accordance with elasticity theory, occurs *instantaneously* at $t = 0$. If it is assumed that the rock mass is constrained so as to be unable to deform in the lateral direction, the only nonzero displacement is the vertical displacement, w, which depends only on the vertical coordinate, z. Consequently, the only nonzero strain is ε_{zz}. The full stress and strain tensors for undrained uniaxial strain are given by (7.64):

$$\tau_{zz}^{o} = P, \quad \tau_{xx}^{o} = \tau_{yy}^{o} = \nu P/(1-\nu), \tag{7.95}$$

$$\varepsilon_{zz}^{o} = P/(\lambda + 2G + \alpha^2 M), \quad \varepsilon_{xx}^{o} = \varepsilon_{yy}^{o} = 0, \tag{7.96}$$

where the superscript o denotes the initial, undrained response. All shear stresses and shear strains are zero. The initial induced pore pressure is found from (7.65) to be

$$P_{\mathrm{p}}^{o} = \frac{\alpha M}{(\lambda + 2G + \alpha^2 M)} P. \tag{7.97}$$

If the layer is assumed to be resting on a rigid, impermeable substrate, then (7.96) can be integrated to yield the instantaneous vertical displacement:

$$w^o = \frac{P}{(\lambda + 2G + \alpha^2 M)}(z - h). \tag{7.98}$$

During the ensuing consolidation process, the displacement and pore pressure fields must satisfy (7.66) and (7.82). Under the assumption of uniaxial strain, the only nonzero component of vector equation (7.66) is

$$(\lambda + 2G)\frac{\partial^2 w(z,t)}{\partial z^2} = -\alpha \frac{\partial P_p(z,t)}{\partial z}. \tag{7.99}$$

Integrating once with respect to z yields

$$(\lambda + 2G)\frac{\partial w(z,t)}{\partial z} + \alpha P_p(z,t) = g(t), \tag{7.100}$$

where the integration constant g may be a function of time. But from (7.45), the left-hand side of (7.100) is seen to be $\tau_{zz}(z,t)$, so (7.100) is equivalent to $\tau_{zz}(z,t) = g(t)$. Evaluation of this equation at $z = 0$, where $\tau_{zz} = P =$ constant, shows that $g(t) = P$, which then implies that

$$\tau_{zz}(z,t) = P, \tag{7.101}$$

for all values of z and t.

Since $\varepsilon_{zz} = \partial w/\partial z$, and $g(t) = P$, (7.100) can be written as

$$\varepsilon_{zz}(z,t) = \frac{1}{(\lambda + 2G)}[P - \alpha P_p(z,t)]. \tag{7.102}$$

In uniaxial strain, $\varepsilon_b = \varepsilon_{zz}$, so differentiation of (7.102) with respect to time gives

$$\frac{\partial \varepsilon_b}{\partial t} = \frac{-\alpha}{(\lambda + 2G)}\frac{\partial P_p}{\partial t}. \tag{7.103}$$

Combining (7.103) with (7.82) gives (Green and Wang, 1990)

$$\frac{k}{\mu}\frac{\partial^2 P_p}{\partial z^2} = \left[\frac{1}{M} + \frac{\alpha^2}{(\lambda + 2G)}\right]\frac{\partial P_p}{\partial t} \equiv S\frac{\partial P_p}{\partial t}. \tag{7.104}$$

where the storage coefficient S is defined in (7.89). The pore pressure is therefore governed by a diffusion equation, with diffusivity $D = k/\mu S$; this coefficient is also referred to in this context as the *coefficient of consolidation*, C. The initial condition for this diffusion equation is supplied by (7.97):

$$P_p(z, t = 0) = P_p^o = \frac{\alpha M}{(\lambda + 2G + \alpha^2 M)}P. \tag{7.105}$$

If the base of the layer is impermeable, and fluid is allowed to drain out at the upper surface through a permeable membrane, then the boundary conditions are, using (7.79),

$$P_p(z = 0, t) = 0, \quad \frac{\partial P_p}{\partial z}(z = h, t) = 0. \tag{7.106a,b}$$

The solution to the mathematical problem (7.104)–(7.106) can be taken from the monograph on heat conduction by Carslaw and Jaeger (1959, p. 97), by redefining the origin to be at the top of the layer, rather than the bottom:

$$P_p(z,t) = \frac{\alpha MP}{(\lambda + 2G + \alpha^2 M)} \sum_{n=1,3,\ldots}^{\infty} \frac{4}{n\pi} \sin\left(\frac{n\pi z}{2h}\right) \exp\left(\frac{-n^2\pi^2 kt}{4\mu Sh^2}\right). \quad (7.107)$$

The pore pressure starts at the value given by (7.105), and then decays to zero (Fig. 7.7a). The time required for the excess pore pressure to decay to a negligible value is found by setting the argument of the dominant exponential in (7.107) equal to 5, since $\exp(-5) < 0.01$; this gives an equilibration time of

$$t_{eq} \approx 20\mu Sh^2/\pi^2 k \approx 2\mu Sh^2/k. \quad (7.108)$$

The equilibration time may be as small as a few seconds for a thin, permeable sand layer, whereas it may be in the order of many years for a thick, relatively impermeable clay layer.

The rate of convergence of the series in (7.107) deteriorates as t approaches 0, so for small values of the time, the following equivalent form of the solution is more computationally convenient:

$$\frac{P_p(z,t)}{P_p^o} = 1 - \sum_{n=0}^{\infty} (-1)^n \left\{ \operatorname{erfc}\left[\frac{2nh + z}{(4kt/\mu S)^{1/2}}\right] + \operatorname{erfc}\left[\frac{2(n+1)h - z}{(4kt/\mu S)^{1/2}}\right] \right\}, \quad (7.109)$$

where $\operatorname{erfc}(x)$ is the coerror function (Abramowitz and Stegun, 1970), defined by

$$\operatorname{erfc}(x) \equiv 1 - \operatorname{erf}(x) = \frac{2}{\sqrt{\pi}} \int_x^{\infty} e^{-\eta^2} d\eta. \quad (7.110)$$

The vertical displacement is found by integrating (7.102), using (7.107) for the pore pressure, and bearing in mind that the displacement is zero at the base of

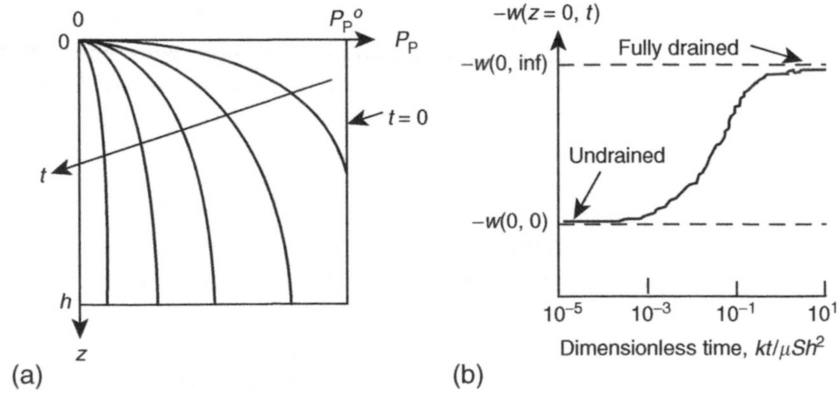

Fig. 7.7 Schematized graphs of the evolution of (a) excess pressure and (b) surface displacement in the one-dimensional consolidation problem.

the layer. This integration process yields

$$w(z,t) = \frac{P}{(\lambda+2G)}\left[(z-h) + \frac{\alpha^2 Mh}{(\lambda+2G+\alpha^2 M)}\sum_{n=1,3,\ldots}^{\infty}\frac{8}{n^2\pi^2}\cos\left(\frac{n\pi z}{2h}\right)\right.$$
$$\left.\times \exp\left(\frac{-n^2\pi^2 kt}{4\mu S h^2}\right)\right]. \quad (7.111)$$

The vertical displacement at the upper surface is given by (Fig. 7.7b)

$$w(0,t) = \frac{-Ph}{(\lambda+2G)}\left[1 - \frac{\alpha^2 M}{(\lambda+2G+\alpha^2 M)}\sum_{n=1,3,\ldots}^{\infty}\frac{8}{n^2\pi^2}\exp\left(\frac{-n^2\pi^2 kt}{4\mu S h^2}\right)\right]. \quad (7.112)$$

Since $1 + 1/3^2 + 1/5^2 + \cdots = \pi^2/8$, the surface displacement is initially equal to $-Ph/(\lambda + 2G + \alpha^2 M)$, which is the value appropriate for *undrained* uniaxial strain, and which agrees with (7.98). As t increases, the exponential terms in (7.112) decay to zero, and the surface displacement equilibrates to the value $-Ph/(\lambda + 2G)$, which corresponds to *drained* uniaxial strain.

Biot (1941) developed a set of equations for poroelastic deformation under the assumption that the pore fluid was incompressible. In his treatment of the consolidation problem, he also assumed a priori that the instantaneous vertical deformation was negligible compared to the equilibrium value of the displacement. Equation (7.112) shows that this second assumption corresponds to the limiting case $M \to \infty$, which physically corresponds to a porous medium composed of incompressible mineral grains, saturated with an incompressible fluid, with all of the compliance of the rock/fluid system being due to the pores and voids. In this case (7.39) shows that $\alpha \to 1$, and (7.104) shows that the consolidation coefficient reduces to $(\lambda + 2G)k/\mu$. The surface displacement (7.112) takes the form

$$w(0,t) = \frac{-Ph}{(\lambda+2G)}\left[1 - \sum_{n=1,3,\ldots}^{\infty}\frac{8}{n^2\pi^2}\exp\left(\frac{-n^2\pi^2(\lambda+2G)kt}{4\mu h^2}\right)\right], \quad (7.113)$$

which varies from 0 at $t = 0$ to $-Ph/(\lambda + 2G)$ as $t \to \infty$.

In Terzaghi's treatment of the one-dimensional consolidation problem (Terzaghi et al., 1996), the governing equation is taken to be

$$\frac{\partial^2 P_p}{\partial z^2} = \frac{\mu m}{k}\frac{\partial P_p}{\partial t}, \quad (7.114)$$

where m is a compliance coefficient that can be identified with $(\lambda + 2G)^{-1}$. The solution found by Terzaghi is essentially equivalent to (7.113). This problem is also discussed in detail by McNamee and Gibson (1960a,b).

A related problem with relevance to certain laboratory permeability measurements is that in which one face of a cylindrical specimen is instantaneously loaded by a fluid at some pressure P, while the lateral surfaces are restrained from expanding, and both the lateral surfaces and the opposing face are impermeable. This fluid loading serves to apply a normal stress at the upper surface of $\tau_{zz}(z = 0, t) = P$, while at the same time maintaining the pore pressure equal to P at that surface. Hence, this problem is equivalent to the consolidation problem solved above, but with the boundary condition (7.106) replaced by $P_{\mathrm{p}}(z = 0, t) = P$. The solution to the pressure-diffusion equation (7.104) is in this case given by

$$P_{\mathrm{p}}(z,t) = P - (P - P_{\mathrm{p}}^{o}) \sum_{n=1,3,\dots}^{\infty} \frac{4}{n\pi} \sin\left(\frac{n\pi z}{2h}\right) \exp\left(\frac{-n^2\pi^2 kt}{4\mu S h^2}\right), \tag{7.115}$$

with the initial induced pore pressure, P_{p}^{o}, again given by (7.105). The pore pressure in this case decays from P_{p}^{o} to P as $t \to \infty$.

Integration of (7.102) gives

$$w(z,t) = \frac{(1-\alpha)P(z-h)}{(\lambda+2G)} - \frac{\alpha(P - P_{\mathrm{p}}^{o})h}{(\lambda+2G)} \sum_{n=1,3,\dots}^{\infty} \frac{8}{n^2\pi^2} \cos\left(\frac{n\pi z}{2h}\right) \times \exp\left(\frac{-n^2\pi^2 kt}{4\mu S h^2}\right). \tag{7.116}$$

The longitudinal surface displacement is

$$w(0,t) = \frac{-(1-\alpha)Ph}{(\lambda+2G)} - \frac{\alpha(P - P_{\mathrm{p}}^{o})h}{(\lambda+2G)} \sum_{n=1,3,\dots}^{\infty} \frac{8}{n^2\pi^2} \exp\left(\frac{-n^2\pi^2 kt}{4\mu S h^2}\right). \tag{7.117}$$

The initial surface displacement is

$$w(z=0,t=0) = \frac{-(1-\alpha)Ph}{(\lambda+2G)} - \frac{\alpha(P - P_{\mathrm{p}}^{o})h}{(\lambda+2G)} = \frac{-(P - \alpha P_{\mathrm{p}}^{o})h}{(\lambda+2G)}, \tag{7.118}$$

whereas the final value of the surface displacement is

$$w(z=0,t\to\infty) = \frac{-(1-\alpha)Ph}{(\lambda+2G)}. \tag{7.119}$$

In practice (for realistic values of the various parameters), the initial induced pore never exceeds the applied pressure P, so it follows from (7.118) and (7.119) that the equilibrium settlement of the loaded surface is *smaller* in magnitude than the initial settlement, that is, the surface *rebounds* after its initial undrained response. The physical reason for this rebound effect is that as time increases, pore fluid diffuses into the rock from the upper surface, causing the rock to expand, in accordance with, say, (7.37). The time evolution of the surface displacement is identical to that shown in Fig. 7.7b for the consolidation problem, if the displacement axis is scaled appropriately so as to satisfy (7.118) and (7.119). Poroelastic problems involving hollow cylinders have been analyzed in detail by Abousleiman and Kanj (2004) and Kanj and Abousleiman (2004).

7.7 Applications of poroelasticity

Although the equations of poroelasticity have been available in substantially the same form as presented in §7.4 and §7.5 since the work of Biot (1941), for many years the only problems solved using the theory were those involving consolidation, which is of more relevance to soils mechanics than to rock mechanics. In the last few decades, however, much progress has been made in applying the equations of poroelasticity to various problems in tectonophysics and, in particular, in petroleum-related rock mechanics. In some specific cases, the results show that the effects of pore pressure on the mechanical deformation is negligible, whereas in other cases the poroelastic analysis reveals important and often unexpected phenomena that are not present in an uncoupled theory. Wang (2000) has given a review of some applications of poroelasticity to geophysical problems. A review of applications of poroelasticity theory to problems in petroleum-related rock mechanics has been given by Cheng et al. (1993).

Roeloffs and Rudnicki (1984) analyzed the induced pore pressures caused by creep along a planar fault in a poroelastic medium. They used a two-dimensional plane strain analysis and modeled the fault as an edge dislocation located along the negative x-axis along which occurs a relative slip of magnitude δ. The tip of the dislocation is assumed to travel to the right at some constant velocity, V. A moving (x, y) coordinate system was used, the origin of which was taken to coincide with the tip of the dislocation (Fig. 7.8). In the limiting case of infinitely rapid slip, the induced pore pressure field was found to be given by

$$P_{\mathrm{p}}(r,\theta) = \frac{BG(1+\nu_{\mathrm{u}})\delta}{3\pi(1-\nu_{\mathrm{u}})}\frac{\sin\theta}{r}, \tag{7.120}$$

whereas a somewhat more complicated pore pressure field was found in the general case of a finite propagation velocity. Using parameters appropriate to the rocks along the San Andreas fault near Hollister, California, and an assumed propagation velocity of 1 km/day, they were able to model with reasonable accuracy the pore pressure changes measured in wells located near the fault.

Segall (1989) used a two-dimensional plane strain poroelastic model to study the possibility of earthquakes being induced by fluid extraction from a subsurface reservoir. The reservoir was assumed to be a slab-shaped permeable layer of thickness h and width $2a$, located at a depth $d \gg h$ below the surface, surrounded by a fluid-saturated but impermeable rock mass (Fig. 7.9). A fluid increment $\Delta\zeta$ was assumed to be removed uniformly from throughout the reservoir, and

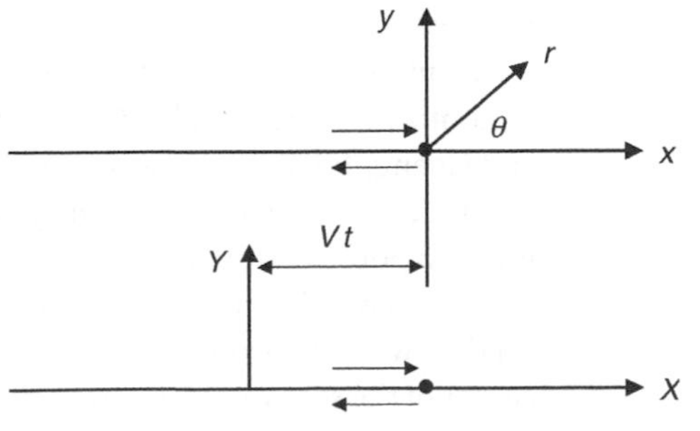

Fig. 7.8 Plane strain shear dislocation, with the relative displacement indicated by small arrows, moving through a poroelastic medium, as used by Roeloffs and Rudnicki (1984) to model a moving fault, showing the stationary coordinate system, (X, Y), and the moving coordinate system, (x, y).

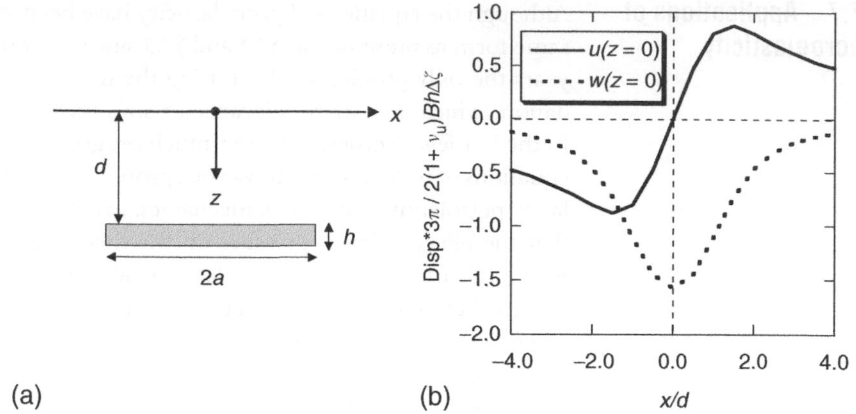

Fig. 7.9 (a) Two-dimensional reservoir of thickness h, width $2a$, and depth d; (b) Normalized vertical (w) and horizontal (u) displacements at the surface, due to withdrawal of fluid increment $\Delta\zeta$ from the reservoir, for the case $d = a$ (Segall, 1989).

the steady-state displacement field was then calculated. The predicted surface displacements were found to be (Fig. 7.9)

$$u(x, z = 0) = \frac{(1 + \nu_u)Bh\Delta\zeta}{3\pi} \ln\left[\frac{d^2 + (x + a)^2}{d^2 + (x - a)^2}\right], \tag{7.121}$$

$$w(x, z = 0) = \frac{2(1 + \nu_u)Bh\Delta\zeta}{3\pi}\left[\tan^{-1}\left(\frac{x - a}{d}\right) - \tan^{-1}\left(\frac{x + a}{d}\right)\right]. \tag{7.122}$$

Recalling the orientation of the axes (Fig. 7.9a) and our sign convention for displacements, we see that points on the surface move down and toward the origin.

The predicted displacements were in rough agreement with those measured above the oil reservoir at Wilmington, California. Moreover, the induced stress changes outside the reservoir were consistent with the type of faulting observed near several oilfields in Europe and North America. In particular, by examining the planes on which the induced shear stress was the largest, Segall correctly predicted that reverse faulting would occur above and below the reservoir, and normal faulting along the flanks. Although the success of this model is clear, by effectively assuming infinite permeability for the reservoir and zero permeability for the surrounding rock, the transient aspects of the poroelastic coupling between deformation and pore pressure are lost; an essentially equivalent analysis could therefore be carried out using a simpler elastic model in which pore fluid extraction acts as a nucleus of hydrostatic strain (Geertsma, 1973).

An interesting poroelastic phenomenon that cannot be predicted by an uncoupled model is the Mandel–Cryer effect (Mandel, 1953; Cryer, 1963; Detournay and Cheng, 1993). Imagine a poroelastic sphere that is instantaneously subjected to a fluid pressure increase of magnitude P along its outer boundary. Initially, an induced pore pressure is generated throughout the sphere, due to the Skempton effect. Eventually, the pore pressure equilibrates back to its initial value as the fluid flows out through the outer surface. However, the pore pressure at the center of the sphere continues to rise for some time after the initial application of the load, before dissipating, for the following reason. Fluid first drains out from the outer surface of the sphere, causing the elastic moduli there to decrease from the undrained to the drained values. The outer edge is then more compliant than the inner core, and so the stress increases within the stiffer inner region (see (8.190)), giving rise to a pore pressure *increase* in the inner core, due to the Skempton effect. A similar phenomenon occurs in a poroelastic layer under uniaxial stress or a poroelastic cylinder pressurized along its curved boundary.

Poroelastic effects also have many implications for boreholes drilled in permeable rocks. Detournay and Cheng (1988) analyze the stresses around a cylindrical borehole in a poroelastic rock mass subjected to a far-field uniform state of biaxial stress, decomposing this stress into an hydrostatic component P and a deviatoric component S (Fig. 7.10). The additional stresses that are induced by the presence of a borehole in a hydrostatic stress field are purely deviatoric (see §8.5), so there are no induced pore pressures in this case, and the poroelastic solution is identical to the elastic solution. However, the *deviatoric* component of the far-field stress does induce an *hydrostatic* component in the stress and strain field around the borehole, thereby bringing poroelastic effects into the problem. Initially, the induced tangential stress concentration around the borehole wall is given by

$$\Delta\tau_{\theta\theta}(a,\theta,t \to 0) = 4S\frac{(1-\nu_u)}{(1-\nu)}\cos 2\theta, \tag{7.123}$$

where a is the radius of the borehole and S is the magnitude of the deviatoric far-field stress. Eventually, after an elapsed time of about $10\mu a^2/Gk$, the stress relaxes to its drained, elastic value of

$$\Delta\tau_{\theta\theta}(a,\theta,t \to \infty) = 4S\cos 2\theta. \tag{7.124}$$

As the undrained Poisson ratio is always larger than the drained Poisson ratio, the stress concentration is initially smaller than it would be in the absence of poroelastic effects (Fig. 7.10). Detournay et al. (1989) use these results to infer that breakdown pressures in permeable formations will be much lower than would be predicted using a purely elastic analysis.

7.8 Thermoelasticity

The theory of thermoelasticity accounts for the effect of changes in temperature on the stresses and displacements in a body. The theory of thermoelasticity is to a great extent mathematically and physically analogous to the theory of poroelasticity, with the temperature playing a role similar to that of the pore pressure. For example, a change in temperature or a change in pore pressure in an isotropic body will each give rise to equal normal strains in three orthogonal directions and no shear strains. Furthermore, both the pore pressure and temperature fields are governed by diffusion-type equations. The analogies actually extend throughout the entire development of the theory, as shown by Geertsma (1957b) and Norris (1992). In practice, however, analysis of thermoelastic problems is usually simpler, because, as will be shown below, the effect of mechanical deformation on the temperature field can usually be ignored, whereas the effect

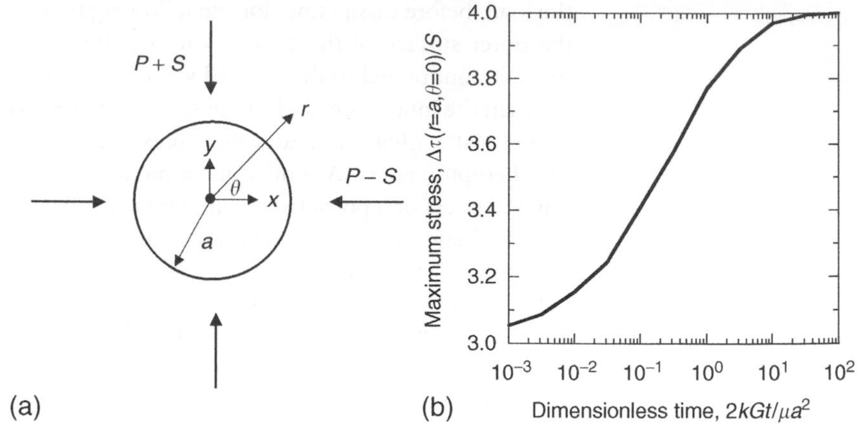

Fig. 7.10
(a) Borehole subjected to a far-field stress composed of a hydrostatic component, P, and a deviatoric component, S. (b) Maximum stress at the borehole wall, as a function of time, for the case $\nu = 0.2$, $\nu_u = 0.4$, $B = 0.8$ (after Detournay et al., 1989).

of mechanical deformation on the pore pressure field in most instances cannot be ignored.

Despite the similarity between the two theories, and the fact that thermoelasticity was developed a full century before poroelasticity (Duhamel, 1833; Biot, 1941), the two theories have traditionally been discussed independently of each other. In this section, the equations of thermoelasticity will be developed in a manner that parallels the derivation given of the poroelasticity equations in §7.4 and §7.5. Connections and analogies between the two theories have been discussed by Geertsma (1957b) and Norris (1992). A combined theory of linearized thermoporoelasticity, in which both thermal effects and pore pressure effects are accounted for, has been developed by, among others, Palciauskas and Domenico (1989), McTigue (1986), and Charlez (1991). The development of fully coupled, nonlinear models of thermohydromechanical behavior is currently being pursued by various research groups (see Stephansson, 1995).

Consider a solid piece of rock that is initially unstressed and at a uniform temperature T_0. This state can be taken as the reference state, in which the strains are by definition taken to be zero. If the temperature is changed to a new value $T > T_0$, the rock will expand. Under the assumption of linearity, this temperature rise will induce strains in the rock that are given by $\boldsymbol{\varepsilon} = -\boldsymbol{\beta}(T - T_0)$, where $\boldsymbol{\beta}$, whose components have dimension [1/°K], is a symmetric second-order tensor known as the *thermal expansivity tensor* (Nye, 1957). Although it is conventional to omit the ° symbol when using the Kelvin scale, we include it so as to avoid confusion with the bulk modulus. As most materials expand upon heating, the components of $\boldsymbol{\beta}$ are defined to be positive numbers, and the minus sign is included to account for the fact that extensional strains are considered to be negative in rock mechanics. If the rock is isotropic, then $\boldsymbol{\beta} = \beta\mathbf{I}$, where the scalar coefficient β is known as the coefficient of (linear) thermal expansion, and the thermal strain is then given by $\boldsymbol{\varepsilon} = -\beta(T - T_0)\mathbf{I}$. Although the coefficient of thermal expansion is more commonly denoted by α, the symbol β will be used here so as to avoid confusion with Biot's effective stress coefficient.

The basic assumption of linear thermoelasticity is that if the rock is subjected to both a temperature change and an applied stress state, then the resulting strain is the sum of the thermal strain and the stress-induced strain, that is,

$$\boldsymbol{\varepsilon} = \frac{1}{2G}\boldsymbol{\tau} - \frac{\nu}{2G(1+\nu)}\text{trace}(\boldsymbol{\tau})\mathbf{I} - \beta(T - T_o)\mathbf{I}, \tag{7.125}$$

which is similar in form to (7.37). To simplify the notation, the difference between the current temperature and the reference temperature can be denoted by ϑ, in which case (7.125) can be written as

$$\boldsymbol{\varepsilon} = \frac{1}{2G}\boldsymbol{\tau} - \frac{\nu}{2G(1+\nu)}\text{trace}(\boldsymbol{\varepsilon})\mathbf{I} - \beta\vartheta\mathbf{I}. \tag{7.126}$$

Taking the trace of both sides of (7.126) yields, in analogy with (7.41),

$$\varepsilon_b = \text{trace}(\boldsymbol{\varepsilon}) = \frac{\tau_m}{K} - 3\beta\vartheta, \tag{7.127}$$

which indicates that 3β is the *volumetric* thermal expansion coefficient. An increase in temperature will cause a negative bulk strain, which is to say that the bulk volume will increase, whereas a decrease in temperature will cause the bulk volume to decrease.

The equations for the stresses in terms of the strains are found by inverting (7.126) with the aid of (7.127):

$$\boldsymbol{\tau} = 2G\boldsymbol{\varepsilon} + \lambda\text{trace}(\boldsymbol{\varepsilon})\mathbf{I} + 3\beta K\vartheta\mathbf{I}, \tag{7.128}$$

which can be written explicitly as

$$\tau_{xx} = 2G\varepsilon_{xx} + \lambda(\varepsilon_{xx} + \varepsilon_{yy} + \varepsilon_{zz}) + 3\beta K\vartheta, \tag{7.129}$$

$$\tau_{yy} = 2G\varepsilon_{yy} + \lambda(\varepsilon_{xx} + \varepsilon_{yy} + \varepsilon_{zz}) + 3\beta K\vartheta, \tag{7.130}$$

$$\tau_{zz} = 2G\varepsilon_{zz} + \lambda(\varepsilon_{xx} + \varepsilon_{yy} + \varepsilon_{zz}) + 3\beta K\vartheta, \tag{7.131}$$

$$\tau_{xy} = 2G\varepsilon_{xy}, \quad \tau_{xz} = 2G\varepsilon_{xz}, \quad \tau_{yz} = 2G\varepsilon_{yz}. \tag{7.132}$$

The last terms on the right-hand sides of (7.129)–(7.131) are often referred to as the *thermal stresses*. However, thermally induced stresses are not caused by temperature changes per se, but rather by the combination of a change in temperature and a mechanical restraint that inhibits free expansion or contraction of the rock. For example, if a homogeneous rock is heated while free to expand, the strain-dependent and temperature-dependent terms in (7.128) cancel out and no "thermal stresses" are induced. On the other hand, if the rock is heated while rigidly clamped at its outer boundary, the normal stresses will be given by the ϑ-dependent terms in (7.129)–(7.131). In the subsurface, rocks are usually constrained to one extent or another, so the thermal stresses will be roughly in the order of $3\beta K\vartheta$. For typical values such as $K \approx 10$ GPa and $\beta \approx 10^{-5}/^\circ$K (Table 7.3), a temperature change of 10°K can induce a thermal stress of about 30 MPa. Temperature changes of this magnitude or greater may be expected to occur around underground radioactive waste canisters (Chan et al., 1982), during injection of cold water into geothermal wells for pressure maintenance (Pruess and Bodvarsson, 1984), or in various natural geothermal processes (Lowell, 1990). In such situations, thermally induced stresses may be of considerable importance.

The thermoelastic version of the Navier equations, which are known as the Duhamel–Neumann equations, are found by combining the stress–strain law (7.128) with the stress equilibrium equations (5.76), and the strain–displacement equations (2.222), to arrive at

$$G\nabla^2\mathbf{u} + (\lambda + G)\nabla(\nabla \cdot \mathbf{u}) = -\mathbf{f} - 3\beta K\nabla\vartheta, \tag{7.133}$$

where $\mathbf{f}$ is the body-force vector per unit volume. The *gradient* of the temperature, multiplied by the thermoelastic parameter $3\beta K$, acts as an additional body force, in a manner analogous to the pore pressure gradient.

The Duhamel–Neumann equations are three equations for the four unknowns $\{\mathbf{u}, \vartheta\}$, and so a fourth equation is needed. Whereas a diffusion-type equation was found for the pore pressure by considering conservation of *mass*, a diffusion-type equation for the temperature is found by considering conservation of *energy*. To do this, we introduce a conductive heat-flux vector, $\mathbf{q}_\mathrm{T}$, with units of $[\mathrm{J/m^2s}]$, or $[\mathrm{W/m^2}]$, which is in many ways analogous to the fluid flux vector, $\mathbf{q}$. The driving force for the conductive heat flux is the temperature gradient, as embodied by Fourier's law (Nye, 1957, p. 195):

$$\mathbf{q}_\mathrm{T} = -\mathbf{k}_\mathrm{T}\nabla\vartheta, \tag{7.134}$$

where $\mathbf{k}_\mathrm{T}$ is the thermal conductivity tensor, with units of W/m°K. The thermal conductivity tensor is a second-order tensor and obeys the usual transformation rules when the coordinate system is rotated. Moreover, it is usually assumed that $\mathbf{k}_\mathrm{T}$ is symmetric, so all of the results concerning principal directions, etc., that apply to the permeability tensor also apply to the thermal conductivity tensor and need not be repeated here. If the rock is isotropic with respect to thermal conductivity, then $\mathbf{k}_\mathrm{T} = k_\mathrm{T}\mathbf{I}$, where the scalar coefficient k_T is referred to as "the thermal conductivity."

The standard diffusion equation for the temperature can be derived from considerations of conservation of energy by ignoring the "strain-dependent" part of the internal energy (recall §5.8) in comparison to the "temperature-dependent" part. The temperature-dependent part of the energy is given by

$$u = u_\mathrm{o} + c_\mathrm{v}\vartheta, \tag{7.135}$$

where u [J/kg] is the energy per unit mass, u_o [J/kg] is some arbitrary reference energy, and c_v [J/kg°K] is the specific heat at constant strain. If the elastic strain energy is neglected, the net flux of heat conducted into any region of rock must equal the rate of change of the nonmechanical part of the internal energy. Following the developments of §7.5, this energy balance can be stated in integral form as

$$\iiint\limits_R \frac{\partial(\rho u)}{\partial t}\mathrm{d}V = -\iint\limits_{\partial R} (\mathbf{q}_\mathrm{T} \cdot \mathbf{n})\mathrm{d}A, \tag{7.136}$$

where the density ρ is needed to convert the energy density *per unit mass* into an energy density *per unit volume*. Use of the divergence theorem gives

$$\rho\frac{\partial u}{\partial t} + \nabla \cdot \mathbf{q}_{\mathrm{T}} = 0, \tag{7.137}$$

where, in keeping with the spirit of a linearized theory, the density is assumed to be constant. Using (7.135) for the internal energy and (7.134) for the conductive heat flux allows (7.137) to be written as

$$\frac{\partial \vartheta}{\partial t} = \frac{k_{\mathrm{T}}}{\rho c_{\mathrm{v}}}\nabla^2\vartheta, \tag{7.138}$$

where $k_{\mathrm{T}}/\rho c_{\mathrm{v}} \equiv D_{\mathrm{T}}$, with units of $[\mathrm{m}^2/\mathrm{s}]$, is the *thermal diffusivity*. Equation (7.138), which governs heat conduction through a rigid medium, is the direct analogue of (7.81) for fluid flow through a rigid porous medium.

If the contribution of the stored strain energy to the overall energy balance is included, an additional component appears on the right-hand side of (7.138), in analogy with (7.82):

$$\frac{\partial \vartheta}{\partial t} = \frac{k_{\mathrm{T}}}{\rho c_{\mathrm{v}}}\nabla^2\vartheta + \frac{3K\beta T_{\mathrm{o}}}{\rho c_{\mathrm{v}}}\frac{\partial \varepsilon_{\mathrm{b}}}{\partial t}. \tag{7.139}$$

The reference temperature T_{o} in (7.139) must be taken as the absolute temperature (Nowacki, 1986, p. 11), which is to say it should be expressed in °K rather than °C. Equations (7.133) and (7.139) supply the four governing equations needed to provide a well-posed mathematical problem for the field variables $\{\mathbf{u}, \vartheta\}$. The diffusion equation for the temperature can also be expressed, in a form analogous to (7.49), by using (7.127) to express the bulk strain in terms of the mean normal stress and the temperature:

$$\frac{\partial \vartheta}{\partial t} = \frac{k_{\mathrm{T}}}{(\rho c_{\mathrm{v}} + 9K\beta^2 T_{\mathrm{o}})}\nabla^2\vartheta + \frac{3\beta T_{\mathrm{o}}}{(\rho c_{\mathrm{v}} + 9K\beta^2 T_{\mathrm{o}})}\frac{\partial \tau_{\mathrm{m}}}{\partial t}. \tag{7.140}$$

A rigorous derivation of the coupled thermomechanical diffusion equation for the temperature, which requires extensive discussion of thermodynamic considerations, is given by Boley and Weiner (1960) and Nowacki (1986). However, the fully coupled form of the temperature equation is rarely used in rock mechanics, because the effect of mechanical stresses or deformations on the temperature is usually small. For example, consider a Berea sandstone undergoing a process in which the temperature change is in the order of 10°K, and the stress change is in the order of 10 MPa. Assuming that the temperature and stress vary at nearly the same rate and using properties from Tables 7.2 and 7.3, it follows that the stress-dependent term on the right-hand side of (7.140) is less than temperature derivative term on the left by about three orders of magnitude. Equation (7.140) will then reduce identically to (7.138) if the following condition is satisfied (McTigue, 1986):

$$\frac{9K\beta^2 T_{\mathrm{o}}}{\rho c_{\mathrm{v}}} \ll 1. \tag{7.141}$$

Examination of the property values in Tables 7.2 and 7.3 shows that the dimensionless ratio in (7.141) is usually less than 0.01.

If the coupling term in (7.140) is neglected, the temperature field can be found by solving the diffusion equation (7.138), without any reference to the stresses or strains. Solutions to the diffusion equation for many geometries and boundary conditions can be found in standard texts such as Crank (1956), Carslaw and Jaeger (1959), and Matthews and Russell (1967). The temperature field can then be used in (7.133) to calculate the pseudo-body-force, after which the displacements, stresses, and strains can be found using any method for solving the elasticity equations with a known body force. Although the temperature field will typically be a function of time, time derivatives of the displacement components do not appear in (7.133), and so t appears merely as a parameter in the equations; the equations of uncoupled thermoelasticity are therefore often referred to as being quasi-static.

Table 7.3 lists the thermal conductivities, thermal expansion coefficients, densities, and heat capacities of various dry rocks and pore fluids. In contrast to the permeability, which varies by many orders of magnitude among the various rock types, the thermal conductivity of a rock is almost always in the range of 1–10 W/m°K (Charlez, 1991, p. 179). This is because the thermal conductivities of most minerals are in this range, and pore structure has a relatively minor effect on thermal conductivity, as compared to, say, permeability. The thermal conductivity of a liquid-saturated rock will be higher than that of the rock under dry conditions – a few percent higher in the case of low-porosity crystalline rocks and as much as factor of two higher for porous rocks such as sandstones (Somerton, 1992, p. 234–5). The thermal expansion coefficients do not vary widely, either, because most rock-forming minerals have similar thermal expansion coefficients, and, in contrast to other rock properties, thermal expansivity is completely unaffected by the presence of cracks and pores (Grimvall, 1986, p. 266). Under drained conditions, the effective thermal expansion coefficient of a fluid-saturated rock is simply equal to that of the dry rock, whereas under undrained conditions it is given by (McTigue, 1986)

$$\beta_u = \beta_m + \phi B(\beta_f - \beta_m), \tag{7.142}$$

where B is the Skempton coefficient, ϕ is the porosity, and the subscripts m and f denote the mineral phase and fluid phase, respectively. The effective density and effective heat capacity of a fluid-saturated rock are exactly given by *volumetrically* weighted averages of the values for the mineral phase and the pore fluid. It follows that the relationship between the density of a rock under the dry and fully saturated conditions is

$$\rho(\text{saturated}) = (1 - \phi)\rho_m + \phi\rho_f = \rho(\text{dry}) + \phi\rho_f. \tag{7.143}$$

Finally, it should be mentioned that thermal properties of rocks, particularly the thermal conductivity, are usually stress-dependent and temperature-dependent; these dependencies are discussed in detail by Somerton (1992). The values in Table 7.3 are generally at 20°C, but at various stress levels, and are intended

Table 7.3 Thermal properties of various rocks and pore fluids.

a – Somerton (1992); b – Giraud and Rousset (1996); c – Arndt et al. (1997); d – McTigue (1986); e – Tuma (1983).

Rock Type	ϕ	β (1/°K)	k_T (W/m°K)	ρc_v (J/m^3°K)	Reference
Berea sandstone (dry)	0.162	1.5×10^{-6}	2.34	1.76×10^6	a
Boom clay (wet)	0.450	3.3×10^{-6}	1.70	2.82×10^6	b
Granodiorite	—	4.7×10^{-5}	2.70	1.90×10^6	c
Halite Salt (wet)	0.001	4.0×10^{-5}	6.62	1.89×10^6	d
Water	—	6.6×10^{-5}	0.60	4.17×10^6	e
Kerosene	—	3.2×10^{-4}	0.13	1.67×10^6	e

to give an indication of the range of values that may be encountered in the subsurface.

A basic thermoelastic problem in rock mechanics is that in which the temperature at a traction-free surface is instantaneously raised by a certain amount ϑ_0, after which it is held constant. The one-dimensional version of this problem is mathematically identical to the pressure-diffusion problem of §7.6. However, the thermal diffusivity is often very low, so in many cases the thermal pulse will not reach any of the other boundaries of the rock mass within elapsed times that are of engineering interest. It is therefore meaningful to consider the case of thermal diffusion into a semi-infinite half-space, rather than into a layer of finite thickness. The solution to (7.138) in this case is (Carslaw and Jaeger, 1959, p. 60)

$$\vartheta(z,t) = \vartheta_0 \mathrm{erfc}[z/(4D_T t)^{1/2}] \equiv \frac{2\vartheta_0}{\sqrt{\pi}} \int_{z/\sqrt{4D_T t}}^{\infty} e^{-\eta^2} d\eta. \tag{7.144}$$

This temperature profile could also be derived from (7.109) and (7.110) by replacing the hydraulic diffusivity with the thermal diffusivity, taking the limit as $h \to \infty$, and appropriately transforming the boundary and initial conditions; since $\mathrm{erfc}(\infty) = 0$, only the first of the coerror function terms remains in the series, yielding (7.144). This solution describes a temperature front of magnitude ϑ_0 diffusing into the rock mass, with a depth of penetration that is approximately equal to $(4D_T t)^{1/2}$. As thermal diffusivities are usually in the order of $10^{-6}\mathrm{m}^2/\mathrm{s}$, the thermal pulse will require a few days to travel 1 m into the rock, about one year to extend 10 m into the rock, and about one hundred years to extend 100 m into the rock.

It can be shown that the displacement field that satisfies the Duhamel–Neumann equations (7.133) and the traction-free boundary condition is (McTigue, 1986)

$$w(z,t) = -\frac{(1+\nu)}{(1-\nu)}\beta \int_0^z \vartheta(\xi,t)d\xi + \frac{(1+\nu)}{(1-\nu)}\beta\vartheta_o\sqrt{\frac{4D_T t}{\pi}}, \tag{7.145}$$

with $\vartheta(z,t)$ given by (7.144), and $u = v = 0$. The only nonzero strain component is found from (7.145) to be

$$\varepsilon_{zz}(z,t) = -\frac{(1+\nu)}{(1-\nu)}\beta\vartheta(z,t), \tag{7.146}$$

after which the nonzero stresses are found from (7.129)–(7.132) to be given by

$$\tau_{xx}(z,t) = \tau_{yy}(z,t) = 3\beta K\frac{(1-2\nu)}{(1-\nu)}\vartheta(z,t). \tag{7.147}$$

The maximum thermally induced stress occurs at the heated surface and is given by

$$\tau_{\max} = \tau_{xx}(0,t) = \tau_{yy}(0,t) = 3\beta K\frac{(1-2\nu)}{(1-\nu)}\vartheta_o. \tag{7.148}$$

The two-dimensional and three-dimensional versions of this problem, in which the heated surface is a cylindrical borehole or a spherical cavity, exhibit similar behavior, although the mathematical details are more complicated, particularly for the cylindrical case. Interestingly, the maximum stress concentration in each case occurs at the heated boundary and is identical to that given by (7.148). If a spherical cavity of radius a has its surface temperature raised by an amount ϑ_o, the tangential normal stresses at the cavity surface will be (Nowacki, 1986, p. 221)

$$\tau_{\theta\theta} = \tau_{\phi\phi} = 3\beta K\frac{(1-2\nu)}{(1-\nu)}\vartheta_o. \tag{7.149}$$

Rehbinder (1985) considered the cases where the temperature of the surface of the spherical cavity increases linearly with time, or oscillates sinusoidally with time, in order to help in the design of underground storage caverns for hot water used in domestic heating.

The thermally induced hoop stress at the wall of a cylindrical borehole or tunnel whose surface is subjected to a temperature change of ϑ_o is given by (Stephens and Voight, 1982)

$$\tau_{\theta\theta} = 3\beta K\frac{(1-2\nu)}{(1-\nu)}\vartheta_o. \tag{7.150}$$

In all three cases, the thermal stresses *within* the heated region of rock are in the order of $3\beta K(T - T_o)$, but are always less than the maximum stress (7.148), which occurs at the heated boundary.

8 Stresses around cavities and excavations

8.1 Introduction

Some of the more important problems in rock mechanics involve the calculation of the stresses and displacements around subsurface cavities and excavations. On a macroscopic scale, calculation of the stresses and displacements around boreholes, tunnels and mine excavations is of paramount importance to petroleum, mining, and civil engineers. On a microscopic scale, the calculation of stresses around small voids and cracks in a rock is a necessary first step in the development of micromechanically based theories of rock deformation and failure (see Chapter 10). As there is no intrinsic "size effect" in classical linear elasticity, or in the classical theories of plastic or viscoelastic behavior, the deformation of both engineering-scale excavations and microscale cracks and pores is governed by the same equations.

Although computational methods such as finite elements (Pande et al., 1990) and boundary elements (Brady and Brown, 2004) are often used nowadays to calculate stresses around excavations of complex shape, analytical solutions for simplified shapes such as cylindrical boreholes or elliptical cracks are extremely useful in elucidating general trends. Analytical solutions have the advantage of clearly displaying the manner in which the results are influenced by parameters such as the Poisson's ratio of the rock or the aspect ratio of a crack. Moreover, there are mathematically complex phenomena, such as stress concentration factors around crack tips, which can reliably be investigated only by analytical means. On the other hand, for the most part, analytical solutions require the assumption of linear elastic behavior, whereas accounting for nonlinear stress–strain behavior poses little additional burden in a fully numerical approach. Hence, the analytical methods presented in this chapter are increasingly being complemented by numerical methods.

In this chapter, we will concentrate on analytical solutions to important problems involving cavities and voids in rock. In two dimensions, powerful general methods exist to solve essentially all such problems, using the theory of complex variables. These methods will be developed in §8.2 and then used to solve for the stresses and displacements around tubular cavities of various cross-sectional shape (§8.5–§8.6, §8.9–§8.11). Problems involving elastic inclusions (§8.8) and circular disks and cylinders (§8.7) can also be solved using the same general approach. Although many techniques have been devised to solve three-dimensional problems in elasticity, these methods are more variegated and complicated. Hence, several important three-dimensional solutions, such as for spherical cavities and penny-shaped cracks, will be presented and discussed with either partial or no details given of the derivation (§8.12–§8.13). A brief discussion of the much more difficult type of problem involving stress-field interactions between nearby cavities is given in §8.14.

8.2 Complex variable method for two-dimensional elasticity problems

Two-dimensional problems in elasticity can be solved using the complex variable method developed by Kolosov (1909) and Muskhelishvili (1963). In this method, the displacements and stresses are represented in terms of two analytic functions of a complex variable. General solutions are readily generated, as any pair of analytic functions automatically leads to displacements and stresses that satisfy the equations of stress equilibrium and Hooke's law; the only nontrivial aspect of the solution procedure is in satisfying the boundary conditions of the specific problem at hand. A full treatment of the theory of functions of a complex variable can be found in standard texts such as Nehari (1961) or Silverman (1967). A brief introduction to those aspects of the theory that are needed for solving elasticity problems is given below.

A complex number z can be represented as $z = x + \mathrm{i}y$, where x and y are real numbers that obey the usual arithmetic laws, and i is an "imaginary" number that has the additional property that $\mathrm{i}^2 = -1$. If z is represented in this form, x is known as the *real part* of z, and y is known as the *imaginary part* of z; symbolically, this is expressed as $x = \mathbf{Re}(z)$ and $y = \mathbf{Im}(z)$. The *complex conjugate* of z is another complex number, $\bar{z}$, that is defined by

$$z = x + \mathrm{i}y, \quad \bar{z} = x - \mathrm{i}y, \tag{8.1}$$

from which it follows that $\mathbf{Re}(z) = (z + \bar{z})/2$, and $\mathbf{Im}(z) = (z - \bar{z})/2\mathrm{i}$.

A complex-valued function of a complex variable can be represented by $\zeta(z)$, where the dependent variable is a complex number that can be written as

$$\zeta = \xi + \mathrm{i}\eta, \tag{8.2}$$

Both ξ and η are real-valued functions of the two real variables x and y. The function $\zeta(z)$ is said to be *analytic* in a certain region of the complex plane if it is continuous throughout that region and has a derivative at every point in the region. The derivative, denoted by $\mathrm{d}\zeta/\mathrm{d}z \equiv \zeta'(z)$, is also a complex number and so can temporarily be written as $\zeta'(z) = a + \mathrm{i}b$. The differential of ζ can therefore be expressed as

$$\mathrm{d}\zeta = \zeta'(z)\mathrm{d}z = (a + \mathrm{i}b)(\mathrm{d}x + \mathrm{i}\mathrm{d}y) = (a\mathrm{d}x - b\mathrm{d}y) + \mathrm{i}(b\mathrm{d}x + a\mathrm{d}y), \tag{8.3}$$

but can also, by virtue of (8.2), be written as

$$\mathrm{d}\zeta = \mathrm{d}\xi + \mathrm{i}\mathrm{d}\eta = \left(\frac{\partial\xi}{\partial x}\mathrm{d}x + \frac{\partial\xi}{\partial y}\mathrm{d}y\right) + \mathrm{i}\left(\frac{\partial\eta}{\partial x}\mathrm{d}x + \frac{\partial\eta}{\partial y}\mathrm{d}y\right). \tag{8.4}$$

Comparison of (8.3) and (8.4) reveals that the two functions ξ and η must satisfy the so-called Cauchy–Riemann equations,

$$\frac{\partial \xi}{\partial x} = \frac{\partial \eta}{\partial y}, \quad \frac{\partial \xi}{\partial y} = -\frac{\partial \eta}{\partial x}. \tag{8.5}$$

Hence, $\zeta'(z)$ can be written in any of the following four equivalent forms:

$$\zeta'(z) = \frac{\partial \xi}{\partial x} - \mathrm{i}\frac{\partial \xi}{\partial y} = \frac{\partial \eta}{\partial y} + \mathrm{i}\frac{\partial \eta}{\partial x} = \frac{\partial \xi}{\partial x} + \mathrm{i}\frac{\partial \eta}{\partial x} = \frac{\partial \eta}{\partial y} - \mathrm{i}\frac{\partial \xi}{\partial y}. \tag{8.6}$$

If $\zeta(z)$ is analytic, it follows from (8.5) that both the real and imaginary parts of ζ satisfy Laplace's equation,

$$\nabla^2\xi(x,y) = 0, \quad \nabla^2\eta(x,y) = 0, \tag{8.7}$$

where the Laplacian operator ∇^2 is defined by

$$\nabla^2 = \frac{\partial^2}{\partial x^2} + \frac{\partial^2}{\partial y^2}. \tag{8.8}$$

Functions that satisfy Laplace's equation are called *harmonic functions*, and two harmonic functions that are related through the Cauchy–Riemann equations (8.5) are called *conjugate harmonic functions*, not to be confused with the unrelated concept of a *complex conjugate* defined above in (8.1). It can be proven that if ξ is any harmonic function, (8.5) can always be integrated to find its harmonic conjugate function, η, and vice versa; moreover, the resulting function $\zeta = \xi + \mathrm{i}\eta$ will be an analytic function of z.

It was demonstrated in §5.7 that two-dimensional elasticity problems can be solved in terms of an Airy stress function, U, which satisfies the biharmonic equation,

$$\nabla^2(\nabla^2 U) = 0. \tag{8.9}$$

In the absence of a body force, the stresses are given by

$$\tau_{xx} = \frac{\partial^2 U}{\partial y^2}, \quad \tau_{yy} = \frac{\partial^2 U}{\partial x^2}, \quad \tau_{xy} = -\frac{\partial^2 U}{\partial x \partial y}. \tag{8.10}$$

We now demonstrate that the Airy stress function for a particular problem can always be expressed in terms of two analytic functions of a complex variable (Timoshenko and Goodier, 1970). First, let

$$P = \tau_{xx} + \tau_{yy} = \frac{\partial^2 U}{\partial y^2} + \frac{\partial^2 U}{\partial x^2} = \nabla^2 U, \tag{8.11}$$

so that, by (8.9), P is an harmonic function. Then, as remarked above, its harmonic conjugate, Q, can in principle be found, and the function

$$f(z) = P + \mathrm{i}Q \tag{8.12}$$

will be analytic. The function found by integrating $f(z)$ along any contour that lies within a region in which $f(z)$ is analytic will itself be analytic (Silverman, 1967), so we can define another analytic function, $\phi(z)$, by

$$\phi(z) = \frac{1}{4}\int f(z)\mathrm{d}z \equiv p + \mathrm{i}q, \tag{8.13}$$

where p and q are the real and imaginary parts of $\phi(z)$. It follows from (8.6), (8.12), and (8.13) that

$$\phi'(z) = \frac{\partial p}{\partial x} + \mathrm{i}\frac{\partial q}{\partial x} = \frac{1}{4}f(z) = \frac{1}{4}(P + \mathrm{i}Q). \tag{8.14}$$

Equating the real and imaginary parts of (8.14) and making use of (8.5), gives

$$\frac{1}{4}P = \frac{\partial p}{\partial x} = \frac{\partial q}{\partial y}, \quad \frac{1}{4}Q = \frac{\partial q}{\partial x} = -\frac{\partial p}{\partial y}. \tag{8.15}$$

Next, note that the function $p_1 = U - px - qy$ is harmonic, since

$$\nabla^2(U - px - qy) = \nabla^2 U - x\nabla^2 p - 2\frac{\partial p}{\partial x} - y\nabla^2 q - 2\frac{\partial q}{\partial y} = 0, \tag{8.16}$$

in which we have used (8.7), (8.11), and (8.15). It follows that p_1 is the real part of an analytic function, which we call $\chi(z)$. Noting that

$$\mathbf{Re}\{\bar{z}\phi(z)\} = \mathbf{Re}\{(x - \mathrm{i}y)(p + \mathrm{i}q)\} = px + qy, \tag{8.17}$$

it follows that

$$\begin{aligned} U &= p_1 + px + qy = \mathbf{Re}\{\chi(z)\} + \mathbf{Re}\{\bar{z}\phi(z)\} \\ &= \frac{1}{2}\{\bar{z}\phi(z) + z\overline{\phi(z)} + \chi(z) + \overline{\chi(z)}\}, \end{aligned} \tag{8.18}$$

where $\overline{\phi(z)} = p - \mathrm{i}q$, etc. We have therefore shown that the Airy stress function U can always be expressed in terms of two analytic functions, $\phi(z)$ and $\chi(z)$.

By differentiating (8.18), it can be shown that the derivatives of U can be expressed in terms of ϕ and χ, as follows:

$$2\frac{\partial U}{\partial x} = \phi(z) + \bar{z}\phi'(z) + \overline{\phi(z)} + z\overline{\phi'(z)} + \chi'(z) + \overline{\chi'(z)}, \tag{8.19}$$

$$2\frac{\partial U}{\partial y} = -\mathrm{i}\phi(z) + \mathrm{i}\bar{z}\phi'(z) + \mathrm{i}\overline{\phi(z)} - \mathrm{i}z\overline{\phi'(z)} + \mathrm{i}\chi'(z) - \mathrm{i}\overline{\chi'(z)}, \tag{8.20}$$

$$2\frac{\partial^2 U}{\partial x^2} = 2\phi'(z) + \bar{z}\phi''(z) + 2\overline{\phi'(z)} + z\overline{\phi''(z)} + \chi''(z) + \overline{\chi''(z)}, \tag{8.21}$$

$$2\frac{\partial^2 U}{\partial y^2} = 2\phi'(z) - \bar{z}\phi''(z) + 2\overline{\phi'(z)} - z\overline{\phi''(z)} - \chi''(z) - \overline{\chi''(z)}, \tag{8.22}$$

$$2\frac{\partial^2 U}{\partial x \partial y} = \mathrm{i}\bar{z}\phi''(z) - \mathrm{i}z\overline{\phi''(z)} + \mathrm{i}\chi''(z) - \mathrm{i}\overline{\chi''(z)}, \tag{8.23}$$

from which the stresses follow directly, as indicated by (8.10).

The strain components are related to the stresses according to (5.46)–(5.47):

$$8G\varepsilon_{xx} = (\kappa + 1)\tau_{xx} + (\kappa - 3)\tau_{yy}, \tag{8.24}$$

$$8G\varepsilon_{yy} = (\kappa + 1)\tau_{yy} + (\kappa - 3)\tau_{xx}, \tag{8.25}$$

where $\kappa = 3 - 4\nu$ for plane strain, and $\kappa = (3 - \nu)/(1 + \nu)$ for plane stress. These relations can also be written as

$$2G\frac{\partial u}{\partial x} = 2G\varepsilon_{xx} = -\tau_{yy} + \frac{1}{4}(\kappa + 1)(\tau_{xx} + \tau_{yy}), \tag{8.26}$$

$$2G\frac{\partial v}{\partial y} = 2G\varepsilon_{yy} = -\tau_{xx} + \frac{1}{4}(\kappa + 1)(\tau_{xx} + \tau_{yy}). \tag{8.27}$$

Making use of (8.10), (8.11), and (8.15), we can rewrite (8.26) and (8.27) as

$$2G\frac{\partial u}{\partial x} = -\frac{\partial^2 U}{\partial x^2} + (\kappa + 1)\frac{\partial p}{\partial x}, \tag{8.28}$$

$$2G\frac{\partial v}{\partial y} = -\frac{\partial^2 U}{\partial y^2} + (\kappa + 1)\frac{\partial q}{\partial y}. \tag{8.29}$$

Integrating (8.28) with respect to x and (8.29) with respect to y, gives

$$2Gu = -\frac{\partial U}{\partial x} + (\kappa + 1)p + g(y), \tag{8.30}$$

$$2Gv = -\frac{\partial U}{\partial y} + (\kappa + 1)q + h(x), \tag{8.31}$$

where g and h are (as yet) unknown functions of y and x, respectively.

To study these two functions, recall that $2G\varepsilon_{xy} = \tau_{xy}$, in which case (8.10) gives

$$2G\left(\frac{\partial u}{\partial y} + \frac{\partial v}{\partial x}\right) = -2\frac{\partial^2 U}{\partial x \partial y}. \tag{8.32}$$

But differentiation of (8.30) and (8.31), and use of (8.15), gives

$$2G\left(\frac{\partial u}{\partial y} + \frac{\partial v}{\partial x}\right) = -2\frac{\partial^2 U}{\partial x \partial y} + g'(y) + h'(x). \tag{8.33}$$

Comparison of (8.32) and (8.33) shows that g and h must satisfy the equation

$$g'(y) + h'(x) = 0, \tag{8.34}$$

the general solution to which is

$$g(y) = \omega y + a, \quad h(x) = -\omega x + b, \tag{8.35}$$

where ω, a, and b are constants. The portion of the total displacement vector that is represented by g and h therefore corresponds to a rigid-body motion and has no stresses associated with it. Ignoring this rigid-body motion, the displacement vector can be written as a complex number, using (8.13), (8.19), (8.20), (8.30), and (8.31):

$$2G(u+\mathrm{i}v) = -\left(\frac{\partial U}{\partial x} + \mathrm{i}\frac{\partial U}{\partial y}\right) + (\kappa+1)(p+\mathrm{i}q) = \kappa\phi(z) - z\overline{\phi'(z)} - \overline{\chi'(z)}. \tag{8.36}$$

Although the additional rigid-body displacement represented by (8.35) has no effect on the stresses, it is not correct to say that it can be ignored. When finding the complex potentials ϕ and χ, care must be taken to insure that the resulting displacement does not contain a spurious rigid-body rotation. In solving the problem of a cavity in an infinite rock mass, the desired solution is usually the one that has no rotation at infinity. In some cases, this will require that an additional rigid-body displacement be added to the displacement that was found by considering only the stress boundary conditions. For example, Maugis (1992) pointed out that the solution given by Stevenson (1945) for an elliptical crack in a biaxial stress field actually contains an unwanted rigid-body rotation component; the correct solution is given in §8.9.

Noting that only $\chi'(z)$ appears in the expressions for the complex displacement and the stresses, rather than the function $\chi(z)$ itself, we can define a new function

$$\psi(z) = \chi'(z), \tag{8.37}$$

allowing (8.36) to be rewritten in a slightly simpler form as

$$2G(u+\mathrm{i}v) = \kappa\phi(z) - z\overline{\phi'(z)} - \overline{\psi(z)}. \tag{8.38}$$

For the case discussed in §5.3, in which there is a uniform strain in the longitudinal direction, we must add a term $-2G\nu\varepsilon$ to the right-hand sides of (8.28) and (8.29), and set $\kappa = 3-4\nu$. Integration of (8.28) and (8.29) then leads, in this case, to

$$2G(u+\mathrm{i}v) = (3-4\nu)\phi(z) - z\overline{\phi'(z)} - \overline{\psi(z)} - 2G\nu\varepsilon z. \tag{8.39}$$

The equations for the change in the displacement and stress components due to a rotation of the coordinate axes take a particularly simple form when written in terms of complex variables. Recalling that $\mathrm{e}^{\mathrm{i}\theta} = \cos\theta + \mathrm{i}\sin\theta$, (2.225) can be written as

$$u' + \mathrm{i}v' = (u+\mathrm{i}v)\mathrm{e}^{-\mathrm{i}\theta}. \tag{8.40}$$

Similarly, the equations (2.25)–(2.27) for the transformation of the stress components can be written as

$$\tau_{y'y'} - \tau_{x'x'} + 2\mathrm{i}\tau_{x'y'} = (\tau_{yy} - \tau_{xx} + 2\mathrm{i}\tau_{xy})\mathrm{e}^{2\mathrm{i}\theta}, \tag{8.41}$$

$$\tau_{y'y'} + \tau_{x'x'} = \tau_{yy} + \tau_{xx}. \tag{8.42}$$

In the special case in which the new coordinate system is a polar coordinate system, with the r direction rotated by an angle θ from the x-axis, we have

$$\tau_{\theta\theta} - \tau_{rr} + 2\mathrm{i}\tau_{r\theta} = (\tau_{yy} - \tau_{xx} + 2\mathrm{i}\tau_{xy})\mathrm{e}^{2\mathrm{i}\theta}, \tag{8.43}$$

$$\tau_{\theta\theta} + \tau_{rr} = \tau_{yy} + \tau_{xx}. \tag{8.44}$$

Subtracting (8.41) from (8.42) gives

$$2(\tau_{x'x'} - \mathrm{i}\tau_{x'y'}) = \tau_{yy} + \tau_{xx} - (\tau_{yy} - \tau_{xx} + 2\mathrm{i}\tau_{xy})\mathrm{e}^{2\mathrm{i}\theta}. \tag{8.45}$$

This relation may be used to express the traction boundary conditions along a surface whose outward unit normal vector is rotated by an angle θ from the x-axis. If the normal and shear components of the traction on this surface are N and T, then (8.45) immediately gives

$$2(N - \mathrm{i}T) = \tau_{yy} + \tau_{xx} - (\tau_{yy} - \tau_{xx} + 2\mathrm{i}\tau_{xy})\mathrm{e}^{2\mathrm{i}\theta}. \tag{8.46}$$

The combination of stress components that appear in (8.46) can be expressed in terms of the complex potentials by making use of (8.10), (8.19)–(8.23), and (8.37):

$$\tau_{yy} + \tau_{xx} = \frac{\partial^2 U}{\partial x^2} + \frac{\partial^2 U}{\partial y^2} = 2[\phi'(z) + \overline{\phi'(z)}] = 4\mathbf{Re}\{\phi'(z)\}, \tag{8.47}$$

$$\tau_{yy} - \tau_{xx} + 2\mathrm{i}\tau_{xy} = \frac{\partial^2 U}{\partial x^2} - \frac{\partial^2 U}{\partial y^2} - 2\mathrm{i}\frac{\partial^2 U}{\partial x \partial y} = 2[\bar{z}\phi''(z) + \psi'(z)]. \tag{8.48}$$

Another useful relationship can be found by combining (8.19) and (8.20):

$$\frac{\partial U}{\partial x} + \mathrm{i}\frac{\partial U}{\partial y} = \phi(z) + z\overline{\phi'(z)} + \overline{\psi(z)}. \tag{8.49}$$

8.3 Homogeneous state of stress

In the previous section, solutions to elasticity problems were presented in terms of arbitrary analytic functions of the complex variable z. Many important problems can be solved by taking the complex potentials to be polynomials in either z or $1/z$. For example, a uniform state of stress and strain can be found by taking the two potentials to be linear functions of z:

$$\phi(z) = cz, \quad \psi(z) = dz, \tag{8.50}$$

where c and d are constants that may be complex. From (8.47) and (8.48), the stresses that follow from (8.50) are seen to be

$$\tau_{xx} - \tau_{yy} = 4\mathbf{Re}\{c\}, \tag{8.51}$$

$$\tau_{xx} - \tau_{yy} + 2\mathrm{i}\tau_{xy} = 2[\bar{z}\phi''(z) + \psi'(z)] = 2d. \tag{8.52}$$

As the imaginary part of c does not affect the stresses, we may take c to be purely real. The imaginary part of d, however, determines the shear component, τ_{xy}, and cannot be ignored.

As c and d are constants, (8.51) and (8.52) show that (8.50) represents a uniform state of stress. If the principal stresses are denoted, as usual, by σ_1 and σ_2, with the direction of σ_1 rotated by an angle β from the x-axis, then it follows from (8.41) and (8.42) that

$$\sigma_1 + \sigma_2 = \tau_{xx} + \tau_{yy} = 4c, \tag{8.53}$$

$$\sigma_2 - \sigma_1 = (\tau_{yy} - \tau_{xx} + 2\mathrm{i}\tau_{xy})\mathrm{e}^{2\mathrm{i}\beta} = 2d\mathrm{e}^{2\mathrm{i}\beta}, \tag{8.54}$$

These relations can be inverted to yield

$$c = \frac{1}{4}(\sigma_1 + \sigma_2), \quad d = \frac{1}{2}(\sigma_2 - \sigma_1)\mathrm{e}^{-2\mathrm{i}\beta}. \tag{8.55}$$

In other words, the uniform stress state consisting of principal stress σ_1 rotated by angle β from the x-axis and principal stress σ_2 rotated by angle β from the y-axis is represented by the complex potentials

$$\phi(z) = \frac{1}{4}(\sigma_1 + \sigma_2)z, \quad \psi(z) = \frac{1}{2}(\sigma_2 - \sigma_1)z\mathrm{e}^{-2\mathrm{i}\beta}. \tag{8.56}$$

For the case of hydrostatic stress, $\sigma_1 = \sigma_2$, and so $\phi(z) = \sigma_1 z/2, \psi(z) = 0$. For uniaxial stress in the x direction, $\sigma_2 = 0$ and $\beta = 0$, so the potentials are $\phi(z) = \sigma_1 z/4$, and $\psi(z) = -\sigma_1 z/2$. A state of pure shear, τ_{xy}, can be generated by taking $\sigma_2 = -\sigma_1$ and $\beta = \pi/4$, which corresponds to $\phi(z) = 0$, $\psi(z) = \mathrm{i}\sigma_1 z$.

The displacements generated by (8.56) are found from (8.38) to be

$$2G(u + \mathrm{i}v) = \frac{1}{4}(\kappa - 1)(\sigma_1 + \sigma_2)z - \frac{1}{2}(\sigma_2 - \sigma_1)\bar{z}\mathrm{e}^{2\mathrm{i}\beta}. \tag{8.57}$$

As we have taken c to be real, the displacement components are

$$2Gu = \frac{1}{4}(\kappa - 1)(\sigma_1 + \sigma_2)x - \frac{1}{2}(\sigma_2 - \sigma_1)\mathbf{Re}\{\bar{z}\mathrm{e}^{2\mathrm{i}\beta}\}, \tag{8.58}$$

$$2Gv = \frac{1}{4}(\kappa - 1)(\sigma_1 + \sigma_2)y - \frac{1}{2}(\sigma_2 - \sigma_1)\mathbf{Im}\{\bar{z}\mathrm{e}^{2\mathrm{i}\beta}\}. \tag{8.59}$$

If x and y are the principal stress directions, then $\beta = 0$, and the displacements are

$$8Gu = [(\kappa + 1)\sigma_1 + (\kappa - 3)\sigma_2]x, \quad 8Gv = [(\kappa + 1)\sigma_2 + (\kappa - 3)\sigma_1]y, \tag{8.60}$$

in agreement with (5.46) and (5.47). The strains are given by

$$8G\varepsilon_1 = (\kappa + 1)\sigma_1 + (\kappa - 3)\sigma_2, \quad 8G\varepsilon_2 = (\kappa + 1)\sigma_2 + (\kappa - 3)\sigma_1. \tag{8.61}$$

Now consider points that initially lie on a circle of radius ρ, centered at $\{\bar{x}, \bar{y}\}$, which are described by

$$x = \bar{x} + \rho\cos\alpha, \quad y = \bar{y} + \rho\sin\alpha. \tag{8.62}$$

After the deformation, these points will be displaced to new positions given by

$$x' = (\bar{x} + \rho\cos\alpha)(1 - \varepsilon_1), \quad y' = (\bar{y} + \rho\sin\alpha)(1 - \varepsilon_2), \tag{8.63}$$

which describe an ellipse centered at $\{\bar{x}(1-\varepsilon_1), \bar{y}(1-\varepsilon_2)\}$, with semiaxes $\rho(1-\varepsilon_1)$ and $\rho(1-\varepsilon_2)$.

If the x and y directions are again taken to be the principal directions, the displacement in polar coordinates is found from (8.57) and (8.40) to be given in complex form by

$$2G(u_r + iu_\theta) = \frac{1}{4}(\kappa - 1)(\sigma_1 + \sigma_2)r + \frac{1}{2}(\sigma_1 - \sigma_2)re^{-2i\theta}, \tag{8.64}$$

which corresponds in component form to

$$8Gu_r = (\kappa - 1)(\sigma_1 + \sigma_2)r + 2(\sigma_1 - \sigma_2)r\cos 2\theta, \tag{8.65}$$

$$8Gu_\theta = -2(\sigma_1 - \sigma_2)r\sin 2\theta. \tag{8.66}$$

If the maximum principal stress is rotated by an angle β from the x direction, as in (8.56), then θ must be replaced by $\theta - \beta$ in (8.65) and (8.66).

Taking c to be a real constant and d to be an arbitrary complex constant in (8.50) therefore leads to a state of uniform (homogeneous) stress and strain. If we consider c to be purely imaginary and d to be zero, that is,

$$\phi(z) = i\delta z, \quad \psi(z) = 0, \tag{8.67}$$

where δ is real, then (8.51) and (8.52) show that the stresses are zero, and (8.38) gives

$$2G(u + iv) = (\kappa + 1)i\delta z, \tag{8.68}$$

which in component form is

$$2Gu = -(\kappa + 1)\delta y, \quad 2Gv = (\kappa + 1)\delta x. \tag{8.69}$$

The strains that follow from (8.69) are all zero, and the rotation is given by

$$\omega_{xy} = \frac{1}{2}\left(\frac{\partial v}{\partial x} - \frac{\partial u}{\partial y}\right) = \frac{1}{4G}[(\kappa + 1)\delta + (\kappa + 1)\delta] = \frac{(\kappa + 1)\delta}{2G}. \tag{8.70}$$

Rigid-body rotation by a small angle ω can therefore be generated by choosing $\delta = 2G\omega/(\kappa + 1)$ in (8.67).

Another interesting state of stress and displacement that can be generated from a simple choice of the potential functions is found by taking ϕ or ψ to be logarithmic. Consider the case

$$\phi(z) = 0, \quad \psi(z) = c\ln z. \tag{8.71}$$

But $z = re^{i\theta}$, so $\ln z = \ln(re^{i\theta}) = \ln r + \ln(e^{i\theta}) = \ln r + i\theta$. From (8.38), the displacement associated with these potentials is therefore given by

$$2G(u + iv) = -c\ln r + ic\theta, \tag{8.72}$$

which is to say, $u = -c\ln r/2G$, and $v = c\theta/2G$. If we start at a specific point z and make a loop around the origin, θ will increase by 2π, and so v will increase by $v = c\pi/G$. Hence, the potentials (8.71) do not represent a continuous displacement field and in general are not admissible. The same would be the case if ϕ contained a logarithmic term. Such logarithmic terms can, however, be used to represent dislocations (Hirth and Lothe, 1992).

8.4 Pressurized hollow cylinder

The problem of a circular hole in an infinite rock mass, with a uniform state of stress at infinity, is probably the most important single problem in rock mechanics. The general case of two unequal far-field principal stresses is treated in §8.5. In this section, we discuss the special case of hydrostatic stress at infinity. To generate this solution, we start with the case of a hollow cylinder subjected to pressure at both its inner and outer boundaries; the solution for the infinite region outside a circular hole is then found by letting the outer radius of the cylinder become infinite. The hollow cylinder problem is of interest in its own right in the context of laboratory testing.

Consider a hollow circular cylinder defined by $a < r < b$. The boundary conditions at the inner and outer surfaces are

$$\tau_{rr}(a) = P_{\rm i}, \quad \tau_{rr}(b) = P_{\rm o}. \tag{8.73}$$

We take the complex potentials to be of the form

$$\phi(z) = cz, \quad \psi(z) = d/z, \tag{8.74}$$

where c and d are constants. The imaginary components of these constants would lead to shear stresses and rotations, so for the present problem, in which there is radial symmetry, we can take c and d to be real. The displacement vector follows from (8.74) and (8.38):

$$2G(u + {\rm i}v) = (\kappa - 1)cz - d/\bar{z}. \tag{8.75}$$

If we use polar coordinates, with $z = r{\rm e}^{{\rm i}\theta}$, then the complex displacement vector (8.75) can be written as

$$2G(u + {\rm i}v) = (\kappa - 1)cr{\rm e}^{{\rm i}\theta} - d{\rm e}^{{\rm i}\theta}/r. \tag{8.76}$$

This equation expresses the Cartesian components of the displacement in terms of the polar variables r and θ; the displacement components (u_r, u_θ) are found from (8.76) by using (8.40):

$$2G(u_r + {\rm i}u_\theta) = 2G(u + {\rm i}v){\rm e}^{-{\rm i}\theta} = (\kappa - 1)cr - d/r. \tag{8.77}$$

The explicit component form of (8.77) is

$$2Gu_r = (\kappa - 1)cr - d/r, \quad 2Gu_\theta = 0. \tag{8.78}$$

The stresses are found from (8.74), (8.47), and (8.48):

$$\tau_{rr} + \tau_{\theta\theta} = \tau_{xx} + \tau_{yy} = 4\mathbf{Re}\{\phi'(z)\} = 4c, \tag{8.79}$$

$$\tau_{yy} - \tau_{xx} + 2{\rm i}\tau_{xy} = 2[\bar{z}\phi''(z) + \psi'(z)] = -2d/z^2 = -2d{\rm e}^{-2{\rm i}\theta}/r^2. \tag{8.80}$$

Use of (8.43) yields

$$\tau_{\theta\theta} - \tau_{rr} + 2{\rm i}\tau_{r\theta} = (\tau_{yy} - \tau_{xx} + 2{\rm i}\tau_{xy}){\rm e}^{2{\rm i}\theta} = -2d/r^2. \tag{8.81}$$

Separating out the real and imaginary parts of (8.81) gives

$$\tau_{\theta\theta} - \tau_{rr} = -2d/r^2, \quad \tau_{r\theta} = 0. \tag{8.82}$$

Imposition of the boundary conditions (8.73) allows (8.79) and (8.82) to be solved for

$$c = \frac{b^2 P_o - a^2 P_i}{2(b^2 - a^2)}, \quad d = \frac{a^2 b^2 (P_i - P_o)}{(b^2 - a^2)}. \tag{8.83}$$

The full solution is then, from (8.78), (8.79), (8.82), and (8.83),

$$2Gu_r = (\kappa - 1)\frac{(b^2 P_o - a^2 P_i)r}{2(b^2 - a^2)} - \frac{a^2 b^2 (P_i - P_o)}{(b^2 - a^2)r}, \tag{8.84}$$

$$\tau_{rr} = \frac{(b^2 P_o - a^2 P_i)}{(b^2 - a^2)} + \frac{a^2 b^2 (P_i - P_o)}{(b^2 - a^2)r^2}, \tag{8.85}$$

$$\tau_{\theta\theta} = \frac{(b^2 P_o - a^2 P_i)}{(b^2 - a^2)} - \frac{a^2 b^2 (P_i - P_o)}{(b^2 - a^2)r^2}, \tag{8.86}$$

where $(\kappa - 1) = 2(1 - 2\nu)$ for the case of plane strain. The stresses from (8.85) and (8.86) are plotted in Fig. 8.1 for the case of a hollow cylinder having $b = 2a$.

From (8.82) and (8.83), we see that the magnitude of the difference between the two principal stresses decreases monotonically from its maximum value of

$$|\tau_{rr} - \tau_{\theta\theta}| = \frac{2|P_o - P_i|b^2}{(b^2 - a^2)} \quad \text{at } r = a, \tag{8.87}$$

to its minimum value of

$$|\tau_{rr} - \tau_{\theta\theta}| = \frac{2|P_o - P_i|a^2}{(b^2 - a^2)} \quad \text{at } r = b. \tag{8.88}$$

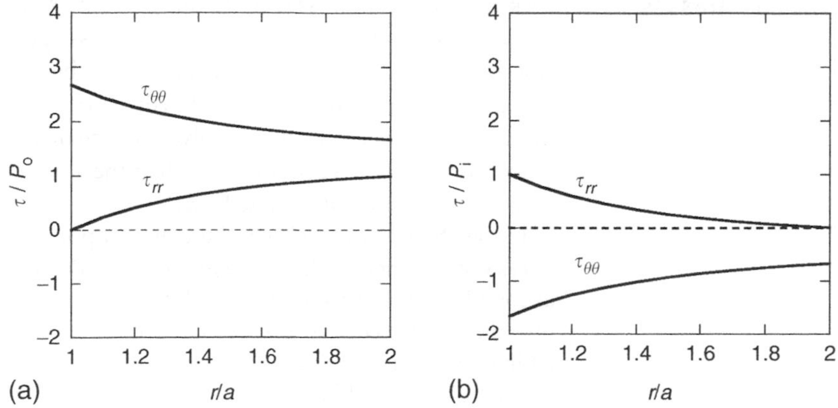

Fig. 8.1 Stress distribution in a hollow cylinder with $b = 2a$, for (a) the case of an external pressure only and (b) the case of an internal pressure only.

The total elastic strain energy stored in the hollow cylinder of length L can be found from the work–energy principle, (5.160), in the form

$$W = \frac{1}{2}P_o(2\pi bL)u_r(b) - \frac{1}{2}P_i(2\pi aL)u_r(a)$$
$$= \frac{\pi L}{2G(b^2 - a^2)}\{(1-2\nu)(P_o b^2 - P_i a^2)^2 + (P_o - P_i)^2 b^2 a^2\}. \quad (8.89)$$

The solution for a circular hole in an infinite rock mass with a far-field hydrostatic stress P_o and an internal pressure P_i is found from (8.84)–(8.86) by letting $b \to \infty$. The results are, for the case of plane strain,

$$2Gu_r(r) = (1-2\nu)P_o r - (P_i - P_o)(a^2/r), \quad (8.90)$$
$$\tau_{rr}(r) = P_o - (P_o - P_i)(a/r)^2, \quad (8.91)$$
$$\tau_{\theta\theta}(r) = P_o - (P_o - P_i)(a/r)^2. \quad (8.92)$$

If a borehole of radius a is loaded by a hydrostatic pressure P, (8.90) shows that the resulting displacement of the borehole wall will be $u_r(a) = -Pa/2G$; this equation is useful in estimating G from borehole measurements.

According to (8.91) and (8.92), the perturbations to the *in situ* stress field due to the presence of the hole are localized to within a few radii of the hole. For example, when $r = 10a$, the magnitude of the second term in (8.92) has decayed to 1 percent of the value it had at the borehole wall, where $r = a$. Hence, the "far-field" stress P_o in (8.90)–(8.92) actually denotes the stress that, in the absence of the hole, would exist in a region around the hole whose extent was about $10a$. The solution given above can therefore be used for a hole in a nonuniform stress field, as long as the length scale of the variations of the *in situ* stress field is greater than about ten times the hole radius.

8.5 Circular hole in a rock mass with given far-field principal stresses

The problem of calculating the displacements and stresses outside a circular hole in an infinite elastic solid, with a uniform state of stress far from the hole, was first solved by the German engineer Kirsch in 1898. As most holes drilled through rock are of circular cross section, this problem is of immense importance in rock engineering. The effect of a hydrostatic stress applied to the surface of the hole was found in §8.4, so we need only consider the case of a nonzero far-field stress and zero pressure in the borehole. Furthermore, since we can utilize the principle of superposition, we need only solve the case of a single nonzero principal stress at infinity. It will be convenient to align the x-axis with this principal stress, the value of which we will denote by σ_1^∞.

In the absence of the hole, the complex potentials associated with a uniaxial stress state aligned with the x-axis would be $\phi(z) = \sigma_1^\infty z/4$, $\psi(z) = -\sigma_1^\infty z/2$, as found in §8.3. Although this solution gives the correct far-field stress, it will give incorrect, nonzero tractions at the borehole wall. We must therefore find additional terms in the potentials that will cancel out these unwanted tractions, but not give any additional stresses at infinity. The only such terms which may be of use and which will lead to stresses that vary as functions of 2θ, are the following:

$$\phi(z) = \frac{1}{4}\sigma_1^\infty\left(z + \frac{A}{z}\right), \quad \psi(z) = -\frac{1}{2}\sigma_1^\infty\left(z + \frac{B}{z} + \frac{C}{z^3}\right), \tag{8.93}$$

where A, B, and C are real constants. The derivatives that will be needed for subsequent calculation of displacements and stresses are

$$\phi'(z) = \frac{1}{4}\sigma_1^\infty\left(1 - \frac{A}{z^2}\right), \quad \phi''(z) = \frac{\sigma_1^\infty A}{2z^3},$$
$$\psi'(z) = -\frac{1}{2}\sigma_1^\infty\left(1 - \frac{B}{z^2} - \frac{3C}{z^4}\right). \tag{8.94}$$

Working in polar coordinates, $z = re^{i\theta}$, it follows from (8.47) that

$$\tau_{rr} + \tau_{\theta\theta} = 4\mathbf{Re}\{\phi'(z)\} = \sigma_1^\infty\mathbf{Re}\{1 - Ar^{-2}e^{-2i\theta}\} = \sigma_1^\infty(1 - Ar^{-2}\cos 2\theta). \tag{8.95}$$

Next, from (8.48),

$$\begin{aligned}\tau_{yy} - \tau_{xx} + 2i\tau_{xy} &= 2[\bar{z}\phi''(z) + \psi'(z)] = A\sigma_1^\infty\bar{z}/z^3 - \sigma_1^\infty(1 - B/z^2 - 3C/z^4) \\ &= A\sigma_1^\infty r^{-2}e^{-4i\theta} - \sigma_1^\infty(1 - Br^{-2}e^{-2i\theta} - 3Cr^{-4}e^{-4i\theta}).\end{aligned} \tag{8.96}$$

We now use (8.43) to find

$$\begin{aligned}\tau_{\theta\theta} - \tau_{rr} + 2i\tau_{r\theta} &= (\tau_{yy} - \tau_{xx} + 2i\tau_{xy})e^{2i\theta} \\ &= \sigma_1^\infty[Br^{-2} - e^{2i\theta} + (Ar^{-2} + 3Cr^{-4})e^{-2i\theta}].\end{aligned} \tag{8.97}$$

The real part of (8.97) gives

$$\tau_{\theta\theta} - \tau_{rr} = \sigma_1^\infty[Br^{-2} - (1 - Ar^{-2} - 3Cr^{-4})\cos 2\theta], \tag{8.98}$$

after which (8.95) and (8.98) can be solved to give

$$\tau_{\theta\theta} = \frac{1}{2}\sigma_1^\infty[1 + Br^{-2} + (3Cr^{-4} - 1)\cos 2\theta], \tag{8.99}$$

$$\tau_{rr} = \frac{1}{2}\sigma_1^\infty[1 - Br^{-2} + (1 - 2Ar^{-2} - 3Cr^{-4})\cos 2\theta]. \tag{8.100}$$

The imaginary part of (8.97) gives

$$\tau_{r\theta} = -\frac{1}{2}\sigma_1^\infty[(1 + Ar^{-2} + 3Cr^{-4})\sin 2\theta]. \tag{8.101}$$

In order for the hole boundary to be traction-free, both (8.100) and (8.101) must vanish at $r = a$; this requires that $\{A, B, C\}$ satisfy the following equations:

$$1 - Ba^{-2} = 0, \quad 1 - 2Aa^{-2} - 3Ca^{-4} = 0, \quad 1 + Aa^{-2} + 3Ca^{-4} = 0, \tag{8.102}$$

the solution to which is

$$A = 2a^2, \quad B = a^2, \quad C = -a^4. \tag{8.103}$$

The full expressions for the stresses follow from (8.99)–(8.101) and (8.103):

$$\tau_{\theta\theta} = \frac{1}{2}\sigma_1^\infty\left[1 + \left(\frac{a}{r}\right)^2\right] - \frac{1}{2}\sigma_1^\infty\left[1 + 3\left(\frac{a}{r}\right)^4\right]\cos 2\theta, \tag{8.104}$$

$$\tau_{rr} = \frac{1}{2}\sigma_1^\infty\left[1 - \left(\frac{a}{r}\right)^2\right] + \frac{1}{2}\sigma_1^\infty\left[1 - 4\left(\frac{a}{r}\right)^2 + 3\left(\frac{a}{r}\right)^4\right]\cos 2\theta, \tag{8.105}$$

$$\tau_{r\theta} = -\frac{1}{2}\sigma_1^\infty\left[1 + 2\left(\frac{a}{r}\right)^2 - 3\left(\frac{a}{r}\right)^4\right]\sin 2\theta. \tag{8.106}$$

The perturbations in the stress field caused by the presence of the hole die away at least as fast as $(a/r)^2$ and are therefore negligible at distances greater than $10a$ from the borehole. Hence, when applying formulae such as (8.104)–(8.106) to a newly drilled or excavated hole, the "far-field" stresses that should be used are those that initially act in the region that lies within a few radii of the (eventual) hole; the far-field principal stresses that act many kilometers from the hole, which might seem to better satisfy the definition of "stresses at infinity," in fact have no relevance to this problem.

At the surface of the hole, the hoop stress $\tau_{\theta\theta}$ varies with θ according to

$$\tau_{\theta\theta}(a, \theta) = \sigma_1^\infty[1 - 2\cos 2\theta], \tag{8.107}$$

and therefore varies from a stress of $-\sigma_1^\infty$ (i.e., tensile) when $\theta = 0$ or π, to a compressive stress of $3\sigma_1^\infty$ when $\theta = \pi/2$ or $3\pi/2$ or $3\pi/2$ (Fig. 8.2a). The mean normal stress is given, from (8.104) and (8.105), by

$$\tau_{\rm m} = \frac{1}{2}(\tau_{\theta\theta} + \tau_{rr}) = \sigma_1^\infty\left[1 - 2\left(\frac{a}{r}\right)^2\cos 2\theta\right], \tag{8.108}$$

so that the mean normal stress is tensile within the region defined by (Fig. 8.2b)

$$(r/a)^2 < 2\cos 2\theta. \tag{8.109}$$

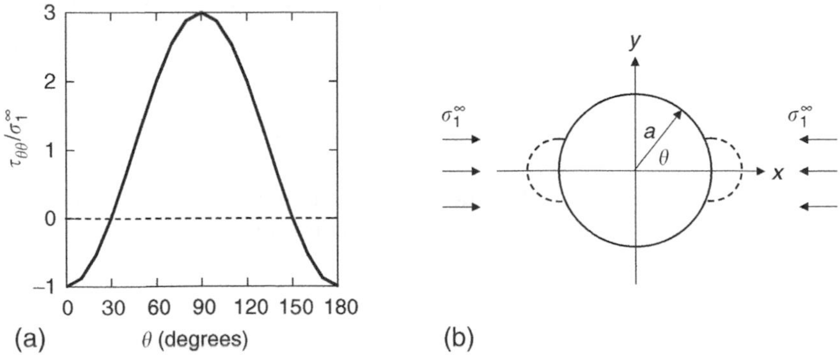

Fig. 8.2 (a) Hoop stress at borehole wall as a function of angle of rotation from direction of far-field stress. (b) Small circular arcs show outer boundaries of regions within which the mean normal stress is tensile.

The boundaries that separate the regions of negative and positive mean normal stress are circles of radius $0.414a$, centered on the points where the x-axis intersects the borehole.

The stresses due to a second far-field principal normal stress, σ_2^∞, that acts in a direction perpendicular to the direction of σ_1^∞, can be found from the solution given above by replacing θ with $\theta + (\pi/2)$, which is equivalent to replacing 2θ with $2\theta + \pi$:

$$\tau_{\theta\theta} = \frac{1}{2}\sigma_2^\infty\left[1 + \left(\frac{a}{r}\right)^2\right] + \frac{1}{2}\sigma_2^\infty\left[1 + 3\left(\frac{a}{r}\right)^4\right]\cos 2\theta, \tag{8.110}$$

$$\tau_{rr} = \frac{1}{2}\sigma_2^\infty\left[1 - \left(\frac{a}{r}\right)^2\right] - \frac{1}{2}\sigma_2^\infty\left[1 - 4\left(\frac{a}{r}\right)^2 + 3\left(\frac{a}{r}\right)^4\right]\cos 2\theta, \tag{8.111}$$

$$\tau_{r\theta} = \frac{1}{2}\sigma_2^\infty\left[1 + 2\left(\frac{a}{r}\right)^2 - 3\left(\frac{a}{r}\right)^4\right]\sin 2\theta. \tag{8.112}$$

The full state of stress around a circular hole in an infinite elastic rock mass with far-field stress σ_1^∞ acting in the x direction and far-field stress σ_2^∞ acting in the y direction is then found by superposing (8.104)–(8.106) and (8.110)–(8.112):

$$\tau_{\theta\theta} = \frac{1}{2}(\sigma_1^\infty + \sigma_2^\infty)\left[1 + \left(\frac{a}{r}\right)^2\right] - \frac{1}{2}(\sigma_1^\infty - \sigma_2^\infty)\left[1 + 3\left(\frac{a}{r}\right)^4\right]\cos 2\theta, \tag{8.113}$$

$$\tau_{rr} = \frac{1}{2}(\sigma_1^\infty + \sigma_2^\infty)\left[1 - \left(\frac{a}{r}\right)^2\right] + \frac{1}{2}(\sigma_1^\infty - \sigma_2^\infty)\left[1 - 4\left(\frac{a}{r}\right)^2 + 3\left(\frac{a}{r}\right)^4\right]\cos 2\theta, \tag{8.114}$$

$$\tau_{r\theta} = -\frac{1}{2}(\sigma_1^\infty - \sigma_2^\infty)\left[1 + 2\left(\frac{a}{r}\right)^2 - 3\left(\frac{a}{r}\right)^4\right]\sin 2\theta. \tag{8.115}$$

As it is conventional to assume that $\sigma_1^\infty \geq \sigma_2^\infty$, in the above equations, it is implicit that the x-axis is aligned with the direction of the *maximum* principal normal stress.

At the surface of the hole, the hoop stress $\tau_{\theta\theta}$ varies with θ according to

$$\tau_{\theta\theta}(a, \theta) = (\sigma_1^\infty + \sigma_2^\infty) - 2(\sigma_1^\infty - \sigma_2^\infty)\cos 2\theta, \tag{8.116}$$

and therefore varies from a minimum value of $3\sigma_2^\infty - \sigma_1^\infty$ when $\theta = 0$, or π, to a maximum value of $3\sigma_1^\infty - \sigma_2^\infty$ when $\theta = \pi/2$ or $3\pi/2$. If $3\sigma_2^\infty > \sigma_1^\infty$, which is equivalent to the condition $\sigma_1^\infty/\sigma_2^\infty < 3$ if both principal normal far-field stresses are compressive, then the hoop stress is positive (compressive) at all points on the boundary of the hole. If this condition is not satisfied, then $\tau_{\theta\theta}(a, \theta)$ is negative (tensile) within the two arcs defined by $\cos 2\theta > (\sigma_1^\infty + \sigma_2^\infty)/2(\sigma_1^\infty - \sigma_2^\infty)$.

If there is also a normal traction of magnitude P acting along the cavity wall, then we must superpose additional stresses that are found from (8.91) and (8.92), with $P_i = P$ and $P_o = 0$; the results are

$$\tau_{\theta\theta} = \frac{1}{2}(\sigma_1^\infty + \sigma_2^\infty)\left[1 + \left(\frac{a}{r}\right)^2\right] - P\left(\frac{a}{r}\right)^2 - \frac{1}{2}(\sigma_1^\infty - \sigma_2^\infty)\left[1 + 3\left(\frac{a}{r}\right)^4\right]\cos 2\theta, \tag{8.117}$$

$$\tau_{rr} = \frac{1}{2}(\sigma_1^\infty + \sigma_2^\infty)\left[1 - \left(\frac{a}{r}\right)^2\right] + P\left(\frac{a}{r}\right)^2 + \frac{1}{2}(\sigma_1^\infty - \sigma_2^\infty)\left[1 - 4\left(\frac{a}{r}\right)^2 + 3\left(\frac{a}{r}\right)^4\right]\cos 2\theta, \tag{8.118}$$

$$\tau_{r\theta} = -\frac{1}{2}(\sigma_1^\infty - \sigma_2^\infty)\left[1 + 2\left(\frac{a}{r}\right)^2 - 3\left(\frac{a}{r}\right)^4\right]\sin 2\theta. \tag{8.119}$$

In this most general case, the hoop stress at the cavity wall varies from a minimum value of $3\sigma_2^\infty - \sigma_1^\infty - P$ when $\theta = 0$ or π, to a maximum value of $3\sigma_1^\infty - \sigma_2^\infty - P$ when $\theta = \pi/2$ or $3\pi/2$. A region of tensile hoop stresses will exist if

$$P > 3\sigma_2^\infty - \sigma_1^\infty; \tag{8.120}$$

this is the simplest criterion for hydraulic fracturing of a formation due to internal pressure in the borehole (Hubbert and Willis, 1957).

The displacements that are associated with the complex potentials given in (8.93) are found from (8.93) and (8.38) to be

$$\frac{8G(u + iv)}{\sigma_1^\infty} = \kappa(re^{i\theta} + Ar^{-1}e^{-i\theta}) - (re^{i\theta} - Ar^{-1}e^{3i\theta}) + 2(re^{-i\theta} + Br^{-1}e^{i\theta} + Cr^{-3}e^{3i\theta}). \tag{8.121}$$

According to (8.40), the (r, θ) components of the displacement are found from (8.121) by multiplying by $e^{-i\theta}$, leading to

$$\frac{8G(u_r + iu_\theta)}{\sigma_1^\infty} = \kappa(r + Ar^{-1}e^{-2i\theta}) - (r - Ar^{-1}e^{2i\theta}) + 2(re^{-2i\theta} + Br^{-1} + Cr^{-3}e^{2i\theta}). \tag{8.122}$$

Using the values of $\{A, B, C\}$ from (8.103), we find

$$\frac{8G(u_r + iu_\theta)}{a\sigma_1^\infty} = \kappa\left[\frac{r}{a} + 2\left(\frac{a}{r}\right)e^{-2i\theta}\right] - \left[\frac{r}{a} - 2\left(\frac{a}{r}\right)e^{2i\theta}\right] + 2\left[\left(\frac{r}{a}\right)e^{-2i\theta} + \left(\frac{a}{r}\right) - \left(\frac{a}{r}\right)^3 e^{2i\theta}\right]. \tag{8.123}$$

Separating out the real and imaginary parts of (8.123) yields

$$\frac{8Gu_r}{a\sigma_1^\infty} = \left[(\kappa-1)\left(\frac{r}{a}\right)+2\left(\frac{a}{r}\right)\right]+2\left[\left(\frac{r}{a}\right)+(\kappa+1)\left(\frac{a}{r}\right)-\left(\frac{a}{r}\right)^3\right]\cos 2\theta, \tag{8.124}$$

$$\frac{8Gu_\theta}{a\sigma_1^\infty} = -2\left[\left(\frac{r}{a}\right)+(\kappa-1)\left(\frac{a}{r}\right)+\left(\frac{a}{r}\right)^3\right]\sin 2\theta, \tag{8.125}$$

where $\kappa = 3 - 4\nu$ for plane strain, and $\kappa = (3-\nu)/(1+\nu)$ for plane stress.

In the more general case, discussed in §5.3, in which there is an axial strain ε in the longitudinal direction, we must, according to (8.39) and (8.40), put $\kappa = 3 - 4\nu$, and add $-\varepsilon\nu r$ to the radial displacement, yielding

$$\frac{4Gu_r}{a\sigma_1^\infty} = \left[(1-2\nu)\left(\frac{r}{a}\right)+\left(\frac{a}{r}\right)\right]+\left[\left(\frac{r}{a}\right)+4(1-\nu)\left(\frac{a}{r}\right)-\left(\frac{a}{r}\right)^3\right] \times \cos 2\theta - \frac{4G\nu\varepsilon r}{a\sigma_1^\infty}. \tag{8.126}$$

The change in the radius of the hole is found by evaluating (8.126) at $r = a$:

$$\frac{u_r(a)}{a} = \frac{(1-\nu)\sigma_1^\infty}{2G}[1+2\cos 2\theta] - \nu\varepsilon = \frac{(1-\nu^2)\sigma_1^\infty}{E}[1+2\cos 2\theta] - \nu\varepsilon. \tag{8.127}$$

If there were an additional principal stress σ_2^∞ acting in the y direction, then the displacement at the borehole wall would be given by

$$\frac{u_r(a)}{a} = \frac{(1-\nu^2)}{E}[(\sigma_1^\infty+\sigma_2^\infty)+2(\sigma_1^\infty-\sigma_2^\infty)\cos 2\theta] - \nu\varepsilon. \tag{8.128}$$

We can express this result solely in terms of the far-field stresses by recalling from (5.26) that $E\varepsilon = \sigma_z^\infty - \nu(\sigma_1^\infty + \sigma_2^\infty)$, in which case we find

$$u_r(a) = \frac{a}{E}[(\sigma_1^\infty+\sigma_2^\infty-\nu\sigma_z^\infty)+2(1-\nu^2)(\sigma_1^\infty-\sigma_2^\infty)\cos 2\theta]. \tag{8.129}$$

This equation has relevance to the theory of many stress-measurement devices operated from boreholes.

8.6 Stresses applied to a circular hole in an infinite rock mass

We now address the problem of a circular hole of radius a in an infinite rock mass, with an arbitrary distribution of tractions along the hole boundary. As the problem of a traction-free hole with given far-field stresses has been solved in §8.5, we now only need to consider the case in which the far-field stresses are zero. We first recall that, if $f(\theta)$ is a piecewise continuous real-valued function defined over the range $-\pi < \theta < \pi$, it can be represented by a Fourier trigonometric series as follows (Tolstov, 1976):

$$f(\theta) = a_0 + \sum_{n=1}^{\infty}(a_n \cos n\theta + b_n \sin n\theta), \tag{8.130}$$

where the Fourier coefficients are given by

$$a_0 = \frac{1}{2\pi}\int_{-\pi}^{\pi} f(\theta)\mathrm{d}\theta, \quad a_n = \frac{1}{\pi}\int_{-\pi}^{\pi} f(\theta)\cos n\theta \mathrm{d}\theta, \quad b_n = \frac{1}{\pi}\int_{-\pi}^{\pi} f(\theta)\sin n\theta \mathrm{d}\theta. \tag{8.131}$$

Recalling that $\mathrm{e}^{\mathrm{i}n\theta} = \cos n\theta + \mathrm{i}\sin n\theta$, (8.130) can be written as

$$\begin{aligned} f(\theta) &= a_0 + \frac{1}{2}\sum_{n=1}^{\infty} a_n(\mathrm{e}^{\mathrm{i}n\theta} + \mathrm{e}^{-\mathrm{i}n\theta}) - \mathrm{i}b_n(\mathrm{e}^{\mathrm{i}n\theta} - \mathrm{e}^{-\mathrm{i}n\theta}) \\ &= a_0 + \frac{1}{2}\sum_{n=1}^{\infty}\left[(a_n - \mathrm{i}b_n)\mathrm{e}^{\mathrm{i}n\theta} + (a_n + \mathrm{i}b_n)\mathrm{e}^{-\mathrm{i}n\theta}\right] = \sum_{n=-\infty}^{\infty} \alpha_n \mathrm{e}^{\mathrm{i}n\theta}, \end{aligned} \tag{8.132}$$

where the complex Fourier coefficients α_n are related to the coefficients a_n and b_n by

$$\alpha_0 = a_0, \quad \alpha_n = \frac{1}{2}(a_n - \mathrm{i}b_n), \quad \alpha_{-n} = \frac{1}{2}(a_n + \mathrm{i}b_n). \tag{8.133}$$

In particular, the normal and shear tractions that act along the surface $r = a$ can be expressed as

$$N - \mathrm{i}T = \sum_{n=-\infty}^{\infty} A_n \mathrm{e}^{\mathrm{i}n\theta}. \tag{8.134}$$

This equation supplies a boundary condition that must be satisfied by the stresses.

If it is desired only to calculate the stresses, and not the displacements, as often is the case, then only the functions $\phi'(z)$, $\phi''(z)$ and $\psi'(z)$ will appear in (8.47) and (8.48). The most general forms that will yield stresses that are bounded at infinity are power series that contain no positive powers of z:

$$\phi'(z) = \sum_{n=0}^{\infty} c_n z^{-n}, \quad \psi'(z) = \sum_{n=0}^{\infty} d_n z^{-n}. \tag{8.135}$$

However, the terms c_0 and d_0 were shown in (8.55) to correspond to the far-field stresses and so will not be needed here. Furthermore, the $n = 1$ terms in (8.135) correspond to logarithmic terms in ϕ and ψ, and therefore, according to (8.72), do not represent continuous displacement fields. The series in (8.135) can therefore be taken to commence with $n = 2$.

According to (8.46)–(8.48), the surface tractions can be expressed in terms of the complex potentials as follows:

$$N - \mathrm{i}T = \phi'(z) + \overline{\phi'(z)} - [\bar{z}\phi''(z) + \psi'(z)]\mathrm{e}^{2\mathrm{i}\theta}, \tag{8.136}$$

with $z = ae^{i\theta}$ along the boundary. Inserting (8.134) and (8.135) into boundary condition (8.136) gives

$$\sum_{n=-\infty}^{\infty} A_n e^{in\theta} = \sum_{n=2}^{\infty}(c_n a^{-n} e^{-in\theta} + \bar{c}_n a^{-n} e^{in\theta})$$
$$+ \left[ae^{-i\theta}\sum_{n=2}^{\infty} nc_n a^{-(n+1)} e^{-i(n+1)\theta} - \sum_{n=2}^{\infty} d_n a^{-n} e^{-in\theta}\right] e^{2i\theta}$$
$$= \sum_{n=0}^{\infty}(n+1)c_n a^{-n} e^{-in\theta} + \bar{c}_n a^{-n} e^{in\theta} - d_{n+2} a^{-(n+2)} e^{-in\theta}, \tag{8.137}$$

where we have used the fact that $c_0 = c_1 = 0$ to allow the series on the right to start with $n = 0$. Equating like powers of $e^{in\theta} = 0$ on both sides of (8.137) yields

$$A_n = \bar{c}_n a^{-n}, \quad n \geq 1; \quad A_0 = -d_2 a^{-2};$$
$$A_{-n} = (n+1)c_n a^{-n} - d_{n+2} a^{-(n+2)}, \quad n \geq 1, \tag{8.138}$$

which can be inverted to give the unknown coefficients $\{c_n, d_n\}$ in terms of the coefficients A_n that are known from the boundary tractions:

$$c_n = \overline{A}_n a^n, \quad n \geq 1; \quad d_2 = -A_0 a^2;$$
$$d_{n+2} = -A_{-n} a^{n+2} + (n+1)c_n a^2, \quad n \geq 1. \tag{8.139}$$

In the simple case of a uniform traction of magnitude P acting on the surface of the hole, the only nonzero coefficient in (8.134) is $A_0 = P$, in which case (8.139) shows that the only nonzero coefficient in (8.135) is $d_2 = -Pa^2$, thereby regenerating the solution given by (8.74) and (8.83): $\psi(z) = -Pa^2/z$.

Now consider the case in which a normal traction of magnitude P is applied over the two symmetrical arcs, $-\theta_o < \theta < \theta_o$ and $\pi - \theta_o < \theta < \pi + \theta_o$, as in Fig. 8.3a. From (8.131), the Fourier coefficients for these boundary tractions are found to be

$$a_0 = 2\theta_o P/\pi, \quad a_n = (2P/n\pi)[1 + (-1)^n] \sin n\theta_o, \quad b_n = 0. \tag{8.140}$$

As the only nonzero coefficients are the a_n for even values of n, the coefficients in the complex form (8.134) of the Fourier series can be written as

$$A_0 = 2\theta_o P/\pi, \quad A_{2m} = A_{-2m} = (P/m\pi)\sin 2m\theta_o, \quad A_{2m+1} = A_{-(2m+1)} = 0, \tag{8.141}$$

in which, as in the remainder of this section, the index m takes on the values $1, 2, 3, \ldots$. The coefficients in (8.135) are found from (8.139) to be given by

$$c_{2m} = (Pa^{2m}/m\pi)\sin 2m\theta_o, \tag{8.142}$$

$$d_2 = -2P\theta_o a^2/\pi, \quad d_{2m+2} = (2Pa^{2m+2}/\pi)\sin 2m\theta_o. \tag{8.143}$$

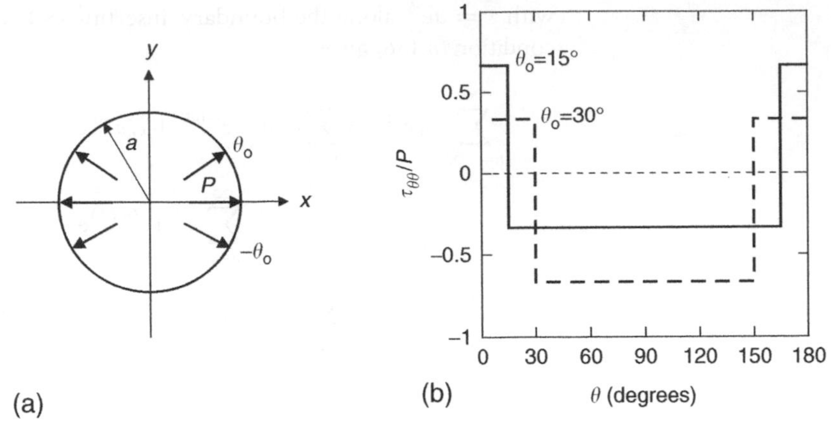

Fig. 8.3 (a) Circular hole loaded with pressure P over two symmetrical arcs. (b) Tangential normal stress along the boundary of the hole.

Inserting (8.142) and (8.143) in (8.135) gives

$$\phi'(z) = \frac{P}{\pi}\sum_{m=1}^{\infty}\frac{\sin 2m\theta_o}{m}\left(\frac{a}{z}\right)^{2m}, \tag{8.144}$$

$$\psi'(z) = -\frac{2P\theta_o}{\pi}\left(\frac{a}{z}\right)^2 + \frac{2P}{\pi}\sum_{m=1}^{\infty}\sin 2m\theta_o\left(\frac{a}{z}\right)^{2m+2}. \tag{8.145}$$

The stresses are found by substituting (8.144) and (8.145) into (8.43), (8.44), (8.47), and (8.48):

$$\tau_{\theta\theta} + \tau_{rr} = \frac{4P}{\pi}\sum_{m=1}^{\infty}\frac{\sin 2m\theta_o \cos 2m\theta}{m}\left(\frac{a}{r}\right)^{2m}, \tag{8.146}$$

$$\tau_{\theta\theta} - \tau_{rr} = -\frac{4P\theta_o}{\pi}\left(\frac{a}{r}\right)^2 - \frac{4P}{\pi}\left[1-\left(\frac{a}{r}\right)^2\right]\sum_{m=1}^{\infty}\sin 2m\theta_o \cos 2m\theta\left(\frac{a}{r}\right)^{2m}, \tag{8.147}$$

$$\tau_{r\theta} = \frac{2P}{\pi}\left[1-\left(\frac{a}{r}\right)^2\right]\sum_{m=1}^{\infty}\sin 2m\theta_o \sin 2m\theta\left(\frac{a}{r}\right)^{2m}. \tag{8.148}$$

At the surface of the hole, the stresses are

$$\tau_{rr}(a) = \frac{2P\theta_o}{\pi} + \frac{2P}{\pi}\sum_{m=1}^{\infty}\frac{\sin 2m\theta_o \cos 2m\theta}{m}, \tag{8.149}$$

$$\tau_{\theta\theta}(a) = -\frac{2P\theta_o}{\pi} + \frac{2P}{\pi}\sum_{m=1}^{\infty}\frac{\sin 2m\theta_o \cos 2m\theta}{m}, \tag{8.150}$$

and $\tau_{r\theta}(a) = 0$. The right-hand side of (8.149) must sum to P in the loaded region and to 0 in the unloaded region. Hence, comparison of (8.149) and (8.150) shows that $\tau_{\theta\theta} = P - (4\theta_o P/\pi)$ along the loaded portion of the hole boundary and $\tau_{\theta\theta} = -4\theta_o P/\pi$ along the unloaded region (Fig. 8.3b).

The displacements are found from (8.144), (8.145), (8.38), and (8.40):

$$\frac{2Gu_r(a)}{a} = -\frac{2P\theta_o}{\pi} - \frac{P}{\pi}\sum_{m=1}^{\infty}\frac{1}{m}\left\{\frac{\kappa}{2m-1} + \frac{1}{2m+1}\right\}\sin 2m\theta_o \cos 2m\theta. \tag{8.151}$$

At $\theta = 0$ or π, which is to say at the midpoint of either of the two loaded segments of the borehole wall, the results of Bromwich (1949, §121) can be used to show that

$$\frac{4\pi Gu_r}{aP} = -4\theta_o - 2(\kappa+1)\sin\theta_o \ln\cot(\theta_o/2) - (\kappa-1)[\pi\cos\theta_o + 2\theta_o - \pi]. \tag{8.152}$$

Likewise, at $\theta = \pi/2$ or $3\pi/2$, the midpoints of the unloaded portions of the borehole wall, the radial displacement is given by

$$\frac{4\pi Gu_r}{aP} = 2(\kappa+1)[\cos\theta_o \ln(\sec\theta_o + \tan\theta_o) - \theta_o] + (\kappa-1)\pi\sin\theta_o. \tag{8.153}$$

If a borehole is stressed in as in Fig. 8.3a, the displacement of the borehole wall in the direction of loading and the direction perpendicular to the loading can be used, in conjunction with (8.152) and (8.153), to determine the elastic moduli of the rock formation. Applications of this solution to borehole measurements are discussed in more detail by Jaeger and Cook (1964). General expressions for both u_r and u_θ at the borehole wall, for arbitrary values of θ, are given by Bray (1987).

8.7 Stresses applied to the surface of a solid cylinder

The problem of a solid cylinder loaded by tractions acting along its outer boundary, in conditions of either plane strain or plane stress, may be treated using the same general approach as for the problem of a circular hole in an infinite rock mass. In this case, however, in order for the stresses to be bounded at the center of the cylinder, the complex potentials may contain only positive powers of z:

$$\phi'(z) = \sum_{n=0}^{\infty} c_n z^n, \quad \psi'(z) = \sum_{n=0}^{\infty} d_n z^n. \tag{8.154}$$

Consider now the case of a cylinder loaded by a pressure P over two symmetric arcs, $-\theta_o < \theta < \theta_o$ and $\pi - \theta_o < \theta < \pi + \theta_o$, as in Fig. 8.4a. This type of loading can be applied by the use of curved jacks, as discussed by Jaeger and Cook (1964). The complex traction vector along the outer boundary of the cylinder can be expressed as

$$N - \mathrm{i}T = A_0 + \sum_{m=-\infty}^{\infty} A_{2m}\mathrm{e}^{2\mathrm{i}m\theta}, \tag{8.155}$$

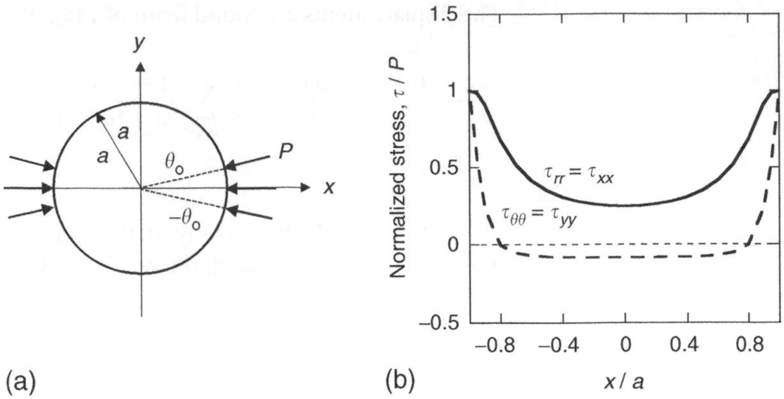

Fig. 8.4 (a) Diametral compression, over two symmetrical arcs, of a circular cylinder of radius a. (b) Stresses along the loaded diameter, for the case of $2\theta_o = 15°$, normalized against P.

with A_0 and A_{2m} given by (8.141). Proceeding as in §8.6, we find that

$$\phi'(z) = \frac{P\theta_o}{\pi} + \frac{P}{\pi}\sum_{m=1}^{\infty}\left(\frac{z}{a}\right)^{2m}\frac{\sin 2m\theta_o}{m}, \tag{8.156}$$

$$\psi'(z) = -\frac{2P}{\pi}\sum_{m=0}^{\infty}\left(\frac{z}{a}\right)^{2m}\sin 2(m+1)\theta_o. \tag{8.157}$$

The stresses are found by substituting (8.156) and (8.157) into (8.43), (8.44), (8.47), and (8.48):

$$\tau_{rr} = \frac{2P\theta_o}{\pi} + \frac{2P}{\pi}\sum_{m=1}^{\infty}\left(\frac{r}{a}\right)^{2m}\left[\left(\frac{r}{a}\right)^{-2} - \left(1 - \frac{1}{m}\right)\right]\sin 2m\theta_o \cos 2m\theta, \tag{8.158}$$

$$\tau_{\theta\theta} = \frac{2P\theta_o}{\pi} - \frac{2P}{\pi}\sum_{m=1}^{\infty}\left(\frac{r}{a}\right)^{2m}\left[\left(\frac{r}{a}\right)^{-2} - \left(1 + \frac{1}{m}\right)\right]\sin 2m\theta_o \cos 2m\theta, \tag{8.159}$$

$$\tau_{r\theta} = \frac{2P}{\pi}\left[1 - \left(\frac{r}{a}\right)^{-2}\right]\sum_{m=1}^{\infty}\left(\frac{r}{a}\right)^{2m}\sin 2m\theta_o \sin 2m\theta. \tag{8.160}$$

Along the x-axis, where $\theta = 0$, the series in (8.158) and (8.159) can be summed in closed form to give (Hondros, 1959)

$$\tau_{rr}(\theta = 0) = \frac{2P}{\pi}\left\{\frac{(1-\rho^2)\sin 2\theta_o}{(1 - 2\rho^2\cos 2\theta_o + \rho^4)} + \arctan\left[\frac{(1+\rho^2)}{(1-\rho^2)}\tan\theta_o\right]\right\}, \tag{8.161}$$

$$\tau_{\theta\theta}(\theta = 0) = -\frac{2P}{\pi}\left\{\frac{(1-\rho^2)\sin 2\theta_o}{(1-2\rho^2\cos 2\theta_o+\rho^4)} - \arctan\left[\frac{(1+\rho^2)}{(1-\rho^2)}\tan\theta_o\right]\right\}, \tag{8.162}$$

in which $\rho = r/a$. Along the y-axis, perpendicular to the loading, where $\theta = \pi/2$, the stresses are

$$\tau_{rr}(\theta = \pi/2) = -\frac{2P}{\pi}\left\{\frac{(1-\rho^2)\sin 2\theta_0}{(1+2\rho^2\cos 2\theta_0+\rho^4)} - \arctan\left[\frac{(1-\rho^2)}{(1+\rho^2)}\tan\theta_0\right]\right\}, \tag{8.163}$$

$$\tau_{\theta\theta}(\theta = \pi/2) = \frac{2P}{\pi}\left\{\frac{(1-\rho^2)\sin 2\theta_0}{(1+2\rho^2\cos 2\theta_0+\rho^4)} + \arctan\left[\frac{(1-\rho^2)}{(1+\rho^2)}\tan\theta_0\right]\right\}. \tag{8.164}$$

The shear stresses are zero along both axes, due to the symmetry of the problem.

The solution for the limiting case of a line load of magnitude W per unit length of cylinder, applied at $\theta = 0$ and $\theta = \pi$, can be found by letting $P \to \infty$ and $\theta_0 \to 0$ in such a way that the resultant load, $W = 2P\theta_0 a$, remains constant. In this case, the stresses along the x-axis are readily found from (8.161) and (8.162) to be

$$\tau_{rr}(\theta = 0) = \frac{W(3+\rho^2)}{\pi a(1-\rho^2)}, \quad \tau_{\theta\theta}(\theta = 0) = -\frac{W}{\pi a}, \tag{8.165}$$

which show that a uniform *tensile* stress of magnitude $-W/\pi a$ acts along this diameter. The other principal stress is compressive and varies from $3W/\pi a$ at the center to infinity at the points where the loading is applied. These infinite stress concentrations are artifacts of the idealization that the load is applied over a vanishingly small arc; for small but nonzero values of θ_0, the other stress is large but finite. The two stresses along the loaded diameter are shown in Fig. 8.4b for the case where $2\theta_0 = 15°$. The state of stress resulting from this type of line-loading is approximated by the so-called Brazilian indirect tension test which is used in laboratory testing to create a tensile stress within a rock specimen; see Jaeger and Hoskins (1966a) and §6.7.

The displacement at the outer edge of the cylinder is found from (8.156), (8.157), (8.38), and (8.40) to be given by

$$\frac{2\pi G u_r(a)}{Pa} = \theta_0(\kappa - 1) + \sum_{m=1}^{\infty}\frac{1}{m}\left(\frac{\kappa}{2m+1} + \frac{1}{2m-1}\right)\sin 2m\theta_0 \cos 2m\theta. \tag{8.166}$$

When $\theta = 0$ or $\pi/2$, the series in (8.166) can be summed using the results of Bromwich (1949, §121). Along the y-axis, where $\theta = \pi/2$, the radial displacement in the limit of small values of θ_0 is given by

$$u_r(a) = -[2(\kappa+1) - \pi(\kappa-1)]W/8\pi G, \tag{8.167}$$

where $W = 2P\theta_0 a$ is the resultant force acting along each of the two loaded arcs.

The problem of a circular cylinder acted upon by several line loads has been discussed by Michell (1900, 1902) and Jaeger (1967). The case of a hollow circular ring loaded over symmetric arcs along either its inner or outer surface has been treated by Ripperger and Davids (1947) and Jaeger and Hoskins (1966b). Hobbs (1965) considered the effect of having an inner hole that was eccentrically located with respect to the outer surface of the cylinder.

8.8 Inclusions in an infinite region

Consider a circular inclusion of radius a, having elastic moduli $\{G_i, \nu_i\}$, in an infinite rock mass whose elastic moduli are $\{G, \nu\}$. The inclusion is assumed to be perfectly welded to the surrounding rock, in which case all tractions and displacements will be continuous at the interface. If the rock mass is subjected to a far-field stress σ_1^∞, which we can assume to act along the x-axis, the stress in the inclusion will in general differ from the far-field stress state. Roughly speaking, the stress will be higher in the inclusion if it is stiffer than the surrounding rock and will be lower if the inclusion is less stiff. The precise relation between the far-field stresses and the stresses induced in the inclusion will be derived below in detail, both for the intrinsic interest of the problem and as an example of the use of the complex variable method to solve more complicated problems.

Following the discussion of the stresses around a circular hole given in §8.5, the complex potentials in the exterior region will be assumed to have the form

$$\phi(z) = \frac{1}{4}\sigma_1^\infty\left(z + \frac{Aa^2}{z}\right), \quad \psi(z) = -\frac{1}{2}\sigma_1^\infty\left(z + \frac{Ba^2}{z} + \frac{Ca^4}{z^3}\right), \tag{8.168}$$

where the terms involving a are introduced so as to simplify the eventual expressions for the constants $\{A, B, C\}$. This form of the potentials ensures that the far-field stress state is one of compression of magnitude σ_1^∞ in the x direction. Inside the inclusion, the potentials will be of the form

$$\phi_i(z) = \frac{1}{4}\sigma_1^\infty\left(A_i z + \frac{B_i z^3}{a^2}\right), \quad \psi_i(z) = -\frac{1}{2}\sigma_1^\infty C_i z, \tag{8.169}$$

which is the only form that yields finite displacements and stresses at the origin and which generates stresses that have the same angular variation, that is, $\cos 2\theta$ and $\sin 2\theta$, as do the far-field stresses.

The six as-yet unknown constants are found by requiring that the tractions and displacements be continuous across the interface at $r = a$. The displacements derived from (8.168) at $r = a$ are

$$8G(u_r + iu_\theta)/a\sigma_1^\infty = (\kappa - 1) + 2B + (\kappa A + 2)e^{-2i\theta} + (A + 2C)e^{2i\theta}, \tag{8.170}$$

whereas those derived from (8.169) are

$$8G_i(u_r + iu_\theta)/a\sigma_1^\infty = A_0(\kappa_i - 1) + \kappa_i B_i e^{-2i\theta} + (2C_i - 3B_i)e^{-2i\theta}. \tag{8.171}$$

Equating the coefficients of the various θ-dependent terms in (8.170) and (8.171) gives

$$e^0: \beta[(\kappa - 1) + 2B] = A_i(\kappa_i - 1), \tag{8.172}$$

$$e^{2i\theta}: \beta(A + 2C) = \kappa_i B_i, \tag{8.173}$$

$$e^{-2i\theta}: \beta(\kappa A + 2) = 2C_i - 3B_i, \tag{8.174}$$

in which the stiffness ratio β is defined as $\beta = G_i/G$.

The complex traction vector at the interface that can be derived from (8.168) using (8.46)–(8.48) is

$$4(N - iT)/\sigma_1^\infty = (2 - 2B) - (3A + 6C)e^{-2i\theta} + (2 - A)e^{2i\theta}, \tag{8.175}$$

whereas the traction vector at $r = a$ that follows from (8.169) is

$$4(N - iT)/\sigma_1^\infty = 2A_i + 3B_i e^{-2i\theta} + (2C_i - 3B_i)e^{2i\theta}, \tag{8.176}$$

Equating the coefficients in (8.175) and (8.176) leads to

$$e^0: 1 - B = A_i, \tag{8.177}$$

$$e^{2i\theta}: 2 - A = 2C_i - 3B_i, \tag{8.178}$$

$$e^{-2i\theta}: A + 2C = -B_i, \tag{8.179}$$

Comparison of (8.173) and (8.179) shows that $B_i = 0$, after which all the coefficients are readily found:

$$A_i = \frac{\beta(\kappa + 1)}{2\beta + (\kappa_i - 1)}, \quad B_i = 0, \quad C_i = \frac{\beta(\kappa + 1)}{\beta\kappa + 1}, \tag{8.180}$$

$$A = \frac{2(1 - \beta)}{\beta\kappa + 1}, \quad B = \frac{(\kappa_i - 1) - \beta(\kappa - 1)}{2\beta + (\kappa_i - 1)}, \quad C = \frac{\beta - 1}{\beta\kappa + 1}. \tag{8.181}$$

The potentials inside the inclusion are therefore

$$\phi_i(z) = \frac{1}{4}\sigma_1^\infty \left[\frac{\beta(\kappa + 1)}{2\beta + (\kappa_i - 1)}\right] z, \quad \psi_i(z) = -\frac{1}{2}\sigma_1^\infty \left[\frac{\beta(\kappa + 1)}{\beta\kappa + 1}\right] z. \tag{8.182}$$

From (8.47) and (8.48), the stress state inside the inclusion is found to be

$$\tau_{xx} = \frac{\sigma_1^\infty}{2}\left[\frac{\beta(\kappa + 1)}{2\beta + (\kappa_i - 1)} + \frac{\beta(\kappa + 1)}{\beta\kappa + 1}\right], \tag{8.183}$$

$$\tau_{yy} = \frac{\sigma_1^\infty}{2}\left[\frac{\beta(\kappa + 1)}{2\beta + (\kappa_i - 1)} - \frac{\beta(\kappa + 1)}{\beta\kappa + 1}\right]. \tag{8.184}$$

If the far-field principal stresses are σ_1^∞ and σ_2^∞, then the principle stresses inside the inclusion would be given by

$$\sigma_1^i = \frac{[\beta(\kappa+2)+\kappa_i]\beta(\kappa+1)}{2(2\beta+\kappa_i-1)(\beta\kappa+1)}\sigma_1^\infty + \frac{[\beta(\kappa-2)-(\kappa_i-2)]\beta(\kappa+1)}{2(2\beta+\kappa_i-1)(\beta\kappa+1)}\sigma_2^\infty, \tag{8.185}$$

$$\sigma_2^i = \frac{[\beta(\kappa-2)-(\kappa_i-2)]\beta(\kappa+1)}{2(2\beta+\kappa_i-1)(\beta\kappa+1)}\sigma_1^\infty + \frac{[\beta(\kappa+2)+\kappa_i]\beta(\kappa+1)}{2(2\beta+\kappa_i-1)(\beta\kappa+1)}\sigma_2^\infty, \tag{8.186}$$

with σ_1^i acting in the same direction as σ_1^∞, and similarly for σ_2^i and σ_2^∞. These results allow the far-field stresses to be determined in terms of the stresses that are induced in the inclusion. The stresses in the region outside the inclusion will be

$$\tau_{\theta\theta} = \frac{1}{2}(\sigma_1^\infty + \sigma_2^\infty)\left[1 + B\left(\frac{a}{r}\right)^2\right] - \frac{1}{2}(\sigma_1^\infty - \sigma_2^\infty)\left[1 - 3C\left(\frac{a}{r}\right)^4\right]\cos 2\theta, \tag{8.187}$$

$$\tau_{rr} = \frac{1}{2}(\sigma_1^\infty + \sigma_2^\infty)\left[1 - B\left(\frac{a}{r}\right)^2\right] + \frac{1}{2}(\sigma_1^\infty - \sigma_2^\infty)\left[1 - 2A\left(\frac{a}{r}\right)^2 - 3C\left(\frac{a}{r}\right)^4\right]\cos 2\theta, \tag{8.188}$$

$$\tau_{r\theta} = -\frac{1}{2}(\sigma_1^\infty - \sigma_2^\infty)\left[1 + A\left(\frac{a}{r}\right)^2 + 3C\left(\frac{a}{r}\right)^4\right]\sin 2\theta, \tag{8.189}$$

with $\{A, B, C\}$ given by (8.181). Although the presence of the inclusion gives rise to a stress-field perturbation that varies rapidly in the region immediately surrounding the inclusion, as shown by (8.187)–(8.189), the inclusion itself is in a state of homogenous stress, as shown by (8.185) and (8.186). The property of having a uniform state of stress within an inclusion subjected to given far-field stresses is shared in two dimensions by all ellipses, of which the circular inclusion is a special case, and in three dimensions by all ellipsoidal inclusions, of which spheroids and spheres are special cases (Eshelby, 1957). Although it is generally thought that only ellipsoidal inclusions (or degenerate cases thereof) have this property, a general proof has not yet been given (Lubarda and Markenscoff, 1998).

As an example of the degree to which the stresses inside the inclusion differ from the far-field stresses, consider a single far-field stress σ_1^∞ in a rock mass having $\nu = 0.25$. If the inclusion also has a Poisson ratio of 0.25, then $\kappa = \kappa_i = 2$, and (8.183) or (8.185) reduce to

$$\sigma_1^i = \frac{3\beta}{2\beta + 1}\sigma_1^\infty. \tag{8.190}$$

The stress inside the inclusion goes to zero when the inclusion is much more compliant than the surrounding rock and reaches $1.5\sigma_1^\infty$ when the inclusion is much stiffer than the surrounding rock.

The solution for circular inclusions was first presented by Sezawa and Nishimura (1931) and Goodier (1933); the derivation presented above using complex variables was given by Muskhelishvili (1963). Donnell (1941) discussed elliptical inclusions that are welded to the surrounding rock material, as in the case discussed above. Kouris et al. (1986) discussed the circular inclusion with a sliding interface, which is assumed to be able to transmit only normal tractions, but no shear tractions. Karihaloo and Viswanathan (1985) treated an elliptical inclusion that has partially debonded from the matrix. In three dimensions, the spherical inclusion was analysed by Goodier (1933), the spheroidal inclusion by Edwards (1951), and the ellipsoidal inclusion by Robinson (1951) and Eshelby (1957). Mura et al. (1985) treated the sliding ellipsoidal inclusion. Qu (1993) considered an ellipsoidal inclusion surrounded by a weakened interfacial layer that was modeled as an elastic spring. Lutz and Zimmerman (1996) examined the problem in which the elastic moduli vary as a function of distance from the outer edge of a spherical inclusion, as would be the case if the inclusion were surrounded by a region of damaged rock. Additional references for elastic inclusion problems are given by Mura (1987).

8.9 Elliptical hole in an infinite rock mass

Consider a transformation defined by

$$z = \omega(\zeta), \tag{8.191}$$

where ω is an analytic function. This transformation maps points $\zeta = \xi + i\eta$ in the ζ-plane into points $z = x + iy$ in the z-plane (Fig. 8.5). In particular, the point $P' = (\xi_0, \eta_0)$ in the ζ-plane is mapped by (8.191) into the point P in the z-plane. The line $\eta = \eta_0$ in the ζ-plane is mapped into the curve PA in the z-plane. The local slope of curve PA at point P can be calculated by starting from

$$dx + idy = dz = \omega'(\zeta)d\zeta = \omega'(\zeta)[d\xi + id\eta] = Me^{i\delta}[d\xi + id\eta], \tag{8.192}$$

where $Me^{i\delta}$ is the polar form of the complex number $\omega'(\zeta)$. The angle δ can be expressed in terms of $\omega'(\zeta)$ as follows:

$$\omega'(\zeta)/\overline{\omega'(\zeta)} = Me^{i\delta}/Me^{-i\delta} = e^{2i\delta}, \tag{8.193}$$

Along the line $\eta = \eta_0 =$ constant, we have $d\eta = 0$, so (8.192) gives

$$dx + idy = Me^{i\delta}d\xi = M(\cos\delta + i\sin\delta)d\xi = M\cos\delta d\xi + iM\sin\delta d\xi, \tag{8.194}$$

which shows that the slope of the curve PA at point P is given by

$$(dy/dx)_{\eta=\eta_0} = M\sin\delta d\xi/M\cos\delta d\xi = \tan\delta. \tag{8.195}$$

Similarly, the line $\xi = \xi_0 =$ constant is transformed into the curve PB in the ζ-plane, whose slope at point P is given by

$$(dy/dx)_{\xi=\xi_0} = -1/\tan\delta. \tag{8.196}$$

The product of the local slopes of the curves PA and PB at point P is -1, which shows that the two orthogonal lines $\xi = \xi_0$ and $\eta = \eta_0$ in the ζ-plane correspond

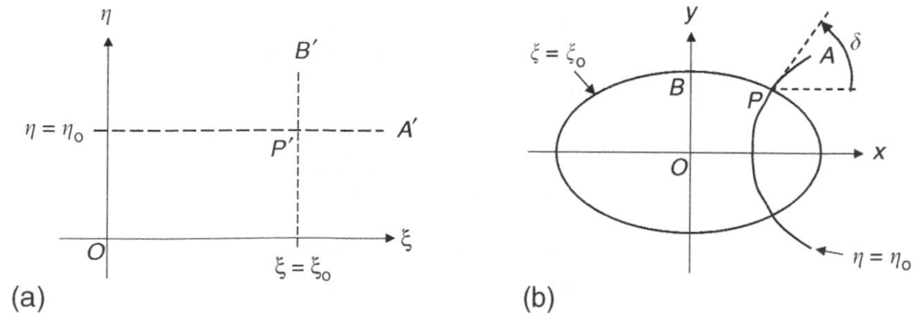

Fig. 8.5 (a) A line of constant η and a line of constant ξ in the $\zeta = (\eta, \xi)$ plane. (b) Image, under mapping (8.201), of these two lines in the $z = (x, y)$ plane.

to two curves PA and PB in the z-plane that are *locally orthogonal*. Hence, the variables (ξ, η) not only represent Cartesian coordinates in the ζ-plane, but can also be thought of as orthogonal coordinates that identify points in the z-plane, as in Fig. 8.5b.

Using (8.193), (8.38), and (8.40), the displacements in the directions of these new coordinates ξ and η can be expressed as

$$2G(u_\xi + \mathrm{i}u_\eta) = [\kappa\phi(z) - z\overline{\phi'(z)} - \overline{\psi(z)}][\overline{\omega'(\zeta)}/\omega'(\zeta)]^{1/2}, \tag{8.197}$$

in which the derivative $\phi'(z)$ is calculated as

$$\phi'(z) = \frac{\mathrm{d}\phi}{\mathrm{d}z} = \frac{\mathrm{d}\phi}{\mathrm{d}\zeta}\frac{\mathrm{d}\zeta}{\mathrm{d}z} = \frac{1}{\omega'(\zeta)}\frac{\mathrm{d}\phi}{\mathrm{d}\zeta}. \tag{8.198}$$

Likewise, the stresses in the (ξ, η) coordinate system can be expressed, using (8.193), (8.41), (8.47), and (8.48), as

$$\tau_{\xi\xi} + \tau_{\eta\eta} = 2[\phi'(z) + \overline{\phi'(z)}], \tag{8.199}$$

$$\tau_{\eta\eta} - \tau_{\xi\xi} + 2\mathrm{i}\tau_{\xi\eta} = 2[\bar{z}\phi''(z) + \psi'(z)][\omega'(\zeta)/\overline{\omega'(\zeta)}]. \tag{8.200}$$

Consider now the specific mapping

$$z = x + \mathrm{i}y = c\cosh\zeta = c\cosh(\xi + \mathrm{i}\eta), \tag{8.201}$$

which in component form is equivalent to

$$x = c\cosh\xi\cos\eta, \quad y = c\sinh\xi\sin\eta. \tag{8.202}$$

From (8.202), we see that the curve in the z-plane that corresponds to $\xi = \xi_o$ satisfies the equation

$$\frac{x^2}{(c\cosh\xi_o)^2} + \frac{y^2}{(c\sinh\xi_o)^2} = 1, \tag{8.203}$$

and is therefore an ellipse (Fig. 8.5b) having semiaxes

$$a = c\cosh\xi_o, \quad b = c\sinh\xi_o. \tag{8.204}$$

The semiaxes are related through $a^2 - b^2 = c^2$. The ratio of the smaller to the larger semiaxis, $b/a = \tanh\xi_o$, is often referred to as the *aspect ratio* of the hole. In the limiting case of $\xi_o = 0$, the ellipse becomes a thin slit extending from $-c \leq x \leq c$. The curve in the z-plane that corresponds to $\eta = \eta_o$ satisfies the equation

$$\frac{x^2}{(c\cos\eta_o)^2} - \frac{y^2}{(c\sin\eta_o)^2} = 1, \tag{8.205}$$

and is therefore a hyperbola. The ellipses and hyperbolae corresponding to constant values of ξ and η are confocal, with foci at $(x = \pm c, y = 0)$.

Now consider an infinite rock mass containing a traction-free elliptical hole, the boundary of which corresponds to $\xi = \xi_0$, so that the region outside the hole corresponds to $\xi > \xi_0$. A far-field stress σ^{∞} acts in a direction rotated from the major axis Ox of the ellipse by an angle β (Fig. 8.6a). This problem was first solved by Inglis (1913) using Airy stress functions. The complex potentials for this problem were found by Stevenson (1945) using elliptical coordinates and by Muskhelishvili (1963), who used a conformal mapping approach (see §8.11). Maugis (1992) pointed out that Stevenson's solution yields the correct stresses but contains an unwanted rotation at infinity. The potentials that satisfy the stress boundary conditions while giving rise to no rotation at infinity are (Maugis, 1992)

$$4\phi(z) = \sigma^{\infty}c[e^{2(\xi_0+i\beta)}\cosh\zeta + \{1 - e^{2(\xi_0+i\beta)}\}\sinh\zeta], \tag{8.206}$$

$$4\psi(z) = -\sigma^{\infty}c[\cosh 2\xi_0 - \cos 2\beta + e^{2\xi_0}\sinh 2(\zeta - \xi_0 - i\beta)]/\sinh\zeta. \tag{8.207}$$

The full state of stress and displacement outside the cavity was discussed in detail by Maugis (1992), who gave an extensive bibliography. Pollard (1973a) also discussed the stresses and displacements in some detail and gave the following brief review of various geological and geomechanical applications of this solution. Brace (1960) and Hoek and Bieniawski (1965) used this solution as the basis of a theory of fracture of brittle rocks. Walsh (1965a,b,c) used these results to analyse the effect of elliptical cracks on the elastic behavior of rocks. Anderson (1951) and Williams (1959) used the flat elliptical crack as a model of faults in the Earth's crust. Anderson (1937) and Pollard (1973b) used the internally pressurized elliptical crack as a model of sheet intrusions, whereas McLain (1968) and Sun (1969) used this solution to study hydraulically induced fractures. Each of these works has given rise to a large number of studies based on the solution for an elliptical hole in a stressed elastic rock mass.

The most important quantity in this problem is the tangential stress at the surface of the hole, that is, $\tau_{\eta\eta}$ at $\xi = \xi_0$. Since the normal stress $\tau_{\xi\xi}$ is zero at the hole, the tangential stress is found by substituting (8.206) into (8.199). We

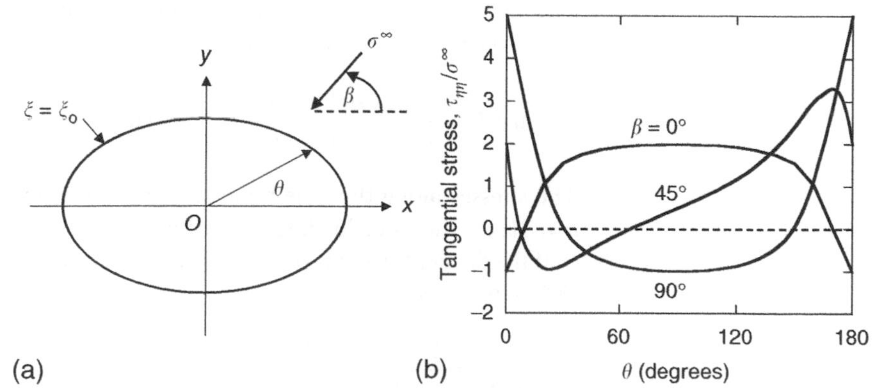

Fig. 8.6 (a) Elliptic hole subject to a far-field stress. (b) Variation of tangential stress, $\tau_{\eta\eta}$, for a hole with $a = 2b$, for three values of β.

first use the chain rule to calculate

$$\phi'(z) = \frac{d\phi}{d\zeta}\frac{d\zeta}{dz} = \frac{d\phi}{d\zeta} \bigg/ \frac{dz}{d\zeta}$$
$$= \frac{1}{4}\sigma^{\infty} c\{e^{2(\xi_0 + i\beta)} \sinh \zeta + [1 - e^{2(\xi_0+i\beta)}] \cosh \zeta\}/c \sinh \zeta$$
$$= \frac{1}{4}\sigma^{\infty}\{e^{2(\xi_0+i\beta)} + [1 - e^{2(\xi_0+i\beta)}] \coth \zeta\}. \tag{8.208}$$

At the hole boundary, $\zeta = \xi_0 + i\eta$, so from (8.199) the tangential stress along the boundary of the hole is given by

$$\tau_{\eta\eta} = 4\mathbf{Re}\{\phi'(z)\} = \sigma^{\infty} e^{2\xi_0} \cos 2\beta + \sigma^{\infty}\mathbf{Re}\{[1 - e^{2(\xi_0+i\beta)}] \coth(\xi_0 + i\eta)\}. \tag{8.209}$$

But $\mathbf{Re}\{fg\} = \mathbf{Re}\{f\}\mathbf{Re}\{g\} - \mathbf{Im}\{f\}\mathbf{Im}\{g\}$, so

$$\frac{\tau_{\eta\eta}}{\sigma^{\infty}} = e^{2\xi_0}\cos 2\beta + (1 - e^{2\xi_0}\cos 2\beta)\mathbf{Re}\{\coth(\xi_0 + i\eta)\}$$
$$- e^{2\xi_0}\sin 2\beta \mathbf{Im}\{\coth(\xi_0 + i\eta)\}$$
$$= e^{2\xi_0}\cos 2\beta + \frac{(1 - e^{2\xi_0}\cos 2\beta)\sinh 2\xi_0}{\cosh 2\xi_0 - \cos 2\eta} - \frac{e^{2\xi_0}\sin 2\beta \sin 2\eta}{\cosh 2\xi_0 - \cos 2\eta}$$
$$= e^{2\xi_0}\cos 2\beta + \frac{\sinh 2\xi_0}{\cosh 2\xi_0 - \cos 2\eta} - \frac{e^{2\xi_0}(\sin 2\beta \sin 2\eta + \cos 2\beta \sinh 2\xi_0)}{\cosh 2\xi_0 - \cos 2\eta}$$
$$= \frac{e^{2\xi_0}\cos 2\beta(\cosh 2\xi_0 - \cos 2\eta) + \sinh 2\xi_0 - e^{2\xi_0}(\sin 2\beta \sin 2\eta + \cos 2\beta \sinh 2\xi_0)}{\cosh 2\xi_0 - \cos 2\eta}$$
$$= \frac{\sinh 2\xi_0 + \cos 2\beta - e^{2\xi_0}\cos 2(\beta - \eta)}{\cosh 2\xi_0 - \cos 2\eta}$$
$$= \frac{2ab + (a^2 - b^2)\cos 2\beta - (a+b)^2 \cos 2(\beta - \eta)}{(a^2 + b^2) - (a^2 - b^2)\cos 2\eta}, \tag{8.210}$$

where in the last step we have used (8.204) to express the result in terms of a and b. To evaluate (8.210), we note that on the hole boundary, where $\xi = \xi_0$, the elliptical coordinate η is related to the polar coordinate θ by

$$\tan\theta = \frac{y}{x} = \frac{c \sinh \xi_0 \sin \eta}{c \cosh \xi_0 \cos \eta} = (b/a) \tan \eta. \tag{8.211}$$

The stresses along the hole boundary are shown in Fig. 8.6b for a hole having $a = 2b$, for the three loading cases of $\beta = \{0°, 45°, 90°\}$.

If the far-field stress is parallel to the major axis of the hole, then $\beta = 0$ and (8.210) reduces to

$$\frac{\tau_{\eta\eta}}{\sigma^{\infty}} = \frac{2ab + (a^2 - b^2) - (a + b)^2 \cos 2\eta}{(a^2 + b^2) - (a^2 - b^2) \cos 2\eta}. \tag{8.212}$$

At the end A of the major axis, $\theta = \eta = 0$, and (8.212) shows that the tangential stress is given by

$$\tau_{\eta\eta}(A) = \sigma^{\infty}, \tag{8.213}$$

regardless of the aspect ratio of the hole. At the end B of the minor axis, $\theta = \eta = \pi/2$, and (8.212) shows that

$$\tau_{\eta\eta}(B) = \sigma^{\infty}[1 + (2b/a)]. \tag{8.214}$$

Hence, the stress concentration factor at point B will be unity for a thin crack-like hole and increase to 3 as $b/a \rightarrow 1$ and the hole becomes a circle, in agreement with the results of §8.5.

If the far-field stress is parallel to the minor axis of the hole, then $\beta = \pi/2$, and (8.210) shows that

$$\tau_{\eta\eta}(A) = \sigma^{\infty}[1 + (2a/b)], \quad \tau_{\eta\eta}(B) = -\sigma^{\infty}. \tag{8.215}$$

Loading perpendicular to the main axis of a very thin crack ($b \ll a$) will therefore give rise to a high compressive stress concentration factor at the ends of the major axis.

If there are two principle stresses at infinity, σ_1^{∞} acting at an angle β to the x-axis and σ_2^{∞} acting at an angle $\beta + (\pi/2)$, then (8.210) generalizes to

$$\tau_{\eta\eta} = \frac{2ab(\sigma_1^{\infty} + \sigma_2^{\infty}) + (\sigma_1^{\infty} - \sigma_2^{\infty})[(a^2 - b^2)\cos 2\beta - (a+b)^2 \cos 2(\beta - \eta)]}{(a^2 + b^2) - (a^2 - b^2)\cos 2\eta}. \tag{8.216}$$

In the limit as $\xi_o \rightarrow \infty$, the aspect ratio $b/a = \tanh \xi_o \rightarrow 1$, and the ellipse degenerates into a circle. Choosing $\beta = 0$ in (8.216) so as to align the maximum principle stress with the x-axis, as in §8.5, we find that (8.216) reduces to (8.116), as it should.

If the far-field stress state is hydrostatic, then $\sigma_1^{\infty} = \sigma_2^{\infty} = \sigma^{\infty}$, and (8.216) reduces to

$$\tau_{\eta\eta} = \frac{4ab\sigma^{\infty}}{(a^2 + b^2) - (a^2 - b^2)\cos 2\eta}. \tag{8.217}$$

The tangential boundary stress along an elliptical hole subjected to an internal pressure of magnitude p, with no far-field stress, can be found by starting with a uniform stress of magnitude p and subtracting off the stress state given by (8.217), with $\sigma^{\infty} = p$, to yield

$$\tau_{\eta\eta} = p\left[1 - \frac{4ab}{(a^2 + b^2) - (a^2 - b^2)\cos 2\eta}\right]. \tag{8.218}$$

The foregoing equations give the stresses only along the boundary of the hole. The stresses and displacements for arbitrary points within the rock mass are given by Maugis (1992). The equations simplify greatly for the important case of a thin elliptical crack, which corresponds to the limit of $\xi_o \to 0$. For a uniaxial far-field stress σ^∞ acting at an angle β to a crack that lies along the x-axis, the full stress field is found by setting $\xi_o = 0$ in (8.199), (8.200), (8.206), and (8.207):

$$\frac{\tau_{\xi\xi} + \tau_{\eta\eta}}{\sigma^\infty} = \cos 2\beta + \frac{(1 - \cos 2\beta)\sinh 2\xi}{\cosh 2\xi - \cos 2\eta} - \frac{\sin 2\beta \sin 2\eta}{\cosh 2\xi - \cos 2\eta}, \tag{8.219}$$

$$\frac{\tau_{\xi\xi} - \tau_{\eta\eta}}{\sigma^\infty} = \frac{\cosh 2\xi \cos 2(\beta - \eta)}{\cosh 2\xi - \cos 2\eta} - \frac{(1 - \cos 2\beta)(1 - \cos 2\eta)\sinh 2\xi}{(\cosh 2\xi - \cos 2\eta)^2} + \frac{\cos 2(\beta - \eta) - \cosh 2\xi \cos 2\beta - \cosh 2\xi \sin 2\beta \sin 2\eta}{(\cosh 2\xi - \cos 2\eta)^2}, \tag{8.220}$$

$$\frac{2\tau_{\xi\eta}}{\sigma^\infty} = \frac{\sinh 2\xi \sin 2(\beta - \eta)}{\cosh 2\xi - \cos 2\eta} - \frac{(1 - \cos 2\beta)(1 - \cosh 2\xi)\sin 2\eta}{(\cosh 2\xi - \cos 2\eta)^2} - \frac{(1 - \cos 2\eta)\sinh 2\xi \sin 2\beta}{(\cosh 2\xi - \cos 2\eta)^2}. \tag{8.221}$$

The solution for the case of pure shear of magnitude τ^∞ directed parallel to the crack can be constructed by superposing the solutions for $\sigma_1^\infty = \tau^\infty$ at $\beta = \pi/4$, and $\sigma_2^\infty = -\tau^\infty$ at $\beta = 3\pi/4$, to yield

$$\frac{\tau_{\xi\xi}}{\tau^\infty} = \frac{(\cosh 2\xi - 1)\sin 2\eta}{\cosh 2\xi - \cos 2\eta} - \frac{(\cosh 2\xi - 1)\sin 2\eta}{(\cosh 2\xi - \cos 2\eta)^2}, \tag{8.222}$$

$$\frac{\tau_{\eta\eta}}{\tau^\infty} = -\frac{(\cosh 2\xi + 1)\sin 2\eta}{\cosh 2\xi + \cos 2\eta} + \frac{(\cosh 2\xi - 1)\sin 2\eta}{(\cosh 2\xi - \cos 2\eta)^2}, \tag{8.223}$$

$$\frac{\tau_{\eta\xi}}{\tau^\infty} = \frac{\sinh 2\xi \cos 2\eta}{\cosh 2\xi - \cos 2\eta} - \frac{(1 - \cos 2\eta)\sinh 2\xi}{(\cosh 2\xi - \cos 2\eta)^2}. \tag{8.224}$$

The displacements are more readily calculated in the z-coordinate system than in the ζ-coordinate system. We start by recalling (8.38):

$$2G(u + iv) = \kappa\phi(z) - z\overline{\phi'(z)} + \overline{\psi(z)}. \tag{8.225}$$

The most important case is that of a far-field stress $\sigma_\perp^\infty$ acting perpendicular to the crack. With $\beta = \pi/2$ and $\xi_o = 0$ to represent a thin crack, the complex potentials (8.206)–(8.208) reduce to

$$\phi(z) = -\frac{c}{4}\sigma_\perp^\infty(\cosh\zeta - 2\sinh\zeta), \tag{8.226}$$

$$\psi(z) = -\frac{c}{4}\sigma_\perp^\infty(2 - \sinh 2\zeta)/\sinh\zeta. \tag{8.227}$$

$$\phi'(z) = -\frac{1}{4}\sigma_\perp^\infty(1 - 2\coth\zeta). \tag{8.228}$$

The complex displacement becomes

$$\frac{8G(u + iv)}{c\sigma_\perp^\infty} = -\kappa\cosh\zeta + 2\kappa\sinh\zeta + (1 - 2\coth\bar{\zeta})\cosh\zeta + (2 - \sinh 2\bar{\zeta})/\sinh\bar{\zeta}. \tag{8.229}$$

The displacement at the surface of the crack is found by setting $\xi = \xi_0 = 0$, in which case $\zeta = i\eta$, $\cosh\zeta = \cos\eta = x/c$, and the normal displacement of the crack face is found to be given by

$$v = \frac{(\kappa + 1)\sigma_\perp^\infty}{4G}(a^2 - x^2)^{1/2}. \tag{8.230}$$

where we have made use of the fact that, for a thin crack, the focal distance c coincides with the half-length, a.

The expressions (8.219)–(8.230) give the leading-order terms for the stresses and displacements. For a real crack with a finite value of ξ_0, additional terms of order ξ_0 and higher would appear in the full expressions, but they would vanish as $\xi_0 \to 0$. Now consider a crack having a small but finite value of ξ_0. From (8.204), we see that $b/a = \tanh\xi_0 \approx \xi_0$, so ξ_0 essentially represents the aspect ratio of the crack in its unstressed state. From (8.203), the initial shape of the crack is seen to be described by

$$y^i = \xi_0(a^2 - x^2)^{1/2}. \tag{8.231}$$

Comparison of (8.230) and (8.231) shows that the displacement of the crack face at any location x is proportional to the initial aperture of the crack at that location, implying that as the far-field stress increases, the crack remains elliptical as it closes. The crack will be fully closed along its entire length when $v = y^i$, which occurs when

$$\sigma_\perp^\infty = \frac{4G\xi_0}{\kappa + 1}. \tag{8.232}$$

For plane strain, $\kappa + 1 = 4(1 - \nu)$, and so the pressure required to close a crack of initial aspect ratio ξ_0 is

$$\sigma_\perp^\infty(\text{closure}) = \frac{G\xi_0}{1 - \nu}. \tag{8.233}$$

Shear moduli of rocks are on the order of 10 GPa, so a crack of initial aspect ratio 0.001 will close up under a stress of about 10 MPa, or 1500 psi, for example.

If a far-field stress of magnitude $\sigma_\parallel^\infty$ is aligned parallel to the crack, which is to say with $\beta = 0$, an analysis similar to that given above would reveal that there is no normal displacement along the crack faces, aside from terms of order ξ_0. Hence, the closure of a thin crack is due entirely to the component of the far-field stress that is aligned perpendicular to it. So, if the far-field principal stresses were σ_2^∞ acting at angle β to the crack, and σ_1^∞ at $\beta + (\pi/2)$, condition (8.233) would be replaced by

$$\sigma_1^\infty \cos^2\beta + \sigma_2^\infty \sin^2\beta = \frac{G\xi_0}{1 - \nu}. \tag{8.234}$$

8.10 Stresses near a crack tip

To study the behavior of the stresses in the vicinity of the crack tip, we first recall from (8.202) that the point $P = (x, y)$ is related to the elliptical coordinates (ξ, η) by

$$x = c\cosh\xi\cos\eta, \quad y = c\sinh\xi\sin\eta. \tag{8.235}$$

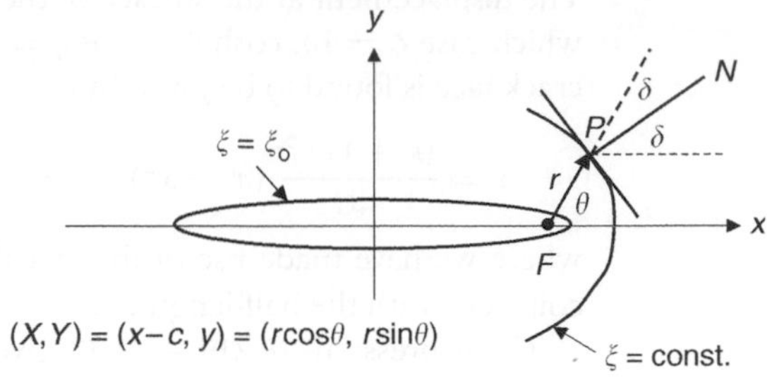

Fig. 8.7 Thin elliptic crack, defined by $\xi = \xi_0$, with the (x, y) coordinate system centered at the center of the crack, and the Cartesian (X, Y) and polar (r, θ) coordinate systems centered at the focal point, F.

We now set up a new (X, Y) coordinate system centered on the focal point $(c, 0)$, defined so that $X = x-c$ and $Y = y$ (Fig. 8.7). For an infinitely thin crack, $\xi_0 = 0$, and both ξ and η will be small in the vicinity of the crack tip, so neglecting terms higher than the second power in (8.235) gives

$$X = x - c = c(\xi^2 - \eta^2)/2, \quad Y = y = c\xi\eta. \tag{8.236}$$

The polar coordinates centered at the focal point can be expressed as

$$X = r\cos\theta, \quad Y = r\sin\theta. \tag{8.237}$$

These relations can be inverted to give

$$\xi = (r/c)^{1/2}(1 + \cos\theta)^{1/2} = (2r/c)^{1/2}\cos(\theta/2), \tag{8.238}$$

$$\eta = (r/c)^{1/2}(1 - \cos\theta)^{1/2} = (2r/c)^{1/2}\sin(\theta/2). \tag{8.239}$$

The term that appears in the denominator of (8.219)–(8.221) takes the form

$$\cosh 2\xi - \cos 2\eta \approx 2(\xi^2 + \eta^2) = 4r/c. \tag{8.240}$$

The stresses in the vicinity of the tip of a thin crack of half-length c, subject to a remote stress of magnitude σ^∞ acting at an inclined angle β, can be found by substituting (8.238)–(8.240) into (8.219)–(8.221). Making use of the fact that ξ and η are small, and neglecting terms of order $r^{1/2}$ or higher in r, we find

$$(\tau_{\xi\xi} - \tau_{\eta\eta})/\sigma^\infty = \cos 2\beta + (c/2r)^{1/2}[2\cos(\theta/2)\sin^2\beta - \sin 2\beta\sin(\theta/2)], \tag{8.241}$$

$$(\tau_{\xi\xi} - \tau_{\eta\eta})/\sigma^\infty = \cos 2\beta + (c/2r)^{1/2}[\sin 2\beta\sin(\theta/2) - \sin^2\beta\sin(\theta/2)\sin(\theta)], \tag{8.242}$$

$$2\tau_{\xi\eta}/\sigma^\infty = (c/2r)^{1/2}[(1/2)\sin 2\beta\sin(\theta/2)(2 - \sin\theta) + \sin^2\beta\cos(\theta/2)\sin(\theta)]. \tag{8.243}$$

Although these are expressed as functions of the polar coordinates r and θ, they nevertheless are the stresses in the (ξ, η) coordinate system. These stresses can be transformed into the (x, y) coordinate system using (8.41) and (8.42), or, more specifically, (8.200) and (8.201). With reference to Fig. 8.7, the line *PN* is orthogonal to the curve of constant ξ and therefore lies in the direction of increasing ξ. The (ξ,η) coordinate system is therefore rotated by angle δ from the (x,y) coordinate system. From (8.193), (8.201), and making use of the smallness of ξ and η, we find

$$e^{2i\delta}=\omega'(\zeta)/\overline{\omega'(\zeta)}=\frac{\sinh(\xi+i\eta)}{\sinh(\xi-i\eta)}\approx\frac{\xi+i\eta}{\xi-i\eta}\cdot\frac{\xi+i\eta}{\xi+i\eta}=\frac{(\xi^2-\eta^2)+i2\xi\eta}{\xi^2+\eta^2}. \tag{8.244}$$

Now using (8.238) and (8.239) to express ξ and η in terms of r and θ, we find

$$e^{2i\delta}=\frac{\cos^2(\theta/2)-\sin^2(\theta/2)+2i\cos(\theta/2)\sin(\theta/2)}{\cos^2(\theta/2)+\sin^2(\theta/2)}=\cos\theta+i\sin\theta=e^{i\theta}, \tag{8.245}$$

which is to say, $\delta=\theta/2$. Hence, in the vicinity of the crack tip, the (r,θ) coordinate system is rotated counterclockwise by angle θ from the (x,y) coordinate system, whereas the (ξ,η) coordinate system is rotated counterclockwise from the (x,y) coordinate system by $\theta/2$.

Using (8.41) and (8.42), we can now calculate the stress components in both the Cartesian and the polar coordinate systems centered on the crack tip:

$$\tau_{rr}=\sigma^{\infty}\cos 2\beta\cos^2\theta+2\sigma^{\infty}(c/8r)^{1/2}\cos(\theta/2)[1+\sin^2(\theta/2)]\sin^2\beta$$
$$+\sigma^{\infty}(c/8r)^{1/2}\sin(\theta/2)[1-3\sin^2(\theta/2)]\sin^2\beta, \tag{8.246}$$

$$\tau_{\theta\theta}=\sigma^{\infty}\cos 2\beta\sin^2\theta+2\sigma^{\infty}(c/8r)^{1/2}\cos^3(\theta/2)\sin^2\beta$$
$$-3\sigma^{\infty}(c/8r)^{1/2}\sin(\theta/2)\cos^2(\theta/2)\sin 2\beta, \tag{8.247}$$

$$\tau_{r\theta}=-\sigma^{\infty}\cos 2\beta\sin\theta\cos\theta+2\sigma^{\infty}(c/8r)^{1/2}\cos^2(\theta/2)\sin(\theta/2)\sin^2\beta$$
$$+\sigma^{\infty}(c/8r)^{1/2}\cos(\theta/2)[1-3\sin^2(\theta/2)]\sin 2\beta, \tag{8.248}$$

$$\tau_{xx}=\sigma^{\infty}\cos 2\beta+2\sigma^{\infty}(c/8r)^{1/2}\cos(\theta/2)[1-\sin(\theta/2)\sin(3\theta/2)]\sin^2\beta$$
$$-\sigma^{\infty}(c/8r)^{1/2}\sin(\theta/2)[2+\cos(\theta/2)\cos(3\theta/2)]\sin 2\beta, \tag{8.249}$$

$$\tau_{yy}=2\sigma^{\infty}(c/8r)^{1/2}\cos(\theta/2)[1+\sin(\theta/2)\sin(3\theta/2)]\sin^2\beta$$
$$+\sigma^{\infty}(c/8r)^{1/2}\sin(\theta/2)\cos(\theta/2)\cos(3\theta/2)\sin 2\beta, \tag{8.250}$$

$$\tau_{xy}=2\sigma^{\infty}(c/8r)^{1/2}\sin(\theta/2)\cos(\theta/2)\cos(3\theta/2)\sin^2\beta$$
$$+\sigma^{\infty}(c/8r)^{1/2}\cos(\theta/2)[1+\sin(\theta/2)\sin(3\theta/2)]\sin 2\beta. \tag{8.251}$$

Note that the nonsingular terms appearing in (8.246)–(8.249), which have been the subject of some discussion and controversy (Maugis, 1992), are *not* simply equal to the stresses that would exist in the absence of the crack. Maugis (2000, p. 159) gives the next terms in the series expansions, which are of order $r^{1/2}$. The leading-order terms in the expressions for the displacements near the crack tip, which are of order $r^{1/2}$, are also given by Maugis.

The stresses, and hence the strains, are seen to be unbounded near the crack tip, with singularities proportional to $(r/c)^{-1/2}$. However, the above solution was developed within the context of linear elasticity, which is founded upon the assumption that the strains are much less than unity. The mathematical singularities would disappear if it were acknowledged that real cracks have finite, nonzero aspect ratios. Nevertheless, these expressions for the stresses near a crack tip, in particular in the equivalent forms given below, play important roles in the theory of *linear elastic fracture mechanics*.

Each of the singular terms in (8.246)–(8.251) involves either $\sin^2\beta$ or $\sin 2\beta$, so a far-field stress oriented parallel to the crack does not give rise to stress singularities. With reference to a far-field stress tensor expressed in the (x,y) coordinate system, therefore, only the stresses τ_{yy}^{∞} and τ_{xy}^{∞} give rise to singular stresses near the crack tip. The singular part of the crack-tip stress field caused by a far-field stress τ_{yy}^{∞}, acting perpendicular to the crack face, is found from (8.246)–(8.248) by setting $\beta = \pi/2$. In this case, $\sin^2\beta = 1$ and $\sin 2\beta = 0$, and, after replacing τ_{yy}^{∞} with the more descriptive notation $\sigma_{\perp}^{\infty}$, we find

$$\tau_{rr} = \sigma_{\perp}^{\infty}(c/2r)^{1/2}\cos(\theta/2)[1+\sin^2(\theta/2)], \tag{8.252}$$

$$\tau_{\theta\theta} = \sigma_{\perp}^{\infty}(c/2r)^{1/2}\cos^3(\theta/2), \tag{8.253}$$

$$\tau_{r\theta} = \sigma_{\perp}^{\infty}(c/2r)^{1/2}\cos^2(\theta/2)\sin(\theta/2), \tag{8.254}$$

$$\tau_{xx} = \sigma_{\perp}^{\infty}(c/2r)^{1/2}\cos(\theta/2)[1-\sin(\theta/2)\sin(3\theta/2)], \tag{8.255}$$

$$\tau_{yy} = \sigma_{\perp}^{\infty}(c/2r)^{1/2}\cos(\theta/2)[1+\sin(\theta/2)\sin(3\theta/2)], \tag{8.256}$$

$$\tau_{xy} = \sigma_{\perp}^{\infty}(c/2r)^{1/2}\sin(\theta/2)\cos(\theta/2)\cos(3\theta/2). \tag{8.257}$$

These expressions are often written, following Irwin (1958), as

$$\tau_{rr} = K_{\rm I}(1/2\pi r)^{1/2}\cos(\theta/2)[1+\sin^2(\theta/2)], \tag{8.258}$$

$$\tau_{\theta\theta} = K_{\rm I}(1/2\pi r)^{1/2}\cos^3(\theta/2), \tag{8.259}$$

$$\tau_{r\theta} = K_{\rm I}(1/2\pi r)^{1/2}\cos^2(\theta/2)\sin(\theta/2), \tag{8.260}$$

$$\tau_{xx} = K_{\rm I}(1/2\pi r)^{1/2}\cos(\theta/2)[1-\sin(\theta/2)\sin(3\theta/2)], \tag{8.261}$$

$$\tau_{yy} = K_{\rm I}(1/2\pi r)^{1/2}\cos(\theta/2)[1+\sin(\theta/2)\sin(3\theta/2)], \tag{8.262}$$

$$\tau_{xy} = K_{\rm I}(1/2\pi r)^{1/2}\sin(\theta/2)\cos(\theta/2)\cos(3\theta/2). \tag{8.263}$$

where $K_{\rm I} = \sigma_{\perp}^{\infty}(\pi c)^{1/2}$ is the *mode I stress intensity factor*. The three modes of deformation at a crack tip are illustrated in Fig. 8.8, with mode I the *crack-opening mode*, mode II the *sliding mode*, and mode III the *tearing mode*.

The stresses arising near the crack tip due to a far-field state of pure shear in the x–y plane can be found by superimposing the solution due to a principal stress of magnitude τ^{∞} at an angle $\beta = \pi/4$, and that due to a principal stress

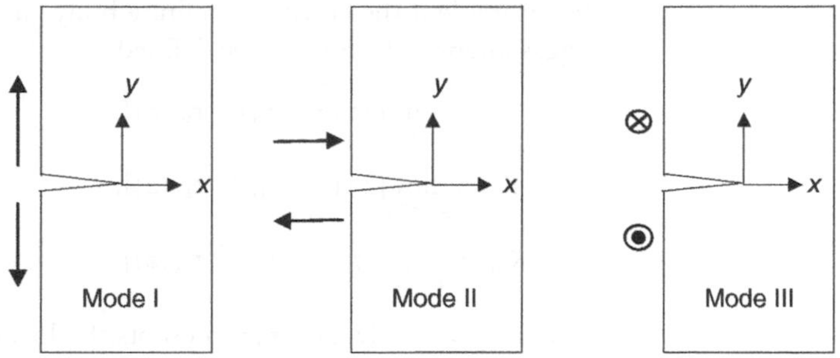

Fig. 8.8 The canonical crack-tip deformation modes: (I) opening mode, (II) sliding mode, and (III) tearing mode, in which top half of figure moves into page, bottom half moves out of page.

of magnitude $-\tau^{\infty}$ at an angle $\beta = 3\pi/4$:

$$\tau_{rr} = \tau^{\infty}(c/2r)^{1/2}\sin(\theta/2)[1 - 3\sin^2(\theta/2)], \tag{8.264}$$

$$\tau_{\theta\theta} = -3\tau^{\infty}(c/2r)^{1/2}\sin(\theta/2)\cos^2(\theta/2), \tag{8.265}$$

$$\tau_{r\theta} = \tau^{\infty}(c/2r)^{1/2}\cos(\theta/2)[1 - 3\sin^2(\theta/2)], \tag{8.266}$$

$$\tau_{xx} = -\tau^{\infty}(c/2r)^{1/2}\sin(\theta/2)[2 + \cos(\theta/2)\cos(3\theta/2)], \tag{8.267}$$

$$\tau_{yy} = \tau^{\infty}(c/2r)^{1/2}\sin(\theta/2)\cos(\theta/2)\cos(3\theta/2), \tag{8.268}$$

$$\tau_{xy} = \tau^{\infty}(c/2r)^{1/2}\cos(\theta/2)[1 + \sin(\theta/2)\sin(3\theta/2)], \tag{8.269}$$

These stresses can be written in terms of the mode II stress intensity factor, $K_{\rm II} = \tau^{\infty}(\pi c)^{1/2}$, as follows:

$$\tau_{rr} = K_{\rm II}(1/2\pi r)^{1/2}\sin(\theta/2)[1 - 3\sin^2(\theta/2)], \tag{8.270}$$

$$\tau_{\theta\theta} = -3K_{\rm II}(1/2\pi r)^{1/2}\sin(\theta/2)\cos^2(\theta/2), \tag{8.271}$$

$$\tau_{r\theta} = K_{\rm II}(1/2\pi r)^{1/2}\cos(\theta/2)[1 - 3\sin^2(\theta/2)], \tag{8.272}$$

$$\tau_{xx} = -K_{\rm II}(1/2\pi r)^{1/2}\sin(\theta/2)[2 + \cos(\theta/2)\cos(3\theta/2)], \tag{8.273}$$

$$\tau_{yy} = K_{\rm II}(1/2\pi r)^{1/2}\sin(\theta/2)\cos(\theta/2)\cos(3\theta/2), \tag{8.274}$$

$$\tau_{xy} = K_{\rm II}(1/2\pi r)^{1/2}\cos(\theta/2)[1 + \sin(\theta/2)\sin(3\theta/2)]. \tag{8.275}$$

The third case, that of a far-field out-of-plane shear stress τ_{yz}^{∞}, gives rise to the following singular stresses near the crack tip (Lardner, 1974, p. 160; Parton and Morozov, 1978, p. 30):

$$\tau_{xz} = -\tau_{yz}^{\infty}(c/2r)^{1/2}\sin(\theta/2) = -K_{\rm III}(1/2\pi r)^{1/2}\sin(\theta/2), \tag{8.276}$$

$$\tau_{yz} = \tau_{yz}^{\infty}(c/2r)^{1/2}\cos(\theta/2) = K_{\rm III}(1/2\pi r)^{1/2}\cos(\theta/2), \tag{8.277}$$

where $K_{\rm III} = \tau_{yz}^{\infty}(\pi c)^{1/2}$ is the mode III stress intensity factor for a single crack in an infinite body. In the context of antiplane loading, the subscript z represents the third, out-of-plane Cartesian coordinate and not the complex variable $z = x + iy$.

The stress intensity factors presented above are relevant to the tip of a single crack in an infinite body, subjected to far-field stresses at infinity. More generally, for example if the crack is in a finite body, or if multiple cracks are present, the stress intensity factors can be defined by

$$K_{\rm I} = \lim_{z\to z_0} [2\pi(z - z_0)^{1/2}\tau_{yy}(z)], \tag{8.278}$$

$$K_{\rm II} = \lim_{z\to z_0} [2\pi(z - z_0)^{1/2}\tau_{xy}(z)], \tag{8.279}$$

$$K_{\rm III} = \lim_{z\to z_0} [2\pi(z - z_0)^{1/2}\tau_{yy}(z)], \tag{8.280}$$

where $z = x + iy$, and z_0 represents the location of the crack tip, which in the present case is $z_0 = c + i0 = c$.

The solution to the problem of a thin elliptical crack subjected to a fluid pressure p along its internal boundary is obtained by starting with the state of uniform hydrostatic stress of magnitude p throughout the body and subtracting off the stresses due to far-field stresses $\tau_{xx}^{\infty} = \tau_{yy}^{\infty} = p$. Of these three stress fields, only the one due to τ_{yy}^{∞} gives rise to singular stresses near the crack tip. Hence, the singular component of the stress field due to internal pressure p is given by (8.252)–(8.257), with τ_{yy}^{∞} replaced by $-p$, or, equivalently, by (8.258)–(8.263) with $K_1 = -p(\pi c)^{1/2}$.

8.11 Nearly rectangular hole

The most powerful technique for finding the stresses and displacements around two-dimensional holes is that of conformal mapping, in which the region outside the hole in the z-plane is transformed into the region outside (or inside) the unit circle in the ζ-plane, through a mapping $z = \omega(\zeta)$. The problem is then solved in the ζ-plane, although the boundary conditions tend to take on a more complicated form in the transformed plane than in the physical z-plane. This method is described in detail in the monographs of Muskhelishvili (1963), Savin (1961), and England (1971). Full exploitation of this method requires knowledge of the various integral theorems of complex analysis, which are, however, beyond the scope of the present discussion.

The Weierstrass theorem assures that the region outside any reasonably well-behaved hole shape can be conformally mapped into the interior of a circle. In particular, if the hole is a polygon, the Schwarz–Christoffel method explicitly yields $z = \omega(\zeta)$ in the form of a convergent infinite series. For a square aligned with its sides parallel to the x- and y-axes, the mapping function is

$$z = \omega(\zeta) = a\left[\zeta^{-1} + \frac{1}{6}\zeta^3 + \frac{1}{56}\zeta^7 + \frac{1}{176}\zeta^{11} + \cdots\right]. \tag{8.281}$$

If only the first two terms of the mapping function are retained, then the unit circle $\zeta = \exp(i\alpha)$ is mapped into the curve given by

$$x = a\left(\cos\alpha - \frac{1}{6}\cos 3\alpha\right), \quad y = -a\left(\sin\alpha + \frac{1}{6}\sin 3\alpha\right), \tag{8.282}$$

which is a quasi-square with rounded corners (Fig. 8.9a). Since $x = 5a/6$ when $\alpha = 0$, the "side" of the square is of length $5a/3$. The radius of curvature at the corners is approximately $a/10$, or about 0.06 times the length of the side.

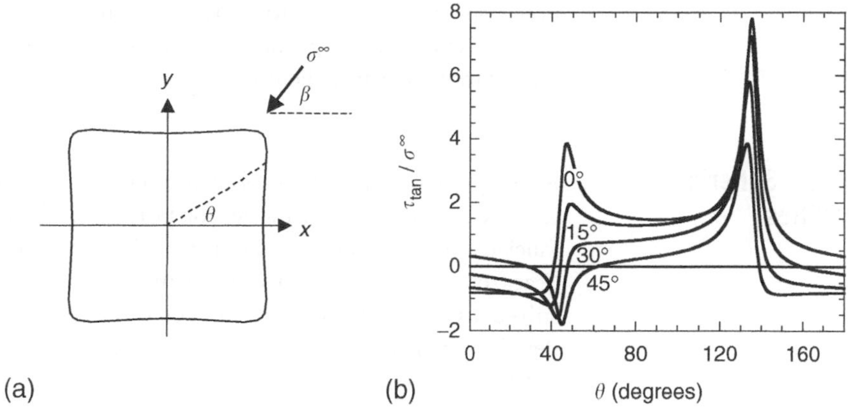

Fig. 8.9 (a) Nearly rectangular hole, described by (8.282), subjected to a far-field stress inclined at an angle β. (b) Normalized tangential stress at the hole boundary; numbers near curves refer to the stress inclination angle, β.

For the case of a far-field stress σ^{∞} acting at an angle β to the x-axis (Fig. 8.9a), Savin (1961) found the following complex stress potentials:

$$\phi(\zeta) = \sigma^{\infty} a \left[\frac{1}{4}\zeta^{-1} + \left(\frac{3}{7}\cos 2\beta + \frac{3}{5}\mathrm{i}\sin 2\beta \right)\zeta + \frac{1}{24}\zeta^{3} \right], \tag{8.283}$$

$$\psi(\zeta) = -\sigma^{\infty} a \left[\frac{(1/2)\mathrm{e}^{-2\mathrm{i}\beta}\zeta^{-1} + \{13\zeta - 26((3/7)\cos 2\beta + (3/5)\mathrm{i}\sin 2\beta)\zeta^{3}\}}{12(2+\zeta^{4})} \right]. \tag{8.284}$$

from which the stresses and displacements can be found using (8.36), (8.47), and (8.48). In particular, the tangential stress at the point on the hole boundary that corresponds to the point $\zeta = \exp(\mathrm{i}\alpha)$ is given by

$$\begin{aligned} \tau_{\tan} &= 4\mathbf{Re}\{\phi'(z)\} = 4\mathbf{Re}\{\phi'(\zeta)/\omega'(\zeta)\}_{\zeta=\exp(\mathrm{i}\alpha)} \\ &= \sigma^{\infty}\mathbf{Re}\left\{ \frac{[70\zeta^{-2} - (120\cos 2\beta + 168\mathrm{i}\sin 2\beta) - 35\zeta^{2}]}{35(2\zeta^{2} + \zeta^{2})} \right\}_{\zeta=\exp(\mathrm{i}\alpha)} \\ &= \sigma^{\infty}\frac{(105 - 360\cos 2\beta\cos 2\alpha + 168\sin 2\beta\sin 2\alpha)}{35(5 + 4\cos 4\alpha)}. \end{aligned} \tag{8.285}$$

Figure 8.9b shows the variation of the tangential stress at the hole boundary, as a function of the polar angle $\theta = \tan^{-1}(y/x)$, for various values of the stress inclination angle, β. The maximum stresses occur close to the "corners" of the hole, where $\theta = 45°$ and $135°$, and can be seen to exceed the maximum value of $3\sigma^{\infty}$ that occurs on the boundary of a circular hole. If more terms are retained in the mapping function, the hole approximates a square more closely; consequently, the radius of curvature at the corners decreases and the maximum tangential stresses increase.

Savin (1961) and Gerçek (1988) give detailed results for rectangular holes of various aspect ratios. Berry (1960a), Berry and Sales (1961,1962), Salamon (1964) and Ryder and Officer (1964) have discussed rectangular holes with reference to underground coal mines, in many cases taking the elastic anisotropy of the ground into account. Gerçek (1997) and Exadaktylos and Stavropoulou (2002) used conformal mapping to analyse the stresses around holes with shapes commonly used for underground tunnels, such as semicircles and a hole with an arched roof and parabolic floor.

8.12 Spherical cavities

Although man-made excavations in rock are rarely if ever spherical, the spherical cavity is useful as a model for pores in sedimentary rocks (Guéguen and Palciauskas, 1994) and for fluid inclusions that occur in some igneous rocks (Lacazette, 1990). Elasticity problems involving spherical geometries can be solved using the Papkovich–Neuber displacement function formulation, the details of which can be found in Sokolnikoff (1956, Chapter 6) or Soutas–Little (1999, Chapter 14).

Consider a spherical cavity of radius a in an infinite elastic rock mass, with the center of the cavity taken as the origin of a Cartesian coordinate system. In general, the far-field stress state will consist of three orthogonal principal stresses, and the full solution will be the appropriate superposition of three solutions for the case of a single far-field principal stress. Without loss of generality, this far-field stress can be taken to be aligned with the z-axis, in which case the resulting state of displacement and stress in the rock mass will be axisymmetric about the z-axis. The solution for the case of three principal stresses can be obtained by appropriate superposition of the solution derived below.

The appropriate boundary conditions for this problem are that

$$\text{as } r \to \infty, \quad \tau_{zz} \to \tau_{zz}^{\infty} \equiv T, \quad \text{all other stresses} \to 0, \tag{8.286}$$

and that all tractions must vanish at $r = a$. To solve this problem, consider first the stress state in which the only nonzero stress component is the uniform stress $\tau_{zz} = T$. We now invoke a standard spherical coordinate system, in which θ is the polar angular coordinate in the x–y plane, and ϕ is the angle of rotation from the z-axis. In terms of this spherical coordinate system, stress state (8.286) corresponds to the following nonzero stress components (Soutas-Little 1999, p. 407):

$$\tau_{rr} = T\cos^2\phi, \quad \tau_{\phi\phi} = T\sin^2\phi, \quad \tau_{r\phi} = -T\sin\phi\cos\phi. \tag{8.287}$$

This stress state satisfies the boundary conditions at infinity, but gives unwanted, nonzero values for the tractions τ_{rr} and $\tau_{r\phi}$ at the cavity surface. Hence, we need to superpose a stress state that cancels out these tractions at the surface $r = a$ and which vanishes at infinity. For axisymmetric elasticity problems in the region exterior to a spherical cavity, the most general solution that gives zero stresses at infinity is (Soutas-Little, 1999, pp. 396–7)

$$u_r = \sum_{n=0}^{\infty}[C_n n(n+3-4\nu)r^{-n} - D_n(n+1)r^{-(n+2)}]P_n(\cos\phi), \tag{8.288}$$

$$u_\phi = \sum_{n=0}^{\infty}[C_n n(-n+4-4\nu)r^{-n} + D_n r^{-(n+2)}]\mathrm{d}P_n(\cos\phi)/\mathrm{d}\phi, \tag{8.289}$$

$$\tau_{rr} = 2G\sum_{n=0}^{\infty}[-C_n n(n^2+3n-2\nu)r^{-(n+1)}$$
$$+ D_n(n+1)(n+2)r^{-(n+3)}]P_n(\cos\phi), \quad (8.290)$$

$$\tau_{r\phi} = 2G\sum_{n=0}^{\infty}[C_n(n^2-2+2\nu)r^{-(n+1)} - D_n(n+2)r^{-(n+3)}]\mathrm{d}P_n(\cos\phi)/\mathrm{d}\phi, \quad (8.291)$$

$$\tau_{\phi\phi} = 2G\sum_{n=0}^{\infty}[C_n(n^2-2n-1+2\nu)r^{-(n+1)} - D_n(n+1)^2r^{-(n+3)}]P_n(\cos\phi)$$
$$+ 2G\sum_{n=0}^{\infty}[C_n(n-4+4\nu)r^{-(n+1)} - D_n r^{-(n+3)}]\cot\phi[\mathrm{d}P_n(\cos\phi)/\mathrm{d}\phi], \quad (8.292)$$

$$\tau_{\theta\theta} = 2G\sum_{n=0}^{\infty}[C_n n(n+3-4n\nu-2\nu)r^{-(n+1)} - D_n(n+1)r^{-(n+3)}]P_n(\cos\phi)$$
$$- 2G\sum_{n=0}^{\infty}[C_n(n-4+4\nu)r^{-(n+1)} - D_n r^{-(n+3)}]\cot\phi[\mathrm{d}P_n(\cos\phi)/\mathrm{d}\phi], \quad (8.293)$$

where P_n is the Legendre polynomial of order n, and the other stress and displacement components are zero. The first three Legendre polynomials are given by

$$P_0(x) = 1, \quad P_1(x) = x, \quad P_2(x) = (3x^2-1)/2. \quad (8.294)$$

Although they will not be needed in the present problem, the remaining Legendre polynomials can be defined in terms of the following recursion relation:

$$(n+1)P_{n+1}(x) = (2n+1)xP_n(x) - nP_{n-1}(x). \quad (8.295)$$

The coefficients C_n and D_n must be chosen so that the tractions τ_{rr} and $\tau_{r\phi}$ given by (8.290) and (8.291) cancel out those given by (8.287) at the cavity surface. Expressing these tractions in terms of Legendre polynomials, we find that the stresses given by (8.290)–(8.293) must satisfy the following boundary conditions at $r = a$:

$$\tau_{rr}(r=a) = -T\cos^2\phi = -(T/3)[1+2P_2(\cos\phi)], \quad (8.296)$$
$$\tau_{r\phi}(r=a) = T\sin\phi\cos\phi = -(T/3)[\mathrm{d}P_2(\cos\phi)/\mathrm{d}\phi]. \quad (8.297)$$

Evaluating the stresses given by (8.290) and (8.291) at $r = a$ and equating them term-by-term to those given by (8.296) and (8.297), yields the following three equations:

$$4GD_0a^{-3} = -T/3, \tag{8.298}$$

$$4G[-2C_2(5-\nu)a^{-3} + 6D_2a^{-5}] = -2T/3, \tag{8.299}$$

$$4G[C_2(1+\nu)a^{-3} - 2D_2a^{-5}] = -T/3, \tag{8.300}$$

the solution to which is (Goodier, 1933)

$$C_2 = 5Ta^3/12G(7-5\nu), \quad D_0 = -Ta^3/12G, \quad D_2 = Ta^5/2G(7-5\nu). \tag{8.301}$$

The complete state of stress is then given by the superposition of the stresses given by (8.287), (8.288)–(8.293), and (8.301).

The uniaxial stress state in the rock mass, as given by (8.286) or (8.287), is perturbed by the presence of the cavity. As for the case of two-dimensional cavities, this effect dies off with distance from the cavity and is greatest at the cavity surface. The nonzero stresses at the cavity surface, found by setting $r = a$ in (8.287), (8.288)–(8.293), and (8.296), are

$$\frac{\tau_{\phi\phi}(a)}{T} = \frac{27-15\nu}{2(7-5\nu)} - \frac{15}{(7-5\nu)}\cos^2\phi, \tag{8.302}$$

$$\frac{\tau_{\theta\theta}(a)}{T} = \frac{-3(1-5\nu)}{2(7-5\nu)} - \frac{15\nu}{(7-5\nu)}\cos^2\phi. \tag{8.303}$$

The maximum value of $\tau_{\phi\phi}$ occurs along the equatorial line $\phi = \pi/2$ (i.e., $z = 0$) and takes the value $2T$ for a rock with $\nu = 0.2$. For this value of ν, the stress component $\tau_{\theta\theta}$ varies between 0 and $-T$ and attains its extreme value of $-0.5T$ at the north ($\phi = 0$) and south ($\phi = \pi$) poles of the sphere. The stress concentration factor along the equator varies within a narrow range from 1.93, when $\nu = 0$, to 2.17, when $\nu = 0.5$. This is in contrast to the stress concentration at the boundary of a two-dimensional circular cavity under far-field uniaxial tension, which has the value of 3 for any value of the Poisson ratio. As a general rule, stress concentrations around three-dimensional cavities are less severe than those around two-dimensional cavities and die off more rapidly with distance from the cavity surface.

The problem of a spherical pore subjected to a pore pressure of magnitude P acting over its surface, with zero stresses at infinity, can be solved by using only the P_0 term in the general solution. From (8.288)–(8.293), the P_0 term gives

$$u_r = -D_0/r^2, \quad u_\phi = u_\theta = 0, \tag{8.304}$$

$$\tau_{rr} = 4GD_0/r^3, \quad \tau_{\phi\phi} = \tau_{\theta\theta} = -2GD_0/r^3. \tag{8.305}$$

The boundary condition $\tau_{rr}(a) = 0$ implies that $D_0 = Pa^3/4G$. Hence, the solution for a pressurized spherical pore is

$$u_r = -Pa^3/4Gr^2, \quad \tau_{rr} = P(a/r)^3, \quad \tau_{\phi\phi} = \tau_{\theta\theta} = -(P/2)(a/r)^3. \tag{8.306}$$

The displacement dies off as r^{-2}, and the stresses decay as r^{-3}. This is in contrast to a pressurized two-dimensional circular pore, for which the displacement varies as r^{-1} and the stresses as r^{-2}.

As the radial displacement at the pore wall is $-Pa/4G$, the total volume increase of the pressurized pore is $P\pi a^3/G$. The "pore compressibility" C_{pp} of a spherical cavity, defined in §7.2 as the fractional derivative of the pore volume with respect to the pore pressure, is therefore equal to $3/4G$. The sphere is the least compressible of all possible pore shapes (Zimmerman, 1991).

8.13 Penny-shaped cracks

The most general three-dimensional cavity (or inclusion) shape that is amenable to analytical treatment is the ellipsoid, which can be thought of as a sphere that has been stretched by possibly different amounts in three mutually orthogonal directions. An ellipsoidal surface centered at the origin can therefore be described by

$$(x/a)^2 + (y/b)^2 + (z/c)^2 = 1, \tag{8.307}$$

where $\{a, b, c\}$ are the semiaxes in the $\{x, y, z\}$ directions, respectively. The problem of an ellipsoidal cavity in an infinite matrix, subjected to a far-field stress state whose principal directions are aligned with the axes of the ellipsoid, was solved by Sadowsky and Sternberg (1949) using ellipsoidal coordinates. This analysis was extended to elastic ellipsoidal inclusions by Robinson (1951). Eshelby (1957) used methods of potential theory to treat the case of arbitrary orientation of the ellipsoid with respect to the principal stresses. Extensive discussion of ellipsoidal elastic inclusions, with cavities as a special case, has been given by Mura (1987).

If all three axes are of equal length, $\{a = b = c\}$, the ellipsoid degenerates into a sphere, which was treated in §8.12. If two axes are of equal length, $\{a = b \neq c\}$, then the ellipsoid degenerates into a spheroid, which can be thought of as being formed by revolving an ellipse about one of its axes of symmetry. In the case $\{a = b > c\}$, the spheroid is referred to as *oblate*, whereas if $\{a = b < c\}$, it is *prolate*. In the limit of $c \gg a$, the prolate spheroid becomes a cylinder, which was discussed in §8.5–§8.7. The remaining limiting case, $c \ll a$, represents a thin, "penny-shaped" crack. This shape is of great interest in rock mechanics as a model for microcracks and hydraulic fractures. Stresses and displacements around penny-shaped cracks can be studied by examining the limiting case of solutions for thin oblate spheroids (Eshelby, 1957) or can be studied by using methods in which the crack is assumed to be infinitely thin *before* the equations are solved (Sack, 1946; Sneddon, 1946).

Sneddon (1946) solved the problem of an infinitely thin crack subjected to uniform normal traction p applied to its faces, such as would be applied by pore fluid. The case of a traction-free crack in a body subjected to a far-field tension perpendicular to the crack plane can be treated using superposition, as in §8.12. The displacement of the crack surface in the direction normal to its plane is given by

$$w = \frac{2(1-\nu)pa}{\pi G}\sqrt{1-(r/a)^2} = w_{\max}\sqrt{1-(r/a)^2}, \tag{8.308}$$

where a is the radius of the crack in its (x, y) plane. The crack-opening displacement is greatest at the center of the crack and zero at its edges. If the crack is initially flat, application of a pore pressure of magnitude p will enlarge it into a spheroid whose semimajor axis equals a, and semiminor axis is w_{max}.

Conversely, if the crack initially is a thin oblate spheroid with $c = \alpha a$ and is subjected to a far-field compressive stress $\sigma_{\perp}^{\infty}$, it will completely close up when the magnitude of the displacement at its center reaches c. Setting w_{max} in (8.308) equal to $c = \alpha a$ and replacing p with $\sigma_{\perp}^{\infty}$ yields the closing pressure of a penny-shaped crack of initial aspect ratio α:

$$\sigma_{\perp}^{\infty}(\text{closure}) = \frac{\pi G\alpha}{2(1-\nu)}. \tag{8.309}$$

The closing pressure of a three-dimensional penny-shaped crack exceeds that of a two-dimensional elliptical crack having the same initial aspect ratio, as given by (8.233), by a factor of $\pi/2$. Note that applying Sneddon's results to an initially oblate crack involves errors on the order of α, arising from the fact that in his solution, the tractions are applied along the x–y plane, whereas the initial crack surface deviates slightly from that plane.

The excess elastic strain energy due to a pressurized crack can be calculated from (5.158), by noting that in the present case $\mathbf{p} \cdot \mathbf{u} = pw$, and $\mathbf{F} = 0$, yielding

$$\begin{aligned}\mathcal{E}_{\text{hydrostatic}} &= \frac{1}{2}\int_{2A} pw\,dA = \int_0^{2\pi}\int_0^a \frac{2(1-\nu)p^2 a}{\pi G}\sqrt{1-(r/a)^2}\,r\,dr\,d\theta \\ &= \frac{4(1-\nu)p^2 a^3}{3G},\end{aligned} \tag{8.310}$$

where the additional factor of 2 arises from considering both faces of the crack. As the initial surface of the crack consists only of faces normal to the z-axis, the state of loading considered by Sneddon also applies to a thin crack loaded by hydrostatic pressure over its entire surface. Superposition arguments then show that (8.310) also represents the excess energy caused by *far-field* hydrostatic loading of magnitude p. It also follows that no additional elastic strain energy arises if the far-field principal stresses are parallel to the plane of the crack (Sack, 1946). To within an error of order α, the strain energy of a pressurized crack is independent of the initial aspect ratio, as can be verified by solving the problem for a spheroid of arbitrary aspect ratio and then taking the limit as the aspect ratio vanishes (Zimmerman, 1985b).

Segedin (1950) solved the problem of an infinitely thin penny-shaped crack whose faces are subjected to uniform shearing tractions. As for the case of normal loading, superposition can be used to find the solution for a traction-free crack subject to far-field shear (Fig. 8.10a). On the crack surface, the only nonzero displacement component, which is the one in the direction of shear, is given by

$$u = \pm\frac{4(1-\nu)Sa}{\pi G(2-\nu)}\sqrt{1-(r/a)^2} = \pm u_{\max}\sqrt{1-(r/a)^2}, \tag{8.311}$$

where S is the magnitude of the shear traction, and the different signs apply to the two surfaces of the crack. The excess elastic strain energy can be found by

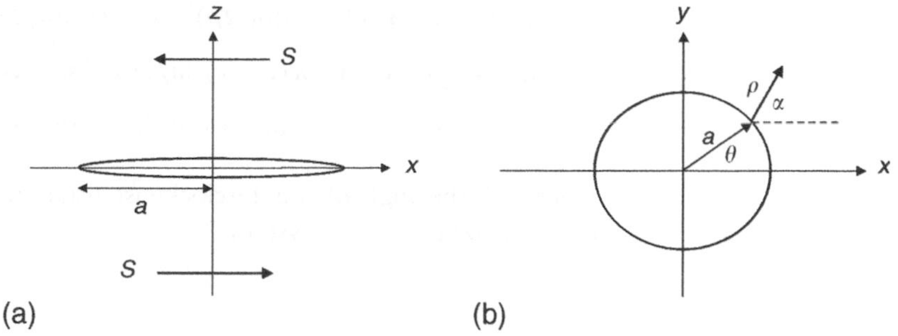

Fig. 8.10
(a) Penny-shaped crack lying in the x-y plane, subject to far-field shear stress. (b) Coordinate system used for stresses near the edge of the crack.

evaluating (5.158), noting that in the present case, $\mathbf{p} \cdot \mathbf{u} = Su$ and $\mathbf{F} = 0$:

$$\mathcal{E}_{\text{shear}} = \frac{1}{2} \int_{2A} Su dA = \int_0^{2\pi} \int_0^a \frac{4(1-\nu)S^2 a}{\pi G(2-\nu)} \sqrt{1-(r/a)^2} r dr d\theta$$

$$= \frac{8(1-\nu)S^2 a^3}{3G(2-\nu)}. \tag{8.312}$$

Comparison of (8.310) and (8.312) shows that the excess energies associated with shear and normal loading of a thin crack are quite similar and coincide if $\nu = 0$.

The stresses in the vicinity of the edge of a penny-shaped crack, which is to say in the vicinity of the points located at $\{r = a, z = 0\}$, can be expressed in forms analogous to those given in §8.10 for two-dimensional cracks (Kassir and Sih, 1975, Chapter 1). For a crack pressurized by a pore pressure p, the dominant terms in the expressions for the stress components are

$$\tau_{\rho\rho} = (p/2\pi)(a/2\rho)^{1/2}[5\cos(\alpha/2) - \cos(3\alpha/2)], \tag{8.313}$$

$$\tau_{\alpha\alpha} = (p/2\pi)(a/2\rho)^{1/2}[3\cos(\alpha/2) + \cos(3\alpha/2)], \tag{8.314}$$

$$\tau_{\rho\alpha} = (p/2\pi)(a/2\rho)^{1/2}[\sin(\alpha/2) + \sin(3\alpha/2)], \tag{8.315}$$

$$\tau_{zz} = (4\nu p/\pi)(a/2\rho)^{1/2}\cos(\alpha/2), \tag{8.316}$$

$$\tau_{\rho z} = \tau_{\alpha z} - 0, \tag{8.317}$$

where (ρ, α) are the local polar coordinates in the (x, y) plane, centered on a point on the edge of the crack (Fig. 8.10b).

If the crack faces are subjected to a shear traction of magnitude S, acting in the x direction, say, then the stresses in the vicinity of the edge of the crack will be, to leading order, given by

$$\tau_{\rho\rho} = [S/(2-\nu)\pi](a/2\rho)^{1/2}\cos\theta[3\sin(3\alpha/2) - 5\sin(\alpha/2)], \tag{8.318}$$

$$\tau_{\alpha\alpha} = -[S/(2-\nu)\pi](a/2\rho)^{1/2}\cos\theta[3\sin(3\alpha/2) + 3\sin(\alpha/2)], \tag{8.319}$$

$$\tau_{\rho\alpha} = [S/(2-\nu)\pi](a/2\rho)^{1/2}\cos\theta[\cos(3\alpha/2) + (1/3)\cos(\alpha/2)], \tag{8.320}$$

$$\tau_{zz} = [8\nu S/3\pi(2-\nu)](a/2\rho)^{1/2}\cos\theta\sin(\alpha/2), \tag{8.321}$$

$$\tau_{\alpha z} = -[4S(1-\nu)/\pi(2-\nu)](a/2\rho)^{1/2}\sin\theta\cos(\alpha/2), \tag{8.322}$$

$$\tau_{\rho z} = -[4S(1-\nu)/\pi(2-\nu)](a/2\rho)^{1/2}\sin\theta\sin(\alpha/2), \tag{8.323}$$

where θ is the angle of counterclockwise rotation from the x-axis to the origin of the (ρ, α) coordinate system.

8.14 Interactions between nearby cavities

The analysis of the stresses around a cavity, be it a microscopic pore or a macroscopic excavation, is facilitated by assuming that the cavity is located within an infinite rock mass. If the cavity is not located too close to any neighboring cavities or other boundaries, such as the ground surface, this assumption is reasonable. Roughly speaking, the nearest distance to another cavity or other type of boundary should be at least three times the characteristic dimension of the cavity in order for this assumption to be acceptable.

Exact solutions of problems involving multiple cavities or inclusions, or cavities/inclusions located near a free surface, are difficult to obtain. One important such problem that has been solved in closed form is that of two nearby circular holes in an infinite region, subjected to far-field tension that is directed either parallel or perpendicular to the line joining the centers of the two holes (Fig. 8.11a). Ling (1948) solved this problem using bipolar coordinates, in the form of infinite series of trigonometric and hyperbolic functions.

In the case of longitudinal far-field tension parallel to the line connecting the centers of the two holes (σ_{long} in Fig. 8.11a), the second hole has a shielding effect, and the maximum stress concentration, which occurs at point B, is less than the value of 3 that obtains for an isolated hole (Fig. 8.11b). This shielding effect is enhanced as the holes become closer, and in the limiting case of two circular holes that are barely touching, the maximum hoop stress is equal to $2.569\sigma_{\text{long}}$. When the applied far-field stress is transverse to the line connecting the two hole centers (σ_{tran} in Fig. 8.11a), the maximum stress concentration, which occurs at the point (A) nearest to the other hole, is enhanced by the presence of the second hole and becomes infinite as the two holes come into contact.

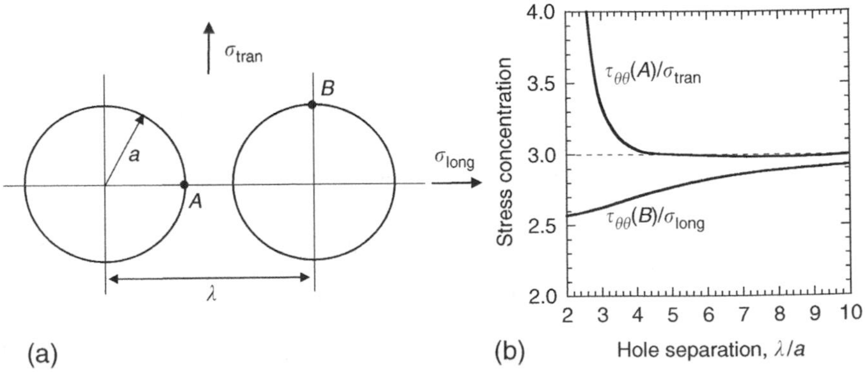

Fig. 8.11 (a) Two nearby holes subjected to longitudinal and transverse far-field stresses. (b) Maximum stress concentrations at the hole boundaries (Ling, 1948).

The most interesting case is that of transverse loading. Asymptotic analyses of Ling's solution, carried out by Zimmerman (1988) and Callias and Markenscoff (1993), reveal that as the holes become closer, the hoop stress at point A is asymptotically of the form

$$\frac{\tau_{\theta\theta}(A)}{\sigma_{\text{tran}}} \approx \frac{1.94}{\sqrt{(\lambda/2a)-1}}, \tag{8.324}$$

where λ is the distance between the two hole centers and a is the radius of either hole. Several seminumerical analyses of this problem (Duan et al., 1986; Tsukrov and Kachanov, 1997), however, have yielded values in the range 2.05–2.13 for the numerical constant in (8.324).

The stresses around a pressurized circular hole near a straight, traction-free boundary were found by Jeffery (1920) and Mindlin (1948) using bipolar coordinates. Verruijt (1998) used conformal mapping for this problem and was also able to find a closed-form expression for the displacements. Callias and Markenscoff (1989) presented an asymptotic analysis of this problem in the limit as the hole approaches the free surface. As in the problem of two nearby holes, the stress singularity was found to be inversely proportional to the square root of the distance between the hole and the free surface.

Howland (1935) investigated an infinite elastic region containing an infinite row of equally sized and equally spaced circular holes and derived an approximate analytic solution for the case when the holes are not too closely spaced. In accordance with Ling's solution for the two-hole problem, he found a stress enhancement effect when the far-field principal stress was perpendicular to the line joining the centers of the holes and a shielding effect when the loading was parallel to this line. Collins (1962a,b) developed approximate analytical solutions for the problem of two pressurized, penny-shaped circular cracks that were either lying in the same plane or directly above one another in parallel planes. His analysis focused on the "pore compressibility" of the cracks, and he found that the compressibility of the coplanar cracks was greater than that of an isolated crack, whereas the compressibility of the parallel cracks was less than that of an isolated crack. Kachanov (1987, 1994) developed a general approximation method for treating interactions between two or more cracks, in either two or three dimensions.

Sternberg and Sadowsky (1952) used a spherical dipolar coordinate system to solve the problem of two nearby spherical cavities in an infinite elastic medium, subjected to uniform pressure along the cavity surfaces, or far-field tension either parallel or perpendicular to the line joining the centers of the two cavities. Eubanks (1954) analyzed the problem of a hemispherical cavity at the surface of an infinite half-space that is subjected to uniaxial tension in a direction parallel to the free surface. McTigue (1987) developed an approximate analytical solution for a pressurized spherical cavity near a free surface, as a model for the inflation of a magma body.

9 Inelastic behavior

9.1 Introduction

In Chapter 5, the theory of elasticity was discussed in some detail. A basic assumption of that theory is that the strain in an elastic body at a given time depends only on the stress acting on that body at that time. In Chapter 8, elasticity theory was used to find the stresses and displacements in the vicinity of cavities and excavations in rock. The theory of elasticity was extended in Chapter 7 to account for the effects of pore pressure and temperature, but still in the context of *elastic* behavior, which is to say that no explicit time dependence or stress-path dependence was considered. An implicit assumption of these elastic-type constitutive models is that the rock undergoes no internal, microscopic degradation as it deforms. This fact implies that the rock will return to its initial state if the stresses acting on it are removed.

Under many circumstances, rock will deform irreversibly. Roughly speaking, if the relationship between stress and strain does not explicitly depend on the rate of deformation, then the inelastic behavior is referred to as *plasticity*, whereas behavior that is explicitly time-dependent is referred to as *viscoelasticity*. Another distinguishing characteristic of plastic behavior is that a finite *yield stress* must be exceeded before irreversible plastic deformation occurs. The basic concepts of *plasticity* and *yield* are presented in §9.2. The stress state around a circular opening in a rock mass that deforms plastically is analyzed in §9.3 and §9.4. In §9.5, we briefly describe the classical theory of *perfectly plastic* behavior, and in §9.6, we use this theory to study plastic flow of a rock being squeezed between two flat surfaces. The phenomenon of hardening, in which the yield stress continues to increase as the rock deforms, is introduced in §9.7. The phenomenon of *creep*, in which the rock continues to deform under the action of constant loads, is introduced in §9.8. The mathematical theory of this type of behavior, known as *viscoelasticity*, is discussed in §9.9 and §9.10, and the solutions to some simple problems of viscoelastic behavior are presented in §9.11.

9.2 Plasticity and yield

The complete stress–strain curve of a rock under uniaxial compression was discussed in detail in Chapter 4. A typical such curve was shown schematically in Fig. 4.3. To solve problems that are more complicated than uniaxial compression, this stress–strain behavior must be represented by a mathematical function(s).

For rocks that are loaded below the yield stress, the stress–strain behavior is typically idealized as being a linear relationship between stress and strain, as indicated in Fig. 4.2a and (4.1). For rocks that are loaded past the yield point, it is necessary to represent the stress–strain behavior in the region *BC* of Fig. 4.2a mathematically.

One simple idealized stress–strain curve that captures the fact that the strain continues to increase without a proportional increase in the stress is that shown in Fig. 9.1a, in which, after the yield stress σ_0 is reached, the stress–strain curve continues to rise but with a shallower slope. The stress σ_0 is known as the *yield stress under uniaxial loading*. The behavior shown in Fig. 9.1a, the further discussion of which is deferred until §9.5, is known as *plastic behavior with strain hardening*. A yet simpler idealization, which lends itself to the solution of important problems such as the deformation of rock around tunnels and cavities, is that shown in Fig. 9.1b, in which the stress remains at σ_0 as the strain continues to increase. A material that obeys a stress–strain law such as that shown in Fig. 9.1b is known as *elastic-perfectly plastic*. A further simplification, in which the elastic strains are neglected entirely and assumed to be negligible compared to the plastic strains, is that of a *rigid-perfectly plastic* material (Fig. 9.1c).

From the point of view of thermodynamics or materials science, the defining feature of the yield stress is that if the load is removed on the portion *OA* of the curve shown in Fig. 9.1a or b, the material will unload along the same curve. However, if the load is removed at some point in the segment *AB*, the material will unload elastically along a curve that is parallel to the original elastic loading curve *OA*. These unloading curves are shown as dashed lines. This phenomenon is of great importance for, say, machine parts or engineered structural elements, which may well undergo many cycles of loading and unloading. Such loading cycles are rare in rock mechanics, however. Hence, in rock mechanics the *thermodynamic irreversibility* of plastic deformation is often not emphasized, whereas the nonlinear character of the stress–strain behavior is.

Under uniaxial deformation, a rock yields if the stress reaches some value σ_0 that is characteristic of that particular rock. Yield stresses are easily measured under uniaxial conditions. The issue then arises of generalizing this result to more complicated stress states. A *yield criterion* is a relationship among the stresses that determines whether or not the material will yield. For an isotropic material, the yield criterion should be a function of the three principal stresses, or equivalently, of the three invariants of the stress.

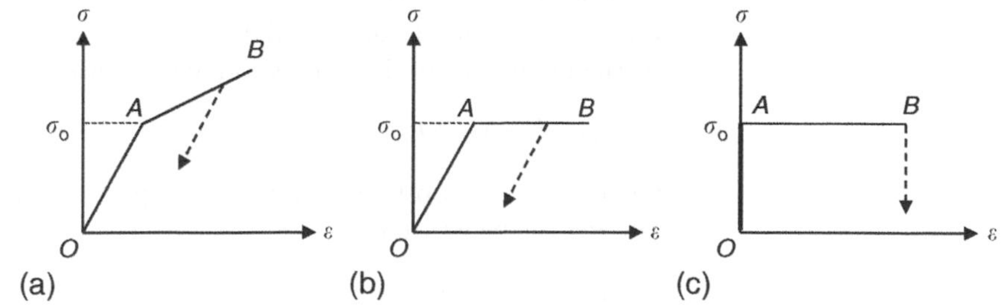

Fig. 9.1 Idealized uniaxial stress–strain curves: (a) elastic–plastic with strain hardening, (b) elastic-perfectly plastic, and (c) rigid-perfectly plastic. Unloading curves are shown as dashed lines.

The simplest and oldest such yield criterion is that proposed by Tresca (1864), who assumed that yield would occur if the maximum shear stress acting on any plane inside the rock reaches some critical value, τ_o. In terms of the three principal stresses, this criterion would be written as $(\sigma_1 - \sigma_3)/2 = \tau_o$. Consideration of the special case of uniaxial stress shows that $\sigma_o/2 = \tau_o$. Hence, the Tresca criterion is

$$\sigma_1 - \sigma_3 = s_1 - s_3 = \sigma_o, \tag{9.1}$$

where the s_i are the principal stress deviations defined in §2.8.

Many yield criteria that have been proposed follow the scheme discussed in §4.8 for brittle failure criteria, in which failure is controlled by J_2, the second invariant of the deviatoric stress, and possibly also by I_1, the mean normal stress. The simplest such yield criterion, that of von Mises (1913), is obtained by assuming that the rock yields when J_2 reaches some critical value. Using the various expressions for J_2 given in §2.8, the von Mises criterion can be written as (von Mises, 1913; Hencky, 1924)

$$J_2 = \sigma_o^2/3, \tag{9.2}$$

$$J_2 = [(\sigma_1 - \sigma_2)^2 + (\sigma_2 - \sigma_3)^2 + (\sigma_3 - \sigma_1)^2]/6 = \sigma_o^2/3, \tag{9.3}$$

$$s_1^2 + s_2^2 + s_3^2 = 2\sigma_o^2/3, \tag{9.4}$$

$$\tau_{oct} = \sqrt{2}\sigma_o/3, \tag{9.5}$$

$$\mathcal{E}_d = \sigma_o^2/6G, \tag{9.6}$$

where σ_o is again the uniaxial yield stress, and $\mathcal{E}_d$ is the distortional strain energy defined in (5.152).

Under conditions of pure shear, such as the stress state $\{\sigma_1, \sigma_2, \sigma_3\} = \{\tau, -\tau, 0\}$, the von Mises criterion predicts that failure will occur when $\tau = \sigma_o/\sqrt{3}$. Thus it is in contrast to the Tresca criterion, which predicts that failure under pure shear occurs when $\tau = \sigma_o/2$. Hence, if the parameter σ_o is chosen so that these criteria agree with experimental data under uniaxial compression, at most one of them will give the correct failure load under pure shear.

The von Mises criterion was originally developed for metals and does not account for the experimental observation that the yield stress of most rocks increases with increasing mean normal stress. For example, note that an *arbitrary* hydrostatic stress increment can be added to an existing stress state without changing the form of criterion (9.3). To account for the strengthening effect of the mean normal stress, a term that depends on I_1 can be included in the yield criterion. Recall that the von Mises criterion (9.2) can be written as $(J_2)^{1/2} = \sigma_o/\sqrt{3} =$ constant. Drucker and Prager (1952) added a "strengthening" term on the right-hand side of this expression that is linear in I_1:

$$(J_2)^{1/2} = a + bI_1, \tag{9.7}$$

where a and b are constants. In terms of the principal stresses, the Drucker–Prager criterion can be written as

$$(\sigma_1 - \sigma_2)^2 + (\sigma_2 - \sigma_3)^2 + (\sigma_3 - \sigma_1)^2 = 6b^2[\sigma_1 + \sigma_2 + \sigma_3 + (a/b)]^2. \tag{9.8}$$

Fitting this criterion to uniaxial compressive strength data shows that $a = (1/\sqrt{3} - b)\sigma_o$, in which case we can put $\alpha = (1/b\sqrt{3}) - 1$ and write (9.8) in the form

$$(\sigma_1 - \sigma_2)^2 + (\sigma_2 - \sigma_3)^2 + (\sigma_3 - \sigma_1)^2 = \frac{2}{(1+\alpha)^2}(\sigma_1 + \sigma_2 + \sigma_3 + \alpha\sigma_o)^2, \quad (9.9)$$

where α is a dimensionless parameter.

Under standard $\{\sigma_1 > \sigma_2 = \sigma_3\}$ triaxial compression, (9.9) takes the form (Fjaer et al., 1992)

$$\sigma_1 = \sigma_o + \frac{\alpha + 3}{\alpha}\sigma_3, \quad (9.10)$$

whereas under $\{\sigma_1 = \sigma_2 > \sigma_3\}$ "triaxial extension," it takes the form

$$\sigma_1 = \frac{\alpha}{\alpha - 1}\sigma_o + \frac{\alpha + 2}{\alpha - 1}\sigma_3. \quad (9.11)$$

Physical plausible values of α are obviously restricted to the range $\alpha > 1$. It follows that the coefficients of both σ_o and σ_3 in (9.11) are greater than the respective coefficients in (9.10). Hence, in $\{\sigma_1, \sigma_3\}$ space, the yield curve for $\{\sigma_1 = \sigma_2 > \sigma_3\}$ loading always lies above the curve for $\{\sigma_1 = \sigma_2 > \sigma_3\}$ loading, illustrating the strengthening effect of the intermediate principal stress.

9.3 Elastic–plastic hollow cylinder

In general, solving problems of a body undergoing plastic deformation requires a full stress–strain law that governs the evolution of the plastic strain. These laws are discussed in §9.4. However, there are certain problems that, because of their symmetry and the monotonicity of the loading, are "statically determinate," in which case the stress state can be found without consideration of the strains. One such problem is that of a pressurized hollow cylinder composed of a rock that obeys the Tresca yield criterion.

Consider a hollow cylinder with inner radius a and outer radius b, under plane strain conditions, with a uniform pressure applied to its outer surface, $r = b$. If this pressure is slowly increased from 0 to some value P_o, at first the cylinder will everywhere be in the elastic regime, and the stresses will be given by (8.85) and (8.86). The maximum shear stress in the cylinder will be $P_o b^2/(b^2 - a^2)$ and will occur at $r = a$. (Note that for any value of Poisson's ratio, we will have $\tau_{\theta\theta} > \tau_{zz} > \tau_{rr}$ at $r = a$.) When the maximum shear stress reaches σ_o, the rock will yield at $r = a$. As P_o increases further, the yielded zone will grow radially outward, and the cylinder will consist of an inner annular region that has yielded and an outer annulus that is still in its elastic state.

Let $r = \rho$ denote the elastic–plastic boundary (Fig. 9.2a). The rock in the region $a < r < \rho$ will have yielded, and so the stresses there will satisfy the Tresca yield criterion, (9.1):

$$\tau_{\theta\theta} - \tau_{rr} = \sigma_o. \quad (9.12)$$

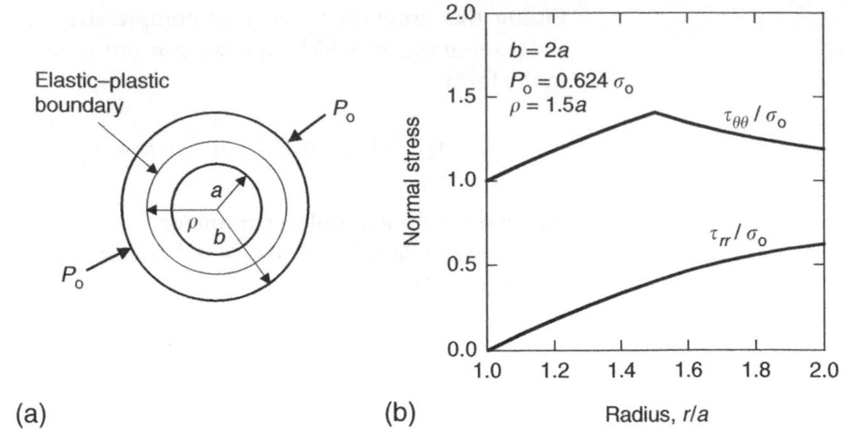

Fig. 9.2 (a) Hollow cylinder subjected to an external pressure. (b) Stresses for the case $b = 2a$, when $P_0 = 0.624\sigma_o$.

The stresses must also satisfy the stress equilibrium equation, (5.108), which takes the form

$$\frac{d\tau_{rr}}{dr} + \frac{\tau_{rr} - \tau_{\theta\theta}}{r} = 0. \tag{9.13}$$

Substitution of (9.12) into (9.13) gives

$$\frac{d\tau_{rr}}{dr} = \frac{\sigma_0}{r}, \tag{9.14}$$

which can be integrated to give

$$\tau_{rr} = \sigma_0 \ln r + C, \tag{9.15}$$

where C is a constant. If the inner surface is traction free, then τ_{rr} must vanish at $r = a$, which shows that $C = -\sigma_0 \ln a$. Hence, the stresses in the plastic region $a < r < \rho$ are

$$\tau_{rr} = \sigma_0 \ln(r/a), \quad \tau_{\theta\theta} = \sigma_0[\ln(r/a) + 1]. \tag{9.16}$$

In the elastic region, $\rho < r < b$, the stresses must be of the form given by (8.79) and (8.82):

$$\tau_{rr} = A - \frac{B}{r^2}, \quad \tau_{\theta\theta} = A + \frac{B}{r^2}, \tag{9.17}$$

where A and B are constants. Three boundary and/or interface conditions are needed to determine the constants $\{A, B, \rho\}$. At the elastic–plastic boundary, $r = \rho$, the elastic stress state (9.17) must be on the verge of yielding. Requiring the stresses given by (9.17) to satisfy the yield condition (9.12) at $r = \rho$ gives $B = \rho^2\sigma_0/2$. Next, requiring the radial stress to be continuous at $r = \rho$ gives

$$A - \frac{B}{\rho^2} = A - \frac{\sigma_0}{2} = \sigma_0 \ln(\rho/a), \tag{9.18}$$

which can be solved for

$$A = \frac{\sigma_o}{2}[1 + 2\ln(\rho/a)]. \tag{9.19}$$

Last, imposing the condition that $\tau_{rr} = P_o$ at $r = b$ gives

$$\tau_{rr}(b) = A - \frac{B}{b^2} = \frac{\sigma_o}{2}[1 + 2\ln(\rho/a)] - \frac{\rho^2\sigma_o}{2b^2} = P_o, \tag{9.20}$$

which can be rearranged to give the following relation between ρ and P_o:

$$[1 + 2\ln(\rho/a) - (\rho/b)^2] = \frac{2P_o}{\sigma_o}. \tag{9.21}$$

The stresses in the elastic region $\rho < r < b$ are given by (Fig. 9.2b)

$$\tau_{rr} = \frac{\sigma_o}{2}[1 + 2\ln(\rho/a) - (\rho/r)^2], \quad \tau_{\theta\theta} = \frac{\sigma_o}{2}[1 + 2\ln(\rho/a) + (\rho/r)^2]. \tag{9.22}$$

The rock first yields at $r = a$ when $P_o = \sigma_o[1-(a/b)^2]/2$. For the limiting case of a borehole or tunnel in an infinite rock mass, this will occur when $P_o = \sigma_o/2$. For thin-walled cylinders, with $a \approx b$, the critical pressure needed to cause yielding can be a small fraction of σ_o. When the external pressure reaches the value $P_o = \sigma_o \ln(b/a)$, the entire cylinder will have yielded. For thin cylinders, this will occur at an external pressure that is only slightly larger than the pressure needed to cause initial yielding. For example, if $b/a = 1.2$, initial yielding will occur when $P_o = 0.153\sigma_o$, and the entire cylinder will be in the plastic state when $P_o = 0.182\sigma_o$. For a hole in an infinite rock mass, the plastic zone will never encompass the entire rock mass, as the pressure required for this to happen grows logarithmically with b.

9.4 Circular hole in an elastic – brittle – plastic rock mass

Consider now a material that obeys a Coulomb-type failure criterion, as given by (4.13):

$$\sigma_1 = C_o + q\sigma_3, \tag{9.23}$$

where $C_o = 2S_o$, $\tan\beta$ is the unconfined compressive strength, S_o is the cohesion, $q = \tan^2\beta = (1 + \sin\phi)/(1 - \sin\phi)$, and ϕ is angle of internal friction. In the special case $q = 1$, this reduces to the Tresca criterion. Assume that the rock mass is initially in a state of hydrostatic stress P_o, and then a circular hole of radius a is drilled into the rock, so that the stress at $r = a$ is reduced to some value P_i. For sufficiently small values of P_o, the rock will be in its elastic state, and the stresses will be given by (8.91) and (8.92). As discussed in §9.3, for larger values of P_o, the rock will fail within some annular region surrounding the borehole. The problem again will be to determine the location of the elastic–plastic boundary, and the stresses in both the yielded and intact regions.

In the failed zone, the stresses will obey (9.23). If $P_o > P_i$, which is the case of practical interest, then the maximum principal stress at the borehole wall will be

$\tau_{\theta\theta}$, and the minimum principal stress will be τ_{rr}. Substitution of (9.23) into the radial stress equilibrium equation (9.13) then gives

$$\frac{d\tau_{rr}}{dr} = \frac{C_o + (q-1)\tau_{rr}}{r}. \tag{9.24}$$

Solving this equation, subject to the condition that $\tau_{rr} = P_i$ when $r = a$, gives

$$\tau_{rr} = \frac{C_o}{1-q} + \left(P_i - \frac{C_o}{1-q}\right)\left(\frac{r}{a}\right)^{q-1}, \tag{9.25}$$

after which (9.23) shows that

$$\tau_{\theta\theta} = \frac{C_o}{1-q} + q\left(P_i - \frac{C_o}{1-q}\right)\left(\frac{r}{a}\right)^{q-1}. \tag{9.26}$$

These stresses will obtain throughout the yielded zone, $a < r < \rho$. In the elastic zone, $r > \rho$, the stresses must have the form given by (9.17). Imposition of the far-field boundary conditions shows that the constant A must equal P_o, which leaves

$$\tau_{rr} = P_o - \frac{B}{r^2}, \quad \tau_{\theta\theta} = P_o + \frac{B}{r^2}. \tag{9.27}$$

At the elastic–plastic boundary, $r = \rho$, the stresses given by (9.27) must satisfy (9.23). This gives

$$B = \rho^2[C_o + P_o(q-1)]/(q+1). \tag{9.28}$$

Last, continuity of τ_{rr} at $r = \rho$ gives, from (9.25), (9.27) and (9.28),

$$\frac{\rho}{a} = \left\{\frac{2[P_o(q-1) + C_o]}{[P_i(q-1) + C_o](q+1)}\right\}^{1/(q-1)}, \tag{9.29}$$

thereby completing the solution.

As an example, consider a rock whose angle of internal friction is 60°, which corresponds to $q = 3$. Then (9.29) gives

$$\frac{\rho}{a} = \left(\frac{C_o + 2P_o}{2C_o + 4P_i}\right)^{1/2}. \tag{9.30}$$

In the yielded region, $a < r < \rho$, (9.25) and (9.26) reduce to

$$\tau_{rr} = [P_i + (C_o/2)](r/a)^2 - (C_o/2), \quad \tau_{\theta\theta} = 3[P_i + (C_o/2)](r/a)^2 - (C_o/2). \tag{9.31}$$

In the elastic region, $r > \rho$, the stresses are, from (9.27), (9.28), and (9.29),

$$\tau_{rr} = P_o - \frac{1}{2}[P_o + (C_o/2)](\rho/r)^2, \quad \tau_{\theta\theta} = P_o + \frac{1}{2}[P_o + (C_o/2)](\rho/r)^2. \tag{9.32}$$

These stresses are plotted in Fig. 9.3a for the case, $P_o = C_o$, $P_i = 0$.

In the special case of a material with no cohesion, the radius of the yielded zone is given by $\rho = a(P_o/2P_i)^{1/2}$, and expressions (9.31) and (9.32) reduce further to

$$\tau_{rr} = P_i(r/a)^2, \quad \tau_{\theta\theta} = 3P_i(r/a)^2, \quad \text{for } a < r < \rho, \tag{9.33}$$

$$\tau_{rr} = P_o\left[1 - \frac{P_o}{4P_i}\left(\frac{a}{r}\right)^2\right], \quad \tau_{\theta\theta} = P_o\left[1 + \frac{P_o}{4P_i}\left(\frac{a}{r}\right)^2\right], \quad \text{for } r > \rho. \tag{9.34}$$

The preceding analysis is most relevant to soils. A more realistic assumption for rock is that the intact material obeys failure law (9.23) with parameters $\{C_o', q'\}$, but once the material has yielded, it obeys the Coulomb failure law with a different set of parameters, $\{C_o, q\}$, where typically $C_o \ll C_o'$. The analysis given above will continue to hold through (9.27). But in the present case, at $r = \rho$, the elastic stresses given by (9.27) must satisfy the failure criterion with parameters appropriate to the intact rock, that is, $\{C_o', q'\}$. Hence, (9.28) is replaced by

$$B = \rho^2[C_o' + P_o(q' - 1)]/(q' + 1). \tag{9.35}$$

Continuity of τ_{rr} at $r = \rho$ now gives

$$\frac{\rho}{a} = \left\{\frac{(2P_o - C_o')(1 - q) - C_o(1 + q')}{[P_i(1 - q) - C_o](1 + q')}\right\}^{1/(q-1)} \tag{9.36}$$

in place of (9.29). Although stresses such as τ_{rr} must be continuous across any surface of constant r, there is no such requirement for $\tau_{\theta\theta}$, and indeed $\tau_{\theta\theta}$ is discontinuous at $r = \rho$ (Fig. 9.3b).

As an example, suppose that $C_o' \gg C_o = 0$, $q' = q = 3$, in which case (9.35) and (9.36) become

$$B = \rho^2(C_o' + 2P_o)/4, \quad \rho = a[(2P_o - C_o')/4P_i]^{1/2}. \tag{9.37}$$

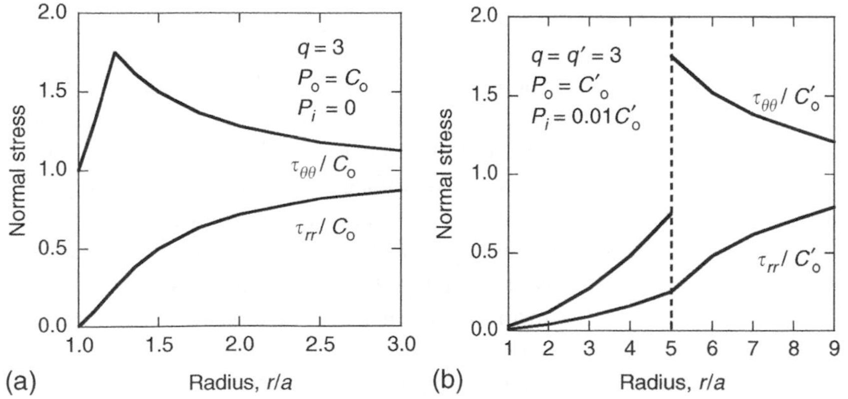

Fig. 9.3 Stresses around a circular hole in an infinite rock mass: (a) Entire rock mass assumed to be in a Coulomb-type stress state; (b) Coulomb-type region assumed to be surrounded by intact rock (see text for details).

The stresses are given by

$$\tau_{rr} = P_i(r/a)^2, \quad \tau_{\theta\theta} = 3P_i(r/a)^2, \quad \text{for } a < r < \rho, \tag{9.38}$$

$$\tau_{rr} = P_o - \frac{(2P_o + C_o')}{4}\left(\frac{\rho}{r}\right)^2, \quad \tau_{\theta\theta} = P_o + \frac{(2P_o + C_o')}{4}\left(\frac{\rho}{r}\right)^2, \quad \text{for } r > \rho. \tag{9.39}$$

Now consider the specific case in which $P_o = C_o'$. According to the analysis given in §9.3, this far-field stress is twice that which would cause failure at a traction-free cavity wall in an elastic–plastic rock mass. From (9.37), the boundary of the yielded zone will be given by

$$\rho = a(C_o'/4P_i)^{1/2}, \tag{9.40}$$

so that if, for example, $P_i = C_o'/100$, then $\rho = 5a$. The stresses in this case are shown in Fig. 9.3b, which illustrates the aforementioned discontinuity in $\tau_{\theta\theta}$.

Ladanyi (1967) analyzed the problem of the expansion of a cylindrical or spherical cavity in a rock mass that is linearly elastic before failure and obeys a generalized Griffith failure criterion. Sharan (2003) gave a closed-form solution for the stresses and displacements around a circular opening in a medium that obeys the Hoek-Brown failure criterion. Detournay and Fairhurst (1987) gave an elasto-plastic analysis of a circular tunnel subject to nonhydrostatic far-field stresses. A large number of solutions for the stresses and displacements around cylindrical and spherical cavities in elasto-plastic and viscoelastic media have been derived and compiled by Yu (2000).

9.5 Perfectly plastic behavior

The theory of perfect/ideal plasticity, which provides a good model for the inelastic behavior of many metals, has been developed in great detail. Among the classic monographs on this topic are those of Hill (1950) and Prager (1959). The applicability of this theory to rocks is restricted to situations of very high stresses and temperatures, as occurs in the deep crust.

As discussed in §9.2, a body will first begin to plastically deform when the stresses become large enough to satisfy the yield criterion. According to the von Mises yield criterion, this will occur when

$$J_2 = [(\sigma_1 - \sigma_2)^2 + (\sigma_2 - \sigma_3)^2 + (\sigma_3 - \sigma_1)^2]/6 = \sigma_o^2/3, \tag{9.41}$$

where σ_o is the uniaxial yield stress. This criterion defines a surface in $\{\sigma_1, \sigma_2, \sigma_3\}$ stress space that is known as the *yield surface*. The region "interior" to the yield surface, for which $J_2 < \sigma_o^2/3$, is an elastic region, whereas stress states lying on the yield surface correspond to states of plastic deformation. (Stress states exterior to the yield surface cannot correspond to states of static equilibrium. For example, if a body that obeys an elasto-plastic stress–strain relationship such as shown in Fig. 9.1b is subjected to a uniaxial stress in excess of σ_o, it will deform uncontrollably, in an unstable manner. In a three-dimensional context, if a region of rock is loaded to the yield point, the stresses will redistribute themselves so that other regions of rock accommodate the excess load.)

The basic assumption of the theory of perfect (or *ideal*) plasticity is that the yield surface (in stress space) does not deform as the rock mass itself deforms. Once the stresses reach the yield surface, further changes in stress may either bring the stress state back inside the yield surface (i.e., unloading, which occurs elastically, as in Fig. 9.1b) or may move the stress state to another point along the yield surface. Assume that the loads are applied monotonically, so that unloading into the elastic regime does not occur. In this case, the externally applied loads can be parameterized by a scalar parameter, which we denote by λ (not related to the Lamé parameter). The condition that a further stress change leads to additional *plastic* deformation is that

$$dJ_2/d\lambda = 0, \tag{9.42}$$

which is equivalent to guaranteeing that the stress state remains on the yield surface. This provides a constraint equation that must hold between the three principal stresses during plastic deformation.

Provided that the strains are small enough to be considered infinitesimal, as in Chapter 2, it is permissible to assume that the total strain consists of the sum of an elastic strain, which is related to the stresses by Hooke's law, and a plastic strain. In many practical problems involving plastic deformation, the elastic strain can be neglected in comparison with the plastic strain. The question then arises of how the plastic strains evolve in response to changes in the stress.

The assumption usually made in classical plasticity theory (Drucker, 1950) is that the plastic strain increment is normal to the yield surface. Such a rule is known as an *associated flow rule*, because the plastic strain increment is associated with the yield function. If the yield surface is defined by some function of the form $J_2(\sigma) = 0$, as can be achieved for the von Mises criterion by moving the constant term to the left-hand side of (9.41), the normal to this surface is given by the gradient of the function $J_2(\sigma)$. The derivatives of J_2 with respect to the principal stresses are given by

$$\frac{\partial J_2}{\partial \sigma_1} = \frac{2(\sigma_1-\sigma_2)+2(\sigma_1-\sigma_3)}{6} = \frac{3\sigma_1-(\sigma_1+\sigma_2+\sigma_3)}{3} = \sigma_1-\sigma_{\mathrm{m}} = s_1, \tag{9.43}$$

where s_1 is the first principal stress deviation, and similarly for the other two principal stresses. Hence, $\nabla J_2 = (s_1, s_2, s_3)$, and so the flow rule associated with the von Mises yield criterion is

$$(de_1, de_2, de_3) = (s_1, s_2, s_3)d\lambda, \tag{9.44}$$

where (e_1, e_2, e_3) are the three principal deviatoric strains, and $d\lambda$ is some scalar increment. No generality is lost by identifying this increment with the parameter λ in (9.42). According to the concept of an associated flow rule, then, the *direction* of the strain increment is determined by the existing stress state. The direction of the strain increment is not controlled by the stress increment, the direction of which is in fact constrained by (9.42) to lie *along* the yield surface, rather than *normal* to it. The *magnitude* of the strain increment, which is proportional to $d\lambda$, must be determined from other aspects of the problem, such as the boundary conditions. Note that the parameter λ may, and in general will, vary with both space and time; it is not a material constant.

A kinematic assumption that is usually made in classical plasticity is that plastic deformation occurs without any change in volume, in which case

$$d\varepsilon_{xx} + d\varepsilon_{yy} + d\varepsilon_{zz} = 0. \tag{9.45}$$

This assumption is valid in materials for which the plastic deformation is due to the motion of dislocations through a crystal lattice, but is not true when plastic deformation is due to, say, the crushing of grains in a sedimentary rock. If (9.45) holds, then the increments in the deviatoric strains are equivalent to the increments in the strains, that is, $(de_1, de_2, de_3) = (d\varepsilon_1, d\varepsilon_2, d\varepsilon_3)$. In terms of an arbitrary coordinate system, the plastic flow equations (9.44) then take the form (Mendelson, 1968, p. 101)

$$d\varepsilon_{xx} = s_{xx}d\lambda = \left[\frac{2}{3}\tau_{xx} - \frac{1}{3}(\tau_{yy} + \tau_{zz})\right]d\lambda, \quad d\varepsilon_{xy} = \tau_{xy}d\lambda, \tag{9.46}$$

and similarly for the other four independent stress components.

The stress state must at all times also satisfy the three equations of stress equilibrium, (5.78)–(5.80). In conjunction with (9.41), (9.42), (9.45), and (9.46), this yields the twelve equations that are needed in order to solve for the six unknown stresses and six unknown strains.

The development of the equations given above for a rigid-perfectly plastic material is due primarily to Saint Venant (1870), Lévy (1871), and von Mises (1913). In the event that the elastic strains are not negligible compared to the plastic strains, Prandtl (1925) and Reuss (1930) suggested that the plastic strain increment would again be given by (9.45) and (9.46), and that the elastic strain increment would be given by an incremental version of Hooke's law for an incompressible material, that is,

$$d\varepsilon^e_{xx} = \tau_{xx}/2G, \quad d\varepsilon^e_{xy} = \tau_{xy}/2G, \quad \text{etc.}, \tag{9.47}$$

where the superscript "e" denotes "elastic." The total strain is assumed to be the sum of the elastic strain, given by (9.47), and the plastic strain, given by (9.46).

We now return to the rigid-plastic case and consider a two-dimensional plane strain deformation in the (x, y) plane. The z direction will necessarily be a direction of principal strain, say the ε_2 direction. Then by (9.44), we see that $s_2 = 0$, after which (2.155) shows that

$$\sigma_2 = \tfrac{1}{2}(\sigma_1 + \sigma_3). \tag{9.48}$$

Combining (9.48) with the yield criterion (9.41) shows that

$$\sigma_1 - \sigma_3 = 2k, \quad k = \sigma_0/\sqrt{3}. \tag{9.49}$$

We write this equation in terms of the parameter k so that, by taking $k = \sigma_0/2$, the analysis can also be applied to a material that obeys the Tresca yield criterion.

In terms of the stress components referred to in the x–y coordinate system, (2.37)–(2.38) show that (9.49) can be written as

$$\tau_{xy}^2 + \frac{1}{4}(\tau_{xx} - \tau_{yy})^2 = k^2. \tag{9.50}$$

In plane strain, and in the absence of body forces, the equations of stress equilibrium, (5.78)–(5.80), take the form

$$\frac{\partial \tau_{xx}}{\partial x} + \frac{\partial \tau_{xy}}{\partial y} = 0, \tag{9.51}$$

$$\frac{\partial \tau_{xy}}{\partial x} + \frac{\partial \tau_{yy}}{\partial y} = 0, \tag{9.52}$$

which can be combined into the single differential equation,

$$\frac{\partial^2 \tau_{xy}}{\partial x^2} - \frac{\partial^2 \tau_{xy}}{\partial y^2} = \pm \frac{\partial^2 (\tau_{xx} - \tau_{yy})}{\partial x \partial y}. \tag{9.53}$$

Inserting (9.50) into (9.53) yields

$$\frac{\partial^2 \tau_{xy}}{\partial x^2} - \frac{\partial^2 \tau_{xy}}{\partial y^2} = \pm 2 \frac{\partial^2}{\partial x \partial y}(k^2 - \tau_{xy}^2)^{1/2}, \tag{9.54}$$

which is a single differential equation for τ_{xy}. If (9.54) can be solved for τ_{xy}, then the two normal stresses follow from (9.51) and (9.52) by direct integration. The constants of integration that arise are found by requiring the stresses to satisfy (9.50).

Alternatively, the stress equilibrium equations (9.51) and (9.52) can be solved by introducing the Airy stress function, as defined in §5.7. In this case, (9.50) becomes

$$\left(\frac{\partial^2 U}{\partial x^2} - \frac{\partial^2 U}{\partial y^2}\right)^2 + 4\left(\frac{\partial^2 U}{\partial x \partial y}\right)^2 = 4k^2, \tag{9.55}$$

thus reducing the problem to a differential equation for U. This equation can be solved using the complex variable methods introduced in Chapter 8 (Annin, 1988).

9.6 Flow between flat surfaces

Consider the problem of the extrusion of a plastic material between two rigid plates that are slowly moving together (Hill, 1950, pp. 226–36). The two rigid plates are located at $y = \pm a$, and the plastic material occupies the region $(0 < x < L, -a < y < a)$. If the plates are very rough, it is reasonable to assume that material has yielded at the interface with these plates, in which case τ_{xy} has its maximum possible value. Hence, if the material is slipping past them to the right, the boundary conditions for the shear stress along the top and bottom surfaces will be

$$\tau_{xy}(x, a) = k, \quad \tau_{xy}(x, -a) = -k, \tag{9.56}$$

A solution to this problem can be obtained by the following semi-inverse method. The boundary conditions (9.56) suggest the following expression for the shear stress distribution within the body:

$$\tau_{xy}(x,y) = ky/a. \tag{9.57}$$

This expression satisfies (9.54). With this form for τ_{xy}, (9.50) and (9.51) can be integrated to give

$$\tau_{xx} = f(y) - kx/a, \quad \tau_{yy} = g(x). \tag{9.58}$$

The requirement that the stresses satisfy (9.50) gives

$$f(y) - g(x) = -kx/a \pm 2k[1 - (y/a)^2]^{1/2}, \tag{9.59}$$

from which it follows that

$$f(y) = P \pm 2k[1 - (y/a)^2]^{1/2}, \quad g(x) = P - kx/a, \tag{9.60}$$

where P is a constant. Choosing the negative sign, the stresses are

$$\tau_{xx} = P - 2k[1 - (y/a)^2]^{1/2} - kx/a, \tag{9.61}$$

$$\tau_{yy} = P - kx/a, \tag{9.62}$$

$$\tau_{xy} = ky/a, \tag{9.63}$$

According to this solution, the rigid plates apply a compressive stress τ_{yy} to the rock that decreases linearly from P at the left edge of the system, to $P - k(L/a)$ at the right edge. The normal stress $\tau_{xx}(0, y)$ at the "inlet" is also compressive and varies parabolically from P at the upper and lower corners to $P - 2k$ at the midplane. Superimposed on this is an additional stress gradient $\partial\tau_{xx}/\partial x = -k/a$, which represents a driving force that helps to extrude the rock to the right.

This problem is statically determinate, in the sense that the stresses can be found without any need to consider the deformation of the rock. To find the displacements, assume that the plates move toward each other with speed V_0. The vertical displacement at the upper surface will be $v(x, a) = V_0 t$, and at the lower surface $v(x, -a) = -V_0 t$. This suggests the following vertical displacement field:

$$v(x,y) = yV_0 t/a. \tag{9.64}$$

The vertical normal strain will then be given by

$$\varepsilon_{yy} = \frac{\partial v}{\partial y} = \frac{V_0 t}{a}, \tag{9.65}$$

and the increment in the normal vertical strain will be

$$d\varepsilon_{yy} = V_0 dt/a. \tag{9.66}$$

By (9.46), we see that $d\varepsilon_{xx}/d\varepsilon_{yy} = s_{xx}/s_{yy}$. From (9.61) and (9.62), the deviatoric stresses are

$$s_{xx} = -s_{yy} = -k[1 - (y/a)^2]^{1/2}, \tag{9.67}$$

from which it follows that

$$d\varepsilon_{xx} = -V_0 dt/a. \tag{9.68}$$

Integration with respect to time, bearing in mind that there is no strain at the start of the process, gives

$$\varepsilon_{xx} = -V_0 t/a. \tag{9.69}$$

Recalling that $\varepsilon_{xx} = \partial u/\partial x$, integration with respect to x gives

$$u(x, y) = -V_0 tx/a + h(y). \tag{9.70}$$

The function $h(y)$ is found by considering the shear strain. Again by (9.46) we see that $d\varepsilon_{xy} = d\varepsilon_{yy}(s_{xy}/s_{yy})$. It follows from (9.63), (9.66), and (9.67) that

$$d\varepsilon_{xy} = \frac{yV_0/a}{(a^2 - y^2)^{1/2}} dt. \tag{9.71}$$

Integration with respect to time gives

$$\varepsilon_{xy} = \frac{yV_0 t/a}{(a^2 - y^2)^{1/2}}. \tag{9.72}$$

But from the definition of the shear strain, using (9.64) and (9.70),

$$\varepsilon_{xy} = \frac{1}{2}\left(\frac{\partial u}{\partial y} + \frac{\partial v}{\partial x}\right) = \frac{1}{2}h'(y). \tag{9.73}$$

Equating (9.72) and (9.73), and integrating, gives

$$h(y) = -2V_0 t[1 - (y/a)^2]^{1/2}. \tag{9.74}$$

Finally, from (9.64), (9.70), and (9.74), the displacement field is

$$u = -V_0 t\{(x/a) + 2[1 - (y/a)^2]^{1/2}\}, \quad v = yV_0 t/a, \tag{9.75}$$

and the velocity field is

$$\dot{u} = -V_0 x/a - 2V_0[1 - (y/a)^2]^{1/2}, \quad \dot{v} = V_0 y/a. \tag{9.76}$$

The rock material moves vertically toward the midplane and horizontally toward the right, with a horizontal velocity component profile that varies parabolically at each cross-sectional location x.

The choice of the positive sign in (9.59) would correspond to the case in which the plates are forced apart by material being squeezed in (intruded), rather than extruded.

A salient feature of this solution is that the stresses, (9.61)–(9.63), are *not* influenced by the velocity V_0 at which the horizontal surfaces approach one another. This is in accordance with the fact that plasticity is not inherently rate-dependent. Note also that there was no need to calculate the value of the parameter λ; its role in equations such as (9.46) is merely to show that, at any value of $\mathbf{x}$ and t, the ratio of each deviatoric strain increment to its corresponding deviatoric stress will have the same value.

The solution described above has been used and extended by Nye (1951) to study the downhill flow of glaciers. Evison (1960) used it to study the growth of continents by plastic extrusion of their deeper layers.

9.7 Flow rules and hardening

We now examine some implications of an associated flow rule, following the discussion given by Fjaer et al. (1992, pp. 68–71). Consider a Coulomb yield criterion (§4.5), which in $\{\sigma_1, \sigma_2, \sigma_3\}$ space can be represented by (Fig. 9.4a)

$$f(\sigma_1, \sigma_2, \sigma_3) = \sigma_1 - \sigma_3 \tan\alpha - \sigma_0 = 0, \tag{9.77}$$

where σ_0 is the uniaxial yield stress, $\tan\alpha = (1 + \sin\phi)/(1 - \sin\phi)$, and ϕ is the angle of internal friction. The normal vector to the yield surface is

$$\nabla f = (1, 0, -\tan\alpha). \tag{9.78}$$

According to the associated flow rule, the plastic strain increments will be normal to the yield surface. In analogy with (9.44), these increments will therefore be

$$\mathrm{d}\varepsilon_1^p = \mathrm{d}\lambda, \quad \mathrm{d}\varepsilon_2^p = 0, \quad \mathrm{d}\varepsilon_3^p = -\tan\alpha\,\mathrm{d}\lambda, \tag{9.79}$$

and the total volumetric plastic strain increment will be

$$\mathrm{d}\varepsilon_v^p = (1 - \tan\alpha)\mathrm{d}\lambda = (1 - \tan\alpha)\mathrm{d}\varepsilon_1^p = \frac{-2\sin\phi}{1 - \sin\phi}\mathrm{d}\varepsilon_1^p. \tag{9.80}$$

For the typical case in which ϕ lies between 0 and 90°, $\mathrm{d}\varepsilon_v^p$ and $\mathrm{d}\varepsilon_1^p$ will have different signs. Hence, if the rock is compressed in the direction of the maximum principal stress, the total volume will in fact *increase*. This phenomenon, known as *dilatancy* (Cook, 1970), is observed in many rocks, but usually not to the degree predicted by (9.80).

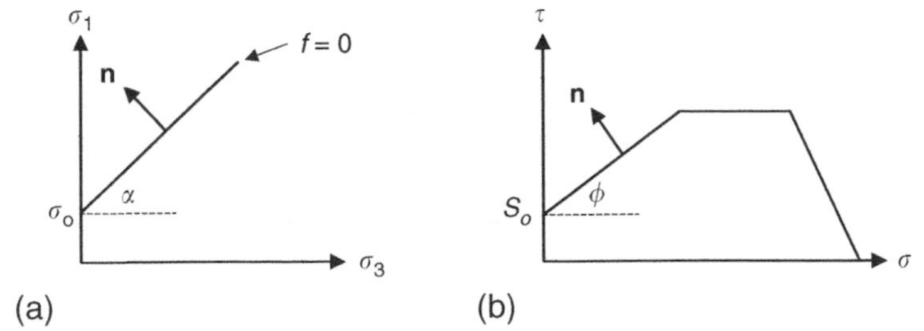

Fig. 9.4 (a) Coulomb failure curve in $\{\sigma_1, \sigma_3\}$ space, showing the normal vector $\mathbf{n} = \nabla f$. (b) Generalized Coulomb failure curve f, in $\{\sigma, \tau\}$ space, showing regions of positive, zero, and negative friction angle.

The amount of dilatancy exhibited by a rock as it fails can be accounted for by relaxing the assumption that the plastic strain is normal to the yield surface. Instead, we can assume that the plastic strain is normal to a different surface, known as the *plastic potential,* which does not coincide with the yield surface. This is an example of a *nonassociated flow rule,* since the flow rule is not associated with the yield function. A simple choice for the plastic potential of a Coulomb-type material would be a function of the same form as (9.77), but with a different parameter φ in place of ϕ:

$$g(\sigma_1, \sigma_2, \sigma_3) = \sigma_1 - \frac{1 + \sin\psi}{1 - \sin\psi}\sigma_3 - \sigma_0 = 0, \tag{9.81}$$

where φ is the *dilatancy angle.* According to (9.80), with φ in place of ϕ, the material will be dilatant if φ is positive and contractant if φ is negative.

The Coulomb failure criterion implies that the compressive strength increases monotonically with the lateral confining stress. In reality, at sufficiently high stresses, the failure curve in $\{\sigma, \tau\}$ space will begin to slope downward and eventually reach the σ-axis. The point at which the failure curve reaches the σ-axis represents failure under purely hydrostatic conditions. Such a failure mode will occur in sedimentary rocks due to grain crushing, which is caused by local stress deviations that are due to the microinhomogeneity of the rock. Hence, the full failure curve can be represented (approximately) on a Mohr diagram by three straight-line segments, having positive, zero, and negative slopes, respectively (Fig. 9.4b).

We return now to the case of the associated flow rule. According to (9.80), the region of positive slope of the $\tau(\sigma)$ failure curve corresponds to dilatant behavior, whereas the region in which there is a negative angle of internal friction corresponds to a rock that contracts volumetrically as it fails. A horizontal region of the failure curve, such as corresponds to the Tresca failure criterion, (9.1), indicates inelastic deformation that occurs without any change in volume.

If the yield stress parameter σ_0 that appears in a yield function such as (9.77) or (9.2) does not change its value during plastic deformation, this yield function will represent elastic-perfectly plastic behavior, as illustrated in Fig. 9.1b. In unconfined uniaxial compression, a rock that behaves in a perfectly plastic manner will not be able to support a stress greater than σ_0. A more realistic model is that shown in Fig. 9.1a, in which the uniaxial stress–strain curve continues to rise after plastic deformation has begun. For cases of monotonic loading, this behavior may seem to be indistinguishable from nonlinear elastic behavior. The fundamental difference appears only during unloading. In a nonlinear elastic material, the unloading curve would coincide with the loading curve, whereas a plastic material would unload elastically along a curve parallel to, but offset from, the original elastic portion of the loading curve. If the compressive stresses acting on the rock are increased again, additional plastic deformation will not commence until the rock again reaches the region *AB* of the stress–strain curve, which will clearly occur at a stress greater than the original value of σ_0.

The region *AB* in Fig. 9.1a is known as the *hardening* region, as it corresponds to an increase in the yield stress parameter σ_0. A full specification of the equations of plasticity in this regime requires a *hardening rule* that governs the evolution of the σ_0 parameter. In order to avoid confusion between the current value of the yield stress and the initial yield stress, it may be convenient to denote the current value by $\kappa\sigma_0$, where κ is known as the *hardening parameter*. This parameter is usually assumed to be a monotonically increasing function of either the total plastic strain or the total plastic work. The plastic work is defined in the same way that the stored elastic strain energy was defined in §5.8 except that only the plastic strains appear in the integrand. These two assumptions concerning the evolution of the hardening parameter are sometimes referred to as describing *strain hardening* and *work hardening*, respectively (Mendelson, 1968, p. 107).

9.8 Creep

The types of stress–strain behavior discussed in previous chapters have for the most part not explicitly depended upon time. According to the elastic model of deformation described in detail in Chapters 5 and 8, the strain at some location $\mathbf{x}$ and some time t depends only on the stress at those same values of $\mathbf{x}$ and t. In the plastic models discussed in previous sections of this chapter, although the strain in a given piece of rock may depend on the stress history (or *stress path*) undergone by that rock, there is no explicit time dependence. More specifically, although the plastic strains may depend on the path that has been traversed in stress space, they do not depend in any way on the *rate* at which the stress path has been traversed. In poroelasticity, as described in Chapter 7, a time dependence is introduced into problems in a "macroscopic" sense, due to the fact that pore pressure diffuses through the rock at a finite rate. Nevertheless, at any given location $\mathbf{x}$ and time t the strain is a function only of the stress and pore pressure at $\mathbf{x}$ and t; the poroelastic stress–strain law itself contains no explicit time, rate, or path dependence.

All of the constitutive models mentioned above are of course idealized approximations to actual rock behavior. There are many situations in which the explicit "time dependence" of rock behavior must be taken into account. Examples of situations in which these effects are important include the long-term deformation of pillars in coal or halite mines (Prasad and Kejriwal, 1984), the closure of cavities in rock salt (Munson, 1997), and the slow "diffusion" of stress between faults that may trigger earthquakes (Huc and Main, 2003). The type of behavior in which the strain continues to evolve under the imposition of a constant stress, or vice versa, is known as *creep*. Figure 9.5a shows the strains measured by Griggs (1940) during uniaxial compression of an alabaster sample immersed in water. At any given value of the compressive stress, the strain continues to increase with time, as the stress is held constant. The rate of creep increases with increasing compressive stress. In general, there is an initial period during which the strain increases rapidly but at a decreasing rate (i.e., the curve is concave down). This is followed by a period of time in which the strain increases at more or less a constant rate (linear curve). For the two highest confining stresses, 25 MPa and 20.5 MPa, after a certain time the strain begins to increase at an ever-increasing rate (concave-upward curve), eventually causing complete collapse of the sample.

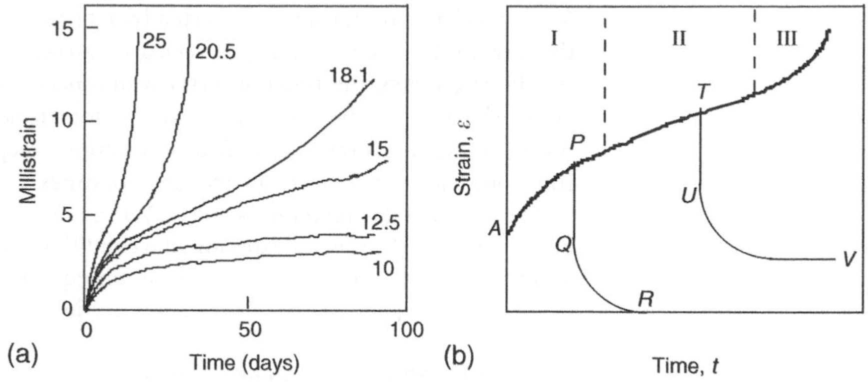

Fig. 9.5
(a) Compressive strains measured by Griggs (1940) on water-saturated alabaster samples subjected to different (constant) values of uniaxial confining stress, indicated in units of megaPascals on each curve. (b) Idealized creep curve showing primary, secondary, and tertiary creep and two unloading paths.

Measurements such as those of Griggs have led to the construction of the idealized one-dimensional creep curve shown in Fig. 9.5b. If a stress is applied instantaneously to a sample, it will immediately give rise to an elastic strain, indicated by A in the figure. Thereafter, the strain continues to increase, albeit at an ever-decreasing rate. This regime, denoted as I, is referred to as *primary* or *transient creep*. This is followed by regime II of *steady-state creep*, in which the strain increases linearly with time. Finally, there may follow regime III of *tertiary creep*, in which the strain increases at an increasing rate, until failure occurs.

The three regimes have different characters that extend beyond their differences with regards to the second derivatives of the strain vs. time curves. If the applied stress is suddenly reduced to zero during primary creep, the strain will relax back to zero, along a path such as PQR. However, if the stress is removed during steady-state creep, the strain will relax to some nonzero value along a path such as TUV, leaving a residual, permanent strain.

Creep strain curves such as that shown in Fig. 9.5b can in principle be represented by an equation of the form

$$\varepsilon = \varepsilon_e + \varepsilon_1(t) + Vt + \varepsilon_3(t), \tag{9.82}$$

where ε_e is the instantaneous elastic strain, $\varepsilon_1(t)$ is the transient creep, Vt is the steady-state creep, and $\varepsilon_3(t)$ is the tertiary or accelerating creep.

Various expressions have been proposed to represent the transient creep term, including

$$\varepsilon_1(t) = At^n, \quad 0 < n < 1 \quad \text{(Cottrell, 1952)}, \tag{9.83}$$

$$\varepsilon_1(t) = A \ln t \quad \text{(Griggs, 1939)}, \tag{9.84}$$

$$\varepsilon_1(t) = A \ln(1 + at) \quad \text{(Lomnitz, 1956)}, \tag{9.85}$$

$$\varepsilon_1(t) = A[(1 + at)^n - 1] \quad \text{(Jeffreys, 1958)}, \tag{9.86}$$

where A, n, and a are constants having the appropriate dimensions. These functions can often fit strain data over certain limited periods of time, but in general do not have appropriate behavior at either small or large values of t. For example, (9.83) and (9.84) give infinite strain *rates* as $t \to 0$. Furthermore, none of these forms yield a smooth transition to steady-state creep as t increases. In order for the strain rate given by (9.82) to approach a constant value in the regime of steady-state creep, the function $\varepsilon_1(t)$ would need to asymptotically approach zero, which is not the case for any of the four functions given above. These limitations demonstrate the difficulty of fitting creep data with mathematical functions that are valid over a wide range of times.

The steady-state creep rate V in (9.82) is not a constitutive parameter per se, but depends on the specific experimental conditions and also on the temperature and the stress. Some of the more common equations that have been used to represent the dependence of V on stress include

$$V = V_o \exp(\sigma/\sigma_o) \quad \text{(Ludwik, 1909)}, \tag{9.87}$$

$$V = V_o \sinh(\sigma/\sigma_o) \quad \text{(Nadai, 1938)}, \tag{9.88}$$

$$V = V_o(\sigma/\sigma_o)^n \quad \text{(Robertson, 1964)}, \tag{9.89}$$

in which V_o is a characteristic strain rate having units of $[1/\text{s}]$, σ_o is a characteristic stress, and n is a dimensionless exponent. Robertson (1964) collected data on a wide range of rocks and found values of n that ranged from 1 to 8.

On a microscopic scale, some types of creep are related to the diffusive motion of defects or dislocations (Evans and Kohlstedt, 1995). These atomic-level processes are thermally activated, in which case the characteristic strain rate varies with temperature according to an Arrhenius-type relation of the form

$$V_o = V_o^{\infty} \exp(-Q/RT), \tag{9.90}$$

where Q is the free energy of activation, with units of $[\text{J/mol}]$, T is the absolute temperature, with units of $[^\circ\text{K}]$, R is the gas constant, with units of $[\text{J/mol}^\circ\text{K}]$, and V_o^{∞} is a hypothetical strain rate at "infinite" temperature.

There exist other creep mechanisms related to the motion of dislocations, however, that lead to strain rates that vary according to (T/T_o) or $(T/T_o)^{-1}$, where T_o is some characteristic temperature. In general, several different creep mechanisms, each of which varies with temperature in a different way, may coexist in a given rock leading to quite complex behavior. These mechanisms are reviewed by Evans and Dresen (1991) and Evans and Kohlstedt (1995).

Of the pioneering laboratory measurements of the creep of rocks, the following short review can be given. Phillips (1931) used bending tests to study the creep of sediments in coal measures. Pomeroy (1956) and Price (1964) studied the creep of coal under bending. Michelson (1917) and Lomnitz (1956) used torsion to study creep of igneous rocks. Uniaxial compression has been used to study the creep of granite, marble and slate (Evans and Wood, 1937), granite (Matsushima, 1960), and sediments (Nishihara, 1958; Hardy, 1959; Price, 1964). Griggs (1936,1939,1940) and le Comte (1965) studied the creep of rock salt under triaxial compression.

An extensive discussion of the rheology of geological materials, and implications for plate tectonics, mantle convection, and other aspects of geodynamics has been given by Ranalli (1995).

9.9 Simple rheological models

In order to solve problems that are more general than the uniaxial compression discussed in the previous section, it is necessary to have stress–strain relations that hold for all types of boundary conditions and geometries. The study of time-dependent stress–strain behavior is known as *rheology*, from the Greek rhei ($\rho\varepsilon\iota$), meaning *flow*, and logos ($\lambda o\gamma o\sigma$), meaning *study*. Standard references on the rheology of materials include Eirich (1956) and Reiner (1971). More specific discussion of rheological models for rocks can be found in Critescu (1989) and Critescu and Hunsche (1997).

The simplest rheological stress–strain relations that attempt to reflect the type of time dependence discussed in the previous section can be constructed using simple mechanical conceptual models based on springs and dashpots. These models are discussed in detail by Bland (1960) and used therein to solve various boundary-value problems. The spring element, shown in Fig. 9.6a, represents an elastic Hookean material in which the stress and strain are related according to Hooke's law,

$$\sigma = k\varepsilon. \tag{9.91}$$

In these models, an analogy is made between the variables of *force* and *displacement* for the spring, and *stress* and *strain* in the solid material. These models can be discussed most simply in terms of uniaxial compression, but can be extended to other types of loading, such as shear. For this reason, the symbol k will be used for the constant of proportionality between stress and strain in a Hookean element; it can represent Young's modulus or the shear modulus, depending on the context. For this type of substance, the stress and strain are related, at any given time, by (9.91), where σ and ε are the instantaneous stress and strain, respectively. Hence, (9.91) shows that the stress will instantaneously rise to $k\varepsilon_0$ if the strain is instantaneously raised from 0 to ε_0 and that the strain will immediately take the value σ_0/k if a stress of σ_0 is instantaneously imposed.

The second basic element used in constructing simple rheological models is the dashpot (Fig. 9.6b), which represents a Newtonian viscous substance that obeys a stress–strain relation of the form

$$\sigma = \eta(\mathrm{d}\varepsilon/\mathrm{d}t) \equiv \eta\dot{\varepsilon}, \tag{9.92}$$

where η is a constant with units of [Pa s], and the overdot indicates the derivative with respect to time. If an instantaneous stress σ_0 is imposed on this element,

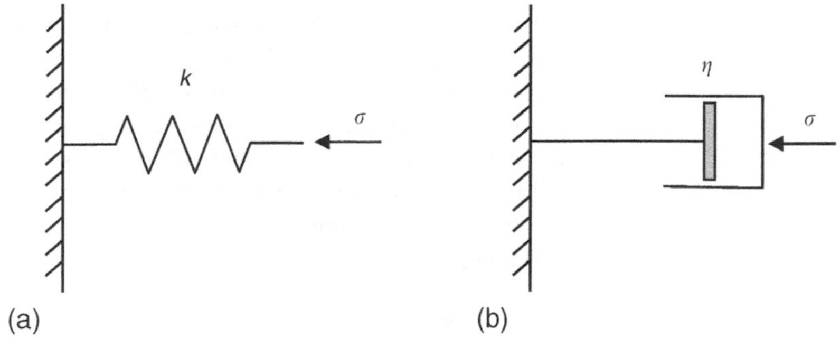

Fig. 9.6 (a) Hookean elastic element, for which $\sigma = k\varepsilon$. (b) Newtonian viscous element, for which $\sigma = \eta(\mathrm{d}\varepsilon/\mathrm{d}t)$.

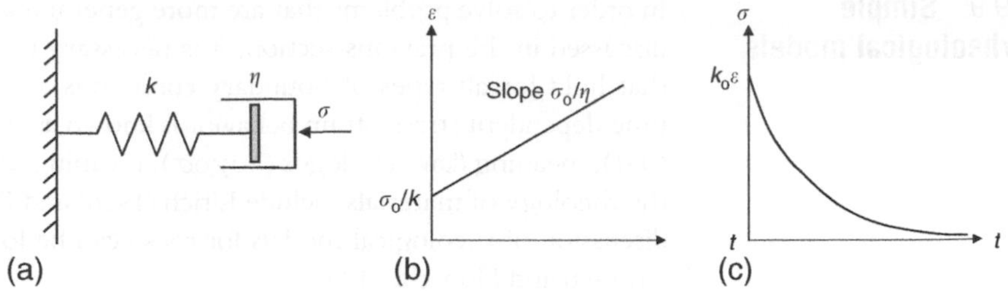

Fig. 9.7
(a) Mechanical model of a Maxwell substance. (b) Response of a Maxwell substance to an instantaneously applied stress. (c) Response of a Maxwell substance to an instantaneously applied strain.

starting from a stress-free and strain-free state, (9.92) can be integrated to show that the strain will grow linearly with time according to

$$\varepsilon = \sigma_o t/\eta. \tag{9.93}$$

As the strain in a Newtonian element is in general equal to the time-integral of the stress, it is not possible to impose an instantaneous jump in the strain, as this would require an infinite stress, which is not realistic.

More complicated types of behavior can be modeled by connecting springs and dashpots together in various series and parallel combinations. A spring and dashpot connected together in series represent a *Maxwell*, or *elasto-viscous substance* (Fig. 9.7a). As the spring and dashpot are both assumed to be massless, the load (stress) carried by each of these elements must be the same at all times. The total displacement, however, will be the sum of the displacement in the spring and the dashpot. Hence, from (9.91) and (9.92), the total strain is governed by the following first-order differential equation:

$$\dot{\varepsilon} = \dot{\varepsilon}_{\text{spring}} + \dot{\varepsilon}_{\text{dashpot}} = (\dot{\sigma}/k) + (\sigma/\eta). \tag{9.94}$$

If the system is initially unstrained and unstressed and a stress σ_0 is instantaneously imposed at $t = 0$, (9.94) can be integrated to yield (Fig. 9.7b)

$$\varepsilon = (\sigma_0/k) + (\sigma_0 t/\eta). \tag{9.95}$$

A Maxwell material therefore exhibits an instantaneous elastic response with stiffness k and a long-term viscous response with viscosity η. The Maxwell substance has been used as a simple model for the Earth's mantle (Carey, 1953).

Now imagine that a Maxwell substance is subjected to an instantaneous jump in the strain, from 0 to ε_0, which is held constant thereafter. As mentioned above, the dashpot cannot undergo a jump in strain, so the jump in strain must initially be accommodated entirely by the spring. Hence, immediately after the imposition of the strain, the stress in the spring will be $k\varepsilon_0$. This stress serves as the initial condition for the differential equation (9.94), which, strictly speaking, must be applied at $t = 0^+$, where 0^+ denotes some infinitesimally small positive value. For subsequent times, the strain is constant, so the left-hand side of (9.94) is zero. Bearing in mind that the "initial stress" is $k\varepsilon_0$, (9.94) can be solved to yield (Fig. 9.7c)

$$\sigma = k\varepsilon_0 e^{-kt/\eta}. \tag{9.96}$$

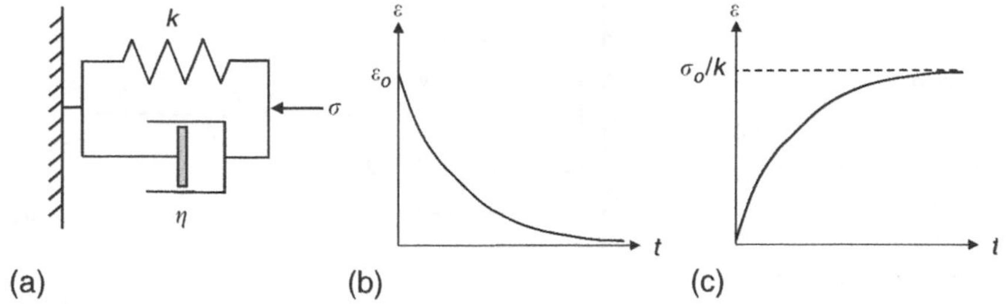

Fig. 9.8
(a) Mechanical model of a Kelvin substance. (b) Response of a Kelvin substance to an instantaneously applied strain. (c) Response of a Kelvin substance to an instantaneously applied stress.

The stress therefore decays down, or *relaxes*, to zero, with a time constant given by $\tau = \eta/k$.

A spring and dashpot connected in parallel form a *Kelvin,* or *firmo-viscous,* substance (Fig. 9.8a). In this case, the total stress will be the sum of the stress carried by the spring and the stress carried by the dashpot:

$$\sigma = \sigma_{\text{spring}} + \sigma_{\text{dashpot}} = k\varepsilon + \eta\dot{\varepsilon}. \tag{9.97}$$

When instantaneous jumps in the stress or strain are imposed on these spring and dashpot systems, the governing differential equation will have a discontinuous forcing function, necessitating the sort of ad hoc solution procedure followed above for the Maxwell substance. Solutions can also be generated systematically using the Laplace transform formalism described in §9.10. In the remainder of this section, the solutions will be written down without derivation.

Suppose that the system is compressed so that the strain is ε_0 when $t = 0$, after which the stress is instantaneously released. The governing differential equation (9.97) then takes the form

$$k\varepsilon + \eta\dot{\varepsilon} = 0. \tag{9.98}$$

subject to the initial condition that ε_0 when $t = 0$. The solution is

$$\varepsilon = \varepsilon_0 e^{-kt/\eta}, \tag{9.99}$$

which shows that the strain decays to zero exponentially, again with a time constant given by $\tau = \eta/k$ (Fig. 9.8b).

If a stress σ_0 is suddenly applied at $t = 0$ to a system that is initially unstrained, the governing equation (9.97) takes the form

$$k\varepsilon + \eta\dot{\varepsilon} = \sigma_0. \tag{9.100}$$

The solution in this case is

$$\varepsilon = (\sigma_0/k)[1 - e^{-kt/\eta}], \tag{9.101}$$

The strain increases asymptotically from 0 to its final, steady-state (elastic) value of σ_0/k, with the time constant $\tau = \eta/k$ that is characteristic of a Kelvin substance (Fig. 9.8c).

The Kelvin model is deficient as a model for creep behavior, as it does not exhibit an instantaneous strain (see Fig. 9.5b). The simplest model that exhibits

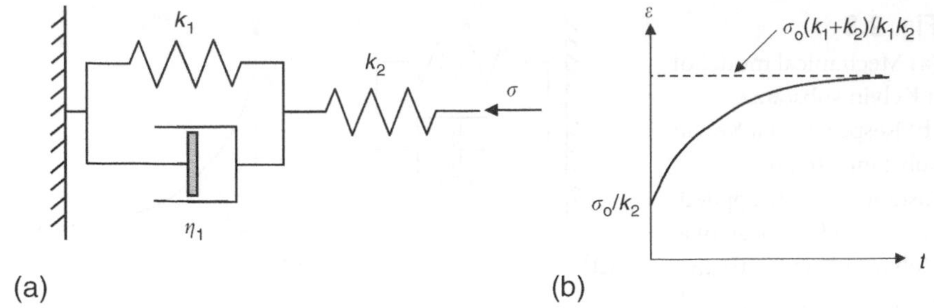

Fig. 9.9
(a) Mechanical model of a generalized Kelvin substance and (b) its response to an instantaneously applied stress.

both an instantaneous strain and a late-time elastic strain is the *generalized Kelvin model*, which consists of a Kelvin element having parameters $\{k_1, \eta_1\}$, arranged in series with a spring k_2 (Fig. 9.9a). As the stress must be the same in both the spring and the Kelvin element, whereas the total strain is the sum of the two strains, it follows that

$$\sigma = \eta_1\dot{\varepsilon}_1 + k_1\varepsilon_1 = k_2\varepsilon_2, \quad \varepsilon = \varepsilon_1 + \varepsilon_2. \tag{9.102}$$

Eliminating the individual strains ε_1 and ε_2 from these equations yields the following single differential equation that governs the behavior of a generalized Kelvin substance:

$$\eta_1\dot{\sigma} + (k_1 + k_1)\sigma = k_2(\eta_1\dot{\varepsilon} + k_1\varepsilon). \tag{9.103}$$

If a stress σ_0 is suddenly applied at $t = 0$ to a generalized Kelvin material that is initially unstrained, the solution for the resulting strain is

$$\varepsilon = \frac{\sigma_0}{k_2} + \frac{\sigma_0}{k_1}(1 - e^{-k_1 t/\eta_1}). \tag{9.104}$$

This model shows an instantaneous strain of σ_0/k_2 and an asymptotic elastic strain of $\sigma_0(k_2 + k_1)/k_2k_1$, approached exponentially with a time constant of η_1/k_1 (Fig. 9.9b).

A model that exhibits instantaneous strain, transient creep, and steady-state creep can be constructed by placing a Kelvin element with parameters $\{\eta_1, k_1\}$ in series with a Maxwell element having parameters $\{\eta_2, k_2\}$, as shown in Fig. 9.10a. The resulting model is known as the *Burgers substance*. The governing equation for this type of material is found by making use of (9.94) for the Maxwell element, (9.97) for the Kelvin element, along with the facts that the stress in each element will be the same, whereas the total strain will be the sum of the two individual strains:

$$\dot{\varepsilon}_2 = (\dot{\sigma}/k_2) + (\sigma/\eta_2), \quad \sigma = \eta_2\dot{\varepsilon}_1 + k_1\varepsilon_1, \quad \varepsilon = \varepsilon_1 + \varepsilon_2. \tag{9.105}$$

Eliminating ε_1 and ε_2 from this equation yields

$$\eta_1\ddot{\varepsilon} + k_1\dot{\varepsilon} = (\eta_1/k_2)\ddot{\sigma} + [1 + (k_1/k_2) + (\eta_1/\eta_2)]\dot{\sigma} + (k_1/\eta_2)\sigma. \tag{9.106}$$

If a Burgers substance that is initially unstrained is subjected to an instantaneous stress of σ_0, the resulting strain will be given by (Fig. 9.10b)

$$\varepsilon = \frac{\sigma_0}{k_2} + \frac{\sigma_0}{k_1}[1 - e^{k_1 t/\eta_1}] + \frac{\sigma_0}{\eta_2}t. \tag{9.107}$$

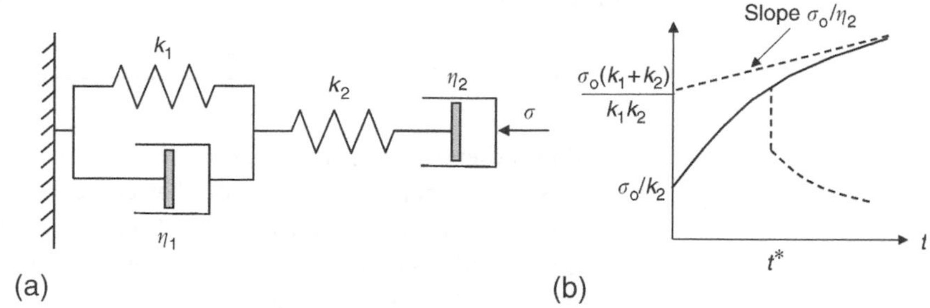

Fig. 9.10 (a) Mechanical model of a Burgers substance and (b) its response to an instantaneously applied stress. The dotted curve shows the strain response when the stress is released at t^*.

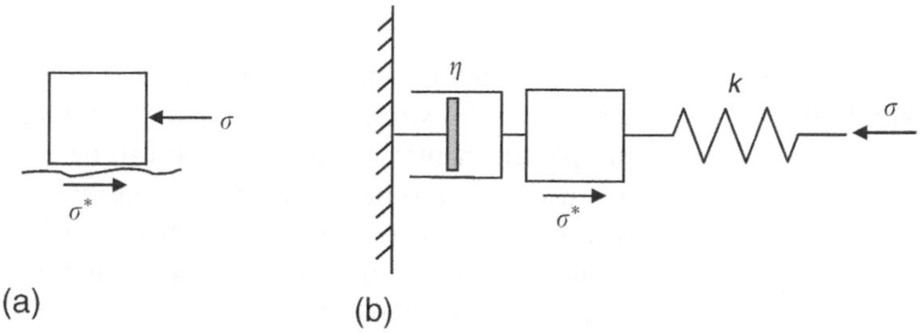

Fig. 9.11 Mechanical models of (a) Saint Venant substance and (b) Bingham substance. The rough surface supplies a resistive frictional force to the block that cannot exceed σ^*.

The three terms on the right side of (9.107) represent the instantaneous strain, the transient creep, and the steady-state creep. If the stress is then released suddenly at some time t^*, the strain drops instantaneously by an amount σ_0/k_2 and then continues to decrease exponentially with the same time constant of η_1/k_1; see the dotted line in Fig. 9.10b. The strain never returns to zero, however, but approaches a permanent residual value of $\sigma_0 t^*/\eta_2$.

Each of the models described above is governed by a linear differential equation, which allows them to be treated using methods such as the Laplace transforms discussed in §9.10. Nonlinear rheological models, which are mathematically more difficult to treat, can be formulated by starting with the *Saint Venant model* of Fig. 9.11a. This model can be represented by a block of mass m placed on a rough, frictional surface. If the friction coefficient between the mass and the surface is μ and the contact area is A, the block will not move until the applied stress σ reaches the value $mg\mu/A$, which can be denoted by σ^*. Hence, the strain will be zero if $\sigma < \sigma^*$ and will be indeterminate if $\sigma > \sigma^*$.

If the Saint Venant element is placed in series with a spring and a dashpot, as in Fig. 9.11b, the resulting model represents a *Bingham substance*. For applied stresses less than σ^*, the block will not move, and the displacement will be confined to the elastic spring. The strain in the elastic spring will be σ/k. If a stress $\sigma_0 > \sigma^*$ is instantaneously applied, the block will move, and this motion will be resisted by the frictional stress σ^* applied to the block by the rough surface. A force balance on the block reveals that the stress transmitted to the dashpot will be $\sigma_0 - \sigma^*$. From (9.93), the strain in the dashpot will be $(\sigma_0 - \sigma^*)t/\eta$. As the dashpot and block are assumed to be rigidly coupled, the total strain of the system will be the sum of the strain in the spring and in the dashpot. Hence, the response of a

Burgers material to an instantaneous stress σ_o is given by

$$\varepsilon = \frac{\sigma_o}{k} \quad \text{if } \sigma_o < \sigma^*, \quad \varepsilon = \frac{\sigma_o}{k} + \frac{(\sigma_o - \sigma^*)}{\eta} t \quad \text{if } \sigma_o > \sigma^*. \tag{9.108}$$

Many other nonlinear rheological models have been proposed. Attewell (1962) proposed a model for porous rock that contains a dashpot of variable viscosity. Price (1964) proposed a Bingham-Voigt model in which the spring of the Bingham model is replaced by a generalized Kelvin element. The resulting material behaves like a generalized Kelvin substance for $\sigma < \sigma^*$ and like a Maxwell substance for greater stresses and at large times.

9.10 Theory of viscoelasticity

The one-dimensional models discussed in §9.9 can be generalized to allow for more complex stress–strain behavior that does not necessarily correspond to any simple mechanical model and also can be extended to three dimensions. First, we use the concept of a *differential operator* to generalize the one-dimensional stress–strain laws of §9.9. The symbol D will be used to represent the operation of "differentiation with respect to time". This type of operator operates on functions of time. For example, $D\sigma \equiv d\sigma/dt$, $D^2\sigma \equiv d^2\sigma/dt^2$, etc. If σ is a function of $\mathbf{x}$ and t, the operator D is interpreted as representing partial differentiation with respect to time. For example, a rheological law such as (9.106) can be written using differential operator notation as

$$(\eta_1 D^2 + k_1 D)\varepsilon = \{(\eta_1/k_2)D^2 + [1 + (k_1/k_2) + (\eta_1/\eta_2)]D + (k_1/\eta_2)\}\sigma. \tag{9.109}$$

An obvious generalization of this notation would be

$$g(D)\varepsilon = f(D)\sigma, \tag{9.110}$$

where f and g are polynomials in D.

Differential equations of the form (9.110) can be solved using the method of Laplace transforms. This method will be described very briefly below. More details can be found in most texts on applied mathematics and specifically in the monographs by Carslaw and Jaeger (1949) and Churchill (1972). Applications to viscoelasticity are given by Christensen (1982) and Lakes (1999).

If u is a function of time, the *Laplace transform of u with respect to time*, denoted by $\hat{u}$, is defined by

$$\hat{u}(p) \equiv L\{u(t)\} \equiv \int_0^{\infty} u(t)e^{-pt}dt, \tag{9.111}$$

where p is some complex parameter whose real part is sufficiently large that the integral in (9.109) converges. (Although a general theory of Laplace transforms requires p to be complex, for the present purposes p can be interpreted as a real parameter.) The transformed function $\hat{u}$ is itself a function of p. Other common notations for the Laplace parameter include s and λ. If u is a function of both space and time, that is, $u(\mathbf{x}, t)$, then its Laplace transform with respect to time will be a function of $\mathbf{x}$ and p. In the following discussion the dependence of the functions u and $\hat{u}$ on $\mathbf{x}$ will be suppressed, for simplicity of notation.

The Laplace transform of many simple functions can be found by direct integration of (9.111). For example,

$$u(t) = 1, \quad \hat{u}(p) = 1/p, \tag{9.112}$$

$$u(t) = t, \quad \hat{u}(p) = 1/p^2, \tag{9.113}$$

$$u(t) = \mathrm{e}^{-at}, \quad \hat{u}(p) = 1/(p + a). \tag{9.114}$$

In order to use Laplace transforms to solve differential equations such as (9.109) or (9.110), it will be necessary to take the Laplace transform of the time derivatives of a function. It follows from applying integration by parts to definition (9.111) that

$$L\{\mathrm{d}u(t)/\mathrm{d}t\} \equiv L\{\mathrm{D}u(t)\} \equiv \int_0^\infty \frac{\mathrm{d}u(t)}{\mathrm{d}t} \mathrm{e}^{-pt} \mathrm{d}t = p\hat{u}(p) - u(0), \tag{9.115}$$

where $u(0)$ is the initial condition of u in the time domain. If this initial condition is zero, as will be the case for a system initially in its undisturbed state, then (9.115) shows that the Laplace transform of $\mathrm{d}u/\mathrm{d}t$ is found by first taking the transform of u and then multiplying by p. If all the time derivatives of u also vanish at $t = 0$ and $f(\mathrm{D})$ is any polynomial-type differential operator, then repeated integration by parts shows that

$$L\{f(\mathrm{D})u\} \equiv \int_0^\infty [f(\mathrm{D})u] \mathrm{e}^{-pt} \mathrm{d}t = f(p)\hat{u}. \tag{9.116}$$

Hence, for systems with zero initial conditions, a linear differential equation for $u(t)$ is replaced, under the Laplace transform operation, by an algebraic equation for $\hat{u}(p)$. Nonzero initial conditions pose no particular additional difficulties, but their treatment is omitted here for the sake of brevity.

Linear differential equations are solved using Laplace transforms by the following procedure. First, the differential equation, say (9.110), is transformed into the Laplace domain, yielding an algebraic equation for $\hat{u}$. This algebraic equation is readily solved for the function $\hat{u}(p)$. This function is then inverted back into the time domain to yield the desired function $u(t)$. This last step can in principle be accomplished using an inversion integral in the complex plane (Churchill, 1972) that is similar in form to (9.111), or by one of many numerical inversion algorithms (Lakes, 1999). In many cases the inversion can be accomplished by consulting published tables of known Laplace transforms.

The procedure can be illustrated by the following example. In differential operator notation, the governing equation for a Kelvin material, (9.97), can be written as

$$(\eta \mathrm{D} + k)\varepsilon = \sigma. \tag{9.117}$$

According to rule (9.116), the Laplace transform of this equation is

$$(\eta p + k)\hat{\varepsilon} = \hat{\sigma}, \tag{9.118}$$

which can be solved for

$$\hat{\varepsilon} = \hat{\sigma}/(\eta p + k), \tag{9.119}$$

If the applied stress is some constant σ_o, then the Laplace transform of the stress can be found from (9.111) and (9.112):

$$\hat{\sigma}(p) = L\{\sigma(t)\} = L\{\sigma_o\} = \int_0^\infty \sigma_o e^{-pt} dt = \frac{\sigma_o}{p}. \tag{9.120}$$

Hence, the transform of the strain is given by

$$\hat{\varepsilon}(p) = \frac{\sigma_o}{p(\eta p + k)} = \frac{\sigma_o}{k}\left[\frac{1}{p} - \frac{1}{p + (k/\eta)}\right]. \tag{9.121}$$

The inverse Laplace transform of (9.121) can be found from any table of $u(t) \leftrightarrow \hat{u}(p)$ pairs, such as (9.112)–(9.114):

$$\varepsilon(t) = \frac{\sigma_o}{k}[1 - e^{-kt/\eta}], \tag{9.122}$$

which agrees with (9.101).

To extend the treatment to three dimensions, first recall the decomposition of Hooke's law into isotropic and deviatoric parts, (5.62):

$$\tau_m = 3K\varepsilon_m, \quad \mathbf{s} = 2G\mathbf{e}, \tag{9.123}$$

where τ_m is the mean normal stress, ε_m is the mean normal strain, $\mathbf{s}$ is the deviatoric stress tensor, (2.151), and $\mathbf{e}$ is the deviatoric strain tensor, (2.234). In light of (9.110), the obvious viscoelastic generalization of (9.123) would be

$$f_1(D)\tau_m = 3g_1(D)\varepsilon_m, \quad f(D)\mathbf{s} = 2g(D)\mathbf{e}. \tag{9.124}$$

A common assumption is that the rock behaves elastically in hydrostatic compression, and that viscoelastic effects occur only in shear, in which case the first equation in (9.124) reduces to the first equation in (9.123).

The Laplace transforms of the constitutive laws (9.124) are

$$f_1(p)\hat{\tau}_m = 3g_1(p)\hat{\varepsilon}_m, \quad f(p)\hat{\mathbf{s}} = 2g(p)\hat{\mathbf{e}}, \tag{9.125}$$

which are identical in form to (9.123), with K replaced by $g_1(p)/f_1(p)$ and G replaced by $g(p)/f(p)$. Suppose now that we know the solution to a certain problem in elasticity in which some known set of stresses are applied to the body. If we take this solution and replace K with $g_1(p)/f_1(p)$ and G with $g(p)/f(p)$, then we automatically have the Laplace transform of the solution for the associated viscoelastic problem in which the same stresses are applied at $t = 0$ to an initially undisturbed body. The actual stresses and strains as functions of time can then be found by inverting the Laplace transforms. This procedure will be illustrated in the next section.

9.11 Some simple viscoelastic problems

The procedure described in §9.10 for generating viscoelastic solutions from elastic solutions will now be illustrated with a few simple examples. In each case, the rock is assumed to behave elastically under hydrostatic compression but viscoelastically under deviatoric stresses.

Consider a uniaxial stress S instantaneously applied to a column of rock that behaves as a Kelvin substance under shear loadings. We align the x-axis with the direction of loading. Comparison of (9.97) and (9.125) shows that $f_1(p) = 1$, $g_1(p) = K, f(p) = 1$, and $g(p) = \eta p + G$, where we identify the generic stiffness k with the shear modulus, G. The constitutive laws in the Laplace domain, (9.125), take the form

$$\hat{\tau}_{\rm m} = 3K\hat{\varepsilon}_{\rm m}, \quad \hat{\mathbf{s}} = 2(\eta p + G)\hat{\mathbf{e}}. \tag{9.126}$$

In component form, for example, $\hat{s}_{xx} = 2(\eta p + G)\hat{e}_{xx}$, etc. Under uniaxial stress, $\tau_{xx} = S$, and $\tau_{yy} = \tau_{zz} = 0$, in which case $\tau_{\rm m} = S/3$ and $s_{xx} = 2S/3$. In the Laplace domain,

$$\hat{\tau}_{\rm m} = S/3p, \quad \hat{s}_{xx} = 2S/3p. \tag{9.127}$$

Solving (9.126) and (9.127) gives

$$\hat{\varepsilon}_{\rm m} = \frac{S}{9Kp}, \quad \hat{e}_{xx} = \frac{S}{3p(\eta p + G)} = \frac{S}{3G}\left[\frac{1}{p} - \frac{1}{p + (G/\eta)}\right]. \tag{9.128}$$

These functions can be inverted back into the time domain using (9.112) and (9.114):

$$\varepsilon_{\rm m} = \frac{S}{9K}, \quad e_{xx} = \frac{S}{3G}(1 - {\rm e}^{-Gt/\eta}). \tag{9.129}$$

Recalling the definition of deviatoric strain, $e_{xx} = \varepsilon_{xx} + \varepsilon_{\rm m}$, we see that the strain in the longitudinal direction is given by

$$\varepsilon_{xx} = \frac{S}{9K} + \frac{S}{3G}(1 - {\rm e}^{-Gt/\eta}). \tag{9.130}$$

Comparison of (9.130) with (9.104) shows that a rock that is elastic in hydrostatic compression and Kelvin-like in distortion will follow a generalized Kelvin law under uniaxial compression.

Consider now a material that is elastic in hydrostatic compression and behaves like a generalized Kelvin substance in shear. Comparison of (9.103) and (9.125) shows that, in this case,

$$f(p) = \eta_1 p + k_1 + k_2, \quad g(p) = k_2(\eta_1 p + k_1). \tag{9.131}$$

Now consider a pore pressure P applied to the wall of a circular borehole of radius a. In the elastic case, the radial displacement is given by (8.90):

$$u = -Pa^2/2Gr. \tag{9.132}$$

In the viscoelastic case, we must replace P with its Laplace transform, P/p, and replace G with $g(p)/f(p)$, so that in the Laplace domain the radial displacement is

$$\hat{u}(r,p) = \frac{-Pa^2(\eta_1 p + k_1 + k_2)}{2k_2(\eta_1 p + k_1)pr} = \frac{-Pa^2}{2k_2 k_1 r}\left[\frac{k_1 + k_2}{p} - \frac{k_2}{p + (k_1/\eta_1)}\right]. \quad (9.133)$$

Inversion, using (9.112) and (9.114), gives

$$u(r,t) = -\frac{Pa^2(k_1 + k_2)}{2k_2 k_1 r} + \frac{Pa^2}{2k_1 r}e^{-k_1 t/\eta_1}. \quad (9.134)$$

The rock mass surrounding the borehole will undergo an instantaneous displacement of $-Pa^2/2k_2 r$. The steady-state displacement will be $-Pa^2(k_1+k_2)/2k_2k_1r$, approached with a time constant of η_1/k_1.

Last, consider a rock that behaves elastically under hydrostatic loading and as a Maxwell substance under deviatoric loading. From (9.94), $f(p) = (p/G) + (1/\eta)$ and $g(p) = p$, where we again identify k with G. If this rock is subjected to uniaxial strain in the z direction, from (5.33), the elastic stresses would be

$$\tau_{zz} = S, \quad \tau_{xx} = \tau_{yy} = \frac{\nu}{1-\nu}S = \frac{3K - 2G}{3K + 4G}S. \quad (9.135)$$

For the viscoelastic problem, we replace S with S/p and G with $g(p)/f(p)$, so that in the Laplace domain the stresses are

$$\hat{\tau}_{zz} = \frac{S}{p},$$

$$\hat{\tau}_{xx} = \hat{\tau}_{yy} = \frac{S[3K(\eta p + G) - 2G\eta p]}{p[3K(\eta p + G) + 4G\eta p]} = S\left[\frac{1}{p} - \frac{6G\eta}{(3K + 4G)\eta p + 3KG}\right]. \quad (9.136)$$

It follows upon inversion that in the time domain the stresses are

$$\tau_{zz} = S, \quad \tau_{xx} = \tau_{yy} = S\left[1 - \frac{6G}{(3K + 4G)}e^{-3KGt/(3K+4G)\eta}\right]. \quad (9.137)$$

As time increases, the stress state approaches a hydrostatic stress of magnitude S, with a time constant of $(3K + 4G)\eta/3KG$.

Ladanyi (1993) gives an extensive review of the application of the theories of creep and viscoelasticity to the time-dependent deformation of rock around underground tunnels and excavations.

10 Micromechanical models

10.1 Introduction

In most of the discussion in previous chapters, "rock" has been thought of as a homogeneous material that can be characterized by macroscopic parameters such as density, elastic moduli, etc., that are uniform over regions at least as large as a laboratory specimen. In most engineering calculations, it is convenient, and often practically necessary, to treat a rock mass as if it were homogeneous on the scale of a borehole, or tunnel, for example. In reality, rocks are quite inhomogeneous on length scales below the centimeter scale, due to pores, microcracks, grain boundaries, etc. Engineering and geophysical calculations cannot explicitly account for inhomogeneity on this "microscale," Nevertheless, consideration of the effect of grains, pores, and cracks on the macroscopic deformation of a rock can shed much light on the constitutive behavior, in the sense of supplying estimates of the elastic moduli and yield strength, for example. The field in which microscale calculations are carried out in order to gain a better understanding of macroscopic rock behavior is known as *micromechanics*.

In this chapter, we discuss some micromechanical models for the elastic and inelastic behavior of rocks. Methods for estimating the effective elastic moduli of a rock that consists of a heterogeneous assemblage of different minerals are presented in §10.2. In §10.3, we discuss the effect of pore structure on the porous rock compressibility coefficients that were defined in §7.2. The nonlinearity in the stress–strain curve that results from crack closure is discussed in §10.4. A brief review of some of the numerous schemes that have been proposed for estimating the effective elastic moduli of porous and cracked rocks is presented in §10.5. In §10.6, we discuss the effects of sliding friction along crack faces and the hysteresis that this friction causes in the stress–strain behavior. The Griffith crack model for rock failure and the concept of the Griffith locus are presented in §10.7. The basic concepts of linear elastic fracture mechanics are briefly discussed in §10.8. Finally, the use of the Griffith crack model to develop a failure criterion in terms of the maximum and minimum principal stresses is presented in §10.9.

10.2 Effective moduli of heterogeneous rocks

Consider first a hypothetical rock that is homogeneous, consisting of a single mineral, and without any cracks or pores. If a piece of this rock having volume V is subjected to a uniform hydrostatic pressure P over its entire outer boundary, a uniform volumetric strain of magnitude P/K will be induced throughout the sample and the resulting volume decrease will be given by $\Delta V = PV/K$.

Now consider a rock that consists of an assemblage of N different minerals, each having its own value of the bulk modulus, K_i, where $i = 1, 2, 3, \ldots, N$. If a piece of this rock, having initial volume V, is subjected to an external hydrostatic stress of magnitude P, the resulting strain inside the rock will not be uniform. In general, the stiffer minerals will deform less than the more compliant minerals, although the precise amount of deformation in each of the solid components of the rock will depend on the geometric configuration of the various minerals. For example, a compliant clay particle that is sitting inside a pore in a sandstone will undergo much less deformation than would a clay particle that is wedged between two sand grains. Nevertheless, this rock will undergo some overall volume decrease, ΔV. From a purely macroscopic point of view, an "effective" bulk modulus K_{eff} can be defined by the relation $K_{\text{eff}} = PV/\Delta V$. This effective bulk modulus can be interpreted as the bulk modulus of a hypothetical homogeneous rock that would undergo the same mean volumetric strain as does the actual heterogeneous rock.

Precise calculation of K_{eff} would require exact knowledge of the microstructure of the rock, which in practice is never available. However, if the volume fractions χ_i and bulk moduli K_i of the various mineral components of the rock are known, the methods of Reuss and Voigt can be used to provide estimates of K_{eff}.

Reuss (1929) made the assumption that the *stresses* would be uniform throughout the heterogeneous rock. In reality this cannot be exactly true, as the different strains in the various mineral grains would then lead to displacement discontinuities at the grain boundaries. If the stress within each mineral were indeed a hydrostatic compression P, then the volume change of each component would be $\Delta V_i = PV_i/K_i$. The total volume change is the sum of the volume changes of the individual minerals, so

$$\Delta V = \sum_{i=1}^{N} \Delta V_i = \sum_{i=1}^{N} \frac{PV_i}{K_i} = \sum_{i=1}^{N} \frac{P\chi_i V}{K_i} = PV \sum_{i=1}^{N} \frac{\chi_i}{K_i}. \tag{10.1}$$

Using the definition $K_{\text{eff}} = PV/\Delta V$,

$$K_{\text{eff}}^{\text{Reuss}} = \frac{PV}{\Delta V} = \left[\sum_{i=1}^{N} \frac{\chi_i}{K_i} \right]^{-1}. \tag{10.2}$$

The Reuss effective bulk modulus is therefore the *weighted harmonic mean* of the individual bulk moduli. In terms of the compressibilities,

$$C_{\text{eff}}^{\text{Reuss}} \equiv 1/K_{\text{eff}}^{\text{Reuss}} = \sum_{i=1}^{N} \frac{\chi_i}{K_i} = \sum_{i=1}^{N} \chi_i C_i, \tag{10.3}$$

and so the Reuss effective compressibility is the *weighted arithmetic mean* of the individual mineral compressibilities.

Voigt (1889), on the other hand, made the assumption that the *volumetric strains* are uniform throughout the heterogeneous body. Again, this cannot be precisely true, as equality of strains implies that the stresses in each mineral phase are different, and so the resulting stress field would be discontinuous, and would not satisfy the stress equilibrium equations. Under the assumption of equal volumetric strains, the mean normal stress in each component would be given by $\sigma_{m,i} = \varepsilon_v K_i$. The average value of the mean normal stress is simply the volumetric average of the mean normal stress in each component:

$$\langle \sigma_m \rangle = \sum_{i=1}^{N} \chi_i \sigma_{m,i} = \sum_{i=1}^{N} \chi_i \varepsilon_v K_i = \varepsilon_v \sum_{i=1}^{N} \chi_i K_i. \tag{10.4}$$

Making the obvious identification of the average value of the mean normal stress with the applied stress P, it follows from the definition $K_{\text{eff}} = PV/\Delta V = P/\varepsilon_v$ that

$$K_{\text{eff}}^{\text{Voigt}} = \frac{P}{\varepsilon_v} = \frac{\langle \sigma_m \rangle}{\varepsilon_v} = \sum_{i=1}^{N} \chi_i K_i. \tag{10.5}$$

Voigt's estimate of the effective bulk modulus is therefore simply the weighted arithmetic mean of the individual bulk moduli. The Voigt estimate of the effective compressibility is

$$C_{\text{eff}}^{\text{Voigt}} \equiv 1/K_{\text{eff}}^{\text{Voigt}} = \left[\sum_{i=1}^{N} \chi_i K_l\right]^{-1} = \left[\sum_{i=1}^{N} \frac{\chi_i}{C_i}\right]^{-1}. \tag{10.6}$$

It follows from elementary algebraic considerations that the Voigt estimate of K_{eff} will always exceed the Reuss estimate. More specifically, however, Hill (1952) used strain energy arguments to prove that the Voigt and Reuss estimates are rigorous upper and lower bounds on the true value of the effective bulk modulus, that is,

$$\left[\sum_{i=1}^{N} \frac{\chi_i}{K_i}\right]^{-1} = K_{\text{eff}}^{\text{Reuss}} \leq K_{\text{eff}} \leq K_{\text{eff}}^{\text{Voigt}} = \sum_{i=1}^{N} \chi_i K_i. \tag{10.7}$$

Hill's proof does not in any way depend upon there being a sufficiently large number of grains so as to have a "statistically meaningful sample," nor on the different minerals being randomly located. The bounds (10.7) apply equally well to a rock specimen that contains a single foreign mineral inclusion, for example.

Hill proposed using the average of the Voigt and Reuss bounds to find a best estimate of K_{eff}. The resulting value is known as the Voigt–Reuss–Hill estimate of the effective modulus:

$$K_{\text{eff}}^{\text{VRH}} = \frac{1}{2}\left[K_{\text{eff}}^{\text{Reuss}} + K_{\text{eff}}^{\text{Voigt}}\right]. \tag{10.8}$$

Although there is no particular justification for assuming that K_{eff} will lie midway between the two bounds, Hill's assumption has the advantage of giving an estimate of the effective modulus that a priori will be guaranteed to have the minimum possible error.

The Voigt, Reuss, and Hill arguments can also be applied to the estimation of the effective *shear* modulus of a heterogeneous rock. In this case,

$$\left[\sum_{i=1}^{N}\frac{\chi_i}{G_i}\right]^{-1} = G_{\text{eff}}^{\text{Reuss}} \leq G_{\text{eff}} \leq G_{\text{eff}}^{\text{Voigt}} = \sum_{i=1}^{N}\chi_i G_i, \tag{10.9}$$

$$G_{\text{eff}}^{\text{VRH}} = \frac{1}{2}\left[G_{\text{eff}}^{\text{Reuss}} + G_{\text{eff}}^{\text{Voigt}}\right]. \tag{10.10}$$

Similar equations are often written for the Young's modulus. However, in general the arithmetic and harmonic means of the individual Young's moduli will *not* necessarily provide bounds on E_{eff} (Grimvall, 1986, p. 261). Instead, bounds on E_{eff} can be obtained from the bounds on K_{eff} and G_{eff} by using the identity $1/E = 1/(3G) + 1/(9K)$.

Values for the elastic moduli of various rock-forming minerals have been compiled by Clark (1966), Simmons and Wang (1971), and Mavko et al. (1998). Bulk moduli values of common minerals range from about 36–38 GPa for quartz, 63–77 GPa for calcite, 130 GPa for olivine, and up to about 253 GPa for corundum. Shear moduli range from about 28–32 GPa for calcite, 44–46 GPa for quartz, 80 GPa for olivine, and up to about 162 GPa for corundum. Hence, the range of values of K and G observed in common minerals span a range of less than one order of magnitude.

The relative lack of variability of the elastic moduli of different minerals causes the Voigt and Reuss bounds to usually be fairly close together. In many cases, the difference between the Voigt and Reuss estimates will be within the experimental uncertainty of the moduli values of the individual minerals.

Brace (1965) measured the bulk moduli of several crystalline rocks at pressures up to 900 MPa. At such high pressures, it can be assumed that any cracks that may have been present at low pressures will be closed. As a representative example, consider the granite from Stone Mountain, Georgia, which was composed of 42 percent plagioclase, 30 percent quartz, 24 percent microcline, and 4 percent mica. In its unstressed state, it had a density of 2631 kg/m^3 and a microcrack porosity of 0.3 percent. At 900 MPa, the individual mineral bulk moduli assumed by Brace were 62.7 GPa for plagioclase, 44.5 GPa for quartz, 60.0 GPa for microcline, and 50.1 GPa for mica. From (10.5), the Voigt estimate is 56.1 GPa, the Reuss estimate is 54.8 GPa, and the Voigt–Reuss–Hill estimate is 55.5 GPa. The measured bulk modulus was 56.8 GPa. Considering that the mineral compositions were reported only to within the nearest percent, the Voigt–Reuss–Hill average agrees with the measured value of the effective bulk modulus as nearly as one could expect. Similar results were found for the other rocks.

10.3 Effect of pores on compressibility

In §10.2, it was shown that the effective bulk or shear modulus of a solid rock can usually be accurately estimated if its mineralogical composition is known. But rocks typically contain some cracks or pores, which cause the moduli to decrease below the values that would be obtained if the rock were nonporous. The Voigt and Reuss bounds could be applied to a porous rock, by treating the pore space as an additional component that has $K_i = G_i = 0$. However, the resulting Reuss bound will be zero, and the spread between the two bounds will be too large for the Voigt–Reuss–Hill average to serve as a useful estimate. Hence, other methods must be used to account for the effect of voids. Moreover, whereas the Voigt and Reuss analyses took no account of the microstructure of the rock, the influence that voids have on the effective moduli depends very strongly on the shape of the voids.

A convenient basis for studying the effect of void shape on the bulk modulus of a rock is to consider the stored elastic strain energy (Eshelby, 1957). According to (5.148), if a homogeneous elastic body of volume V_b and bulk modulus K_m is subjected to a uniform hydrostatic stress P over its outer surface, the elastic strain energy stored in the body will be $P^2V_b/2K_m$. Imagine now that pores are introduced into this body while maintaining the hydrostatic boundary conditions. The total elastic strain energy will change by some amount $\Delta\mathcal{E}_{\text{hydro}}$, to a new value $P^2V_b/2K_m + \Delta\mathcal{E}_{\text{hydro}}$. The effective bulk modulus K can be defined so that the strain energy stored in the porous body is equal to that which would be stored in a solid body having bulk modulus K:

$$\frac{P^2V_b}{2K} = \frac{P^2V_b}{2K_m} + \Delta\mathcal{E}_{\text{hydro}}. \tag{10.11}$$

We drop the subscript eff, for simplicity, and in recognition of the fact that the bulk modulus value that would be used in geological or engineering calculations is actually the effective bulk modulus of a material that is heterogeneous and porous at the microscale.

The definition of the effective bulk modulus of a heterogeneous medium in terms of the stored strain energy is equivalent to the definition given in §10.2 in terms of the mean stresses and strains (Markov, 2000). For a porous rock, the effective bulk modulus is the inverse of the parameter C_{bc} of §7.2. The equivalence between the "energy" definition and the "strain" definition of C_{bc} can be demonstrated explicitly, as follows. Consider a solid piece of rock under hydrostatic stress. First, imagine that the pores are carved out of the body, while the appropriate stresses are maintained at the pore boundaries so as not to allow the pore surfaces to relax. The body will lose an amount of strain energy equal to that which was stored in the carved-out material, that is, $P^2V_p/2K_m = P^2V_pC_m/2$. Now imagine that the stresses at the pore boundaries are slowly relaxed to zero. As the pore surfaces relax, the pore volume will change by $-C_{pp}V_pP$, and the body will perform an amount of work against this stress that is given by $C_{pp}V_pP^2/2$. This term also represents energy that is *lost* by the body. Finally, as the fictitious pore stresses are relaxed, the *outer* boundary of the body will contract, leading to a bulk volume change of $-C_{bp}V_bP$. Since the pressure on the outer boundary remains unchanged at P during this process, the external stress will perform an amount of work *on* the body that is equal to $C_{bp}V_bP^2$. The total energy change due to the presence of the pores is therefore

$$\Delta\mathcal{E}_{\text{hydrostatic}} = -(P^2V_pC_m/2) - (P^2V_pC_{pp}/2) + (P^2V_bC_{bp}). \tag{10.12}$$

Using (7.14), $C_{bp} = \phi C_{pc}$, and (7.15), $C_{pp} = C_{pc} - C_m$, this can be written as

$$\Delta\mathcal{E}_{\text{hydrostatic}} = P^2V_pC_{pc}/2. \tag{10.13}$$

Combining (10.11) and (10.13), and factoring out $P^2 V_b/2$, we recover (7.16):

$$1/K = C_{bc} = C_m + \phi C_{pc}. \tag{10.14}$$

Hence, the calculation of the effective bulk modulus reduces to the calculation of the pore compressibility, C_{pc}.

The parameter C_{pc} represents the compressibility of the *entire* pore space of the rock. Although pores are often interconnected and form a continuous structure, most methods of relating compressibility to pore structure assume that the pores exist as isolated voids. The compressibility C_{pc} of an isolated void can be found by solving the elasticity problem of a traction-free single void in an infinite rock mass, subjected to hydrostatic stress at infinity.

One commonly used model of a pore is that of a cylindrical tube of circular or elliptical cross section. This model has been used to study the attenuation of elastic waves in fluid-filled porous media (Biot, 1956a,b), and to study the effect of porosity and stress on permeability (Bernabe et al., 1982; Sisavath et al., 2000). The problem of a tubular elliptical cavity in an infinite elastic body, subjected to a hydrostatic stress P at infinity, was first solved by Inglis (1913). If a is the semimajor axis and b is the semiminor axis, the normal component of the displacement at the surface of the hole, under plane stress conditions, is

$$u_n = \frac{P}{hE_m}[(a^2 + b^2) - (a^2 - b^2)\cos 2\beta], \tag{10.15}$$

where E_m is the Young's modulus of the intact rock, h is the metric coefficient of the elliptical coordinate system, and β is the angular coordinate. The change in the cross-sectional area of the hole is found by integrating the normal displacement along the perimeter of the hole (Walsh et al., 1965):

$$\Delta A = \int_0^{2\pi} u_n(hd\beta)$$
$$= \int_0^{2\pi} \frac{P}{hE_m}[(a^2 + b^2) - (a^2 - b^2)\cos 2\beta]hd\beta = \frac{2\pi(a^2 + b^2)P}{E_m}. \tag{10.16}$$

The initial area of the hole is πab, so the plane stress pore compressibility is

$$C_{pc} = \frac{\Delta A}{A^i P} = \frac{2\pi(a^2 + b^2)P}{E_m \pi abP} = \frac{2}{E_m}\left(\frac{a}{b} + \frac{b}{a}\right) = \frac{2}{E_m}\left(\alpha + \frac{1}{\alpha}\right), \tag{10.17}$$

where $\alpha = b/a \leq 1$ is the aspect ratio of the hole.

This expression is appropriate for plane stress conditions, but must be modified to render it applicable to tubular pores in rocks, which will be more nearly under conditions of plane strain. This is done by expressing (10.17) in terms of G and ν, and then replacing ν with $\nu/(1 - \nu)$. Since $E = 2G(1 + \nu)$, the plain strain version of (10.17) is (Fig. 10.1a)

$$C_{pc}(\text{ellipse}) = \frac{1 - \nu_m}{G_m}\left(\alpha + \frac{1}{\alpha}\right). \tag{10.18}$$

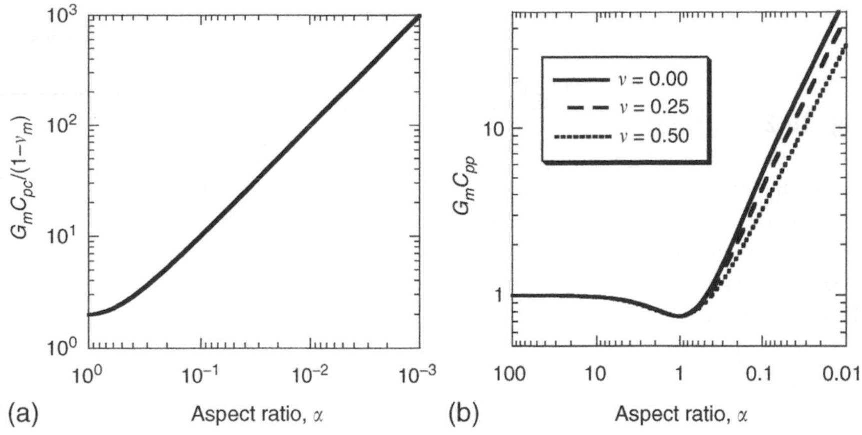

Fig. 10.1 Pore compressibility of (a) two-dimensional elliptical voids, and (b) three-dimensional spheroidal voids, as functions of the aspect ratio.

For circular pores, $\alpha = 1$, and the pore compressibility reduces to

$$C_{pc}(\text{circle}) = 2(1 - \nu_m)/G_m. \tag{10.19}$$

For small deviations from circularity, (10.18) can be expanded in terms of the eccentricity, $e = 1 - \alpha$, to yield

$$C_{pc} = \frac{(1 - \nu_m)}{G_m}\left[(1 - e) + \frac{1}{1 - e}\right] = \frac{2(1 - \nu_m)}{G_m}\left[1 + \frac{e^2}{2} + \cdots\right], \tag{10.20}$$

which shows that slight deviations from circularity have little effect on the pore compressibility. The circular hole is the stiffest of all ellipses, and the compressibility increases as the aspect ratio decreases. For crack-like pores of small aspect ratio, (10.18) reduces to

$$C_{pc}(\text{2D crack}) = (1 - \nu_m)/\alpha G_m, \tag{10.21}$$

which is consistent with the results of §8.9, where the initial aspect ratio was denoted by ξ_o. The effect of elliptical pores on elastic properties has been discussed in detail by Kachanov (1994).

The complex variable methods of Chapter 8 can be used to study the compressibility of two-dimensional pores (Savin, 1961; Zimmerman, 1986; Jasiuk, 1995). One simple family of mapping functions that can represent a wide variety of pore shapes is the hypotrochoid,

$$z = \omega(\zeta) = \frac{1}{\zeta} + m\zeta^n, \tag{10.22}$$

in which n is a positive integer, and m is a real number that must satisfy $0 \le m < 1/n$ in order for the mapping to be single-valued. A hypotrochoid of exponent n has $n + 1$ "corners," which become more acute as m increases. As $m \to 0$, the hole rounds off into a circle, whereas as $m \to 1/n$, the corners become cusp-like. A family of such curves is shown in Fig. 10.2a for $n = 3$, with each curve labeled

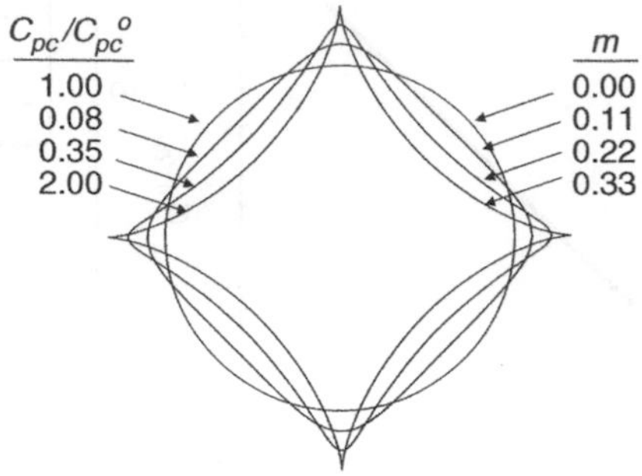

Fig. 10.2 (a) Pore compressibility C_{pc} of several hypotrochoidal holes of $n = 3$, normalized with respect to that of a circular hole. (b) Dimensionless normalized pore compressibility β_{pc} for various two-dimensional pore shapes; α is the aspect ratio.

	Shape	$\dfrac{G_m C_{pc}}{2(1-\nu_m)}$	$\dfrac{\Pi 2}{4\pi A^i}$	β_{pc}
	○	1.00	1.00	1.00
	△	1.58	1.65	0.96
	□	1.19	1.27	0.93
	▭	0.393/α	0.318/α	1.23
	⬭	0.500/α	0.4.5/α	1.23

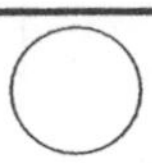

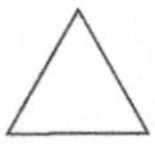

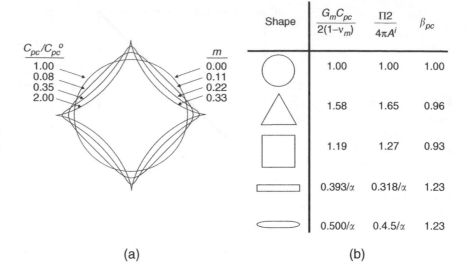

C_{pc}/C_{pc}^o

1.00	
0.08	
0.35	
2.00	

m

0.00	
0.11	
0.22	
0.33	

(a)   (b)

by its value of m, as well as by its pore compressibility, which is given by (Mavko 1980; Zimmerman, 1986)

$$C_{pc}(\text{hypotrochoid}) = \frac{2(1 - \nu_m)}{G_m}\left[\frac{1 + nm^2}{1 - nm^2}\right]. \tag{10.23}$$

A four-sided hypotrochoid could be used as a simplified two-dimensional model for the pore shape changes that occur as a result of the diagenetic alteration of a sandstone. When $m = 1/3$, the pore resembles a cross section of the space between a set of four adjacent spherical grains. As m decreases to zero, the shape mimics the evolution of a pore as the rock undergoes diagenesis. This model has been used by Vernik (1997) to study the compaction of sandstones, and by O'Donnell and Steif (1989) to study the sintering of powdered metals.

When $m = 2/n(n + 1)$, the hypotrochoid represents the first two terms of the mapping function for an equilateral polygon of $n + 1$ sides (Savin, 1961) and resembles a polygon with slightly concave sides and rounded corners. From (10.23), the pore compressibility can be expressed as a function of the number of sides of the pore, s:

$$C_{pc}(\text{quasi-polygon of } s \text{ sides}) = \frac{2(1 - \nu_m)}{G_m}\left[\frac{s^2(s - 1) + 4}{s^2(s - 1) - 4}\right]. \tag{10.24}$$

The quasi-triangular pore is 57 percent more compressible than a circular pore whereas the quasi-square is only 18 percent more compressible. The compressibility of these quasi-polygonal pores decreases rapidly as the number of sides increases and the pore becomes more circular. Consideration of more terms in the exact mapping functions of the polygons alters the predicted pore compressibilities by no more than 1–2 percent (Jasiuk, 1995).

For many different two-dimensional hole shapes, the pore compressibility can be approximately expressed in terms of the perimeter Π and initial area A of the hole (Zimmerman, 1986; Tsukrov and Novak, 2002). We start by expressing the pore compressibility as

$$C_{pc} = \frac{2(1-\nu_m)}{G_m}\frac{\Pi^2}{4\pi A}\beta_{pc}, \tag{10.25}$$

where the first term on the right is the compressibility of a circular pore, the second term is a dimensionless geometric factor that equals unity for a circle and thereby measures departures from circularity, and β_{pc} is dimensionless normalized pore compressibility. By construction, $\beta_{pc} = 1$ for a circular pore. For the quasi-polygons obtained by setting $m = 2/n(n+1)$, β_{pc} always lies in the range 0.95–1.00. For the infinitely thin elliptical slit that is obtained by setting $n = 1$ and letting $m \to 1$, $\beta_{pc} = 1.23$. The smallest value of β_{pc} for the entire family of hypotrochoids is 0.617; this occurs when $m = 1/n$ and $n \to \infty$, which corresponds to a near-circular pore that contains an infinite number of small cusps of vanishing amplitude. The parameter β_{pc} does not depart drastically from unity (Fig. 10.2b), so it is approximately true that

$$C_{pc} \approx \frac{2(1-\nu_m)}{G_m}\frac{\Pi^2}{4\pi A}. \tag{10.26}$$

The only three-dimensional pore shape that is amenable to exact analytical treatment is the ellipsoid (Sadowsky and Sternberg, 1949; Eshelby, 1957). In its various forms, the ellipsoid can represent a variety of shapes, such as spheres, cylinders, and thin cracks. An expression for the pore compressibility of an ellipsoidal pore with three axes of unequal lengths is implicitly contained in the results of Eshelby, although the result is expressed in terms of unwieldy elliptical integrals. However, the pore compressibility of an ellipsoid depends mainly on the ratio of the minimum axis to the intermediate axis; the value of the maximum axis has a minor effect. For this reason, and to avoid using two aspect ratios to characterize a pore, three-dimensional pores are typically modeled not as ellipsoids, but as spheroids, which are degenerate ellipsoids having two axes of equal length.

A spheroid can be formed by revolving an ellipse about one of its axes of symmetry. Revolution about the minor axis generates an *oblate* spheroid, whereas revolution about the major axis produces a *prolate* spheroid. The aspect ratio can be defined as the ratio of the length of the unequal axis to the length of one of the two equal axes. By this definition, prolate spheroids have $\alpha > 1$, while oblate spheroids have $\alpha < 1$. As $\alpha \to \infty$, the prolate spheroid becomes a long needle-like cylinder, and as $\alpha \to 0$ it becomes a thin "penny-shaped" crack. A spheroid having $\alpha = 1$ is a sphere.

Zimmerman (1985b) used the general solution of the elasticity equations in prolate spheroidal coordinates (Edwards, 1951), and chose the arbitrary coefficients so as to give a hydrostatic stress at infinity, and a traction-free cavity wall. Integration of the normal component of the displacement over the surface of the cavity yields the pore compressibility in the following form, which is most conveniently expressed in terms of C_{pp} rather than C_{pc} (Fig. 10.1b):

$$C_{pp} = \frac{2(1-2\nu_m)(1+2R)-(1+3R)\{1-2(1-2\nu_m)R-[3\alpha^2/(\alpha^2-1)]\}}{4G_m\{(1+3R)[\alpha^2/(\alpha^2-1)]-(1+R)(\nu_m+\nu_m R+R)\}},$$

$$R = \frac{1}{\alpha^2-1} + \frac{\alpha}{2(\alpha^2-1)^{3/2}} \ln \frac{\alpha - \sqrt{\alpha^2-1}}{\alpha + \sqrt{\alpha^2-1}}. \tag{10.27}$$

The minimum value of the pore compressibility of a prolate spheroid occurs when $\alpha = 1$, where it has the value

$$C_{pp}(\text{sphere}) = \frac{3}{4G_m}, \tag{10.28}$$

as was derived explicitly for a sphere in §8.12. At the other end of the aspect ratio spectrum, the needle-like pore has $C_{pp} = 1/G_m$. The values of C_{pp} for all prolate spheroids lie between $3/4G_m$ and $1/G_m$, and are nearly independent of Poisson's ratio. The variation of C_{pp} with ν_m is less than 1 percent and is within thickness of the curves shown in Fig. 10.1b.

The fact that the compressibility of a prolate spheroidal void is very insensitive to the Poisson ratio of the surrounding rock can be exploited to find a simple approximation to C_{pp}. Setting $\nu_m = 1/2$ in (10.27), the following asymptotic expression can be found:

$$G_m C_{pp}(\text{prolatespheroid}) \approx \frac{2\alpha^2+1}{2\alpha^2+\ln(4\alpha^2)} \quad \text{as} \quad \alpha \to \infty, \tag{10.29}$$

which is accurate to within 1 percent for all $\alpha > 3$ and to within 2 percent for all $\alpha > 2$. A simpler approximation to the compressibility of a prolate spheroid, which is accurate to within 2 percent for all $\alpha > 1$, is

$$G_m C_{pp} \approx 1 - 0.25e^{-(\alpha-1)/3}. \tag{10.30}$$

Whereas (10.29) is an asymptotic expansion of (10.27), (10.30) is merely a convenient curve-fit.

The results for a prolate spheroid can be transformed into a form applicable to oblate spheroids by making a simple change of variables (Edwards, 1951); the result is (Fig. 10.2b)

$$C_{pp} = \frac{(1+3R)[1-2(1-2\nu_m)R+3\alpha^2]-2(1-2\nu_m)(1+2R)}{4G_m[(1+3R)\alpha^2+(1+R)(\nu_m+\nu_m R+R)]},$$

$$R = \frac{-1}{1-\alpha^2} + \frac{\alpha}{(1-\alpha^2)^{3/2}} \arcsin\sqrt{1-\alpha^2}. \tag{10.31}$$

Although the expression for the pore compressibility of a three-dimensional oblate spheroidal pore is more complicated than that for the two-dimensional elliptical pore, the variation of compressibility with aspect ratio is very similar (compare Figs. 10.1a,b). In both cases the slope of the C_{pp} curve is zero near $\alpha = 1$, and C_{pp} grows like $1/\alpha$ as $\alpha \to 0$.

An asymptotic expression for the compressibility of very thin penny-shaped cracks, also known as "circular cracks" because of their circular planform, can be derived from (10.31) by considering small values of α (Zimmerman, 1985b):

$$C_{pc}(\text{penny-shaped crack}) = \frac{2(1-\nu_m)}{\pi\alpha G_m}. \tag{10.32}$$

Comparison of (10.21) and (10.32) shows that the three-dimensional penny-shaped crack is stiffer than the two-dimensional crack by a factor of $\pi/2$ (Walsh, 1965a).

The pore compressibility expressions presented above were derived for isolated pores and are valid for pores that are sufficiently distant from neighboring pores that stress-field interactions between the pores are negligible (see §8.14). Consider a rock containing a dilute concentration of spherical pores. From (10.28), the compressibility C_{pp} of each pore can be approximated by $3/4G_m$. Combining this with (10.14) and using the identity $E = 2G(1+\nu)$ gives the following expression for the compressibility of a rock containing a small amount of spherical porosity:

$$C_{bc} = C_m\left[1 + \frac{3(1-\nu_m)}{2(1-2\nu_m)}\phi\right]. \tag{10.33}$$

As typical values of the Poisson ratio are in the range of 0.1–0.3, a 10 percent volume concentration of spherical pores will increase the bulk compressibility by about 20 percent.

The effect of cracks on the bulk compressibility is not conveniently expressed in terms of porosity. Consider a rock containing N penny-shaped cracks within a total bulk volume V_b, each having semimajor axis a and semiminor axis $b = \alpha a$. The porosity will be

$$\phi = \frac{N}{V_b}\frac{4\pi a^2 b}{3} = \frac{4\pi N a^3\alpha}{3V_b} = \frac{4\pi\alpha}{3}\Gamma, \tag{10.34}$$

where $\Gamma = Na^3/V_b$ is a dimensionless crack-density parameter. Insertion of (10.32) and (10.34) into (10.14), and using the identity $3K(1-2\nu) = 2G(1+\nu)$, yields (Walsh, 1965a)

$$C_{bc} = C_m\left[1 + \frac{16(1-\nu_m^2)}{9(1-2\nu_m)}\Gamma\right]. \tag{10.35}$$

Cracks of different aspect ratios can be combined into a single value of Γ, because the aspect ratio cancels out of the calculations that lead to (10.35).

Note from (10.34) that $4\pi\Gamma/3$ is the porosity that would exist if the cracks were replaced by circumscribed spherical pores. Hence, the increase in bulk compressibility is not proportional to the crack porosity, but is essentially proportional to the porosity that would exist if the cracks were replaced by circumscribed spherical pores.

10.4 Crack closure and elastic nonlinearity

Many crystalline rocks contain crack-like voids that are very thin in one direction. These voids have a strong influence on the mechanical and transport properties, despite the fact that total crack porosity may be very small. One important property of crack-like voids is that they can be closed under sufficiently large differential pressures, at which point they cease to contribute to the bulk compressibility. Since different cracks close at different pressures, the result is a nonlinearity in the elastic stress–strain curve of a rock. Although sedimentary rocks sometimes do not have microcracks per se, as opposed to fractures on a larger scale, imperfectly bonded grain boundaries can also be modeled as "crack-like" voids.

The initial nonlinear elastic portion of the stress–strain curve, such as the region *OA* in Fig. 4.3, has often been modeled by assuming that the unstressed rock contains a distribution of two or three-dimensional elliptical/spheroidal cracks of various aspect ratios (Walsh and Decker, 1966; Morlier, 1971; Cheng and Toksöz, 1979; Seeburger and Nur, 1984; Zimmerman, 1991). As shown in §8.9 and §8.13, elliptical (2D) or spheroidal (3D) cracks close up by an amount that is proportional to the confining pressure. Hence, the pore compressibility, when defined with respect to the initial volume, is constant. For these cracks, $\Delta A/A^i = -C_{pc}\Delta P_c$, so the pressure at which the crack fully closes is found by setting $\Delta A = -A^i$, leading to

$$P_c(\text{closing}) \equiv P^* = 1/C_{pc}. \tag{10.36}$$

It follows from (10.32) that the closing pressure of a three-dimensional penny-shaped crack of *initial* aspect ratio α is

$$P^*(\text{penny-shaped crack}) = \frac{\pi\alpha G_m}{2(1-\nu_m)} = \frac{3\pi\alpha(1-2\nu_m)}{4(1-\nu_m^2)C_m}. \tag{10.37}$$

It follows from the discussion given in §8.13 that pore pressure increments have an equal but opposite effect on cracks as do increments in the confining pressure. Hence, crack closure is actually a function of the differential pressure, $P_d = P_c - P_p$, although if the pore pressure is constant, the crack closure depends only on P_c.

If the cracks have cusp-like shapes, the closure is a nonlinear function of the confining stress (Mavko and Nur, 1978). However, the pressure at which the crack becomes fully closed generally obeys an equation very similar to (10.37), except for slight differences in the numerical constant (Zimmerman, 1991).

Morlier (1971) devised the following method for relating the distribution of initial aspect ratios to the shape of the stress–strain curve. For rocks with low concentrations of penny-shaped cracks, (10.35) gives

$$C_{bc} = C_m + \frac{16(1-\nu_m^2)}{9(1-2\nu_m)}C_m\Gamma. \tag{10.38}$$

Any additional equidimensional pores will be nonclosable under typical elastic stresses and so will contribute an additional term to the bulk compressibility that is independent of the stress. For the present purposes, such contributions to C_{bc}

can be incorporated into the first term C_m on the right side of (10.38), and do not affect the analysis.

Assume that the confining pressure is increased while the pore pressure is held constant. To simplify the notation, we drop all subscripts on P. At any given pressure, the parameter Γ in (10.38) must refer to the density of *open* cracks. The cracks that are still open at a confining pressure P will be those cracks whose closing pressures are greater than P. According to (10.37), these are the cracks that had an initial aspect ratio greater than $4(1-\nu_m^2)C_m P/3\pi(1-2\nu_m)$. If we think of α as identifying, at any given pressure, the aspect ratio of those cracks that are just at the verge of closing, then we can say $\mathrm{d}\alpha/\mathrm{d}P = 4(1-\nu_m^2)C_m/3\pi(1-2\nu_m)$.

Differentiation of (10.38) with respect to P, and use of the chain rule, gives

$$\frac{\mathrm{d}C_{bc}}{\mathrm{d}P} = \frac{\mathrm{d}C_{bc}}{\mathrm{d}\Gamma}\frac{\mathrm{d}\Gamma}{\mathrm{d}\alpha}\frac{\mathrm{d}\alpha}{\mathrm{d}P} = \frac{4\pi}{3}\left[\frac{4(1-\nu_m^2)C_m}{3\pi(1-2\nu_m)}\right]^2\frac{\mathrm{d}\Gamma}{\mathrm{d}\alpha}. \tag{10.39}$$

Since $\Gamma(\alpha)$ represents the number of cracks whose initial aspect ratios are *greater* than a, the aspect ratio distribution function, $\gamma(\alpha)$, is given by

$$\gamma(\alpha) = -\frac{\mathrm{d}\Gamma}{\mathrm{d}\alpha} = \frac{-3}{4\pi}\left[\frac{3\pi(1-2\nu_m)}{4(1-\nu_m^2)C_m}\right]^2\left[\frac{\mathrm{d}C_{bc}}{\mathrm{d}P}\right]_{P=P^*}, \tag{10.40}$$

where the derivative $\mathrm{d}C_{bc}/\mathrm{d}P$ must be evaluated at the crack closing pressure P^*, given by (10.37). Aside from the multiplicative constant, the aspect ratio distribution function is the derivative of the compressibility curve, which is to say, the *second* derivative of the stress–strain curve. As differentiation is a numerically unstable operation, small amounts of noise in the stress–strain data can lead to large fluctuations in the computed aspect ratio distribution; this can be avoided by fitting smooth curves to the stress–strain data.

Compressibility data can often be fit by exponentially decreasing functions of the form (Wyble, 1958; Zimmerman, 1991)

$$C_{bc} = C_{bc}^{\infty} + (C_{bc}^{i} - C_{bc}^{\infty})\mathrm{e}^{-P/\hat{P}}, \tag{10.41}$$

where the superscript i denotes the initial (zero stress) value, the superscript ∞ denotes the value at high stresses, and $\hat{P}$ is a characteristic (crack-closing) pressure. For this type of stress–strain curve, (10.40) gives

$$\gamma(\alpha) = \left[\frac{9(1-2\nu_m)}{16(1-\nu_m^2)C_m}\right]\frac{(C_{bc}^{i} - C_{bc}^{\infty})}{\hat{\alpha}}\mathrm{e}^{-\alpha/\hat{\alpha}}, \tag{10.42}$$

where $\hat{\alpha}$ is related to $\hat{P}$ by (10.37).

By definition, the "crack density" whose initial aspect ratio lies between α and $\alpha + \mathrm{d}\alpha$ is given by $\mathrm{d}\Gamma = \gamma(\alpha)\mathrm{d}\alpha$. If all microcracks in the rock are assumed to have the same radius a, then (10.34) shows that the porosity contained within this range of α is $\mathrm{d}\phi = (4\pi\alpha/3)\mathrm{d}\Gamma$. Hence, the distribution function for the porosity, denoted by $c(\alpha)$, is

$$c(\alpha) = \frac{\mathrm{d}\phi}{\mathrm{d}\alpha} = \frac{4\pi\alpha}{3}\gamma(\alpha) = \frac{3\pi(1-2\nu_m)}{4(1-\nu_m^2)}\frac{(C_{bc}^{i} - C_{bc}^{\infty})}{C_m}\frac{\alpha}{\hat{\alpha}}\mathrm{e}^{-\alpha/\hat{\alpha}}. \tag{10.43}$$

The function $c(\alpha)$ is also known as the *aspect ratio distribution function*. The particular form given by (10.43) rises almost linearly from the origin, has a maximum at $\alpha = \hat{\alpha}$, and then effectively decays to zero when α reaches about $5\hat{\alpha}$. The total crack porosity is found by integrating $c(\alpha)$ over all values of α, from 0 to 1. However, $c(\alpha)$ decays so rapidly that the range of the integral can be extended to infinity:

$$\phi_{\text{crack}} = \int_0^\infty c(\alpha)\mathrm{d}\alpha = \int_0^\infty \frac{3\pi(1-2\nu_m)}{4(1-\nu_m^2)} \frac{(C_{bc}^i - C_{bc}^\infty)}{C_m} \frac{\alpha}{\hat{\alpha}} \mathrm{e}^{-\alpha/\hat{\alpha}} \mathrm{d}\alpha$$

$$= \frac{3\pi(1-2\nu_m)}{4(1-\nu_m^2)} \frac{(C_{bc}^i - C_{bc}^\infty)}{C_m} \hat{\alpha} = (C_{bc}^i - C_{bc}^\infty)\hat{P}, \qquad (10.44)$$

where in the last step (10.37) is used to relate $\hat{\alpha}$ to $\hat{P}$.

Zimmerman (1991) fit measured compressibilities of three consolidated sandstones, Boise, Berea, and Bandera, to functions of the form (10.41). Table 10.1 shows the fitted parameters in (10.41), the Voigt–Reuss–Hill estimates of the intact rock parameters $\{C_m, \nu_m\}$, and the estimate of total crack porosity computed from (10.44). The compressibility curves and the aspect ratio distribution functions are shown in Fig. 10.3.

Morlier's method assumes that each crack behaves as an isolated void in an otherwise infinite, intact rock. As explained in §10.5, stress-field interactions between nearby cracks lead to a pore compressibility that increases with crack density. Consequently, the bulk compressibility will increase with Γ in a nonlinear manner, more rapidly than predicted by (10.38). Accounting for these effects

Table 10.1 Compressibility parameters for three sandstones (Zimmerman, 1991).

Sandstone	C_{bc}^∞ (1/GPa)	C_{bc}^i (1/GPa)	C_m (1/GPa)	ν_m	$\hat{P}$ (MPa)	ϕ_{crack}
Bandera	0.082	0.617	0.0226	0.210	8.33	0.0044
Berea	0.105	0.740	0.0222	0.218	4.74	0.0030
Boise	0.095	0.374	0.0251	0.188	7.01	0.0019

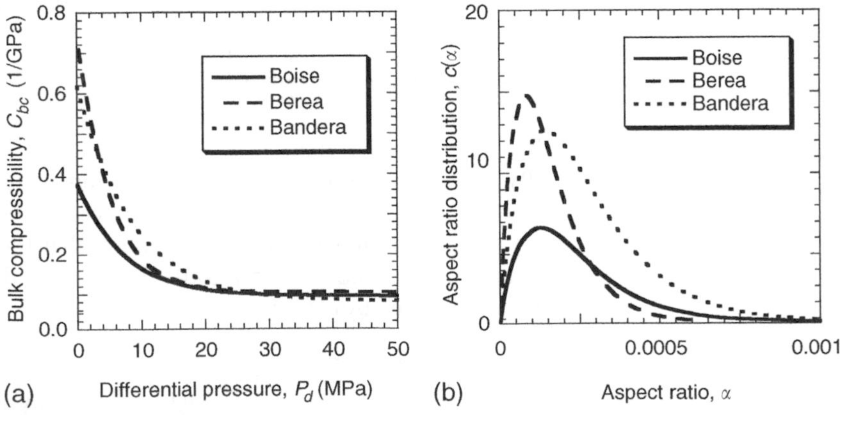

Fig. 10.3 (a) Compressibility curves of several consolidated sandstones, and (b) the associated aspect ratio distribution functions, as computed by Morlier's method (Zimmerman, 1991).

(Cheng and Toksöz, 1979; Zimmerman, 1991) will lead to aspect ratio distributions that are shifted to lower values of α and to estimated crack porosities that are smaller than that given by (10.44).

10.5 Effective medium theories

The compressibility of pores of different shapes was discussed in §10.3 under the assumption that a given pore was surrounded by an infinite expanse of intact rock. However, the effect of nearby pores can be shown to increase the compressibility of a given pore, by the following argument. Imagine a rock permeated with various pores, only one of which is pressurized by the pore fluid. The expansion of this pore will be resisted not by an infinite expanse of nonporous rock, but by the porous rock, whose effective elastic moduli will be lower than that of the intact rock. Hence, the pressurized pore will expand by a greater amount than it would if it were imbedded in intact rock. The pore compressibility of a given pore will therefore be increased by the presence of nearby pores. This phenomenon is referred to as being due to "pore–pore interactions," although the effect is one of interaction of the stress fields around each pore, rather than physical intersection or coalescence of the voids.

Since the compressibility of an isolated pore can be calculated by solving the suitable elastostatic boundary-value problem, it might be thought that this same approach could be used in the case of multiple pores. Unfortunately, exact solutions to elasticity problems involving multiple cavities are very difficult to obtain. Some solutions exist for bodies containing a pair of pores, such as two spherical cavities (Sternberg and Sadowsky, 1952; Willis and Bullough, 1969; Fond et al., 2001; Chalon and Montheillet, 2003) or two cylindrical cavities (Ling, 1948; Zimmerman, 1988). However, it is not clear that these two-pore solutions are useful in finding the effective moduli of a body that contains, say, 20–30 percent spherical pores by volume, such as might be used to model the behavior of a sandstone (see Chen and Acrivos, 1978).

The most fruitful approaches to the problem of predicting the compressibility of a body containing a finite concentration of pores have been those approximate methods that avoid solving multipore interaction problems. Numerous such methods have been proposed in the fields of rock physics, ceramics, and materials science. These methods are all applicable to the more general problem of predicting the effective elastic moduli of heterogeneous, multicomponent materials (Christensen, 1991; Nemat-Nasser and Hori, 1993; Milton, 2002).

Most of these methods can be discussed within the general formalism of the energy approach discussed in §10.3. The effective bulk modulus of a porous rock can be defined by

$$\frac{P^2 V_b}{2K} = \frac{P^2 V_b}{2K_m} + \Delta\mathcal{E}_{\text{hydro}}, \tag{10.45}$$

where $\Delta\mathcal{E}_{\text{hydro}}$ is the excess elastic strain energy that would be stored in a rock of volume V_b if the pores were introduced into the initially solid rock while maintaining a hydrostatic confining pressure P. The effective shear modulus can be defined in a similar way, with the rock assumed to be subjected to a shear stress of magnitude S:

$$\frac{S^2 V_b}{2G} = \frac{S^2 V_b}{2G_m} + \Delta\mathcal{E}_{\text{shear}}. \tag{10.46}$$

Utilization of (10.45) and (10.46) requires the estimation of the energy perturbation terms for hydrostatic and shear loading. The simplest approach is to neglect pore–pore interactions, and calculate the energy change due to each pore as if it were an isolated void in an infinite intact rock, and then sum up this energy for all the pores, as in §10.3. The two energy perturbations, $\Delta\mathcal{E}_{\text{hydro}}$ and $\Delta\mathcal{E}_{\text{shear}}$, will depend on the moduli of the intact rock, $\{K_m, G_m\}$, and will be proportional to the porosity. The equations for the two effective moduli will be uncoupled, and can be solved explicitly for the effective moduli K and G.

For example, the energy terms for an isolated spherical pore are given by (Nemat-Nasser and Hori, 1993)

$$\Delta\mathcal{E}_{\text{hydrostatic}} = \frac{P^2 V_b}{2K_m}\left[\frac{3K_m + 4G_m}{4G_m}\right]\phi, \tag{10.47}$$

$$\Delta\mathcal{E}_{\text{shear}} = \frac{P^2 V_b}{2G_m}\left[\frac{15K_m + 20G_m}{9K_m + 8G_m}\right]\phi. \tag{10.48}$$

Expression (10.47) is consistent with the expression that could be obtained by combining (10.13) and (10.33). If the energy terms (10.47) and (10.48) are used in (10.45) and (10.46), the effective moduli are predicted to be

$$\frac{K}{K_m} = \left[1 + \frac{3K_m + 4G_m}{4G_m}\phi\right]^{-1} = \left[1 + \frac{3(1-\nu_m)}{2(1-2\nu_m)}\phi\right]^{-1}, \tag{10.49}$$

$$\frac{G}{G_m} = \left[1 + \frac{15K_m + 20G_m}{9K_m + 8G_m}\phi\right]^{-1} = \left[1 + \frac{15(1-\nu_m)}{(7-5\nu_m)}\phi\right]^{-1}, \tag{10.50}$$

As these predictions ignore pore–pore interactions, they are correct only to first-order in ϕ and increasingly overestimate the moduli as the porosity increases. This overestimation is also clear from the fact that they predict finite elastic moduli when the porosity reaches 100 percent (Fig. 10.4).

In the so-called "self-consistent" scheme of Hill (1965) and Budiansky (1965), the excess strain energy due to each single pore is calculated by assuming that it is introduced into a homogeneous medium that has the elastic properties of the actual porous material. This leads to the same functional forms for the two energy terms as does the "no-interaction" approach, except that $\{K_m, G_m\}$ are replaced by $\{K, G\}$. In general, (10.45) and (10.46) become two coupled nonlinear algebraic equations that require numerical solution. For spherical pores, the self-consistent method yields

$$\frac{1}{K} = \frac{1}{K_m} + \frac{1}{K}\left[\frac{3K + 4G}{4G}\right]\phi, \tag{10.51}$$

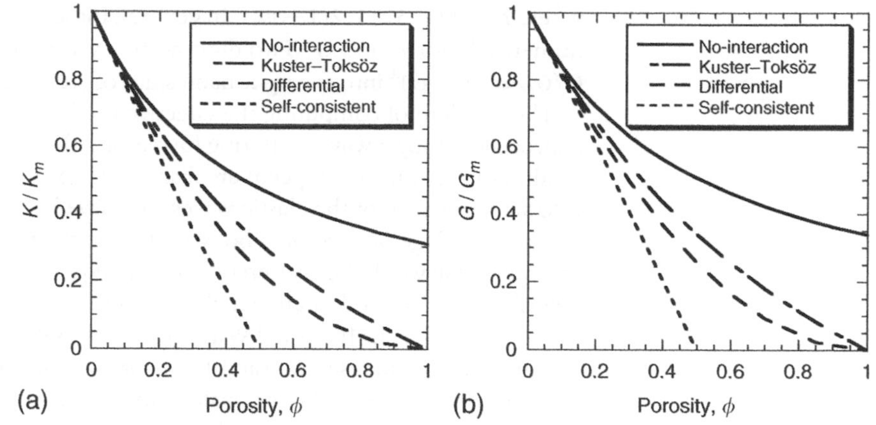

Fig. 10.4 Elastic moduli of a rock containing dry, randomly distributed spherical pores, according to various effective medium theories. The Poisson ratio of the intact rock is taken to be 0.25.

$$\frac{1}{G} = \frac{1}{G_m} + \frac{1}{G}\left[\frac{15K + 20G}{9K + 8G}\right]\phi. \tag{10.52}$$

This method yields much lower effective moduli than does the no-interaction method, and predicts that the moduli vanish at some finite porosity (Fig. 10.4).

Bruner (1976) suggested that the self-consistent method implicitly takes interactions between pairs of pores into account twice, since the "typical" pore is assumed to be imbedded in an effective medium whose elastic moduli already reflect in part the interactions between this typical pore and all other pores. To avoid this double-counting, one can introduce the pores into the rock sequentially, with pore $n + 1$ considered to be placed into a homogeneous medium which has the effective elastic properties of the body with n pores, etc. In this way, pore $n + 1$ feels the effect of pore n, but not vice versa. Each new pore is assumed to be randomly placed, and so if the "current" porosity is ϕ, the next pore "replaces" intact rock with probability $1 - \phi$, and replaces existing pore space with probability ϕ (McLaughlin, 1977). In the limit in which each new addition of pores is infinitesimal, this method gives two coupled ordinary differential equations for the two effective moduli.

In the case of spherical pores, these equations are

$$-\frac{(1-\phi)}{K}\frac{dK}{d\phi} = \frac{3K + 4G}{4G}, \tag{10.53}$$

$$-\frac{(1-\phi)}{G}\frac{dG}{d\phi} = \frac{15K + 20G}{9K + 8G}. \tag{10.54}$$

The initial conditions are that $K = K_m$ and $G = G_m$ when $\phi = 0$. These equations can be integrated to give the following implicit expressions for the effective moduli (Norris, 1985):

$$\frac{G}{G_m} = (1-\phi)^2\left[\frac{1 + \beta(G/G_m)^{3/5}}{1 + \beta}\right]^{1/3}, \tag{10.55}$$

$$\frac{K}{K_m} = \frac{G}{G_m}\left[\frac{1 + 2\beta}{1 + 2\beta(G/G_m)^{3/5}}\right], \tag{10.56}$$

where $\beta = (1 - 5\nu_m)/2(1 + \nu_m)$. In general, these two equations must be solved numerically for K and G. Two-digit accuracy can be achieved by substituting $G/G_m = (1 - \phi)^2$ into the right-hand sides of (10.55) and (10.56).

The problem of computing the effective moduli of a porous rock can also be approached using a wave-scattering formalism; the relationships between elastic moduli and seismic wave speeds are discussed in §11.4. Kuster and Toksöz (1974) calculated the sum of the elastic waves that have been scattered once from each of an assemblage of pores in a body with moduli $\{K_m, G_m\}$, and equated this to the wave that would be scattered from an "equivalent homogeneous spherical inclusion" whose moduli are equal to the effective moduli of the porous rock, $\{K, G\}$. Toksöz et al. (1976) and Wilkens et al. (1986) used this approach to model seismic velocities in reservoir sandstones, and Zimmerman and King (1986) used it to study the effect of the ice/water ratio on seismic velocities in permafrost. For a rock containing dry spherical pores, the method of Kuster and Toksöz gives (Fig. 10.4)

$$\frac{K}{K_m} = \frac{1 - \phi}{1 + (3K_m/4G_m)\phi} = \frac{1 - \phi}{1 + [(1 + \nu_m)/2(1 - 2\nu_m)]\phi}, \tag{10.57}$$

$$\frac{G}{G_m} = \frac{1-\phi}{1+[(6K_m+12G_m)/(9K_m+8G_m)]\phi} = \frac{1-\phi}{1+[2(4-5\nu_m)/(7-5\nu_m)]\phi}. \tag{10.58}$$

The predicted moduli lie between those of the no-interaction method and the differential method and vanish at a porosity of 100 percent. The Kuster–Toksöz predictions for a rock containing spherical pores also coincide *exactly* with the upper bounds of Hashin and Shtrikman (1961), which are valid regardless of the geometry of the pores.

The predictions of these four methods are plotted in Fig. 10.4, for $\nu_m = 0.25$. Each approach predicts similar, although in general not identical, behavior for G as it does for K. Consider the curves for the effective bulk modulus as a function of spherical porosity. If expanded in Taylor series in ϕ, all four methods agree to first-order, but give different values for the higher-order coefficients. The predictions diverge from each other markedly for porosities greater than about 0.10. As an indication of the validity of these approaches, consider the suite of porous glass specimens that were fabricated by Walsh et al. (1965) to have pores that were as nearly spherical as possible. The measured bulk moduli, for porosities ranging from 0.05–0.70, generally fell about midway between the predictions of the Kuster–Toksöz and the differential schemes (Zimmerman, 1991, p. 120).

Each of the sets of predicted effective moduli takes on a particularly simple form when $K_m = 4G_m/3$, which corresponds to $\nu_m = 0.2$. In this case, the no-interaction method gives $K/K_m = G/G_m = (1 + 2\phi)^{-1}$, the self-consistent method gives $K/K_m = G/G_m = 1 - 2\phi$, the differential method gives $K/K_m = G/G_m = (1 - \phi)^2$, and the method of Kuster and Toksöz gives $K/K_m = G/G_m = (1 - \phi)/(1 + \phi)$.

In order to apply the effective moduli theories to a body with penny-shaped cracks, the two strain energy perturbation terms are needed. The energy perturbation $\Delta\mathcal{E}_{hydro}$ for a single penny-shaped crack of radius a under hydrostatic loading is given by (8.310), or by (10.13) and (10.32):

$$\Delta\mathcal{E}_{hydro} = \frac{4(1-\nu_m)a^3P^2}{3G_m} = \frac{8(1-\nu_m^2)a^3P^2}{9(1-2\nu_m)K_m}. \tag{10.59}$$

The strain energy perturbation $\Delta\mathcal{E}_{shear}$ for an isolated crack subjected to a shear stress of magnitude S will depend on the orientation of the shear stress with respect to the crack plane; this issue does not arise for hydrostatic loading, nor does it arise for shear loading in the case of a spherical pore. If the crack lies in, say, the x–y plane, then the strain energy due to a far-field shear stress $\tau_{xy} = S$ is given by (8.312). If the orientations of the cracks within the rock are randomly distributed, the strain energy must be averaged over all possible angles of inclination with respect to the direction of shear. In the general case in which the crack is inclined to the shear, there will be both a hydrostatic and a shear component to the far-field stress and to the strain energy. After performing the averaging process, the average excess strain energy per crack, for a random distribution of crack planes, is found to be

$$\Delta\mathcal{E}_{shear(random\ orientation)} = \frac{16(1-\nu_m)(5-\nu_m)a^3S^2}{45(2-\nu_m)G_m}. \tag{10.60}$$

According to the no-interaction scheme, the effective elastic moduli of a randomly/isotropically cracked body can be found by inserting the energy perturbations (10.59) and (10.60) into the general expressions (10.45) and (10.46), to yield

$$\frac{K}{K_m} = \left[1 + \frac{16(1-\nu_m^2)}{9(1-2\nu_m)}\Gamma\right]^{-1}, \tag{10.61}$$

$$\frac{G}{G_m} = \left[1 + \frac{32(1-\nu_m)(5-\nu_m)}{45(2-\nu_m)}\Gamma\right]^{-1}. \tag{10.62}$$

The predictions of the self-consistent method are also found by inserting (10.59) and (10.60) into (10.45) and (10.46), but with $\{K, G, \nu\}$ used in the excess energy terms. This leads to the following implicit expressions for K and G (O'Connell and Budiansky, 1974):

$$\frac{K}{K_m} = 1 - \frac{16(1-\nu^2)}{9(1-2\nu)}\Gamma, \tag{10.63}$$

$$\frac{G}{G_m} = 1 - \frac{32(1-\nu)(5-\nu)}{45(2-\nu)}\Gamma. \tag{10.64}$$

These equations can be partially inverted by using $3K(1-2\nu) = 2G(1+\nu)$ to eliminate G and K, to arrive at

$$\Gamma = \frac{45(\nu_m-\nu)(2-\nu)}{16(1-\nu^2)(10\nu_m-3\nu_m\nu-\nu)}. \tag{10.65}$$

After (10.65) is solved numerically for ν as a function of Γ, this value of ν can be used in (10.63) and (10.64) to find K and G. The curves of K and G as functions of Γ are somewhat nonlinear, but the effective Young's modulus that follows from (10.63) and (10.64) is very nearly a linear function of crack density. For all $0 < \nu_m < 1/2$, the following expression is accurate to within 1 percent:

$$\frac{E}{E_m} = 1 - \frac{16}{9}\Gamma. \tag{10.66}$$

The equations of the differential scheme, for a rock containing randomly distributed and oriented cracks, are (Salganik, 1973)

$$\frac{1}{K}\frac{dK}{d\Gamma} = -\frac{16(1-\nu^2)}{9(1-2\nu)}, \tag{10.67}$$

$$\frac{1}{G}\frac{dG}{d\Gamma} = -\frac{32(1-\nu)(5-\nu)}{45(2-\nu)}. \tag{10.68}$$

Using the initial conditions that $K = K_m$ and $G = G_m$ when $\Gamma = 0$, these equations can be integrated to yield (Zimmerman, 1985c)

$$e^{\Gamma} = \left(\frac{3-\nu}{3-\nu_m}\right)^{5/128}\left(\frac{1-\nu}{1-\nu_m}\right)^{30/128}\left(\frac{1+\nu}{1+\nu_m}\right)^{45/128}\left(\frac{\nu}{\nu_m}\right)^{-80/128}, \tag{10.69}$$

$$\frac{K}{K_m} = \left(\frac{\nu}{\nu_m}\right)^{10/9}\left(\frac{3-\nu}{3-\nu_m}\right)^{-1/9}\left(\frac{1-2\nu}{1-2\nu_m}\right)^{-1}. \tag{10.70}$$

A simple and accurate approximate solution to (10.67) and (10.68) is (Bruner, 1976):

$$\frac{E}{E_m} = e^{-16\Gamma/9}, \quad \frac{\nu}{\nu_m} = e^{-8\Gamma/5}. \tag{10.71}$$

Kuster and Toksöz (1974) presented equations for the effective moduli of a body containing oblate spheroidal pores of arbitrary aspect ratio, but did not explicitly consider the limiting case of infinitely thin cracks. In the limit, as the aspect ratio goes to zero, the effective moduli predicted by this method are found to be

$$\frac{K}{K_m} = \frac{1 - [32(1+\nu_m)/27]\Gamma}{1 + [16(1+\nu_m)^2/27(1-2\nu_m)]\Gamma}, \tag{10.72}$$

$$\frac{G}{G_m} = \frac{1 - [32(5-\nu_m)(7-5\nu_m)/675(2-\nu_m)]\Gamma}{1 + [64(5-\nu_m)(4-5\nu_m)/675(2-\nu_m)]\Gamma}. \tag{10.73}$$

The various predictions for the effective moduli of a rock containing a random distribution of cracks are plotted in Fig. 10.5. Note that the relative positions of the curves are not quite the same as for spherical pores; for cracks, the Kuster–Toksöz model predicts lower moduli than does the differential scheme. All four methods agree to first-order in crack density, but begin to diverge appreciably

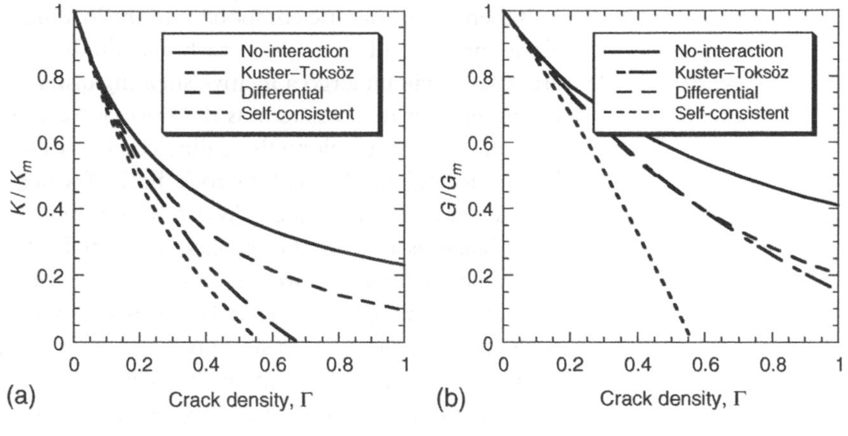

Fig. 10.5 Elastic moduli of a rock containing dry, randomly distributed and randomly oriented penny-shaped cracks, according to various effective medium theories. The Poisson ratio of the intact rock is taken to be 0.25.

for crack densities greater than about 0.15. The no-interaction method predicts that the moduli decay to zero roughly as $1/\Gamma$, whereas the differential method predicts a faster, exponential decay. Nevertheless, both of these methods predict substantial moduli values for crack densities as high as 1.0. The self-consistent method, on the other hand, predicts that the moduli both vanish at a crack density of 0.5625. The Kuster–Toksöz method also predicts that the moduli vanish at some finite crack density, although this critical density differs for K and G, and varies with the Poisson ratio of the intact rock.

All three energy-based methods predict that as the moduli decay to zero due to an increase in crack density, the effective Poisson's ratio goes to zero. However, as the Kuster–Toksöz method predicts that K decays faster than G, there is a range of crack densities for which this method predicts positive values for both of the elastic moduli, but a *negative* value for Poisson's ratio. For example, if $\nu_m = 0.25$, negative values of ν are predicted for $0.420 < \Gamma < 0.675$.

Budiansky and O'Connell (1976) showed that the results for cracks with circular planforms can be applied to cracks having elliptical planforms, if the crack density is defined as $\Gamma = 2NA^2/\pi PV_b$, where A is the area of the crack in its plane, and P is its perimeter. Various special cases of two and three-dimensional bodies containing systems of aligned cracks are discussed by Hashin (1988), Kachanov (1994), Nemat-Nasser and Hori (1993), and Mavko et al. (1998). The effective moduli of a body containing randomly oriented needle-shaped pores are discussed by Berryman (1995).

10.6 Sliding crack friction and hysteresis

As discussed in §10.4, an increase in the hydrostatic stress will cause cracks to close up and consequently cause the bulk modulus to increase. If the applied load is purely hydrostatic, and if crack–crack stress-field interactions are ignored, the deformation of each crack face will be purely normal to the plane of the crack, and there will be no relative tangential displacement between the two crack faces. In this case, the issue of friction along the two contacting crack faces is irrelevant. Under hydrostatic compression, a crack that is closed by stress acts as if it were welded shut.

If there is a deviatoric component to the loading, however, the traction acting along the plane of any closed crack will also have a *shear* component. The two crack faces will undergo a relative shearing displacement if the shear traction τ is greater than $\mu\sigma$, where σ is the normal traction and μ is the coefficient of friction. The friction along the sliding crack faces will influence the shear modulus and Young's modulus of the rock, both of which quantify the behavior of the rock under deviatoric loading. Walsh (1965b) first analyzed this situation, using a hybrid analysis in which the cracks were treated as three-dimensional objects, but the excess energy terms were calculated under the two-dimensional plane stress or plane strain approximations. For consistency with the case in which there is no frictional sliding, and to simplify some of the calculations needed when averaging over all cracks orientations, we will analyze this problem under the plane stress assumption. Although this approximation is strictly applicable only to thin plates, the main purpose of our analysis is to give a qualitative understanding of the effect that cracks have on a rock undergoing uniaxial compression.

Consider a rock specimen of length L, width b, and thickness t, subjected to a uniaxial tension T in the longitudinal (y) direction. In complete analogy with the discussion of the effective bulk and shear moduli given in §10.5, the effective Young's modulus of this specimen can be defined by

$$\frac{T^2 V_b}{2E} = \frac{T^2 V_b}{2E_m} + \Delta\mathcal{E}_{\text{uniaxial}}, \tag{10.74}$$

where $\Delta\mathcal{E}_{\text{uniaxial}}$ is the excess energy due to the presence of the cracks. In contrast to the calculations of §10.5, in this case the excess energy will consist of an elastic stored strain energy, plus an *inelastic* term that represents the energy dissipated through frictional sliding.

To calculate these energy terms, first consider a single isolated crack of length $2c$, lying in a plane whose outward unit normal vector is oriented at some angle β to the direction of loading. The crack is assumed to pass through the entire thickness t of the specimen, and we assume for now that the crack is *not* closed. In a coordinate system oriented with the crack, the stress components are

$$\tau_{x'x'} = T\sin^2\beta, \quad \tau_{y'y'} = T\cos^2\beta, \quad \tau_{x'y'} = \tau_{y'x'} = (T/2)\sin 2\beta. \tag{10.75}$$

As discussed in §8.9, if the crack is assumed to be infinitely thin, the stress component acting parallel to the crack, $\tau_{x'x'}$, causes no relative displacement of the crack faces, and so gives rise to no additional strain energy. The strain energy due to the normal stress $\tau_{y'y'}$ can be calculated by the following argument. The component $\tau_{x'x'}$ gives rise to no strain energy, so the strain energy due to a uniaxial stress normal to the crack must be the same as that due to *hydrostatic* loading. In general, as shown in §10.3, a hydrostatic stress P gives rise to an excess strain energy of $P^2 V_p C_{pc}/2$. The pore compressibility C_{pc} of a thin elliptical crack of aspect ratio α under plane stress conditions is $2/\alpha E_m$, by (10.17), and its pore volume is $\pi c^2 t\alpha$. Hence, the excess elastic strain energy due to a crack under a far-field hydrostatic stress P is $P^2\pi c^2 t/E_m$. Replacing P with $\tau_{y'y'}$ from (10.75) shows that the elastic strain energy due solely to the normal stress component is

$$\Delta\mathcal{E}_{\text{uniaxial loading, resolved normal stress}} = \frac{T^2\pi c^2 t \cos^4\beta}{E_m}. \tag{10.76}$$

If the rock contains N cracks of length $2c$, whose angles of inclination are randomly distributed, the total energy contribution due to compression of the cracks can be found by averaging (10.76) over the full range $-\pi/2 \le \beta \le \pi/2$. The average value of $\cos^4\beta$ is $3/8$, so the total contribution to the elastic strain energy due solely to crack compression is

$$\Delta\mathcal{E}_{\text{uniaxial loading, resolved normal stress}} = \frac{3T^2\pi Nc^2 t}{8E_m}. \tag{10.77}$$

Starr (1928) showed that the excess strain energy of a crack due to shear loading of magnitude S is given by the same expression as for hydrostatic loading, with P replaced by S. This is consistent with the results of §8.13, where it was shown that the algebraic form of the excess energy of a three-dimensional penny-shaped crack under shear loading coincides with that for hydrostatic loading, when the Poisson ratio is zero. In the present case of plane stress, neither of the two energy terms has any dependence on Poisson's ratio, and the expressions coincide in all cases. With the resolved shear stress given by (10.75), the elastic strain energy of a single crack oriented at angle β is

$$\Delta\mathcal{E}_{\text{uniaxial loading, resolved shear stress}} = \frac{T^2\pi c^2 t \sin^2 2\beta}{4E_m}. \tag{10.78}$$

The average value of $\sin^2 2\beta$ is $1/2$, so the total strain energy due to the resolved shear stress, for a rock containing N randomly oriented cracks, is

$$\Delta\mathcal{E}_{\text{uniaxial loading, resolved shear stress}} = \frac{T^2 N\pi c^2 t}{8E_m}. \tag{10.79}$$

The total excess strain energy of the cracked rock under uniaxial tension T is the sum of (10.77) and (10.79):

$$\Delta\mathcal{E}_{\text{uniaxial loading}} = \frac{T^2 N\pi c^2 t}{2E_m}. \tag{10.80}$$

Using this result in (10.74) gives the following expression for the effective Young's modulus of a two-dimensional randomly cracked body containing open cracks (Bristow, 1960):

$$\frac{E}{E_m} = \frac{1}{1+\pi\Gamma_2}, \tag{10.81}$$

where $\Gamma_2 = Nc^2/A$ is the two-dimensional crack density parameter, and $A = V_b/t = bL$ is the bulk area of the sample. This estimate (10.81) neglects stress field interactions between nearby cracks. Application of the self-consistent effective medium theory would yield $E/E_m = 1 - \pi\Gamma_2$, whereas the differential scheme would yield $E/E_m = \exp(-\pi\Gamma_2)$.

Now consider the same rock, again under uniaxial compression, but assume that all of the cracks are *closed*. Consider first a single crack having orientation angle β. The effective Young's modulus is again defined by (10.74). As the crack is already closed, and the resolved stress normal to the crack plane is compressive, there can be no additional normal displacement of the crack faces. Hence, the strain energy term (10.77) will not be present. The strain energy term (10.78), which is due to the resolved shear stress along the crack plane, can only arise if the crack faces are able to slide past each other. If the coefficient of friction along the crack faces is μ, sliding will occur only if the resolved shear stress τ, which is given by $\tau_{x'y'}$ in (10.75), exceeds $\mu\sigma$, where the normal traction σ is given by $\tau_{y'y'}$ in (10.75). Hence, the effective shear stress that is available to cause displacements around the crack face is given by

$$\tau_{\text{eff}} = |\tau| - \mu\sigma. \tag{10.82}$$

The excess elastic strain energy due to the crack is then given by (10.78), with the effective shear stress in place of the resolved shear stress:

$$\Delta\mathcal{E} = \tau_{\text{eff}}^2 \pi c^2 t / E_m. \tag{10.83}$$

This energy is stored within the rock in the vicinity of the crack, in the form of elastic strain energy.

Superposition arguments, such as those used in §7.2, in which the strains due to a remote shear stress τ are decomposed into the difference between those caused by a shear stress τ at infinity and a shear traction τ along the crack faces, minus those caused by shear tractions along the face of the crack alone, show that the stored elastic strain energy is also equal to one-half of the product of the resolved shear traction along the crack face, multiplied by the mean displacement of the crack face in the direction of that traction. But this expression is precisely equal to the energy that is dissipated due to frictional sliding along the crack faces, except that in this case, although the displacement is controlled by τ_{eff}, the work is done only by the shear traction $\mu\sigma$ that resists the sliding. Hence, the energy dissipated by frictional sliding is given by

$$\Delta W = (\mu\sigma/\tau_{\text{eff}})\Delta\mathcal{E} = \mu\sigma\tau_{\text{eff}}\pi c^2 t / E_m. \tag{10.84}$$

This energy is dissipated as heat and is not stored in the rock as recoverable elastic strain energy.

Including both the stored elastic strain energy (10.83) and the frictionally dissipated energy (10.84) on the right side of (10.74), the expression for the effective Young's modulus becomes, after using (10.82),

$$\frac{T^2 bLt}{2E} = \frac{T^2 bLt}{2E_m} + \frac{(\tau^2 - \mu\sigma|\tau|)\pi c^2 t}{E_m}. \tag{10.85}$$

With the shear and normal tractions on the crack face given by (10.75), the effective Young's modulus for the rock containing this single closed crack is therefore given by

$$\frac{1}{E} = \frac{1}{E_m}\left[1 + \frac{(\sin^2 2\beta - 2\mu\cos^2\beta|\sin 2\beta|)\pi c^2}{2A}\right]. \tag{10.86}$$

If the rock contains a random distribution of cracks, the effective modulus can be found by averaging the bracketed term in (10.86) over all cracks for which sliding occurs. These are the cracks for which $|\tau| > \mu\sigma$, which from (10.75) can be seen to be those cracks for which $\tan^{-1}\mu < |\beta| < \pi/2$. The resulting expression for the effective modulus is

$$\frac{1}{E} = \frac{1}{E_m}\left[1 + \frac{(1+\mu^2)\tan^{-1}(1/\mu) - \mu}{2(1+\mu^2)}\Gamma_2\right], \tag{10.87}$$

where $\Gamma_2 = Nc^2/A$ is again the total two-dimensional crack density. If the friction coefficient is zero, (10.87) reduces to $E/E_m = 1/[1 + (\pi/4)\Gamma_2]$, which differs from expression (10.81) for open cracks, because, even in the absence of friction, closed cracks are prevented from undergoing compressive deformation normal to their planes. As μ increases, (10.87) reduces to $E = E_m$. This limit is essentially reached for $\mu > 2$, although such high coefficients of friction may be unrealistic.

The effective bulk modulus K of a rock containing closed cracks will equal the crack-free bulk modulus, K_m. It follows that $K_m = E_m/3(1-2\nu_m) = E/3(1-2\nu)$, and so the effective Poisson ratio ν is given by

$$(1 - 2\nu) = (1 - 2\nu_m)E/E_m. \tag{10.88}$$

Closed cracks that slide against friction therefore cause the Poisson ratio to increase, in contrast to open cracks, which cause it to decrease; *cf.* (10.71).

To study the hysteretic effect of closed cracks, consider again a rock containing a single crack oriented at some angle β, such that $\tan^{-1}\mu < |\beta| < \pi/2$. If the uniaxial compressive stress T is increased from zero, the rock will deform with an elastic modulus given by (10.86). Now imagine that loading ceases when $T = T'$, at which point the stresses on the crack face are given by τ', σ', and τ'_{eff}. Relative motion between the crack faces is possible only if $|\tau| > \mu\sigma$. As soon as T begins to decrease (i.e., unloading), the direction of the frictional resistance reverses, so that the shear stress resulting from the deformation of the crack must overcome both the frictional resistance and the shear stress due to the load, before reverse sliding can commence. The shear stress due to the deformation of the crack is equal in magnitude, but opposite in sign, to the effective shear stress that produced this deformation. Therefore, the condition for reverse sliding can be expressed as

$$\tau'_{\text{eff}} \geq |\tau| + \mu\sigma. \tag{10.89}$$

At the conclusion of the loading cycle, the conditions were

$$|\tau'| = \tau'_{\text{eff}} + \mu\sigma'. \tag{10.90}$$

Denoting the conditions at which reverse crack sliding first occurs by τ'', etc., we find from (10.89) and (10.90) that

$$|\tau'| + \mu\sigma' = |\tau''_{\text{eff}}| - \mu\sigma''. \tag{10.91}$$

Using (10.75) for the shear and normal stresses, along with elementary trigonometric identities, gives the following condition for the onset of reverse sliding:

$$T'' = T'(\tan\beta - \mu)/(\tan\beta + \mu). \tag{10.92}$$

As a crack which underwent sliding during loading must have had tan $> \mu$, (10.92) shows that reverse sliding will begin at some compressive stress T'', where $0 < T'' < T'$.

A loading and unloading cycle of a rock containing a single closed crack, subject to uniaxial compression at an angle β to the crack normal vector, is shown in Fig. 10.6a. Initial loading along *OA* occurs with an effective modulus $E < E_m$, given by (10.86). Unloading (path *AB*) from some maximum load T' first occurs with the modulus E_m of the uncracked material (10.75). When the applied compressive stress is reduced to T'', as given by (10.92), reverse sliding begins on the crack, and the rock specimen continues to deform along *BO* with a modulus E'' that is lower than the initial loading modulus, E. If the applied load returns to zero, the strain again vanishes.

If the initial unloading line is extended down to the strain axis, as shown in the dashed line, then the triangular area *ADC* represents the work that would have been stored at maximum load if the rock were uncracked. The area between this line *ABC* and the actual loading line *OA* represents the sum of the excess strain energy due to the crack, and the work done against friction, during loading. The elastic portion of this work, area *OBC*, is recovered during unloading, so that the net work done by the load during the entire loading cycle is given by the area of the triangular region *OABO*. As the total relative displacement of the crack faces must be the same (in magnitude) during loading and unloading, the ratio of energy dissipation due to friction during loading and unloading is in the ratio of T'/T''.

The uniaxial stress–strain curve for a rock containing cracks having a distribution of aspect ratios is shown schematically in Fig. 10.6b. Initially, the modulus is given by an expression similar to (10.81) for a body with open cracks. As the load increases, cracks close, and eventually the modulus is given by an expression such as (10.87) for a body containing closed, sliding cracks. Unloading begins

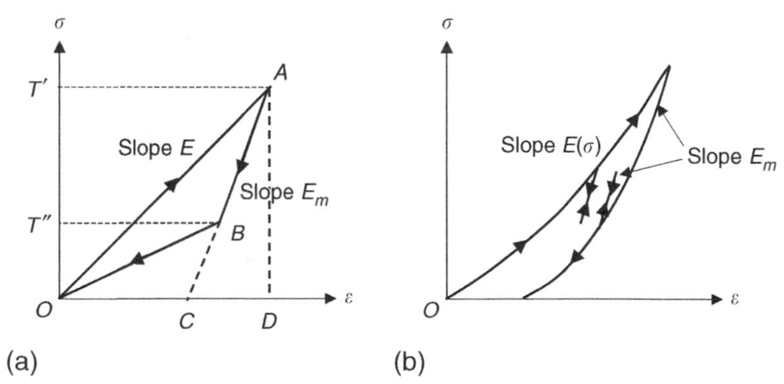

Fig. 10.6 (a) Uniaxial stress–strain diagram of a rock containing a single closed crack, loaded and unloaded along path *OABO*. (b) Uniaxial stress–strain diagram for a rock containing a distribution of cracks at different orientations, with some lateral confining stress (see text for details).

with the intrinsic elastic modulus, E_m. But as the load decreases, cracks with successively smaller values of β begin reverse sliding, according to (10.92), and so the modulus continually decreases as the stress and strain return to zero. If there were a lateral confining stress in addition to the axial load, this would cause an additional frictional resistance along the crack faces, which would tend to inhibit reverse sliding. In this case, the axial strain would not return to zero when the axial load has been completely removed, and some residual strain would remain in the rock.

10.7 Griffith cracks and the Griffith locus

The analysis given thus far has assumed that cracks of a given fixed length are present in the rock. Griffith (1920,1924) used thermodynamic arguments to find a necessary criterion for a crack to grow due to an applied load. A key ingredient of his analysis is the recognition that, as a crack grows, energy is needed to create the new surface area. Another aspect of a thermodynamic approach is that it becomes necessary to consider, in addition to the rock itself, the agency that supplies the load to the rock.

Consider again a thin rock specimen of length L, width b, and thickness t, containing a thin crack of length $2c$ lying perpendicular to the side of length L. Imagine that a tensile load T is applied to this rock by a hanging mass m connected to the rock by a cord that passes over a frictionless pulley (Fig. 10.7a). As the rock slab is assumed to be very thin, we can use plane stress analysis for the deformation of the rock. This system is assumed to be under equilibrium, with the crack having length c. The total energy of this thermodynamic system consists of the elastic strain energy in the rock, the surface energy of the crack, and the gravitational potential energy of the mass.

Now imagine that the crack grows from half-length c to some new half-length, $c + \delta c$, while the load is maintained constant. According to Griffith, this change will only be thermodynamically possible if extension of the crack allows the total energy of the system to *decrease*. (Alternatively, as in §6.5, one could include an additional term in the energy balance, representing energy that is available to

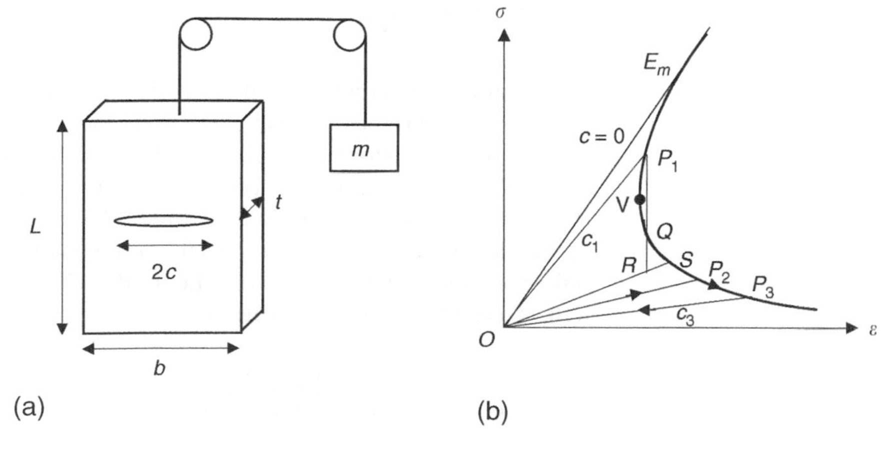

Fig. 10.7 (a) Crack extended under constant applied load, used in the derivation of the Griffith criterion. (b) Stress–strain diagram for a rock containing cracks of initial length c, showing the initial linear behavior, and the Griffith locus (see text for details.)

cause unstable crack growth with its associated seismic energy, and look for the condition such that this excess energy is *positive*.) With all of the other parameters fixed and c treated as a variable, we will see below that the total energy is an increasing function of c for c less than some critical value c^* and a decreasing function of c for $c > c^*$. Hence, the critical crack half-length can be found by maximizing the total energy as a function of c. The condition is therefore found by solving the equation

$$\left(\frac{\partial \mathcal{E}_{\text{total}}}{\partial c}\right)_T = \frac{d\mathcal{E}_{\text{elastic}}}{dc} + \frac{d\mathcal{E}_{\text{surface}}}{dc} + \frac{d\mathcal{E}_{\text{gravitational}}}{dc} = 0. \tag{10.93}$$

The elastic strain energy stored in the rock is given, from (10.74), by $T^2bLt/2E$, where E is the *effective* Young's modulus of the rock. Within the context of small-strain elasticity, this energy does not vary with L, so it is convenient to use the unstrained length, L_0, in which case the elastic energy can be written as $T^2bL_0t/2E$. The effective modulus of a rock containing a single crack lying normal to the loading direction is found by setting $\beta = 0$ in (10.74) and (10.76), to yield $E = E_m/[1 + (2\pi c^2/Lb)]$. If the initial crack is assumed to be small compared to the specimen dimensions, this relation can be approximated by $E = E_m[1 - (2\pi c^2/Lb)]$. Hence, the first derivative on the right side of (10.93) can be calculated as follows:

$$\begin{aligned}\frac{d\mathcal{E}_{\text{elastic}}}{dc} &= \frac{d\mathcal{E}_{\text{elastic}}}{dE}\frac{dE}{dc} = \left(\frac{-T^2bL_0t}{2E^2}\right)\left(\frac{-4\pi E_m c}{bL_0}\right) \\ &= \frac{2\pi T^2 tE_m c}{E^2} \approx \frac{2\pi T^2 tc}{E_m},\end{aligned} \tag{10.94}$$

where in the last step, we use the fact that E is only slightly less than E_m.

Next, consider the surface energy term. The surface energy of any cavity having surface area A is equal to γA, where γ is the surface energy per unit area. Hence, the surface area of a crack of length $2c$ that passes through the entire thickness t of the rock slab is $4\gamma ct$. The second derivative term on the right of (10.93) is therefore simply given by

$$\frac{d\mathcal{E}_{\text{surface}}}{dc} = 4\gamma t. \tag{10.95}$$

The gravitational potential energy of the mass that supplies the load to the rock is mgh, where h is the elevation of the mass above some datum. This elevation h is related to the deformation of the rock by the fact that if the rock expands by some amount δL, the mass will descend by the same amount. The instantaneous length L of the specimen is given by $L = L_0(1 + \varepsilon) = L_0[1 + (T/E)]$, where it is convenient here to abandon the usual "compression = positive" convention and take the tensile strain as positive. The third derivative term on the right side of (10.94) can therefore be calculated as

$$\frac{d\mathcal{E}_{\text{grav}}}{dc} = \left(\frac{d\mathcal{E}_{\text{grav}}}{dh}\right)\left(\frac{dh}{dL}\right)\left(\frac{dL}{dE}\right)\left(\frac{dE}{dc}\right)$$

$$= (mg)\,(-1)\left(\frac{-L_0 T}{E^2}\right)\left(\frac{-E_m 4\pi c}{bL_0}\right)$$
$$= -\frac{mgE_m 4\pi cT}{bE^2} \approx -\frac{4\pi cTmg}{bE_m}. \tag{10.96}$$

The applied load tensile stress T is given by $T = mg/bt$, so (10.96) takes the form

$$\frac{d\mathcal{E}_{\text{grav}}}{dc} = -\frac{4\pi cT^2 bt}{bE_m} = -\frac{4\pi cT^2 t}{E_m}. \tag{10.97}$$

Combining (10.93), (10.94), (10.95), and (10.97) yields

$$\frac{2\pi T^2 tc}{E_m} + 4\gamma t - \frac{4\pi T^2 tc}{E_m} = 4\gamma t - \frac{2\pi T^2 tc}{E_m} = 0. \tag{10.98}$$

This derivative of the total energy of the system is positive for small values of c and negative for large values of c. Equation (10.98) can be solved for the critical crack length c^* above which cracks are thermodynamically free to grow at fixed tensile stress T:

$$c^* = \frac{2\gamma E_m}{\pi T^2}. \tag{10.99}$$

Alternatively, we can consider that the load increases quasi-statically and solve (10.98) for the tensile stress T^* at which a crack of initial length c will be able to grow:

$$T^* = \sqrt{\frac{2\gamma E_m}{\pi c}}. \tag{10.100}$$

A similar analysis for plane strain conditions yields the criterion

$$T^* = \sqrt{\frac{2\gamma\, E_m}{\pi(1 - \nu_m^2)c}}, \tag{10.101}$$

whereas for a three-dimensional penny-shaped crack, the result is

$$T^* = \sqrt{\frac{\pi\gamma\, E_m}{4(1 - \nu_m^2)c}}. \tag{10.102}$$

These three expressions differ only by multiplicative factors on the order of unity. They each show that small cracks will require relatively large stresses to allow crack growth, whereas large cracks can grow under the application of smaller stresses.

Griffith's criterion (10.100) is a *necessary* condition that must be satisfied for crack growth to be thermodynamically possible, but in itself is not *sufficient* to cause growth to occur. The sufficient condition presumably is that the tensile stress at the crack tip must be large enough to break the atomic bonds at the crack tip. But as the stress concentration at the tip of an elliptical crack of small aspect ratio is quite large (§8.10), this latter condition will in practice always be satisfied if Griffith's criterion is satisfied.

A similar analysis can be carried out under the assumption that the rock is subject to fixed displacement rather than fixed stress at its outer boundary. The elastic energy stored in the rock is again equal to $T^2bL_0t/2E$. Expressed in terms of the tensile strain, the elastic energy is $bL_0tE\varepsilon^2/2$. The surface energy of the crack is again $4\gamma ct$. As the strain is assumed to be constant, the external loads applied to the rock do not undergo any motion, and consequently the energy of the loading agency does not change. The derivative of the total energy can therefore be calculated as follows:

$$\left(\frac{\partial \mathcal{E}_{\text{total}}}{\partial c}\right)_\varepsilon = \frac{d\mathcal{E}_{\text{elastic}}}{dc} + \frac{d\mathcal{E}_{\text{surface}}}{dc} = \left(\frac{d\mathcal{E}_{\text{elastic}}}{dE}\right)\left(\frac{dE}{dc}\right) + \frac{d\mathcal{E}_{\text{surface}}}{dc}$$

$$= \left(\frac{bL_0t\varepsilon^2}{2}\right)\left(\frac{-4\pi cE_m}{bL_0}\right) + 4\gamma t = -2\pi tcE_m\varepsilon^2 + 4\gamma t. \quad (10.103)$$

Setting this derivative to zero yields $\varepsilon^2 = 2\gamma/\pi E_m c$. Reexpressing this condition in terms of stress rather than strain, and using the approximation that $E \approx E_m$ for a small crack, leads again to a critical stress value of $T^* = \sqrt{2\gamma E_m/\pi c}$, which agrees with the result (10.100) obtained under the assumption of constant load.

Now imagine that this rock slab contains N isolated horizontal cracks, each of half-length c. If this slab is subjected to a slowly increasing tensile stress, T, it will at first deform as a linear elastic body (Fig. 10.7) with an effective modulus given by

$$E = E_m/[1 + (2\pi Nc^2/Lb)]. \quad (10.104)$$

When the applied stress reaches the critical value T^*, as given by (10.100), the cracks will begin to extend. The relationship between the stress and the strain at the point of incipient crack extension can be found by using (10.100) and (10.104) to eliminate c from the expression $T = E\varepsilon$, to yield

$$\varepsilon^* = \frac{T^*}{E_m} + \frac{8\gamma^2NE_m}{\pi Lb(T^*)^3}. \quad (10.105)$$

This curve, known as the Griffith locus (Berry, 1960b; Cook, 1965), demarcates in stress–strain space the boundary of the region of linear elastic behavior (Fig. 10.7).

A rock that is filled with small cracks, say of half-length c_1, will initially deform with a modulus only slightly less than E_m and will intersect the Griffith locus at a point such as P_1. For larger values of the initial crack length, say $c_2 > c_1$, the initial slope will be lower, and the stress–strain curve will intersect the Griffith locus at a point farther along, such as P_2. If the rock is deformed slowly, under conditions of controlled strain, its path in stress–strain space will be able to follow the Griffith locus, with increasing strain and decreasing stress, from P_2 out to, say, P_3. Unloading from P_3 would occur along a straight line, back to the origin, with a slope given by the effective modulus that corresponds to the new crack length, c_3. The area enclosed by OP_2P_3O represents the work that was done to extend the crack from c_2 to c_3.

This type of stable, controlled deformation of the rock would not be possible if the Griffith locus were first reached at a point such as P_1, however, since in this region crack extension along the locus is associated with a decrease in *both* the stress and the strain. The critical crack length c_c that separates these two branches of the Griffith locus corresponds to the point V (large dot in Fig. 10.7b) at which $d\varepsilon^*/dT^* = 0$, which, from (10.105) and (10.100), is found to be given by $c_c = N\sqrt{bL/6}$. Under conditions of controlled strain, the stress–strain state of the rock would drop down from P_1 to R, corresponding to some crack length $c_R > c_c > c_1$, after which it could proceed along the linear stress–strain curve from R to S, etc. Berry (1960b) showed that point R, and hence the crack length c_R, is determined by the condition that the area bounded by P_1VQP_1 is equal to that bounded by $QRSQ$.

10.8 Linear elastic fracture mechanics

The thermodynamic ideas of the previous section can be combined with the concept of stress intensity factors, defined in §8.10, to express the conditions for crack extension in terms of a critical value of the stress intensity factor at the crack tip. These ideas lead to the theory of *linear elastic fracture mechanics*, which is described in a large number of monographs (Lawn and Wilshaw, 1975; Broek, 1986; Kanninen and Popelar, 1986; Lawn, 1993). A brief introduction to the main ideas of this theory now follows.

It was shown in §10.7 that a crack can extend only if the energy released by this process is at least large enough to supply the surface energy required to form the new surface area of the crack faces. To be concrete, consider again a thin plate of thickness t, under conditions of plane stress. For a thin crack of length $2c$ to extend by an amount δc in each direction, energy in the amount $\delta \mathcal{E}_{\text{surf}} = 4\gamma t\delta c$ must be supplied, where γ is the surface energy per unit area. If the far-field stress is held constant, this energy is supplied by the decrease in the potential energy of the loading system, as shown in (10.97). If the far-field displacement is held constant, this energy is supplied by a decrease in the stored elastic strain energy of the body, as shown in (10.103). In either case, the energy released by the combined rock and loading agency is equal to

$$\frac{\delta \mathcal{E}_{\text{released}}}{\delta c} = \frac{2\pi T^2 tc}{E}, \tag{10.106}$$

where T is the applied tensile stress, and E is the Young's modulus of the rock. Equating these two terms leads to the criterion (10.100), which states that the crack can extend if T reaches the critical value given by

$$T_c = \sqrt{\frac{2\gamma E}{\pi c}}. \tag{10.107}$$

These results, expressed here in terms of the far-field load, can also be expressed in terms of the conditions that exist at the crack tip. Using definition (8.262) for the mode I (tensile) stress intensity factor, $K_{\text{I}} = T\sqrt{\pi c}$, the energy released by the extension of the crack becomes

$$\frac{\delta \mathcal{E}_{\text{released}}}{\delta c} = \frac{2K_{\text{I}}^2 t}{E}. \tag{10.108}$$

Half of this energy can be identified as the energy released as the right tip of the crack extends to the right and the other half attributed to the extension of the left tip. It is conventional to then define the *energy release rate*, $\mathcal{G}$, as the energy released per unit thickness of the plate, as *one* of the crack tips extends; this definition allows the concept to apply to situations such as a half-crack that intersects a free boundary or emanates from a circular void, for example. Hence, the energy release rate for an isolated crack under mode I loading in plane stress conditions is

$$\mathcal{G} \equiv \frac{\delta \mathcal{E}_{\text{released}}}{\delta c} = \frac{K_{\text{I}}^2}{E}. \tag{10.109}$$

The necessary thermodynamic criterion for crack extension, (10.107), can be expressed in terms of a critical *stress intensity factor*, K_{Ic}, or a *critical value of the energy release rate*, $\mathcal{G}_c$, by substituting $K_{\text{I}} = T\sqrt{\pi c}$ into (10.107):

$$K_{\text{Ic}} = \sqrt{2\gamma E}, \quad \mathcal{G}_c = \frac{K_{\text{Ic}}^2}{E} = 2\gamma. \tag{10.110}$$

Hence, in order for the crack to grow, the stress intensity factor K_{I} must reach the critical value K_{Ic}, or, equivalently, the energy release rate $\mathcal{G}$ must reach the critical value $\mathcal{G}_c = K_{\text{Ic}}^2/E$. According to Griffith's analysis, the critical energy release rate is simply equal to twice the surface energy (since two new crack face elements are created). Irwin (1958) extended Griffith's concept by pointing out that in many materials, as a crack grows, energy must also be expended to create a damaged zone of irreversible, plastic deformation ahead of the crack tip. In rock, this zone may consist of crushed grains, microcracking, etc. Irwin proposed to add these energy terms to the surface energy term, calling the new term Γ, the *fracture toughness*. This has the effect of replacing γ with Γ in criterion (10.110).

Although the excess elastic strain energy associated with a crack is of course localized around the crack, it is by no means entirely located near the crack *tip*. Hence, it may seem peculiar that the energy release accompanying crack extension can be related to a parameter, namely the stress intensity factor, which has relevance only *at* the crack tip. The following derivation of (10.109) shows more explicitly why $\mathcal{G}$ can be expressed so succinctly in terms of K_{I}.

As mentioned above, the energy release rate does not depend on whether the boundary conditions for the plate are constant traction or constant displacement. So, for simplicity we consider constant displacement boundary conditions. In this case, it is clear that no work is done by the loading agency at the outer boundary of the rock. Furthermore, as the faces of the existing crack are traction-free, no work is done there. Hence, the energy released at the right edge of the crack, per unit thickness in the third direction, can be calculated from the following

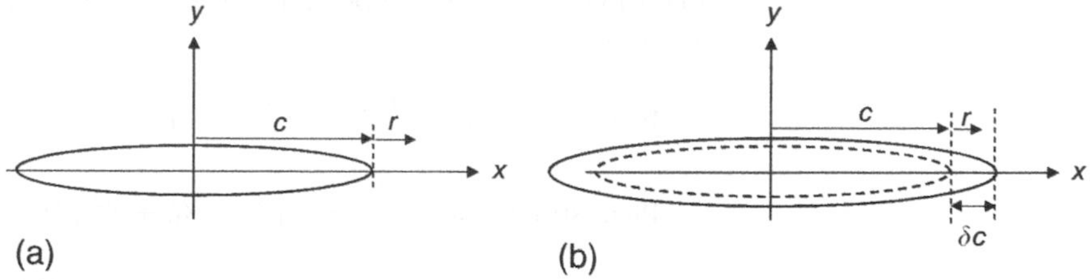

Fig. 10.8 Crack geometry and definition of variables, (a) before crack extension, and (b) after crack extension. In (b), the dotted ellipse represents the crack shape before extension, and the solid ellipse represents the crack after it has grown by an amount δc at each edge.

integral (Lawn and Wilshaw, 1975, p. 56):

$$\mathcal{G} = 2 \int_{c}^{c+\delta c} \frac{1}{2} \tau_{yy} v \, dx, \tag{10.111}$$

where the factor of 2 outside the integral accounts for the upper and lower faces of the newly created portion of the crack.

The traction τ_{yy} used in (10.111) must be that which exists ahead of the crack tip, before the stresses are relieved. This traction is calculated from (8.262), with $\theta = 0$, and $x = c + r$ (Fig. 10.8a), that is, $\tau_{yy} = K_{\mathrm{I}}/(2\pi r)^{1/2}$. The displacement v in (10.111) must be the displacement undergone by the element of rock that initially lies on the x-axis, but now forms part of the newly exposed crack face (Fig. 10.8b). This displacement can, for plane stress, be calculated from (8.230), with c replaced by the new coordinate of the right crack tip, $c + \delta c$. This gives $v = 2T[(c + \delta c)^2 - x^2]^{1/2}/E$. Using the relation $x = c + r$ and the definition $K_{\mathrm{I}} = T\sqrt{\pi c}$, (10.111) takes the form

$$\mathcal{G} = \frac{2K_{\mathrm{I}}^2}{\pi E \sqrt{2c}} \int_{c}^{c+\delta c} \frac{[(c+\delta c)^2 - (c+r)^2]^{1/2}}{\sqrt{r}} \mathrm{d}x. \tag{10.112}$$

Both δc and r are very small compared to c, so the numerator of the integrand can be expressed as $[2c(\delta c - r)]^{1/2}$. Next, we change the variable of integration from x to r, noting that $\mathrm{d}x = \mathrm{d}r$ and r varies from 0 to δc, to arrive at

$$\mathcal{G} = \frac{2K_{\mathrm{I}}^2}{\pi E} \int_{0}^{\delta c} \frac{\sqrt{\delta c - r}}{\sqrt{r}} \mathrm{d}r. \tag{10.113}$$

Finally, we let $r = (\delta c) \sin^2 \vartheta$, which transforms (10.113) into

$$\mathcal{G} = \frac{2K_{\mathrm{I}}^2}{\pi E} \int_{0}^{\pi/2} \frac{\cos \vartheta}{\sin \vartheta} 2 \sin \vartheta \cos \vartheta \, \mathrm{d}\vartheta = \frac{4K_{\mathrm{I}}^2}{\pi E} \int_{0}^{\pi/2} \cos^2 \vartheta \, \mathrm{d}\vartheta = \frac{K_{\mathrm{I}}^2}{E}, \tag{10.114}$$

which agrees with (10.109).

A similar analysis of the other two crack growth modes leads to the following expressions for the energy release rate, in which the energy release rates for the three modes are taken to be additive (Lawn and Wilshaw, 1975, p. 57; Maugis, 2000, p. 184):

$$\text{Plane stress:}\quad \mathcal{G} = \frac{1}{E}[K_{\mathrm{I}}^2 + K_{\mathrm{II}}^2 + (1+\nu)K_{\mathrm{III}}^2], \tag{10.115}$$

$$\text{Plane strain:}\quad \mathcal{G} = \frac{(1-\nu^2)}{E}[K_{\mathrm{I}}^2 + K_{\mathrm{II}}^2 + K_{\mathrm{III}}^2/(1-\nu)]. \tag{10.116}$$

These results assume that the crack grows by extending itself in its own plane, that is, along the x-axis in Fig. 10.8, regardless of the mode of loading to which it is subjected. Although expressions (10.115) and (10.116) are quoted in many texts and monographs, it must be noted that there is little if any evidence that cracks in isotropic rocks extend in their own plane under shear loading (see Scholz, 1990, p. 26). In fact, there is evidence that shear loading will create new microcracks along the boundary of the existing crack, which then propagate in mode I (Brace and Bombolakis, 1963; Cox and Scholz, 1988).

According to the formalism of linear elastic fracture mechanics, the problem of crack growth therefore involves two main aspects: calculating the stress intensity factor, which can be achieved by solving the elasticity equations for the given crack geometry and loading conditions, and having knowledge of the critical energy release rate for the rock in question, which can be obtained experimentally. Calculated stress intensity factors for many crack geometries have been compiled by Tada et al. (2000), among others. Aside from a geometry-dependent multiplicative factor, usually on the order of unity, the stress intensity factor is typically equal to the applied far-field stress divided by the square root of a suitable crack dimension. Measured values of the critical stress intensity factor $K_{\mathrm{I}c}$ and critical energy release rate $\mathcal{G}_c$ for various rocks and minerals have been compiled by Atkinson and Meredith (1987). Values of $K_{\mathrm{I}c}$ at room temperature vary from about 0.1 MPa $\mathrm{m}^{1/2}$ for coal, up to about 3.5 MPa $\mathrm{m}^{1/2}$ for harder rocks such as granite or dunite.

10.9 Griffith theory of failure

The theory of fracture mechanics briefly outlined in the previous section applies to cracked bodies subjected to tensile loading. Griffith (1924) used the conceptual model of a two-dimensional rock containing a collection of randomly oriented thin elliptical cracks to derive a failure criterion that applies under tensile or compressive loads. This simple model leads to a curved, nonlinear failure surface in Mohr space that is in several respects more realistic than the linear Coulomb law. The only requirements of this derivation are the expressions for the stresses along the boundary of the crack, which were obtained in §8.9, and the assumption that a crack will extend in an unstable manner when the maximum *tensile* stress at any point along the crack boundary reaches some value that is characteristic of the rock. Note that although both the following analysis and the energy-related considerations of §10.8 are due to Griffith, the present model for failure under compressive loads is fundamentally different from that discussed in the previous sections, in that it assumes a "critical stress" criterion for crack propagation, rather than a criterion based on energy release.

Consider, as in §8.9, a flat elliptical crack, with the coordinate system aligned so that the x-axis lies along the major axis of the ellipse. The semiaxes of the crack are, from (8.204),

$$a = c \cosh \xi_o, \quad b = c \sinh \xi_o. \tag{10.117}$$

Let the minor principal stresses σ_2 be oriented at an angle β to the x-axis, and the major principal stress σ_1 be oriented at an angle $(\pi/2) + \beta$ to the x-axis. In the x–y coordinate system, these far-field stresses are

$$\tau_{xx} = \sigma_1 \sin^2 \beta + \sigma_2 \cos^2 \beta, \quad \tau_{yy} = \sigma_1 \cos^2 \beta + \sigma_2 \sin^2 \beta, \tag{10.118}$$

$$\tau_{xy} = -\frac{1}{2}(\sigma_1 - \sigma_2) \sin 2\beta. \tag{10.119}$$

From (8.210), the tangential stress along the edge of the hole is

$$\tau_{\eta\eta} = \frac{(\sigma_1 + \sigma_2) \sinh 2\xi_o + (\sigma_1 - \sigma_2)[e^{2\xi_o} \cos 2(\beta - \eta) - \cos 2\beta]}{\cosh 2\xi_o - \cos 2\eta}, \tag{10.120}$$

or, in terms of the stresses (10.118) and (10.119) in the coordinate system aligned with the crack,

$$\tau_{\eta\eta} = \frac{2\tau_{yy} \sinh 2\xi_o + 2\tau_{xy}[(1 + \sinh 2\xi_o) \cot 2\beta - e^{2\xi_o} \cos 2(\beta - \eta) \operatorname{cosec} 2\beta]}{\cosh 2\xi_o - \cos 2\eta}. \tag{10.121}$$

For thin cracks, ξ_o will be small. The maximum tensile stress is expected to occur near the sharp tip of the crack, where η is small. Hence, we expand (10.121) for small values of ξ_o and η, to arrive at

$$\tau_{\eta\eta} = \frac{2(\xi_o \tau_{yy} - \eta \tau_{xy})}{\xi_o^2 - \eta^2}. \tag{10.122}$$

For a given crack in a given stress field, we now calculate the location along the crack boundary at which this stress attains its smallest (i.e., most tensile) value. Differentiating (10.122) with respect to η, we find that $d\tau_{\eta\eta}/d\eta = 0$ implies

$$\tau_{xy}\eta^2 - (2\xi_o \tau_{yy})\eta - \xi_o^2 \tau_{xy} = 0, \tag{10.123}$$

which is a quadratic equation for η. The two solutions are

$$\eta = \xi_o[\tau_{yy} \pm (\tau_{yy}^2 + \tau_{xy}^2)^{1/2}]/\tau_{xy}. \tag{10.124}$$

Substituting this value of η into (10.122) gives the two extreme values of the tangential stress:

$$\xi_o \tau_{\eta\eta} = \tau_{yy} \mp (\tau_{yy}^2 + \tau_{xy}^2)^{1/2}. \tag{10.125}$$

The negative sign in (10.125) corresponds to a tensile stress, so the greatest tensile stress along the crack boundary is given by

$$\begin{aligned}\xi_o \tau_{\eta\eta} &= \tau_{yy} - (\tau_{yy}^2 + \tau_{xy}^2)^{1/2} \\ &= (\sigma_1 \cos^2\beta + \sigma_2 \sin^2\beta) - (\sigma_1^2 \cos^2\beta + \sigma_2^2 \sin^2\beta)^{1/2}, \end{aligned} \tag{10.126}$$

and occurs when

$$\begin{aligned}\frac{\eta}{\xi_o} &= \frac{\tau_{yy} + (\tau_{yy}^2 + \tau_{xy}^2)^{1/2}}{\tau_{xy}} \\ &= \frac{2[(\sigma_1 \cos^2\beta + \sigma_2 \sin^2\beta) + (\sigma_1^2 \cos^2\beta + \sigma_2^2 \sin^2\beta)^{1/2}]}{(\sigma_2 - \sigma_1)\sin 2\beta}. \end{aligned} \tag{10.127}$$

Next, we determine the orientation of the crack that will give the largest (in magnitude) value of this tensile stress. Differentiating (10.127) with respect to β gives

$$\xi_o \frac{d\tau_{\eta\eta}}{d\beta} = \left[2(\sigma_2 - \sigma_1) + \frac{(\sigma_1^2 - \sigma_2^2)}{(\sigma_1^2 \cos^2\beta + \sigma_2^2 \sin^2\beta)^{1/2}}\right] \sin\beta \cos\beta. \tag{10.128}$$

This derivative will be zero if $\beta = 0$, $\beta = \pi/2$, or

$$\cos 2\beta = -(\sigma_1 - \sigma_2)/2(\sigma_1 + \sigma_2). \tag{10.129}$$

This inclined orientation will exist only if the right-hand side of (10.129) lies between -1 and $+1$, which requires that

$$\sigma_1 + 3\sigma_2 > 0 \text{ and } \sigma_2 + 3\sigma_1 > 0. \tag{10.130}$$

But $\sigma_1 > \sigma_2$ by definition, so the second inequality will hold whenever the first inequality holds. Assuming that the stresses satisfy $\sigma_1 + 3\sigma_2 > 0$, substitution of (10.129) into (10.126) gives the maximum tensile stress as

$$\tau_{\eta\eta} = \frac{-(\sigma_1 - \sigma_2)^2}{4(\sigma_1 + \sigma_2)\xi_o}. \tag{10.131}$$

Griffith then assumed that the crack would extend when this maximum tensile stress reaches some critical value that is characteristic of the rock. Furthermore, this crack extension is identified with "failure" of the rock. In this case, (10.131) provides a failure criterion in terms of the two principal stresses. However, it contains the aspect ratio parameter ξ_o, which will in practice be difficult to estimate. But both ξ_o and the critical (local) tensile stress can be eliminated from the failure criterion, as follows. Consider a combination of far-field stresses that do *not* satisfy condition (10.130). For this to be true, σ_2 must be negative, and it follows from (10.126) and (10.127), or directly from (10.120), that the greatest tensile stress along the crack surface occurs when $\beta = \pi/2$, and has the value

$$\tau_{\eta\eta}^{\text{crit}} = 2\sigma_2/\xi_o. \tag{10.132}$$

Uniaxial tension, in which $\sigma_1 = 0$ and $\sigma_2 < 0$, is one particular case for which (10.130) is not satisfied, and so (10.132) holds. The value of σ_2 at failure is, by definition, $-T_o$, so (10.132) shows that $\xi_o \tau_{\eta\eta}^{\text{crit}} = -2T_o$. Substitution of this result into (10.131), for the "predominantly compressive" stress regimes defined by $\sigma_1 + 3\sigma_2 > 0$, allows (10.131) to be expressed in a form that does not involve the parameters ξ_o or $\tau_{\eta\eta}^{\text{crit}}$:

$$(\sigma_1 - \sigma_2)^2 = 8T_o(\sigma_1 + \sigma_2), \quad \text{if } \sigma_1 + 3\sigma_2 > 0, \tag{10.133}$$

which, along with

$$\sigma_2 = -T_o, \quad \text{if } \sigma_1 + 3\sigma_2 < 0, \tag{10.134}$$

constitutes Griffith's criterion for failure. The only adjustable parameter that appears in this criterion is the uniaxial tensile strength, T_o.

This criterion is represented in the (σ_1, σ_2) plane (Fig. 10.9a) by the segment AC of the line $\sigma_2 = -T_o$ that extends from $\sigma_1 = -T_o$ to $\sigma_1 = 3T_o$ and the portion CDE of the parabola (10.133) that connects with the straight line at C. It can be shown from (10.133) that this composite curve has a continuous slope, as the parabola is indeed horizontal at C. The complete parabola (10.133) passes through the origin and is symmetric with respect to the line $\sigma_2 = \sigma_1$. Values above and to the left of this line have no physical meaning, since by definition $\sigma_1 \geq \sigma_2$.

For states of uniaxial compression, $\sigma_2 = 0$, and (10.133) shows that failure occurs at a value $\sigma_2 = 8T_o$, which by definition is the uniaxial compressive strength. This value of $C_o/T_o = 8$ for the ratio of compressive strength to tensile strength is reasonable, but somewhat lower than is observed for most rocks.

To display Griffith's criterion on a Mohr diagram, we first write (10.133) and (10.134) in terms of the mean normal stress, $\sigma_m = (\sigma_1 + \sigma_2)/2$, and the maximum shear stress, $\tau_m = (\sigma_1 - \sigma_2)/2$:

$$\tau_m^2 = 4T_o\sigma_m, \quad \text{if } 2\sigma_m > \tau_m, \tag{10.135}$$

$$\tau_m = \sigma_m + T_o, \quad \text{if } 2\sigma_m < \tau_m. \tag{10.136}$$

The point (σ_m, τ_m) represents the top of a Mohr's circle. The locus of these circle tops consists of dashed line AB with a slope of unity for $-T_o < \sigma_m < T_o$, and

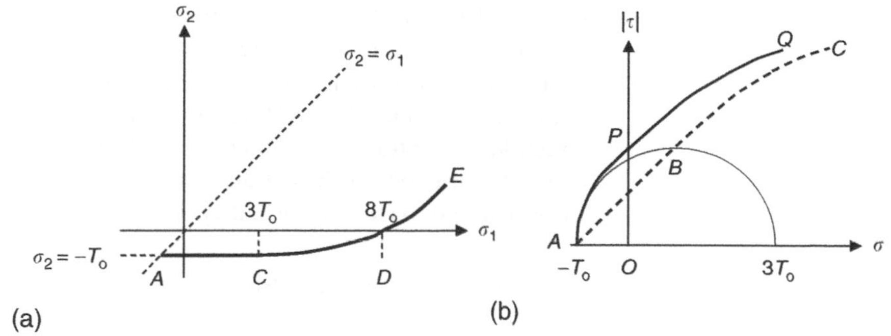

Fig. 10.9 Griffith's failure criterion (a) in principal stress space and (b) on a Mohr diagram.

the portion BC of the parabola (10.135) for $\sigma_m > T_o$ (Fig. 10.9b). The Mohr's circles whose tops lie on line AB all meet at A, where $\sigma_2 = -T_o$. Hence, A is the terminus of the Mohr envelope.

To find the Mohr envelope corresponding to the circles whose tops lie on parabola BC, we must find the envelope of the family of circles having centers located at $(\sigma_m, 0)$, and having radii τ_m, where σ_m and τ_m are related by (10.135). The equation of these circles can be written as

$$f(\sigma_m) = (\sigma - \sigma_m)^2 + \tau^2 - \tau_m^2 = (\sigma - \sigma_m)^2 + \tau^2 - 4T_o\sigma_m = 0, \qquad (10.137)$$

The envelope of this family of circles is found by eliminating σ_m between (10.137) and the equation for $\partial f/\partial\sigma_m = 0$, which is

$$\partial f/\partial\sigma_m = \sigma - \sigma_m + 2T_o = 0. \qquad (10.138)$$

The resulting equation for the Mohr envelope is

$$\tau^2 = 4T_o(\sigma + T_o), \qquad (10.139)$$

which is the parabola APQ.

Murrell (1963) extended Griffith's theory of failure into the true-triaxial domain in the following ad hoc manner. In this exercise, it is convenient to ignore the restriction $\sigma_1 \geq \sigma_2 \geq \sigma_3$ and consider the entire stress space. First, the parabolic failure surface in (σ_1, σ_2) space becomes a paraboloid of revolution in $(\sigma_1, \sigma_2, \sigma_3)$ space, with its axis taken to be the line $\sigma_1 = \sigma_2 = \sigma_3$. Next, since the two-dimensional Griffith criterion also included portions of the lines $\sigma_1 = \sigma_2 = -T_o$, in three dimensions the paraboloid is assumed terminate at the pyramid of the three mutually perpendicular planes $\sigma_1 = \sigma_2 = \sigma_3 = -T_o$. Lastly, in analogy with the fact that the Griffith parabola is tangent to the line $\sigma_2 = -T_o$, Murrell required that the paraboloid of revolution be tangent to the planes $\sigma_2 = \sigma_3 = -T_o$. These criteria uniquely determine the failure surface, which can be expressed (Murrell, 1963) as

$$(\sigma_2 - \sigma_3)^2 + (\sigma_3 - \sigma_1)^2 + (\sigma_1 - \sigma_2)^2 = 24T_o(\sigma_1 + \sigma_2 + \sigma_3), \qquad (10.140)$$

or, in terms of octahedral stresses,

$$\tau_{\text{oct}}^2 = 8T_o\sigma_{\text{oct}}. \qquad (10.141)$$

It follows from (10.140) that Murrell's theory predicts that the uniaxial compressive strength is given by $C_o = 12T_o$, which is to say that the ratio of compressive strength to tensile strength is predicted to have the value 12.

McClintock and Walsh (1962) extended Griffith's model, within a two-dimensional context, by including the effect of friction along the crack faces. If μ is the friction coefficient along the crack faces and σ_c is the compressive stress required to close a crack, they derive the following failure criterion:

$$\sigma_1[(1 + \mu^2)^{1/2} - \mu] - \sigma_2[(1 + \mu^2)^{1/2} + \mu] = 4T_o[1 + (\sigma_c/T_o)]^{1/2} - 2\mu\sigma_c. \qquad (10.142)$$

If σ_c is small enough to be neglected, this reduces to

$$\sigma_1[(1+\mu^2)^{1/2}-\mu]-\sigma_2[(1+\mu^2)^{1/2}+\mu]=4T_o, \qquad (10.143)$$

which is essentially the Coulomb criterion, (4.16), with S_o identified with $2T_o$.

Theories such as those of Griffith and Murrell provide a useful framework for qualitatively understanding the manner in which failure depends on the state of stress. However, it must be acknowledged (Scholz, 1990; Paterson and Wong, 2005) that the identification of rock *failure* with the extension of a *single, isolated crack* is a gross oversimplification of the complex processes involved in rock deformation.

Computational methods are obviously well suited for studying the process of rock degradation and failure, without requiring simplistic models such as that of a single crack in an infinite, homogeneous elastic medium. Paterson and Wong (2005, Chapter 6) and Yuan and Harrison (2006) have reviewed the numerous stochastic/statistical approaches that have been taken, from which the following small sampling is extracted. Scholz (1968) imagined that a rock sample could be divided into a number of small elements, within each of which the local stress and local strength were distributed about their mean values. Individual elements were assumed to fail when the local stress reaches the local strength, and this failed region propagates to an adjacent element if that element has the appropriate combination of higher stress and/or lower strength. As the stress is increased incrementally, the entire rock specimen may eventually fail, as the local regions of failed rock percolate throughout the body. Allègre et al. (1982) considered an array of eight cubical elements, each of which is either intact or already "failed," with probabilities p and $1-p$, respectively. By considering all possible topological combinations, they calculated the probability of the 8-cube array to be in the failed state, and then grouped this macrocube with seven adjacent macrocubes, continuing the process (the so-called "renormalization group" approach) to find that the entire sample will be in a macroscopic "failed" state if $p > 0.896$. Madden (1983) extended this approach by relating the probability of failure p to the local crack density. Lockner and Madden (1991) assumed that each failed element contained a microcrack and accounted for possible crack closure and frictional sliding. Their model was able to predict the development of dilatancy and shear localization.

Blair and Cook (1998) developed a lattice-based model in which the local elements are allowed to fail by tensile cracking if the local tensile stress exceeds the local tensile strength, which was assumed to be distributed either uniformly over a certain range, or bimodally. They found that the locations of cracked elements are initially random, but eventually these failed elements coalesce to form a macroscopic fracture, consistent with experimental observation (Fig. 4.8). Reuschlé (1998) used a similar network model to study the influence of heterogeneity and loading conditions on the fracture process. These types of network models can also be used to study other inelastic deformation processes, such as unstable fault slip (Hazzard et al., 2002), or the compaction bands that occur in some sedimentary rocks (Katsman et al., 2005).

Tang et al. (2000) developed a finite element code in which the individual rock elements are represented by an elastic–brittle constitutive model with zero residual strength. The elastic modulus and strength of the individual elements were assumed to follow a Weibull (1951) distribution. Their model was able to predict many of the features of rock deformation, both pre- and post-failure. Fang and Harrison (2002) developed a similar model that also accounted for the strengthening effect that confining pressure has on the individual elements, and their model was able to show the transition between brittle and ductile behavior as the confining pressure is increased. Yuan and Harrison (2005) extended this model to the study of hydromechanical coupling by incorporating a relationship between local damage and local hydraulic conductivity.

11 Wave propagation in rocks

11.1 Introduction

It is usually assumed that rock is at rest under the action of static stresses, and most problems in rock mechanics are treated as problems of statics. However, there are a number of important situations in which the stresses are of a dynamic nature and the propagation of these stresses through the rock as a *wave* must be considered. Such situations may arise naturally, for example, in earthquakes. Dynamic stresses in rock may also be the result of man-made activities, such as explosive blasting. In other cases they are inadvertent consequences of human activities, such as the rockbursts that occur in underground mines due to the redistribution of stresses caused by the excavations. The amplitudes of the dynamic stresses are usually small compared to the compressive strength of the rock, except perhaps in the immediate vicinity of the source, and the time of application is generally short. In such situations, the resulting stresses and displacements can be analyzed using the dynamic theory of linear elasticity. Waves traveling through rock, governed by the laws of linear elasticity, are known as *seismic waves*.

Seismic waves originating from earthquakes are used to locate the earthquake foci and to study the mechanism at the source (Aki and Richards, 1980). Seismic waves from man-made explosive sources are used to study the near-surface structure of the earth, particularly for the purpose of locating minerals and hydrocarbons (Sheriff and Geldart, 1995). In mining, man-made seismic waves are used to assess the quality of the rock around excavations (Gibowicz and Kijko, 1994; Falls and Young, 1998).

Much of this chapter is concerned with elastic wave propagation, in which the stress–strain behavior is governed by the equations of linear elasticity. The basic theory of one-dimensional elastic wave propagation, including reflection and refraction at an interface, is treated in §11.2. The basic theory and relevant definitions for harmonic waves are discussed in §11.3. The propagation of harmonic elastic waves in unbounded, three-dimensional regions is discussed in §11.4. Reflection and refraction of such waves at an interface between two rock types are treated in §11.5, and harmonic waves propagating along an interface between two rock layers or along a free surface are discussed in §11.6. Transient waves traveling unidirectionally, or radially from a spherical source, are treated briefly in §11.7. The effect of pore fluids on the propagation of waves in porous rock is discussed in §11.8. Various mechanisms of attenuation, in which mechanical energy is lost as the wave travels through the rock, are discussed in §11.9. Lastly, inelastic waves are briefly discussed in §11.10.

11.2 One-dimensional elastic wave propagation

The propagation of elastic waves through rock is governed by the three-dimensional equations of elasticity, (5.76), with the inertia terms retained on the right-hand side. If these equations are supplemented with Hooke's law, for example, in the form given by (5.18)–(5.21) for isotropic rocks, they form a complete set of equations whose solutions describe transient propagation of stress/strain waves through a rock. Although elastic waves often travel in a unidirectional manner, the coupling between the different normal stresses and strains due to the Poisson effect causes the motion to never be truly one-dimensional in a mathematical sense. Hence, the analysis of elastic waves becomes mathematically complex. However, many of the concepts of elastic wave propagation can be understood within the context of a simplified, one-dimensional model. This model, which is developed and discussed below, actually applies rigorously to waves propagating along a thin elastic bar.

Imagine a thin elastic rod of a given cross-sectional shape that is uniform along the length of the rod. The axial coordinate is x. The precise shape of the cross section is not relevant to the following analysis, but it may be convenient to think of it as circular, with radius a. If the rod is thin, it seems reasonable to assume that the stress τ_{xx} will not vary over the cross section; hence, τ_{xx} varies only with x and t. (This assumption is true for waves whose wavelengths are greater than about ten times the radius of the rod; Graff, 1975, p. 471). If the rod is acted upon only by longitudinal forces in the x direction, and its outer boundary is traction-free, then the only nonzero stresses within the rod will be τ_{xx}. A force balance in the x direction taken on the infinitesimal segment of rod between x and $x + \Delta x$ then yields

$$[\tau_{xx}(x,t) - \tau_{xx}(x+\Delta x,t)]A = -\rho(A\Delta x)\frac{\partial^2 u}{\partial t^2}, \tag{11.1}$$

where the – sign appears on the right because, as explained in §2.10, the displacement u is reckoned positive if the particle displaces in the negative x direction. Dividing both sides of (11.1) by $A\Delta x$, and taking the limit as $\Delta x \to 0$, yields

$$\frac{\partial \tau_{xx}}{\partial x} = \rho\frac{\partial^2 u}{\partial t^2}. \tag{11.2}$$

This equation is independent of the constitutive equation that applies to the rock. Under the assumption of linear elastic behavior, and noting that there are no forces acting in the directions perpendicular to the x-axis, then $\tau_{xx} = E\varepsilon_{xx} = E(\partial u/\partial x)$, and (11.2) becomes

$$\frac{E}{\rho}\frac{\partial^2 u}{\partial x^2} = \frac{\partial^2 u}{\partial t^2}. \tag{11.3}$$

Differentiation of both sides of (11.3) with respect to x, and invoking the fact that the various partial differentiation operators commute with each other, shows that both the strain ε_{xx} and the stress τ_{xx} also satisfy this same differential equation:

$$\frac{E}{\rho}\frac{\partial^2 \varepsilon_{xx}}{\partial x^2} = \frac{\partial^2 \varepsilon_{xx}}{\partial t^2}, \quad \frac{E}{\rho}\frac{\partial^2 \tau_{xx}}{\partial x^2} = \frac{\partial^2 \tau_{xx}}{\partial t^2}. \tag{11.4}$$

Equation (11.3) is the one-dimensional wave equation. It describes disturbances that propagate along the bar, in either the $+x$ or $-x$ direction, at a speed c that is given by

$$c = (E/\rho)^{1/2}. \tag{11.5}$$

The wave-like nature of the solutions to (11.3) is most easily seen by utilizing the following analysis, first given by the French mathematician and philosopher Jean d'Alembert in 1747. Consider any differentiable function f of one variable, and let the argument of f be the new variable $\eta = x - ct$, where c is given by (11.5). Use of the chain rule gives

$$\frac{\partial f}{\partial x} = \frac{df}{d\eta}\frac{\partial \eta}{\partial x} = \frac{df}{d\eta}, \quad \text{so } \frac{\partial^2 f}{\partial x^2} = \frac{d^2 f}{d\eta^2}, \tag{11.6}$$

$$\frac{\partial f}{\partial t} = \frac{df}{d\eta}\frac{\partial \eta}{\partial t} = -c\frac{df}{d\eta}, \quad \text{so } \frac{\partial^2 f}{\partial t^2} = c^2\frac{d^2 f}{d\eta^2} = \frac{E}{\rho}\frac{d^2 f}{d\eta^2}, \tag{11.7}$$

which shows that the function $f(x - ct)$ satisfies (11.3). Hence, any differentiable function that depends on the two variables x and t only through the combination $x - ct$ will be a solution to the wave equation (11.3).

The function $f(x - ct)$, which according to (11.3) and (11.4) may stand for displacement, stress or strain, represents a disturbance that moves to the right at speed c. To be concrete, consider the peak of the disturbance shown in Fig. 11.1, which we take to be located at $x = x_0$ at time $t = 0$. The magnitude of the disturbance at this peak is given by $f(\eta = x_0 - c0) = f(x_0)$. At some time $t > 0$ later, this peak will move to a location at which the variable η has the same value as it did at $t = 0$, that is, $x' - ct = x_0$, or

$$x' = x_0 + ct, \quad \text{so } \left(\frac{\partial x}{\partial t}\right)_{\eta=\text{constant}} = c. \tag{11.8}$$

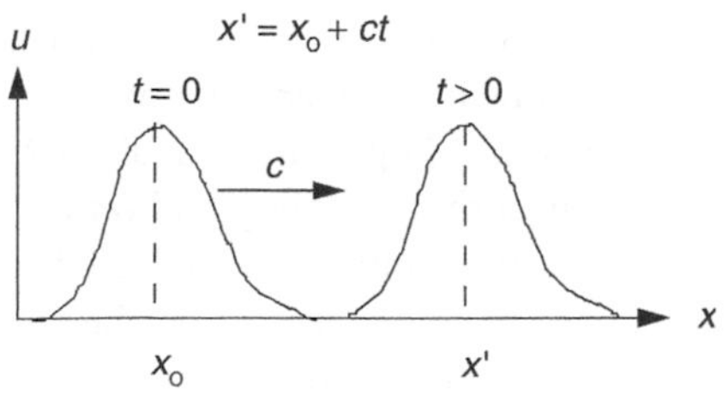

Fig. 11.1 Elastic disturbance moving to the right at velocity c. During a time increment t the pulse moves to the right by a distance ct, without altering its shape.

This argument holds for any point on the wave, not just the peak, so the solution $f(x - ct)$ represents a disturbance propagating, without distortion, to the right at speed c. The waveform shown in Fig. 11.1 can be interpreted as the graph of the displacement as a function of x, for a fixed value of t, but it can also be thought of as the time variation (with a scaling factor c) of the displacement at a fixed location x. The variable η is sometimes called the *phase* of the wave and c is the *phase velocity*.

An identical analysis, using the variable $\xi = x + ct$, shows that $g(x + ct)$ is also a solution to the wave equation for an arbitrary twice-differentiable function g and represents a disturbance moving, without distortion, to the *left* at speed c. It can be shown (Bers et al., 1964) that the general solution to (11.3) is given by

$$u(x, t) = f(x - ct) + g(x + ct), \tag{11.9}$$

in the sense that any solution to (11.3) can be written in this form.

Imagine a thin elastic bar extending infinitely far in both directions. At time $t = 0$, assume that the displacement and velocity of each point along the bar are each given by some known function, that is,

$$u(x, t = 0) = U(x), \quad \frac{\partial u}{\partial t}(x, t = 0) = V(x). \tag{11.10}$$

Equation (11.3), along with the two initial conditions given in (11.10), forms a well-posed initial value problem whose solution can be shown to be (Pearson, 1959, p. 179; Fetter and Walecka, 1980, p. 213)

$$u(x, t) = \frac{1}{2}[U(x - ct) + U(x + ct)] + \frac{1}{2c}\int_{x-ct}^{x+ct} V(s)\,ds. \tag{11.11}$$

Hence, the initial disturbance $U(x)$ splits into two parts, half propagating to the left and half to the right. The influence of the initial velocity $V(x)$ on the resulting wave, which is given by the integral term, is not as easily visualized, although it can also be written in terms of left-traveling and right-traveling waves by defining a function H that is the indefinite integral of V (Graff, 1975, p. 15):

$$\int_{x-ct}^{x+ct} V(s)\,ds \equiv H(x + ct) - H(x - ct). \tag{11.12}$$

The foregoing analysis applies to waves traveling in an infinite, unbounded bar. Such waves will, in principle, travel indefinitely, without changing shape. But whenever an elastic wave traveling through a given medium 1 reaches a boundary with another medium 2, the waveform will be altered. A "transmitted wave" will pass into medium 2, and, in general, a "reflected wave" will reflect off of the boundary and return into medium 1 (Fig. 11.2a). The amplitudes of the transmitted and reflected waves, relative to that of the incident wave, can be shown to depend on the elastodynamic properties of the two media in the following manner (Bedford and Drumheller, 1994, pp. 62–64).

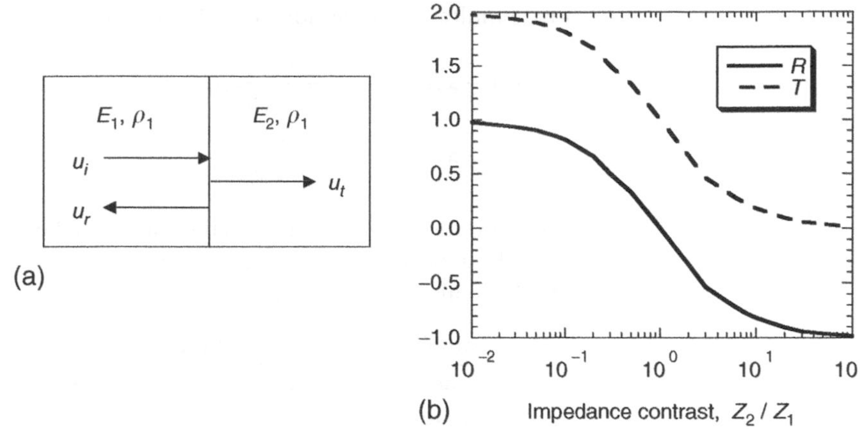

Fig. 11.2 (a) Incident wave, transmitted wave, and reflected wave at a welded interface; (b) Reflection and transmission coefficients as functions of the impedance contrast, from (11.18) and (11.19).

Consider an incident wave $u_i(x-c_1t)$, traveling to the right though medium 1, which occupies the region $x < 0$. The transmitted wave $u_t(x - c_2t)$ travels to the right through medium 2 and the reflected wave $u_r(x + c_1t)$ travels to the left through medium 1. At the interface between the two media, $x = 0$, the displacements and tractions must be continuous; these are the so-called "welded interface" boundary conditions. If the interface is a fracture or fault, different boundary conditions may be appropriate (see §12.7). The variable $x-ct$ can be multiplied by any constant k without altering the fact that $u(x - ct)$ satisfies the wave equation, so without loss of generality we can take $k = -1/c$ and thereby denote the solutions as $u(t - x/c)$. At the interface, $x = 0$, and so the displacement due to the incident wave is $u_i(t)$, for example. As usual, all displacements are considered positive if the motion is in the *negative x* direction, regardless of the direction of propagation of the wave. The condition of continuity of the displacement at the interface then can be expressed as

$$u_i(t) + u_r(t) = u_t(t). \tag{11.13}$$

The stress is related to the displacement by $\tau_{xx} = E\varepsilon_{xx} = E(\partial u/\partial x) = \rho c^2(\partial u/\partial x)$. By the chain rule, $\partial u/\partial x = -(1/c)u'$ for a right-traveling wave and $\partial u/\partial x = (1/c)u'$ for a left-traveling wave, where the prime denotes differentiation with respect to the argument of the function. The condition of stress continuity at the interface therefore can be written as

$$-\rho_1c_1u_i'(t) + \rho_1c_1u_r'(t) = -\rho_2c_2u_t'(t). \tag{11.14}$$

Integration of (11.14) yields

$$-\rho_1c_1u_i(t) + \rho_1c_1u_r(t) = -\rho_2c_2u_t(t) + B, \tag{11.15}$$

where B is a constant of integration. If the incoming wave is of finite duration, then after a sufficiently long time has elapsed, the stresses associated with each of the three waves must be zero, implying that the integration constant must be zero. Simultaneous solution of (11.13) and (11.15) yields

$$u_t(t - x/c_2) = \frac{2\rho_1 c_1}{\rho_1 c_1 + \rho_2 c_2} u_i(t - x/c_1), \tag{11.16}$$

$$u_r(t + x/c_1) = \frac{\rho_1 c_1 - \rho_2 c_2}{\rho_1 c_1 + \rho_2 c_2} u_i(t - x/c_1). \tag{11.17}$$

The product of the density and the wave speed of a material, $\rho c = (\rho E)^{1/2}$, is called the *acoustic impedance* and is usually denoted by Z. The results (11.16) and (11.17) can also be expressed as

$$T = \frac{\text{amplitude of transmitted wave}}{\text{amplitude of incident wave}} = \frac{2Z_1}{Z_1 + Z_2}, \tag{11.18}$$

$$R = \frac{\text{amplitude of reflected wave}}{\text{amplitude of incident wave}} = \frac{Z_1 - Z_2}{Z_1 + Z_2}. \tag{11.19}$$

According to the sign convention used for displacements, $u < 0$ corresponds to compression and $u > 0$ corresponds to extension. If $Z_2 < Z_1$, the transmitted wave has the same "sense" as the incident wave (i.e., compression or rarefaction) and a larger amplitude; the reflected wave has the same sense as the incident wave but a smaller amplitude. If $Z_2 > Z_1$, the transmitted wave has the same sense as the incident wave but a smaller amplitude, whereas the reflected wave has the opposite sense and a smaller amplitude. In this case, therefore, a compression wave would be reflected as a wave of rarefaction and vice versa. The reflection and transmission coefficients R and T are plotted in Fig. 11.2b as functions of the impedance ratio. If these coefficients are defined in terms of stresses rather than displacements, the expressions and curves are somewhat different. If the displacements are reckoned relative to the direction of wave propagation, rather than with respect to a coordinate system fixed in space, the expression for T in (11.18) should be multiplied by -1 (Mavko et al., 1998, p. 58).

The special case of a wave impinging on a free surface can be obtained by letting $\rho_2, E_2 \to 0$, in which case $Z_2 \to 0$. The reflected wave then has the same sense and same magnitude as the incident wave, whereas the transmitted wave has the same sense but twice the magnitude of the incident wave. The prediction of a transmitted wave traveling through a medium with zero stiffness and zero density may seem paradoxical, but it must be remembered that media with zero density and stiffness do not exist. The above result should be thought of as an asymptotic result that holds in the limit in which the impedance of medium 2 is very small relative to that of medium 1.

The other extreme case is a wave impinging on an interface with a medium of infinitely large impedance. If $Z_2 \to \infty$, (11.18) shows that there will be no transmitted wave, and (11.19) shows that the incident wave will be totally reflected back into medium 1 but with the opposite sense. Hence, a compressive wave will be reflected as a rarefaction wave and vice versa. A third interesting case is when the two media have identical values of the acoustic impedance. In this case, the wave will be fully transmitted into medium 2, with no change in amplitude, and there will be no reflected wave.

An elastic wave carries energy with it, in the form of both elastic strain energy and kinetic energy. For a wave described by $u(x - ct)$, the strain is

$$\varepsilon_{xx} = \frac{\partial u}{\partial x} = f'(x - ct). \tag{11.20}$$

The stress is then $\tau_{xx} = Ef'(x - ct)$, and so from (5.142), the elastic strain energy density is

$$\mathcal{E} = \frac{1}{2}\tau_{xx}\varepsilon_{xx} = \frac{1}{2}\rho c^2 (f')^2, \tag{11.21}$$

where (11.5) has been used to write $E = \rho c^2$. The local particle velocity of the rock is *not* equal to the phase velocity of the wave, c, but is found by the chain rule to be given by

$$\dot{u} = \frac{\partial u}{\partial t} = -cf'(x - ct). \tag{11.22}$$

The kinetic energy density per unit volume is therefore

$$\mathcal{K} = \frac{1}{2}\rho(\dot{u})^2 = \frac{1}{2}\rho c^2 (f')^2, \tag{11.23}$$

and is exactly equal, at each location x and at each time t, to the elastic strain energy density. The total energy density contained in the wave is

$$\mathcal{T} = \mathcal{K} + \mathcal{E} = \rho c^2 (f')^2. \tag{11.24}$$

The total energy contained in any region of the bar could be found by multiplying (11.24) by the cross-sectional area A and integrating along the length.

Comparison of (11.20) and (11.22) shows that the particle velocity is related to the phase velocity by $\dot{u} = -c\varepsilon_{xx}$ for a right-traveling wave; the analogous relationship is $\dot{u} = c\varepsilon_{xx}$ for a left-traveling wave. As the strain must by necessity be very small in order for the theory of linear elasticity to apply, it follows that the particle velocity in an elastic wave is but a small fraction of the phase velocity, which is the speed at which the "wave" travels. Use of Hooke's law and (11.5) in the expression $\dot{u} = c\varepsilon_{xx}$ shows that the particle velocity can also be expressed as

$$|\dot{u}| = \tau_{xx}/\rho c. \tag{11.25}$$

showing that the acoustic impedance can also be interpreted as the coefficient that relates the stress to the particle velocity.

11.3 Harmonic waves and group velocity

The analysis in §11.2 treated the general case of a wave of arbitrary form. But the most important type of wave, and that which is used as the basis for most mathematical analyses, is a *harmonic* wave in which the displacement (and hence also the strain and the stress) oscillates in a sinusoidal manner. The reason for the importance of harmonic waves is that, by use of Fourier's theorem, a wave of arbitrary time-variation can be decomposed into a combination of harmonic waves of different frequencies, each with its own amplitude. In this section, we discuss some of the properties of harmonic waves, using the nomenclature of longitudinal wave propagation in a thin elastic bar, as in §11.2.

Consider a displacement described by

$$u(x,t) = U_0\mathbf{Re}\{e^{ik(x-ct)}\} = U_0\cos[k(x-ct)] \equiv U_0\cos(kx-\omega t), \qquad (11.26)$$

where $\omega = kc$. The latter form $U_0\cos(kx - \omega t)$ is convenient, although it obscures the fact that k and ω are not independent but are related by $\omega/k = c$. Although it is less ambiguous to use only the real part of the exponential as the displacement, it is usually simpler to carry out mathematical manipulations using the complex exponential form and then to take the real part at the end. As the argument of an exponential must be dimensionless, the parameter k, called the *wave number*, must have dimensions of [1/L]. The parameter ω therefore has dimensions of [1/T].

At a fixed value of t, the wavelength λ of this displacement wave is determined by the condition $k(x+\lambda) = kx + 2\pi$, which shows that $\lambda = 2\pi/k$; the wave number is therefore a "spatial frequency". Similarly, at a fixed location, the period T of the wave is determined by the condition $\omega(t+T) = \omega t + 2\pi$, so $T = 2\pi/\omega$. The period T represents the number of "seconds per cycle", so $1/T$ would be the number of "cycles per second", and hence $\omega = 2\pi/T$ represents the number of "radians per second", that is, the *angular frequency*. It is often more natural to refer to the frequency in terms of the number of "cycles per second". This frequency, the so-called *cyclic frequency*, or simply "the frequency", is usually denoted by either f or ν, and is related to ω by $\nu = \omega/2\pi = 1/T$. The units of "cycles per second" are also known as *Hz*, in honor of the nineteenth century German physicist and acoustician Heinrich Hertz. A few of the more useful of the various relations between the parameters of a harmonic wave are

$$\omega = kc, \quad \omega = 2\pi f, \quad T = 2\pi/\omega, \quad \lambda = 2\pi/k, \quad \lambda = 2\pi c/\omega, \quad \lambda = c/f. \qquad (11.27)$$

The strain and stress associated with a harmonic displacement wave are also sinusoids having the same frequency as the displacement, but 1/4-cycle out of phase with it:

$$\varepsilon_{xx} = \partial u/\partial x = -kU_0\sin(kx-\omega t), \qquad (11.28)$$

$$\tau_{xx} = E\varepsilon_{xx} = -kEU_0\sin(kx-\omega t). \qquad (11.29)$$

From (5.142), the stored elastic strain energy density (per unit volume) is given by

$$\mathcal{E} = \frac{1}{2}\varepsilon_{xx}\tau_{xx} = \frac{k^2EU_0^2}{2}\sin^2(kx-\omega t). \qquad (11.30)$$

Using the relations $k = \omega/c$ and $c^2 = E/\rho$, this can be written as

$$\mathcal{E} = \frac{\rho\omega^2U_0^2}{2}\sin^2(kx-\omega t). \qquad (11.31)$$

The kinetic energy density is

$$\mathcal{K} = \frac{1}{2}\rho(\dot{u})^2 = \frac{\rho\omega^2 U_o^2}{2}\sin^2(kx - \omega t), \tag{11.32}$$

and so the total energy density is

$$\mathcal{T} = \mathcal{K} + \mathcal{E} = \rho\omega^2 U_o^2 \sin^2(kx - \omega t). \tag{11.33}$$

At any location x, the energy density varies in time between 0 and $\rho\omega^2 U_o^2$, with an average value of $\rho\omega^2 U_o^2/2$. Likewise, at any time t, the energy density varies in space between 0 and $\rho\omega^2 U_o^2$, with an average value of $\rho\omega^2 U_o^2/2$. At all times, the total energy contained in the wave is equally partitioned between elastic strain energy and kinetic energy, as was shown in a more general context in §11.2.

The time-averaged flux of energy through a given cross section located at x can be calculated as follows. Consider the total energy contained in the region between $x - \lambda$ and x:

$$\Delta\mathcal{T} = \int_{x-\lambda}^{x} \mathcal{T} A dx = \int_{x-\lambda}^{x} \rho\omega^2 U_o^2 \sin^2(kx - \omega t) A dx = \frac{\rho\omega^2 U_o^2 A\lambda}{2}. \tag{11.34}$$

The entire wave travels to the right at speed c, so all of the energy in the region between $x - \lambda$ and x at time t will pass through the plane at x within an elapsed time given by $\Delta t = \lambda/c$. The time-averaged *power flux*, which is the rate at which energy flows past location x, per unit area, is equal to

$$\mathcal{P} = \frac{\Delta\mathcal{T}}{A\Delta t} = \frac{\rho c\omega^2 U_o^2}{2}. \tag{11.35}$$

As discussed in §11.2, the actual velocity of the material particles is not the same as the velocity at which the "wave" propagates. The particle velocity is found from (11.26) by differentiation:

$$\dot{u} = \partial u/\partial t = \omega U_o \sin(kx - \omega t) = ckU_o \sin(kx - \omega t). \tag{11.36}$$

Comparison of (11.36) and (11.28) shows that, in accord with the general results of §11.2, $\dot{u} = -c\varepsilon_{xx}$, and so $|\dot{u}|/c \ll 1$.

Although the theory of wave propagation along a thin elastic bar was presented in §11.2 in part to provide a simplified, one-dimensional context in which to develop the basic ideas of elastic wave propagation, it has great importance in its own right, because most laboratory measurements of wave propagation in rocks are made on cylindrical core samples. According to this simplified low-frequency, long-wavelength theory, the wave speed c is independent of the wavelength or wave number. This is also the case for wave propagation in three-dimensional, unbounded elastic media (see §11.4). Waves for which the speed is independent of wavelength are called *nondispersive*. This term refers to the fact that, since the various frequency components of the wave each travel at the same speed, the waveform retains its shape as it travels through the medium.

The fully three-dimensional theory of wave propagation in an elastic bar (Graff, 1975, pp. 468–74) predicts that the wave speed actually depends on wavelength and hence on frequency. The wave speed asymptotically approaches $(E/\rho)^{1/2}$ as the wavelength becomes infinite and decreases as the wave number increases. Moreover, there are additional, higher "modes", each with its own complicated $c(k)$ relationship, that correspond to motions in which the stress is not of the same sign across a given cross section at any given time. It is generally thought that these higher modes are not excited in most laboratory measurements. Nevertheless, the wave speed in an elastic bar, as would be measured in the laboratory, does vary with wave number. Wave speed will vary with wave number whenever an elastic wave travels through a "waveguide"; for example, when a wave travels along a layer of rock that is bounded above and below by strata having different elastic properties. Waves whose speed varies with frequency are called "dispersive", because, as the different components of the total wave travel at different speeds, the wave form will *not* retain its overall shape as it moves through the medium. Waves that travel through viscoelastic media also exhibit dispersion, although in this case not for geometrical reasons but because the waves lose energy as they travel.

The phenomenon of dispersion leads to the concept of the *group velocity* of a wave, which was first analyzed in the following manner by the British mathematical physicist George Stokes in 1876. Consider a wave that consists of two harmonic components, with the same amplitude but slightly different wave numbers, k_1 and $k_2 = k_1 + \Delta k$, and slightly different frequencies, ω_1 and $\omega_2 = \omega_1 + \Delta\omega$:

$$u = U_o \cos(k_1 x - \omega_1 t) + U_o \cos(k_2 x - \omega_2 t). \tag{11.37}$$

Using standard trigonometric identities, this can be written as

$$u = 2U_o \cos\left\{\frac{1}{2}(k_1 + k_2)x - \frac{1}{2}(\omega_1 + \omega_2)t\right\} \cos\left\{\frac{1}{2}\Delta kx - \frac{1}{2}\Delta\omega t\right\}. \tag{11.38}$$

Now denote the mean wave number by k and the mean frequency by ω, so that (11.38) can be written as

$$u = 2U_o \cos\left\{k(x - \frac{\omega}{k}t)\right\} \cos\left\{\frac{1}{2}\Delta k(x - \frac{\Delta\omega}{\Delta k}t)\right\}. \tag{11.39}$$

The first cosine term is a high-frequency "carrier" wave that travels at a speed $c = \omega/k$. Since k_1 and k_2 each differs only slightly from k, and ω_1 and ω_2 each differs only slightly from ω, both of the two individual waves in (11.37) travel essentially at velocity c, the so-called *phase velocity*. However, (11.38) shows that these two waves combine in such a way that the wave traveling at velocity c is *modulated* by the second cosine term, which represents a low-frequency wave that travels at a speed given by $c_g = \Delta\omega/\Delta k$, the so-called *group velocity*.

This modulated wave is illustrated schematically in Fig. 11.3. If one focuses on the detailed motion of the medium, one indeed observes the carrier wave traveling at speed c. But if one ignores the detailed motion and focuses attention

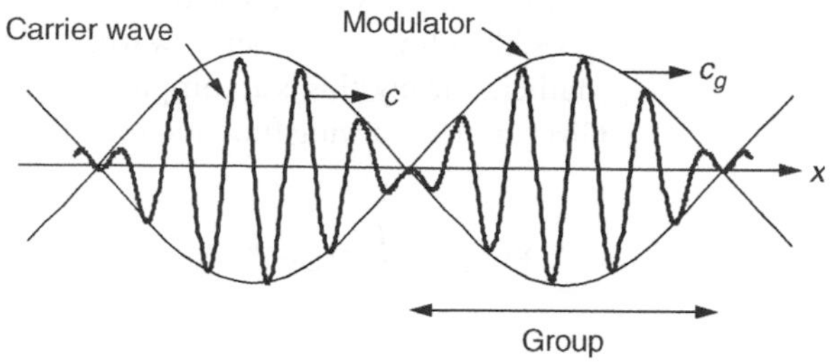

Fig. 11.3 Superposition of two waves of slightly differing frequencies, which combine to yield a high-frequency carrier wave traveling at the phase velocity c, modulated by a low-frequency modulator wave traveling at the group velocity, c_g, as described by (11.37)–(11.39).

on the "macroscopic" motion, one observes the envelope moving forward at speed c_g. If $c_g > c$, the individual wavelets seem to appear at the front of the group and disappear at the rear, as they are overtaken by the modulator wave. If $c_g < c$, the individual wavelets seem to appear at the rear of the group, travel forward through it, and disappear at the front (Brillouin, 1960). An easily observable example of this phenomenon is the radially diverging wave pattern produced by dropping a small object into a still lake or pond. In this case $c_g > c$, and the individual ripples start at the outer edge of the ring and eventually disappear at the inner edge.

In the more general context in which a wavepacket may contain a range of frequencies, the group velocity is defined by $c_g = d\omega/dk$. Using the relationships given in (11.27), the group velocity can also be expressed as

$$c_g = \frac{d\omega}{dk} = c + k\frac{dc}{dk} = c - \lambda\frac{dc}{d\lambda}. \tag{11.40}$$

Another useful form of (11.40) is

$$\frac{1}{c_g} = \frac{1}{c} - \frac{\omega}{c^2}\frac{dc}{d\omega}, \tag{11.41}$$

which is expressed in terms of the *slowness*, $1/c$.

Dispersion is termed *normal* if the group velocity decreases with increasing frequency (Bourbié et al., 1987, p. 111) and is termed *anomalous* or *inverse* if c_g increases with increasing frequency. Dispersion that is due to geometrical effects, such as the dispersion of waves in an elastic layer, is typically of the normal type, whereas dispersion that is due to viscoelastic or other energy-dissipative effects is usually anomalous (Mavko et al., 1998, p. 55).

For a nondispersive wave, $c_g = d\omega/dk = c =$ constant and the group velocity coincides with the phase velocity. In this case, the energy travels with velocity c. But for dispersive waves in an elastic medium, it can be shown by a lengthy mathematical argument (Achenbach, 1973, pp. 211–15) that the energy actually travels through the medium at the group velocity, c_g. The group velocity is therefore also equal to the velocity of energy propagation. Waves in inelastic media are also dispersive, but they lose energy as they propagate, and consequently the relationship between the various velocities is not so simple or meaningful (Mavko et al., 1998, pp. 56–7).

An arbitrary wave can be thought of as being composed of a superposition of a (possibly infinite) number of waves, each with its own frequency and amplitude. Mathematically, this is accomplished by the Fourier transform (Bracewell, 1986). Given a time-varying function $f(t)$, its Fourier transform $F(\omega)$ can be defined as

$$F(\omega) = \int_{-\infty}^{\infty} f(t)e^{-i\omega t}dt. \tag{11.42}$$

Other notations for the Fourier transform of $f(t)$ are $f^*(\omega)$ or $\hat{f}(\omega)$. The mathematical conditions under which the integral in (11.42) exists are discussed by Bracewell (1986); in practice, a Fourier transform exists for all waveforms arising in wave propagation in rocks, in the laboratory or the field. The function $F(\omega)$ can be thought of as representing that portion of the total wave $f(t)$ that has frequency ω. This is seen from the Fourier inversion integral, in which the function $f(t)$ is represented as a superposition of the various "components" $F(\omega)$:

$$f(t) = \frac{1}{2\pi}\int_{-\infty}^{\infty} F(\omega)e^{i\omega t}d\omega. \tag{11.43}$$

A more symmetric form of these relationships is obtained by utilizing the cyclic frequency, $\omega = 2\pi\nu$. A simple change of variables applied to (11.42) and (11.43) shows that the Fourier transform can also be defined as follows:

$$F(\nu) = \int_{-\infty}^{\infty} f(t)e^{-2\pi i\nu t}dt, \quad f(t) = \int_{-\infty}^{\infty} F(\nu)e^{2\pi i\nu t}d\nu. \tag{11.44}$$

Fourier transforms are useful in solving wave propagation problems (Miklowitz, 1978) and are ubiquitous in the analysis of seismic data (Berkhout, 1987).

11.4 Elastic waves in unbounded media

Elastic waves travel at the speed $c = (E/\rho)^{1/2}$ only in the special case of a long-wavelength disturbance traveling along an elastic bar of constant cross section. Waves that travel in three-dimensional, unbounded, isotropic elastic media can be studied by starting with the full three-dimensional equations of motion, (5.90), with the inertia term included:

$$(\lambda + G)\nabla(\nabla \cdot \mathbf{u}) + G\nabla^2\mathbf{u} = \rho\ddot{\mathbf{u}}. \tag{11.45}$$

Consider a planar wave traveling at speed c along direction $\mathbf{n}$, which can be represented by

$$\mathbf{u} = f(\mathbf{x}\cdot\mathbf{n} - ct)\mathbf{d}, \tag{11.46}$$

where $\mathbf{d}$, the particle displacement vector, is taken to be a constant. In the one-dimensional theory of §11.2, the particle displacement must necessarily be in the same direction as the wave motion, but this need not be true in three dimensions.

As $\mathbf{x} \cdot \mathbf{n}$ represents the projection of the vector $\mathbf{x}$ onto the direction $\mathbf{n}$, the phase $\eta = \mathbf{x} \cdot \mathbf{n} - ct$ has the same value for all points $\mathbf{x}$ that lie on a given plane perpendicular to $\mathbf{n}$. So for simplicity, consider a point lying on the vector $\mathbf{n}$, that is, $\mathbf{x} = \zeta\mathbf{n}$. Along direction $\mathbf{n}$, the phase is equal to $\eta = (\zeta\mathbf{n}) \cdot \mathbf{n} - ct = \zeta - ct$, since $\mathbf{n}$ is a unit vector. The velocity at which the wavefront moves along $\mathbf{n}$ is then given, as in (11.8), by

$$\mathbf{v}(\textit{wavefront}) = \left(\frac{\partial \mathbf{x}}{\partial t}\right)_\eta = \left(\frac{\partial(\zeta\mathbf{n})}{\partial t}\right)_\eta = \left(\frac{\partial \zeta}{\partial t}\right)_\eta \mathbf{n} = c\mathbf{n}. \tag{11.47}$$

Hence, the wave does indeed propagate in the $\mathbf{n}$ direction, and the parameter c that appears in (11.46) is the phase velocity of this disturbance.

The time derivatives of $\mathbf{u}$ are found by applying the chain rule to (11.46):

$$\dot{\mathbf{u}} = -cf'(\mathbf{x} \cdot \mathbf{n} - ct)\mathbf{d}, \quad \ddot{\mathbf{u}} = c^2 f''(\mathbf{x} \cdot \mathbf{n} - ct)\mathbf{d}. \tag{11.48}$$

The phase of this wave is explicitly given by $\eta = xn_x + yn_y + zn_z - ct$, so the spatial derivatives of $\mathbf{u}$ are, for example,

$$\frac{\partial \mathbf{u}}{\partial x} = f'(\mathbf{x} \cdot \mathbf{n} - ct)\mathbf{d}\frac{\partial \eta}{\partial x} = f'(\mathbf{x} \cdot \mathbf{n} - ct)\mathbf{d}n_x. \tag{11.49}$$

Hence, it follows that

$$\nabla \cdot \mathbf{u} = f'(\eta)\mathbf{d} \cdot \mathbf{n}, \tag{11.50}$$

$$\nabla(\nabla \cdot \mathbf{u}) = f''(\eta)(\mathbf{d} \cdot \mathbf{n})\mathbf{n}, \tag{11.51}$$

$$\nabla^2 \mathbf{u} = f''(\eta)\mathbf{d}, \tag{11.52}$$

and (11.45) reduces to

$$(\lambda + G)f''(\eta)(\mathbf{d} \cdot \mathbf{n})\mathbf{n} + Gf''(\eta)\mathbf{d} = \rho c^2 f''(\eta)\mathbf{d}. \tag{11.53}$$

Aside from the trivial case $f''(\eta) = 0$, which leads to either a rigid-body motion or a state of uniform strain that is independent of time, (11.53) is equivalent to

$$(\lambda + G)(\mathbf{d} \cdot \mathbf{n})\mathbf{n} + (G - \rho c^2)\mathbf{d} = 0. \tag{11.54}$$

If the particle motion is perpendicular to the direction of wave propagation, then $\mathbf{d} \cdot \mathbf{n} = 0$, and (11.54) can only be satisfied for nonzero $\mathbf{d}$ if

$$c \equiv c_T = \sqrt{G/\rho}, \tag{11.55}$$

where c_T is the velocity of *transverse waves*, for which the particle motion is transverse to the direction of wave propagation. The direction of propagation $\mathbf{n}$ is completely arbitrary, as is the amplitude of the particle displacement (aside from the requirement to maintain the small-strain approximation). Hence, transverse plane waves can travel in any direction of an isotropic elastic medium but can only travel at a speed given by (11.55). As this speed is independent of the frequency or wavelength of the disturbances, such waves are nondispersive.

If, on the other hand, $\mathbf{d}$ is *parallel* rather than *perpendicular* to the direction of wave propagation, then $\mathbf{d} = d\mathbf{n}$, where d is a scalar, and (11.54) reduces to

$$(\lambda + 2G - \rho c^2)d\mathbf{n} = 0, \tag{11.56}$$

which can only be satisfied by nonzero particle displacement d if

$$c \equiv c_L = \sqrt{(\lambda + 2G)/\rho}. \tag{11.57}$$

Longitudinal plane waves, in which the particle velocity is in the same direction as the wave propagation, can therefore travel through an isotropic elastic medium in any direction but only at a velocity given by (11.57). These longitudinal waves are also nondispersive because c_L does not vary with frequency.

Use of relations (5.13)–(5.15) allows the longitudinal wave speed to be written in the following forms:

$$c_L = \sqrt{\frac{K + (4G/3)}{\rho}} = \sqrt{\frac{(1-\nu)E}{(1+\nu)(1-2\nu)\rho}} = \sqrt{\frac{2(1-\nu)G}{(1-2\nu)\rho}}, \tag{11.58}$$

from which it follows that the ratio of the two wave speeds is

$$\frac{c_L}{c_T} = \sqrt{\frac{2(1-\nu)}{(1-2\nu)}}. \tag{11.59}$$

In any elastic medium, longitudinal waves always travel faster than transverse waves. The ratio of the two wave speeds is $\sqrt{2} \approx 1.41$ when $\nu = 0$, increases with increasing ν, and becomes unbounded when $\nu \to 0.5$.

Without loss of generality, in an isotropic medium the direction of propagation $\mathbf{n}$ can be taken to be the x-axis and the direction of the particle displacement vector $\mathbf{d}$ to be the y-axis. The displacement vector for a transverse wave then has the form $\mathbf{u} = f(x - c_T t)d_y\mathbf{e_y}$, from which it follows that the only nonzero strain component is $\varepsilon_{xy} = f'(x - c_T t)d_y/2$ and the nonzero stress is $\tau_{xy} = Gf'(x - c_T t)d_y$. A transverse wave is therefore a *shear wave*, consistent with the fact that the shear modulus is the only elastic modulus that affects c_T.

Similarly, by proper alignment of the coordinate system, a longitudinal wave can be represented by $\mathbf{u} = f(x - c_T t)d_x\mathbf{e_x}$, for which the only nonzero strain component is $\varepsilon_{xx} = f'(x - c_T t)d_x$. Hence, a longitudinal wave is a wave of *uniaxial strain*, consistent with the fact that, according to (5.33), $\lambda + 2G$ is the uniaxial strain modulus. Due to the Poisson effect, longitudinal waves are *not* waves of uniaxial stress, as the normal stresses on planes perpendicular to the direction of wave propagation must satisfy $\tau_{yy} = \tau_{zz} = \nu\tau_{xx}/(1-\nu)$ in order to maintain a state of uniaxial strain. In contrast to this situation, when a long-wavelength longitudinal wave travels along a thin bar, as in §11.2, the *stress* is essentially uniaxial and the two lateral strains are nonzero.

In geophysics, the transverse wave velocity is usually denoted by V_s and the longitudinal velocity by V_p. The subscript s can be thought of as signifying a *shear wave*, or it can be thought of as signifying a *secondary wave*, as these waves arrive at a receiver later than the faster-moving longitudinal waves. The subscript p can similarly be thought of as standing for *primary wave* or *pressure wave*. These two types of waves are often referred to as P-waves and S-waves.

The two wave velocities depend on the elastic moduli and density of the rock, which in turn depend not only on mineral composition, pore structure, fluid properties (see §11.8), but also vary with stress, temperature, etc. For example, an increase in confining stress tends to close up cracks and grain boundary pores, thereby increasing the elastic moduli and the wave speeds. Given the great variability in rock properties, even within the same rock type, it is consequently difficult, and not very meaningful, to cite specific values for specific rocks. Table 11.1, adapted from Bourbié et al. (1987), gives ranges of representative values for several types of rock.

A useful mathematical tool for the analysis of elastodynamic problems is the Helmholtz decomposition of the displacement vector into the gradient of a scalar potential, φ, plus the curl of a divergence-free vector potential, $\boldsymbol{\Psi}$ (Sternberg, 1960):

$$\mathbf{u} = \text{div}\,\varphi + \text{curl}\boldsymbol{\Psi} = \nabla\varphi + \nabla \times \boldsymbol{\Psi}. \tag{11.60}$$

The displacement corresponding to $\boldsymbol{\Psi}$ is divergence-free, and hence has no volumetric strain, and is therefore a state of pure shear. To prove

Table 11.1 Ranges of wave speeds and densities of various rock types.

Rock type	V_p (m/s)	V_s (m/s)	ρ (kg/m^3)
Vegetal soil	300–700	100–300	1700–2400
Dry sands	400–1200	100–500	1500–1700
Wet sands	1500–2000	400–600	1900–2100
Saturated shales and clays	1100–2500	200–800	2000–2400
Marls	2000–3000	750–1500	2100–2600
Saturated shale / sand sections	1500–2200	500–750	2100–2400
Porous saturated sandstones	2000–3500	800–1800	2100–2400
Limestones	3500–6000	2000–3300	2400–2700
Chalk	2300–2600	1100–1300	1800–2300
Salt	4500–5500	2500–3100	2100–2300
Anhydrite	4000–5500	2200–3100	2900–3000
Dolomite	3500–6500	1900–3600	2500–2900
Granite	4500–6000	2500–3300	2500–2700
Basalt	5000–6000	2800–3400	2700–3100
Gneiss	4400–5200	2700–3200	2500–2700
Coal	2200–2700	1000–1400	1300–1800
Water	1450–1500	—	1000
Ice	3400–3800	1700–1900	900
Oil	1200–1250	—	600–900

this, consider

$$\mathbf{u} = \operatorname{curl}\boldsymbol{\Psi} \equiv -2\operatorname{asym}(\nabla\boldsymbol{\Psi}) = (\nabla\boldsymbol{\Psi})^{\mathrm{T}} - \nabla\boldsymbol{\Psi}$$
$$= \left[\left(\frac{\partial\psi_z}{\partial y} - \frac{\partial\psi_y}{\partial z}\right), \left(\frac{\partial\psi_x}{\partial z} - \frac{\partial\psi_z}{\partial x}\right), \left(\frac{\partial\psi_y}{\partial x} - \frac{\partial\psi_x}{\partial y}\right)\right], \tag{11.61}$$

from which it readily follows that

$$\varepsilon_v = \nabla\cdot\mathbf{u} = \left(\frac{\partial^2\psi_z}{\partial x\partial y} - \frac{\partial^2\psi_z}{\partial y\partial x}\right) + \left(\frac{\partial^2\psi_y}{\partial z\partial x} - \frac{\partial^2\psi_y}{\partial x\partial z}\right) + \left(\frac{\partial^2\psi_x}{\partial y\partial z} - \frac{\partial^2\psi_x}{\partial z\partial y}\right) = 0, \tag{11.62}$$

in which case (11.45) reduces to

$$G\nabla^2\mathbf{u} = \rho\ddot{\mathbf{u}}. \tag{11.63}$$

Equation (11.63) represents three uncoupled wave equations, one for each of the displacement components, each with wave speed $c_T = (G/\rho)^{1/2}$:

$$\nabla^2 u = \frac{1}{c_T^2}\frac{\partial^2 u}{\partial t^2}, \quad \nabla^2 v = \frac{1}{c_T^2}\frac{\partial^2 v}{\partial t^2}, \quad \nabla^2 w = \frac{1}{c_T^2}\frac{\partial^2 w}{\partial t^2}. \tag{11.64a–c}$$

Taking the partial derivative of (11.64a) with respect to y and adding it to the partial derivative of (11.64b) with respect to x and similarly for the other two pairs of equations that can be chosen from (11.64), shows that each of the components of the strain tensor also satisfies this same wave equation:

$$\nabla^2\varepsilon_{xy} = \frac{1}{c_T^2}\frac{\partial^2\varepsilon_{xy}}{\partial t^2}, \quad \nabla^2\varepsilon_{xz} = \frac{1}{c_T^2}\frac{\partial^2\varepsilon_{xz}}{\partial t^2}, \quad \nabla^2\varepsilon_{yz} = \frac{1}{c_T^2}\frac{\partial^2\varepsilon_{yz}}{\partial t^2}. \tag{11.65}$$

Alternatively, subtracting the partial derivative of (11.64a) with respect to y from the partial derivative of (11.64b) with respect to x, etc., which is essentially equivalent to applying the curl operator to (11.45), shows that the three independent components of the rotation tensor also satisfy this same wave equation:

$$\nabla^2\omega_{xy} = \frac{1}{c_T^2}\frac{\partial^2\omega_{xy}}{\partial t^2}, \quad \nabla^2\omega_{xz} = \frac{1}{c_T^2}\frac{\partial^2\omega_{xz}}{\partial t^2}, \quad \nabla^2\omega_{yz} = \frac{1}{c_T^2}\frac{\partial^2\omega_{yz}}{\partial t^2}. \tag{11.66}$$

Hence, "rotation" is also propagated through the medium at speed c_T. Consequently, shear waves are also sometimes called *rotational waves*.

The other part of the decomposed displacement vector, $\mathbf{u} = \nabla\varphi$, corresponds to a deformation in which there is *no* rotation. For example,

$$\omega_{xy} = \frac{1}{2}\left(\frac{\partial u}{\partial y} - \frac{\partial v}{\partial x}\right) = \frac{1}{2}\left(\frac{\partial}{\partial y}\left[\frac{\partial\varphi}{\partial x}\right] - \frac{\partial}{\partial x}\left[\frac{\partial\varphi}{\partial y}\right]\right) = 0, \tag{11.67}$$

and similarly for ω_{xz} and ω_{yz}. In this case the term $\nabla(\nabla\cdot\mathbf{u})$ in (11.45) reduces to $\nabla^2\mathbf{u}$, because, for example,

$$\frac{\partial(\nabla\cdot\mathbf{u})}{\partial x} = \frac{\partial}{\partial x}\left(\frac{\partial u}{\partial x} + \frac{\partial v}{\partial y} + \frac{\partial w}{\partial z}\right) = \frac{\partial^2 u}{\partial x^2} + \frac{\partial}{\partial y}\left(\frac{\partial v}{\partial x}\right) + \frac{\partial}{\partial z}\left(\frac{\partial w}{\partial x}\right)$$
$$= \frac{\partial^2 u}{\partial x^2} + \frac{\partial}{\partial y}\left(\frac{\partial u}{\partial y}\right) + \frac{\partial}{\partial z}\left(\frac{\partial u}{\partial z}\right) = \nabla^2 u, \tag{11.68}$$

and similarly for the other two components of $\nabla(\nabla \cdot \mathbf{u})$, in which case (11.45) takes the form

$$(\lambda + 2G)\nabla^2 \mathbf{u} = \rho \ddot{\mathbf{u}}. \tag{11.69}$$

This is equivalent to the following three uncoupled scalar wave equations:

$$\nabla^2 u = \frac{1}{c_L^2}\frac{\partial^2 u}{\partial t^2}, \quad \nabla^2 v = \frac{1}{c_L^2}\frac{\partial^2 v}{\partial t^2}, \quad \nabla^2 w = \frac{1}{c_L^2}\frac{\partial^2 w}{\partial t^2}. \tag{11.70a–c}$$

Hence, irrotational waves travel at the longitudinal wave velocity, c_L; equivalently, longitudinal waves are associated with irrotational motions.

Finally, differentiating (11.70a) with respect to x, (11.70b) with respect to y, and (11.70c) with respect to z, and adding the results, which is equivalent to taking the divergence of (11.69), yields

$$\nabla^2(\nabla \cdot \mathbf{u}) = \nabla^2 \varepsilon_v = \frac{1}{c_L^2}\frac{\partial^2 \varepsilon_v}{\partial t^2}. \tag{11.71}$$

Hence, the bulk volumetric strain also propagates at the speed c_L.

More generally, the scalar potential φ satisfies the wave equation with wave speed c_L. This is proven by substituting $\nabla \cdot \mathbf{u} = \nabla \cdot (\nabla\varphi) = \nabla^2\varphi$ into (11.71) and noting that the operator ∇^2 commutes with the time derivatives, leading to

$$\nabla^2 \varphi = \frac{1}{c_L^2}\frac{\partial^2 \varphi}{\partial t^2}. \tag{11.72}$$

Similarly, it can be shown that each of the three Cartesian components of the vector potential $\boldsymbol{\Psi}$ satisfy the wave equation with wave speed c_T:

$$\nabla^2 \psi_x = \frac{1}{c_T^2}\frac{\partial^2 \psi_x}{\partial t^2}, \quad \nabla^2 \psi_y = \frac{1}{c_T^2}\frac{\partial^2 \psi_y}{\partial t^2}, \quad \nabla^2 \psi_z = \frac{1}{c_T^2}\frac{\partial^2 \psi_z}{\partial t^2}. \tag{11.73}$$

The full explicit relationships between the displacements and the potentials are

$$u = \frac{\partial \varphi}{\partial x} + \frac{\partial \psi_z}{\partial y} - \frac{\partial \psi_y}{\partial z}, \quad v = \frac{\partial \varphi}{\partial y} + \frac{\partial \psi_x}{\partial z} - \frac{\partial \psi_z}{\partial x}, \quad w = \frac{\partial \varphi}{\partial z} + \frac{\partial \psi_y}{\partial x} - \frac{\partial \psi_x}{\partial y}. \tag{11.74}$$

11.5 Reflection and refraction of waves at an interface

In §11.2, the transmission of a wave across an interface between two possibly different elastic media was studied in the one-dimensional case. It was seen that, in general, some portion of the energy is transmitted through to the second medium and the remaining portion is reflected back into the medium from which the wave came. The amplitudes of the reflected and transmitted waves depended on the ratios of the acoustic impedances of the two media, where the acoustic impedance is the product of the density and the wave speed.

When an elastic wave impinges upon an interface between two rock types at an oblique angle, the behavior is more complicated. This problem was apparently first studied by Knott (1899), although the resulting equations for the reflection and transmission coefficients are often attributed to Zoeppritz (1919). Other early studies were made by Jeffreys (1926), Muskat and Meres (1940) and Ott (1942). Detailed results are given by Ewing et al. (1957) and Brekhovskikh (1980). In general, regardless of whether the impinging wave is a shear wave or a compressional wave, four waves are created: a shear and a compressional wave are transmitted (refracted) across the interface, and a shear and a compressional wave are reflected back into the first medium. The amplitudes of these four waves and the angles which they make with the interface will depend not only on the acoustic impedances of the two media but also on the angle of incidence of the incident wave.

An exception to this general behavior occurs when the incident wave is a shear wave whose displacement vector is parallel to the interface. Such waves are called "SH" waves, referring to the fact that if the interface is horizontal, the displacement of such a wave will lie in the horizontal plane. Although interfaces between two rock types need not be horizontal, waves in which the displacement has no component normal to the interface are known as SH-waves. In this case, the transmitted and reflected waves will both be of the SH type and no compressional waves will be generated. As this is the simplest case, it will be presented first, in detail.

Consider two semi-infinite half-spaces, with their interface coinciding with the x–y plane. The region $z < 0$ is labeled with superscript 1 and the region $z > 0$ is labeled with superscript 2. A plane shear wave propagates through medium 1, with its direction of propagation ($\mathbf{n}$ in the notation of §11.4) lying in the x–z plane, making an angle θ_0 with the z-axis (Fig. 11.4a). The particle displacement is in the y direction; this motion therefore represents an "SH" wave. The only nonzero displacement component of the incoming wave is

$$v^{(0)} = A_0 \exp[ik_0(x \sin\theta_0 + z\cos\theta_0 - c_{T1}t)]. \tag{11.75}$$

It is traditional (and simpler) to use complex numbers to represent the displacements, stresses, etc., when solving such problems, bearing in mind that only the

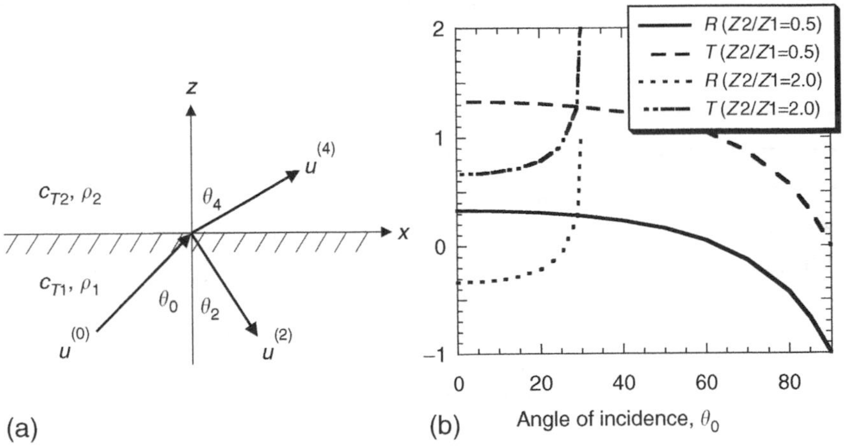

Fig. 11.4 (a) Incident, transmitted, and reflected SH-wave impinging on a plane interface; (b) Reflection and transmission coefficients as a function of θ_0, for the case of $\rho_1 = \rho_2$.

real component has physical significance. The wave that is reflected back into medium 1 is taken to be of the form

$$v^{(2)} = A_2 \exp[ik_2(x \sin\theta_2 - z\cos\theta_2 - c_{T1}t)], \tag{11.76}$$

and the wave that is refracted into medium 2 is taken to be of the form

$$v^{(4)} = A_4 \exp[ik_4(x \sin\theta_4 + z\cos\theta_4 - c_{T2}t)]. \tag{11.77}$$

The term "refracted" is used here instead of "transmitted", to emphasize that, in general, the wave changes its velocity of propagation as well as its direction of propagation upon entering the second medium.

The total displacement in medium 1 will be the sum of the displacement from the incoming wave and the reflected wave. In general, all three components of the total displacement vector must be continuous across a "welded" interface between two rock types. If the rocks are separated by a fracture or other type of mechanical discontinuity, different boundary conditions are appropriate, as described in §12.7. For welded interfaces, the traction vector acting on the interface must also be continuous. As the unit normal vector of the interface is $\mathbf{e_z}$, the traction vector along the interface has components $\{\tau_{zx}, \tau_{zy}, \tau_{zz}\}$. In general, there are six continuity conditions, corresponding to the three displacement components and three traction components. In the present case, the only nonzero displacement component is v and the only nonzero stress component that represents a traction along the interface is τ_{zy}.

From (11.75)–(11.77), the condition for continuity of v across the interface $z = 0$ takes the form

$$\begin{aligned} &A_0 \exp[ik_0(x \sin\theta_0 - c_{T1}t)] + A_2 \exp[ik_2(x \sin\theta_2 - c_{T1}t)] \\ &\quad = A_4 \exp[ik_4(x \sin\theta_4 - c_{T2}t)]. \end{aligned} \tag{11.78}$$

As this equation must hold for all values of x, the factors multiplying x in each of the three exponential terms must be identical, that is,

$$k_0 \sin\theta_0 = k_2 \sin\theta_2 = k_4 \sin\theta_4. \tag{11.79}$$

The same must be true for the time-dependent terms, so

$$k_0 c_{T1} = k_2 c_{T1} = k_4 c_{T2}. \tag{11.80}$$

Solving (11.79) and (11.80) for the angles and wave numbers of the reflected and refracted waves gives

$$k_2 = k_0, \quad k_4 = (c_{T1}/c_{T2})k_0, \quad \theta_2 = \theta_0, \quad \sin\theta_4 = (c_{T2}/c_{T1})\sin\theta_0, \tag{11.81}$$

after which (11.78) reduces to

$$A_0 + A_2 = A_4. \tag{11.82}$$

Equation (11.81) shows that the angle of the reflected wave is always equal to the angle of the incident wave, whereas the refracted wave is "bent" toward the normal when medium 2 is "slower" than medium 1 and bent away from the normal if medium 2 is faster than medium 1. These facts, as implied by (11.81), are equivalent to Snell's law of optics.

The stress τ_{zy} that corresponds to a displacement of the form (11.75)–(11.77) is, suppressing the subscripts,

$$\tau_{zy} = \pm\, ik \cos\theta G A \exp[ik(x \sin\theta \pm z \cos\theta - c_T t)], \tag{11.83}$$

where the + sign is used for the incident and refracted wave, and the − sign is used for the reflected wave. From (11.83) and (11.81), the condition of continuity of τ_{zy} across the interface takes the form

$$\cos\theta_0 G_1 A_0 - \cos\theta_0 G_1 A_2 = (c_{T1}/c_{T2}) \cos\theta_4 G_2 A_4. \tag{11.84}$$

Solving (11.82) and (11.84) for the amplitudes of the reflected and transmitted waves gives

$$\frac{A_2}{A_0} = \frac{\rho_1 c_{T1} \cos\theta_0 - \rho_2 c_{T2} \cos\theta_4}{\rho_1 c_{T1} \cos\theta_0 + \rho_2 c_{T2} \cos\theta_4}, \tag{11.85}$$

$$\frac{A_4}{A_0} = \frac{2\rho_1 c_{T1} \cos\theta_0}{\rho_1 c_{T1} \cos\theta_0 + \rho_2 c_{T2} \cos\theta_4}. \tag{11.86}$$

The product of the density and shear wave velocity is the shear wave impedance (see §11.2), and so the *displacement* reflection and transmission coefficients, $R = A_2/A_0$ and $T = A_4/A_0$, depend on the shear wave impedance ratio, Z_{T2}/Z_{T1}, and the angle of incidence of the incoming wave, θ_0. However, these coefficients also depend on the angle of the transmitted wave, θ_4, which in turn depends on the ratio of wave speeds. Hence, the reflection and transmission coefficients depend on two material property ratios, impedance and wave speed. From (11.83), it is clear that if the reflection and transmission coefficients were defined in terms of the amplitudes of the stresses, rather than the displacements, they would have somewhat different forms (Daehnke and Rossmanith, 1997).

The wavelength or frequency of the incident wave does not appear in the expressions for the reflection and transmission coefficients, so these expressions hold for arbitrary superposition of waves of different frequencies.

The reflection and transmission ratios are plotted in Fig. 11.4b, for two different impedance ratios. As there is somewhat more variability in wave speed than in density between different rocks (see Table 11.1), for the purposes of illustration and discussion, the densities of the two media are assumed equal, in which case the impedance ratio coincides with the velocity ratio. Aside from cases in which one of the media is a fluid, impedance ratios typically lie within the range shown in Fig. 11.4b, namely 0.5–2.0. The following observations can be made:

1 If the incoming wave is normal to the interface, then $\theta_0 = \theta_2 = \theta_4 = 0$, and it is seen that the reflection and transmission coefficients reduce to those given in (11.18) and (11.19) and Fig. 11.2b for the one-dimensional wave model.
2 If the second medium has zero shear impedance, such as occurs when a wave impinges on an interface between rock and fluid, the reflection coefficient is

unity. Although the transmission coefficient approaches 2, this coefficient refers to the amplitudes, not the energies. If either the wave speed or the density goes to zero, (11.23) shows that no energy will be transmitted across the interface; the energy is entirely reflected back into medium 1. This apparent paradox of having a nonzero transmission coefficient can be eliminated by utilizing the *stress* transmission coefficient, which would vanish in this case.

3 There is usually one particular angle of incidence for which there is no reflected wave. This angle is found by simultaneously solving, from (11.81), $\sin\theta_4 = (c_{T2}/c_{T1})\sin\theta_0$ and, from (11.85), $\cos\theta_4 = (Z_{T1}/Z_{T2})\cos\theta_0$. For the case in which the rocks on either side of the interface have the same density, this occurs for $\sin\theta_0 = [1 + (c_{T2}/c_{T1})^2]^{-1/2}$.

4 If $(c_{T2}/c_{T1})\sin\theta_0 > 1$, which can only occur if $c_{T2} > c_{T1}$, Snell's law yields an imaginary value for $\cos\theta_0$. The transmitted wave then has the form (Miklowitz, 1978, p. 184)

$$v^{(4)} = A_4 \exp(-bz)\exp[ik_4(x\sin\theta_4 - c_{T2}t)], \tag{11.87}$$

where $b = k_0[(c_{T2}/c_{T1})^2\sin^2\theta_0 - 1]^{1/2}$. Rather than representing a wave that propagates into medium 2, this displacement propagates only in the x direction, parallel to the interface, with an amplitude (in medium 2) that decays exponentially with distance from the interface. Hence, this wave carries no energy into medium 2, and all of the incoming energy is reflected back into medium 1 with the reflected wave. This situation is referred to as *total internal reflection*.

If the incident wave is either an SV-wave or a P-wave, with particle motion in the x–z plane, the conditions of continuity for the displacement and tractions across the interface will, in general, only be satisfied by a combination of a pair of reflected P and SV-waves, and a pair of refracted P and SV-waves. An incident P-wave propagating toward the interface at an angle θ to the z-axis can be represented by

$$\begin{aligned}\mathbf{u}^0 = \{u^{(0)}, v^{(0)}, w^{(0)}\} &= A_0 \exp[ik_0(x\sin\theta_0 + z\cos\theta_0 - c_{L1}t)]\\ &\quad\times \{\sin\theta_0, 0, \cos\theta_0\}.\end{aligned} \tag{11.88}$$

The reflected P-wave is denoted by 1, the "reflected" SV-wave by 2, the transmitted P-wave by 3, and the transmitted SV-wave by 4, as follows:

$$\begin{aligned}\mathbf{u}^1 = \{u^{(1)}, v^{(1)}, w^{(1)}\} &= A_1 \exp[ik_1(x\sin\theta_1 - z\cos\theta_1 - c_{L1}t)]\\ &\quad\times \{\sin\theta_1, 0, -\cos\theta_1\}.\end{aligned} \tag{11.89}$$

$$\begin{aligned}\mathbf{u}^2 = \{u^{(2)}, v^{(2)}, w^{(2)}\} &= A_2 \exp[ik_2(x\sin\theta_2 - z\cos\theta_2 - c_{T1}t)]\\ &\quad\times \{\sin\theta_2, 0, -\cos\theta_2\}.\end{aligned} \tag{11.90}$$

$$\begin{aligned}\mathbf{u}^3 = \{u^{(3)}, v^{(3)}, w^{(3)}\} &= A_3 \exp[ik_3(x\sin\theta_3 + z\cos\theta_3 - c_{L2}t)]\\ &\quad\times \{\sin\theta_3, 0, \cos\theta_3\}.\end{aligned} \tag{11.91}$$

$$\begin{aligned}\mathbf{u}^4 = \{u^{(4)}, v^{(4)}, w^{(4)}\} &= A_4 \exp[ik_4(x\sin\theta_4 + z\cos\theta_4 - c_{T2}t)]\\ &\quad\times \{\sin\theta_4, 0, \cos\theta_4\}.\end{aligned} \tag{11.92}$$

Matching the phases of the five waves along the interface $z = 0$ leads to the following equations that define the four angles $\{\theta_1, \theta_2, \theta_3, \theta_4\}$ and four wave numbers $\{k_1, k_2, k_3, k_4\}$:

$$k_0 \sin\theta_0 = k_1 \sin\theta_1 = k_2 \sin\theta_2 = k_3 \sin\theta_3 = k_4 \sin\theta_4, \tag{11.93}$$

$$k_0 c_{L1} = k_1 c_{L1} = k_2 c_{T1} = k_3 c_{L2} = k_4 c_{T2}. \tag{11.94}$$

A set of four algebraic equations for the amplitudes of the four waves generated by the incident wave is found by imposing continuity conditions on the nonzero displacement components u and w, and the nonzero traction components τ_{zz} and τ_{zx}. The resulting equations are (Achenbach, 1973, p. 186)

$$\begin{bmatrix} -\sin\theta_1 & -\cos\theta_2 & \sin\theta_3 & -\cos\theta_4 \\ \cos\theta_1 & -\sin\theta_2 & \cos\theta_3 & \sin\theta_4 \\ \sin 2\theta_1 & \dfrac{c_{L1}}{c_{T1}}\cos 2\theta_2 & \dfrac{G_2}{G_1}\dfrac{c_{L1}}{c_{L2}}\sin 2\theta_3 & -\dfrac{G_2}{G_1}\dfrac{c_{L1}}{c_{T2}}\cos 2\theta_4 \\ -\cos 2\theta_2 & \dfrac{c_{T1}}{c_{L1}}\sin 2\theta_2 & \dfrac{G_2}{G_1}\dfrac{c_{L2}}{c_{L1}}\left(\dfrac{c_{T1}}{c_{T2}}\right)^2 \cos 2\theta_4 & \dfrac{G_2}{G_1}\dfrac{c_{T1}}{c_{T2}}\dfrac{c_{T1}}{c_{L1}}\sin 2\theta_4 \end{bmatrix} \begin{bmatrix} A_1 \\ A_2 \\ A_3 \\ A_4 \end{bmatrix}$$

$$= A_0 \begin{bmatrix} \sin\theta_0 \\ \cos\theta_0 \\ \sin 2\theta_0 \\ \cos 2\theta_2 \end{bmatrix} \tag{11.95}$$

Closed-form solutions to these equations have been given by Ewing et al. (1957) and extensive numerical tables have been generated by Muskat and Meres (1940). The results display complicated and often nonmonotonic behavior, and, as mentioned by the latter authors, "no simple physical interpretation or explanation can be given for the manifold variations of the coefficients with the parameters". This difficulty is somewhat mitigated by the fact that numerical solutions to (11.95) can be generated in a straightforward manner using matrix inversion algorithms.

One important special case, which can be solved and interpreted easily, is that of a P-wave impinging on a traction-free surface (Fig. 11.5a). In general, both a P-wave and an SV-wave will be reflected back into the rock. This case is obtained from (11.95) by setting $A_3 = A_4 = 0$, so as to ignore "transmitted" waves, and also ignoring the first two equations in (11.95), because the "continuity of displacement" boundary conditions are not relevant at a free surface. The third and fourth equations in (11.95), representing the traction-free boundary conditions, remain relevant. These two equations, corresponding to the lower-left 2×2 submatrix in (11.95), can be solved for

$$\frac{A_1}{A_0} = \frac{\sin 2\theta_0 \sin 2\theta_2 - (c_L/c_T)^2 \cos^2 2\theta_2}{\sin 2\theta_0 \sin 2\theta_2 + (c_L/c_T)^2 \cos^2 2\theta_2}, \tag{11.96}$$

$$\frac{A_2}{A_0} = \frac{2(c_L/c_T)\sin 2\theta_0 \cos 2\theta_2}{\sin 2\theta_0 \sin 2\theta_2 + (c_L/c_T)^2 \cos^2 2\theta_2}. \tag{11.97}$$

These coefficients are shown in Fig. 11.5b for two values of Poisson's ratio. When $\nu < 0.26$, there will be two angles of incidence for which the amplitude

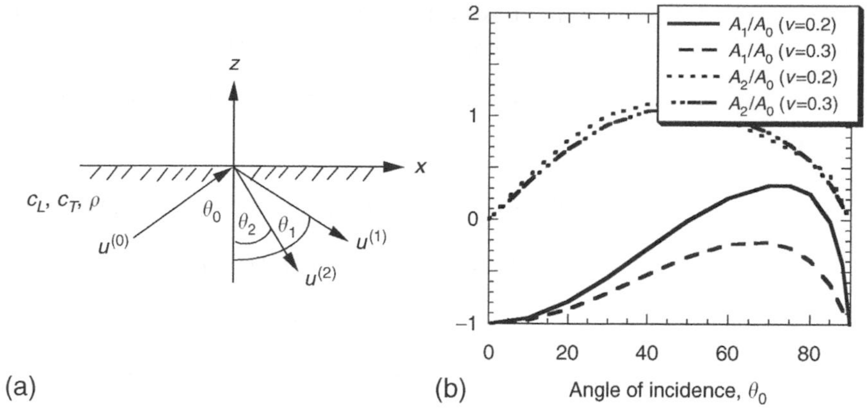

Fig. 11.5 (a) Incident P-wave and reflected P and SV-waves at a free surface. (b) Amplitude ratios for the reflected P-wave, A_1/A_0, and the reflected SV-wave, A_2/A_0.

of the reflected P-wave is zero. In this case, known as mode conversion, all of the incident energy is reflected as an SV-wave. The two angles at which mode conversion occurs are equal to 38° and 90° when $\nu = 0$, and coalesce to 68° when $\nu = 0.26$ (Arenberg, 1948).

11.6 Surface and interface waves

Although waves that propagate through an unbounded elastic medium must travel at either the longitudinal/compressional wave speed c_L or the transverse/shear wave speed c_T, this is not the case for *bounded* media. For example, long-wavelength disturbances can travel along a thin elastic bar at the speed $c = (E/\rho)^{1/2}$, as discussed in §11.2. Whenever there is a free surface or a boundary with another medium, other types of waves become possible. The most famous and important of these is the Rayleigh (1885) surface wave, which travels along a planar free surface of an elastic medium.

Consider a semi-infinite half-space whose surface is defined by $z = 0$, with z increasing into the medium. Without loss of generality, consider a plane wave traveling in the x direction parallel to the free surface with a displacement vector that lies in the x–z plane with no y-component. Searching for waves that are essentially confined near the free surface, consider the following displacement potentials:

$$\varphi = Ae^{-qz}e^{ik(x-ct)}, \quad \psi_y = Be^{-sz}e^{ik(x-ct)}. \tag{11.98}$$

Substitution of these potentials into the wave equations (11.72) and (11.73) yields

$$q = k(1 - c^2/c_L^2)^{1/2}, \quad s = k(1 - c^2/c_T^2)^{1/2}, \tag{11.99}$$

where the negative roots are discarded, since both q and s must be positive for the displacements to decay with increasing z. Furthermore, as imaginary values of q or s would lead to displacements that oscillate with z rather than of decay with increasing z, (11.99) immediately shows that the as-yet-unknown wave speed must satisfy the constraint $c < c_T < c_L$.

The amplitudes $\{A, B\}$ and the phase velocity c must be chosen so that the tractions vanish on the free surface, $z = 0$. As $\mathbf{n} = -\mathbf{e_z}$ on this surface, the stresses that must vanish are $\{\tau_{zx}, \tau_{zy}, \tau_{zz}\}$. From (11.74), the two nonzero displacements associated with (11.98) are

$$u = [ikAe^{-qz} + sBe^{-sz}]e^{ik(x-ct)}, \quad w = [-qAe^{-qz} + ikBe^{-sz}]e^{ik(x-ct)}. \tag{11.100}$$

The strains in the x–z plane are therefore given by

$$\varepsilon_{xx} = [-k^2Ae^{-qz} + iksBe^{-sz}]e^{ik(x-ct)}, \tag{11.101}$$

$$\varepsilon_{zz} = [q^2Ae^{-qz} - iksBe^{-sz}]e^{ik(x-ct)}, \tag{11.102}$$

$$2\varepsilon_{xz} = -[2ie^{-qz} + (s^2 + k^2)Be^{-sz}]e^{ik(x-ct)}. \tag{11.103}$$

Using (5.18)–(5.21), (11.55)–(11.57), and (11.99), the nonzero stresses are

$$\tau_{xx} = -\rho k^2\{[(c_L^2 - 2c_T^2)(c^2/c_L^2) + 2c_T^2]Ae^{-qz} - 2c_T(c_T^2 - c^2)^{1/2}iBe^{-sz}\}e^{ik(x-ct)}, \tag{11.104}$$

$$\tau_{zz} = -\rho k^2[(c^2 - 2c_T^2)Ae^{-qz} + 2c_T(c_T^2 - c^2)^{1/2}ie^{-sz}]e^{ik(x-ct)}, \tag{11.105}$$

$$\tau_{xz} = -\rho k^2[2(c_T^2/c_L)(c_L^2 - c^2)^{1/2}ie^{-qz} + (2c_T^2 - c^2)Be^{-sz}]e^{ik(x-ct)}. \tag{11.106}$$

Setting the two stresses given by (11.105) and (11.106) to zero on the surface $z = 0$ leads to

$$(c^2 - 2c_T^2)A + 2c_T(c_T^2 - c^2)^{1/2}iB = 0, \tag{11.107}$$

$$2(c_T^2/c_L)(c_L^2 - c^2)^{1/2}iA + (2c_T^2 - c^2)B = 0. \tag{11.108}$$

This pair of equations can have nonzero solutions for $\{A, B\}$ only if its determinant vanishes. This condition can be written as

$$c_L^2(2c_T^2 - c^2)^4 - 16c_T^6(c_L^2 - c^2)(c_T^2 - c^2) = 0. \tag{11.109}$$

According to (11.59), $(c_T/c_L)^2 = (1 - 2\nu)/2(1 - \nu)$, so (11.56) can also be written as

$$(2 - r^2)^4 - 16(1 - r^2)(1 - \alpha^2 r^2) = 0, \quad r = c/c_T, \quad \alpha^2 = (c_T/c_L)^2 = \frac{(1 - 2\nu)}{2(1 - \nu)}. \tag{11.110}$$

Expanding out (11.110) and factoring out the extraneous double root at $r^2 = 0$ gives the following cubic equation for r^2:

$$r^6 - 8r^4 + 8(3 - 2\alpha^2)r^2 - 16(1 - \alpha^2) = 0. \tag{11.111}$$

Equation (11.111) has exactly one real root satisfying $0 < c < c_T$ (Achenbach, 1973, pp. 189–91), which is the condition that is needed to ensure that (11.100) actually represents a wave propagating in the x direction with displacements that decay away from the free surface. This root is called c_R, the speed of the Rayleigh wave. Several exact expressions have been given for the relevant root of (11.109), although they are cumbersome to compute (Rahman and Barber, 1995; Nkemzi, 1997; Mavko et al., 1998). The following approximation is accurate to within 0.03 percent for all $0 < \nu < 0.5$ (Fig. 11.6):

$$\frac{c_R}{c_T} = 0.874032 + 0.200396\nu - 0.0756704\nu^2. \tag{11.112}$$

Expression (11.112) is more accurate than the often-cited approximation of Viktorov (1967), $c_R/c_T = (0.87 + 1.12\nu)/(1 + \nu)$, which gives errors as large as 0.5 percent for some values of ν.

Having calculated c_R from (11.109) or (11.111), the ratio B/A can be found from (11.107), after which the displacements (11.100) can be rewritten as

$$u = \mathrm{i}kA\left\{\mathrm{e}^{-(1-\alpha^2r^2)^{1/2}kz} - [1-(r^2/2)]\mathrm{e}^{-(1-r^2)^{1/2}kz}\right\}\mathrm{e}^{\mathrm{i}k(x-ct)}, \tag{11.113}$$

$$w = -kA\left\{(1-\alpha^2r^2)^{1/2}\mathrm{e}^{-(1-\alpha^2r^2)^{1/2}kz} - \frac{[1-(r^2/2)]}{(1-r^2)^{1/2}}\mathrm{e}^{-(1-r^2)^{1/2}kz}\right\}\mathrm{e}^{\mathrm{i}k(x-ct)}. \tag{11.114}$$

The actual displacements correspond to the real parts of (11.113) and (11.114). Taking A to be purely real yields (Bedford and Drumheller, 1994, p. 109)

$$u = -kA\left\{\mathrm{e}^{-(1-\alpha^2r^2)^{1/2}kz} - [1-(r^2/2)]\mathrm{e}^{-(1-r^2)^{1/2}kz}\right\}\sin k(x-c_Rt), \tag{11.115}$$

$$w = -kA\left\{(1-\alpha^2r^2)^{1/2}\mathrm{e}^{-(1-\alpha^2r^2)^{1/2}kz} - \frac{[1-(r^2/2)]}{(1-r^2)^{1/2}}\mathrm{e}^{-(1-r^2)^{1/2}kz}\right\} \times \cos k(x-c_Rt), \tag{11.116}$$

where c_R is given implicitly by (11.110) and $r = c_R/c_T$. The wave number k is arbitrary, implying that Rayleigh waves may travel at any frequency. Likewise,

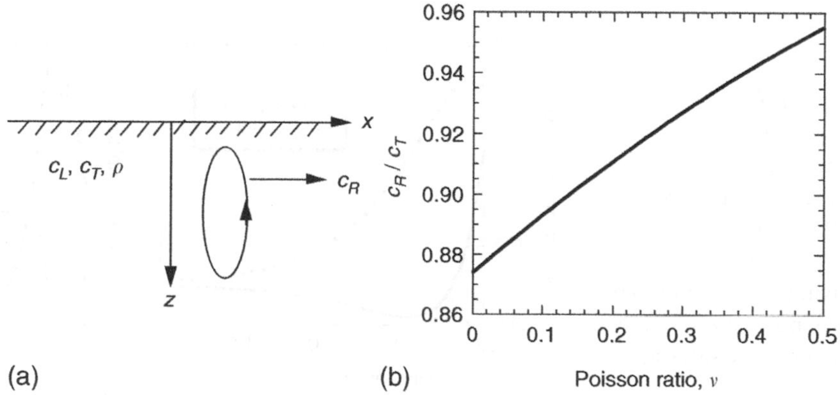

Fig. 11.6 (a) Particle velocity (ellipse) of a Rayleigh wave traveling to right at speed c_R; (b) Phase velocity of Rayleigh waves as a function of the Poisson ratio.

A is arbitrary, so Rayleigh waves may have arbitrary amplitude, apart from the restriction that the strains be sufficiently small that the linearized theory applies. The parameters r and α depend only on the Poisson ratio, and $kz = 2\pi z/\lambda$, so for any value of ν the amplitudes of the two displacement components vary only with the dimensionless depth, z/λ.

At any location (x, z), the particles move in elliptical trajectories, with the aspect ratio determined by the ratio of the two bracketed terms in (11.115) and (11.116). The magnitude of the displacement normal to the free surface, w, is nonzero at all depths, whereas the displacement parallel to the free surface, u, changes sign once as the depth increases. The critical depth at which u vanishes varies with ν, but is roughly equal to one-fifth of the wavelength. The two displacement amplitudes are plotted in Fig. 11.7a, normalized against the normal component of the displacement at the surface, $w_0 \equiv w(z = 0)$, for the case $\nu = 0.25$. The amplitudes of the stresses $\{\tau_{xx}, \tau_{zz}, \tau_{xz}\}$ are shown in Fig. 11.7b, normalized against the surface value of τ_{xx}. All displacement and stress components are negligible at depths greater than about two wavelengths. The curves would be qualitatively similar for different values of ν.

Many other types of waves can propagate along free surfaces or along interfaces between different media. Lamb waves are waves of plane strain that can propagate along a plate having traction-free upper and lower surfaces, with displacement components in both the direction of propagation along the plate and perpendicular to the plane of the plate. These waves, which are discussed in depth by Viktorov (1967) and Miklowitz (1978), are highly dispersive. At wavelengths greater than about ten times the plate thickness, the phase and group velocity of the Lamb wave approach the value

$$c_{\text{plate}} = \sqrt{E/\rho(1 - \nu^2)}, \tag{11.117}$$

which generally satisfies $c_T < c_{\text{plate}} < c_L$. The phase velocity decreases with decreasing wavelength, approaching the Rayleigh wave speed for wavelengths less than about half the plate thickness.

An analysis similar to that given for Rayleigh waves shows that surface waves in a half-space cannot have a displacement in the y direction, that is to say, parallel to the free surface but perpendicular to the direction of propagation,

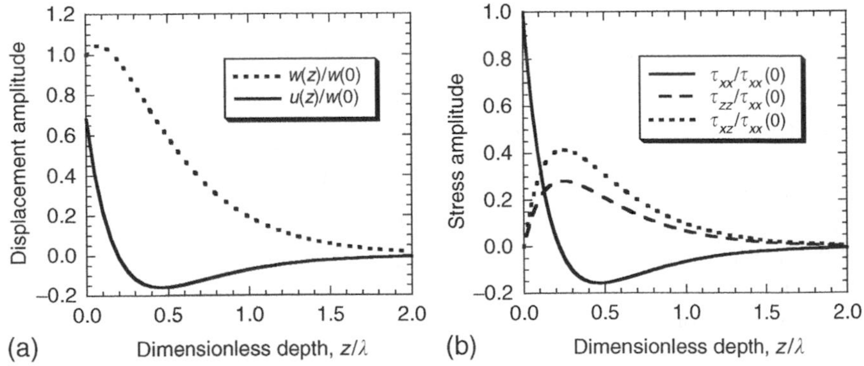

Fig. 11.7 (a) Displacement amplitudes and (b) stress amplitudes for a Rayleigh wave traveling in a rock having a Poisson ratio of 0.25, normalized to the values at the surface, $z = 0$.

because such waves cannot satisfy the traction-free boundary condition at $z = 0$. However, surface-localized SH-waves are observed to accompany earthquakes. Love (1911) explained this by showing that SH surface waves could propagate if the half-space is overlaid by a surface layer that has a shear wave speed less than that of the medium below. Love waves are dispersive; the velocity of the lowest mode is less than the velocity of shear waves in the sublayer, approaching the substrate shear velocity at very long wavelengths.

Stoneley (1924) studied waves that propagate along the planar interface between two elastic half-spaces. He found that such waves could exist only in a small range of the parameter space defined by the density ratio and the shear wave velocity ratio of the two media (Miklowitz, 1978, p. 168). For two media with nearly equal densities, as would be the case for most rocks, Stoneley waves can only propagate if the shear wave speeds of the two media are nearly equal. The speed of the Stoneley wave is then slightly less than the shear wave speeds in either of the two adjacent media.

A special case of a Stoneley wave is that of a wave traveling along the interface between an elastic solid and a fluid; these are also known as Scholte waves. The most important examples of these waves for engineering purposes are the "borehole waves" that travel along a fluid-filled borehole (Paillet and Cheng, 1991; Fjaer et al., 1992, chapter 8). The most important borehole mode travels at a speed less than that of the compressional wave speed in the fluid, approaching c_L(fluid) for "fast" formations, defined as those for which c_T(rock) $\gg c_L$(fluid). The energy in these borehole waves is to a great extent localized within the fluid-filled borehole.

11.7 Transient waves

The previous two sections considered sinusoidally varying "steady-state" waves, such as are used in seismic exploration. In rock mechanics, transient waves, such as are produced by man-made explosions or by naturally occurring rockbursts, are frequently of interest. As the simplest example of a transient wave, consider an isotropic half-space $x > 0$, subjected to a spatially uniform, time varying normal traction

$$\tau_{xx}(x = 0, t) = p(t), \tag{11.118}$$

where $p(t)$ is some function that satisfies $p(t) = 0$ if $t < 0$. By the symmetry of this problem, the only nonzero displacement will be u, and it will vary only with x and t. Consequently, the resulting motion $u(x, t)$ will be irrotational and will be governed by (11.70a):

$$\frac{\partial^2 u}{\partial x^2} = \frac{1}{c_L^2}\frac{\partial^2 u}{\partial t^2}. \tag{11.119}$$

From (11.9), the general solution to this equation is

$$u(x, t) = f(x - c_L t) + g(x + c_L t). \tag{11.120}$$

It can be shown rigorously that $g = 0$ in this problem (Achenbach, 1973, p. 23), although this is clear on physical grounds, since f represents a wave that travels *into* the half-space, whereas g would represent a wave that starts within the half-space and travels *toward* the surface.

As in §11.2, it is convenient to write (11.119) in the equivalent form $u(x,t) = f(t - x/c_L)$. The normal strain in the x direction is then given by

$$\varepsilon_{xx} = \frac{\partial u}{\partial x} = -\frac{1}{c_L} f'(t - x/c_L), \tag{11.121}$$

and the normal stress in the x direction is given by

$$\tau_{xx} = (\lambda + 2G)\varepsilon_{xx} = -\frac{(\lambda + 2G)}{c_L} f'(t - x/c_L). \tag{11.122}$$

Boundary condition (11.118) implies that

$$-\frac{(\lambda + 2G)}{c_L} f'(t) = p(t), \tag{11.123}$$

which can be integrated to yield

$$f(t) = -\frac{c_L}{\lambda + 2G} \int_0^t p(s) ds + A, \tag{11.124}$$

where A is a constant of integration. Recalling that $u(x,t) = f(t - x/c_L)$ and $\lambda + 2G = \rho c_L^2$, we find

$$u(t - x/c_L) = -\frac{1}{\rho c_L} \int_0^{t - x/c_L} p(s) ds + A. \tag{11.125}$$

The displacement must vanish throughout the medium when $t < 0$, so A must be zero. Therefore,

$$u(x,t) = -\frac{1}{\rho c_L} \int_0^{t - x/c_L} p(s) ds, \tag{11.126}$$

from which it follows that the stress is given by

$$\tau_{xx}(x,t) = (\lambda + 2G)\frac{\partial u}{\partial x} = -(\lambda + 2G)\left(\frac{-1}{\rho c_L^2}\right) p(t - x/c_L) = p(t - x/c_L). \tag{11.127}$$

The stress at any location x is zero until t reaches the value $t = x/c_L$, corresponding to the arrival of the wavefront. After this time, the stress at x follows the imposed traction history at the surface, with a time delay of x/c_L.

The displacement at any location x is zero for $t < x/c_L$, because the integrand in (11.126) vanishes until this time. For $t > x/c_L$, the displacement increases in proportion to the time integral of the surface traction. The particle velocity can be found by differentiating (11.126):

$$\dot{u}(x,t) = \frac{\partial u(x,t)}{\partial t} = -\frac{1}{\rho c_L} p(t - x/c_L), \tag{11.128}$$

which shows that the acoustic impedance ρc_L can be interpreted as the coefficient that relates the particle velocity to the traction at a wavefront. The minus sign in (11.128) is an artifact of the sign convention used for displacements; a compressive traction at the free surface does indeed cause the rock to move in the positive x direction, into the half-space, as would be expected.

If the imposed surface traction is of finite duration, say Δt, then $p(t) = 0$ for $t > \Delta t$. After the wave has passed a given location x, a steady-state residual displacement will remain, given by (Rinehart, 1975, pp. 32–3)

$$u(x, t > \Delta t + x/c_L) = -\frac{1}{\rho c_L} \int_0^{\Delta t - x/c_L} p(s)ds. \tag{11.129}$$

Whereas the above planar wave travels unaltered through the rock mass, transient waves that emanate from cylindrical or spherical cavities have an entirely different character. Consider, for example, a normal traction $p(t)$ acting against the surface of a spherical cavity of radius r_0, in an infinite rock mass. The resultant displacement will have spherical symmetry, and in spherical coordinates, the only nonzero displacement component will be $u(r)$. The stress–displacement equations for spherically symmetric deformations are, from (5.117)–(5.120),

$$\tau_{rr} = (\lambda + 2G)\frac{\partial u}{\partial r} + 2\lambda\frac{u}{r}, \qquad \tau_{\theta\theta} = \lambda\frac{\partial u}{\partial r} + 2(\lambda + G)\frac{u}{r}, \tag{11.130}$$

where $\tau_{\theta\theta} = \tau_{\phi\phi}$ are the normal stresses in the two directions perpendicular to r. The only nontrivial equation of motion is given by (5.122), with the inertia term included on the right-hand side:

$$\frac{\partial \tau_{rr}}{\partial r} + \frac{2(\tau_{rr} - \tau_{\theta\theta})}{r} = \rho\frac{\partial^2 u}{\partial t^2}. \tag{11.131}$$

Substituting (11.130) into (11.131) yields

$$\frac{\partial^2 u}{\partial r^2} + \frac{2}{r}\frac{\partial u}{\partial r} - \frac{2u}{r^2} = \frac{1}{c_L^2}\frac{\partial^2 u}{\partial t^2}. \tag{11.132}$$

This equation can be put into the standard form of a wave equation by defining, in analogy with the treatment in §11.4, a scalar potential φ, such that

$$u = \frac{\partial \varphi}{\partial r}. \tag{11.133}$$

Substitution of (11.133) into (11.132) shows that the displacement u will satisfy the equation of motion, provided that φ atisfies the following equation:

$$\frac{\partial^2 (r\varphi)}{\partial r^2} = \frac{1}{c_L^2}\frac{\partial^2 (r\varphi)}{\partial t^2}. \tag{11.134}$$

Hence, the displacement component u does *not* satisfy a wave equation, but the product $r\varphi$ does satisfy a wave equation, with wave speed c_L. The general solution therefore can be written as

$$r\varphi(r,t) = f(r - c_L t) + g(r + c_L t). \tag{11.135}$$

The function g can again be discarded, as it represents a wave moving toward the cavity.

It is easily verified by use of the chain rule that if $f(\eta)$ is a solution to the wave equation, with $\eta = r - c_L t$, then $f(a\eta + b)$ will also be a solution, for any constants a and b. Choosing $a = -1/c_L$ and $b = r_0/c_L$ allows the solution to be written in the following equivalent but more convenient form:

$$\varphi(r,t) = \frac{1}{r} f[t - (r - r_0)/c_L] \equiv \frac{1}{r} f(s). \tag{11.136}$$

The variable $s = t - (r - r_0)/c_L$ represents the time that has elapsed since the wavefront first arrived at location r.

The displacement and stress fields associated with (11.136) follow from (11.130) and (11.133):

$$u(r,t) = -\frac{f(s)}{r^2} - \frac{f'(s)}{c_L r}, \tag{11.137}$$

$$\tau_{rr}(r,t) = \frac{(\lambda + 2G) f''(s)}{c_L^2 \quad r} + \frac{4G f'(s)}{c_L \quad r^2} + 4G\frac{f(s)}{r^3}, \tag{11.138}$$

$$\tau_{\theta\theta}(r,t) = \tau_{\phi\phi}(r,t) = \frac{\lambda f''(s)}{c_L^2 \quad r} - \frac{2G f'(s)}{c_L \quad r^2} - 2G\frac{f(s)}{r^3}, \tag{11.139}$$

and all shear stress components in the spherical coordinate system are zero. Applying the condition that the normal traction given by (11.138) must equal $p(t)$ at the cavity surface, where $s = t$, leads to

$$f''(s) + \frac{4c_T^2}{r_0 c_L} f'(s) + \frac{4c_T^2}{r_0^2} f(s) = \frac{r_0 c_T^2}{G} p(s), \tag{11.140}$$

The solution to this problem, for arbitrary $p(t)$, is given by Achenbach (1973, pp. 130–1) in the form of a convolution integral. For the case of a constant pressure $p(t) = p_0$ applied instantaneously at $t = 0$, we proceed as follows. The general solution to (11.140) is

$$f(s) = (A\cos\omega s + B\sin\omega s)e^{-\zeta s} + \frac{r_0^3 p_0}{4G}, \tag{11.141}$$

$$\text{where} \quad \zeta = \frac{2c_T^2}{r_0 c_L}, \qquad \omega = \frac{2c_T(c_L^2 - c_T^2)^{1/2}}{r_0 c_L}, \tag{11.142}$$

and $\zeta/\omega = (1 - 2\nu)^{1/2}$. If the rock mass is at rest before the application of the surface tractions on the cavity wall, the initial conditions that must be satisfied are $f(0) = f'(0) = 0$. Imposition of these conditions leads to

$$f(s) = \frac{r_o^3 p_o}{4G}\left[1 - \left(\cos\omega s + \frac{\zeta}{\omega}\sin\omega s\right)e^{-\zeta s}\right], \tag{11.143}$$

with $s = t - (r - r_o)/c_L$ and the implicit understanding that $f = 0$ for $s < 0$. Finally, insertion of (11.143) into (11.137) yields the displacement in the form (Fig. 11.8a)

$$u(r,t) = -\frac{r_o^3 p_o}{4Gr^2}\left\{1 + \left[\left(\frac{2r}{r_o} - 1\right)\frac{\zeta}{\omega}\sin\omega s - \cos\omega s\right]e^{-\zeta s}\right\}H(s), \tag{11.144}$$

where $H(s)$ is the Heaviside unit-step function, defined to be zero for $s < 0$ and 1 for $s > 0$. The solution given by Graff (1975, p. 298) has an erroneous − sign in front of the square-bracketed term, causing the displacements shown on p.299 to exhibit unrealistic *instantaneous* jumps at the wavefront. The displacements plotted by Sharpe (1942) and repeated by Rinehart (1975, p. 47) were computed by ignoring the steady-state component of (11.144), on the grounds that it is of order $(r_o/r)^2$ and presumably negligible compared to the transient terms of order (r_o/r); these graphs therefore erroneously show the displacement to stabilize at zero, rather than at the steady-state value.

The radial normal stress is found from (11.138) and (11.143) to be, for $s = t - (r - r_o)/c_L > 0$,

$$\frac{\tau_{rr}}{p_o} = \left(\frac{r_o}{r}\right)^3 + \left\{\left[\left(\frac{r_o}{r}\right) - \left(\frac{r_o}{r}\right)^3\right]\cos\omega s - \left[\left(\frac{r_o}{r}\right) - 2\left(\frac{r_o}{r}\right)^2 + \left(\frac{r_o}{r}\right)^3\right]\frac{\zeta}{\omega}\sin\omega s\right\}e^{-\zeta s}, \tag{11.145}$$

where again $\zeta/\omega = (1 - 2\nu)^{1/2}$.

At any fixed location r, the displacement, stresses, and strains will be zero until the arrival of the wavefront, which occurs at $t = (r - r_o)/c_L$. The character of the wave at, and immediately behind, the wavefront is found by fixing r and

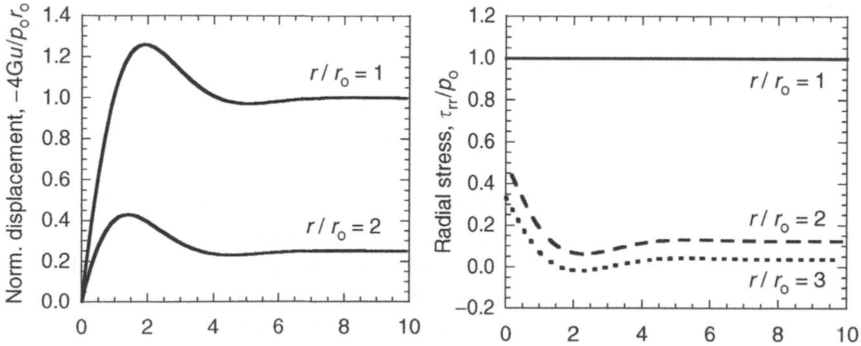

Fig. 11.8 (a) Radial displacement and (b) radial stress due to a step-function traction applied at the surface of a spherical cavity in an infinite rock mass.

taking the limit as s approaches 0^+. The results are

$$-u(r, \text{small } s) \approx \frac{p_o c_T^2}{G c_L}\left(\frac{r_o}{r}\right) s = \frac{p_o}{\rho c_L}\left(\frac{r_o}{r}\right) s, \tag{11.146}$$

$$-\dot{u}(r, \text{small } s) \approx \frac{p_o}{\rho c_L}\left(\frac{r_o}{r}\right), \tag{11.147}$$

$$\tau_{rr}(r, \text{small } s) \approx p_o\left(\frac{r_o}{r}\right), \tag{11.148}$$

$$\tau_{\theta\theta}(r, \text{small } s) = \tau_{\phi\phi}(r, \text{small } s) \approx \frac{\nu}{1-\nu} p_o\left(\frac{r_o}{r}\right), \tag{11.149}$$

where it is implicit that each of these equations has a Heaviside unit-step function on the right side, so that all terms vanish for $s < 0$. The normal stress at location r jumps abruptly to $p_o(r_o/r)$ at $t = (r - r_o)/c_L$ and then decays in a highly damped oscillatory manner, eventually stabilizing at the steady-state value, $p_o(r_o/r)^3$ (Fig. 11.8b). The displacement grows linearly with time after the arrival of the front, but its magnitude also scales as r_o/r. This $1/r$ decay of the stress and displacement is a general property of spherical wavefronts (Miklowitz, 1978, p. 77). The relationship between particle velocity and normal stress, $\tau_{rr} = -\rho c_L \dot{u}$, is the same as for plane waves.

As $t \to \infty$ for a fixed value of r, the displacement approaches the value

$$u(r, t \to \infty) = -\frac{r_o^3 p_o}{4Gr^2}, \tag{11.150}$$

which is precisely the displacement given by the static theory, (8.306). After the initial rise at $t = (r - r_o)/c_L$, the displacement oscillates due to the sinusoidal term in (11.144), and decays in time due to the exponential term, eventually stabilizing at the static value at long times. Since the exponential term becomes negligible when $\zeta s \approx 4$, it can be shown from (11.142) and (11.144) that the oscillations will die out when

$$c_L s/r_o \approx 2(c_L/c_T)^2 = 4(1-\nu)/(1-2\nu). \tag{11.151}$$

For a rock with a Poisson ratio of 0.25, the right-hand side of (11.151) equals 6. Since the distance traveled by the wavefront during an elapsed time s is $c_L s$, the oscillations at any location r will die out after the wave has traveled a distance equivalent to an additional six cavity radii.

The solution for a pressure pulse applied to the surface of a cylindrical cavity is much more mathematically complicated, involving Bessel functions (Selberg, 1952; Miklowitz, 1978, pp. 282–90). However, the general behavior is qualitatively similar to that of the spherical cavity. The wave arrives at location r at time $t = (r - r_o)/c_L$, at which time the stresses rise abruptly and the displacement grows linearly with elapsed time. However, the magnitudes of these quantities decay spatially as $(r_o/r)^{1/2}$, rather than as (r_o/r), as occurred for spherical waves.

Near the wavefront, for $t > (r - r_0)/c_L$,

$$u(r,t) \approx -\frac{p_0}{\rho c_L}\left(\frac{r_0}{r}\right)^{1/2}[t-(r-r_0)/c_L], \tag{11.152}$$

$$\dot{u}(r,t) \approx -\frac{p_0}{\rho c_L}\left(\frac{r_0}{r}\right)^{1/2}, \tag{11.153}$$

$$\tau_{rr}(r,t) \approx p_0\left(\frac{r_0}{r}\right)^{1/2}, \tag{11.154}$$

$$\tau_{\theta\theta}(r,t) = \tau_{\phi\phi}(r,t) \approx \frac{\nu}{1-\nu}p_0\left(\frac{r_0}{r}\right)^{1/2}. \tag{11.155}$$

Eventually, the transient part of the wave dies away, leaving the static displacement, as given by (8.90):

$$u(r,t\to\infty) = -\frac{r_0^2 p_0}{2Gr}. \tag{11.156}$$

11.8 Effects of fluid saturation

The theory of wave propagation presented in previous sections of this chapter applies to a homogeneous, single-phase elastic material. When the void space of a porous or fractured rock is saturated with a fluid, the rock can no longer be considered to be a homogeneous, single-phase material. Several approaches, having differing degrees of rigor and complexity, have been used to account for the effect of fluid saturation on seismic velocities.

Consider a rock with porosity ϕ, saturated with a fluid having density ρ_f and bulk modulus K_f; the shear modulus of a fluid is zero. The mineral grains have effective moduli $\{K_m, G_m\}$, which can be accurately estimated from the mineral composition using the Voigt-Reuss-Hill average of §10.2, and effective density ρ_m, which is exactly given by the volumetric average of the densities of the individual mineral components. Wyllie et al. (1956) approximated the travel time of a wave passing through this rock by a volume-weighted average of the travel times through a layer of solid rock and a layer of pore fluid. This model, known as the *Wyllie time-average*, leads to

$$\frac{1}{V_p} = \frac{\phi}{V_{pf}} + \frac{1-\phi}{V_{pm}}, \tag{11.157}$$

where $V_{pf} = (K_f/\rho_f)^{1/2}$ and $V_{pm} = [(3K_m + 4G_m)/3\rho_m]^{1/2}$ are the compressional wave speeds of the fluid and mineral phase and V_p is the compressional wave speed of the actual fluid-saturated rock. This model usually underestimates wave speeds, but has some validity for cemented and consolidated sandstones, particularly at high pressures, when the crack-like pores have been shut. This approach cannot be applied to shear waves, as the shear wave speed in a fluid is effectively zero.

The next level of sophistication is to treat the fluid-saturated rock as an "effective elastic medium". In this approach, the effective moduli and density are calculated in the static limit and the wave speeds are then calculated from (11.55) and (11.57). Let the rock have effective moduli $\{K_d, G_d\}$ under dry/drained conditions. Addition of a pore fluid cannot increase the shear stiffness, so the shear modulus of the fluid-saturated rock must be G_d. In the "quasi-static" limit of low frequencies, the effective bulk modulus of the fluid-saturated rock is assumed to be given by the Gassmann equation for the "undrained" bulk modulus, (7.27), rewritten in (11.158) in terms of bulk moduli rather than compressibilities. Hence, the effective moduli of the fluid-saturated rock are taken to be

$$K_u = \frac{\phi(1/K_f - 1/K_m) + (1/K_d - 1/K_m)}{\phi(1/K_f - 1/K_m)/K_d + (1/K_d - 1/K_m)/K_m}, \quad G_u = G_d. \tag{11.158}$$

The Gassmann equation assumes that the pressure in the pore fluid is uniform throughout the rock. If the wave is of sufficiently low frequency, the pore pressure will have time to locally equilibrate within the time needed for the stress pulse to pass through a region of the rock, in accordance with Gassmann's assumption. The effective density in the effective medium approach is the volumetrically weighted average of the mineral and fluid densities:

$$\rho = (1 - \phi)\rho_m + \phi\rho_f. \tag{11.159}$$

The wave speeds in the fluid-saturated rock are then given by

$$V_p(\text{low } \omega) = \{[3K_u + (4/3)G_u]/\rho\}^{1/2}, \quad V_s(\text{low } \omega) = (G_u/\rho)^{1/2}, \tag{11.160}$$

where the effective moduli and density are given by (11.158) and (11.159).

Saturating the pore space with fluid will increase the density and leave the shear modulus unchanged, leading to a slightly lower shear wave velocity. For example, in a sandstone of 20 percent porosity, with mineral density 2.65 g/cm^3, saturating the pore space with water having density 1.00 g/cm^3 will lower the shear wave speed by about 6 percent. The increase in bulk modulus usually overshadows the increase in density, so that compressional wave speeds are greater under saturated conditions than under dry conditions. The Gassmann approach usually works very well at seismic frequencies, below about 100 Hz.

At higher frequencies, in the logging (10 kHz) or laboratory ultrasonic (1 MHz) range, the pore fluid does not have sufficient time to redistribute itself so as to locally equilibrate the pore pressure. In this regime, the fluid-saturated rock cannot be treated as an effective single-phase continuum. Rather, the motion of the fluid, as distinguished from that of the solid phase, must somehow be accounted for. Biot (1956a,b) developed a theory based on the model of pores being long, cylindrical tubes, that allowed for macroscopic flow of the fluid phases. This theory reduces to the Gassmann approach at low frequencies, but predicts higher P- and S-wave velocities in the high-frequency limit. At intermediate frequencies, the velocities are given by complicated expressions that involve the parameters appearing in (11.158) and (11.159), along with a "tortuosity" parameter and a characteristic grain size parameter (Berryman, 1980).

Although Biot's theory works reasonably well for very porous, high-permeability sediments (Stoll, 1989), it does not give accurate results for consolidated rocks. Wave propagation at logging or acoustic frequencies in rocks seems to involve small-scale "squirt-like" fluid flow that occurs on the length scale of individual pores and cracks (Mavko and Nur, 1975), not necessarily in the direction of propagation of the wave, as is implicitly assumed in Biot's model. Mavko and Jizba (1991) developed a procedure for estimating the wave speeds over the complete range of frequencies, which required knowledge of the wave speeds in the dry rock at a given value of the effective stress, and in the high-stress limit, when all crack-like pores are closed. The computational procedure is summarized by Mavko et al. (1998, pp. 186–9).

11.9 Attenuation

When a wave travels through an elastic medium, the total energy contained in the wave, which was shown in §11.2 and §11.3 to be partitioned between elastic strain energy and kinetic energy, is conserved. A *plane* elastic wave will propagate without any change in its amplitude. For waves that spread out radially, such as those emanating from spherical cavities or cylindrical boreholes, the amplitude will decrease, because a finite amount of energy is spread out over a wavefront having ever-increasing area. This type of amplitude decay is known as *geometric attenuation* and is not associated with any loss of overall kinetic energy.

However, rocks do not behave entirely elastically under transient conditions. There are numerous mechanisms which cause the kinetic energy of seismic waves to be transformed into internal energy. This energy is not lost, but rather serves to raise the temperature of the rock slightly. But from a purely mechanical point of view, this energy appears to be "lost" or "dissipated".

The attenuation of a plane wave can be introduced by following the development in §11.2 for one-dimensional wave propagation, but using a simple viscoelastic constitutive model, such as the Kelvin-Voigt model of §9.9:

$$\tau_{xx} = E\varepsilon_{xx} + \eta\dot{\varepsilon}_{xx}, \tag{11.161}$$

where η is a viscosity-like parameter. This constitutive model is often interpreted as an elastic element (spring) in parallel with a viscous element (dashpot). If we consider a process in which the strain varies according to $\varepsilon_{xx} = \varepsilon_0 e^{i\omega t}$, the stress can be written as

$$\tau_{xx} = (E + i\omega\eta)\varepsilon_0 e^{i\omega t} = M_R + iM_I\varepsilon_{xx}, \tag{11.162}$$

which shows that E can be interpreted as the real part of the complex modulus and $\omega\eta$ as the imaginary part.

Substituting (11.161) into the governing equation (11.2) yields

$$E\frac{\partial^2 u}{\partial x^2} + \eta\frac{\partial^3 u}{\partial t \partial x^2} = \rho\frac{\partial^2 u}{\partial t^2}. \tag{11.163}$$

Now consider a plane wave described by

$$u(x,t) = U_0 \exp\{i[(k_R + (ik_I))x - \omega t]\}, \tag{11.164}$$

where we take the wave number to have both a real and an imaginary part. Substitution of (11.164) into (11.163) yields the requirement that

$$(k_R + ik_I)^2(E - i\omega\eta) = \rho\omega^2, \tag{11.165}$$

the solution to which is (Kolsky, 1963, p. 117)

$$k_R = \left[\frac{\rho E \omega^2}{2(E^2 + \eta^2\omega^2)} \left\{ \left(\frac{E^2 + \eta^2\omega^2}{E^2} \right)^{1/2} + 1 \right\} \right]^{1/2}, \tag{11.166}$$

$$k_I = \left[\frac{\rho E \omega^2}{2(E^2 + \eta^2\omega^2)} \left\{ \left(\frac{E^2 + \eta^2\omega^2}{E^2} \right)^{1/2} - 1 \right\} \right]^{1/2}. \tag{11.167}$$

The positive root for k_R is chosen so that the wave propagates to the right, whereas the positive root must be chosen for k_I so as not to yield a wave whose amplitude grows as it propagates.

For a nonmolten rock, the elastic part of the stress would be expected to dominate the viscous part, which is to say η must in some sense be small. Expanding (11.166) and (11.167) for small values of η gives

$$k_R = \frac{\omega}{c_0}\left[1 - \frac{3}{8}\left(\frac{\eta\omega}{E}\right)^2\right], \quad k_I = \frac{\eta\omega^2}{2Ec_0}, \tag{11.168}$$

where $c_0 = (E/\rho)^{1/2}$ is the velocity that the elastic wave would have in the absence of any dissipative mechanisms. The actual velocity, $c = \omega/k_R$, varies with frequency, as it must for any dissipative medium, as required by the Kramers-Kronig relations (Mavko et al., 1998, pp. 75–7). However, to first-order in η, the wave speed is unaffected by a small amount of viscous damping and is given by $c = (E/\rho)^{1/2}$. Using this further simplification, the wave (11.163) can be expressed as

$$u(x,t) = U_0 \exp(-k_I x) \exp\{i[(\omega/c_0)x - \omega t]\}, \tag{11.169}$$

where k_I is given by (11.168). Thus, the wave travels at velocity c_0 but with an amplitude that decays exponentially with distance. This represents actual viscous attenuation, rather than the geometrical attenuation found in spherical or cylindrical waves.

According to this model, the attenuation seems to increase with the square of the frequency. However, there are various mechanisms in rocks that give rise to viscous-like behavior, and each has, in effect, its own dependence of η on frequency. Thus, each mechanism predicts a frequency-dependence of attenuation that will reflect both the ω^2 term from (11.168) and the frequency-dependence of η, usually giving rise to an exponent that differs from 2. Before discussing these dissipative mechanisms, we discuss several standard definitions that are used to quantify attenuation.

The imaginary part of the wavenumber, k_I, is also denoted by α, the *attenuation coefficient*. Its inverse, $1/k_I$, is the length over which the amplitude will decay by a factor of $1/e \approx 0.37$. The mechanical energy (kinetic plus elastic strain energy) contained in a sinusoidal plane wave is proportional to the square of the amplitude, according to (11.33), so the fractional loss of energy over one wavelength is

$$\frac{\Delta\mathcal{T}}{\mathcal{T}} = \frac{\exp(-2\alpha x) - \exp(-2\alpha\{x+\lambda\})}{\exp(-2\alpha x)} = 1 - \exp(-2\alpha\lambda) \approx 2\alpha\lambda. \tag{11.170}$$

The quality factor Q is defined in terms of this fractional energy loss as follows:

$$\frac{1}{Q} \equiv \frac{\Delta \mathcal{T}}{2\pi \mathcal{T}} = \frac{2\alpha\lambda}{2\pi} = \frac{2\alpha c}{\omega}, \qquad (11.171)$$

where the last step makes use of the relation $\lambda = 2\pi c/\omega$. Substituting k_{I} from (11.168) into (11.171), and recalling that $\omega\eta = M_{\mathrm{I}}$ and $E = M_{\mathrm{R}}$ shows that Q can also be expressed as

$$\frac{1}{Q} = \frac{2\alpha c}{\omega} = \frac{2c\eta\omega^2}{2Ec\omega} = \frac{\eta\omega}{E} = \frac{M_{\mathrm{I}}}{M_{\mathrm{R}}}. \qquad (11.172)$$

It can also be shown that, if α is small, $1/Q$ is equal to the phase shift (in radians) between the stress and the strain, under sinusoidal oscillations such as described in (11.162). Another parameter occasionally used to quantify attenuation in rocks is the *logarithmic decrement*, defined by $\delta = \pi/Q$.

Although α and Q contain the same information, α essentially measures the energy loss per distance traveled by the wave, whereas $1/Q$ measures the energy loss per wave cycle. Hence, as seen in (11.172), they will vary with frequency in different ways, a fact that should be remembered when viewing such graphs.

Versions of the relations (11.170)–(11.172) that do not require the assumption of small attenuation are given by Bourbié et al. (1987, p. 113). Relations similar to those described above for waves propagating along a thin bar can be derived for bulk P- and S-waves in terms of the real and imaginary parts of K and G, and the P- and S-wave quality factors (Winkler and Nur, 1979).

Toksöz and Johnston (1981) have collected many of the seminal papers on wave attenuation in rocks and have provided several summary/overview chapters. Bourbié et al. (1987) present much data on attenuation measurements and also review the various mechanisms. Measured values of Q for P-waves in various rocks are shown in Table 11.2. Porous rocks such as sandstones and limestones tend to have Q values in the range of 10–100, whereas igneous and metamorphic

Table 11.2 P-wave quality factors of several rocks.

Rock	Condition	*f* (Hz)	*Q*	Source
Tennessee marble	Dry	$0\text{–}2 \times 10^4$	480	Wyllie et al. (1962)
Quincy granite	Air dry	$2\text{–}45 \times 10^2$	125	Birch and Bancroft (1938)
Solenhofen limestone	Air dry	$3\text{–}15 \times 10^6$	112	Peselnick and Zeitz (1959)
Amherst sandstone	Oven dry	$1\text{–}13 \times 10^3$	52	Born (1941)
Pierre shale	*in situ*	$5\text{–}45 \times 10^1$	32	McDonal et al. (1958)
Berea sandstone	Brine sat.	$2\text{–}8 \times 10^5$	10	Toksöz et al. (1979)

rocks will be in the range 100–1000 (Bradley and Fort, 1966). As an approximation, it is usually assumed that the attenuation arising from different mechanisms, as quantified by Q^{-1}, is additive. Individual mechanisms that give rise to values of $Q > 1000$ are therefore negligible in comparison with other, more dominant, mechanisms.

Walsh (1966) developed a model in which attenuation is due to sliding friction along the faces of closed elliptical cracks. The model predicts $Q^{-1} = f(\mu, E/E_{\mathrm{m}})\varpi$, where μ is the coefficient of sliding friction along the crack faces, f is a dimensionless function whose values are in the order of 0.1, and ϖ is the crack-density parameter for those cracks whose faces are barely touching. There will be no frictional sliding along faces of open cracks, and the small stresses associated with seismic waves will be insufficient to cause sliding on crack faces that are tightly closed. Savage (1969) argued that there are unlikely to be a sufficient number of "barely closed" cracks to yield appreciable values of Q^{-1}. Mavko (1979) considering a tapered crack, a portion of whose two faces will always be in contact, and for typical parameter values found $Q^{-1} \approx \varepsilon/\bar{\alpha}_i$, where ε is the incremental strain associated with the wave, and $\bar{\alpha}_i$ is some appropriate mean value of the initial crack aspect ratio. For values such as $\bar{\alpha}_i \approx 10^{-3}$, this attenuation may be appreciable under laboratory conditions but would be negligible at the strains encountered in seismic waves, which would typically be $<10^{-6}$, except very close to the source (Winkler et al., 1979).

Savage (1966) analyzed attenuation arising from the conversion of mechanical energy into internal energy due to the coupling between strain and heat flow (§7.8). Strain rises and falls sharply in the vicinity of cracks or pores, causing local heat flow, leading to wave attenuation described by

$$Q^{-1} = \left(\frac{\beta^2 K T_{\mathrm{o}}}{\rho c_{\mathrm{v}}}\right) g(\nu)\phi F(\omega), \tag{11.173}$$

where β is the *volumetric* thermal expansion coefficient, K is the bulk modulus, T_{o} is the ambient temperature, ρ is the rock density, c_{v} is the specific heat, ϕ is the porosity, g is a dimensionless function of the Poisson ratio (taking on a slightly different form depending on whether the wave is compressional or shear), and F is a dimensionless function of frequency. The term in parenthesis in (11.173) is the thermoelastic coupling parameter that measures the extent to which strain induces heat flow and is on the order of 10^{-3}–10^{-4} for most rocks (Zimmerman, 2000). The function F is greatest at a characteristic frequency of about $\omega^* = k_{\mathrm{T}}/\rho c_{\mathrm{v}} a^2$, where a is the pore/crack size and k_{T} is the thermal conductivity; it increases with ω for lower frequencies and drops off as ω^{-1} at higher frequencies. At the critical frequency, the diffusion length of the induced heat flux is in the order of the pore/crack size. By assuming a range of crack sizes so as to smear out the function F over a range of frequencies, Savage found Q values of about 225 for longitudinal waves and 350 for transverse waves in granite, about twice as high as the measured values. This mechanism therefore seems to be important for cracked igneous rocks and can lead to attenuations as large as about $Q^{-1} \approx 0.01$.

The Biot theory of wave propagation in fluid-saturated rocks predicts attenuation due to the viscous drag exerted by the pore walls on the pore fluid. At frequencies below the Biot critical frequency $\omega^* = \phi\mu/k\rho_f$, which for most rock/fluid systems exceeds 1 MHz, the theory predicts (Schon, 1996, p. 252)

$$Q^{-1} = \frac{g\rho_f^2 k\omega}{2\rho\mu}, \tag{11.174}$$

where ρ_f is the density of the fluid, ρ is the density of the fluid-saturated rock, k is the permeability, and g is a numerical constant that equals 1 for shear waves and is less than 1 for compressional waves. The inverse relation between attenuation and viscosity predicted by (11.174) is in contradiction to most measured data (Jones and Nur, 1983). Mochizuki (1982) made measurements on a Massilon sandstone and found attenuations much greater than those predicted by Biot theory. For consolidated rocks with permeabilities below about 1 Darcy, Biot attenuation appears to be negligible at seismic and logging frequencies ($<10^4$ Hz). Biot theory has been more successful when applied to highly permeable sediments, for which (11.174) yields appropriately high values of $1/Q$ (Stoll, 1989).

Whereas the Biot attenuation mechanism is based on "global flow" of the pore fluid, Mavko and Nur (1975), Murphy et al. (1984) and others have modeled the attenuation arising from the local flow of fluid squirting out of compliant cracks that are compressed by the passing elastic wave. This "squirt" flow is local, and is not necessarily aligned with the direction of wave propagation. Instead, it occurs locally in directions determined by the geometry of crack and pore intersections. The attenuation predicted by squirt flow models goes to zero at low frequencies, since in the limit of zero frequency, there can be no viscous attenuation, and also goes to zero at very high frequencies, at which the fluid does not have sufficient time to move from one crack to a neighboring crack within one period of the wave. The attenuation is peaked about a critical frequency that is roughly given by (Sams et al., 1997)

$$\omega^* \approx K\alpha^3/\mu, \tag{11.175}$$

where K is the bulk modulus of the rock and α is the aspect ratio of the cracks. Sams et al. (1997) measured wave speeds and attenuations on a finely layered sequence of limestones, sandstones, siltstones, and mudstones in northeast England, over a range of methods/frequencies spanning vertical seismic profiling (30–280 Hz), crosshole surveys (0.2–2.3 kHz), sonic logging (8–24 kHz), and laboratory measurements (300–900 kHz), and found a bell-shaped curve for $1/Q$ vs. $\ln\omega$ that fit very well to the squirt-flow model.

One mechanism that produces "attenuation" without any conversion of mechanical energy into internal energy is elastic wave scattering. When an elastic wave impinges upon an inhomogeneity, such as a pore, a crack, a fracture, etc., the inhomogeneity causes some portion of the energy to be scattered in directions other than the direction of propagation of the incident wave (Sato and Fehler, 1998), thereby decreasing the amplitude of the original pulse. Yamakawa

(1962) calculated the scattering from isolated spherical pores of radius a and found

$$Q^{-1} = g\phi(\omega a/c)^3 = g\phi(ka)^3, \tag{11.176}$$

where ϕ is the porosity and g is a dimensionless parameter of order 1 whose precise value depends on the moduli and densities of the rock and pore fluid. Other pore shapes lead to the same general form, with a being some characteristic dimension such as crack length, but with different values of g. Yamakawa's result applies asymptotically for wavelengths much larger than the inclusion size, the so-called *Rayleigh scattering* limit. As "typical" values of the parameters will be $a \approx 100\,\mu\text{m}$ and $c \approx 5000\,\text{m/s}$, attenuation due to Rayleigh scattering will become appreciable only for frequencies greater than about 1 MHz, that is, perhaps important at laboratory frequencies but negligible at seismic frequencies.

11.10 Inelastic waves

The waves discussed in previous sections of this chapter were elastic waves, in which the strains are sufficiently small that the additional stresses and strains associated with the wave can be assumed to obey Hooke's law. If the stresses or strains are sufficiently large, the rock will cease to be elastic and inelastic waves may propagate. Two types of inelastic wave that are of occasional significance in rock mechanics are *plastic waves* and *shock waves*. A brief discussion of plastic waves is given in this section, based on the analysis of von Kármán and Duwez (1950), followed by a brief analysis of shock waves, based on the analysis given by Kolsky (1963, pp. 178–82). More extensive treatments of waves in inelastic media have been given by Nowacki (1978) and Drumheller (1998).

A plane wave traveling through a rock in the x direction is governed by

$$\rho\frac{\partial^2 u}{\partial t^2} = \frac{\partial \tau_{xx}}{\partial x}, \tag{11.177}$$

which can also be written as

$$\rho\frac{\partial^2 u}{\partial t^2} = \frac{d\sigma}{d\varepsilon}\frac{\partial \varepsilon}{\partial x} = \frac{d\sigma}{d\varepsilon}\frac{\partial^2 u}{\partial x^2}, \tag{11.178}$$

where σ is written for τ_{xx} and ε for ε_{xx}. For a wave traveling along a thin rod, the local slope of the stress–strain curve is $\partial\sigma/\partial\varepsilon = E(\varepsilon)$, where E must be interpreted as the tangent modulus, rather than the secant modulus. The rock is assumed to behave in a linear elastic manner up to some value of the strain, beyond which the strain is a nonlinear, monotonically increasing function of stress. For a plane wave traveling through bulk rock, the relevant modulus would be $\lambda + 2G$; using E for simplicity of notation, (11.178) can be written as

$$\frac{\partial^2 u}{\partial t^2} = \frac{E(\varepsilon)}{\rho}\frac{\partial^2 u}{\partial x^2}. \tag{11.179}$$

A general solution of this equation for an arbitrary nonconstant function $E(\varepsilon)$ is not available. However, consider the specific problem of a semi-infinite body that occupies the region $x < 0$, and is initially at rest, with a constant velocity V applied at face $x = 0$, starting at $t = 0$. Expecting the strain to propagate into the body at some constant speed ξ, we look for solutions to (11.179) of the form $\varepsilon = \varepsilon(\xi)$, where $\xi = x/t$. Substitution into (11.179) gives

$$u = \int_{-\infty}^{x} \frac{\partial u}{\partial x'} \mathrm{d}x' = \int_{-\infty}^{x} \varepsilon(\xi') \mathrm{d}x' = t \int_{-\infty}^{\xi} \varepsilon(\xi') \mathrm{d}\xi' \tag{11.180}$$

where use has been made of the fact that the displacement must vanish as $x \to -\infty$. Differentiation, bearing in mind that $(\partial\xi/\partial t)_x = -x/t^2 = -\xi/t$ and $(\partial\xi/\partial x)_t = 1/t$, yields

$$\frac{\partial^2 u}{\partial t^2} = \frac{\xi^2}{t}\varepsilon'(\xi), \quad \frac{\partial^2 u}{\partial x^2} = \frac{1}{t}\varepsilon'(\xi), \tag{11.181}$$

after which substitution into (11.179) gives

$$\left(\xi^2 - \frac{E}{\rho}\right)\varepsilon'(\xi) = 0. \tag{11.182}$$

This equation will be satisfied if either $\varepsilon'(\xi) = 0$ or $\rho\xi^2 = E$. This first choice represents a uniform strain and corresponds to the solution

$$u(x,t) = V[t + (x/c_1)], \tag{11.183}$$

where c_1 is a constant that will be seen to correspond to the plastic wave speed. The strain associated with (11.183) is constant and equal to $\varepsilon_1 = V/c_1$. The second choice leads to the solution

$$E(\varepsilon) = \rho(x^2/t^2). \tag{11.184}$$

Since E is a known function of ε, (11.184) represents a solution in which ε is an implicit function of the similarity variable, $\xi = x/t$. Along with (11.183) and (11.184), a third solution to (11.179) is provided by $\varepsilon = 0$.

The full solution to the problem is constructed by piecing together these three solutions. For $-c_1 < \xi < 0$, that is, $|x| < c_1 t$, the displacement is given by (11.183), and the strain is constant and equal to ε_1. For $-c_0 < \xi < -c_1$, that is, $c_1 t < |x| < c_0 t$, where c_0 is the elastic wave speed, the strain is given implicitly by (11.184). Finally, for $\xi < -c_0$, that is, $|x| > c_0 t$, the displacement and strain are both zero.

The elastic wave speed c_0 is given by $[E(0)/\rho]^{1/2}$, where $E(0)$ is the tangent modulus at small strains, when the rock is in the elastic regime. The plastic wave speed c_1 is found by setting $x = 0$ in (11.180), which leads to

$$\frac{u(x=0,t)}{t} = V = \int_{-\infty}^{0} \varepsilon(\xi') \mathrm{d}\xi'. \tag{11.185}$$

Integration by parts gives

$$V = \int_{-\infty}^{0} \varepsilon \, d\xi = \varepsilon\xi\Big]_{-\infty}^{0} - \int_{0}^{\varepsilon_1} \xi \, d\varepsilon = -\int_{0}^{\varepsilon_1} \xi \, d\varepsilon = \int_{0}^{\varepsilon_1} [E(\varepsilon)/\rho]^{1/2} d\varepsilon, \quad (11.186)$$

where use has been made of the fact that ε vanishes at the lower limit of integration, ξ vanishes at the upper limit of integration, and the negative square root is chosen to agree with the fact that $\xi < 0$. Equation (11.186) gives $\varepsilon_1 = V/c_1$ implicitly as a function of V, thus completing the solution. Since the integrand is nonnegative, the strain ε_1 is a monotonically increasing function of the impact velocity, V.

Since $V = c_1\varepsilon_1$, (11.186) shows that c_1 can be interpreted as the mean value of the "local" wave speed, $[E(\varepsilon)/\rho]^{1/2}$, averaged over all strains from 0 to ε_1. As the tangent modulus is usually a decreasing function of strain, the plastic wave speed, c_1, will be smaller than the elastic wave speed, c_0. If the strain remains within the linear elastic range, then $E(\varepsilon) = E$ is constant and (11.186) gives $V = c_0\varepsilon_1$. The stress would then be given by

$$\sigma = E\varepsilon_1 = \rho c_0^2(V/c_0) = \rho c_0 V = \rho c_0 \dot{u}, \quad (11.187)$$

which is consistent with (11.25).

The strain, for the case in which the strain exceeds the elastic limit, is shown schematically as a function of ξ in Fig. 11.9. Ahead of the elastic wavefront, the strain is always at its undisturbed value, zero. Behind the elastic wavefront, but ahead of the plastic wavefront, the strain is variable, as each strain-increase from ε to $\varepsilon + \delta\varepsilon$ propagates with a velocity $[E(\varepsilon)/\rho]^{1/2}$ corresponding to the strain ε. Last, behind the plastic wavefront, the strain is constant and equal to ε_1.

For most rocks, the modulus decreases with increasing strain, so the plastic wave has a velocity *less* than that of an elastic wave in the same material. However, at very large values of the confining stress, the stiffness of a rock may increase with increasing strain (Bridgman, 1931). In this case, large strains propagate faster than small strains, and so a strain pulse tends to acquire a steep front, giving rise to a *shock wave* that propagates *faster* than an elastic wave of small amplitude. A shock wave can be analyzed as a traveling plane of discontinuity, with different values of the density, pressure, etc., ahead of and behind the front.

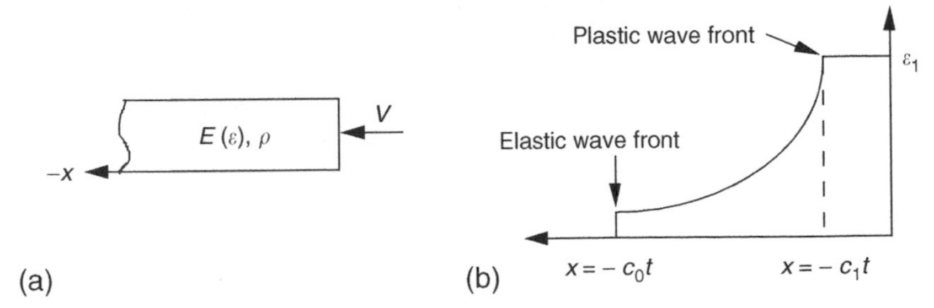

Fig. 11.9 (a) Velocity V imposed at the free end of a semi-infinite elastic–plastic rod and (b) the resulting strain pulse.

Consider a plane shock wave traveling at constant velocity c_s through a region of rock. The rock is undisturbed ahead of the shock front, and behind the front the particle velocity, $-\dot{u}$, the pressure, p, and the specific volume, $\upsilon \equiv 1/\rho$, are each assumed to be constant. Immediately behind the shock front is a thin disturbed "transition" zone. Now choose a coordinate system that travels with the shock front, and designate the region ahead of the front by subscript 1 and the region behind the front with subscript 2. By conservation of mass, the rates $\dot{m}$ at which mass enters and leaves the shock zone (per unit area and unit time) must be equal, so

$$\dot{m} = \dot{u}_1/\upsilon_1 = \dot{u}_2/\upsilon_2. \tag{11.188}$$

Equating the rate of change in momentum across the shock zone to the net force acting on the rock in that zone yields

$$\dot{m}(\dot{u}_1 - \dot{u}_2) = p_2 - p_1. \tag{11.189}$$

The equation for conservation of energy across the shock zone can be written as

$$p_2\dot{u}_2 - p_1\dot{u}_1 = \dot{m}[(\dot{u}_1^2 - \dot{u}_2^2)/2 + \Delta w]. \tag{11.190}$$

where Δw is the rate at which mechanical energy is transformed into internal energy, per unit mass.

The velocity of propagation of the shock front through the rock is equal to the velocity of the undisturbed region relative to this front, but with the opposite sign. So, from (11.188) and (11.189), the velocity of the shock front, relative to a coordinate system fixed in space, is given by

$$c_s = -\dot{u}_1 = -\upsilon_1[(p_2 - p_1)/(\upsilon_1 - \upsilon_2)]^{1/2}. \tag{11.191}$$

The particle velocity behind the shock zone is

$$-\dot{u} = \dot{u}_2 - \dot{u}_1 = -[(p_2 - p_1)(\upsilon_1 - \upsilon_2)]^{1/2}, \tag{11.192}$$

and the rate of increase in specific internal energy across the shock zone is

$$\Delta w = \frac{1}{2}[(p_2 + p_1)(\upsilon_1 - \upsilon_2)]. \tag{11.193}$$

Equations (11.188)–(11.190) are the Rankine conditions for a shock in a perfect gas, and (11.193) is known as the Hugoniot relation. For small pressure differences, the shock speed reduces to

$$c_s \approx [\upsilon^2 \Delta p/\Delta\upsilon]^{1/2} = \left[\frac{1/\rho}{(1/\upsilon)\,(\partial\upsilon/\partial p)}\right]^{1/2} = \left[\frac{K}{\rho}\right]^{1/2}, \tag{11.194}$$

which is the same as the speed of elastic waves in a fluid-like medium having no shear rigidity.

Shock waves are created in rock masses during blasting processes associated with excavations and quarrying. Wu et al. (1998) studied the propagation of blast-induced shock waves in a jointed rock mass. Natale et al. (1998) studied shock waves propagating in hyperthermal fluid-pressurized regions, as a model for understanding volcanic systems. Shock waves are generated by the impact of meteorites on the earth, moon, or other planets, and so the analysis of shock behavior is therefore an important component of impact studies (Bjork, 1961; Kieffer and Simonds, 1980; Asphaug and Melosh, 1993; Koeberl and Henkel, 2005). Data on Hugoniot curves for many rock types have been collected by Ahrens and Johnson (1995). Laboratory methods for measuring Hugoniot data using air guns have been described in detail by Furnish (1993) and Shang et al. (2000).

12 Hydromechanical behavior of fractures

12.1 Introduction

To a great extent, it is the nearly ubiquitous presence of fractures that makes the mechanical behavior of rock masses different from that of most engineering materials. These fractures also cause the behavior of *rock masses* to differ from that of small laboratory-sized rock samples. Most laboratory tests on rock samples are conducted on specimens that are "intact," and so, by definition, do not contain fractures. But almost all rock masses contain fractures on a scale larger than that of laboratory samples, with typical fracture spacings that range from tens of centimeters to tens of meters. These fractures have a controlling influence on the mechanical behavior of rock masses, since existing fractures provide planes of weakness on which further deformation can more readily occur. Fractures also often provide the major conduits through which fluids can flow. The field-scale permeability of a fractured rock mass may be many orders of magnitude larger than the permeability that would be measured on an intact core-scale specimen from the same field.

The hydromechanical behavior of rock fractures can be studied on the scale of a single fracture and also on the scale of a fractured rock mass that contains many fractures. Obviously, the behavior of single fractures must be thoroughly understood before the behavior of fractured rock masses can be understood. The mechanical, hydraulic, and seismic behaviors of a single rock fracture are now fairly well understood. Each of these properties depends almost exclusively on the geometry of the fracture void space, which is discussed in §12.2. The normal stiffness of a fracture is defined and discussed in §12.3, and the shear stiffness is treated in §12.4. The hydraulic transmissivity of single rock fractures is examined in §12.5. Coupling between the mechanical and hydraulic properties of a fracture is treated in §12.6. The influence of a fracture on seismic wave propagation, both across and along the fracture, is discussed in §12.7. Finally, §12.8 discusses attempts that have been made to relate the properties of single fractures to the *macroscopic* properties of *fractured rock masses*.

12.2 Geometry of rock fractures

An idealized rock fracture or joint consists of two nominally planar, rough surfaces. The surfaces are typically in contact with each other at some locations, but separated at others. The distance of separation, usually measured perpendicular to the nominal fracture plane, is known as the *aperture*. If the fracture has undergone substantial shear, it is usually classified as a fault; otherwise, it is denoted as a *joint* (Mandl, 2000). The space between the two rock surfaces may be clean, or may contain (i) fault gouge that has been produced by the shearing of the two faces of rock, (ii) clay minerals, (iii) mineral coatings that have been precipitated from flowing pore fluids, or (iv) microbial films. The genesis of faults and fractures in rocks is discussed at length by Mandl (2000). This chapter will describe the hydraulic, mechanical, and seismic behavior of existing fractures, rather than the generation of new fractures or the growth of existing ones. We begin with a discussion of various mathematical concepts and definitions that are used to characterize fracture surfaces and apertures, focusing on clean fractures that contain no infilling or coating material.

Consider a nominally planar fracture that lies in the x–y plane. Fracture surfaces are typically well-correlated at very large wavelengths, so that, even if the fracture has waviness at large scales, a nominal fracture plane can usually be defined locally. Two parallel reference planes can be drawn, one inside the lower region of rock, the other inside the upper region (Fig. 12.1). The distance between these two planes is denoted by d. The lower rock surface is then described by a "surface height" function $z_1(x, y)$ and the upper surface by the function $z_2(x, y)$. The aperture, defined as the distance between the opposing rock surfaces, measured perpendicular to the two reference planes, is then given by

$$h(x, y) = d - z_1(x, y) - z_2(x, y). \tag{12.1}$$

In principle, if the two surface profiles were known, the aperture would be known exactly, through (12.1). Moreover, all relevant hydromechanical properties of the fracture, such as its hydraulic transmissivity, shear and normal stiffnesses, etc., could *in principle* be found from the geometry, by solving the relevant solid or fluid-mechanical problem. But this detailed geometric information is usually not known, and moreover, solution of the problem of elastic (or plastic) deformation of the contacting surface, or the problem of fluid flow through the fracture's void space, is currently not computationally feasible for realistic fracture profiles. Hence, current practice is to try to characterize the fracture in terms of a small number of statistical parameters and to develop theories that relate the properties of the fracture to this set of parameters. In doing so, the fracture profiles and aperture fields are often treated as random variables, and the actual fracture is viewed as one stochastic realization of a random process that

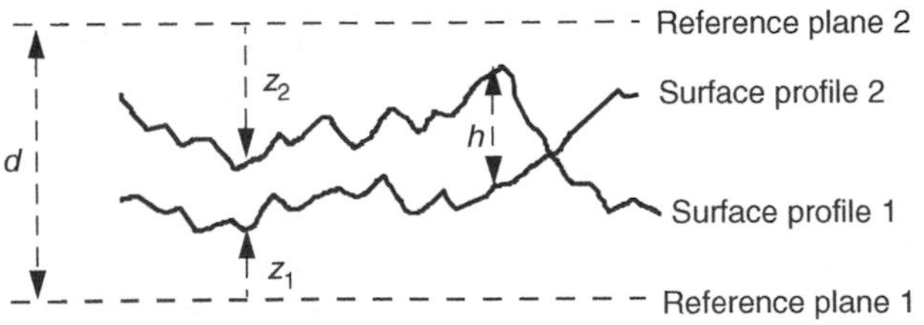

Fig. 12.1 Two rough fracture surface profiles, separated by an aperture h, along with the two reference planes, separated by a distance d.

has certain statistical properties. If the statistical properties of each realization of a stochastic process are the same, the process is said to be *ergodic* (Lanaro, 2000). In this case, statistical parameters such as the mean and variance can be calculated from a single realization.

The most basic statistical property of a "random variable" such as one of the surface profile functions $z(x, y)$ is the mean, defined by

$$\mu_z = \lim_{A\to\infty} \frac{1}{A} \iint_A z(x, y)\mathrm{d}x\mathrm{d}y \equiv E\{z\}, \tag{12.2}$$

where A is the nominal area of the fracture in the x–y plane. With reference to a particular transect of the fracture, say at a fixed value of y, we could define

$$\mu_z = \lim_{L\to\infty} \frac{1}{L} \int_0^L z(x, y)\mathrm{d}x. \tag{12.3}$$

If the statistical properties of a function are invariant with respect to translation of the origin, the process is called *homogeneous* or *stationary* (Adler and Thovert, 1999). In this case, the degree of correlation between the value of z at one location x and at another location displaced from x by an amount ξ, can be quantified by the *autocovariance function*,

$$\mathrm{cov}_z(\xi) = E\{[z(x) - \mu_z][z(x + \xi) - \mu_z]\} = E\{z(x)z(x + \xi)\} - \mu_z^2. \tag{12.4}$$

Evaluation of the autocovariance function at a lag distance of $\xi = 0$ yields the *variance*,

$$\sigma_z^2 \equiv \mathrm{cov}_z(\xi = 0) = E\{z^2(x)\} - \mu_z^2, \tag{12.5}$$

the square-root of which is the *standard deviation*, σ_z.

With regards to a surface defined over a region of the two-dimensional x–y plane, rather than a linear transect, the autocovariance can be defined as a function of the vector $\boldsymbol{\xi}$, as follows (Adler and Thovert, 1999)

$$\mathrm{cov}_z(\boldsymbol{\xi}) = E\{[z(\mathbf{x}) - \mu_z][z(\mathbf{x} + \boldsymbol{\xi}) - \mu_z]\} = E\{z(\mathbf{x})z(\mathbf{x} + \boldsymbol{\xi})\} - \mu_z^2, \tag{12.6}$$

where $\mathbf{x} = (x, y)$ and $\boldsymbol{\xi} = (\zeta, \eta)$. If the surface is isotropic, the autocovariance will depend only on the length of the lag vector, $|\boldsymbol{\xi}| = (\zeta^2 + \eta^2)^{1/2}$. In this case, no generality is lost by putting $\eta = 0$. For simplicity of notation, isotropy will be assumed henceforth, in which case x and ξ can be treated as one-dimensional variables.

Another measure of spatial correlation is the *variogram* function, $\gamma_z(\xi)$, defined by

$$\gamma_z(\xi) = E\{[z(x + \xi) - z(x)]^2\}. \tag{12.7}$$

Expansion of the term inside the brackets, and comparison with (12.4) and (12.5), shows that

$$\mathrm{cov}_z(\xi) = \sigma_z^2 - \frac{1}{2}\gamma_z(\xi), \tag{12.8}$$

where the term $(1/2)\gamma_z(\xi)$ is referred to as the *semivariogram*. The average slope of a surface z between two locations x and $x+\xi$ is given by $[z(x+\xi)-z(x)]/\xi$. The variance of the average slope is, by (12.5), given by

$$\sigma^2_{\text{slope}}(\xi) = E\{[z(x+\xi)-z(x)]^2/\xi^2\} - \mu^2_{\text{slope}} = \frac{\gamma_z(\xi)}{\xi^2}, \tag{12.9}$$

where the last step makes use of definition (12.7), and the fact that the mean value of the average slope must vanish, by appropriate choice of the reference plane. The variogram is therefore closely related to the variance of the mean value of the surface slope taken over the lag distance.

From definition (12.7), the semivariogram should vanish at $\xi = 0$, although in practice, this is often obscured by an inability to make measurements at sufficiently small scales. At sufficiently large lag distances, a fracture surface will usually become uncorrelated, in which case its autocovariance goes to zero, and the semivariogram approaches the variance. The power spectrum of $z(x)$ can then be defined as the Fourier transform of its autocovariance function:

$$G_z(k) = \frac{1}{2\pi}\int_{-\infty}^{\infty} \text{cov}_z(\xi)e^{-ik\xi}\,d\xi, \tag{12.10}$$

where $k = 2\pi/\lambda$ is the wavenumber and λ is the wavelength.

Two common models for the autocovariance are the exponential and the Gaussian models:

$$\text{cov}_z(\xi) = \sigma_z^2 \exp(-|\xi|/\xi_0), \quad \text{cov}_z(\xi) = \sigma_z^2 \exp[-(\xi/\xi_0)^2]. \tag{12.11}$$

For an exponential autocovariance, the surface is effectively uncorrelated at distances greater than about $4\xi_0$, whereas for the Gaussian model, the correlation is negligible for $\xi > 2\xi_0$. The *correlation length*, for which several different definitions can be given, is the distance beyond which the correlation between $z(x)$ and $z(x+\xi)$ is negligible. For exponential or Gaussian autocovariances, the parameter ξ_0 gives an indication of the correlation length. From (12.10) and (12.11), the power spectra of the exponential and Gaussian models are, respectively,

$$\text{exponential:}\quad G_z(k) = \frac{\sigma_z^2}{\pi}\frac{(1/\xi_0)}{(1/\xi_0)^2+k^2}, \tag{12.12}$$

$$\text{Gaussian:}\quad G_z(k) = \frac{\xi_0\sigma_z^2}{2\sqrt{\pi}}\exp(-k^2\xi_0^2/4). \tag{12.13}$$

A profile $z(x)$ is said to be *self-affine* if $z(\lambda x) = \lambda^H z(x)$ for some constant H, which is known as the *Hurst exponent*. A profile is *statistically self-affine* if $z(x)$ is statistically similar to $\lambda^{-H}z(\lambda x)$. A self-affine profile has a power spectrum of the form

$$G_z(k) = Ck^{-\alpha}, \tag{12.14}$$

where $\alpha = 2H + 1$ (Adler and Thovert, 1999, pp. 44–6). Such power spectra have been observed for profiles of fractures in crystalline and sedimentary rocks, for bedding plane surfaces, and for frictional wear surfaces (Brown and Scholz, 1985a; Power and Tullis, 1991). In practice, a power law can only apply between a lower limit of $k_{\min} = 2\pi/L$, where L is the length of the profile, and an upper limit of $k_{\max} = 2\pi/l$, where l is the distance along the x-axis between successive measurements (i.e., the sampling interval).

12.3 Normal stiffness of rock fractures

If a rock core containing a through-going fracture that is aligned more or less perpendicular to the axis of the core is tested under uniaxial compression, the length change measured between the two end-plates will consist of two contributions: the deformation of the intact rock and an excess deformation, δ, that can be attributed to the fracture (Goodman, 1976). This excess deformation is called the *joint closure* and is defined to be a nonnegative number that increases as the joint compresses. If the initial length of the specimen is L and the normal stress is σ, the incremental change in the overall length of the core can be expressed as

$$dL = dL_r - d\delta = -\frac{L}{E_r}d\sigma - \frac{1}{\kappa_n}d\sigma, \tag{12.15}$$

where E_r is the Young's modulus of the intact rock, and κ_n, with dimensions of [Pa/m], is the normal stiffness of the fracture. An apparent Young's modulus of the fractured rock, E_{fr}, could be defined, but it would not be a meaningful property of the rock, as its value would depend on the length of the specimen, that is,

$$\frac{1}{E_{fr}} \equiv -\frac{1}{L}\frac{dL}{d\sigma} = \frac{1}{E_r} + \frac{1}{L\kappa_n}. \tag{12.16}$$

Goodman (1976) made measurements of joint closure as a function of stress on artificially induced fractures by measuring the displacement across the total length of an intact sample and then repeating the measurement across the core after it had been fractured. Joint closure measurements were made for mated joints in which the two halves of the core were placed in the same relative position that they occupied before the core was fractured, and on nonmated joints in which the two surfaces were rotated from their initial positions relative to one another. The unmated surfaces allowed much greater joint closure and had much lower joint stiffness (Fig. 12.2). The joint closure is a highly nonlinear function of stress and levels off to some asymptotic value at high values of the confining stress. Goodman related the joint closure to the stress through the following empirical relation:

$$\sigma = \sigma_o\left[1 + \left(\frac{\delta}{\delta_m - \delta}\right)^t\right], \quad \text{for } \sigma \geq \sigma_o. \tag{12.17}$$

where σ_o is some initial, low "seating stress," t is a dimensionless empirical exponent, and δ_m is the maximum possible joint closure, approached asymptotically as the stress increases.

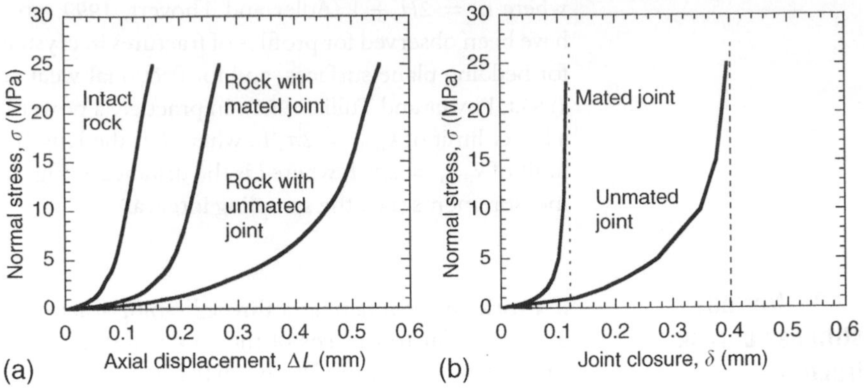

Fig. 12.2 Measurements made by Goodman (1976, p. 172) of joint closure on a granodiorite specimen: (a) axial displacement of intact core, core with mated joint, and core with unmated joint; (b) joint closure, computed by subtracting displacement for intact specimen from displacement of jointed specimen.

Bandis et al. (1983) made extensive measurements of joint closure on a variety of natural, unfilled joints in dolerite, limestone, siltstone, and sandstone and found that cycles of loading and unloading exhibited hysteresis and permanent set that diminished rapidly with successive cycles. Barton et al. (1985) later suggested that the hysteresis was a laboratory artifact and that *in situ* fractures probably behave in a manner similar to the third or fourth loading cycle. Bandis et al. (1983) fit the joint closure with functions of the form

$$\sigma = \frac{\kappa_o \delta}{1 - (\delta/\delta_m)} = \frac{\kappa_o \delta_m \delta}{\delta_m - \delta}, \tag{12.18}$$

where κ_o is an empirical parameter. The joint closure is related to the normal stress by

$$\delta = \left(\frac{\sigma}{\sigma + \kappa_o \delta_m}\right) \delta_m, \tag{12.19}$$

The normal stiffness of the fracture is given by

$$\kappa_n = \frac{d\sigma}{d\delta} = \frac{\kappa_o}{(1 - \delta/\delta_m)^2}, \tag{12.20}$$

which shows that κ_o is the normal stiffness at low confining stress. The function proposed by Goodman reduces to (12.18) when $t = 1$ and $\sigma \gg \sigma_o$.

Many aspects of the normal closure of an initially mated fracture can be qualitatively explained by the conceptual model developed by Myer (2000), in which a fracture is represented by a collection of collinear elliptical cracks (Fig. 12.3). The cracks have length $2a$, the spacing between the centers of adjacent cracks is 2λ, the fractional contact area is $c = 1 - (a/\lambda)$, and the cracks can have an arbitrary distribution of initial aspect ratios. From the elasticity solution of Sneddon and Lowengrub (1969), the incremental joint closure due to a small increase in normal stress is

$$\delta = \frac{4\lambda(1-\nu)\sigma}{\pi G} \ln \sec\left(\frac{\pi a}{2\lambda}\right) = \frac{4(1-\nu)a\sigma}{\pi G(1-c)} \ln \sec\left[\frac{\pi}{2}(1-c)\right]. \tag{12.21}$$

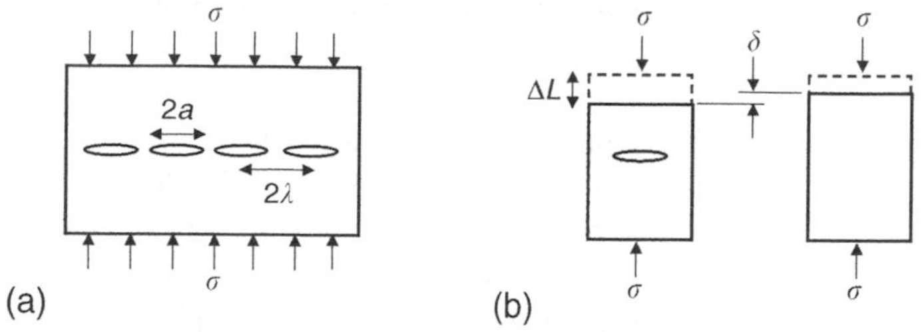

Fig. 12.3 (a) Schematic model of a fracture as an array of two-dimensional cracks of length $2a$ and spacing 2λ. (b) Unit cell of fractured and intact rock, showing definition of δ (Myer, 2000).

The normal compliance of the joint is given by

$$\frac{1}{\kappa_n} = \frac{d\delta}{d\sigma} = \frac{4(1-\nu)a}{\pi G(1-c)} \ln \sec\left[\frac{\pi}{2}(1-c)\right]. \tag{12.22}$$

At low stresses, the fractional contact area is small and the compliance will be large. As the normal stress increases, those cracks with smaller aspect ratio close up. Although this disturbs the periodicity of the array, it can be modeled approximately by assuming that a (the half-length of the open cracks) remains the same, but λ (the mean spacing between adjacent cracks) increases, leading to an increase in c, and a consequent decrease in joint compliance. Expanding (12.22) for small values of $(1-c)$ shows that as c increases,

$$\delta \approx \frac{\pi a(1-\nu)(1-c)}{2G}\sigma, \quad \frac{1}{\kappa_n} \approx \frac{\pi a(1-\nu)(1-c)}{2G} \tag{12.23}$$

So, as the contact area increases, the compliance goes to zero, and the joint stiffness becomes infinite, in accordance with experimental observations. This model also indicates a size-dependence, in that (other factors, such as c, being equal), smaller "crack" sizes a lead to stiffer fractures.

Pyrak-Nolte et al. (1987) made Wood's metal casts of the void space of a natural granitic fracture under various normal stresses, at 3, 33, and 85 MPa. Myer (2000) took transects of these casts and found that as the normal stress increases, in addition to complete closure of some "cracks," the rock faces occasionally come into contact at isolated points within existing cracks, creating two cracks with half-lengths less than a. Hence, as the normal stress increases, the contact area c increases and the mean crack length a decreases. According to (12.22) and (12.23), both the increase in c and the decrease in a lead to higher joint stiffness.

For fracture surfaces that are unmated, perhaps as a result of previous shear displacement, Bandis et al. (1983) found that the normal stress could be fit with an equation of the form

$$\ln(\sigma/\sigma_o) = J\delta, \tag{12.24}$$

where σ_o is an initial, small stress level at which the joint closure is taken to be zero and J is a constant with dimensions of $1/L$. The normal stiffness associated with this stress-closure relationship is

$$\kappa_n = \frac{d\sigma}{d\delta} = J\sigma, \tag{12.25}$$

which increases linearly with stress.

Lee and Harrison (2001) developed a method to relate the parameters that appear in the empirical equations of Goodman (12.17) and Bandis et al. (12.18), (12.24) to the statistical properties of the fracture surfaces. They used the conceptual model proposed by Hopkins (1990), in which the asperities are treated as columns with circular cross-sections. The Boussinesq solution (§13.5) was used to calculate the deformation under and around the regions of circular contact between the two fracture surfaces, and Hooke's law for a column was used to find the deformation of the asperity itself. By appropriate choices of the initial contact area, mean aperture, and correlation length, their model could be made to agree with the various empirical equations (12.17), (12.18), and (12.24).

Another conceptual model for the normal stiffness of a rock fracture is to treat the fracture surface as a rough elastic surface and use Hertzian contact theory (Timoshenko and Goodier, 1970, pp. 409–16) to analyze the deformation of the contacting asperities. Greenwood and Williamson (1966) considered a single, rough elastic surface whose asperities each have radius of curvature R, with a distribution of peak heights $\phi(Z^*)$, where the height Z of an asperity is measured relative to a reference plane that is parallel to the nominal fracture plane and can conveniently be located entirely within the rock (i.e., below the lowest troughs of the fracture surface). A value of Z^* is associated with each local peak, of which there are assumed to be η per unit area of fracture in the undeformed (zero stress) state. The height of the highest peak, measured from the reference plane, is initially equal to d_o (Fig. 12.4a).

If such a surface is pressed against a smooth elastic surface of area A, the density of contacts is given by

$$n = N/A = \eta \int_{d_o-\delta}^{\infty} \phi(Z^*)dZ^*. \tag{12.26}$$

As the distribution function $\phi(Z^*)$ vanishes for $Z^* > d_o$, by construction, the contact density is zero when the joint closure δ is zero. In the hypothetical situation in which all asperities were pressed flat against the upper flat surface, δ would equal d_o, so the integral in (12.26) would approach unity, and the fraction of asperities in contact, n/η, would reach unity. The fractional contact area of asperities is given by

$$c = A_{\text{contact}}/A = \pi R\eta \int_{d_o-\delta}^{\infty} (Z^* - d_o + \delta)\phi(Z^*)dZ^*, \tag{12.27}$$

and the average normal stress acting over the surface is

$$\sigma = \frac{4}{3}\eta R^{1/2}E' \int_{d_o-\delta}^{\infty} (Z^* - d_o + \delta)^{3/2}\phi(Z^*)dZ^*, \tag{12.28}$$

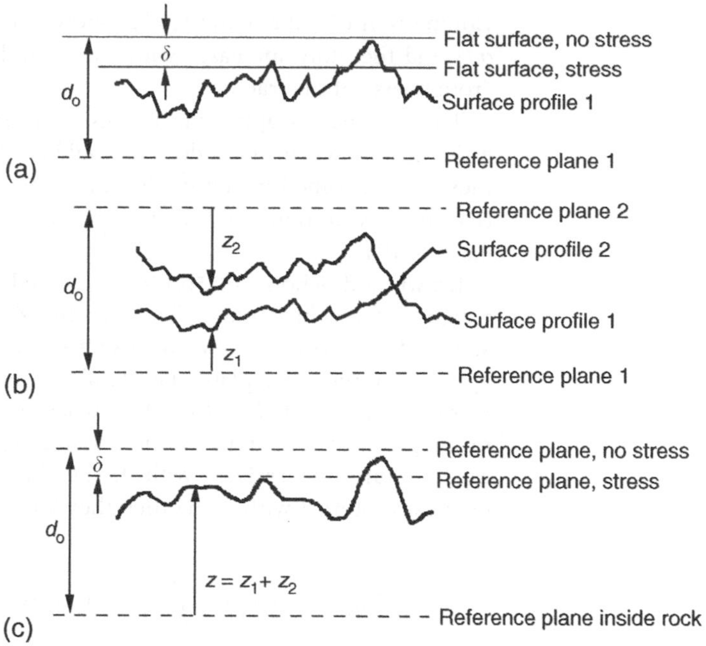

Fig. 12.4 (a) Single rough profile in contact with a smooth surface; (b) two rough surfaces; (c) composite profile (Cook, 1992).

where the reduced elastic modulus E' is defined by

$$\frac{1}{E'} = \frac{1-\nu_1^2}{E_1} + \frac{1-\nu_2^2}{E_2}, \tag{12.29}$$

and the subscripts 1, 2 denote the properties of the rough and smooth surface, respectively.

Swan (1983) measured the topography of ten different surfaces of Offerdale slate and showed that the peak heights of asperities followed a Gaussian distribution. Greenwood and Williamson (1966) showed that the upper quartile of a Gaussian distribution could be approximated by an exponential distribution of the form

$$\phi(Z^*) = \frac{1}{s}\exp(-Z^*/s), \tag{12.30}$$

where s is the mean, as well as the standard deviation, of the exponential distribution. Equations (12.26)–(12.28) lead in this case to

$$\ln\{\sigma/[(\pi Rs)^{1/2} sE'\eta]\} = (\delta - d_o)/s, \tag{12.31}$$

which has the same form as the empirical relation found by Bandis et al. (1983) for fractures with unmated surfaces. Comparison of (12.24) and (12.31) shows that the model of Swan and Greenwood and Williamson predicts

$$J = 1/s, \quad \sigma_o = (\pi Rs)^{1/2}(sE'\eta)\exp(-d_o/s). \tag{12.32}$$

Comparison of (12.25) and (12.32) shows that the normal stiffness is equal to σ/s and therefore increases with stress, and is inversely proportional to the "roughness" of the fracture.

The parameters appearing in this expression for σ_0 would be difficult to estimate in practice, and indeed R would not typically be the same for all asperities, as is assumed in the model. However, Olsson and Brown (1993) noted that, for a wide range of fractures, σ_0 varies in the relatively narrow range of 0.2–0.6 MPa.

Brown and Scholz (1985b,1986) extended this model to the closure of *two* rough surfaces in contact. The variable Z^* was redefined to represent the summed heights of the two opposing surfaces, each measured relative to the appropriate reference plane (Fig. 12.4b,c), and the effective radius of curvature was taken as $R = R_1R_2/(R_1 + R_2)$, where R_1 and R_2 are the radii of curvature of the pair of contacting asperities. Assuming that the radii of curvature of the asperities are uncorrelated with the heights and that nearby asperities do not elastically interact with each other, they found

$$\sigma = \frac{4}{3}\eta\langle R^{1/2}\rangle\langle E'\rangle\langle\psi\rangle \int_{d_o-\delta}^{\infty} (Z^* - d_o + \delta)^{3/2}\phi(Z^*)\mathrm{d}Z^*, \tag{12.33}$$

where the brackets denote the mean value taken over all contacting asperities, and ψ is a "tangential stress correction factor" whose mean value is very close to unity.

If the shear stress within a particular asperity becomes sufficiently large, the asperity will yield and undergo irreversible plastic deformation. Two contacting spherical asperities will begin to yield when the displacement (at that particular contact point) reaches a critical value given by (Greenwood and Williamson, 1966; Brown and Scholz, 1986)

$$\delta_p \approx CR\left(\frac{H}{E'}\right)^2, \tag{12.34}$$

where R is the effective radius of curvature, C is a dimensionless constant in the order of unity, and H is the indentation hardness of the rock mineral. For crystalline plasticity, $H \approx 3Y$, where Y is the yield stress. As the fracture compresses, the highest asperities will be plastically flattened first. The fractional area of *plastic* contact is given by

$$A_{\text{plastic}}/A = \pi R\eta \int_{d_o-\delta+\delta_p}^{\infty} (Z^* - d_o + \delta)\phi(Z^*)\mathrm{d}Z^* \tag{12.35}$$

Greenwood and Williamson (1966) suggested that plastic deformation becomes nonnegligible when the ratio of plastic contact area to total contact area, that is, the ratio of the integrals in (12.27) and (12.35), reaches about 2–10 percent.

12.4 Behavior of rock fractures under shear

If a fracture is located in a rock mass with a given ambient state of stress, the traction acting across the fracture plane can be resolved into a normal component and a shear component. The normal traction gives rise to a normal closure of the fracture, as described in §12.3. The shear component of the traction causes the two rock faces to undergo a relative deformation parallel to the nominal fracture plane, referred to as a *shear deformation*. However, a tangential traction also typically causes the mean aperture to *increase*, in which case the fracture is said to *dilate*. Dilation arises because the asperities of one fracture surface must by necessity ride up in order to move past those of the other surface. Hence, shear deformation of a fracture is inherently a coupled process in which both normal and shear displacement occur.

Displacement parallel to the nominal fracture plane is called the *shear displacement* and is usually denoted by Δu (Fig. 12.5a). The displacement in the direction perpendicular to the fracture plane is known in this context as dilation and is denoted by Δv. Shear displacement is reckoned positive if it is in the direction of the applied shear stress, whereas the dilation is positive if the two fracture surfaces move apart from each other. A typical but idealized curve for the shear displacement as a function of shear stress, as would be measured under conditions of constant normal stress, is shown in Fig. 12.5b.

The shear stress first increases in a manner that is nearly proportional to the shear displacement. The slope of this line is the *shear stiffness*, κ_s. During this phase of the deformation, the two fracture surfaces ride over each other's asperities, causing dilation of the fracture, but little degradation to the surfaces (Gentier et al., 2000). A peak shear stress τ_p is eventually reached, corresponding to the point at which the asperities begin to shear off, causing irreversible damage to the fracture surfaces. This peak shear stress is also known as the *shear strength* of the fracture. The displacement at the peak shear stress is known as the *peak displacement*, u_p.

If the fracture continues to be deformed under conditions of controlled shear displacement, the peak shear stress will be followed by an unstable softening regime, during which the shear stress decreases to a value known as the *residual shear stress*, τ_r. During this phase the asperities continue to be crushed and

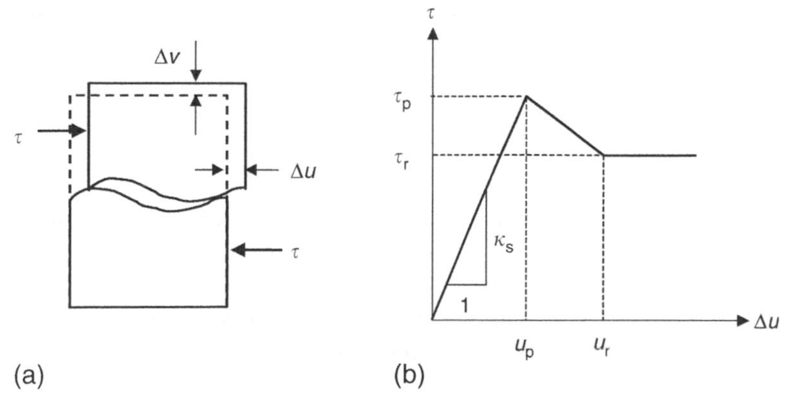

Fig. 12.5 (a) Schematic diagram of a fracture sheared under constant normal stress (Goodman, 1989, p. 163). (b) Shear stress as a function of shear displacement (Goodman, 1976, p. 174).

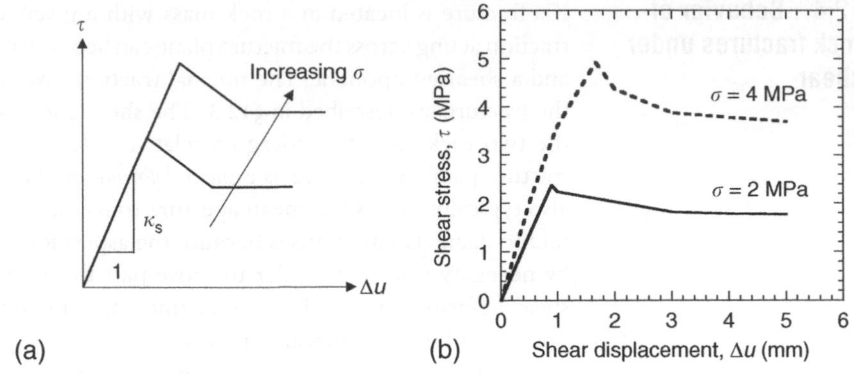

Fig. 12.6 (a) Effect of normal stress σ on the relationship between shear stress and shear displacement (Goodman, 1976, p. 177). (b) Measurements made by Olsson and Barton (2001) on a granite fracture from Äspö in Sweden.

sheared off, the fractional contact area between the two surfaces increases, and the dilation continues but at a decreased rate. The level of displacement at which the shear stress first reaches its residual value is known as the *residual displacement*, u_r.

The behavior of a fracture under shear depends very strongly on the normal stress acting across the fracture. A highly schematic view of the manner in which the relationship between τ and Δu varies with normal stress is shown in Fig. 12.6a (Goodman, 1976, p. 177). In this model, the shear stiffness is independent of the normal stress but both the peak shear stress and the residual shear strength increase with increasing normal stress. This is roughly consistent with the experimental measurements made by Olsson and Barton (2001) on a granite fracture taken from Äspö in Sweden (Fig. 12.6b).

The variation of peak shear stress as a function of normal stress is called the *shear strength curve*. Patton (1966) found the following bilinear function for τ_p as a function of σ (Fig. 12.7):

$$\text{for } \sigma < \sigma_T: \quad \tau_p = \sigma \tan(\phi_b + i),$$
$$\text{for } \sigma > \sigma_T: \quad \tau_p = C_J + \sigma \tan \phi_r. \tag{12.36a,b}$$

At low normal stresses, shear deformation is assumed to take place predominantly by asperities sliding over each other. At higher normal stresses, the fracture possesses a cohesion C_J that is due to the inherent shear strength of the asperities and has an effective angle of internal friction of $\phi_r < \phi_b + i$. Trigonometric considerations show that the parameters in (12.36) are related by $\tan(\phi_b + i) - \tan\phi_r = C_J/\sigma_T$. Jaeger (1971) proposed the following continuous function,

$$\tau_p = (1 - e^{-\sigma/\sigma^*})C_J + \sigma \tan \phi_r, \tag{12.37}$$

which asymptotically approaches (12.36a) and (12.36b) for small and large normal stresses, respectively. The parameter σ^* is a transition stress that roughly demarcates the two regimes but is not numerically identical to the parameter σ_T. The peak shear stresses measured by Gentier et al. (2000) on cement replicas of a fracture in Guéret granite show qualitative agreement with this type of model (Fig. 12.7b).

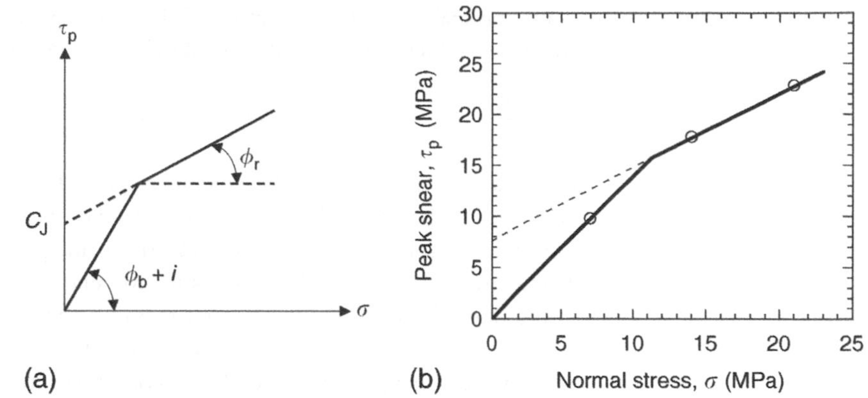

Fig. 12.7 (a) Bilinear model for the peak shear strength of a joint; parameters defined as in (12.36). (b) Peak shear stresses measured on cement replicas of a fracture in Guéret granite (Gentier et al., 2000) in the direction labeled by them as $-30°$ and fit to a curve of the form given by (12.36).

A simple mechanical model of two flat surfaces that have an intrinsic (or "basic") friction angle of ϕ_b, and whose interface is inclined by an angle i from the nominal fracture plane, leads directly to (12.36a). Recognizing that this model is oversimplified, but that the coefficient i in (12.36a) must depend on the roughness of the fracture, Barton (1973) correlated i to the *joint roughness coefficient* (*JRC*), an empirical measure of roughness whose value is estimated by comparing a fracture surface profile with standard profiles that have been assigned roughness values ranging from 1–20. Examination of data from fractures in various sedimentary, igneous, and metamorphic rocks led Barton to the correlation

$$i = JRC \log_{10}(JCS/\sigma), \tag{12.38}$$

where *JCS* is the *joint compressive strength*, which is equal to the unconfined compressive strength of the intact rock for unweathered fracture surfaces, but which has a much lower value for weathered surfaces (Barton and Choubey, 1977). Grasselli and Egger (2003) have attempted to correlate i to objectively quantifiable measures of roughness that can be estimated using optical means.

12.5 Hydraulic transmissivity of rock fractures

In many rock masses, field-scale fluid flow takes place predominantly through joints, faults, or fractures, rather than through the matrix rock itself. In some cases most of the flow may take place through a single such discontinuity, which for simplicity will be referred to as a "fracture," whereas in other cases the flow occurs through an interconnected network of such fractures. Fracture-dominated flow is of importance in many areas of technological interest. Nearly half of all known hydrocarbon reserves are located in naturally fractured formations (Nelson, 1985), as are most geothermal reservoirs (Bodvarsson et al., 1986). Fracture flow is of importance in understanding and predicting the performance of underground radioactive waste repositories (Wu et al., 1999). Indeed, it has become increasingly clear during the past few decades that fracture-dominated flow is the rule, rather than the exception, in much of the subsurface.

On the scale of a single fracture, fluid flow is governed by the Navier-Stokes equations, which can be written as (Batchelor, 1967, pp. 147–50)

$$\frac{\partial \mathbf{u}}{\partial t} + (\mathbf{u} \cdot \nabla)\mathbf{u} = \mathbf{F} - \frac{1}{\rho}\nabla p + \frac{\mu}{\rho}\nabla^2 \mathbf{u}, \tag{12.39}$$

where $\mathbf{u} = (u_x, u_y, u_z)$ is the velocity vector, $\mathbf{F}$ is the body-force vector per unit mass, ρ is the fluid density, μ is the fluid viscosity, and p is the pressure. The Navier-Stokes equations embody the principle of conservation of linear momentum, along with a linear constitutive relation that relates the stress tensor to the rate of deformation. The first term on the left of (12.39) represents the acceleration of a fluid particle due to the fact that, at a fixed point in space, the velocity may change with time. The second term, the "advective acceleration," represents the acceleration that a particle may have, even in a steady-state flow field, by virtue of moving to a location at which there is a different velocity. The forcing terms on the right side represent the applied body force, the pressure gradient, and the viscous forces.

Often, the only appreciable body force is that due to gravity, in which case $\mathbf{F} = -g\mathbf{e_z}$, where $\mathbf{e_z}$ is the unit vector in the upward vertical direction and $g = 9.81 \text{ m/s}^2$. If the density is uniform, gravity can be eliminated from the equations by defining a reduced pressure, $\hat{p} = p + \rho g z$ (Phillips, 1991, p. 26), in which case

$$\mathbf{F} - \frac{1}{\rho}\nabla p = -g\mathbf{e_z} - \frac{1}{\rho}\nabla p = -\frac{1}{\rho}(\nabla p + \rho g\mathbf{e_z}) = -\frac{1}{\rho}\nabla(p + \rho g z) = -\frac{1}{\rho}\nabla\hat{p}. \tag{12.40}$$

Hence, the governing equations can be written without the gravity term, if the pressure is replaced by the reduced pressure. For simplicity of notation, the reduced pressure will henceforth be denoted by p.

In the steady-state, the Navier-Stokes equations then reduce to

$$\mu\nabla^2\mathbf{u} - \rho(\mathbf{u} \cdot \nabla)\mathbf{u} = \nabla p. \tag{12.41}$$

Equation (12.41) represents three equations for the four unknowns: the three velocity components and the pressure. An additional equation to close the system is provided by the principle of conservation of mass, which for an incompressible fluid is equivalent to conservation of volume, and takes the form

$$\text{div}\mathbf{u} = \nabla \cdot \mathbf{u} = \frac{\partial u_x}{\partial x} + \frac{\partial u_y}{\partial y} + \frac{\partial u_z}{\partial z} = 0. \tag{12.42}$$

The compressibility of water is roughly 5×10^{-10}/Pa (Batchelor, 1967, p. 595), so a pressure change of 10 MPa would alter the density by only 0.5 percent; the assumption of incompressibility is therefore reasonable.

The set of four coupled partial differential equations, (12.41) and (12.42), must be augmented by the "no-slip" boundary conditions, which state that at the interface between a solid and a fluid, the velocity of the fluid must equal that of the solid. This implies that at the fracture walls, not only must the normal component of the fluid velocity be zero, but the tangential component must vanish as well.

The simplest conceptual model of a fracture, for hydrological purposes, is that of two smooth, parallel walls separated by a uniform aperture, h. For this geometry, the Navier-Stokes equations can be solved exactly, to yield a velocity profile that is parabolic between the two walls. If the x-axis is aligned with the pressure gradient, the y-axis taken perpendicular to the pressure gradient within the plane of the fracture, and the z-axis taken normal to the fracture plane, with the fracture walls located at $z = \pm h/2$, the solution to (12.41) and (12.42) is (Zimmerman and Bodvarsson, 1996)

$$u_x = -\frac{1}{2\mu}\frac{\partial p}{\partial x}[(h/2)^2 - z^2], \quad u_y = 0, \quad u_z = 0. \tag{12.43}$$

The total volumetric flux, with units of [m^3/s], is found by integrating the velocity:

$$Q_x = w\int_{-h/2}^{+h/2} u_x dz = -\frac{w}{2\mu}\frac{\partial p}{\partial x}\int_{-h/2}^{+h/2}[(h/2)^2 - z^2]dz = -\frac{wh^3}{12\mu}\frac{\partial p}{\partial x}, \tag{12.44}$$

where w is the depth of the fracture in the y direction, normal to the pressure gradient (i.e., into the page in Fig. 12.8a). The term $T = wh^3/12$ is known as the fracture *transmissivity*. As the transmissivity is proportional to the cube of the aperture, this result is known as the "cubic law." The transmissivity is sometimes reckoned per unit length in the y direction, in which case the factor w does not appear.

The result $T = wh^3/12$, which is exact only for smooth-walled fractures of uniform aperture, must be modified to account for roughness and asperity contacts. To do this rigorously requires solution of the Navier-Stokes equations for more realistic geometries. However, in general, the presence of the advective acceleration term $(\mathbf{u}\cdot\nabla)\mathbf{u}$ renders the Navier-Stokes nonlinear and consequently very difficult to solve. An exact solution is obtainable for flow between smooth parallel plates only because the nonlinear term vanishes identically in this case: the velocity vector lies in the x direction and is thus orthogonal to the velocity gradient, which lies in the z direction. In principle, the Navier-Stokes equations could be solved numerically for realistic fracture geometries, but computational difficulties have as yet not allowed this to be achieved. Consequently, the Navier-Stokes equations are usually reduced to more tractable equations, such as the Stokes or Reynolds equations.

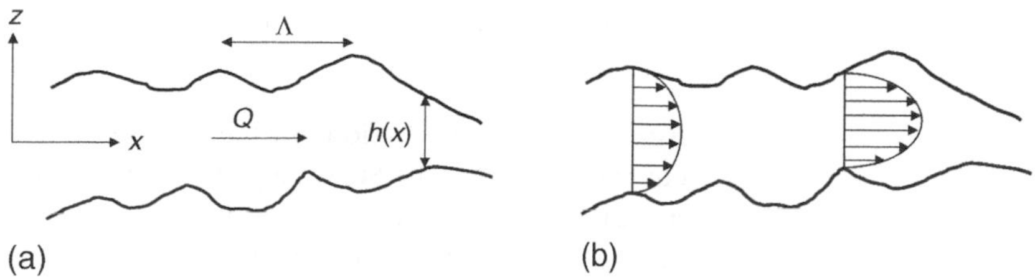

Fig. 12.8
(a) Schematic of a rough-walled fracture. (b) Parabolic velocity profile assumed in the derivation of the Reynolds equation, (12.59).

The Stokes equations derive from the Navier-Stokes equations by neglect of the advective acceleration terms, which is justified if these terms are small compared to the viscous terms. A priori estimates of the magnitudes of the various terms in the steady-state Navier-Stokes equations for flow through a variable-aperture fracture can be achieved as follows (Zimmerman and Bodvarsson, 1996). Let U be a characteristic velocity, such as the mean velocity in the direction of the macroscopic pressure gradient. As the in-plane velocity varies quasi-parabolically from zero at the upper and lower surfaces, to some maximum value in the order of U in the interior, the magnitude of the viscous terms can be estimated as (Fig. 12.8a)

$$|\mu\nabla^2\mathbf{u}| \approx \left|\mu\frac{\partial^2 u_x}{\partial z^2}\right| \approx \frac{\mu U}{h^2}. \tag{12.45}$$

The term h^2 appears because the velocity is differentiated *twice* with respect to the variable z. Since the advective acceleration, or "inertia" terms, contain first derivatives of velocity, their order of magnitude can be estimated as

$$|\rho(\mathbf{u}\cdot\nabla)\mathbf{u}| \approx \frac{\rho U^2}{\Lambda}, \tag{12.46}$$

where Λ is some characteristic dimension in the direction of flow, such as the dominant wavelength of the aperture variation or the mean distance between asperities. The condition for the inertia forces to be negligible compared to the viscous forces is

$$\frac{\rho U^2}{\Lambda} \ll \frac{\mu U}{h^2}, \quad \text{or} \quad Re^* \equiv \frac{\rho U h^2}{\mu\Lambda} \ll 1, \tag{12.47}$$

where the reduced Reynolds number Re^* is the product of the traditional Reynolds number, $\rho Uh/\mu$, and the geometric parameter h/Λ.

If condition (12.47) is satisfied, which necessarily will be the case at sufficiently low velocities, the Navier-Stokes equations reduce to the Stokes equations:

$$\mu\nabla^2\mathbf{u} = \nabla p, \tag{12.48}$$

which can be written in component form as

$$\frac{\partial^2 u_x}{\partial x^2} + \frac{\partial^2 u_x}{\partial y^2} + \frac{\partial^2 u_x}{\partial z^2} = \frac{1}{\mu}\frac{\partial p}{\partial x}, \tag{12.49}$$

$$\frac{\partial^2 u_y}{\partial x^2} + \frac{\partial^2 u_y}{\partial y^2} + \frac{\partial^2 u_y}{\partial z^2} = \frac{1}{\mu}\frac{\partial p}{\partial y}, \tag{12.50}$$

$$\frac{\partial^2 u_z}{\partial x^2} + \frac{\partial^2 u_z}{\partial y^2} + \frac{\partial^2 u_z}{\partial z^2} = \frac{1}{\mu}\frac{\partial p}{\partial z}. \tag{12.51}$$

Again, these three equations must be supplemented by the conservation of mass equation, (12.42). The Stokes equations are linear and consequently somewhat more tractable than the Navier-Stokes equations. More importantly, if the flow is governed by the Stokes equations, the resulting relation between the volumetric flux and pressure gradient will be linear, in analogy with Darcy's law, (7.73). Only under such conditions will the flux be given by $\mathbf{Q} = -(T/\mu)\nabla p$, where the transmissivity T is independent of the pressure gradient, and where $\mathbf{Q}$ and ∇p each lie within the fracture plane.

Various approaches, including more precise order-of-magnitude estimates that account for typical values of the parameter h/Λ (Oron and Berkowitz, 1998; Zimmerman and Yeo, 2000), numerical simulations of flow through simulated fracture apertures (Skjetne et al., 1999), and perturbation solutions for flow between a smooth wall and a sinusoidal wall (Hasegawa and Izuchi, 1983), each show that the relationship between flux and pressure gradient is nearly linear for Reynolds numbers less than about 10. For higher Reynolds numbers, a nonlinear relationship of the form

$$|\nabla p| = \frac{\mu|\mathbf{Q}|}{T} + \beta\,|\mathbf{Q}|^2. \tag{12.52}$$

is observed. At a given flowrate, an additional "non-Darcy" pressure drop is added to the Darcian pressure drop. Reynolds numbers greater than 10 are difficult to avoid in laboratory experiments, if the flowrates and pressure drops are to be large enough to measure with sufficient accuracy (Witherspoon et al., 1980; Yeo et al., 1998). However, in most subsurface flow situations, the nonlinear term is negligible. This nonlinearity is not necessarily due to turbulence, which occurs only at much higher Reynolds numbers. The nonlinearity observed at values of *Re* in the range 10–100 is due to the effects of curvature of the streamlines (Phillips, 1991, p. 28) and occurs in the laminar flow regime.

Brown et al. (1995) and Mourzenko et al. (1995) have solved the Stokes equations numerically for a few simulated fracture profiles, but use of the Stokes equation for studying fracture flow is not yet common and is by no means computationally straightforward. Typically, the Stokes equations are reduced further to the Reynolds lubrication equation, which requires that the variations in aperture occur gradually in the plane of the fracture. The magnitudes of the second derivatives in (12.49) can be estimated as

$$\left|\frac{\partial^2 u_x}{\partial x^2}\right| \approx \left|\frac{\partial^2 u_x}{\partial y^2}\right| = \frac{U}{\Lambda^2}, \quad \left|\frac{\partial^2 u_x}{\partial z^2}\right| = \frac{U}{h^2}, \tag{12.53}$$

and similarly for (12.50). If $(h/\Lambda)^2 \ll 1$, the derivatives within the plane will be negligible compared to the derivative with respect to z, and (12.49), (12.50) can be replaced by

$$\frac{\partial^2 u_x}{\partial z^2} = \frac{1}{\mu}\frac{\partial p}{\partial x}, \quad \frac{\partial^2 u_y}{\partial z^2} = \frac{1}{\mu}\frac{\partial p}{\partial y}. \tag{12.54}$$

Integration of both of these equations with respect to z, bearing in mind the no-slip boundary conditions at the top and bottom walls, $z = h_1$ and $z = -h_2$, yields (Fig. 12.8b)

$$u_x(x,y,z) = \frac{1}{2\mu}\frac{\partial p(x,y)}{\partial x}(z-h_1)(z+h_2), \tag{12.55}$$

$$u_y(x,y,z) = \frac{1}{2\mu}\frac{\partial p(x,y)}{\partial y}(z-h_1)(z+h_2). \tag{12.56}$$

This is essentially the same parabolic velocity profile as occurs for flow between parallel plates, except that the velocity vector is now aligned with the *local* pressure gradient, which is not necessarily collinear with the *global* pressure gradient. Integration across the fracture aperture, using a temporary variable $\zeta = z + h_2$ that represents the vertical distance from the bottom wall, yields

$$h\bar{u}_x(x,y) = \int_{-h_2}^{h_1} u_x(x,y,z)\mathrm{d}z = -\frac{h^3(x,y)}{12\mu}\frac{\partial p(x,y)}{\partial x}, \tag{12.57}$$

$$h\bar{u}_y(x,y) = \int_{-h_2}^{h_1} u_y(x,y,z)\mathrm{d}z = -\frac{h^3(x,y)}{12\mu}\frac{\partial p(x,y)}{\partial y}, \tag{12.58}$$

where $h = h_1 + h_2$ is the total aperture, and the overbar indicates averaging over the z direction.

Equations (12.57) and (12.58) represent an approximate solution to the Stokes equations, but contain an unknown pressure field. A governing equation for the pressure field is found by appealing to the conservation of mass equation, (12.42), which, however, applies to the local velocities, not their integrated values. But $\nabla \cdot \mathbf{u} = 0$, so the integral of $\nabla \cdot \mathbf{u}$ with respect to z must also be zero. Interchanging the order of the divergence and integration operations, which is valid as long as the velocity satisfies the no-slip boundary condition, then shows that the divergence of the z-integrated velocity, $\bar{\mathbf{u}}$, must also vanish. Hence,

$$\frac{\partial (h\bar{u}_x)}{\partial x} + \frac{\partial (h\bar{u}_y)}{\partial y} = 0, \quad \text{so} \quad \frac{\partial}{\partial x}\left[h^3\frac{\partial p}{\partial x}\right] + \frac{\partial}{\partial y}\left[h^3\frac{\partial p}{\partial y}\right] = 0, \tag{12.59}$$

which is the Reynolds (1886) "lubrication" equation.

Insofar as flow through a fracture is accurately represented by the Reynolds equation, the problem of finding the transmissivity of a variable-aperture fracture therefore reduces to the well-studied problem of finding the effective conductivity of a heterogeneous two-dimensional conductivity field, with h^3 playing the role of the conductivity. This problem is conveniently discussed in terms of the "hydraulic aperture," h_{H}, which is defined so that $T = wh_{\mathrm{H}}^3/12$. Using variational principles, it can be shown that the hydraulic aperture is bounded by (Beran, 1968, p. 242)

$$\langle h^{-3}\rangle^{-1} \le h_{\mathrm{H}}^3 \le \langle h^3\rangle, \tag{12.60}$$

where $\langle x\rangle \equiv x_{\mathrm{m}}$ is the arithmetic mean value of the quantity x. The lower bound, the so-called harmonic mean, corresponds to the case in which the aperture varies only in the direction of flow, whereas the upper bound, the arithmetic mean, corresponds to aperture variation only in the direction transverse to the flow (Neuzil and Tracy, 1981; Silliman, 1989). These bounds are theoretically important, because they are among the few results pertaining to flow in rough-walled fractures that are rigorously known, but they are usually too far apart to be quantitatively useful. For instance, it is invariably the case that the hydraulic aperture is less than the mean aperture, that is, $h_{\rm H}^3 \leq \langle h \rangle^3$. But $\langle h \rangle^3 < \langle h^3 \rangle$ for any nonuniform distribution, so the bounds alone are not sufficiently powerful to show that $h_{\rm H}^3 \leq \langle h \rangle^3$.

Elrod (1979) used Fourier transforms to solve the Reynolds equation for a fracture with an aperture having "sinusoidal ripples in two mutually perpendicular directions," and showed that, for the isotropic case,

$$h_{\rm H}^3 = \langle h \rangle^3 \left[1 - 1.5\sigma_{\rm h}^2 / \langle h \rangle^2 + \cdots\right]. \tag{12.61}$$

Zimmerman et al. (1991) considered the case of small regions of unidirectional ripples, which were then randomly assembled, and found results that agreed with (12.61) up to second-order, for both sinusoidal and sawtooth profiles. Furthermore, (12.61) is consistent with the results of Landau and Lifshitz (1960, pp. 45–6), who required only that the aperture field be continuous and differentiable. An alternative expression that agrees with (12.61) up to second-order, but which does not yield unrealistic negative values for large values of the standard deviation, is (Renshaw, 1995)

$$h_{\rm H}^3 = \langle h \rangle^3 \left[1 + \sigma_{\rm h}^2 / \langle h \rangle^2\right]^{-3/2}. \tag{12.62}$$

Dagan (1993) expressed the effective conductivity of a heterogeneous two-dimensional medium in a form that, in the context of fracture flow, can be written as

$$h_{\rm H}^3 = e^{3\langle \ln h \rangle}(1 + a_2\sigma_{\rm Y}^2 + a_4\sigma_{\rm Y}^4 + \cdots) \equiv h_{\rm G}^3(1 + a_2\sigma_{\rm Y}^2 + a_4\sigma_{\rm Y}^4 + \cdots), \tag{12.63}$$

where $Y = \ln h$, $\sigma_{\rm Y}$ is the standard deviation of $\ln h$, and $h_{\rm G} = \exp\langle \ln h \rangle$ is the geometric mean of the aperture distribution. Using a perturbation method and the assumption of a lognormal aperture distribution, Dagan showed that the coefficients a_n vanish at least up to $n = 6$, implying that the geometric mean is a very good approximation for the hydraulic aperture in the lognormal case. Dagan's result agrees with (12.61) up to second order in $\sigma_{\rm Y}$ (Zimmerman and Bodvarsson, 1996).

The predictions of (12.62) compare reasonably well with several numerical simulations and with some laboratory data (Fig. 12.9a). Patir and Cheng (1978) used finite differences to solve the Reynolds equation for flow between two surfaces, the half-apertures of which each obeyed a Gaussian height distribution with linearly decreasing autocorrelation functions. Brown (1987) performed a similar analysis for simulated fractures having fractal roughness profiles. These profiles had fractal dimensions between $D = 2$, which corresponds to a fracture having smooth walls, and $D = 2.5$, which was found by Brown and Scholz (1985a) to

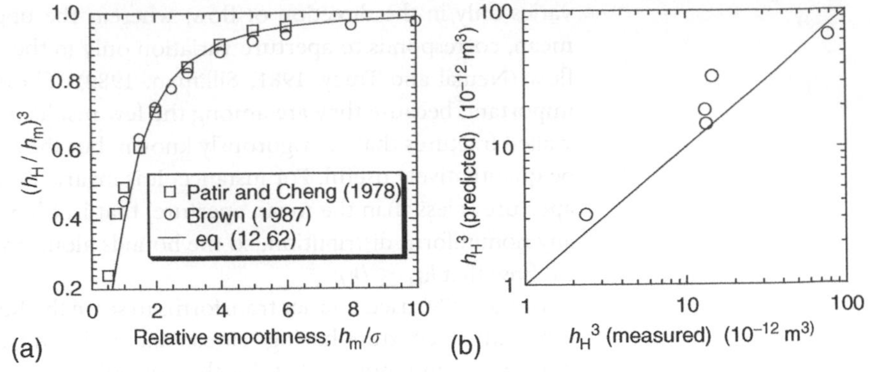

Fig. 12.9
(a) Normalized transmissivity of simulated fractures.
(b) Transmissivities measured by Hakami (1989) on fractures in granite, compared with predictions of (12.62).

correspond to the maximum amount of roughness that occurs in real fractures. The transmissivities computed by Brown were found to essentially depend on $\langle h \rangle$ and σ_h, with little sensitivity to D; the data in Fig. 12.9a are for $D = 2.5$. Fig. 12.9b compares the predictions of (12.62) with values measured by Hakami (1989) in the laboratory on five granite cores from Stripa, Sweden. These five fractures had mean apertures that ranged from 161–464 μm, and relative roughnesses $\sigma/\langle h \rangle$ in the range 0.38–0.75; one other fracture, with a mean aperture of 83 μm, had no measurable transmissivity.

In passing from the Stokes equations to the Reynolds equation, the momentum equation in the z direction, (12.51), is ignored. The z component of the pressure gradient vanishes in the mean, as does the z component of the velocity, but this does not necessarily imply that all of these terms are small *locally*. The error incurred by replacing the Stokes equations by the Reynolds equation is in some sense related to the extent to which the terms in (12.51) are indeed negligible. Visual examination of fracture casts shows that the condition $(h/\Lambda)^2 \ll 1$, which is needed in order for these terms to be negligible, is not always satisfied. The problem is not aperture variation per se, but rather the abruptness with which the aperture varies. Yeo et al. (1998) measured aperture profiles and transmissivities of a fracture in a red Permian sandstone from the North Sea, and solved the Reynolds equation for this fracture using finite elements, and found that the Reynolds equation overpredicted the transmissivity by 40–100 percent, depending on the level of shear displacement. Similar results have been found for artificial fractures (Nicholl et al., 1999), implying that the Reynolds equation may in some cases be an inadequate model for fracture flow.

Fluid flow through fractures is also hindered by the presence of asperity regions at which the opposing fracture walls are in contact and the local aperture is consequently zero. Models such as the geometric mean and the harmonic mean predict zero transmissivity if there is a finite probability of having $h = 0$. A lognormal aperture distribution, on the other hand, does not allow for any regions of zero aperture. These facts suggest using the methods described above for the regions of nonzero aperture and treating the contact regions by other methods (Walsh, 1981; Piggott and Elsworth, 1992).

If an effective hydraulic aperture, call it h_o, can be found for the "open" regions of the fracture, the effect of asperity regions can be modeled by assuming that the fracture consists of regions of aperture $h = h_o$ and regions of aperture $h = 0$. If the flowrate is sufficiently low, that is, $Re^* \ll 1$, and the characteristic in-plane dimension a of the asperity regions is much greater than the aperture, that is, $h_o/a \ll 1$, then flow in the open regions is governed by (12.59), with $h = h_o =$ constant, yielding Laplace's equation,

$$\nabla^2 p = \frac{\partial^2 p}{\partial x^2} + \frac{\partial^2 p}{\partial y^2} = 0. \tag{12.64}$$

Boundary conditions must be prescribed along the contours Γ_i in the (x, y) plane that form the boundaries of the contact regions. As no fluid can enter these regions, the component of the velocity vector normal to these contours must vanish. But the velocity vector is parallel to the pressure gradient, as shown by (12.57) and (12.58), so the boundary conditions for (12.64), along each contour Γ_i, are

$$\frac{\partial p}{\partial n} \equiv (\nabla p) \cdot \mathbf{n} = 0, \tag{12.65}$$

where $\mathbf{n}$ is the outward unit normal vector to Γ_i and n is the coordinate in the direction of $\mathbf{n}$. This mathematical model of flow between two smooth parallel plates, obstructed by cylindrical posts, is known as the Hele-Shaw model (Bear, 1988, pp. 687–92).

Boundary condition (12.65) assures that no flow enters the asperity regions, but the no-flow condition also requires the *tangential* component of the velocity to vanish, that is, $(\nabla p) \cdot \mathbf{t} = 0$, where $\mathbf{t}$ is a unit vector in the (x, y) plane, perpendicular to $\mathbf{n}$. However, it is not possible to impose boundary conditions on both the normal and tangential components of the derivative when solving Laplace's equation (Bers et al., 1964, pp. 152–4). So, solutions to the Hele-Shaw equations typically do not satisfy $(\nabla p) \cdot \mathbf{t} = 0$ and therefore do not account for viscous drag along the sides of the asperities. The relative error induced by this incorrect boundary condition is in the order of h/a (Thompson, 1968; Kumar et al., 1991). Pyrak-Nolte et al. (1987) observed apertures in a fracture in crystalline rock that were in the order of $10^{-4} - 10^{-3}$ m, and asperity dimensions (in the fracture plane) that were in the order of $10^{-1} - 10^{-3}$ m. Gale et al. (1990) measured apertures and asperity dimensions on a natural fracture in a granite from Stripa, Sweden, under a normal stress of 8 MPa and found $h \approx 0.1$ mm, and $a \approx 1.0$ mm. These results imply that viscous drag along the sides of asperities will be negligible compared to the drag along the upper and lower fracture walls, consistent with the assumptions of the Hele-Shaw model.

Walsh (1981) used the solution for potential flow around a single circular obstruction of radius a (Carslaw and Jaeger, 1959, p. 426) to develop the following estimate of the influence of contact area on fracture transmissivity:

$$h_H^3 = h_o^3(1 - c)/(1 + c), \tag{12.66}$$

where c is the fraction of the fracture plane occupied by asperity regions. This expression has been validated numerically (Zimmerman et al., 1992) for asperity concentrations up to 0.25, which covers the range of contact areas that have been observed in real fractures (Witherspoon et al., 1980; Pyrak-Nolte et al., 1987). If the contact regions are randomly oriented ellipses of aspect ratio $\alpha \leq 1$, then (Zimmerman et al., 1992)

$$h_H^3 = h_o^3(1 - \beta c)/(1 + \beta c), \quad \text{where } \beta = (1 + \alpha)^2/4\alpha. \tag{12.67}$$

As the ellipses become more elongated, the factor β increases, and the hydraulic aperture decreases. Although contact areas are not perfectly elliptical, (12.67) can be used if the actual asperity shapes are replaced by "equivalent" ellipses having the same perimeter/area ratios.

12.6 Coupled hydromechanical behavior

As both the mechanical and hydraulic behaviors of rock fractures are controlled to a great extent by the morphology of the fracture surfaces, it is to be expected that the stiffnesses (normal and shear) and transmissivity of a fracture should be related to one another in some way (Cook, 1992; Pyrak-Nolte and Morris, 2000). Although this is undoubtedly true, the relationship is indirect and very complex, and no simple correlations seem to exist between the mechanical and hydraulic properties. This is because hydraulic transmissivity depends primarily on the aperture of the open areas of the fracture, and to a lesser extent on the contact area, whereas normal stiffness depends mainly on the amount and distribution of the contact areas (Hopkins, 2000).

As the normal stress on a fracture increases, the mean aperture decreases, causing the transmissivity to decrease. Due to roughness and asperity contact, however, the change in mean aperture is not exactly equivalent to the joint closure defined in §12.3. Furthermore, although transmissivity is proportional to mean aperture cubed, it also depends on the variance of the aperture, and the amount of contact area, through relations such as (12.62) and (12.67). Consequently, the transmissivity of a fracture that is deforming under a normal load is not always directly proportional to the cube of the mean aperture.

Witherspoon et al. (1980) measured the transmissivity of a tensile fracture in marble, while compressing it under a normal load (Fig. 12.10). At low stresses, the aperture is large and the relative roughness $\sigma/\langle h \rangle$ is low. The fracture therefore approximates the parallel plate model and the transmissivity varies with the cube of the mean aperture (region I). As the stress increases, in the open areas of the fracture the mean aperture will decrease while σ remains nearly the same (Renshaw, 1995), causing the transmissivity to decrease faster than the cube of the mean aperture, as shown by, say, (12.62). In other areas, regions of the fracture will come into contact. This phenomenon also causes an additional decrease in transmissivity; see (12.66) and (12.67). The combination of a decrease in mean aperture, increase in relative roughness, and increase in contact area causes the transmissivity to increase faster than the cube of the mean aperture (region II). In this region, the relation can be approximated by a power law with an exponent

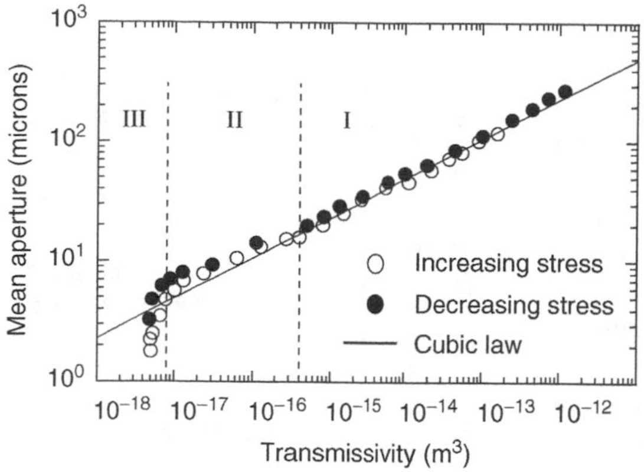

Fig. 12.10 Transmissivity of a marble fracture as a function of mean aperture, as measured by Witherspoon et al. (1980). Data points are from three different loading cycles, although for clarity, not all points are shown.

greater than 3 (Pyrak-Nolte et al., 1987), often as high as 8–10. However, this power law holds over a very narrow range of mean apertures, typically only a factor of two or three.

Sisavath et al. (2003) modeled the transition between regions I and II by considering a fracture with a sinusoidal aperture variation. As the normal stress increases, the two surfaces are assumed to move toward each other, giving a decrease in $\langle h \rangle$ while σ remains constant. Flow was modeled by a perturbation solution of the Stokes equations, including terms up to $(\sigma/\lambda)^2$, where λ is the wavelength. Plausible values of initial roughness and wavelength, such as $\sigma/\langle h \rangle = 0.5$ and $\lambda/\langle h \rangle \approx 5$, lead to slopes of 3–4 at high mean apertures, but abruptly increase to 8–10 as mean aperture decreases.

As normal stress increases further, the mean aperture continues to decrease, but the transmissivity stabilizes at some small but nonzero residual value (region III). Cook (1992) suggested the following qualitative explanation for the existence of a residual transmissivity. Metal castings of the void space of a natural granite fracture under a range of stresses (Pyrak-Nolte et al., 1987) revealed, at high stresses (85 MPa), the existence of large "oceanic" regions of open fracture, connected by tortuous paths through "archipelagic" regions filled with numerous small, closely spaced contact regions. As the stress increases, the oceanic regions, necessarily having very low aspect ratios will continue to deform, leading to a continued decrease in mean aperture. The resistance to flow, however, is controlled by the small tortuous channels that connect the oceanic regions. These channels necessarily have relatively large aspect ratios and are therefore extremely stiff; see §9.3. For example, the channels observed by Pyrak-Nolte et al. (1987) had widths of less than 100 μm (Myer, 2000). The mean aperture at high stresses was in the order of 10 μm, so the aspect ratios of these channels were in the order of 0.1. If the Young's modulus of the intact rock is in the range of 10–100 GPa, the apertures of such channels will decrease by less than 1 percent under an additional stress of a few tens of MegaPascals. Hence, the result is a residual fracture transmissivity that is nearly stress-independent.

As a fracture undergoes shear displacement, on the other hand, the fracture will dilate and the transmissivity will increase. Olsson and Brown (1993)

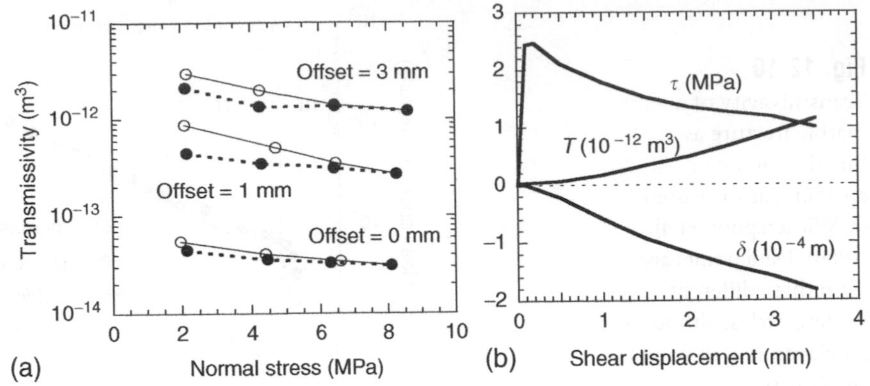

Fig. 12.11 Transmissivity of a fracture in Austin chalk, as measured by Olsson and Brown (1993): (a) Transmissivity as a function of normal stress, for different values of shear offset, with o denoting increasing stress, and • decreasing stress; (b) shear stress τ, transmissivity T, and joint closure δ, as functions of shear displacement, with normal stress held at 4.3 MPa.

measured the hydraulic transmissivity of a fracture in Austin chalk, while either increasing and decreasing the normal stress, keeping the shear displacement constant, or increasing the shear displacement, holding the normal stress constant. The flow geometry was annular-radial, with fluid entering the fracture plane at the inner radius of 24.0 mm and leaving at the outer radius of 60.3 mm. At a fixed value of shear displacement, the transmissivity decreased with increasing normal stress and then increased as the normal stress was removed although some hysteresis is observed (Fig. 12.11a). When the fracture was subjected to 3.5 mm of shear displacement at constant normal stress (equal to 4.3 MPa in Fig. 12.11b), the joint dilated (i.e., negative joint closure) at a rate of about 50 μm per millimeter of shear displacement, and the transmissivity increased by about two orders of magnitude. Qualitatively similar results were found by Yeo et al. (1998) for a red Permian sandstone fracture from the North Sea and by Chen et al. (2000) for a granitic fracture from Olympic Dam mine in Central Australia.

Esaki et al. (1999) subjected an artificially split fracture in a granite from Nangen, Korea to shear displacements up to 20 mm, under various values (1, 5, 10, and 20 MPa) of normal stress. Flow was measured from a central borehole to the outer boundaries of a rectangular fracture plane of dimensions 100 × 120 mm. Transmissivity typically decreased very slightly for the first 0.5–1.0 mm or so of shear, until the peak shear stress was reached. It then increased by about two orders of magnitude as the shear offset increased to about 5 mm. After this point, when the shear stress had reached its residual value, the transmissivity essentially leveled off (Fig. 12.12). When the shear displacement was reversed, the transmissivity decreased, but not to its original value, leaving an excess residual transmissivity at zero shear displacement. This hysteresis was larger at larger values of the normal stress.

12.7 Seismic response of rock fractures

Joints, fractures, and faults also influence the seismic response of rock masses. By introducing an additional compliance into the rock mass, over and above that associated with the adjacent intact rock, they lead to a decrease in the velocities of both compressional and shear waves. They also cause attenuation of

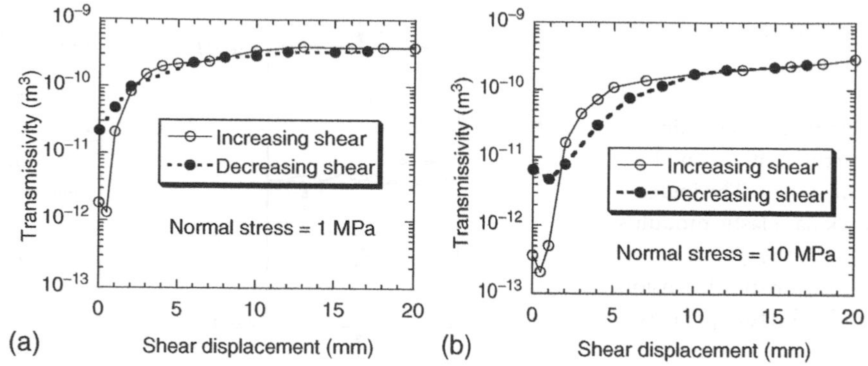

Fig. 12.12 Transmissivity of a granitic joint as a function of shear displacement, as measured by Esaki et al. (1999) under a normal stress of (a) 1 MPa and (b) 10 MPa.

seismic signals, and tend preferentially to filter out high-frequency components of that signal, thereby acting as a low-pass filter. Most of these aspects of the seismic response of individual fractures can be modeled successfully with the displacement discontinuity model. Kendall and Tabor (1971), who referred to it as the "incomplete interface" model, used it to study transmission across an interface at normal incidence. Schoenberg (1980), referring to it as the "linear slip interface" model, extended the analysis to waves impinging on the interface at oblique angles.

When elastic waves impinge on a "welded" interface between two rock layers, the displacements and tractions are assumed to be continuous across the interface (§11.5). In the displacement discontinuity model, the finite "thickness of the fracture," in reality on the order of the mean fracture aperture, is ignored, and the fracture is taken to be an interface of zero thickness separating two regions of intact rock. The condition of continuity of tractions is retained, but the displacements immediately on either side of the interface are assumed to differ by an amount that is proportional to the traction. This condition is consistent with the concept of fracture stiffness, as described in §12.3, as will now be shown using a simplified one-dimensional model.

Consider a rock core of length $2H$, containing a fracture located roughly at its midpoint (Fig. 12.13a), subjected to a normal stress, τ_{zz}. At distances from the fracture that are greater than the characteristic size of the distance between asperities, denoted here by Λ, the longitudinal strain will approach the value $\varepsilon_{zz} = \tau_{zz}/E_r$. The stresses in the vicinity of the nominal fracture plane will be complex and locally heterogeneous, as the void space of the fracture closes up and the asperity contacts deform. This region of thickness 2Λ can be replaced by a homogeneous layer of the same thickness, with an effective Young's modulus E_l chosen so that the overall compression of this homogeneous layer is equal to the average deformation of the actual fracture layer (Fig. 12.13b).

If the origin is fixed to the lower surface of the core, the displacement at the two interfaces between the rock and the thin layer will be given by

$$w(z = H) = \frac{\tau_{zz}H}{E_r}, \quad w(z = H + 2\Lambda) = \frac{\tau_{zz}H}{E_r} + \frac{\tau_{zz}(2\Lambda)}{E_l}. \tag{12.68}$$

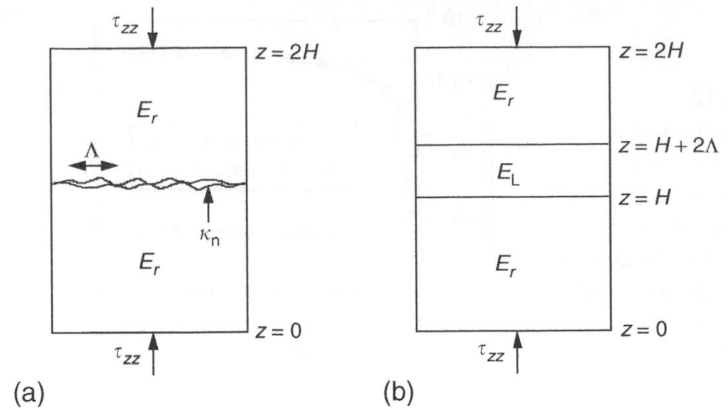

Fig. 12.13 (a) Uniaxial compression of a core of height $2H$, containing a fracture with normal stiffness κ_n. The intact rock has elastic modulus E_r, and the characteristic dimension in the fracture plane is Λ. (b) Fracture zone modeled as a homogeneous layer of thickness 2Λ and elastic modulus E_l.

The difference in the displacements between the upper edge of the layer and the lower edge is

$$w(z = H + 2\Lambda) - w(z = H) = \frac{(2\Lambda)\tau_{zz}}{E_l}. \tag{12.69}$$

For wavelengths that are much greater than the thickness of this layer, that is, $\lambda \gg \Lambda$, it would seem possible to ignore the layer thickness by letting $\Lambda \to 0$, while maintaining the ratio $2\Lambda/E_l$ at the same value, which can be denoted by $1/\kappa_n$, yielding

$$w(H^+) - w(H^-) = \frac{\tau_{zz}}{\kappa_n}. \tag{12.70}$$

The parameter κ_n is a stiffness that has units of Pa/m. Comparison of (12.70) with (12.15) shows that the discontinuity of the displacement across the nominal fracture plane is equivalent to the joint closure, δ.

Angel and Achenbach (1985) solved the problem of elastic wave propagation across a periodic array of cracks, as in Fig. 12.3, and indeed found that applying the displacement discontinuity boundary condition along the nominal fracture plane provides a good approximation to the exact results, as long as $\lambda/(2l) > 5$, that is, for wavelengths greater than about five times the spacing between the centers of two adjacent asperities. (The spacing $2l$ is denoted by 2λ in Fig. 12.3). For asperity spacings that do not exceed 1 mm and wave speeds c on the order of 5000 m/s (see Table 11.1), the displacement discontinuity approximation will be valid for frequencies below the MHz range.

Consider a compressional wave impinging at normal incidence on a joint that is modeled as a displacement discontinuity (Fig. 12.14a). For simplicity, the properties of the rock on either side of the joint are assumed to be the same and the nominal fracture plane is located at $z = 0$. At a given level of the *in situ* stress, the fracture will have a certain normal stiffness κ_n, given by the tangent to the stress-closure curve. If the stress increment associated with the incoming wave is small compared to the *in situ* stress, the fracture stiffness can be assumed to be constant, and the equations of linearized elasticity can be used, with the stresses

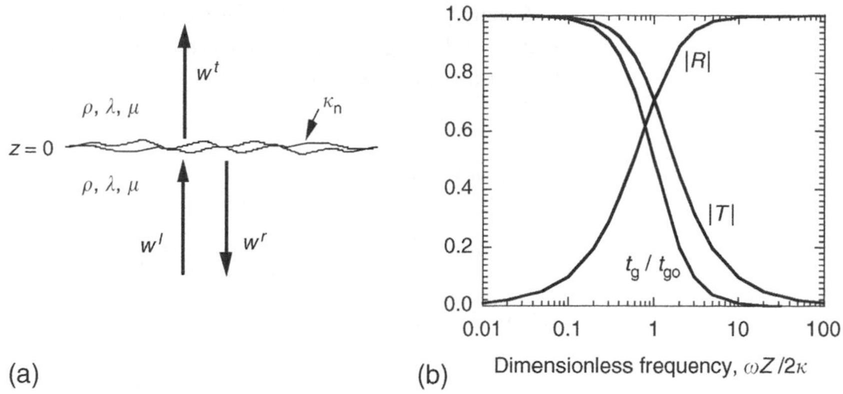

Fig. 12.14 (a) Waves impinging on a fracture at normal incidence. (b) Transmission and reflection coefficients, and group time delay (normalized to its value at zero frequency).

and displacements denoting the incremental values produced by the wave, *not* the *in situ* values.

Using a notation similar to that of §11.5, the incoming wave can be expressed as

$$w^i = A_\mathrm{i} \exp[ik_\mathrm{i}(z - c_\mathrm{L} t)], \tag{12.71}$$

the reflected wave as

$$w^r = A_\mathrm{r} \exp[ik_\mathrm{r}(-z - c_\mathrm{L} t)], \tag{12.72}$$

and the transmitted wave as

$$w^t = A_\mathrm{t} \exp[ik_\mathrm{t}(z - c_\mathrm{L} t)], \tag{12.73}$$

where c_L is the compressional wave velocity in the intact rock on either side of the discontinuity.

The displacement terms are of the form $w = A\exp[ik(\pm z - c_\mathrm{L} t)]$, from which the normal stress follows as $\tau_{zz} = (\lambda + 2\mu)\varepsilon_{zz} = \pm(\lambda + 2\mu)ikA\exp[ik(\pm z - c_\mathrm{L} t)]$. The total stress in the region $z < 0$ will be the sum of the stresses from the incident wave and the reflected wave, whereas the stress in the region $z > 0$ corresponds only to that of the transmitted wave. The condition of continuity of tractions across the interface therefore takes the form

$$k_\mathrm{i} A_\mathrm{i} \exp(-ik_\mathrm{i} c_\mathrm{L} t) - k_\mathrm{r} A_\mathrm{r} \exp(-ik_\mathrm{r} c_\mathrm{L} t) = k_\mathrm{t} A_\mathrm{t} \exp(-ik_\mathrm{t} c_\mathrm{L} t), \tag{12.74}$$

after canceling out the common term $(\lambda + 2\mu)i$ and setting $z = 0$ in the expressions for the stresses. In order for this relationship to hold for all values of time, the wave numbers must be the same for the incident, reflected, and transmitted waves, that is, $k_\mathrm{i} = k_\mathrm{r} = k_\mathrm{t} \equiv k$, in which case (12.74) takes the form

$$A_\mathrm{i} - A_\mathrm{r} = A_\mathrm{t}. \tag{12.75}$$

Likewise, the displacement boundary condition (12.70) takes the form

$$A_\mathrm{i} + A_\mathrm{r} = A_\mathrm{t} - ik[(\lambda + 2\mu)/\kappa_\mathrm{n}]A_\mathrm{t}. \tag{12.76}$$

Solving (12.75) and (12.76) for the amplitudes of the reflected and transmitted waves, and using relations such as $\omega = kc$ and $(\lambda + 2\mu) = \rho c_{\rm L}^2$, gives the reflection and transmission coefficients as functions of the frequency (Schoenberg, 1980):

$$R(\omega) = \frac{A_{\rm r}}{A_{\rm i}} = \frac{-i(\omega Z_{\rm L}/2\kappa_{\rm n})}{1 - i(\omega Z_{\rm L}/2\kappa_{\rm n})} \equiv \frac{-i\varpi}{1 - i\varpi}, \tag{12.77}$$

$$T(\omega) = \frac{A_{\rm t}}{A_{\rm i}} = \frac{1}{1 - i(\omega Z_{\rm L}/2\kappa_{\rm n})} \equiv \frac{1}{1 - i\varpi}, \tag{12.78}$$

where $Z_{\rm L} = \rho c_{\rm L}$ is the seismic impedance of longitudinal (compressional) waves in the intact rock and $\varpi = (\omega Z_{\rm L}/2\kappa_{\rm n})$ is a dimensionless frequency. These expressions are different in form, but equivalent to, those derived by Kendall and Tabor (1971), who defined their reflection and transmission coefficients in terms of stress amplitudes rather than displacement amplitudes. The reflection coefficient given by Pyrak-Nolte et al. (1990) differs from (12.77) in sign, due to the inclusion of a minus sign in their definition of the amplitude of the reflected wave.

As only the real component of the wave has physical significance, taking $A_{\rm i}$ to be purely real gives an incident wave described by

$$w^i = A_{\rm i}\cos[k(z - c_{\rm L}t)]. \tag{12.79}$$

Multiplying numerator and denominator of (12.78) by $1 + i\varpi$, the complex transmission coefficient can be written in Cartesian form as

$$T(\omega) = T_{\rm R}(\omega) + iT_{\rm I}(\omega) = 1/(1 + \varpi^2) + i[\varpi/(1 + \varpi^2)], \tag{12.80}$$

or in polar form as

$$\begin{aligned} T(\omega) &= |T(\omega)|\exp(i\Delta) = (T_{\rm R}^2 + T_{\rm I}^2)^{1/2}\exp[i\arctan(T_{\rm I}/T_{\rm R})] \\ &= (1 + \varpi^2)^{-1/2}\exp(i\arctan\varpi). \end{aligned} \tag{12.81}$$

The transmitted wave then has the complex displacement

$$\begin{aligned} w^t &= A_{\rm i}T(\omega)\exp[ik(z - c_{\rm L}t)] \\ &= A_{\rm i}(1 + \varpi^2)^{-1/2}\exp[ik(z - c_{\rm L}t) + i\arctan\varpi], \end{aligned} \tag{12.82}$$

the real (physical) part of which is

$$w^t = A_{\rm i}(1 + \varpi^2)^{-1/2}\cos[k(z - c_{\rm L}t) + \arctan\varpi]. \tag{12.83}$$

The transmitted wave therefore has an amplitude that differs from that of the incident wave by the multiplicative factor $|T(\omega)| = (1 + \varpi^2)^{-1/2}$, and, since the time term and the phase angle enter (12.83) with opposite signs, lags behind the incident wave by a phase angle of $\Delta_{\rm t} = \arctan\varpi$.

Similarly, the real component of the displacement of the reflected wave is

$$w^r = A_{\rm i}\varpi(1 + \varpi^2)^{-1/2}\cos[-k(z + c_{\rm L}t) + \arctan(-1/\varpi)]. \tag{12.84}$$

The reflected wave has an amplitude that is less than that of the incident wave by the multiplicative factor $|R(\varpi)| = \varpi/(1+\varpi^2)^{1/2}$ and a phase angle of $\Delta_r = \arctan(-1/\varpi)$.

If the stiffness of the interface vanishes, the dimensionless frequency $\varpi = (\omega Z_L/2\kappa_n)$ approaches unity, and (12.83) and (12.84) show that $|R| \to 1$ and $|T| \to 0$, in agreement with the results for normal incidence at a free surface (Fig. 12.14b). As the stiffness of the interface becomes infinite, the dimensionless frequency approaches zero, and (12.83) and (12.84) show that $|R| \to 0$ and $|T| \to 1$, as would be expected for a welded interface between two media having the same properties. For a given interface having a finite stiffness, at low frequencies $|R| \approx \varpi$ and $|T| \approx 1-(\varpi^2/2)$, whereas at high frequencies $|R| \approx 1-(1/2\varpi^2)$ and $|T| \approx 1/\varpi$. At all frequencies, it is the case that $|R|^2 + |T|^2 = 1$, indicating that mechanical energy is conserved during the process of wave transmission across the interface. The reflection and transmission coefficients obey the relation $|R(1/\varpi)| \approx |T(\varpi)|$, which in fact follows from $|R|^2 + |T|^2 = 1$.

Referring to the model of the interface as a thin layer of thickness 2Λ (Fig. 12.13b), the dimensionless frequency can be shown to be equal to π times the ratio of the layer thickness to the wavelength of the incoming wave, that is, $\varpi = \pi\Lambda/\lambda$. Using the row-of-cracks model shown in Fig. 12.3, Angel and Achenbach (1985) found that the displacement discontinuity approach, which smears the effect of the individual cracks/asperities over the entire fracture plane, is valid only for $\lambda > 10l$, where $2l$ is the spacing between centers of adjacent cracks. The layer thickness Λ was defined as the distance from the fracture plane over which the stress, in a static compression test, is perturbed from its mean value; hence, Λ will be in the order of l. These considerations seem to imply that the displacement discontinuity model should be accurate in the range $\varpi < 1$ and perhaps be only qualitatively useful for higher values of the dimensionless frequency.

The time dependence of a monochromatic traveling wave with frequency ω and phase angle Δ is $\cos(kz - \omega t + \Delta)$. Compared to a similar wave having the same frequency but zero phase angle, this wave can be said to lag behind with a "time delay" of $t_d = \Delta/\omega$. Likewise, the *group time delay* of a dispersive wave can be defined as $t_g = d\Delta/d\omega$. From the phase angles shown in (12.83) and (12.84), it follows that the group time delay for both the transmitted and reflected waves is given by (Fig. 12.14b)

$$t_g = \frac{(Z_L/2\kappa_n)}{1+(\omega Z_L/2\kappa_n)^2}. \tag{12.85}$$

The granite sample tested by Pyrak-Nolte et al. (1990) had a longitudinal wave speed of 5200 m/s, a density of 2650 kg/m^3, and a normal fracture stiffness in the order of 10^{13} Pa/m. The group time delays predicted by (12.85) are in the range of 0.5 μs, which are close to the measured values, which ranged from 0.12–0.96 μs.

The analysis given above for normal incidence of a (longitudinal) P-wave upon a fracture modeled as a displacement discontinuity applies in all details to SV- and SH-waves (Pyrak-Nolte et al., 1990), with the normal stiffness κ_n replaced by the

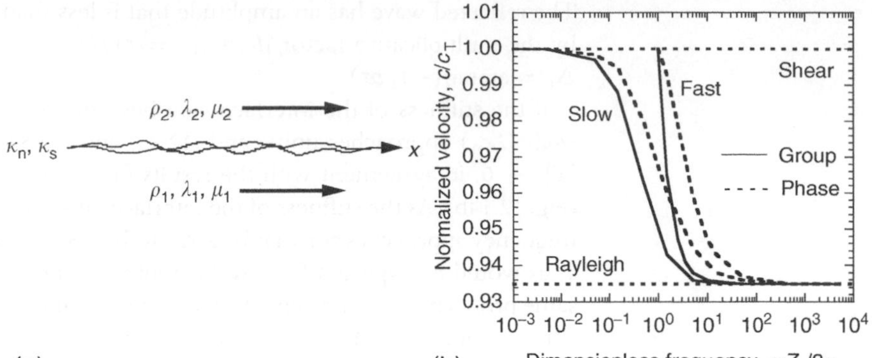

(a) (b)

Fig. 12.15 (a) Interface wave propagating (→) along a fracture. (b) Group and phase velocities, for the case $\rho_1 = \rho_2, \lambda_1 = \lambda_2 = \mu_1 = \mu_2$ (after Pyrak-Nolte and Cook, 1987).

shear stiffness κ_s, and the impedance Z_L replaced by the shear wave impedance Z_T. If the fracture is anisotropic, the shear stiffness values appropriate for the SV- and SH-waves will differ.

Detailed results for incidence of P-, SV- and SH-waves at arbitrary angles to the fracture plane have been given by Gu et al. (1996b). Joints that are filled with viscous fluid can be modeled by assuming both a displacement discontinuity and a velocity discontinuity. The velocity discontinuity boundary condition involves a "specific viscosity" parameter that represents the ratio of the shear stress acting on the fracture to the discontinuity that is produced in the velocity (Pyrak-Nolte et al., 1990).

Pyrak-Nolte and Cook (1987) investigated interface waves that propagate *along* a fracture modeled as a displacement discontinuity; these waves can be thought of as generalized Rayleigh or Stoneley waves. In the special case in which the rock on either side of the interface has the same properties, two types of interface waves can travel along the fracture (Fig. 12.15). At high frequencies (or low values of the fracture stiffness), the two half-spaces become progressively uncoupled, and the speed of both waves approaches that of a Rayleigh wave in the intact rock. Below a certain critical frequency of about $\varpi^* \approx 1$, the faster of these two interface waves cannot propagate. At lower frequencies, the speed of the "slow" interface wave approaches the speed of shear waves in the intact rock. Both interface waves are weakly dispersive, with the group velocity being slightly lower than the phase velocity. If the properties of the rock differ greatly on the two sides of the interface, only the slow interface wave exists. Detailed studies of these waves as functions of rock properties, shear and normal stiffnesses of the fracture, etc., have been given by Pyrak-Nolte and Cook (1987) and Gu et al. (1996a). Laboratory measurements of these fracture interface waves have been reported by Pyrak-Nolte et al. (1992).

12.8 Fractured rock masses

The mechanical, hydraulic, and seismic behavior of single fractures has been described in §12.2–§12.7. In some cases, the behavior of a rock mass may be controlled by a single large fracture or fault. But in many rock masses, fractures are so pervasive that the rock must be viewed as a new type of medium known as a *fractured rock mass*, rather than as a mass of intact rock that happens to contain one or several distinct fractures. A fractured rock mass can often be treated as a continuum whose properties, such as elastic moduli and permeability, are controlled by the density and orientation of the fractures and by the hydromechanical properties of these individual fractures. This situation is analogous to the manner in which, on a much smaller length scale, pores and cracks control the mechanical and seismic properties of "intact" rock (as described in §10.3).

Consider a rock mass that contains three orthogonal sets of uniformly spaced fractures. The fractures that lie in the y–z plane are taken to have spacing S_x, normal stiffness κ_{nx}, and shear stiffness in the y and z directions of κ_{sy} and κ_{sz}, respectively, with similar notation for the other two fracture sets. The intact rock between the fractures is assumed to be isotropic, with moduli $\{E_r, G_r, V_r\}$. Such a medium will be elastically orthotropic, as discussed in §5.10, with at most nine distinct elastic moduli, which depend on the properties of the intact rock and the fractures.

Since the medium is periodic in x with unit cell of length S_x, the effective Young's modulus for deformation in the x direction is given by (12.16), with the spacing S_x in place of the "specimen length" L:

$$\frac{1}{E_x} = \frac{1}{E_r} + \frac{1}{S_x \kappa_{nx}}. \tag{12.86}$$

By adding the deformation of the intact rock to the shear deformation occurring on the fractures, the effective shear modulus in the x–y plane is found to be (Amadei and Savage, 1993)

$$\frac{1}{G_{xy}} = \frac{1}{E_r} + \frac{1}{S_x \kappa_{sx}} + \frac{1}{S_y \kappa_{sy}}. \tag{12.87}$$

If the fractures themselves are assumed not to contribute to the Poisson effect, that is, they contribute no additional strain ε_{xx} due to a stress τ_{xx}, then the compliance coefficient that relates ε_{xx} to τ_{xx} will be the same as for the intact rock. In the Voigt notation of §5.10, this compliance is $s_{12} = -\nu_r/E_r$. As the intact rock is assumed to be isotropic, $s_{12} = s_{13} = s_{23}$. Such a fractured medium therefore has only seven independent moduli or compliances, which in Voigt notation (see (5.184)) are

$$s_{11} = \frac{1}{E_x} = \frac{1}{E_r} + \frac{1}{S_x \kappa_{nx}}, \quad s_{22} = \frac{1}{E_y} = \frac{1}{E_r} + \frac{1}{S_y \kappa_{ny}}, \quad s_{33} = \frac{1}{E_z} = \frac{1}{E_r} + \frac{1}{S_z \kappa_{nz}}, \tag{12.88}$$

$$s_{44} = \frac{1}{G_{yz}} = \frac{1}{G_r} + \frac{1}{S_y \kappa_{sy}} + \frac{1}{S_z \kappa_{sz}}, \tag{12.89}$$

$$s_{55} = \frac{1}{G_{xz}} = \frac{1}{G_r} + \frac{1}{S_x \kappa_{sx}} + \frac{1}{S_z \kappa_{sz}}, \tag{12.90}$$

$$s_{66} = \frac{1}{G_{xy}} = \frac{1}{G_r} + \frac{1}{S_x\kappa_{sx}} + \frac{1}{S_y\kappa_{sy}}, \tag{12.91}$$

$$s_{12} = s_{13} = s_{23} = s_{21} = s_{31} = s_{32} = \frac{-\nu_r}{E_r}. \tag{12.92}$$

If the spacings and fracture properties of the three fracture set are identical, the rock mass will have $s_{11} = s_{22} = s_{33}$ and $s_{44} = s_{55} = s_{66}$, but will not be elastically isotropic, as there will be three independent compliances rather than two. Other special cases such as a rock mass containing only one set of joints can be recovered by setting either the fracture stiffnesses or fracture spacings equal to infinity in two of the directions, say y and z. In this case, the rock mass is transversely isotropic in the y–z plane, and has four independent elastic coefficients (Morland, 1976). Fossum (1985) started with the result for a single set of fractures and, by an averaging procedure, derived an expression for the elastic moduli of a rock mass containing randomly oriented fractures. All of the above-mentioned results ignore dilatancy, that is, the coupling between shear stress and normal deformation, and thus are valid only for small stress and strain increments, for which dilatancy may be negligible.

The strength of a rock that contains a single plane of weakness was discussed in §4.8, following the treatment given by Jaeger (1960). This analysis can also be applied to a rock mass that contains a set of parallel joints, although it does not account for the effect of the intermediate principal stress σ_2, except in the special case where the joints strike in the σ_2 direction (Amadei, 1988). A detailed discussion of the failure of a fractured rock mass, accounting for the effect of three independent principal stresses, has been given by Amadei and Savage (1993).

For wavelengths that are long compared to the spacing between fractures, seismic wave propagation through a fractured rock mass can be modeled as wave propagation through an equivalent elastic medium, with the elastic moduli given by equations such as (12.88)–(12.92). As wave speeds in fractured rocks are typically about 4000 m/s and fracture spacings are in the order of 0.1–10 m, the equivalent continuum model will apply for frequencies in the range below 100 Hz, as are used in low-frequency seismic surveys. Equivalent medium representations for seismic wave propagation in fractured rock have been developed by, among others, Schoenberg and Sayers (1995), Boadu and Long (1996), and Liu et al. (2000).

The conceptual model of a rock mass containing three orthogonal sets of fractures can also be used to derive expressions for the macroscopic permeability. Let the fracture set that lies in the y–z plane have spacing S_x, and hydraulic apertures h_{xy} and h_z in the y and z directions, respectively, with similar notations for the other two sets. Each fracture is assumed to be of infinite extent in its plane. As the transmissivity of the intact rock material between the fractures is usually orders of magnitude less than that of the fracture sets, the intact rock can be assumed to be impermeable. A unit cell of this rock mass would have dimensions $\{S_x, S_y, S_z\}$, and contain one fracture in each of the three orthogonal directions (Fig. 12.16). A pressure gradient in the x direction would give rise to a fluid flux in the x direction, through each of the two fractures that lie in the x–y

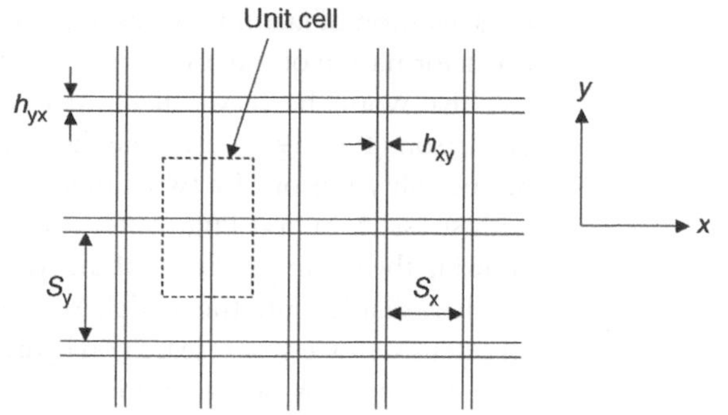

Fig. 12.16 Schematic diagram of a two-dimensional rock mass containing two sets of orthogonal fractures, showing a unit cell used to help calculate the permeability. The notation h_{xy} denotes the hydraulic aperture, for flow in the y direction, of a fracture whose outward unit normal vector is in the x direction.

and x–z plane, according to (12.44):

$$Q_x = -\frac{S_y h_{zx}^3}{12\mu}\frac{\partial p}{\partial x} - \frac{S_z h_{yx}^3}{12\mu}\frac{\partial p}{\partial x}. \tag{12.93}$$

The cross-sectional area of the unit cell in the direction perpendicular to the flux is $S_y S_z$, so the permeability is given by

$$k_x = \frac{-Q_x \mu}{A(\partial p/\partial x)} = \frac{h_{zx}^3}{12S_z} + \frac{h_{yx}^3}{12S_y}, \tag{12.94}$$

and similarly for the other two directions. The permeability is inversely proportional to the fracture spacings, that is, proportional to the fracture frequencies. Such a rock mass will generally be hydraulically anisotropic.

Snow (1969) extended this model to account for fracture sets that are oriented at oblique angles, but which are still of infinite length. Real fractures are of course of finite extent, and so, unlike in the infinite-length fracture model, may or may not intersect other neighboring fractures. Hence, the effective permeability is expected to be less than would be predicted by an infinite-length fracture model. The effect of interconnectivity on the macroscopic permeability of a two-dimensional rock mass was studied by Long and Witherspoon (1985), under the assumption that all fractures were of the same length and same hydraulic aperture. Their analysis was generalized by Hestir and Long (1990), who allowed the fractures to be of different lengths. Charlaix et al. (1987) used percolation theory to study the effective permeability of randomly oriented disk-shaped fractures having a wide distribution of apertures. Wei et al. (1995) conducted numerical experiments of flow through a two-dimensional fracture network, and concluded that since fracture lengths usually exceed fracture spacings by an order of magnitude, corrections for finite fracture lengths would be less than the errors inherent in other aspects of the analysis. The situation in which the matrix permeability is not negligible compared to that of the fractures has been studied by Sayers (1991) for aligned, two-dimensional fractures of finite lengths.

The geometric complexity of actual fracture networks suggests that numerical simulations may play an important role in elucidating the coupled hydromechanical behavior of fractured rock masses. Indeed, numerical simulations permit the consideration of factors such as nonlinear constitutive relations for the mechanical deformation of the fractures, distributions of fracture lengths and apertures, etc., that would be very difficult to account for by analytical means. Zhang and Sanderson (1996,1998) used the distinct element method to study the effective permeability tensor of a two-dimensional fractured rock mass and its variation with stress. Min and Jing (2003) and Min et al. (2004) extended this work by allowing the normal stiffness of the fractures to be nonlinear and by incorporating shear dilation into the model. Rutqvist and Stephansson (2003) have provided an extensive review of research on hydromechanical coupling in fractured rocks, with an emphasis on implications for subsidence, induced seismicity, and the performance of underground radioactive waste repositories.

13 State of stress underground

13.1 Introduction

Knowledge of the *in situ* state of stress in the subsurface is of crucial importance in all areas of rock engineering. In general, the existing state of stress in the subsurface varies as a function of depth, but the manner in which the three principal stresses and their associated directions vary with depth does not have an easily predictable pattern. These stresses will be influenced by the topography, by the tectonic forces, by the constitutive behavior of the rock, and by the local geological history. Whenever an excavation is made for a tunnel, underground repository, underground storage space, etc., the excavation will alter the existing stress state in the rock mass. Depending on the constitutive model used for the rock, the stresses and displacements that are induced in the rock can be calculated by methods such as those discussed in Chapters 8 and 9. Prediction of the resulting final state of stress and deformation in the vicinity of the excavation, and of the stability of the excavation, therefore requires, among other things, knowledge of the preexisting *in situ* stresses. The same is true for boreholes drilled for oil and gas wells, geothermal wells, and water wells.

Various issues related to the state of stress in the subsurface are discussed in this chapter. Some basic and highly simplified conceptual models of the state of stress in the subsurface are presented in §13.2. A discussion of measured data on subsurface stresses, collected throughout the world by various researchers, is given in §13.3. In §13.4 and §13.5, the theory of elasticity is used to develop some solutions for stresses due to surface loads and due to gravity forces in the presence of irregular topography. The use of hydraulic fracturing for estimating the *in situ* stresses is discussed in §13.6. Several other methods for estimating the subsurface stress state, such as the use of flat-jacks and overcoring, are briefly reviewed in §13.7.

A more detailed discussion of the state of stress in the subsurface is given in the monograph by Amadei and Stephansson (1997). Extensive discussion of the ways in which the subsurface stresses influence fault patterns and other geological structures can be found in various books on structural geology, such as Price and Cosgrove (1990), Bayly (1992) and Pollard and Fletcher (2005).

13.2 Simple models for the state of stress in the subsurface

As discussed in Chapter 2, the complete state of stress at any point in a rock can be fully specified by the values of the three principal stresses and their associated directions. At a free surface the shear tractions are zero, and so the normal vector to any unloaded free surface is necessarily a principal stress direction. In regions of flat or nearly flat topography, it therefore seems reasonable that one of the principal stress directions will be vertical. The simplest assumption is that the normal compressive stress at any depth z below the surface must be sufficient to support the weight of the overburden rock. This can be expressed as

$$\tau_{zz} = \int_0^z \rho(z')g(z')\mathrm{d}z'. \tag{13.1}$$

where, in general, the density ρ and the gravitational acceleration g may vary with depth. For the relatively shallow depths that are of engineering interest, say less than a few kilometers, g can be considered constant, and (13.1) gives $\tau_{zz} = \bar{\rho}gz$, where $\bar{\rho}$ is the mean density of the rocks above the depth z.

If the rock is porous or fractured, the mean density that appears in the expression $\tau_{zz} = \bar{\rho}gz$ must be averaged over the mineral grains *and* the pore fluids. For example, if the overburden consists of a porous rock of porosity ϕ and mineral density ρ_{m}, fully saturated with a pore fluid of density ρ_{f}, then the mean density would be given by

$$\bar{\rho} = (1 - \phi)\rho_{\mathrm{m}} + \phi\rho_{\mathrm{f}}. \tag{13.2}$$

In the sequel, we will assume that the density is uniform, and drop the overbar and simply write $\tau_{zz} = \rho gz$.

Expression (13.1) for the vertical stress is almost universally assumed to hold. Nevertheless, it should be borne in mind that (13.1) is valid only in the absence of shear stresses τ_{xz} or τ_{yz}, which could also serve to support the overburden, as can be seen by doing a simple force balance on a prism of rock lying between depth z and the surface. Measurements, discussed in more detail in §13.3, tend to verify (13.1), except occasionally at very shallow depths, of at most a few tens of meters, in areas of irregular surface topography.

The simplest assumption regarding the entire stress state is that the other two principal stresses are also equal to the overburden stress, that is, $\tau_{xx} = \tau_{yy} = \rho gz$. Any state of stress in which all three principal stresses are equal is referred to in mechanics as a *hydrostatic* stress state, as this is the situation that occurs within a static column of water. However, in the rock mechanics context, to avoid confusion with stresses that are indeed due to pore fluids, the stress state $\tau_{xx} = \tau_{yy} = \tau_{zz} = \rho gz$ is referred to as *lithostatic*. This state of stress is also characterized by the absence of deviatoric stresses, which means that no shear stress components will appear in the stress tensor, in any coordinate system.

The assumption of a lithostatic stress state in the subsurface is known as *Heim's rule*. Heim's original argument for this assumption was that if the rock is assumed to behave viscoelastically, according to a Maxwell or Burgers model as described in §9.9, the stress state will eventually approach a lithostatic condition, given sufficient time (Heim, 1878). However, this model neglects the effect of tectonic forces, which will often vary on timescales that are shorter than the relaxation times of the rock. Furthermore, for many near-surface rocks, brittle behavior may be a more realistic assumption than viscoelastic behavior.

Another model that has often been invoked to predict the magnitudes of the other two principal stresses is to assume that the gravitational forces due to the weight of the rock can be instantaneously turned on, after which the rock is constrained against any lateral deformation. This leads to the state of uniaxial strain that has been described in §5.3. Assuming uniform and isotropic rock properties and that the vertical stress is given by $\tau_{zz} = \rho g z$, the resulting state of stress will be

$$\tau_{zz} = \rho g z, \quad \tau_{xx} = \tau_{yy} = \frac{\nu}{1-\nu}\tau_{zz}, \quad \tau_{xy} = \tau_{yz} = \tau_{zx} = 0. \tag{13.3}$$

For a typical value of Poisson's ratio, such as 0.25, this model predicts that the vertical (lithostatic) stress is the maximum principal stress and that other two principal stresses are equal to one-third of the lithostatic value.

Aside from the unphysical assumption that the gravitational forces act only after the rock has been emplaced, the assumption that no lateral strain can occur is clearly an extreme case. The other extreme assumption would be that when gravity begins to act, there are no lateral stresses. This leads to a state of uniaxial stress, also described in §5.3:

$$\tau_{zz} = \rho g z, \quad \tau_{xx} = \tau_{yy} = \tau_{xy} = \tau_{yz} = \tau_{zx} = 0. \tag{13.4}$$

All three of the simple models described above can be written in the following unified form:

$$\tau_{zz} = \rho g z, \quad \tau_{xx} = \tau_{yy} = k\tau_{zz}, \quad \tau_{xy} = \tau_{yz} = \tau_{zx} = 0, \tag{13.5}$$

where k is the lateral stress coefficient. This coefficient equals 1 according to Heim's rule, $\nu/(1-\nu)$ according to the uniaxial strain model, and 0 according to the assumption of unconstrained lateral deformation.

Each of the aforementioned models predicts that the lateral stresses will not exceed the vertical stress. Another related model that can predict values of $k > 1$ is that of McCutchen (1982). In this model, the crust is taken to be a thin, elastically isotropic and homogeneous shell, resting on top of a rigid mantle and deforming due to gravity. The difference between this model and the uniaxial strain model discussed above is that the spherical geometry of the Earth is now taken into account. The boundary conditions are that the radial normal stress vanishes at the surface, and the tangential displacement vanishes at the crust–mantle boundary. If the outer radius of the Earth is a, the radius of the crust/mantle interface is b, solving the equations of elastic equilibrium for this model yields

$$\tau_{rr} = \frac{\rho g a}{4(1-\nu)}\left\{\left[2-(1+\nu)\frac{b}{a}\right]\frac{(1+\nu)+2(1-2\nu)(b/r)^3}{(1+\nu)+2(1-2\nu)(b/a)^3} + (1+\nu)\frac{b}{a} - 2\frac{r}{a}\right\}, \tag{13.6}$$

$$\tau_{\theta\theta} = \tau_{\phi\phi} = \frac{\rho g a}{4(1-\nu)}\left\{\left[2-(1+\nu)\frac{b}{a}\right]\frac{(1+\nu)-(1-2\nu)(b/r)^3}{(1+\nu)+2(1-2\nu)(b/a)^3} + (1+\nu)\frac{b}{a} - (1+2\nu)\frac{r}{a}\right\}, \tag{13.7}$$

where ν is the radial displacement, r is the distance from the center of the Earth, and the depth is given by $z = a - r$.

These expressions for the stresses can be simplified considerably by making use of the fact that the thickness ($t = a - b$) of the crust, which is in the order of a few tens of kilometers, is very small compared to the radius of the Earth, which is about 6.4×10^6 km. Expanding (13.6) and (13.7) in power series of z/a and neglecting terms of order $(t/a)^2 \approx 0.0001$, gives

$$\tau_{rr} = \sigma_{\text{vertical}} = \rho g z, \tag{13.8}$$

$$\tau_{\theta\theta} = \tau_{\phi\phi} = \sigma_{\text{horizontal}} = \frac{t}{4a}\rho g t + \frac{\nu}{1-\nu}\rho g z. \tag{13.9}$$

The vertical and horizontal stresses predicted by this model are shown in Fig. 13.1a, using a mean crustal thickness of 40 km, a mean density of 2700 kg/m^3, and a Poisson's ratio of 0.25. As the crustal thickness is a small fraction of the Earth's radius, the model predicts an essentially linear variation of stresses with depth. The predicted vertical stress gradient is very nearly equal to ρg, and the tangential stress gradient is again essentially equal to $\nu\rho g/(1-\nu)$. However, because of the assumed spherical geometry, the model predicts a *finite* tangential stress at the surface, in contrast with the zero tangential stress predicted by the uniaxial strain model. The predicted stress ratio k decreases nonlinearly with depth, according to

$$k = \frac{\tau_{\phi\phi}}{\tau_{rr}} = \frac{\nu}{1-\nu} + \frac{t^2}{4az}. \tag{13.10}$$

This predicted $1/z$ decrease of k with depth is in qualitative agreement with many field measurements, as can be seen by comparing Figs. 13.1b and 13.2b.

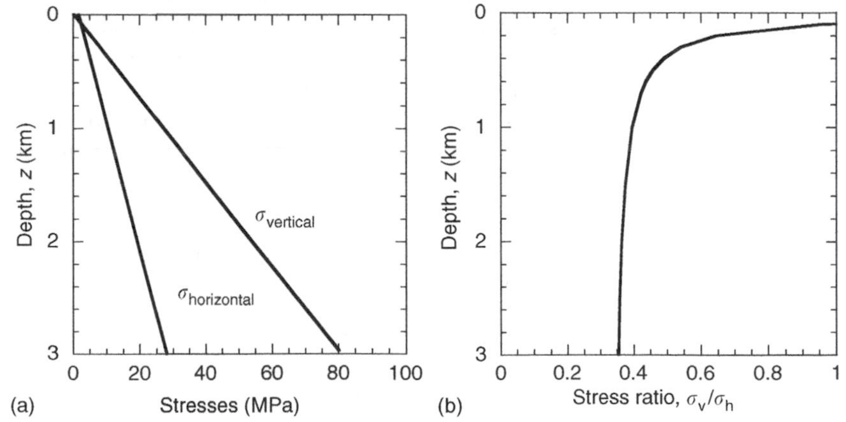

Fig. 13.1 Predictions of the spherical shell model; (a) vertical and horizontal stresses; (b) ratio of horizontal to vertical stress.

Sheorey (1994) extended this model by considering the crust and underlying mantle each to be composed of several layers and by accounting for the geothermal temperature gradient and thermal expansion effects. This model predicts a variation of k with depth that is in better general agreement with the data collected by Brown and Hoek (1978), and predicts that the horizontal stresses at the surface will be about 11 MPa. Horizontal stresses of this magnitude have indeed occasionally been measured within a few tens of meters of the surface (Lo, 1978; Herget, 1987).

13.3 Measured values of subsurface stresses

Subsurface stresses have been measured, or rather have *been inferred from measurements*, at numerous locations throughout the world, at depths down to several kilometers. Some of the methods used to estimate these stresses are discussed in §13.6 and §13.7. Among the more notable reports of subsurface stress measurements are those of Hast (1969), Herget (1974), Haimson and Voight (1977), McGarr and Gay (1978), and Zoback and Zoback (1980). An extensive review of many of these data sets has been given by Amadei and Stephansson (1997).

The most well known compilation of measured subsurface stress values is that of Brown and Hoek (1978). These data are shown in Fig. 13.2, without individual labels denoting the publication source of the data or the geographical locations of the measurements, which can be found in Brown and Hoek (1978). As can be seen in Fig. 13.2a, these data set shows considerable scatter, but nevertheless allow some general trends to be noted. The vertical stresses do indeed cluster around the line $\tau_{zz} = \rho g z$ that corresponds to a mean density of $2700\,\mathrm{kg/m^3}$. Deviations are probably attributable to variations in local topography, as discussed in §13.4, or to local geological inhomogeneities. The assumption that the deviations are due to topographical effects is consistent with the fact that the fractional deviations from the linear trend line decrease with depth.

The ratio of the average value of the two horizontal normal stresses to the vertical stress is plotted in Fig. 13.2b. At depths of 300 m or less, this ratio is found to range between 1–4. At greater depths, the range of k narrows considerably, and below 2000 m, the observed values are generally less than 1. Clearly, neither Heim's rule nor the uniaxial compaction model, (13.3), provides a good

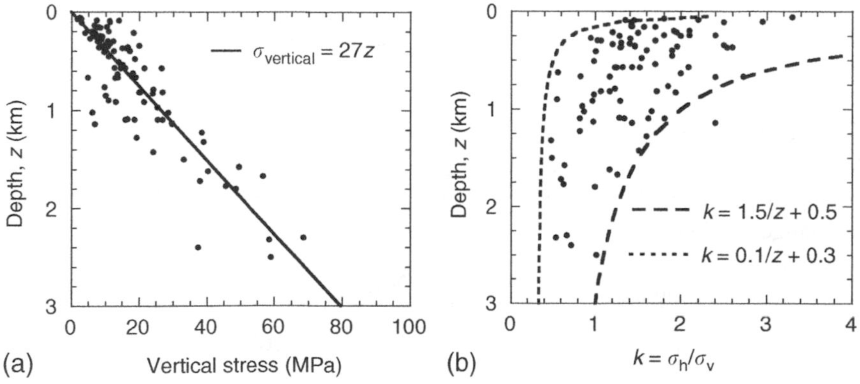

Fig. 13.2 Measured subsurface stress data, collected by Brown and Hoek (1978).

explanation for the data trends, which show a highly nonlinear decrease of the stress ratio *k* with depth.

Brown and Hoek (1978) fitted two envelopes that provide rough bounds to the data:

$$0.3 + \frac{100}{z} \le k \le 0.5 + \frac{1500}{z}, \tag{13.11}$$

where z is taken in meters. Both envelopes are of the same algebraic form, $A + B/z$, as is predicted by the spherical shell model of §13.2. In fact, the asymptotic values of 0.3 and 0.5 correspond to the values predicted by the spherical shell model if the Poisson's ratio varies between 0.23 and 0.33, which is quite reasonable for crustal rocks. The two empirical bounding values of the B coefficient are 100 and 1500. By way of comparison, McCutchen's spherical shell model gives $B = 62.5$, whereas Sheorey's multilayer, nonisothermal extension of McCutchen's model gives $B = 410$.

13.4 Surface loads on a half-space: two-dimensional theory

The complex variable method for solving two-dimensional elasticity problems that was presented in §8.2 can be used to investigate the stresses and displacements in an elastic half-space under the action of surface loads. In order to maintain the standard practice of letting x and y be real variables that represent the two coordinates, and taking $z = x + \mathrm{i}y$ to be a complex variable, in this section, we let x be the vertical coordinate normal to the surface of the half-space $x \ge 0$ (Fig. 13.3a).

The total load applied to the surface between points A and B, per unit length in direction normal to the x–y plane, can be calculated from

$$\int_A^B (N + \mathrm{i}T)\mathrm{d}y = \int_A^B \left[\frac{\partial^2 U}{\partial y^2} - \mathrm{i}\frac{\partial^2 U}{\partial x \partial y}\right]\mathrm{d}y$$

$$= -\mathrm{i}\left[\frac{\partial U}{\partial x} + \mathrm{i}\frac{\partial U}{\partial y}\right]_A^B = -\mathrm{i}\left[\phi(z) + z\overline{\phi'(z)} + \overline{\psi(z)}\right]_A^B, \tag{13.12}$$

where (8.10) and (8.46)–(8.49) have been used. This result will be needed later.

Consider now the functions

$$\phi(z) = C \ln z, \quad \psi(z) = D \ln z, \tag{13.13}$$

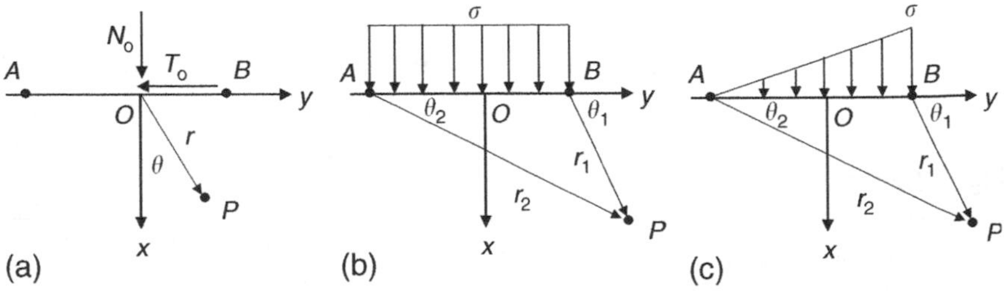

Fig. 13.3 Surface loads applied to a half-space: (a) normal and tangential line loads, (b) uniform normal traction applied over a strip, (c) linearly increasing normal traction applied over a strip.

which have singularities at the origin and so may be expected to correspond to point loads applied there. We now write point A as $ae^{-i\pi/2}$ and B as $be^{i\pi/2}$. Using (13.12), the total loads represented by the functions given in (13.13) are found to be

$$\int_A^B (N + iT)dy = -i\left[C\ln(b/a) + i\pi C + \bar{D}\ln(b/a) - i\pi\bar{D}\right]. \tag{13.14}$$

For this to correspond to a point load having components (N_0, T_0), the complex constants C and D must be chosen so that

$$-i\left[C\ln(b/a) + i\pi C + \bar{D}\ln(b/a) - i\pi\bar{D}\right] = N_0 + iT_0. \tag{13.15}$$

This is achieved by choosing $C = (N_0 + iT_0)/2\pi$ and $D = -(N_0 - iT_0)/2\pi$, in which case we see that

$$\phi(z) = \frac{N_0 + iT_0}{2\pi}\ln z, \quad \psi(z) = -\frac{N_0 - iT_0}{2\pi}\ln z, \tag{13.16}$$

are the complex stress potentials that correspond to a concentrated load (N_0, T_0) per unit length in the third direction, applied to the surface at the origin. As the load acts along the entire line corresponding to $(x = 0, y = 0)$, that is, the entire z-axis, these solutions represent *line loads*. It is convenient to consider normal and tangential loads separately.

13.4.1 Normal line load

Taking $T_0 = 0$ in (13.16) and using (8.43)–(8.44), gives

$$\tau_{rr} + \tau_{\theta\theta} = \frac{2N_0}{\pi}\text{Re}\left(\frac{1}{z}\right) = \frac{2N_0\cos\theta}{\pi r}, \tag{13.17}$$

$$\tau_{\theta\theta} - \tau_{rr} + 2i\tau_{r\theta} = \frac{N_0}{\pi}\left(-\frac{\bar{z}}{z^2} - \frac{1}{z}\right)e^{2i\theta} = -\frac{2N_0\cos\theta}{\pi r}, \tag{13.18}$$

from which the stresses follow as

$$\tau_{rr} = \frac{2N_0\cos\theta}{\pi r}, \quad \tau_{\theta\theta} = \tau_{r\theta} = 0. \tag{13.19}$$

If we consider the tractions acting over the surface of a small semicircle of radius r, centered on the origin, integration shows that the resultant force is indeed a vertical force of magnitude N_0. This problem was first solved by Flamant (1892), by integrating the point load solution of Boussinesq (1878) that is described in §13.5.

In terms of Cartesian coordinates, (13.17) and (13.18) can be expressed as

$$\tau_{xx} + \tau_{yy} = \frac{2N_0 x}{\pi r^2}, \tag{13.20}$$

$$\tau_{yy} - \tau_{xx} + 2i\tau_{xy} = -\frac{N_0}{\pi}\left[\frac{(x - iy)^3}{r^4} + \frac{(x - iy)}{r^2}\right], \tag{13.21}$$

from which the stresses follow as

$$\tau_{xx} = \frac{2N_0 x^3}{\pi r^4}, \quad \tau_{yy} = \frac{2N_0 xy^2}{\pi r^4}, \quad \tau_{xy} = \frac{2N_0 x^2 y}{\pi r^4}. \tag{13.22}$$

The displacement is, by (8.38),

$$2G(u + iv) = \frac{N_0}{2\pi}\left\{4(1-\nu)\ln r - \cos 2\theta + i\left[2(1-2\nu)\theta - \sin 2\theta\right]\right\}, \tag{13.23}$$

so that along the upper loaded surface, the normal displacement is

$$u(x = 0) = \frac{(1-\nu)N_0}{G\pi}\ln y, \tag{13.24}$$

in which the constant term has been neglected, as an arbitrary constant may always be added to the displacement without altering the stresses. This displacement has the peculiar property of becoming unbounded at infinite distances from the line of the applied load. This unrealistic result can be overcome by working with the relative vertical displacement between two points on the surface, which is always finite (Davis and Selvadurai, 1996, p. 143).

13.4.2 *Tangential line load*

Proceeding in the same way for the case of $N_0 = 0$ in (13.16), we find

$$\tau_{xx} = \frac{2T_0 x^2 y}{\pi r^4}, \quad \tau_{yy} = \frac{2T_0 y^3}{\pi r^4}, \quad \tau_{xy} = \frac{2N_0 xy^2}{\pi r^4}, \tag{13.25}$$

$$2G(u + iv) = \frac{T_0}{2\pi}\left\{-2(1-2\nu)\theta - \sin 2\theta + i\left[4(1-\nu)\ln r + \cos 2\theta\right]\right\}. \tag{13.26}$$

The stresses and displacements arising from more complicated loads may be found by integrating the solutions arising from these line loads, with the appropriate weighting factors.

13.4.3 *Normal traction σ applied over a strip of width 2a*

Imagine that this vertical line load, which corresponds to a total force N_0 per unit length in the z direction, is actually distributed over an infinitesimal strip of width dy. The compressive normal traction σ under this load will then be $N_0/\mathrm{d}y$, which shows that the loading should be represented by $N_0 = \sigma\,\mathrm{d}y$. Using superposition, the resultant stresses and displacements are found by integration of (13.22) and (13.23) over the entire loaded region, noting that in the integrand, y must represent the horizontal distance between the point of application of the load and the observation point. For the case (Fig. 13.3b) of a uniform normal stress applied over the strip $-a \leq y \leq a$,

$$\tau_{xx} = \frac{2\sigma x^3}{\pi} \int_{-a}^{a} \frac{dy'}{[x^2 + (y - y')^2]^2} = \frac{\sigma}{\pi} \left[(\theta_1 - \theta_2) - \frac{x(y-a)}{r_1^2} + \frac{x(y+a)}{r_2^2} \right]$$

$$= \frac{\sigma}{\pi} \left[(\theta_1 - \theta_2) - \sin(\theta_1 - \theta_2) \cos(\theta_1 + \theta_2) \right], \tag{13.27}$$

where the angles and radii are as defined in Fig. 13.3b. As partial verification of this result, note that for $y > a$, we have $\theta_1 = \theta_2 = 0$, and (13.27) gives $\tau_{xx} = 0$, whereas for $-a < y < a$, we have $\theta_1 = \pi$ and $\theta_2 = 0$, and (13.27) gives $\tau_{xx} = \sigma$. Similarly, the other stress components are given by

$$\tau_{yy} = \frac{\sigma}{\pi} \left[(\theta_1 - \theta_2) + \frac{x(y-a)}{r_1^2} - \frac{x(y+a)}{r_2^2} \right]$$

$$= \frac{\sigma}{\pi} [(\theta_1 - \theta_2) + \sin(\theta_1 - \theta_2) \cos(\theta_1 + \theta_2)], \tag{13.28}$$

$$\tau_{xy} = \frac{\sigma x^2 (r_2^2 - r_1^2)}{\pi r_1^2 r_2^2} = \frac{\sigma}{\pi} \left[\sin(\theta_1 - \theta_2) \sin(\theta_1 + \theta_2) \right]. \tag{13.29}$$

The normal displacement of the surface is, except for an additive constant,

$$u(x = 0) = \frac{(1 - \nu) N_o}{G\pi} \left[2a + (y - a) \ln |y - a| - (y + a) \ln |y + a| \right]. \tag{13.30}$$

13.4.4 Linearly increasing normal load

Suppose that the normal surface traction increases linearly from 0 at A to σ at B (Fig. 13.3c). In this case, the results are

$$\tau_{xx} = \frac{\sigma}{2\pi} \left\{ \left[1 + (y/a) \right] (\theta_1 - \theta_2) - \sin 2\theta_1 \right\}, \tag{13.31}$$

$$\tau_{yy} = \frac{\sigma}{2\pi} \left\{ \left[1 + (y/a) \right] (\theta_1 - \theta_2) + \sin 2\theta_1 - (2x/a) \ln(r_2/r_1) \right\}, \tag{13.32}$$

$$\tau_{xy} = \frac{\sigma}{2\pi} \left[1 - (x/a)(\theta_1 - \theta_2) + \cos 2\theta_1 \right]. \tag{13.33}$$

Jeffreys (1992) used this solution to estimate the stresses in the crust beneath mountain ranges. In such models, the weight of the mountain is assumed to provide a vertical normal traction that acts on the surface of a flat half-space. The subsurface stresses thus calculated would then be added to the $\rho g x$ term that is due to the weight of the material *below* the nominal $x = 0$ surface. The overall effect is that the subsurface vertical stress will, in general, *not* be equal to that which would be calculated from the lithostatic gradient using the depth below the actual ground surface. Various piecewise linear normal loadings can be modeled by superposition of the two above cases of uniform traction and linearly increasing traction (Jürgenson, 1934). Martel and Muller (2000) used the displacement discontinuity boundary element method to study the stresses below two-dimensional slopes of arbitrary shape and used the computed stresses to determine possible locations of slope failures.

13.4.5 Sinusoidally varying normal load

If the surface is subjected to a normal traction that varies as $\tau_{xx} = \sigma \cos \omega y$, then

$$\tau_{xx} = \frac{2\sigma x^3}{\pi} \int_{-\infty}^{\infty} \frac{\cos \omega y' \mathrm{d}y'}{[x^2 + (y - y')^2]^2} = \sigma(1 + \omega x)\mathrm{e}^{-\omega x} \cos \omega y, \quad (13.34)$$

and similarly,

$$\tau_{yy} = \sigma(1 - \omega x)\mathrm{e}^{-\omega x} \cos \omega y, \quad \tau_{xy} = \sigma \omega x \mathrm{e}^{-\omega x} \sin \omega y. \quad (13.35)$$

13.4.6 Shear traction τ applied over a strip of width 2a

If the surface is subjected to a uniform shear traction of magnitude τ, applied over a strip of width $2a$, as in Fig. 13.3b, the solution is found by integrating the stresses given in (13.25):

$$\tau_{xx} = \frac{\tau}{\pi} \sin(\theta_1 - \theta_2) \sin(\theta_1 + \theta_2), \quad (13.36)$$

$$\tau_{yy} = \frac{\tau}{\pi} \left[2 \ln(r_2/r_1) - \sin(\theta_1 - \theta_2) \sin(\theta_1 + \theta_2)\right], \quad (13.37)$$

$$\tau_{xy} = \frac{\tau}{\pi} \left[(\theta_1 - \theta_2) + \sin(\theta_1 - \theta_2) \cos(\theta_1 + \theta_2)\right]. \quad (13.38)$$

Solutions such as those presented in this section are of great importance in soil mechanics and foundation engineering and are discussed in detail by Poulos and Davis (1974) and Davis and Selvadurai (1996).

13.5 Surface loads on a half-space: three-dimensional theory

In the previous section, the fundamental solution for a line load on a half-space was used to develop solutions for more complicated two-dimensional surface loads. A three-dimensional version of this analysis can be developed by starting with the fundamental solutions for point loads acting perpendicular to the surface (Boussinesq, 1878) or tangential to the surface (Cerruti, 1882). These fundamental solutions can be derived in many ways, none of which are straightforward (Westergaard, 1952); hence, we present them here without derivation.

If a concentrated normal force N is applied at the origin of the half-space $z \geq 0$, the resulting displacements and stresses are

$$u = \frac{N}{4\pi G} \left[\frac{(1 - 2\nu)x}{r(z + r)} - \frac{xz}{r^3}\right], \quad (13.39)$$

$$v = \frac{N}{4\pi G} \left[\frac{(1 - 2\nu)y}{r(z + r)} - \frac{yz}{r^3}\right], \quad (13.40)$$

$$w = -\frac{N}{4\pi G}\left[\frac{2(1-\nu)}{r} + \frac{z^2}{r^3}\right], \tag{13.41}$$

$$\tau_{xx} = \frac{N}{2\pi}\left[\frac{3x^2z}{r^5} + \frac{(1-2\nu)(y^2+z^2)}{r^3(z+r)} - \frac{(1-2\nu)z}{r^3} - \frac{(1-2\nu)x^2}{r^2(z+r)^2}\right], \tag{13.42}$$

$$\tau_{yy} = \frac{N}{2\pi}\left[\frac{3y^2z}{r^5} + \frac{(1-2\nu)(x^2+z^2)}{r^3(z+r)} - \frac{(1-2\nu)z}{r^3} - \frac{(1-2\nu)y^2}{r^2(z+r)^2}\right], \tag{13.43}$$

$$\tau_{zz} = \frac{N}{2\pi}\left[\frac{3z^3}{r^5}\right], \tag{13.44}$$

$$\tau_{xy} = \frac{N}{2\pi}\left[\frac{3xyz}{r^5} - \frac{(1-2\nu)xy(z+2r)}{r^3(z+r)^2}\right], \tag{13.45}$$

$$\tau_{yz} = \frac{N}{2\pi}\left[\frac{3yz^2}{r^5}\right], \tag{13.46}$$

$$\tau_{xz} = \frac{N}{2\pi}\left[\frac{3xz^2}{r^5}\right], \tag{13.47}$$

where r is the distance from the origin to the point (x, y, z).

Now consider the case of an arbitrary distribution of normal tractions applied over some region of the surface. Denoting the normal traction on the surface by σ, the total load applied over a small region of area $d\xi\, d\eta$, located at $(x = \xi, y = \eta, z = 0)$, would be $\sigma(\xi, \eta)d\xi\, d\eta$. The displacements and stresses at an arbitrary point (x, y, z) can then be found by integrating (13.39)–(13.47) over the entire loaded region, with N replaced by $\sigma(\xi, \eta)d\xi\, d\eta$, and x in the integrand replaced by $x - \xi$, etc. For example, the normal component of the displacement of the surface would be, from (13.41),

$$w(x, y, 0) = \frac{-(1-\nu)}{2\pi G}\iint_A \frac{\sigma(\xi, \eta)}{\sqrt{(x-\xi)^2 + (y-\eta)^2}} d\xi\, d\eta, \tag{13.48}$$

where A is the region of the surface over which the loads are applied (Fig. 13.4a).

An important example is the case of a uniform normal traction σ applied over a circular region of radius a. By symmetry, we need only consider the displacements along, say, the x-axis. At the point $(x = b, y = 0, z = 0)$, the normal displacement is

$$w(b, 0, 0) = \frac{-(1-\nu)\sigma}{2\pi G}\iint_A \frac{1}{\sqrt{(x-\xi)^2 + \eta^2}} d\xi\, d\eta. \tag{13.49}$$

Consider now only points within the loaded region, for which $b \le a$. We now change (Fig. 13.4b) to a polar coordinate system on the surface, centered at $(x = b, y = 0)$. The integration in (13.49) must take place over the entire loaded circular region. It can be seen in Fig. 13.4c that as the new radial variable ρ varies along the chord EF and the angle ψ varies from 0 to $\pi/2$, the shaded region is covered. For example, the chord E_1F_1 corresponds to $\psi = 0$, etc. By symmetry, the full integral will be twice the value for this shaded region. Noting that the

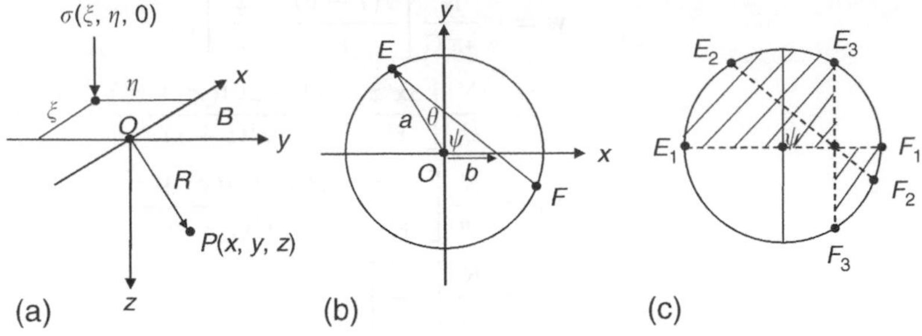

Fig. 13.4 (a) Notation used for stresses and displacements at a generic observation point P, due to surface tractions applied at $(x = \xi, y = \eta, z = 0)$; (b) new variables used to integrate results for the case of uniform load applied over a circular region; (c) illustration of region covered as ψ varies from 0 to $\pi/2$.

denominator in (13.49) is equal to ρ, and that $d\xi\, d\eta$ must be replaced by $\rho d\rho d\psi$ when passing to polar coordinates, we have

$$w(b, 0, 0) = \frac{-(1-\nu)\sigma}{\pi G} \int_0^{\pi/2} d\psi \int_E^F d\rho. \tag{13.50}$$

The integral in ρ is taken over the chord EF, and so its value is $2a\cos\theta$. But $a\sin\theta = b\sin\psi$, so (13.50) becomes

$$w(b, 0, 0) = \frac{-2a(1-\nu)\sigma}{\pi G} \int_0^{\pi/2} \sqrt{1 - (b/a)^2 \sin^2\psi}\, d\psi. \tag{13.51}$$

This is an elliptic integral and, in general, cannot be simplified further or expressed in terms of elementary functions. The maximum vertical displacement occurs at the center of the circle, where $b = 0$ and $w = -a(1-\nu)\sigma/G$. At the outer edge of the loaded region, $b = a$ and $w = -2a(1-\nu)\sigma/\pi G = -0.637a(1-\nu)\sigma/G$. The mean displacement over the loaded region is $-0.848a(1-\nu)\sigma/G$.

Similarly, the vertical displacement of points on the surface outside the loaded circular region, that is, $b > a$, can be expressed in terms of two elliptical integrals (Davis and Selvadurai, 1996, p. 120). Manipulation of these results shows that as b increases, w decays as $1/b$. This should be expected from (13.41), since when $b \gg a$, the displacements must reduce to those that would occur for a point load of magnitude $\sigma\pi a^2$, applied at the origin.

Boussinesq also solved the related problem of indentation of the surface of a half-space by a rigid circular plate of radius a, by using the above superposition process shown in (13.48) and making an educated guess of the appropriate surface traction distribution needed to yield a uniform value of w within the region $r \le a$ (Davis and Selvadurai, 1996, pp. 131–5). If the uniform vertical indentation is $-w_0$, the mean value of the traction over the loaded region is found to be $4Gw_0/\pi a(1-\nu)$. Hence, the constants of proportionality relating the mean vertical displacement and mean normal traction differ by only 8 percent between the cases of uniform surface traction and uniform surface displacement.

The displacements and stresses arising from a tangential point load T acting in the x direction, at the origin of the half-space, are (Cerruti, 1882; Davis and

Selvadurai, 1996, p. 92)

$$u = \frac{T}{4\pi G}\left[\frac{(x^2+r^2)}{r^3} + \frac{(1-2\nu)}{z+r} - \frac{(1-2\nu)x^2}{r(z+r)^2}\right], \tag{13.52}$$

$$v = \frac{T}{4\pi G}\left[\frac{xy}{r^3} - \frac{(1-2\nu)xy}{r(z+r)^2}\right], \tag{13.53}$$

$$w = \frac{T}{4\pi G}\left[\frac{xz}{r^3} + \frac{(1-2\nu)x}{r(z+r)}\right], \tag{13.54}$$

$$\tau_{xx} = -\frac{T}{2\pi}\left[-\frac{3x^3}{r^5} + \frac{(1-2\nu)(r^2-y^2)x}{r^3(z+r)^2} - \frac{2(1-2\nu)ry^2x}{r^3(z+r)^3}\right], \tag{13.55}$$

$$\tau_{yy} = -\frac{T}{2\pi}\left[-\frac{3y^2x}{r^5} + \frac{(1-2\nu)(3r^2-x^2)x}{r^3(z+r)^2} - \frac{2(1-2\nu)rx^3}{r^3(z+r)^3}\right], \tag{13.56}$$

$$\tau_{zz} = \frac{T}{2\pi}\left[\frac{3xz^2}{r^5}\right], \tag{13.57}$$

$$\tau_{xy} = -\frac{T}{2\pi}\left[-\frac{3x^2y}{r^5} + \frac{(1-2\nu)(x^2-r^2)y}{r^3(z+r)^2} + \frac{2(1-2\nu)rx^2y}{r^3(z+r)^3}\right], \tag{13.58}$$

$$\tau_{yz} = \frac{T}{2\pi}\left[\frac{3xyz}{r^5}\right], \tag{13.59}$$

$$\tau_{xz} = \frac{T}{2\pi}\left[\frac{3x^2z}{r^5}\right]. \tag{13.60}$$

As was the case for the normal point load, superposition can be used to generate solutions for problems involving variable shear tractions acting along the surface of a half-space.

13.6 Hydraulic fracturing

One of the more widely used methods for estimating the subsurface stress state is *hydraulic fracturing*. In this procedure, water or drilling mud is pumped into a hydraulically isolated segment of the wellbore, until the induced stresses at the borehole wall are large enough to cause a fracture to open and propagate into the formation. In the petroleum industry, wells are often hydraulically fractured to increase the ability of fluid to flow from the formation into the wellbore. The basic rock mechanics analysis of the hydraulic fracturing process, for the purposes of predicting the wellbore fluid pressure that would be needed to create a fracture in a rock mass under a given state of *in situ* stress, was first given by Hubbert and Willis (1957). Scheidegger (1960) and Fairhurst (1964) pointed out that these same equations could be used in an inverse manner to infer the *in situ* stresses from the pressure data collected during the hydraulic fracturing process. The following discussion will emphasize the use of hydraulic fracturing as a tool for stress estimation, although the same basic equations apply when it is used for the purpose of oilwell stimulation. A detailed discussion of hydraulic fracturing as a well stimulation tool has been given in the monograph edited by Economides and Nolte (2000).

The simplest situation to analyze is that of a vertical borehole in a nonporous, impermeable formation. Assume that, before the borehole is drilled, the three principal stresses are the vertical stress, σ_v, and two horizontal principal stresses, σ_H and σ_h, that are oriented normal to each other in the horizontal plane. By convention, the horizontal stresses are labeled according to $\sigma_H > \sigma_h$. If the pressure of the fluid in the borehole is P_w, then the principal stresses around the borehole wall will be, according to the solution presented in §8.5,

$$\tau_{rr} = P_w, \tag{13.61}$$

$$\tau_{\theta\theta} = \sigma_H + \sigma_h - 2(\sigma_H - \sigma_h)\cos 2\theta - P_w, \tag{13.62}$$

$$\tau_{zz} = \sigma_v + \nu\left[(\sigma_H + \sigma_h) - 2(\sigma_H - \sigma_h)\cos 2\theta\right], \tag{13.63}$$

where θ is the angle of clockwise rotation from the direction of the maximum horizontal stress.

A tensile fracture is assumed to form at the borehole wall and propagate into the formation if any of these principal stresses becomes sufficiently tensile (negative) to equal $-T_0$, where T_0 is the tensile strength of the rock. The most common situation, particularly for deep wells, is for at least one of the horizontal *in situ* principal stresses to be less than the vertical *in situ* stress. In this case the minimum (i.e., most negative) stress value that exists after the borehole is drilled will be the tangential normal stress at $\theta = 0$ and $\theta = \pi$, where its value will be $\tau_{\theta\theta} = 3\sigma_h - \sigma_H - P_w$. The condition for fracturing is found by setting $\tau_{\theta\theta} = -T_0$. As it will be rare for the horizontal stresses to be so anisotropic that $3\sigma_h - \sigma_H < 0$ (Amadei and Stephansson, 1997, p. 26), the criterion for fracture will usually be met only when the wellbore fluid pressure P_w reaches the value

$$P_w = 3\sigma_h - \sigma_H + T_0. \tag{13.64}$$

The pressure at which fracturing first occurs is referred to as the *breakdown pressure*, P_b. Once this pressure is reached, a fracture forms and abruptly propagates into the formation. Fluid from the borehole then rushes into the newly created fracture, causing an instantaneous drop in the wellbore pressure. Hence, the breakdown pressure is readily identified during the fracturing operation as being the *peak pressure* recorded. So, if the tensile strength of the rock can be estimated by other means, (13.64) gives one of the two equations needed to determine the two horizontal principal stresses. Expressed in terms of the breakdown pressure, this equation is

$$P_b = 3\sigma_h - \sigma_H + T_0. \tag{13.65}$$

If the pressure is again increased some time after the initial fracture is created, the fracture will reopen at some *reopening pressure*, P_r. The reopening pressure is sometimes identified with the peak pressure achieved during the repressurization process (Whittaker et al., 1992, p. 378). Amadei and Stephansson (1997, p. 144) identify the reopening pressure as the point in the pressure vs. time curve, just before the peak pressure is reached, at which the rate of pressure increase slows down – presumably indicating that fluid from the borehole has begun to infiltrate into the fracture. As the reopening fracture has no intrinsic tensile strength, application of (13.64) with the tensile strength set to zero gives

$$P_r = 3\sigma_h - \sigma_H. \tag{13.66}$$

Use of the reopening pressure as in (13.66) eliminates the need to estimate T_0. Alternatively, measurement of both the breakdown pressure and the reopening pressure will, in principle, allow the tensile strength of the intact rock to be determined, as comparison of (13.65) and (13.66) shows that (Bredehoeft et al., 1976)

$$T_0 = P_b - P_r. \tag{13.67}$$

Consider now the case in which the rock is porous and saturated with a pore fluid at pressure P_p. If the permeability of the rock is sufficiently low, the drilling fluid will not penetrate into the formation rapidly enough to influence the fracturing process. Hence, the pore pressure P_p and the wellbore pressure P_w must be considered to be independent of each other. According to the empirical effective stress law for failure presented in §4.7, the pore pressure should be subtracted from each principal stress before applying a rock failure criterion. This is equivalent to taking the effective stress coefficient for failure to be unity. If the pore pressure is subtracted from $\tau_{\theta\theta}$ in (13.62), then criterion (13.65) for fracture initiation becomes

$$P_b = 3\sigma_h - \sigma_H + T_0 - P_p. \tag{13.68}$$

A more general model, which should reduce to the two models presented above in the two limiting cases of zero permeability and zero porosity, is that of a porous and permeable *poroelastic* formation that allows the drilling fluid to infiltrate into the rock at the borehole wall (Haimson and Fairhurst, 1967). In this case, the breakdown pressure is given by

$$P_b = \frac{3\sigma_h - \sigma_H + T_0 - 2\eta P_p}{2(1-\eta)}, \tag{13.69}$$

where $\eta = \alpha(1-2\nu)/2(1-\nu)$ and α is the Biot parameter (§7.4). The parameter 2η lies between 0 and 1, and a typical value is about 0.6. Detailed discussions of the stresses around a borehole in a poroelastic formation, including transient effects and their implications for hydraulic fracturing, have been given by Detournay and Cheng (1988) and Detournay et al. (1989).

Schmitt and Zoback (1989) pointed out that there was actually very limited experimental justification for the effective stress coefficient for tensile failure to be taken equal to unity. They proposed that only some fraction, $\beta < 1$, of the pore pressure should be subtracted from the principal stresses when applying a tensile failure criterion. This assumption leads to the following equation for the breakdown pressure in a porous but impermeable rock:

$$P_b = 3\sigma_h - \sigma_H + T_0 - \beta P_p. \tag{13.70}$$

For the poroelastic case, their suggestion leads to the following modification of (13.69):

$$P_b = \frac{3\sigma_h - \sigma_H + T_o - 2\eta P_p}{1 + \beta - 2\eta}. \tag{13.71}$$

Schmitt and Zoback (1989) analyzed the results of some laboratory hydraulic fracturing experiments conducted on a Valders limestone and used (13.71) to back-calculate the value of β, the effective stress coefficient for tensile fracture. Their computed values of β, for different experiments on the same rock, were in the range of about 0.2–0.6, apparently justifying their rejection of the assumption that $\beta = 1$.

Regardless of the model used for the breakdown pressure, knowledge of P_b gives only one equation, which contains both of the horizontal *in situ* principal stresses. A second equation is obtained from the *shut-in pressure*, as described below, although estimation of this pressure is not without difficulties.

After the fracture is either initially opened, or reopened, the pressure will decrease, as drilling fluid flows into the fracture from the borehole and leaks off into the formation from the fracture faces. If pumping is then ceased and the borehole is hydraulically shut in, the pressure sometimes drops abruptly, as the fracture closes up. In these cases, the quasi-stabilized pressure that is observed is identified as the *shut-in pressure*, P_s. In many cases, however, there is no clearly observed pressure plateau. The shut-in pressure may then be identified by the presence of a "knee" in the pressure decline curve, as the rate of pressure decline decreases due to the fact that fluid can no longer leak off into the formation from the (now closed) fracture faces. Numerous data analysis methods have been proposed to identify the shut-in pressure from the pressure vs. time signal (Lee and Haimson, 1989; Tunbridge, 1989; Amadei and Stephansson, 1997; Haimson and Cornet, 2003).

The shut-in pressure is usually interpreted as representing the minimum horizontal *in situ* stress, σ_h, according to the following argument. As explained above, failure is assumed to first occur at $\theta = 0$ and/or $\theta = \pi$ along the borehole wall, when the local stress becomes sufficiently tensile. The fracture is then assumed to penetrate into the formation, in a vertical plane that passes through the borehole axis. Although the stresses near the borehole wall reflect the stress perturbations due to the borehole, once the fracture has penetrated into the formation to a distance of more than three or four borehole radii, it will essentially be subjected to a far-field normal stress σ_h, which acts to close it up, and a pore pressure P_p, which acts to prop it open. Hence, the value of P_p at which the fracture first closes should be equal to σ_h. This pressure is denoted as the shut-in pressure, P_s, in which case we have

$$\sigma_h = P_s. \tag{13.72}$$

After σ_h is found from the shut-in pressure, one of the models for the breakdown pressure, such as (13.65), (13.68), (13.69), (13.70), or (13.71), can be used to estimate the maximum horizontal stress, σ_H.

The discussion given above explains the basic procedure by which the *magnitudes* of the two *in situ* horizontal principal stresses are determined from hydraulic fracturing data. The principal stress *directions* can be estimated if the trace of the fracture along the borehole wall can be located. One method for doing this is to use an impression packer, which consists of an inflatable element wrapped with a soft rubber film (Amadei and Stephansson, 1997, section 4.2.2). When the packer is inflated inside the borehole, the film is extruded into the fracture, leaving a permanent impression that can be recorded after the packer is retrieved from the borehole. According to the conceptual model of hydraulic fracturing presented above, the fracture trace(s) should coincide with the $\theta = 0$ and/or $\theta = \pi$ directions, which is to say, with the direction of σ_H. These traces can also be inferred from borehole televiewer images (Zemanek et al., 1970). In practice, the traces are often irregular and possibly discontinuous. Statistical techniques to identify the fracture direction from the observed traces have been developed by Lee and Haimson (1989). Many other practical aspects of the hydraulically fracturing process are discussed by Amadei and Stephansson (1997) and Economides and Nolte (2000).

13.7 Other stress-measurement methods

Another method for estimating subsurface stresses involves the use of *flat-jacks*. This method is relatively simple and robust, but can only be used to estimate the stresses in the vicinity of preexisting underground excavations or tunnels. A tabular slot is cut into the rock surface, such as at the wall of a tunnel (Fig. 13.5). Before cutting the slot, the rock immediately above and below the slot is instrumented so that any changes in the distance between two points on either side of the slot can be monitored. If the initial stress perpendicular to the slot is compressive, as is usually the case, the two faces will converge toward each other when the slot is cut. Next, a flat hydraulic jack is inserted into the slot and cemented into place. This jack consists of two flat steel plates, welded together around their edges, with a tube to allow pressurized oil to be pumped into the space between the plates (Goodman, 1989, p. 121). The jack is then pressurized, until the distance between the measuring points on either side of the slot returns to its initial value. When this has occurred, it is assumed that the pressure of the fluid in the jack is equal to the compressive stress that acts in the direction normal to the plane of the jack. As the flat-jacks are necessarily installed near the free surface of an excavation, this stress must be interpreted as representing both the *in situ* stresses and the stresses that were induced by the excavation.

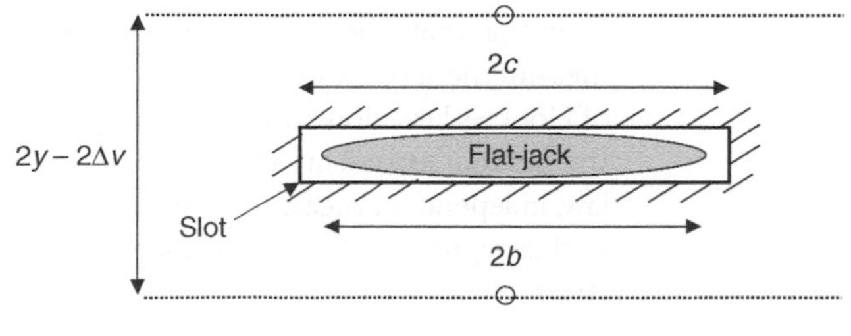

Fig. 13.5 Stress measurement by flat-jacks; view is normal to the free surface of the preexisting excavation.

In practice, the width of the jack, $2b$, will be less than the width of the slot, $2c$. Furthermore, an additional width d at each end of the jack, where the steel plates are welded together, will be inoperative. A simple force balance in the direction normal to the plane of the jack then shows that the normal stress σ will be related to the jack pressure p by

$$\sigma = p(b - d)/c. \tag{13.73}$$

The flat-jack method seems to have been first used by Mayer et al. (1951) and Tincelin (1951). An advantage of this method is that no knowledge of the elastic properties of the rock is required for its application and interpretation. However, the standard interpretation described above implicitly assumes that the rock exhibits *elastic* behavior and neglects the possibility of creep occurring during the process of installing and pressuring the jack. Panek and Stock (1964) have discussed several methods for accounting for possible creep deformation during flat-jack tests.

If the measured deformation across the slot is monitored as a function of the jack pressure, this information can be used to estimate the elastic moduli of the rock. The simplest model would be to treat the slot as a thin ellipse and use the equations of §8.9 for the deformation around an elliptical cavity. The convergence of two points on the rock face that are initially spaced at distances $\pm y$ from the slot plane, along the center line normal to that plane, due to the stress relief caused by the cutting of the slot, would be given by

$$2\Delta v = \frac{2c\sigma}{E}\left\{(1-\nu)[\sqrt{1+(y/c)^2} - (y/c)] + \frac{(1+\nu)}{\sqrt{1+(y/c)^2}}\right\}. \tag{13.74}$$

Measurements of the convergence of two pairs of points initially located at two different values of y will allow both E and ν to be estimated, using (13.74). However, Li and Cornet (2004) numerically modeled the deformation occurring during a flat-jack test in a fractured rock mass at a granite quarry in central France and found that full three-dimensional modeling was needed in order to yield estimates of E that agreed with those inferred from laboratory and seismic tests.

Stress measurement using flat-jacks is similar to hydraulic fracturing in that neither method requires knowledge of the elastic moduli of the rock. There are several other methods of subsurface stress estimation that do require knowledge of the elastic moduli, because they actually measure displacements or strains and rely on Hooke's law to convert the strains to stresses. Many of these methods can collectively be referred to as *relief methods*, as they involve a coring or drilling operation that relieves the stresses in some small region of rock, accompanied by measurement of the deformation that occurs during this stress relief. Application of Hooke's law allows the stress change to be estimated, thereby giving the initial stresses that existed in the rock prior to the stress relief. In order to apply Hooke's law, independent measurements must be made of the elastic moduli of the rock in that region, preferably on the same specimen that had been involved in the stress relief.

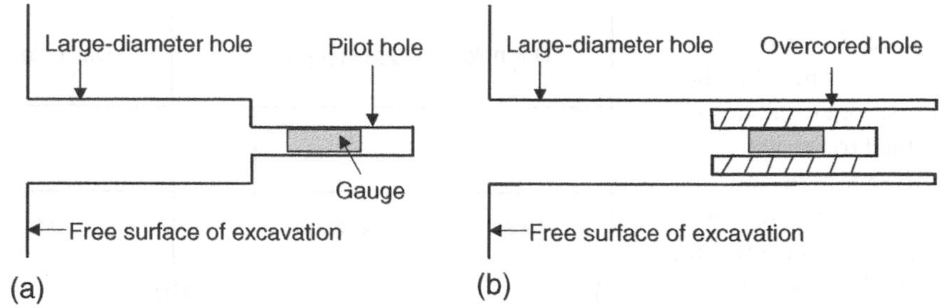

Fig. 13.6 Stress measurement by overcoring, using a device such as the USBM gauge: (a) gauge is inserted into the pilot hole, (b) which is then overcored, creating an annular region (shaded) that is relieved of stress. The resulting deformation is measured by the gauge.

Among the most common stress relief methods are those that are carried out in a borehole and involve *overcoring*. One stress-measurement device that utilizes the concept of overcoring is the US Bureau of Mines stress gauge (Obert et al. 1962; Hooker et al., 1974). The USBM gauge is essentially a cylindrical tool that has three diametrically opposed pairs of pistons protruding from its outer surface, equally spaced around the circumference. These pistons are connected inside the tool to cantilevers whose deflection is measured with strain gauges.

To use the USBM gauge, a small diameter borehole, of roughly the same diameter as the gauge (38 mm), is drilled into the rock. The gauge is inserted into the hole, and the pistons are initially tensioned so as to make good contact with the borehole walls (Fig. 13.6a). This small hole is then overcored with a larger diameter drill bit (typically 150 mm), to a depth extending at least one overcore diameter past the gauge (Fig. 13.6b). The overcoring process will create an annular rock region that is essentially free of stress. As the stresses that had been acting on this annular region are relieved, the ensuing radial deformation of the (inner) borehole is measured by the three sets of cantilevers.

Hooke's law is then used to relate the displacements at the borehole to the change in stress. This requires an analytical expression for the stress state that existed in the annular region before it was overcored, as a function of the initial *in situ* stresses. Analytical solutions have been presented for isotropic media by Leeman (1967) and Hiramatsu and Oka (1968), and for anisotropic media by Becker (1968) and Hooker and Johnson (1969). A very general solution, valid for transversely isotropic or orthotropic media, with an arbitrary inclination between the borehole and the axes of elastic symmetry of the rock, has been presented by Amadei and Stephansson (1997, section 5.4.2).

Overcoring can also be used to measure the strain relief that occurs at the flat bottom surface of a borehole (Fig. 13.7). For example, consider the *CSIR doorstopper cell* developed at the South African Council for Scientific and Industrial Research (Leeman, 1964,1967,1971). This cell consists of a cylindrical plug of silicone rubber, 35 mm in diameter and of roughly the same height, on the bottom surface of which is attached a rosette of three or four strain gauges. The cell is pressed against the flattened and smoothed bottom surface of the borehole and glued into place. The cell is then overcored to relieve the stresses from the region of rock to which the cell is attached. The strains are recorded before and after overcoring. Again, elastic solutions for the local stresses and strains at the bottom of a truncated cylindrical cavity are needed in order to

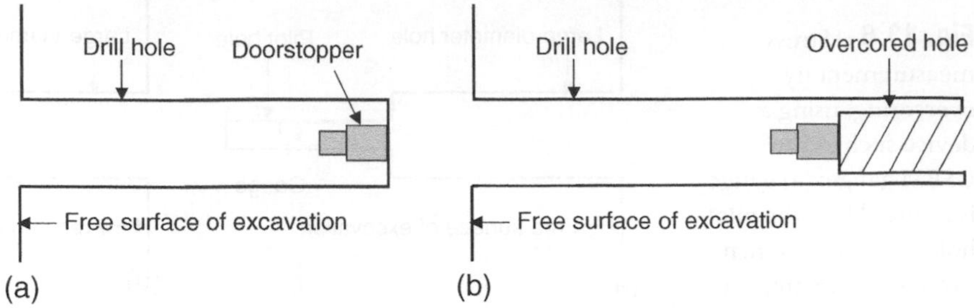

Fig. 13.7 Stress measurement using the doorstopper cell: (a) cell is glued to the flat bottom face of borehole, (b) which is then overcored, creating a cylindrical region (shaded) that is relieved of stress. The resulting strains are measured by the strain gauges mounted on the face of the cell.

relate the measured strains to the stresses. As this geometry does not permit a closed-form analytical solution, the stress–strain relations used in the analysis are usually simple relations fit to the results of numerical simulations (Galle and Wilhiot, 1962; Leeman, 1964; Hoskins, 1967; Bonnechere and Fairhurst, 1968; van Heerden, 1969; Coates and Yu, 1970; Hocking, 1976; Rahn, 1984). Due to the geometry of the doorstopper cell, stresses can only be estimated within the plane normal to the borehole axis, provided that the borehole is aligned with one of the principal stress directions.

Fairhurst (2003) has given an historical review of subsurface stress-measurement methods. Sjöberg et al. (2003) describe a suggested methodology for overcoring measurements, with particular reference to the Borre stress probe developed at the Swedish State Power Board. Worotnicki (1993) provides a detailed analysis of the CSIRO (Australia) hollow inclusion stress cell. Several other stress-measurement devices are described and analyzed by Amadei and Stephansson (1997). Many articles describing recent advances in subsurface stress estimation methods, along with some case studies and four ISRM Suggested Methods for rock stress estimation, are contained in a Special Issue of the *International Journal of Rock Mechanics and Mining Sciences* (Hudson and Cornet, 2003).

14 Geological applications

14.1 Introduction

As remarked in Chapter 1, many of the problems of structural geology are similar to those of engineering rock mechanics, except that they are on a larger scale.

The basic concepts of Mohr–Coulomb failure, as described in Chapter 4, are used in §14.2 to explain the classic Andersonian theory of the types of faulting that may be expected to occur in various subsurface stress regimes. Hubbert and Rubey's classic analysis of low-angle overthrust faulting, which is based on the Mohr–Coulomb failure model and the effective stress principle, is presented in §14.3. In §14.4, elasticity theory is used to develop some models for the stresses around faults. The relationship between the shear stress acting on a fault and the relative displacement of the two opposing fault faces, is discussed for several fault geometries. The mechanics of igneous intrusion is studied in §14.5, again using basic ideas of rock mechanics discussed in previous chapters. Sheet intrusions are analyzed based on the Griffith failure model of §10.9, and some elastic stress solutions from §8.12 are used to study ring dykes and cone sheets. Beam models for crustal deformation and folding are developed in §14.6, based on the elastic beam equations of §6.9 and the viscoelastic stress–strain laws presented in §9.9. Finally, in §14.7, energy considerations from Chapter 5, along with the frictional sliding models of Chapter 3, are used to provide a simple explanation of some aspects of earthquakes.

This chapter, by necessity, touches only briefly on a few geological and geophysical applications of rock mechanics. Further analysis of geological structures, based on the principles of continuum mechanics in general and rock mechanics in particular, can be found in various books on structural geology, such as Price and Cosgrove (1990), Bayly (1992) and Pollard and Fletcher (2005).

14.2 Stresses and faulting

The phenomena of brittle fracture discussed in Chapter 4 on a laboratory scale occur as well on a geological scale. Faults are geological fractures of rock for which there is a relative displacement of the rock on the two opposing faces. They are therefore shear fractures in the sense of §4.4. Griggs and Handin (1960) and others have used the term "fault" for shear fractures on both the laboratory and geological scale.

The surface of the fracture is referred to as the *fault plane* and is specified by its *strike* and *dip*. The strike is the direction of any horizontal line in the fault plane, often measured relative to due north. The dip is the angle between the horizontal plane and the fault plane, measured within a vertical plane that is normal to the fault plane. If the relative motion along the fault plane is in the direction of the strike, the fault is described as a *strike-slip* fault. If the motion is in the direction of the dip, the fault is called a *dip-slip* fault. If the motion is in some other direction, the fault is referred to as *oblique-slip*.

In strike-slip faulting, the observed fault plane is often nearly vertical. If a hypothetical observer stands vertically and faces the plane, the fault is called *right-handed*, or *dextral*, if the rock on the opposite side of the fault has moved to the right, and *left-handed*, or *sinistral*, if the rock on the opposite side of the fault has moved to the left. The terms "wrench fault" and "transcurrent fault" have also been used for strike-slip faults.

A dip-slip fault is referred to as a *normal fault* if its dip angle is greater than 45°, and if the upper surface (the *hanging wall*) has moved downward relative to the lower surface (the *footwall*). If the hanging wall has moved upward relative to the footwall, the fault is described as a *reverse fault*. If the dip is less than 45°, and the hanging wall moves upward relative to the footwall, the fault is a *thrust fault*. Thrust faults with very shallow dips, say less than 10°, are called *overthrust faults*. Normal faults with small dip angles are known as *detachment faults*.

These types of fault were discussed by Anderson (1951) on the basis of the Mohr–Coulomb theory of shear fracture (§4.5) and classified on the basis of the relative magnitudes of the principal stresses. According to this theory, fracture takes place in one or both of a pair of conjugate planes that pass through the direction of the intermediate principal stress and are equally inclined at an angle less than 45° from the direction of the maximum principal stress.

Under the usual assumption that one principal stress is vertical, three cases arise.

14.2.1 Thrust faulting

If the vertical stress is the *least* principal stress, failure may occur on either of two planes such as the one shown in Fig. 14.1a, inclined at an angle $\psi < 45°$ to the horizontal.

14.2.2 Normal faulting

If the vertical stress is the *greatest* principal stress, failure may occur on either of two planes such as the one shown in Fig. 14.1b, inclined at an angle $\psi < 45°$ to the vertical.

14.2.3 Strike-slip faulting

If the vertical stress is the *intermediate* principal stress, failure may occur on either of two vertical planes such as the one shown in Fig. 14.1c, inclined at an angle $\psi < 45°$ to the σ_1 direction.

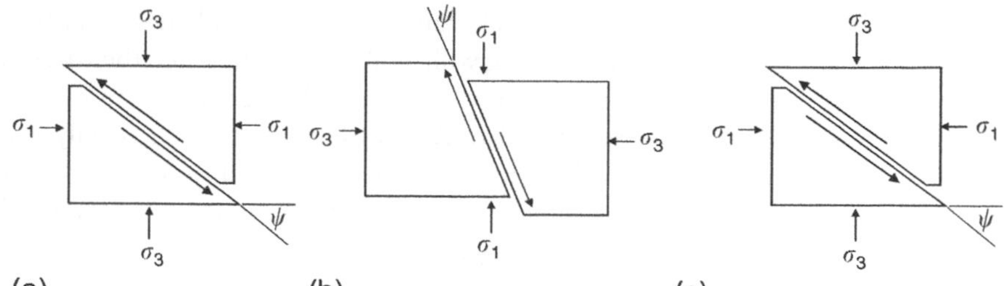

Fig. 14.1 (a) Thrust fault (vertical plane), (b) normal fault (vertical plane), and (c) strike-slip fault (horizontal plane).

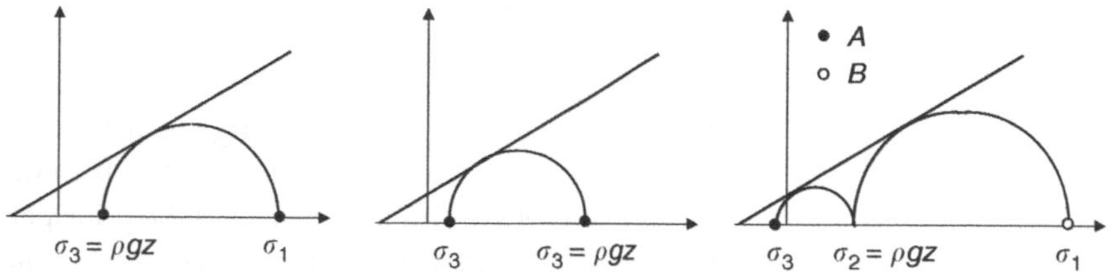

Fig. 14.2 Mohr's circle analysis of (a) thrust faulting, (b) normal faulting, and (c) strike-slip faulting, according to the Mohr–Coulomb failure criterion.

The angle between the fault plane and the direction of maximum principal stress is, by (4.6),

$$\psi = (\pi/2) - \alpha = (\pi/4) - (\phi/2), \tag{14.1}$$

where ϕ is the angle of internal friction. Since values of ϕ between 30° and 55° are commonly measured in laboratory experiments, values of ψ between 17° and 30° might be expected in the field. Sax (1946) observed values of 25°–30° for normal faults, and 20°–25° for thrust faults, in the South Limburg coalfield region of the Netherlands. Price (1962) found similar values in western Wales. Values of ψ in the order of 30° are commonly measured for strike-slip faults (Moody and Hill, 1956; Williams, 1959). Hence, there is quite good agreement between laboratory and geological measurements, based on the simple Mohr–Coulomb theory.

Hubbert (1951) used Mohr diagrams to study the stress ratios that may be expected to cause the different types of faulting. Making only the simple assumption that one of the principal stresses is the vertical lithostatic stress, the three types of faulting discussed above can be represented as in Fig. 14.2. For thrust faulting to occur, the vertical stress is σ_3, and σ_1 must increase until the (σ_1, σ_3) Mohr's circle touches the failure line (Fig. 14.2a). Hence, fairly high values of σ_1, relative to $\sigma_3 = \rho gz$, will be needed to cause thrust faulting.

For normal faulting, the vertical stress will be the greatest principal stress, and so the least principal stress σ_3 must be reduced until the (σ_1, σ_3) Mohr's circle touches the failure line (Fig. 14.2b). If ρgz is less than the unconfined compressive strength of the rock, then the required value of σ_3 would actually be tensile. Hence, it seems that the horizontal stresses will be small in regions in which normal faulting occurs.

For strike-slip faulting, the vertical stress will be the intermediate principal stress. Faulting will, as always, occur when the (σ_1, σ_3) Mohr's circle touches the failure line. If σ_1 is only slightly greater than $\sigma_2 = \rho g z$, this will occur when σ_3 is reduced to A as in Fig. 14.2c. If σ_3 is only slightly less than $\sigma_2 = \rho g z$, this will occur when σ_1 is increased to B. Other stress states sufficient to cause strike-slip faulting will fall between these two sets of bounds on σ_1 and σ_3, that is, $A < \sigma_3 < \rho g z < \sigma_1 < B$.

14.2.4 Oblique-slip faulting

In this case, it is necessary to study in more detail the direction of the shear stress in an arbitrary plane, following the discussion of the stress transformation equations in §2.6. Let line OP in Fig. 14.3 be the normal to the plane, and let the direction cosines of this line relative to the principal stress axes $Oxyz$ be denoted by $\{l, m, n\}$. Line OP can also be specified by its colatitude, θ, and longitude, λ.

Taking new axes Pz' along direction OP, Px' in the plane OPz, with Py' completing the right-handed Cartesian coordinate system, the stresses in the plane $Px'y'$ are

$$\tau_{z'z'} = (\tau_{xx}\cos^2\lambda + \tau_{yy}\sin^2\lambda)\sin^2\theta + \tau_{zz}\cos^2\theta, \tag{14.2}$$

$$\tau' \equiv \tau_{x'z'} = \tfrac{1}{2}(\tau_{xx}\cos^2\lambda + \tau_{yy}\sin^2\lambda - \tau_{zz})\sin 2\theta, \tag{14.3}$$

$$\tau'' \equiv \tau_{y'z'} = -\tfrac{1}{2}(\tau_{xx} - \tau_{yy})\sin\theta\sin 2\lambda. \tag{14.4}$$

Introducing the direction cosines $\{l, m, n\}$ of line OPz', namely

$$l = \sin\theta\cos\lambda, \quad m = \sin\theta\sin\lambda, \quad n = \cos\theta, \tag{14.5}$$

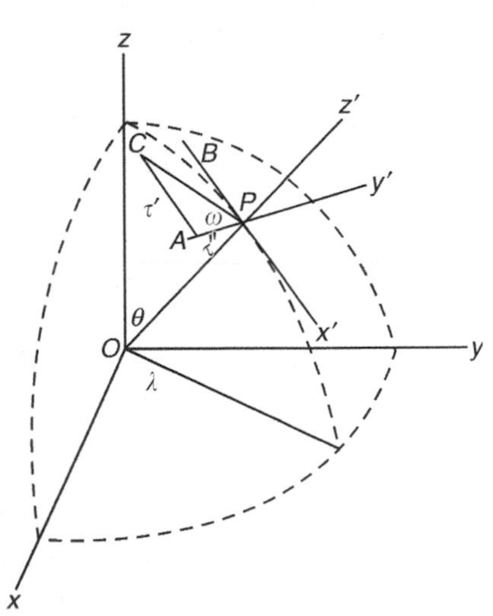

Fig. 14.3 Geometric construction for studying oblique-slip faulting (see text for details).

(14.3) and (14.4) can be written as

$$\tau' = [m^2(\tau_{yy} - \tau_{xx}) - (1 - n^2)(\tau_{zz} - \tau_{xx})]n(1 - n^2)^{-1/2}, \tag{14.6}$$

$$\tau'' = (\tau_{yy} - \tau_{xx})lm(1 - n^2)^{-1/2}. \tag{14.7}$$

The stresses τ' and τ'' are positive in the directions *PA* and *PB*, and the angle ω that the resultant shear stress makes with *PA* is given by

$$\tan\omega = \frac{n}{lm}\left[m^2 - (1 - n^2)\frac{\tau_{zz} - \tau_{xx}}{\tau_{yy} - \tau_{xx}}\right]. \tag{14.8}$$

Suppose now that the principal stress τ_{zz} is vertical, and consider positive values of l, m, and n. If faulting takes place, the direction of slip will be parallel to *PC*, and the system can be characterized by this direction. Now if $\tau_{yy} > \tau_{xx}$, then (14.7) shows that $\tau'' > 0$, and the direction of slip is to the right, so the system is called *dextral*. If $\tau_{yy} < \tau_{xx}$, then $\tau'' < 0$, and the system is called *sinistral*.

If $\tau_{yy} > \tau_{xx} > \tau_{zz}$, it follows from (14.8) that $\tan\omega \to \infty$ if τ_{yy} approaches τ_{xx}, and $\tan\omega \to nm/l$ if τ_{yy} approaches τ_{zz}. Hence, if $\tau_{yy} > \tau_{xx} > \tau_{zz}$, then

$$\tan^{-1}(nm/l) < \omega < 90°, \tag{14.9}$$

and the system is called *dextral thrust*. Similarly, if $\tau_{yy} > \tau_{zz} > \tau_{xx}$, then

$$-\tan^{-1}(ln/m) < \omega < \tan^{-1}(nm/l), \tag{14.10}$$

and the system is called *dextral wrench*. Finally, if $\tau_{zz} > \tau_{yy} > \tau_{xx}$, then

$$-90° < \omega < -\tan^{-1}(ln/m), \tag{14.11}$$

and the system is called *dextral gravity*.

Similarly, there are three sinistral cases, and, allowing for the six cases on which two stresses are equal, Bott (1959) distinguishes twelve "tectonic regimes" corresponding to various combinations of the tectonic stresses and the orientation of the plane.

14.3 Overthrust faulting and sliding under gravity

Hubbert and Rubey (1959,1960,1961) investigated the importance of pore fluid pressure in the mechanics of sliding of large rock masses, with particular reference to low-angle overthrust faulting, along with some related problems of geological significance.

Consider slab *OABC* of height h and length L, sliding along horizontal plane *CB*, on which the coefficient of sliding friction is μ (Fig. 14.4a). The origin is taken at O and the mean density of the fluid-filled rock is ρ, so the vertical stress at depth z is $\tau_{zz} = \rho g z$. The pore fluid density is written as $\rho_f = \lambda\rho$, where λ is some number less than unity, so the pore pressure p at depth z is $p = \lambda\rho g z$. If $\tau'_{zz} = \tau_{zz} - p$ is the effective stress, then

$$\tau_{zz} = \rho g z, \quad p = \lambda\rho g z, \quad \tau'_{zz} = (1 - \lambda)\rho g z. \tag{14.12}$$

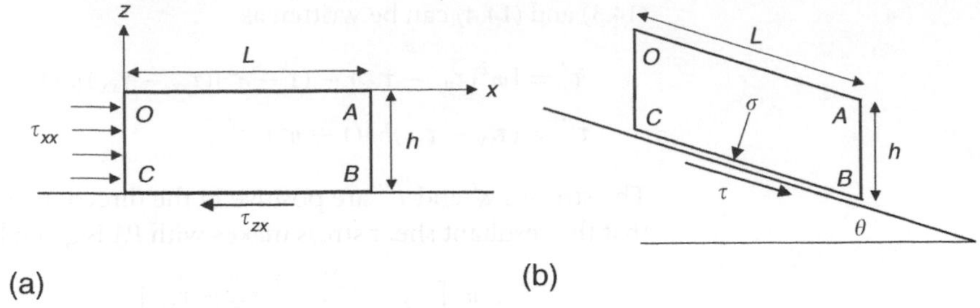

Fig. 14.4 Two scenarios analyzed by Hubbert and Rubey (1959): (a) block being pushed along a horizontal surface; (b) block sliding down a sloped surface.

The slab is now imagined to be pushed to the right by stresses τ_{xx} applied on face *OC*. These stresses are assumed to be the maximum sustainable by the rock, so that the rock is in a state of incipient failure all along *OC*. Writing (4.13) in terms of the effective stresses, we have

$$\tau'_{xx} = C_0 + q\tau'_{zz} = C_0 + q(1 - \lambda)\rho g z, \tag{14.13}$$

where C_0 and $q = \tan^2 \beta$ are the Mohr–Coulomb parameters for the rock mass. It follows that

$$\tau_{xx} = C_0 + \rho g z[q + \lambda(1 - q)]. \tag{14.14}$$

The condition for motion along face *BC* is found by writing (3.6) in terms of effective stresses:

$$\tau_{zx} = S_0 + \mu(1 - \lambda)\rho g h, \tag{14.15}$$

where S_0 is the inherent shear strength of surface *BC*, and it is recalled that, to calculate the effective stress, the pore pressure is subtracted only from the normal stresses.

As the total shear force along *BC* that resists motion will increase with *L*, but the driving force along *OC* is independent of *L*, the problem is to determine the maximum length *L* that can be moved in this manner. This is achieved by noting that, for the slab to be in static equilibrium, the net force acting on it in the *x* direction must be zero, that is,

$$\int_0^h \tau_{xx}(x = 0)\mathrm{d}z = \int_0^L \tau_{zx}(z = -h)\mathrm{d}x. \tag{14.16}$$

Substituting expressions (14.14) and (14.15) for the stresses into this force balance and integrating gives

$$C_0 h + \rho g[q + \lambda(1 - q)]\frac{h^2}{2} = S_0 L + \mu(1 - \lambda)\rho g h L, \tag{14.17}$$

which can be solved to yield

$$L = \frac{2C_0 h + \rho g h^2[q + \lambda(1 - q)]}{2S_0 + 2\mu(1 - \lambda)\rho g h}. \tag{14.18}$$

Hubbert and Rubey (1959) examined some specific cases and showed that the effect of pore fluid pressure, as represented by the coefficient λ, may be to greatly increase the sliding distance L.

Consider now a block $OABC$ of length L and depth h, sitting on a surface CB having slope q, coefficient of sliding friction μ, and inherent shear strength S_0 (Fig. 14.4b). If ρ is the mean density of the fluid-saturated medium and $\lambda\rho$ is the density of the pore fluid, then the shear stress acting along CB is

$$\tau = \rho g h \sin\theta \cos\theta, \qquad (14.19)$$

and the compressive normal stress acting along CB is

$$\sigma = \rho g h \cos^2\theta. \qquad (14.20)$$

Using a variation of the effective stress concept, Hubbert and Rubey (1959) argued that the effective normal stress should not include the fluid density term, in which case

$$\sigma' = \rho(1-\lambda) g h \cos^2\theta. \qquad (14.21)$$

The condition for the block to slide down the slope is then $\tau \geq S_0 + \mu\sigma'$, that is,

$$\rho g h \sin\theta \cos\theta \geq S_0 + \mu\rho(1-\lambda) g h \cos^2\theta. \qquad (14.22)$$

For the simple case in which $S_0 = 0$, sliding will occur if

$$\tan\theta \geq \mu(1-\lambda). \qquad (14.23)$$

Hence, the effect of pore pressure will be to reduce the minimum angle at which sliding may occur. Two slightly more complicated models for this phenomenon have been analyzed by Raleigh and Griggs (1963).

14.4 Stresses around faults

In a Volterra dislocation, a cut is made across a surface in an elastic solid, a shear displacement takes place along this surface, and the two faces are then welded together. The result is a state of stress and strain in the medium, in the absence of any applied external loads. The concept of a Volterra dislocation can be used to develop simple models of the stresses around faults.

Based on some analytical solutions developed by Steketee (1958), Chinnery (1961, 1966a,b) used the concept of dislocations to model a strike-slip fault as a uniform shear displacement U taking place over a rectangular region within the fault plane (Fig. 14.5a). The fault plane is located within the $z < 0$ half-space, $z = 0$ is taken to be a free surface, and the dislocation surface is defined by

$$-L < x < L, \quad -(D+d) < z < -d, \quad y = 0. \qquad (14.24)$$

The shear displacement on the fault surface occurs in the x direction. In general, the integrals that define the displacement field must be evaluated numerically,

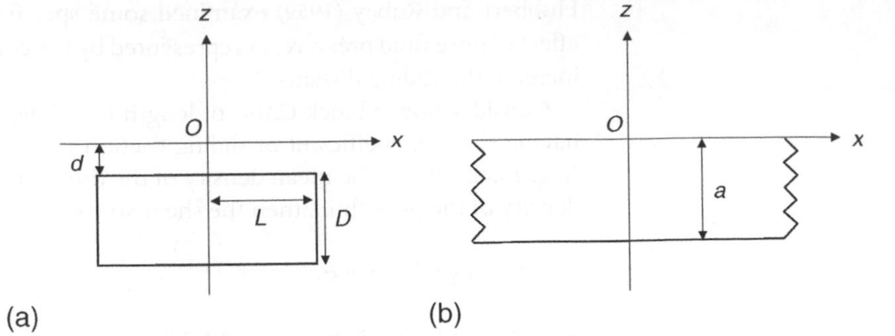

Fig. 14.5 Two dislocation models for a strike-slip fault: (a) Chinnery (1961), (b) Knopoff (1958). Shear displacement along the fault is in the x direction.

but in the limiting case of $L \to \infty$, the displacements components can be explicitly evaluated. The only nonzero displacement is

$$u = \frac{U}{2\pi}\left[\arctan\left(\frac{D+z}{y}\right) + \arctan\left(\frac{D-z}{y}\right) - \arctan\left(\frac{d+z}{y}\right) - \arctan\left(\frac{d-z}{y}\right)\right]. \tag{14.25}$$

It follows that the only nonzero stress components are

$$\tau_{xy} = \frac{GU}{2\pi}\left[\frac{d+z}{y^2+(d+z)^2} + \frac{d-z}{y^2+(d-z)^2} - \frac{D+z}{y^2+(D+z)^2} - \frac{D-z}{y^2+(D-z)^2}\right], \tag{14.26}$$

$$\tau_{xz} = \frac{GUy}{2\pi}\left[\frac{1}{y^2+(D+z)^2} - \frac{1}{y^2+(D-z)^2} - \frac{1}{y^2+(d+z)^2} + \frac{1}{y^2+(d-z)^2}\right]. \tag{14.27}$$

On the free surface $z = 0$, the stress component τ_{xz} vanishes, as it must, and

$$u = \frac{U}{\pi}\left[\arctan\left(\frac{D}{y}\right) - \arctan\left(\frac{d}{y}\right)\right], \tag{14.28}$$

$$\tau_{xy} = \frac{GU}{\pi}\left[\frac{d}{y^2+d^2} - \frac{D}{y^2+D^2}\right]. \tag{14.29}$$

Knopoff (1958) developed exact solutions for an infinitely long, vertical strike-slip fault that extends from the surface down to a depth a (Fig. 14.5b). Imagine first an intact semi-infinite space bounded by the plane $z = 0$, subjected to a uniform shear stress τ_{yx}, with magnitude τ . Now imagine that slip occurs over a fault plane defined by

$$-\infty < x < \infty, \quad -a < z < 0, \quad y = 0. \tag{14.30}$$

This slip causes the magnitude of the shear stress along the fault plane to drop from τ down to 0. The resulting displacement field was found by Knopoff to be given by

$$u = \frac{\tau}{G}\mathbf{Im}\sqrt{(z+\mathrm{i}y)^2 - a^2}, \quad v = 0, \quad w = 0, \tag{14.31}$$

where z and y must each be considered to be real variables. In the plane of the fault, $y = 0$, the displacement in the slip direction is found from (14.31) to be given by

$$u = 0 \quad \text{for } |z| > a, \quad u = \pm\frac{\tau}{G}\sqrt{a^2 - z^2} \quad \text{for } |z| < a, \tag{14.32}$$

which shows that the relative displacement of two points that are initially facing each other across the fault is equal to $2\tau a/G$ at the surface and drops monotonically with depth, vanishing at depth a. Along the free surface, $z = 0$, (14.31) shows that the displacements parallel to the fault plane are $\pm(\tau/G)(y^2 + a^2)^{1/2}$. The nonvanishing stress components are given by

$$\tau_{yx} = \tau\mathbf{Re}\left[\frac{z+\mathrm{i}y}{\sqrt{(z+\mathrm{i}y)^2 - a^2}}\right], \quad \tau_{xz} = \tau\mathbf{Im}\left[\frac{z+\mathrm{i}y}{\sqrt{(z+\mathrm{i}y)^2 - a^2}}\right]. \tag{14.33}$$

The mean value of the relative fault displacement, averaged over the depth of the fault, is

$$\overline{\Delta u} = \frac{1}{a}\int_{-a}^{0}\frac{2\tau}{G}\sqrt{a^2 - z^2}\,\mathrm{d}z = \frac{2\tau a}{G}\int_{0}^{1}\sqrt{1 - s^2}\,\mathrm{d}s = \frac{\pi\tau a}{2G}. \tag{14.34}$$

The mean offset of the fault is therefore related to the drop in shear stress by $\overline{\Delta u} = \pi a\Delta\tau/2G$. Hence, $2G/\pi a$ can be identified as the spring constant k that could be used in the "mass on a spring" model discussed in §3.4.

Another model for the relationship between the shear displacement and stress drop along a fault can be developed by assuming that the displacement occurs over a circular region of the fault, having radius a. The local relative displacement due to a uniform stress change τ over the circular region is given by (8.311):

$$\Delta u = \frac{8(1-\nu)\tau a}{\pi(2-\nu)G}\sqrt{1 - (r/a)^2}, \tag{14.35}$$

where the coordinate system is aligned so that x direction lies in the plane of the crack, parallel to τ. The mean value of the relative displacement is

$$\begin{aligned}\overline{\Delta u} &= \frac{1}{\pi a^2}\int_0^a\int_0^{2\pi}\frac{8(1-\nu)\tau a}{\pi(2-\nu)G}\sqrt{1-(r/a)^2}\,r\mathrm{d}r\mathrm{d}\theta \\ &= \frac{16(1-\nu)\tau}{\pi(2-\nu)Ga}\int_0^a\sqrt{1-(r/a)^2}\,r\mathrm{d}r = \frac{16(1-\nu)a\tau}{\pi(2-\nu)G}\int_0^1\sqrt{1-\rho^2}\,\rho\,\mathrm{d}\rho \\ &= \frac{16(1-\nu)a\tau}{3\pi(2-\nu)G}.\end{aligned} \tag{14.36}$$

In this case, the spring constant k that relates the stress change to the mean shear displacement is $3\pi(2-\nu)G/16(1-\nu)a$. This value is often reported (Scholz, 1990, p. 182) as $7\pi G/16a$, following the common geophysical assumption that $\nu = 1/4$. This result, along with the result $k = 2G/\pi a$ for an infinite fault of depth a, illustrates the general fact that $k = CG/a$, where a is the characteristic dimension (usually the "smallest" dimension) of the slipped region and C is a dimensionless constant on the order of unity.

14.5 Mechanics of intrusion

The primary mechanism of igneous intrusion is thought to be tensile failure under stresses caused by the pressure of magma, and several models for this process have been proposed.

14.5.1 Sheet intrusions

The obvious model for this case is a flat elliptical crack filled with magma at a pressure p (Anderson, 1937). This was discussed in §10.9 in the context of Griffith's failure criterion. Modifying those results to account for pore pressure and using notation appropriate to a three-dimensional stress state, the condition for the crack to propagate becomes

$$\sigma_3 - p + T_0 = 0, \tag{14.37}$$

provided that $\sigma_1 + 3\sigma_3 < 4p$.

Vertical dykes are frequently associated with normal faulting, so it may be assumed that they have been formed under conditions in which the horizontal principal stress σ_h is less than the vertical stress $\rho g z$. In this case, condition (14.37) becomes

$$p = \sigma_h + T_0, \tag{14.38}$$

provided that $\rho g z + 3\sigma_h < 4p$. This condition is likely to be satisfied as p is expected to be on the order of $\rho g z$.

According to the present model, horizontal sheets of magma can propagate if $\sigma_h > \rho g z$, and (14.37) gives

$$p = \rho g z + T_0, \tag{14.39}$$

provided that $\sigma_h + 3\rho g z < 4p$. Pollard and Holzhausen (1979) analyzed this problem in more detail, using Schwarz's iterative approach to develop an approximate solution for the manner in which the traction-free ground surface changes the state of stress along the boundary of the horizontal crack.

If the pressure in the crack is sufficiently larger than the weight of the overburden, the magma can lift the overburden and form a *laccolith*. Kerr and Pollard (1998) studied laccolith formation by modeling the overburden as an elastic plate, using the governing equation for plate bending that is analogous to the equation of beam bending used in §14.6. Zenzri and Keer (2001) extended these models further by allowing the substrate and overburden to have different elastic moduli and considered both elongated intrusions, idealized as plane strain problems, as well as axisymmetric geometries.

14.5.2 Ring dykes and cone sheets

Anderson (1935) discussed the effects caused by pressure in a spherical magma chamber near the Earth's surface. He considered the stresses due to a small, pressurized spherical cavity of radius a, located at depth $d \gg a$ below the surface (Fig. 14.6), that is, at $z = d$, with the z-axis taken to point downward. To solve this problem, he started with the solution to this problem of a pressurized cavity in an infinite medium, as given in (8.304)–(8.306). This solution gives unwanted tractions at the free surface, $z = 0$. As a first step to removing these tractions, Anderson imagined that the half-space was replaced by an infinite medium and considered an "image" cavity located at $z = -d$. In Cartesian coordinates, the displacements due to the superposition of these two pressurized cavities are

$$u_1 = -Pa^3x(r_1^{-3} + r_2^{-3}), \quad v_1 = -Pa^3y(r_1^{-3} + r_2^{-3}),$$
$$w_1 = -Pa^3[(z-d)r_1^{-3} + (z+d)r_2^{-3})], \tag{14.40}$$

where r_1 and r_2 are the lengths of AP and $A'P$ in Fig. 14.6a, that is,

$$r_1 = [x^2 + y^2 + (z-d)^2]^{1/2}, \quad r_2 = [x^2 + y^2 + (z+d)^2]^{1/2}. \tag{14.41}$$

By symmetry, this image cavity cancels out the shear stresses on the plane $z = 0$. However, there is still an unwanted normal stress on the free surface, given by

$$\tau_{zz} = -4GPa^3(r_1^{-3} - 3d^2r_1^{-5}). \tag{14.42}$$

These tractions can be removed by superposing the solution for the half-space $z > 0$, with the negative of these tractions applied at the surface. This solution

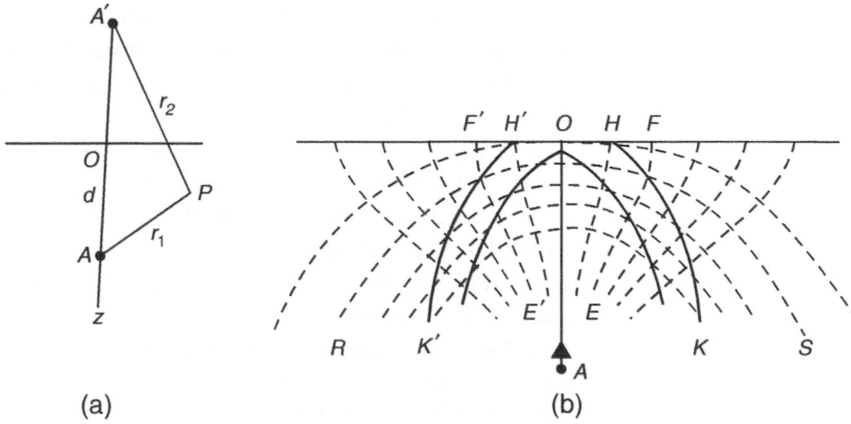

Fig. 14.6 (a) Small pressurized cavity A at depth d below free surface and image cavity A', and (b) the computed stress trajectories.

can be found by appropriate superposition of the fundamental solution for a point load on a half-space that was presented in §13.5. The result is

$$u_2 = 2Pa^3\left[\frac{3xz(z+d)}{r_2^5} - \frac{(1-2\nu)x}{r_2^3}\right], \tag{14.43}$$

$$v_2 = 2Pa^3\left[\frac{3yz(z+d)}{r_2^5} - \frac{(1-2\nu)y}{r_2^3}\right], \tag{14.44}$$

$$w_2 = 2Pa^3\left[\frac{3z(z+d)^2}{r_2^5} + \frac{(1-2\nu)z}{r_2^3} + \frac{2(1-\nu)d}{r_2^3}\right]. \tag{14.45}$$

The complete solution (see also Mindlin, 1936) is then given by $u = u_1 + u_2$, etc., after which the stresses can be found by the usual process of differentiation and application of Hooke's law.

As a second example, Anderson considered a point force of magnitude F, acting vertically upward at point A. In this case, the complete displacement field was found to be

$$u = B\left[\frac{-x(z-d)}{r_1^3} + \frac{(3-4\nu)x(d-z)}{r_2^3} + \frac{4(1-\nu)(1-2\nu)x}{r_2(r_2+z+d)} - \frac{6dxz(z+d)}{r_2^5}\right], \tag{14.46}$$

$$v = B\left[\frac{-y(z-d)}{r_1^3} + \frac{(3-4\nu)y(d-z)}{r_2^3} + \frac{4(1-\nu)(1-2\nu)y}{r_2(r_2+z+d)} - \frac{6dyz(z+d)}{r_2^5}\right], \tag{14.47}$$

$$w = B\left[\begin{array}{l} \dfrac{-(z-d)^2}{r_1^3} + \dfrac{(z+d)^2}{r_2^3} - \dfrac{(3-4\nu)}{r_1} + \dfrac{(3-4\nu)}{r_2} - \dfrac{8(1-\nu)^2}{r_2} \\ \quad - \dfrac{6dz(z+d)^2}{r_2^5} - \dfrac{4(1-\nu)^2 z^2}{r_2^3} - \dfrac{2(3-4\nu)dz}{r_2^3} - \dfrac{2d^2}{r_2^3} \end{array}\right], \tag{14.48}$$

where $B = -F/16\pi G(1-\nu)$.

For geological applications, it is sufficient to consider this latter case of the vertical push at A, the stress trajectories of which are shown in Fig. 14.6b. Of these, the set of type EF, $E'F'$, corresponds to a tensional principal stress and intersects the surface vertically in circles centered at O and having diameter FF'. Thus, a force at A due to upward-thrusting magma would be expected to cause tensional failure on these surfaces and give rise to circular system of dykes, dipping approximately vertically. These are the commonly observed *ring dykes*. The theory for a center of hydrostatic pressure, derived from displacements (14.43)–(14.45), leads to similar conclusions.

Associated with ring dykes, another circular system, *cone sheets*, also occurs. These dip outward from the center. The mechanism suggested for these involves a retreat of magma from the magma chamber. Since this region will now have a density less than the normal value, this effect may be simulated by a negative (i.e., downward) push at A. The stress trajectories will be as for the upward force, but σ_3 will now lie along the set EF and σ_1 in the orthogonal direction. Shear failure may be expected to occur along curves such as HK inclined at an angle ψ

to the trajectories of greatest principal stress, such as RS, where angle ψ will be in the order of 30°. As seen in Fig. 14.6b, these directions, shown in heavy lines, dip outward in the required fashion. Further discussion of these phenomena has been given by Robson and Barr (1964).

14.6 Beam models for crustal folding

The problem of bending of an elastic (or viscoelastic) beam or plate has often been used to develop models for deformation of the lithosphere. Consider a beam AOB of length $2L$ that is freely hinged at its two ends and is subjected to an axial compressive force F (Fig. 14.7). Suppose also that the vertical movement of this beam is resisted by a restraining force applied at it midpoint by an elastic spring of stiffness k. Imagine that the centre point O of the beam has been displaced downward by an amount v_0. The restraining force exerted by the spring will be kv_0, and so reaction forces of $-kv_0/2$ must be supplied at the two hinges.

Taking the origin to be located at the displaced position of the midpoint of the beam, for $0 \leq z \leq L$, the governing equation (6.20) for the deflection of the beam takes the form

$$EI\frac{d^2v}{dz^2} = F(v_0 - v) - \frac{1}{2}kv_0(L - z), \tag{14.49}$$

where E and I are the Young's modulus and moment of inertia of the beam, respectively. By symmetry, only the region $0 \leq z \leq L$ needs to be considered. The general solution of this differential equation is

$$v = A\sin\omega z + B\cos\omega z + v_0 - (k/2F)v_0(L - z), \tag{14.50}$$

where $\omega = \sqrt{F/EI}$. The constants must be chosen so that $v(z = 0) = 0$, which follows from our choice of origin, and also to satisfy the two boundary conditions,

$$v(z = L) = v_0, \quad \frac{dv}{dz}(z = 0) = 0. \tag{14.51}$$

The conditions $v(0) = 0$ and $v'(0) = 0$ lead immediately to

$$B = [(kL/2F) - 1]v_0, \quad A = -(k/2F\omega)v_0, \tag{14.52}$$

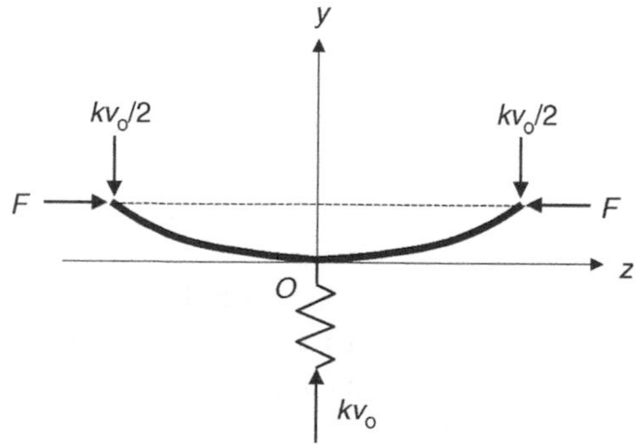

Fig. 14.7 Elastic beam subjected to axial compression and an elastic restraining force at its midpoint.

which reduces (14.50) to the form

$$v = -(kv_0/2F\omega)\sin\omega z + [(kL/2F) - 1]v_0\cos\omega z + v_0 - (kv_0/2F)(L - z). \tag{14.53}$$

Finally, imposition of the condition $v(L) = v_0$ leads to

$$\{-(k/2F\omega)\sin\omega L + [(kL/2F) - 1]\cos\omega L\}\, v_0 = 0. \tag{14.54}$$

This equation could be satisfied by choosing $v_0 = 0$, which corresponds to a beam that has undergone no deflection, but this solution is of no interest in the present context. However, if F takes on a value that causes the bracketed term in (14.54) to vanish, then a solution exists for any value of the amplitude v_0. Such values of F, known as the *eigenvalues* of this boundary-value problem, are therefore the roots of the following equation:

$$\tan\omega L = \omega L\left[1 - \frac{2EI}{kL^3}(\omega L)^2\right]. \tag{14.55}$$

Although this is a transcendental equation that must be solved numerically, there are an infinite number of positive roots for the parameter ωL. Recalling that $\omega = \sqrt{F/EI}$, these values of ωL define the critical values of the axial load F for which nontrivial deflections of the beam can occur. This type of deformation, in which the beam deflects in a direction that is *normal* to the applied load F, is known as *buckling*.

In the special case of no elastic restraint, that is, $k = 0$, (14.55) reduces to $\tan\omega L = -\infty$, the positive roots of which are $\omega L = \pi/2, 3\pi/2$, etc. The most important root is the smallest one, $\omega L = \pi/2$, which leads to a critical axial load of

$$F_{\text{crit}} = \frac{\pi^2 EI}{4L^2}. \tag{14.56}$$

This problem was first analyzed in 1744 by the Swiss mathematician and mechanician, Leonhard Euler, and so (14.56) is known as the *Euler buckling load* for a freely hinged column. In this case, (14.53) reduces to

$$v(z) = v_0[1 - \cos(\pi z/2L)], \tag{14.57}$$

and the deformed configuration of the entire beam is seen to be one half-wavelength of a cosine function; the full wavelength of this configuration would be $4L$. If F increases slowly from some small value, the beam will not deflect as long as F is less than the critical value given by (14.56). When this value is reached, the beam can assume the sinusoidal shape given by (14.57). The amplitude of the deflection, v_0, will be controlled by the other kinematic constraints of the problem.

The case in which the deflection of the beam is resisted by an elastic medium that supplies a restraining force that is proportional to the vertical displacement at all points along the beam may be treated in the same manner (Goldstein, 1926).

The foregoing analysis can be generalized as follows. Consider an infinite beam subjected to an axial force F and a distributed load in the y direction given by $q(z)$. The governing equation (6.20) then takes the form

$$EI\frac{d^2v}{dz^2} = \int_z q(\xi)(\xi - z)d\xi - Fv, \tag{14.58}$$

where the integral gives the moment due to the distributed loads. Differentiating twice with respect to z yields

$$EI\frac{d^4v}{dz^4} + F\frac{d^2v}{dz^2} = -q(z). \tag{14.59}$$

Suppose that the deflection of the beam is resisted by a viscous medium, as described in §9.9, in which case (14.59) takes the form

$$EI\frac{d^4v}{dz^4} + F\frac{d^2v}{dz^2} = -\eta\frac{dv}{dt}. \tag{14.60}$$

We look for solutions to (14.60) that are spatially periodic with wavelength $\lambda = 2\pi/\omega$, and so we take $v(z,t) = V(t)\exp(i\omega z)$. Inserting this form into (14.60) yields

$$\eta\frac{dV}{dt} = \omega^2(F - EI\omega^2)V, \tag{14.61}$$

the solution to which is

$$V(t) = V_o \exp[\omega^2(F - EI\omega^2)t/\eta]. \tag{14.62}$$

For fixed values of the parameters $\{F, E, I\}$, this solution will decay to zero with time for spatial frequencies ω that are greater than $\sqrt{F/EI}$. Hence, if for any reason the beam starts to take on a shape having a spatial frequency greater than this value, the viscous restraint force will tend to force it back into its straight, undeformed configuration. Deformations with spatial frequencies less than this critical value, on the other hand, will tend to grow in time. Of course, at some point, the assumptions of linear elastic behavior and small deflections, which were required for the derivation of (6.20), will cease to hold. But this linear analysis does nevertheless indicate the tendency for deflections having sufficiently long wavelengths to grow in an unstable manner.

This smallest unstable wavelength is given by $\lambda = 2\pi\sqrt{EI/F}$. Recalling that the wavelength of the unstable configuration of the hinged beam of length L was $4L$, we see from (14.56) that this smallest unstable wavelength corresponds exactly to the wavelength of the shortest hinged beam that would be unstable under a given axial load F. This critical value of $\lambda = 2\pi\sqrt{EI/F}$ is known as the *Euler wavelength*, and deformations having any wavelength greater than this are also unstable. But the coefficient $\omega^2(F - EI\omega^2)$ appearing in (14.62) attains its maximum value when $\omega = \sqrt{F/2EI}$, and so in some sense the wavelength $\lambda_d = 2\pi\sqrt{2EI/F}$ is the "most unstable," as configurations having this wavelength will grow fastest. If the beam initially has some arbitrary small deflection, which by Fourier analysis can be decomposed into a superposition of different spatial frequencies, eventually the frequency $\omega_d = \sqrt{F/2EI}$ will dominate, and the beam will assume the shape of a sinusoid of this frequency.

Biot (1961) gave the foregoing analysis as a preliminary to the discussion of the folding of geological systems. He considered next the case of an elastic plate of thickness h, immersed in an infinite medium of viscosity η, and found the dominant wavelength to be given by $\lambda_d = \pi h[E/(1 - \nu^2)F]^{1/2}$, where F is now the in-plane, normal compressive traction applied to the edges of the plate. Finally, he considered a planar viscous sheet of thickness h and viscosity η_s, immersed in a medium of viscosity η, and found $\lambda_d = 2\pi h[\eta_s/6\eta]^{1/3}$. This theory can be extended to several layers and to include the effects of gravity (Biot, 1957,1961,1965; Biot et al., 1961; Currie et al., 1962; Ramberg, 1964). Ramberg (1967) discussed the use of physical scale models to study such problems. A detailed discussion of the folding of geological layers can be found in the monograph by Johnson and Fletcher (1994).

14.7 Earthquake mechanics

There are two outstanding features of earthquakes. First, earthquakes are concentrated into narrow zones that form a network around the Earth. Second, those that occur at depths of less than 60 km account for 75 percent of all energy released by earthquakes. These observations appear to be consistent with the theory of plate tectonics, originally proposed in 1915 by Alfred Wegener as "continental drift" (Hallam, 1973). According to this theory, the crust comprises a number of discrete plates that move across the surface of the Earth in response to viscous flow in the mantle (Fig. 14.8). At locations where the relative motion of these plates is resisted, mechanical instabilities may occur in the form of earthquakes.

It is generally believed that most earthquakes are mechanical instabilities that result from the sudden failure of rock to sustain the shear stresses that act across

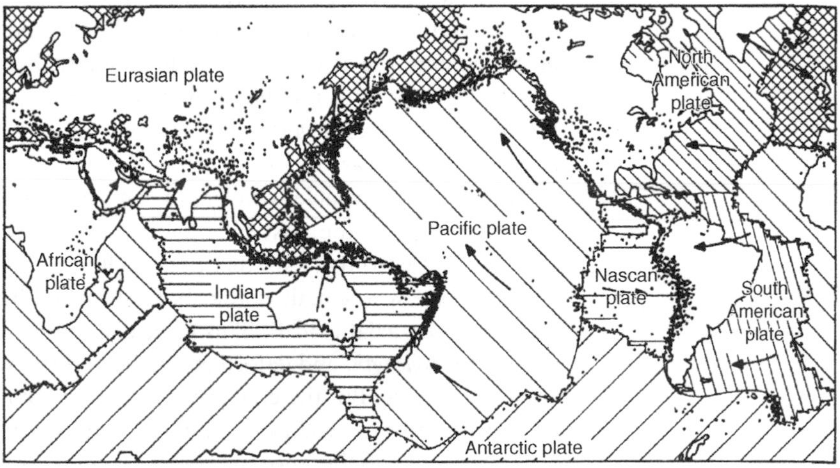

Fig. 14.8 Spatial distribution of earthquakes and their relationship to tectonic plates (after Nur, 1974).

a surface. The surface may be a preexisting fault or a new fracture caused by the failure. If it is assumed as a first approximation that most of the rock around such a surface responds in a linear, elastic manner to the sudden change in stress associated with the earthquake, much of the mechanics of earthquakes can be explained, in simple terms.

From such an assumption, it follows that a linear relationship exists between any change in the value of the shear stress acting along the surface and the relative tangential displacement of the two opposing faces. It is convenient to think in terms of an average displacement and average shear stress, as in §14.4, which for simplicity will be denoted here by u and τ. A schematic representation of this linear relationship is shown in Fig, 14.9, with u taken to be zero and τ equal to τ_1 in the configuration before slip has occurred. The slope of this straight line is determined by the shape and size of the surface and by the elastic moduli of the adjacent rock, as discussed briefly in §14.4. For the present discussion, the value of this slope, that is, the spring constant relating u and τ, is not needed.

Imagine now that slip occurs by an average amount u_2, over an area A. If this slip had occurred quasi-statically, that is, *reversibly* in a thermodynamic sense, the system would have moved down line *bega*, starting at b and ending at g. In this case, the work done by the rock would be given by the area A of the fault, multiplied by the area *obgho* in Fig. 14.9, and this work would also necessarily equal the loss in strain energy of the rock mass adjacent to the fault. The area *obgho* is given by the product of the mean stress and the displacement, so $\Delta\mathcal{E}_{\text{elastic}} = A(\tau_1 + \tau_2)u_2/2$. But the elastic strain energy is a thermodynamic state property, and so its change when the system moves from $u = 0$ to $u = u_2$ cannot depend on the path. Hence, even if the fault displacement occurred in a rapid, irreversible jolt, it must nevertheless be the case that

$$\Delta\mathcal{E}_{\text{elastic}} = A(\tau_1 + \tau_2)u_2/2. \tag{14.63}$$

It is convenient here to ignore signs and bear in mind that (14.63) is a positive quantity that represents the elastic strain energy *released* by the rock mass.

Now consider the more realistic case of a sudden fault displacement, in which the system does not necessarily follow curve *bega* at all times. It seems reasonable to assume that this sudden motion is precipitated, or at least accompanied by,

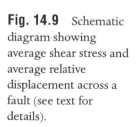

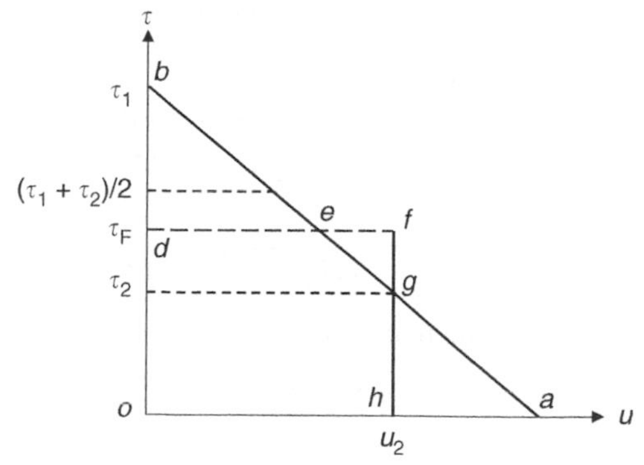

Fig. 14.9 Schematic diagram showing average shear stress and average relative displacement across a fault (see text for details).

a sudden drop in the shear stress, say to some value τ_F. Assume that the stress along the fault remains at τ_F during the displacement. In this case, an amount of energy equal to $A\tau_F u_2$ would be dissipated by friction along the fault. This energy manifests itself as an increase in the temperature of the rock and so is actually converted into internal energy. However, as this energy is no longer either potential energy or mechanical energy of any sort, it is common to refer to it as "lost" or "dissipated." This frictional energy "loss" is proportional to area *dfho*. Hence,

$$\Delta\mathcal{E}_{\text{friction}} = A\tau_F u_2, \tag{14.64}$$

where again we treat this as a positive number and account later for the relative senses of the various terms in the energy balance.

The difference between the elastic strain energy liberated during this process and the frictional energy dissipated, will manifest itself mainly as seismic energy in elastic waves that radiate outward from the fault surface. Assuming only these three terms in the energy balance, the seismic energy released by the earthquake is given by

$$\Delta\mathcal{E}_{\text{seismic}} = \Delta\mathcal{E}_{\text{elastic}} - \Delta\mathcal{E}_{\text{friction}} = A(\tau_1 + \tau_2)u_2/2 - A\tau_F u_2. \tag{14.65}$$

Without considering the details of this process, it follows immediately from (14.65) that the frictional stress τ_F cannot exceed $(\tau_1 + \tau_2)/2$, or else the radiated seismic energy would be negative, which is not physically plausible. Furthermore, the dynamic nature of the slip process would cause the two opposing sides of the fault to overshoot the equilibrium position corresponding to stress τ_F, which is given by *e* in Fig. 14.9; hence, τ_F must not be less than τ_2 (Orowan, 1960).

The seismic energy (14.65) can also be written as

$$\Delta\mathcal{E}_{\text{seismic}} = A(\tau_1 - \tau_2)u_2/2 - A(\tau_F - \tau_2)u_2 = Au_2\Delta\tau/2 - A(\tau_F - \tau_2)u_2, \tag{14.66}$$

where $\Delta\tau = \tau_1 - \tau_2$ is called the *stress drop*. The stress drop is sometimes called *complete* if $\tau_2 = 0$ and *partial* if $\tau_2 > 0$ (Kanamori, 1977). For a given stress drop, the seismic energy will attain its maximum value if $\tau_F = \tau_2$, in which case it takes the value (Orowan, 1960)

$$\Delta\mathcal{E}_{\text{seismic}} = Au_2\Delta\tau/2. \tag{14.67}$$

The *seismic efficiency* η is defined as the ratio of the seismic energy released during the earthquake to the elastic strain energy released by the rock. From (14.63) and (14.66), we find

$$\eta = \frac{\Delta\mathcal{E}_{\text{seismic}}}{\Delta\mathcal{E}_{\text{elastic}}} = \frac{(\tau_1 + \tau_2)/2 - \tau_F}{(\tau_1 + \tau_2)/2} = \frac{\tau_1 + \tau_2 - 2\tau_F}{\tau_1 + \tau_2}. \tag{14.68}$$

The maximum efficiency occurs when $\tau_F = \tau_2$, in which case

$$\eta = \frac{\tau_1 - \tau_2}{\tau_1 + \tau_2}. \tag{14.69}$$

If the stress drop is small, this expression for maximum seismic efficiency reduces to $\eta \approx \Delta\tau/2\tau_1$. Assuming the often-quoted rule of thumb that the stress drop is about one-tenth of the ambient stress (Scholz, 1990, p. 165), the seismic efficiency would be about 0.05. Kanamori and Brodsky (2004) report that stress drops in earthquakes are typically in the range of 1–10 MPa.

A parameter that is often used to quantify the sizes of earthquakes is the *seismic moment*, **M**. This is a second-order tensor defined as

$$\mathbf{M} = AG \begin{bmatrix} 2un_x & un_y + vn_x & un_z + wn_x \\ vn_x + un_y & 2vn_y & vn_z + wn_y \\ wn_x + un_z & wn_y + vn_z & 2wn_z \end{bmatrix}, \tag{14.70}$$

where (u, v, w) are the Cartesian components of the relative displacement of the opposing fault faces and (n_x, n_y, n_z) are the components of the normal vector to the fault plane. For example, a fault geometry such as shown in Fig. 14.5 would have $\mathbf{u} = (u, 0, 0)$ and $\mathbf{n} = (0, 1, 0)$, so the only nonzero components of **M** would be $M_{xy} = M_{yx} = AGu$, and hence AGu can be thought of as the "magnitude" of the seismic moment. If the stress and displacement vary along the fault, M would need to be calculated by integrating the terms in (14.70) over the spatial extent of the slipped region of the fault.

According to the simple model described above, and assuming that $\tau_F = \tau_2$, it follows from (14.67) that the radiated seismic energy is related to the seismic moment by

$$\Delta\mathcal{E}_{\text{seismic}} = M\Delta\tau/2G. \tag{14.71}$$

The value of M varies by many orders of magnitude between different earthquakes, whereas the stress drop seems to vary by barely an order of magnitude and likewise for the shear modulus. Hence, the seismic moment provides a rough estimate of the seismic energy released in an earthquake.

References

Abousleiman, Y. N. and Kanj, M. Y. (2004). The generalized Lamé problem – Part II: Applications in poromechanics, *J. Appl. Mech.*, 71, 180–9.

Abramowitz, M. and Stegun, I. A. (1970). *Handbook of Mathematical Functions*, Dover, New York.

Achenbach, J. D. (1973). *Wave Propagation in Elastic Solids*, North-Holland, Amsterdam.

Adams, F. D. (1912). An experimental contribution to the question of the depth of the zone of flow in the earth's crust, *J. Geol.*, 20, 97–118.

Adler, P. M. and Thovert, J.-F. (1999). *Fractures and Fracture Networks*, Kluwer, Dordrecht.

Ahrens, T. J. and Johnson, M. L. (1995). Shock wave data for rocks, in *Rock Physics and Phase Relations – A Handbook of Physical Constants*, T. J. Ahrens, ed., American Geophysical Union, Washington, pp. 35–44.

Airy, G. B. (1863). On the strains in the interior of beams, *Phil. Trans. Roy. Soc. Lond.*, 153, 49–80.

Aki, K. and Richards, P. G. (1980). *Quantitative Seismology*, W. H. Freeman, San Francisco.

Al-Ajmi, A. and Zimmerman, R. W. (2005). Relation between the Mogi and the Coulomb failure criteria, *Int. J. Rock Mech.*, 42, 431–9.

Allègre, C. J., Le Mouel, J. L., and Provost, A. (1982). Scaling rules in rock fracture and possible implications for earthquake prediction, *Nature*, 297, 47–9.

Almusallam, A. and Taher, S. E.-D. (1995). Three-dimensional Mohr's circle for shear stress components, *J. Eng. Mech.*, 121, 477–81.

Alsayed, M. I. (2002). Utilising the Hoek triaxial cell for multiaxial testing of hollow rock cylinders, *Int. J. Rock Mech.*, 39, 355–66.

Amadei, B. (1983). *Rock Anisotropy and the Theory of Stress Measurements*, Springer-Verlag, New York.

Amadei, B. (1988). Strength of a regularly jointed rock mass under biaxial and axisymmetric loading conditions, *Int. J. Rock Mech.*, 25, 3–13.

Amadei, B. (1996). Importance of anisotropy when estimating and measuring *in situ* stresses in rock, *Int. J. Rock Mech.*, 33, 293–325.

Amadei, B. and Savage, W. Z. (1993). Effect of joints on rock mass strength and deformability, in *Comprehensive Rock Engineering*, J. A. Hudson, ed., Pergamon, Oxford, pp. 331–65.

Amadei, B. and Stephansson, O. (1997). *Rock Stress and its Measurement*, Chapman & Hall, London and New York.

Amadei, B., Savage, W. Z., and Swolfs, H. S. (1987). Gravitational stresses in anisotropic rock masses, *Int. J. Rock Mech.*, 24, 5–14.

Amenzade, I. A. (1979). *Theory of Elasticity*, Mir, Moscow.

Anagnostopoulos, A. (1993). *Geotechnical Engineering of Hard Soils – Soft Rocks*, Balkema, Rotterdam.

Andersen, M. A. and Jones, F. O., Jr. (1985). A comparison of hydrostatic-stress and uniaxial strain pore-volume compressibilities using nonlinear elastic theory, in *Proc. 26th U.S. Symp. Rock Mech.*, E. Ashworth, ed., Balkema, Rotterdam, pp. 403–10.

Anderson, E. M. (1935). The dynamics of the formation of cone-sheets, ring-dykes and caldron-subsidencies, *Proc. Roy. Soc. Edin.*, 56, 128–57.

Anderson, E. M. (1937). The dynamics of sheet intrusion, *Proc. Roy. Soc. Edin.*, 58, 242–51.

Anderson, E. M. (1951). *The Dynamics of Faulting and Dyke Formation with Applications to Britain*, 2nd ed., Oliver & Boyd, Edinburgh.

Anderson, O. L., Schreiber, E., Liebermann, R., and Soga, N. (1968). Some elastic constant data on minerals relevant to geophysics, *Rev. Geophys. Space Phys.*, 6, 491–524.

Andreev, G. E. (1995). *Brittle Failure of Rock Material*, Balkema, Rotterdam.

Angel, Y. C. and Achenbach, J. D. (1985). Reflection and transmission of elastic waves by a periodic array of cracks, *J. Appl. Mech.*, 52, 33–41.

Annin, B. D. (1988). *Elastic Plastic Problems*, Am. Soc. Mech. Eng., New York.

Arenberg, D. L. (1948). Ultrasonic solid delay lines, *J. Acoust. Soc. Am.*, 20, 1–26.

Arndt, J., Bartel, T., Scheuber, E., and Schilling, F. (1997). Thermal and rheological properties of granodioritic rocks from the Central Andes, North Chile, *Tectonophysics*, 271, 75–88.

Asphaug, E. and Melosh, H. J. (1993). The Stickney impact of Phobos: A dynamical model, *Icarus*, 101, 144–64.

Atkinson, B. K. (1979). A fracture mechanics study of subcritical tensile cracking of quartz in wet environments, *Pure Appl. Geophys.*, 117, 1011–24.

Atkinson, B. K. and Meredith, P. G. (1987). The theory of subcritical crack growth with applications to minerals and rocks, in *Fracture Mechanics of Rock*, B. K. Atkinson, ed., Academic Press, London and San Diego, pp. 111–66.

Attewell, P. B. (1962). Composite model to simulate porous rock, *Engineering*, 193, 574–5.

Bandis, S. C., Lumsden, A. C., and Barton, N. R. (1983). Fundamental of rock joint deformation, *Int. J. Rock Mech.*, 20, 249–68.

Barnard, P. R. (1964). Researches into the complete stress-strain curve for concrete, *Mag. Concr. Res.*, 16, 203–10.

Barton, N. (1973). Review of a new shear strength criterion for rock joints, *Eng. Geol.*, 7, 287–332.

Barton, N. and Choubey, V. (1977). The shear strength of rock joints in theory and practice, *Rock Mech.*, 10, 1–54.

Barton, N. R., Bandis, S. C., and Bakhtar, K. (1985). Strength, deformation and conductivity coupling of rock joint deformation, *Int. J. Rock Mech.*, 22, 121–40.

Batchelor, G. K. (1967). *An Introduction to Fluid Dynamics*, Cambridge University Press, Cambridge and New York.

Bayly, B. M. (1992). *Mechanics in Structural Geology*, Springer-Verlag, Berlin and New York.

Bear, J. (1988). *Dynamics of Fluids in Porous Media*, Dover, New York.

Becker, G. F. (1893). Finite homogeneous strain, flow and rupture of rocks, *Bull. Geol. Soc. Am.*, 4, 13–90.

Becker, R. M. (1968). An anisotropic elastic solution for testing stress relief cores, *U. S. Bureau Mines Rep. Invest. 7143*.

Bedford, A. and Drumheller, D. S. (1994). *Introduction to Elastic Wave Propagation*, John Wiley & Sons, New York and Chichester.

Beran, M. J. (1968). *Statistical Continuum Theories*, Wiley Interscience, London and New York.

Berenbaum, R. and Brodie, I. (1959). The tensile strength of coal, *J. Inst. Fuel*, 32, 320–7.

Berkhout, A. J. (1987). *Applied Seismic Wave Theory*, Elsevier, Amsterdam.

Bernabe, Y., Brace, W. F., and Evans, B. (1982). Permeability, porosity, and pore geometry of hot-pressed calcite, *Mech. Mater.*, 1, 173–83.

Bernhard, R. K. (1940). Influence of the elastic constant of testing machines, *ASTM Bull.*, 88, 14–15.

Berry, D. S. (1960a). An elastic treatment of ground movement due to mining. I. Isotropic ground, *J. Mech. Phys. Solids*, 8, 280–92.

Berry, D. S. and Sales, T. W. (1961). An elastic treatment of ground movement due to mining. II. Transversely isotropic ground, *J. Mech. Phys. Solids*, 9, 52–62.

Berry, D. S. and Sales, T. W. (1962). An elastic treatment of ground movement due to mining. III. Three-dimensional problem, transversely isotropic ground, *J. Mech. Phys. Solids*, 10, 73–83.

Berry, J. P. (1960b). Some kinetic consideration of the Griffith criterion for fracture. I. Equations of motion at constant force, *J. Mech. Phys. Solids*, 8, 207–16.

Berryman, J. G. (1980). Confirmation of Biot's theory, *Appl. Phys. Lett.*, 37, 382–84.

Berryman, J. G. (1992). Effective stress for transport properties of inhomogeneous porous rock, *J. Geophys. Res.*, 97, 17, 409–24.

Berryman, J. G. (1995). Mixture theories for rock properties, in *Rock Physics and Phase Relations – A Handbook of Physical Constants*, T. J. Ahrens, ed., American Geophysical Union, Washington, D. C., pp. 205–28.

Bers, L., John, F., and Schechter, M. (1964). *Partial Differential Equations*, Wiley Interscience, New York.

Bieniawski, Z. T., Denkhaus, H. G., and Vogler, U. W. (1969). Failure of fractured rock, *Int. J. Rock Mech.*, 6, 323–41.

Biot, M. A. (1941). General theory of three-dimensional consolidation, *J. Appl. Phys.*, 12, 155–64.

Biot, M. A. (1956a). Theory of propagation of elastic waves in a fluid-saturated porous solid, I. Lower freqency range, *J. Acoust. Soc. Am.*, 28, 168–78.

Biot, M. A. (1956b). Theory of propagation of elastic waves in a fluid-saturated porous solid, II. Higher freqency range, *J. Acoust. Soc. Am.*, 28, 179–91.

Biot, M. A. (1957). Folding instability of a layered visco-elastic medium under compression, *Proc. Roy. Soc. Lond. Ser. A.*, 242, 444–54.

Biot, M. A. (1961). Theory of folding of stratified visco-elastic media and its implication in tectonics and orogenesis, *Bull. Geol. Soc. Am.*, 72, 1595–620.

Biot, M. A. (1965). Theory of viscous buckling and gravity instability of multi-layers with large deformation, *Bull. Geol. Soc. Am.*, 76, 371–8.

Biot, M. A. and Willis, D. G. (1957). The elastic coefficients of the theory of consolidation, *J. Appl. Mech.*, 24, 594–601.

Biot, M. A., Ode, E., and Roever, W. L. (1961). Experimental verification of the folding of stratified visco-elastic media, *Bull. Geol. Soc. Am.*, 72, 1621–32.

Birch, F. and Bancroft, D. (1938). Elasticity and internal friction in a long column of granite, *Bull. Seism. Soc. Am.*, 28, 243–54.

Bjork, R. L. (1961). Analysis of the formation of Meteor Crater, Arizona: A preliminary report, *J. Geophys. Res.*, 66, 3379–87.

Black, A. D., Dearing, H. L., and Dibona, B. G. (1985). Effects of pore pressure and mud filtration on drilling rates in a permeable sandstone, *J. Petrol. Tech.*, 37, 1671–81.

Blair, S. C. and Cook, N. G. W. (1998). Analysis of compressive fracture in rock using statistical techniques: part I, a non-linear rule-based model, *Int. J. Rock Mech.*, 35, 837–48.

Bland, D. R. (1960). *The Theory of Linear Viscoelasticity*, Pergamon, Oxford and New York.

Boadu, F. K. and Long, L. T. (1996). Effects of fractures on seismic-wave velocity and attenuation, *Geophys. J. Int.*, 127, 86–110.

Bodvarsson, G. S., Pruess, K., and Lippmann, M. J. (1986). Modeling of geothermal systems, *J. Petrol. Tech.*, 38, 1007–21.

Böker, R. (1915). Die Mechanik der bleibenden Formänderung in Kristallinisch aufgebauten Körpern [The mechanics of permanent deformation in crystalline bodies], *Ver. dt. Ing. Mitt. Forsch.*, 175, 1–51.

Boley, B. A. and Weiner, J. H. (1960). *Theory of Thermal Stresses*, John Wiley & Sons, New York.

Bonnechere, F. J. and Fairhurst, C. (1968). Determination of the regional stress field from doorstopper measurements, *J. S. Afr. Inst. Min. Metall.*, 69, 520–44.

Born, W. T. (1941). Attenuation constant of earth materials, *Geophysics*, 6, 132–48.

Bott, M. H. P. (1959). The mechanics of oblique slip faulting, *Geol. Mag.*, 96, 109–17.

Bourbié, T., Coussy, O., and Zinszner, B. (1987). *Acoustics of Porous Media*, Gulf Publishing, Houston.

Boussinesq, J. (1878). Équilibre élastique d'un solide isotrope de masse négligeable, soumis à différents poids [Elastic equilibrium of a weightless isotropic solid under various loads], *C. Rend. Acad. Sci. Paris*, 86, 1260–63.

Bowden, F. P. (1954). The friction of non-metallic solids, *J. Inst. Petrol.*, 40, 89–103.

Bowden, F. P. and Tabor, D. (1985). *The Friction and Lubrication of Solids*, Clarendon Press, Oxford.

Brace, W. F. (1960). An extension of Griffith theory of fracture to rocks, *J. Geophys. Res.*, 65, 3477–80.

Brace, W. F. (1964). Brittle fracture of rocks, in *State of Stress in the Earth's Crust*, W. R. Judd, ed., Elsevier, New York, pp. 111–74.

Brace, W. F. (1965). Some new measurements of the linear compressibility of rocks, *J. Geophys. Res.*, 70, 391–8.

Brace, W. F. and Bombolakis, E. G. (1963). A note on brittle crack growth in compression, *J. Geophys. Res.*, 68, 3709–13.

Brace, W. F. and Byerlee, J. D. (1966). Stick-slip as a mechanism for earthquakes, *Science*, 153, 990–2.

Bracewell, R. N. (1986). *The Fourier Transform and its Applications*, McGraw-Hill, New York and London.

Bradley, F. E. (1990). Development of an Airy stress function of general applicability in one, two, or three dimensions, *J. Appl. Phys.*, 67, 225–6.

Bradley, J. J. and Fort, A. N., Jr. (1966). Internal friction in rocks, in *Handbook of Physical Constants*, S. P. Clark, ed., Geological Society of America, New York, pp. 175–93.

Brady, B. H. G. (1979). A direct formulation of the boundary element method of stress analysis for complete plane strain, *Int. J. Rock Mech.*, 16, 235–44.

Brady, B. H. G. and Brown, E. T. (2004). *Rock Mechanics for Underground Mining*, Springer, Berlin and New York.

Brandt, H. (1955). A study of the speed of sound in porous granular media, *J. Appl. Mech.*, 22, 479–86.

Bray, J. (1987). Some applications of elastic theory, in *Analytical and Computational Methods in Engineering Rock Mechanics*, E. T. Brown, ed., Allen & Unwin, London, pp. 32–94.

Bredehoeft, J. D., Wolff, R. G., Keys, W. S., and Shuter, E. (1976). Hydraulic fracturing to determine the regional *in situ* stress field, Piceance Basin, Colorado, *Geol. Soc. Am. Bull.*, 87, 250–58.

Brekhovskikh, L. M. (1980). *Waves in Layered Media*, 2nd ed., Academic Press, New York.

Bridgman, P. (1931). *The Physics of High Pressures*, Bell, London.

Brillouin, L. (1960). *Wave Propagation and Group Velocity*, Academic Press, New York and London.

Bristow, J. R. (1960). Microcracks, and the static and dynamic elastic constants of annealed and heavily cold-worked metals, *Brit. J. Appl. Phys.*, 11, 81–85.

Broek, D. (1986). *Elementary Engineering Fracture Mechanics*, 4th ed., Kluwer, Dordrecht.

Bromwich, T. J. I. (1949). *An Introduction to the Theory of Infinite Series*, Macmillan, London.

Brown, E. T. and Hoek, E. (1978). Trends in relationships between in situ stresses and depth, *Int. J. Rock Mech.*, 15, 211–25.

Brown, R. J. S. and Korringa, J. (1975). On the dependence of the elastic properties of a porous rock on the compressibility of the pore fluid, *Geophysics*, 40, 608–16.

Brown, S. R. (1987). Fluid flow through rock joints: The effect of surface roughness, *J. Geophys. Res.*, 92, 1337–47.

Brown, S. R. and Scholz, C. H. (1985a). Broad bandwidth study of the topography of natural surfaces, *J. Geophys. Res.*, 90, 12, 575–82.

Brown, S. R. and Scholz, C. H. (1985b). Closure of random elastic surfaces in contact, *J. Geophys. Res.*, 90, 5531–45.

Brown, S. R. and Scholz, C. H. (1986). Closure of rock joints, *J. Geophys. Res.*, 91, 4939–48.

Brown, S. R., Stockman, H. W., and Reeves, S. J. (1995). Applicability of the Reynolds equation for modeling fluid flow between rough surfaces, *Geo. Res. Letts.*, 22, 2537–40.

Bruner, W. M. (1976). Comment on Seismic velocities in dry and cracked solids by Richard J. O'Connell and B. Budiansky, *J. Geophys. Res.*, 81, 2573–6.

Budiansky, B. (1965). On the elastic moduli of some heterogeneous materials, *J. Mech. Phys. Solids*, 13, 223–27.

Budiansky, B. and O'Connell, R. J. (1976). Elastic moduli of a cracked solid, *Int. J. Solids Struct.*, 12, 81–97.

Burridge, R. and Knopoff, L. (1967). Model and theoretical seismicity, *Bull. Seism. Soc. Am.*, 57, 341–71.

Byerlee, J. D. (1967). Frictional characteristics of granite under high confining pressure, *J. Geophys. Res.*, 72, 3639–48.

Byerlee, J. D. (1978). Friction of rocks, *Pure Appl. Geophys.*, 116, 615–26.

Callias, C. J. and Markenscoff, X. (1989). Singular asymptotics analysis for the singularity at a hole near a boundary, *Q. Appl. Math.*, 47, 233–45.

Callias, C. J. and Markenscoff, X. (1993). The singularity of the stress field of two nearby holes in a planar elastic medium, *Q. Appl. Math.*, 51, 547–57.

Carey, S. W. (1953). The rheid concept in geotectonics, *J. Geol. Soc. Austral.*, 1, 67–117.

Carpenter, C. B. and Spencer, G. B. (1940). Measurements of compressibility of consolidated oil-bearing sandstones, *Report 3540*, U. S. Bureau of Mines, Denver.

Carroll, M. M. (1979). An effective stress law for anisotropic elastic deformation, *J. Geophys. Res.*, 84, 7510–2.

Carroll, M. M. and Katsube, N. (1983). The role of Terzaghi effective stress in linearly elastic deformation, *J. Energy Resour. Tech.*, 105, 509–11.

Carslaw, H. S. and Jaeger, J. C. (1949). *Operational Methods in Applied Mathematics*, 2nd ed., Oxford University Press, Oxford.

Carslaw, H. S. and Jaeger, J. C. (1959). *Conduction of Heat in Solids*, 2nd ed., Clarendon Press, Oxford.

Carter, N. L., Christie, J. M., and Griggs, D. T. (1964). Experimental deformation and recrystallization of quartz, *J. Geol.*, 72, 687–733.

Cerruti, V. (1882). Recerche intorno all' equilibrio de' corpi elastici isotropi [Research into the equilibrium of an isotropic elastic body], *Atti Acad. Naz. Lincei Rend., Ser. 3a*, 13, 81–122.

Chalon, F. and Montheillet, F. (2003). The interaction of two spherical gas bubbles in an infinite elastic solid, *J. Appl. Mech.*, 70, 789–98.

Chan, T., Hood, M., and Board, M. (1982). Rock properties and their effect on thermally-induced displacements and stresses, *J. Energy Resour. Tech.*, 104, 384–8.

Chang, C. and Haimson, B. (2005). Non-dilatant deformation and failure mechanism in two Long Valley Caldera rocks under true triaxial compression, *Int. J. Rock Mech.*, 42, 402–14.

Charlaix, E., Guyon, E., and Roux, S. (1987). Permeability of a random array of fractures of widely varying apertures, *Transp. Porous Media*, 2, 31–43.

Charlez, P. A. (1991). *Rock Mechanics: Theoretical Fundamentals*, Editions Technip, Paris.

Chau, K. T. (1997). Young's modulus interpreted from compression tests with end friction, *J. Eng. Mech.*, 123, 1–7.

Chen, C.-S., Pan, E., and Amadei, B. (1998). Determination of deformability and tensile strength of anisotropic rock using Brazilian tests, *Int. J. Rock Mech.*, 35, 43–61.

Chen, H.-S. and Acrivos, A. (1978). The effective elastic moduli of composite materials containing spherical inclusions at non-dilute concentrations, *Int. J. Solids Struct.*, 14, 349–64.

Chen, Z., Narayan, S. P., Yang, Z., and Rahman, S. S. (2000). An experimental investigation of hydraulic behaviour of fractures and joints in granitic rock, *Int. J. Rock Mech.*, 37, 1061–71.

Cheng, A. H.-D. (1997). Material coefficients of anisotropic poroelasticity, *Int. J. Rock Mech.*, 34, 199–205.

Cheng, A. H.-D., Abousleiman, Y., and Roegiers, J.-C., (1993). Review of some poroelastic effects in rock mechanics, *Int. J. Rock Mech.*, 30, 1119–26.

Cheng, C. H. and Toksöz, M. N. (1979). Inversion of seismic velocities for the pore aspect ratio spectrum of a rock, *J. Geophys. Res.*, 84, 7533–43.

Chilingarian, G. V. and Wolf, K. H. (1975). *Compaction of Coarse-Grained Sediments*, Vol. 1, Elsevier, Amsterdam.

Chinnery, M. A. (1966a). Secondary faulting. I. Theoretical aspects, *Canad. J. Earth Sci.*, 3, 163–74.

Chinnery, M. A. (1966b). Secondary faulting. II. Geological aspects, *Canad. J. Earth Sci.*, 3, 175–90.

Chou, P. C. and Pagano, N. J. (1992). *Elasticity: Tensor, Dyadic, and Engineering Approaches*, Dover, New York.

Christensen, R. M. (1982). *Theory of Viscoelasticity*, 2nd ed., Academic Press, New York and London.

Christensen, R. M. (1991). *Mechanics of Composite Materials*, Krieger, Malabar, FL.

Churchill, R. V. (1972). *Operational Mathematics*, 3rd ed., McGraw-Hill, New York.

Claesson, J. and Bohloli, B. (2002). Brazilian test: stress field and tensile strength of anisotropic rocks using an analytical solution, *Int. J. Rock Mech.*, 39, 991–1004.

Clark, N. J., Jr. (1966) *Handbook of Physical Constants* (Geol. Soc. Am. Mem. 97), Geological Society of America, New York.

Cleary, M. P. (1978). Elastic and dynamic response regimes of fluid-impregnated solids with diverse microstructures, *Int. J. Solids Struct.*, 14, 795–819.

Coates, D. F. and Yu, Y. S. (1970). A note on the stress concentration at the end of a cylindrical hole, *Int. J. Rock Mech.*, 7, 585–88.

Collins, W. D. (1962a). Some axially symmetric stress distributions in elastic solids containing penny-shaped cracks, *Proc. Roy. Soc. Ser. A.*, 266, 369–86.

Collins, W. D. (1962b). Some coplanar punch and crack problems in three-dimensional elastostatics, *Proc. Roy. Soc. Ser. A.*, 274, 507–28.

Colmenares, L. B. and Zoback, M. D. (2002). A statistical evaluation of intact rock failure criteria constrained by polyaxial test data for five different rocks, *Int. J. Rock Mech.*, 39, 695–729.

Cook, N. G. W. (1962). *A Study of Failure in the Rock Surrounding Underground Excavations*, Ph.D. thesis, University of Witwatersrand, Johannesburg.

Cook, N. G. W. (1965). The failure of rock, *Int. J. Rock Mech.*, 2, 389–403.

Cook, N. G. W. (1970). An experiment proving that dilatancy is a pervasive volumetric property of brittle rock loaded to failure, *Rock Mech.*, 2, 181–8.

Cook, N. G. W. (1992). Natural joints in rock: Mechanical, hydraulic and seismic behaviour and properties under normal stress, *Int. J. Rock Mech.*, 29, 198–223.

Cook, N. G. W. and Hojem, J. P. M. (1966). A rigid 50-ton compression and tension testing machine, *S. Afr. Mech. Eng.*, 16, 89–92.

Cook, N. G. W., Hoek, E., Pretorius, J. P. G., Ortlepp, W. D., and Salamon, M. D. G. (1966). Rock mechanics applied to the study of rockbursts, *J. South Afr. Inst. Min. Metall.*, 66, 436–528.

Cottrell, A. H. (1952). The time laws of creep, *J. Mech. Phys. Solids*, 1, 53–63.

Coulomb, C. A. (1773). Application des règles de maxima et minima à quelques problèmes de statique relatifs à l'Architecture [Application of the rules of maxima and minima to some problems of statics related to architecture], *Acad. Roy. Sci. Mem. Math. Phys.*, 7, 343–82.

Coviello, A., Lagioia, R., and Nova, R. (2005). On the measurement of the tensile strength of soft rocks, *Rock Mech. Rock Eng.*, 38, 251–73.

Cox, S. J. D. and Scholz, C. H. (1988). On the formation and growth of faults: An experimental study, *J. Struct. Geol.*, 10, 413–30.

Crank, J. (1956). *Mathematics of Diffusion*, Clarendon Press, Oxford.

Critescu, N. (1989). *Rock Rheology*, Kluwer, Boston and Dordrecht.

Critescu, N. and Hunsche, U. (1997). *Time Effects in Rock Mechanics*, Wiley, Chichester.

Crouch, S. L. (1970). Experimental determination of volumetric strains in failed rock, *Int. J. Rock Mech.*, 7, 589–603.

Crouch, S. L. (1972). A note on post-failure stress-strain path dependence in norite, *Int. J. Rock Mech.*, 9, 197–204.

Cryer, C. W. (1963). A comparison of the three-dimensional compaction theories of Biot and Terzaghi, *Q. J. Mech. Appl. Math.*, 16, 401–12.

Currie, J. B., Patnode, H. W., and Trump, R. P. (1962). Development of folds in sedimentary strata, *Bull. Geol. Soc. Am.*, 73, 655–74.

Daehnke, A. and Rossmanith, H. P. (1997). Reflection and refraction of plane stress waves at interfaces modelling various rock joints, *Int. J. Blast. Frag.*, 1, 111–231.

Dagan, G. (1993). Higher-order correction for effective permeability of heterogeneous isotropic formations of lognormal conductivity distribution, *Transp. Porous Media*, 12, 279–90.

Davis, L. A. and Gordon, R. B. (1968). Pressure dependence of the plastic flow stress of alkali halide single crystals, *J. Appl. Phys.*, 39, 3885–97.

Davis, R. O. and Selvadurai, A. P. S. (1996). *Elasticity and Geomechanics*, Cambridge University Press, Cambridge.

de Marsily, G. (1986). *Quantitative Hydrogeology*, Academic Press, San Diego.

de Sitter, L. U. (1956). *Structural Geology*, McGraw-Hill, New York.

Detournay, E. and Cheng, A. H.-D. (1988). Poroelastic response of a borehole in a non-hydrostatic stress field, *Int. J. Rock Mech.*, 25, 171–82.

Detournay, E. and Cheng, A. H.-D. (1993). Fundamentals of poroelasticity, in *Comprehensive Rock Engineering*, J. A. Hudson, ed., Pergamon, Oxford, pp. 113–71.

Detournay, E. and Fairhurst, C. (1987). Two-dimensional elastoplastic analysis of a long, cylindrical cavity under non-hydrostatic loading, *Int. J. Rock Mech.*, 24, 197–211.

Detournay, E., Cheng, A. H.-D., Roegiers, J. C., and McLennan, J. D. (1989). Poroelasticity considerations in *in situ* stress determination by hydraulic fracturing, *Int. J. Rock Mech.*, 26, 507–13.

Dewers, T. and Ortoleva, P. J. (1989). Mechanico-chemical coupling in stressed rocks, *Geochem. Cosmochem. Acta*, 53, 1243–58.

Dieterich, J. H. (1972). Time-dependent friction in rocks, *J. Geophys. Res.*, 77, 3690–7.

Dieterich, J. H. (1978). Time-dependent friction and the mechanics of stick-slip, *Pure Appl. Geophys.*, 116, 790–806.

Digby, P. J. (1981). The effective elastic moduli of porous granular rocks, *J. Appl. Mech.*, 48, 803–8.

Dobrynin, V. M. (1962). Effect of overburden stress on some properties of sandstones, *Soc. Petrol. Eng. J.*, 2, 360–6.

Domenico, S. N. (1977). Elastic properties of unconsolidated porous sand reservoirs, *Geophysics*, 42, 1339–68.

Donath, F. A. (1961). Experimental study of shear failure in anisotropic rocks, *Bull. Geol. Soc. Am.*, 72, 985–90.

Donath, F. A. (1966). A triaxial pressure apparatus for testing of consolidated or unconsolidated materials subject to pore pressure, in *Testing Techniques for Rock Mechanics* (ASTM Spec. Tech. Pub. 402), American Society for Testing and Materials, Philadelphia, pp. 41–51.

Donnell, L. H. (1941). Stress concentrations due to elliptical discontinuities in plates under edge forces, in *Theodore von Karman Anniversary Volume*, Calif. Inst. Tech., Pasadena, pp. 293–309.

Dove, P. M. (1995). Geochemical controls on the kinetics of quartz fracture at subcritical tensile stresses, *J. Geophys. Res.*, 100, 22, 349–59.

Drucker, D. C. (1950). Some implications of work-hardening and ideal plasticity, *Q. Appl. Math.*, 7, 411–8.

Drucker, D. C. and Prager, W. (1952). Soil mechanics and plastic analysis of limit design, *Q. Appl. Math.*, 10, 157–65.

Drumheller, D. S. (1998). *Introduction to Wave Propagation in Nonlinear Fluids and Solids*, Cambridge University Press, Cambridge and New York.

Duan, Z. P., Kienzler, R., and Herrmann, G. (1986). An integral equation method and its application to defect mechanics, *J. Mech. Phys. Solids*, 34, 539–61.

Duhamel, J. M. C. (1833). Mémoire sur la méthode générale relative au mouvement de la chaleur dans les corps solides plongés dans les milieux de température variable au cours du temps [Memoir on the general method related to heat transfer in solid bodies immersed in a medium whose temperature varies with time], *J. Ecole Polytech. Paris*, 14, 20–66.

Dullien, F. A. L. (1992). *Porous Media: Fluid Transport and Pore Structure*, 2nd ed., Academic Press, San Diego.

Durelli, A. J., Phillips, E. A., and Tsao, C. H. (1958). *Introduction to the Theoretical and Experimental Analysis of Stress and Strain*, McGraw-Hill, New York.

Economides, M. J. and Nolte, K. G. (2000). *Reservoir Stimulation*, 3rd ed., John Wiley & Sons, Hoboken, N. J. and Chichester.

Edelman, F. (1949). On the compression of a short cylinder between rough end-blocks, *Q. Appl. Math.*, 7, 334–37.

Edmond, J. M. and Paterson, M. S. (1972). Volume changes during the deformation of rocks at high pressures, *Int. J. Rock Mech.*, 9, 161–82.

Edwards, R. H. (1951). Stress concentrations around spheroidal inclusions and cavities, *J. Appl. Mech.*, 18, 19–27.

Eirich, F. R. (1956). *Rheology: Theory and Applications*, 5 vols., Academic Press, New York.

Elrod, H. G. (1979). A general theory for laminar lubrication with Reynolds roughness, *J. Lubr. Tech.*, 101, 8–14.

Engelder, J. T., Logan, J. M., and Handin, J. (1975). The sliding characteristics of sandstone on quartz fault-gauge, *Pure Appl. Geophys.*, 113, 69–86.

England, A. H. (1971). *Complex Variable Methods in Elasticity*, Wiley-Interscience, Chichester.

Esaki, T., Du, S., Mitani, Y., Ikusada, K., and Jing, L. (1999). Development of a shear-flow test apparatus and determination of coupled properties for a single rock joint, *Int. J. Rock Mech.*, 36, 641–50.

Eshelby, J. D. (1957). The determination of the elastic field of an ellipsoidal inclusion, and related problems, *Proc. Roy. Soc. Lond. Ser. A*, 241, 376–96.

Eubanks, R. A. (1954). Stress concentration due to a hemispherical pit at a free surface, *J. Appl. Mech.*, 21, 57–62.

Evans, B. and Dresen, G. (1991). Deformation of Earth materials – six easy peices, rheology of rocks, *Rev. Geophys.*, 29, Part 2, Supp. S, 823–43.

Evans, B. and Kohlstedt, D. L. (1995). Rheology of rocks, in *Rock Physics and Phase Relations – A Handbook of Physical Constants*, T. J. Ahrens, ed., American Geophysical Union, Washington, D. C., pp. 148–65.

Evans, I. (1961). The tensile strength of coal, *Colliery Eng.*, 38, 428–34.

Evans, R. H. and Wood, R. H. (1937). Transverse elasticity of natural stones, *Proc. Leeds Phil. Lit. Soc.*, 3, 340–52.

Evison, F. F. (1960). On the growth of continents by plastic flow under gravity, *Geophys. J. Roy. Astr. Soc.*, 3, 155–90.

Ewing, W. M., Jardetzky, W. S., and Press, F. (1957). *Elastic Waves in Layered Media*, McGraw-Hill, New York.

Ewy, R. T., Ray, P., Bovberg, C. A., Norman, P. D., and Goodman, H. E. (2001). Openhole stability and sanding predictions by 3D extrapolation from hole-collapse tests, *SPE Drill. Complet.*, 16, 243–51.

Exadaktylos, G. E. and Stavropoulou, M. C. (2002). A closed-form solution for stresses and displacements around tunnels, *Int. J. Rock Mech.*, 39, 905–16.

Exadaktylos, G. E., Vardoulakis, I., and Kourkoulis, S. K. (2001a). Influence of nonlinearity and double elasticity on flexure of rock beams – II. Technical theory, *Int. J. Solids Struct.*, 38, 4091–117.

Exadaktylos, G. E., Vardoulakis, I., and Kourkoulis, S. K. (2001b). Influence of nonlinearity and double elasticity on flexure of rock beams – II. Characterization of Dionysos marble, *Int. J. Solids Struct.*, 38, 4119–45.

Fabre, D. and Gustkiewicz, J. (1997). Poroelastic properties of limestones and sandstones under hydrostatic conditions, *Int. J. Rock Mech.*, 34, 127–34.

Fahrenthold, E. P. and Cheatham, J. B. (1986). An approximate rock stress-field for steady flow into production casing, *J. Energy Resour. Tech.*, 108, 116–9.

Fairhurst, C. (1964). Measurement of *in situ* rock stresses, with particular reference to hydraulic fracturing, *Rock Mech. Eng. Geol.*, 2, 129–47.

Fairhurst, C. (2003). Stress estimation in rock: a brief history and review, *Int. J. Rock Mech.*, 40, 957–73.

Falls, S. D. and Young, R. P. (1998). Acoustic emission and ultrasonic-velocity methods used to characterise the excavation disturbance associated with deep tunnels in hard rock, *Tectonophysics*, 289, 1–15.

Fang, Z. and Harrison, J. P. (2002), Application of a local degradation model to the analysis of brittle fracture of laboratory scale rock specimens under triaxial conditions, *Int. J. Rock Mech.*, 39, 459–76.

Fatt, I. (1958). The Biot-Willis elastic coefficients of a sandstone, *J. Appl. Mech.*, 26, 296–7.

Fetter, A. L. and Walecka, J. D. (1980). *Theoretical Mechanics of Particles and Continua*, McGraw-Hill, New York.

Filon, L. N. G. (1902). On the equilibrium of circular cylinders under certain practical systems of load, *Phil. Trans. Roy. Soc. Ser. A*, 198, 147–233.

Filonenko-Borodich, M. (1965). *Theory of Elasticity*, Dover, New York.

Fjaer, E., Holt, R. M., Horsrud, P., Raaen, A. M., and Risnes, R. (1992). *Petroleum Related Rock Mechanics*, Elsevier, Amsterdam.

Flamant, A. A. (1892). Sur la répartition des pressions dans un solide rectangulaire chargé transversalement [On the distribution of stresses in a rectangular solid under transverse load], *C. Rend. Acad. Sci. Paris*, 114, 1465–68.

Fond, C., Riccardi, A., Schirrer, R., and Montheillet, F. (2001). Mechanical interaction between spherical inhomogeneities: An assessment of a method based on the equivalent inclusion, *Eur. J. Mech. A Solids*, 20, 59–75.

Fossum, A. F. (1985). Effective elastic properties for a randomly jointed rock mass, *Int. J. Rock Mech.*, 22, 467–70.

Friedman, M. (1963). Petrofabric analysis of experimentally deformed calcite-cemented sandstones, *J. Geol.*, 71, 12–37.

Friedman, M. (1964). Petrofabric techniques for the determination of principal stress directions in rocks, in *State of Stress in the Earth's Crust*, W. R. Judd, ed., Elsevier, New York, pp. 451–550.

Frocht, M. M. (1941). *Photoelasticity*, Vol. 1, John Wiley & Sons, New York.

Fumi, F. G. (1952a). Physical properties of crystals: The direct-inspection method, *Acta Crystall.*, 5, 44–8.

Fumi, F. G. (1952b). Matter tensors in symmetrical systems, *Il Nuovo Cimento*, 9, 739–55.

Fung, Y. C. (1965). *Foundations of Solid Mechanics*, Prentice-Hall, Englewood Cliffs, N.J.

Furnish, M. D. (1993). Recent advances in methods for measuring the dynamic response of geological materials to 100 GPa, *Int. J. Impact Eng.*, 14, 267–77.

Gale, J., MacLeod, R., and LeMessurier, P. (1990). *Site characterization and validation – Measurement of flowrate, solute velocities and aperture variation in natural fractures as a function of normal and shear stress*, Stripa Project Report 90–11, Swedish Nuclear Fuel and Waste Management Company, Stockholm.

Galle, E. M. and Wilhiot, J. (1962). Stresses around a well bore due to internal pressure and unequal geostatic stresses, *Soc. Petrol. Eng. J.*, 2, 145–55.

Gassmann (1951a). Elastic waves through a packing of spheres, *Geophysics*, 15, 673–85.

Gassmann (1951b). Uber die Elasticität Poröser Medien [On the elasticity of porous media], *Vierteljahrsschrift der Naturforschenden Gesellschaft in Zürich*, 96, 1–23.

Geertsma, J. (1957a). The effect of fluid decline on volumetric changes of porous rocks, *Petrol. Trans. AIME*, 210, 331–40.

Geertsma, J. (1957b). A remark on the analogy between thermoelasticity and the elasticity of saturated porous media, *J. Mech. Phys. Solids*, 6, 13–6.

Geertsma, J. (1973). Land subsidence above compacting oil and gas reservoirs, *J. Petrol. Technol.*, 25, 734–44.

Gentier, S., Riss, J., Archambault, G., Flamand, R., and Hopkins, D. (2000). Influence of fracture geometry on shear behavior, *Int. J. Rock Mech.*, 37, 161–74.

Gerçek, H. (1988). Calculation of elastic boundary stresses for rectangular underground openings, *Mining Sci. Tech.*, 7, 173–82.

Gerçek, H. (1997). An exact solution for stresses around tunnels with conventional shapes, *Int. J. Rock Mech.*, 34, 708.

Gibowicz, S. J. and Kijko, A. (1994). *An Introduction to Mining Seismology*, Academic Press, New York.

Giraud, A. and Rousset, G. (1996). Time-dependent behaviour of deep clays, *Eng. Geol.*, 41, 181–95.

Goldstein, S. (1926). The stability of a strut under thrust when buckling is resisted by a force proportional to the displacement, *Proc. Camb. Phil. Soc.*, 23, 120–9.

Goodier, J. N. (1933). Concentration of stress around spherical and cylindrical inclusions and flaws, *Trans. ASME*, 55, 39–44.

Goodman, R. E. (1976). *Methods of Geological Engineering in Discontinuous Rocks*, West Publishing, New York.

Goodman, R. E. (1989). *Introduction to Rock Mechanics*, 2nd ed., John Wiley & Sons, New York and Chichester.

Goodman, R. E. (1993). *Engineering Geology: Rock in Engineering Construction*, John Wiley & Sons, New York and Chichester.

Graff, K. F. (1975). *Wave Motion in Elastic Solids*, Clarendon Press, Oxford.

Grasselli, G. and Egger, P. (2003). Constitutive law for the shear strength of rock joints based on three-dimensional surface parameters, *Int. J. Rock Mech.*, 40, 25–40.

Green, A. E. and Zerna, W. (1954). *Theoretical Elasticity*, Clarendon Press, Oxford.

Green, D. H. and Wang, H. F. (1986). Fluid pressure response to undrained compression in saturated sedimentary rock, *Geophysics*, 51, 948–56.

Green, D. H. and Wang, H. F. (1990). Specific storage as a poroelastic coefficient, *Water Resour. Res.*, 26, 1631–7.

Greenberg, H. J. and Truell, R. (1948). On a problem in plane strain, *Q. Appl. Math.*, 6, 53–62.

Greenwood, J. A. and Williamson, J. (1966). Contact of nominally flat surfaces, *Proc. Roy. Soc. Lond. Ser. A*, 295, 300–19.

Gresseth, E. W. (1964). Determination of principal stress directions through an analysis of rock joint and fracture orientations, Star Mine, Burke, Idaho, *U. S. Bureau Mines Rep. Invest. 6413*.

Griffith, A. A. (1920). The phenomena of flow and rupture in solids, *Phil. Trans. Roy. Soc. Lond. Ser. A*, 221, 163–98.

Griffith, A. A. (1924). Theory of rupture, in *Proc. 1st Int. Cong. Appl. Mech.*, C. B. Biezano and J. M. Burgers, eds., J. Waltman Jr, Delft, pp. 53–63.

Griggs, D. T. (1936). Deformation of rocks under high confining pressures, *J. Geol.*, 44, 541–77.

Griggs, D. T. (1939). Creep of rocks, *J. Geol.*, 47, 225–51.

Griggs, D. T. (1940). Experimental flow of rocks under conditions favouring recrystallization, *Bull. Geol. Soc. Am.*, 51, 1001–22.

Griggs, D. T. and Handin, J. (1960). Observations on fracture and an hypothesis of earthquakes, in *Rock Deformation* (Geol. Soc. Am. Mem. 79), D. T. Griggs and F. J. Turner, eds., Geological Society of America, New York, pp. 347–64.

Griggs, D. T., Turner, F. J., and Heard, H. C. (1960). Deformation of rocks at 500°C to 800°C, in *Rock Deformation* (Geol. Soc. Am. Mem. 79), Geological Society of America, New York, pp. 39–104.

Grimvall, G. (1986). *Thermophysical Properties of Materials*, North-Holland, Amsterdam.

Gu, B., Nihei, K. T., Myer, L. R., and Pyrak-Nolte, L. J. (1996a). Fracture interface waves, *J. Geophys. Res.*, 101, 827–35.

Gu, B., Suarez-Rivera, R., Nihei, K. T., and Myer, L. R. (1996b). Incidence of plane waves upon a fracture, *J. Geophys. Res.*, 101, 25, 337–46.

Gu, J. and Wong, T.-F. (1991). Effects of loading velocity, stiffness, and inertia on the dynamics of a single degree of freedom spring-slider system, *J. Geophys. Res.*, 96, 21, 677–91.

Gu, J., Rice, J. R., Ruina, A. L., and Tse, S. T. (1984). Slip motion and stability of a single degree of freedom elastic system with rate and state dependent friction, *J. Mech. Phys. Solids*, 32, 167–96.

Guéguen, Y. and Palciauskas, V. (1994). *Introduction to the Physics of Rocks*, Princeton University Press, Princeton.

Gurtin, M. E. (1972). The linear theory of elasticity, in *Handbuch der Physik*, Vol. Via/2, S. Flügge, ed., Springer-Verlag, Berlin, pp. 1–295.

Hadley, K. (1975). Azimuthal variation of dilatancy, *J. Geophys. Res.*, 80, 4845–50.

Haimson, B. and Chang, C. (2000). A new true triaxial cell for testing mechanical properties of rock, and its use to determine rock strength and deformability of Westerly granite, *Int. J. Rock Mech.*, 37, 285–96.

Haimson, B. and Fairhurst, C. (1967). Initiation and extension of hydraulic fractures in rocks, *Soc. Petrol. Eng. J.*, 7, 310–18.

Haimson, B. C. and Cornet, F. H. (2003). Suggested methods for rock stress estimation – Part 3: Hydraulic fracturing (HF) and/or hydraulic testing of pre-existing fractures, *Int. J. Rock Mech.*, 40, 1011–20.

Haimson, B. C. and Song, I. S. (1995). A new borehole failure criterion for estimating *in situ* stress from breakout span, in *Proc. 8th Int. Cong. Rock Mech.*, Vol. 1, T. Fujii, ed., Balkema, Rotterdam, pp. 341–6.

Haimson, B. C. and Voight, B. (1977). Crustal stress in Iceland, *Pure Appl. Geophys.*, 115, 153–90.

Hakami, E. (1989). *Water Flow in Single Rock Joints*, Licentiate diss., Luleå Univ. Tech., Luleå, Sweden.

Hallbauer, D. K., Wagner, H., and Cook, N. G. W. (1973). Some observations concerning the microscopic and mechanical behaviour of quartzite specimens in stiff, triaxial compression tests, *Int. J. Rock Mech.*, 10, 713–26.

Handin, J. (1953). An application of high pressure in geophysics: Experimental rock deformation, *Trans. Am. Soc. Mech. Eng.*, 75, 315–24.

Handin, J. and Fairbairn, H. W. (1955). Experimental deformation of Hasmark dolomite, *Bull. Geol. Soc. Am.*, 66, 1257–73.

Handin, J. and Hager, R. V. (1957). Experimental deformation of sedimentary rocks under confining pressure: Tests at room temperature on dry samples, *Bull. Am. Assoc. Petrol. Geol.*, 41, 1–50.

Handin, J. and Stearns, D. W. (1964). Sliding friction of rock, *Trans. AGU*, 45, 103.

Handin, J., Higgs, D. V., and O'Brien, J. K. (1960). Torsion of Yule marble under confining pressure, in *Rock Deformation* (Geol. Soc. Am. Mem. 79), D. T. Griggs and F. J. Turner, eds., Geological Society of America, New York, pp. 245–74.

Handy, R. L. (1981). Linearizing triaxial test failure envelopes, *Geotech. Test. J.*, 4, 188–91.

Hardy, H. R., Jr. (1959). Time-dependent deformation of geologic materials, *Colo. Sch. Mines Q.*, 54, 135–75.

Hart, D. J. and Wang, H. F. (1995). Laboratory measurements of a complete set of poroelastic moduli for Berea sandstone and Indiana limestone, *J. Geophys. Res.*, 100, 17,741–51.

Hart, D. J. and Wang, H. F. (2001). A single test method for determination of poroelastic constants and flow parameters in rocks with low hydraulic conductivities, *Int. J. Rock Mech.*, 38, 577–83.

Hasegawa, E. and Izuchi, H. (1983). On the steady flow through a channel consisting of an uneven wall and a plane wall, *Bull. Jap. Soc. Mech. Eng.*, 26, 514–20.

Hashin, Z. (1988). The differential scheme and its application to cracked materials, *J. Mech. Phys. Solids*, 36, 719–34.

Hashin, Z. and Shtrikman, S. (1961). Note on a variational approach to the theory of composite elastic materials, *J. Franklin Inst.*, 271, 336–41.

Hast, N. (1969). The state of stress in the upper part of the Earth's crust, *Tectonophysics*, 8, 169–211.

Hazzard, J. F., Collins, D. S., Pettitt, W. S., and Young, R. P. (2002). Simulation of unstable fault slip in granite using a bond-particle model, *Pure Appl. Geophys.*, 159, 221–45.

Heard, H. C. (1960). Transition from brittle to ductile flow in Solenhofen limestone as a function of temperature, confining pressure and interstitial fluid pressure, in *Rock Deformation* (Geol. Soc. Am. Mem. 79), Geological Society of America, New York, pp. 193–226.

Heim, A. (1878). Untersuchungen uber den Mechanismus der Gebirgsbildung [Investigation into the mechanisms of mountain-building], in *Anschluss and die Geologische Monographie der Tödi-Windgälen-Gruppe*, B. Schwabe, Basel.

Helbig, K. (1994). *Foundations of Anisotropy for Exploration Seismics*, Pergamon, New York.

Hencky, H. Z. (1924). Zur Theorie Plastischer Deformationen und der hierdurch im Material hervorgerufenen Nachspannungen [On the theory of plastic deformation and the caused late deformations], *Zeit. Ang. Math. Mech.*, 4, 323–34.

Herget, G. (1974). Ground stress determinations in Canada, *Rock Mech.*, 6, 53–74.

Herget, G. (1987). Stress assumptions for underground excavations in the Canadian shield, *Int. J. Rock Mech.*, 24, 95–97.

Hestir, K. and Long, J. C. S. (1990). Analytical expressions for the permeability of random two-dimensional Poisson fracture networks based on regular lattice percolation and equivalent medium theories, *J. Geophys. Res.*, 95, 21, 565–81.

Hetényi, M. (1950). *Handbook of Experimental Stress Analysis*, John Wiley & Sons, New York.

Hickey, C. J., Spanos, T. J. T., and DeLaCruz, V. (1995). Deformation parameters of permeable media, *Geophys. J. Int.*, 121, 359–70.

Hill, R. (1950). *The Mathematical Theory of Plasticity*, Oxford University Press, Oxford.

Hill, R. (1952). The elastic behaviour of a polycrystalline aggregate, *Proc. Phys. Soc. London*, A65, 349–54.

Hill, R. (1965). A self-consistent mechanics of composite materials, *J. Mech. Phys. Solids*, 13, 213–22.

Hiramatsu, Y. and Oka, Y. (1968). Determination of the stress in rock unaffected by boreholes or drifts from measured strains or deformations, *Int. J. Rock Mech.*, 5, 337–53.

Hirth, J. P. and Lothe, J. (1992). *Theory of Dislocations*, 2nd ed., Krieger, Malabar, FL.

Hobbs, D. W. (1962). The strength of coal under biaxial compression, *Coll. Eng.*, 39, 285–90.

Hobbs, D. W. (1964). The tensile strength of rocks, *Int. J. Rock Mech.*, 1, 385–96.

Hobbs, D. W. (1965). An assessment of a technique for determining the tensile strength of rock, *Brit. J. Appl. Phys.*, 16, 259–68.

Hocking, G. (1976). Three-dimensional elastic stress distribution around the flat end of a cylindrical cavity, *Int. J. Rock Mech.*, 13, 331–37.

Hoek, E. (1964). Fracture of anisotropic rock, *J. South Afr. Inst. Min. Metall.*, 64, 501–18.

Hoek, E. (1990). Estimating Mohr-Coulomb friction and cohesion values from the Hoek-Brown failure criterion, *Int. J. Rock Mech.*, 27, 227–9.

Hoek, E. and Bieniawski, Z. T. (1965). Brittle fracture propagation in rock under compression, *Int. J. Fract. Mech.*, 1, 137–55.

Hoek, E. and Brown, E. T. (1980). *Underground Excavations in Rock*, Institution of Mining and Metallurgy, London.

Hoek, E. and Franklin, J. A. (1968). Simple triaxial cell for field or laboratory testing of rock, *Trans. Inst. Min. Metall.*, 77, A22–26.

Hojem, J. P. M. and Cook, N. G. W. (1968). The design and construction of a triaxial and polyaxial cell for testing rock specimens, *South Afr. Mech. Eng.*, 18, 57–61.

Hojem, J. P. M., Cook, N. G. W., and Heins, C. (1975). A stiff, two meganewton testing machine for measuring the 'work-softening' behaviour of brittle materials, *South Afr. Mech. Eng.*, 25, 250–70.

Hondros, G. (1959). The evaluation of Poisson's ratio and the modulus of materials of a low tensile resistance by the Brazilian (indirect tensile) test with particular reference to concrete, *Austral. J. Appl. Sci.*, 10, 243–64.

Hooker, V. E. and Johnson, C. F. (1969). Near surface horizontal stresses including the effects of rock anisotropy, *U. S. Bureau Mines Rep. Invest. 7224*.

Hooker, V. E., Aggson, J. R., and Bickel, D. L. (1974). Improvements in the three component borehole deformation gage and overcoring techniques, *U. S. Bureau Mines Rep. Invest. 7894*.

Hopkins, D. L. (1990). *The Effect of Surface Roughness on Joint Stiffness, Aperture and Acoustic Wave Propagation*, Ph.D. diss., University of California, Berkeley.

Hopkins, D. L. (2000). The implications of joint deformation in analyzing the properties and behavior of fractured rock masses, underground excavations, and faults, *Int. J. Rock Mech.*, 37, 175–202.

Horn, H. M. and Deere, D. U. (1962). Frictional characteristics of minerals, *Géotechnique*, 12, 319–35.

Hoskins, E. (1967). Investigation of strain relief methods of measuring rock stress, *Int. J. Rock Mech.*, 4, 155–64.

Hoskins, E. R., Jaeger, J. C., and Rosengren, K. J. (1968). A medium scale direct friction experiment, *Int. J. Rock Mech.*, 5, 143–54.

Howland, R. C. (1935). Stresses in a plate containing an infinite row of holes, *Proc. Roy. Soc. Lond. Ser. A*, 148, 471–91.

Hubbert, M. K. (1951). Mechanical basis for certain familiar geological structures, *Bull. Geol. Soc. Am.*, 62, 355–372.

Hubbert, M. K. and Rubey, W. W. (1959, 1960, 1961). Role of fluid pressure in mechanics of overthrust faulting, *Bull. Geol. Soc. Am.*, 70, 111–205; 71, 617–28; 72, 1581–94.

Hubbert, M. K. and Willis, D. G. (1957). Mechanics of hydraulic fracturing, *Petrol. Trans. AIME*, 210, 153–68.

Huc, M. and Main, I. G. (2003). Anomalous stress diffusion in earthquake triggering: Correlation length, time dependence, and directionality, *J. Geophys. Res. Solid Earth*, 108(B7), art. 2324.

Hudson, J. A. and Harrison, J. P. (1997). *Engineering Rock Mechanics: An Introduction to the Principles*, Pergamon, Oxford.

Hudson, J. A., Crouch, S. L., and Fairhurst, C. (1972). Soft, stiff and servo-controlled testing machines: a review with reference to rock failure, *Eng. Geol.*, 6, 155–89.

Hunsche, U. and Albrecht, H. (1990). Results of true triaxial strength tests on rock salt, *Eng. Fract. Mech.*, 35, 867–77.

Inglis, C. E. (1913). Stresses in a plate due to the presence of cracks and sharp corners, *Trans. Inst. Nav. Archit.*, 55, 219–30.

Irwin, G. R. (1958). Fracture, in *Handbuch der Physik*, Vol. 6, S. Flügge, Springer-Verlag, Berlin, pp. 551–90.

Jaeger, J. C. (1943). The effect of absorption of water on the mechanical properties of sandstones, *J. Inst. Eng. Austral.*, 15, 164–6.

Jaeger, J. C. (1959). The frictional properties of joints in rock, *Geof. Pura Appl.*, 43, 148–58.

Jaeger, J. C. (1960). Shear failure of anisotropic rocks, *Geol. Mag.*, 97, 65–72.

Jaeger, J. C. (1961). The cooling of irregularly shaped igneous bodies, *Am. J. Sci.*, 259, 721–34.

Jaeger, J. C. (1963). Extension failures in rocks subject to fluid pressure, *J. Geophys. Res.*, 68, 6066–7.

Jaeger, J. C. (1967). Failure of rocks under tensile conditions, *Int. J. Rock Mech.*, 4, 219–27.

Jaeger, J. C. (1971). Friction of rocks and the stability of rock slopes, *Géotechnique*, 21, 97–134.

Jaeger, J. C. and Cook, N. G. W. (1964). Theory and application of curved jacks for measurement of stresses, in *State of Stress in the Earth's Crust*, W. R. Judd, ed., Elsevier, New York, pp. 381–95.

Jaeger, J. C. and Hoskins, E. R. (1966a). Rock failure under the confined Brazilian test, *J. Geophys. Res.*, 71, 2651–9.

Jaeger, J. C. and Hoskins, E. R. (1966b). Stresses and failure in rings of rock loaded in diametral tension or compression, *Brit. J. Appl. Phys.*, 17, 685–92.

Jasiuk, I. (1995). Cavities vis-a-vis rigid inclusions: Elastic moduli of materials with polygonal inclusions. *Int. J. Solids Struct*, 32, 407–22.

Jeffery, G. B. (1920). Plane stress and plane strain in bipolar coordinates, *Trans. Roy. Soc. Lond. Ser. A*, 221, 265–93.

Jeffreys, H. (1926). The reflection and refraction of elastic waves, *Monthly Notices Roy. Astron. Soc. Geophys. Suppl.*, 1, 321–34.

Jeffreys, H. (1958). A modification of Lomnitz's law of creep in rocks, *Geophys. J. Roy. Astron. Soc.*, 1, 321–34.

Jeffreys, H. (1992). *The Earth*, 4th ed., Cambridge University Press, Cambridge.

Johnson, A. M. and Fletcher, R. C. (1994). *Folding of Viscous Layers*, Columbia University Press, New York.

Jones, T. and Nur, A. (1983). Velocity and attenuation in sandstone at elevated temperatures and pressures, *Geophys. Res. Letts.*, 10, 140–3.

Judd, W. R. (1964). Rock stress, rock mechanics and research, in *State of Stress in the Earth's Crust*, W. R. Judd, ed., Elsevier, New York, pp. 5–51.

Jürgenson, L. (1934). The application of theories of elasticity and plasticity to foundation problems, *J. Boston Soc. Civ. Eng.*, 21, 148–242.

Kachanov, M. (1987). Elastic solids with many cracks: A simple method of analysis, *Int. J. Solids Struct.*, 23, 23–43.

Kachanov, M. (1994). Elastic solids with many cracks and related problems, *Adv. Appl. Mech.*, 30, 259–445.

Kanj, M. Y. and Abousleiman, Y. N. (2004). The generalized Lamé problem – Part I: Coupled poromechanical solutions, *J. Appl. Mech.*, 71, 168–79.

Kanninen, M. F. and Popelar, C. H. (1986). *Advanced Fracture Mechanics*, Oxford University Press, Oxford.

Karihaloo, B. L. and Viswanathan, K. (1985). Elastic field of a partially debonded elliptic inhomogeneity in an elastic matrix (plane-strain), *J. Appl. Mech.*, 52, 835–40.

Kassir, M. K. and Sih, G. C. (1975). *Mechanics of Fracture 2: Three-Dimensional Fracture Problems*, Noordhoff, Leyden.

Katsman, R., Aharonov, E., and Scher, H. (2005). Numerical simulation of compaction bands in high-porosity sedimentary rock, *Mech. Mater.*, 37, 143–62.

Kellogg, O. D. (1970). *Foundations of Potential Theory*, Frederick Ungar, New York.

Kendall, K. and Tabor, D. (1971). An ultrasonic study of the area of contact between stationary and sliding surfaces, *Proc. Roy. Soc. Lond. Ser. A*, 323, 321–40.

Kerr, A. D. and Pollard, D. D. (1998). Toward more realistic formulations for the analysis of laccoliths, *J. Struct. Geol.*, 20, 1783–93.

Kieffer, S. W. and Simonds, C. H. (1980). The roles of volatiles and lithology in the impact cratering process, *Rev. Geophys. Space Phys.*, 18, 143–81.

King, L. V. (1912). On the limiting strength of rocks under conditions of stress existing in the earth's interior, *J. Geol.*, 20, 119–38.

Kirchhoff, G. (1850). Uber das Gleichgewicht und die Bewegung einer elastischen Scheibe [On the equilibrium and motion of an elastic disk], *J. Reine Angew. Math.*, 40, 51–88.

Knopf, E. B. (1957). Petrofabrics in structural geology, *Colo. Sch. Mines Q.*, 52, 99–111.

Knopoff, L. (1958). Energy release in earthquakes, *Geophys. J. Roy. Astron. Soc.*, 1, 44–52.

Knops, R. J. (1958). On the variation of Poisson's ratio in the solution of elastic problems, *Q. J. Mech. Appl. Math.*, 11, 326–50.

Knott, C. G. (1899). Reflection and refraction of elastic waves with seismological applications, *Phil. Mag.* [5], 48, 64–97.

Koeberl, C. and Henkel, H., eds. (2005). *Impact Tectonics*, Springer, Berlin and New York.

Kolosov, G. V. (1909). *On the Application of the Theory of Functions of a Complex Variable to a Plane problem in the Mathematical Theory of Elasticity*, Ph.D. diss., Dorpat University (in Russian).

Kolsky, H. (1963). *Stress Waves in Solids*, Dover, New York.

Kouris, D. A., Tsuchida, E., and Mura, T. (1986). An anomaly of sliding inclusions, *J. Appl. Mech.*, 53, 724–26.

Krim, J. (1996). Atomic-scale origins of friction, *Langmuir*, 12, 4564–6.

Kumar, S., Zimmerman, R. W., and Bodvarsson, G. S. (1991). Permeability of a fracture with cylindrical asperities, *Fluid Dyn. Res.*, 7, 131–7.

Kümpel, H.-J. (1991). Poroelasticity: parameters reviewed, *Geophys. J. Int.*, 105, 783–99.

Kuske, A. and Robertson, G. (1974). *Photoelastic Stress Analysis*, John Wiley & Sons, London.

Kuster, G. T. and Toksöz, M. N. (1974). Velocity and attenuation of seismic waves in two-phase media: Part I. Theoretical formulations, *Geophysics*, 39, 587–606.

Labuz, J. F. and Bridell, J. M. (1993). Reducing frictional constraint in compression testing through lubrication, *Int. J. Rock Mech.*, 30, 451–55.

Lacazette, A. (1990). Application of linear elastic fracture-mechanics to the quantitative-evaluation of fluid-inclusion decrepitation, *Geology*, 18, 782–5.

Lachenbruch, A. H. (1961). Depth and spacing of tension cracks, *J. Geophys. Res.*, 66, 4273–92.

Ladanyi, B. (1967). Expansion of cavities in brittle media, *Int. J. Rock Mech.*, 4, 301–28.

Lakes, R. S. (1987). Foam structures with a negative Poisson's ratio, *Science*, 235, 1038–40.

Lakes, R. S. (1999). *Viscoelastic Solids*, CRC Press, Boca Raton, FL.

Lanaro, F. (2000). A random field model for surface roughness and aperture of rock fractures, *Int. J. Rock Mech.*, 37, 1195–210.

Landau, L. D. and Lifshitz, E. M. (1960). *Electrodynamics of Continuous Media*, Pergamon, New York.

Lang, S. (1971). *Linear Algebra*, 2nd ed., Addison-Wesley, Reading, MA.

Lang, S. (1973). *Calculus of Several Variables*, Addison-Wesley, Reading, MA.

Lardner, R. W. (1974). *Mathematical Theory of Dislocations and Fracture*, University of Toronto Press, Toronto.

Laurent, J., Bouteca, M. J., Sarda, J.-P., and Bary, D. (1993). Pore-pressure influence in the poroelastic behavior of rocks: experimental studies and results, *SPE Form. Eval.*, 8, 117–22.

Lavrov, A. and Vervoort, A. (2002). Theoretical treatment of tangential loading effects on the Brazilian test stress distribution, *Int. J. Rock Mech.*, 39, 275–83.

Lawn, B. R. (1993). *Fracture of Brittle Solids*, 2nd ed., Cambridge University Press, Cambridge and New York.

Lawn, B. R. and Wilshaw, T. R. (1975). *Fracture of Brittle Solids*, 1st ed., Cambridge University Press, Cambridge and New York.

Le Comte, P. (1965). Creep in rock salt, *J. Geol.*, 73, 469–84.

Lee, C.-H. and Farmer, I. (1993). *Fluid Flow in Discontinuous Rocks*, Chapman & Hall, London.

Lee, D. H., Juang, C. H., Chen, J. W., Lin, H. M., and Shieh, W. H. (1999). Stress paths and mechanical behavior of a sandstone in hollow cylinder tests, *Int. J. Rock Mech.*, 36, 857–70.

Lee, M. Y. and Haimson, B. C. (1989). Statistical evaluation of hydraulic fracturing stress measurement parameters, *Int. J. Rock Mech.*, 26, 447–56.

Lee, S. D. and Harrison, J. P. (2001). Empirical parameters for non-linear fracture stiffness from numerical experiments of fracture closure, *Int. J. Rock Mech.*, 38, 721–7.

Leeman, E. R. (1964). The measurement of stress in rock – Parts I, II, and III, *J. South Afr. Inst. Min. Metall.*, 65, 45–114, 254–84.

Leeman, E. R. (1967). The borehole deformation type of rock stress measuring instrument, *Int. J. Rock Mech.*, 4, 23–44.

Leeman, E. R. (1971). The CSIR doorstopper and triaxial rock stress measuring instruments, *Rock Mech.*, 3, 25–50.

Lévy, M. (1871). Mémoire sur les équations différentielles générales des mouvements internes aux corps solides [Memoir on the development of the differential equations of the internal motion in solid bodies], *Compt. Rend.*, 70, 473–80.

Li, L. and Cornet, F. H. (2004). Three dimensional consideration of flat jack tests, *Int. J. Rock Mech.*, 41, paper 1B-10, 403–4.

Ling, C. B. (1948). On the stresses in a plate containing two circular holes, *J. Appl. Phys.*, 19, 77–82.

Liu, E. R., Hudson, J. A., and Pointer, T. (2000). Equivalent medium representation of fractured rock, *J. Geophys. Res.*, 105, 2981–3000.

Lo, K. Y. (1978). Regional distribution of in-situ horizontal stresses in rocks in southern Ontario, *Can. Geotech. J.*, 15, 371–81.

Lockner, D. A. and Madden, T. R. (1991). A multiple-crack model of brittle fracture, 1, non-time-dependent simulations, *J. Geophys. Res.*, 96, 19, 623–42.

Lockner, D. A. and Stanchits, S. A. (2002). Undrained poroelastic response of sandstones to deviatoric stress change, *J. Geophys. Res.*, 107, paper 2353.

Lockner, D. E. (1995). Rock failure, in *Rock Physics and Phase Relations – A Handbook of Physical Constants*, T. J. Ahrens, ed., American Geophysical Union, Washington, D. C., pp. 127–47.

Lomnitz, C. (1956). Creep measurements in igneous rocks, *J. Geol.*, 64, 473–9.

Long, J. C. S. and Witherspoon, P. A. (1985). The relationship of the degree of interconnection to permeability in fracture networks, *J. Geophys. Res.*, 95, 3087–97.

Long, J. C. S., Remer, J. S., Wilson, C. R., and Witherspoon, P. A. (1982). Porous media equivalents for networks of discontinuous fractures, *Water Resour. Res.*, 18, 645–58.

Love, A. E. H. (1911). *Some Problems in Geodynamics*, Cambridge University Press, Cambridge.

Love, A. E. H. (1927). *A Treatise on the Mathematical Theory of Elasticity*, Dover, New York.

Lowell, R. P. (1990). Thermoelasticity and the formation of black smokers, *Geophys. Res. Letts.*, 17, 709–12.

Lubarda, V. and Markenscoff, X. (1998). On the absence of Eshelby property for non-ellipsoidal inclusions, *Int. J. Solids Struct.*, 35, 3405–11.

Ludwik, P. (1909). *Elemente der Teknologischen Mechanik* [Elements of Engineering Mechanics], Springer, Berlin.

Lutz, M. P. and Zimmerman, R. W. (1996). Effect of the interphase zone on the bulk modulus of a particulate composite, *J. Appl. Mech.*, 63, 855–61.

Madden, T. R. (1983). Microcrack connectivity in rocks: a renormalization group approach to the critical phenomena of conduction and failure in crystalline rocks, *J. Geophys. Res.*, 88, 585–92.

Mal, A. K. and Singh, S. J. (1991). *Deformation of Elastic Solids*, Prentice-Hall, Englewood Cliffs, NJ.

Malvern, L. E. (1969). *Introduction to the Mechanics of a Continuous Medium*, Prentice-Hall, Englewood Cliffs, NJ.

Mandel, J. (1953). Consolidation des sols – étude mathématique [Mathematical study of the consolidation of soils], *Géotechnique*, 3, 287–99.

Mandl, G. (2000). *Faulting in Brittle Rocks*, Springer-Verlag, Berlin.

Markov, K. Z. (2000). Elementary micromechanics of heterogeneous media, in *Heterogeneous Media*, K. Markov and L. Preziosi, eds., Birkhäuser, Basel and Boston, pp. 1–162.

Martel, S. J. and Muller, J. R. (2000). A two-dimensional boundary element method for calculating elastic gravitational stresses in slopes, *Pure Appl. Geophys.*, 157, 989–1007.

Matsushima, S. (1960). On the flow and fracture of igneous rocks, *Bull. Disast. Prevent. Inst. Kyoto Univ.*, 36, 1–9.

Matthews, C. S. and Russell, D. G. (1967). *Pressure Buildup and Flow Tests in Wells*, Society of Petroleum Engineers, Dallas.

Maugis, D. (1992). Stresses and displacements around cracks and elliptic cavities – exact solutions, *Eng. Fract. Mech.*, 43, 217–55.

Maugis, D. (2000). *Contact, Adhesion, and Rupture of Elastic Solids*, Springer-Verlag, Berlin.

Mavko, G. (1979). Frictional attenuation: an inherent amplitude dependence, *J. Geophys. Res.*, 84, 4769–75.

Mavko, G. (1980). Velocity and attenuation in partially molten rocks, *J. Geophys. Res.*, 85, 5173–89.

Mavko, G. and Jizba, D. (1991). Estimating grain-scale fluid effects on velocity dispersion in rocks, *Geophysics*, 56, 1940–9.

Mavko, G. and Nur, A. (1975). Melt squirt in the asthenosphere, *J. Geophys. Res.*, 80, 1444–8.

Mavko, G., Mukerji, T., and Dvorkin, J. (1998). *The Rock Physics Handbook*, Cambridge University Press, Cambridge.

Maxwell, J. C. (1870). On reciprocal figures, frames, and diagrams of force, *Trans. Roy. Soc. Edin.*, 26, 1–40.

Mayer, A., Habib, P., and Marchand, R. (1951). Mesure des pressions sur le terrain [In situ measurement of ground stresses], in Conférence Internationale sur les Pressions de Terrain et le Soutènement dans les Chantiers d'Exploitation, *Annales des Mines de Belgique, Institut National de l'Industrie Charbonnière*, Liege, pp. 221–24.

McClintock, F. A. and Walsh, J. B. (1962). Friction on Griffith cracks under pressure, in *Proc. 4th U. S. National Congress of Applied Mechanics*, pp. 1015–21.

McCutchen, W. R. (1982). Some elements of a theory of in situ stress, *Int. J. Rock Mech.*, 19, 201–03.

McDonal, F. J., Angona, F. A., Mills, R. L., Sengbush, R. L., van Nostrand, R. G., and White, J. E. (1958). Attenuation of shear and compressional waves in Pierre shale, *Geophysics*, 23, 421–39.

McGarr, A. and Gay, N. C. (1978). State of stress in the Earth's crust, *Ann. Rev. Earth Planet. Sci.*, 6, 403–36.

McLain, W. C. (1968). Rock mechanics in the disposal of radioactive wastes by hydraulic fracturing, *Rock Mech. Eng. Geol.*, 6, 131–61.

McLaughlin, R. (1977). A study of the differential scheme for composite materials, *Int. J. Eng. Sci.*, 15, 237–44.

McLellan, A. G. (1980). *The Classical Thermodynamics of Deformable Materials*, Cambridge University Press, Cambridge.

McNamee, J. and Gibson, R. E. (1960a). Displacement functions and linear transforms applied to diffusion through porous media, *Q. J. Mech. Appl. Math.*, 13, 98–111.

McNamee, J. and Gibson, R. E. (1960b). Plane strain and axially symmetric problems of the consolidation of a semi-infinite clay stratum, *Q. J. Mech. Appl. Math.*, 13, 210–27.

McTigue, D. F. (1986). Thermoelastic response of fluid-saturated porous rock, *J. Geophys. Res.*, 91, 9533–42.

McTigue, D. F. (1987). Elastic stress and deformation near a finite spherical magma body: resolution of the point source paradox, *J. Geophys. Res.*, 92, 12, 931–40.

Means, W. D. (1976). *Stress and Strain: Main Concepts of Continuum Mechanics for Geologists*, Springer-Verlag, New York.

Means, W. D. and Paterson, M. S. (1966). Experiments on preferred orientation of platy minerals, *Contrib. Min. Petrol.*, 13, 108–33.

Mehrabadi, M. M. and Cowin, S. C. (1990). Eigentensors of linear anisotropic elastic materials, *Q. J. Mech. Appl. Math.*, 43, 15–41.

Mendelson, A. (1968). *Plasticity: Theory and Application*, Macmillan, New York.

Mesri, G., Adachi, K., and Ullrick, C. R. (1976). Pore pressure response in rock to undrained change in all round stress, *Géotechnique*, 26, 317–30.

Michelitsch, T. and Wunderlin, A. (1996). Stress functions and internal-stresses in linear three-dimensional anisotropic elasticity, *Zeit. Phys. B Cond. Matter.*, 100, 53–6.

Michell, J. H. (1900). Elementary distribution of plane stress, *Proc. Lond. Math. Sci.*, 34, 134–42.

Michell, J. H. (1902). The inversion of plane stress, *Proc. Lond. Math. Sci.*, 32, 35–61.

Michelson, A. A. (1917). The laws of elastico-viscous flow, *J. Geol.*, 24, 405–10.

Miklowitz, J. (1978). *The Theory of Elastic Waves and Waveguides*, North-Holland, Amsterdam.

Milton, G. W. (2002). *The Theory of Composites*, Cambridge University Press, Cambridge and New York.

Min, K.-B. and Jing, L. (2003). Numerical determination of the equivalent elastic compliance tensor for fractured rock masses using the distinct element method, *Int. J. Rock Mech.*, 40, 795–816.

Min, K.-B., Rutqvist, J., Tsang, C.-F., and Jing, L. (2004). Stress-dependent permeability of fractured rock masses: A numerical study, *Int. J. Rock Mech.*, 41, 1191–1210.

Mindlin, R. D. (1936). Force at a point in the interior of a semi-infinite solid, *Physics*, 7, 195–202.

Mindlin, R. D. (1948). Stress distribution around a hole near the edge of a plate under tension, *Proc. Soc. Exp. Stress Anal.*, 5, 56–67.

Mochizuki, S. (1982). Attenuation in partially saturated rocks, *J. Geophys. Res.*, 87, 8598–604.

Mogi, K. (1965). Deformation and fracture of rocks under confining pressure (2). Elasticity and plasticity of some rocks, *Bull. Earthquake Res. Inst, Tokyo Univ.*, 43, 349–79.

Mogi, K. (1966). Some precise measurements of fracture stress of rocks under uniform compressive stress, *Rock Mech. Eng. Geol.*, 4, 41–55.

Mogi, K. (1971). Fracture and flow of rocks under high triaxial compression, *J. Geophys. Res.*, 76, 1255–69.

Mohr, O. (1900). Welche Umstände bedingen die Elastizitätsgrenze und den Bruch eines Materials? [What are the conditions for the elastic limit and the fracturing of a material?], *Z. Ver. dt. Ing.*, 44, 1524–30, 1572–7.

Mohr, O. (1914). *Abhandlungen aus dem Gebiete der Technische Mechanik* [Treatise on Topics in Engineering Mechanics], 2nd ed., Ernst und Sohn, Berlin.

Moody, J. D. and Hill, M. J. (1956). Wrench fault tectonics, *Bull. Geol. Soc. Am.*, 67, 1207–46.

Mora, A. and Place, D. (1994). Simulation of the frictional stick-slip instability, *Pure Appl. Geophys.*, 143, 61–87.

Morera, G. (1892). Soluzione generale delle equazioni indefinite dell'equilibrio di un corpo continuo [General indefinite solution of the equations of equilibrium of a continuous body], *Atti Reale Acad. Naz. Lincei*, 5, 137–41.

Morland, L. W. (1976). Elastic anisotropy of regularly jointed media, *Rock Mech.*, 8, 35–48.

Morlier, P. (1971). Description de l'état de fissurisation d'une roche à partir d'essais simples et non-destructifs [Description of the state of rock fracturization through simple non-destructive tests], *Rock Mech.*, 3, 125–38.

Mourzenko, V. V., Thovert, J. F., and Adler, P. M. (1995). Permeability of a single fracture – validity of the Reynolds equation, *J. Phys. II*, 5, 465–82.

Munson, D. E. (1997). Constitutive model of creep in rock salt applied to underground room closure, *Int. J. Rock Mech.*, 34, 233–47.

Mura, T. (1987). *Micromechanics of Defects in Solids*, 2nd ed., M. Nijhoff, Dordrecht.

Mura, T., Jasiuk, I., and Tsuchida, B. (1985). The stress-field of a sliding inclusion, *Int. J. Solids Struct.*, 21, 1165–79.

Murphy, W. F. (1984). Acoustic measurements of partial gas saturation in tight sandstones, *J. Geophys. Res.*, 89, 11,549–59.

Murphy, W. M., Winkler, K. W., and Kleinberg, R. L. (1984). Frame modulus reduction in sedimentary rocks: The effect of adsorption on grain contacts, *Geophys. Res. Lett.*, 11, 805–8.

Murrell, S. A. (1964). Theory of the propagation of elliptical Griffith cracks under various conditions of plane stress or plane strain, *Brit. J. Appl. Phys.*, 15, 1195–223.

Murrell, S. A. (1965). The effect of triaxial stress systems on the strength of rocks at atmospheric temperatures, *Geophys. J. Roy. Astron. Soc.*, 10, 231–81.

Murrell, S. A. F. (1963). A criterion for brittle fracture of rocks and concrete under triaxial stress and the effect of pore pressure on the criterion, in *Proc. 5th Rock Mech. Symp.*, C. Fairhurst, ed., Pergamon Press, Oxford, pp. 563–77.

Murrell, S. A. F. and Chakravarty, S. (1973). Some new rheological experiments on igneous rocks at temperatures up to 1120°C, *Geophys. J. Roy. Astron. Soc.*, 34, 211–50.

Murrell, S. A. F. and Ismail, I. A. H. (1976). The effect of decomposition of hydrous minerals on the mechanical properties of rocks at high pressures and temperatures, *Tectonophysics*, 31, 207–58.

Muskat, M. and Meres, M. W. (1940). Reflection and transmission coefficients for plane waves in elastic media, *Geophysics*, 5, 115–48.

Muskhelishvili, N. I. (1963). *Some Basic Problems of the Mathematical Theory of Elasticity*, 4th ed., Noordhoff, Groningen.

Myer, L. R. (2000). Fractures as collections of cracks, *Int. J. Rock Mech.*, 37, 231–43.

Nadai, A. (1938). The influence of time upon creep, the hyperbolic creep law, in *Stephen Timoshenko 60th Anniversary Volume*, J. M. Lessels, Macmillan, New York, pp. 155–70.

Nadai, A. (1950). *Theory of Flow and Fracture of Solids*, Vol. 1, 2nd ed., McGraw-Hill, New York.

Natale, G., Salusti, E., and Troisi, A. (1998). Rock deformation and fracturing processes due to nonlinear shock waves propagating in hyperthermal fluid pressurized domains, *J. Geophys. Res.*, 103, 15, 325–38.

Nehari, Z. (1961). *Introduction to Complex Analysis*, Allyn and Bacon, Boston.

Nelson, R. A. (1985). *Geologic Analysis of Naturally Fractured Reservoirs*, Gulf Publishing, Houston.

Nemat-Nasser, S. and Hori, M. (1993). *Micromechanics: Overall Properties of Heterogeneous Materials*, North-Holland, Amsterdam.

Neuzil, C. E. and Tracy, J. V. (1981). Flow through fractures, *Water Resour. Res.*, 17, 191–9.

Newman, G. H. (1973). Pore volume compressibility of consolidated, friable, and unconsolidated reservoir rocks under hydrostatic loading, *J. Petrol. Technol.*, 25, 129–34.

Nicholl, M. J., Rajaram, H., Glass, R. J., and Detwiler, R. (1999). Saturated flow in a single fracture: Evaluation of the Reynolds equation in measured aperture fields, *Water Resour. Res.*, 35, 3361–73.

Nishihara, M. (1958). Stress-strain-time relations of rocks, *Doshisha Eng. Rev.*, 8, 32–55, 85–115.

Nkemzi, D. (1997). A new formula for the velocity of Rayleigh waves, *Wave Motion*, 26, 199–205.

Norris, A. (1985). A differential scheme for the effective moduli of composites, *Mech. Mater.*, 4, 1–16.

Norris, A. (1992). On the correspondence between poroelasticity and thermoelasticity, *J. Appl. Phys.*, 71, 1138–41.

Nowacki, W. (1978). *Stress Waves in Non-Elastic Solids*, Pergamon, Oxford.

Nowacki, W. (1986). *Thermoelasticity*, 2nd ed., Pergamon, Oxford.

Nur, A. and Byerlee, J. D. (1971). An exact effective stress law for elastic deformation of rock with fluids, *J. Geophys. Res.*, 76, 6414–9.

Nye, J. F. (1951). The flow of glaciers and ice sheets as a problem in plasticity, *Proc. Roy. Soc. Lond. Ser. A.*, 207, 554–72.

Nye, J. F. (1957). *Physical Properties of Crystals*, Clarendon Press, Oxford.

Obert, L. and Stephenson, D. E. (1965). Stress conditions under which core discing occurs, *Soc. Min. Eng. Trans.*, 232, 227–34.

Obert, L., Merrill, R. H., and Morgan, T. A. (1962). Borehole deformation gauge for determining the stress in mine rock, *U. S. Bureau Mines Rep. Invest. 5978.*

O'Connell, R. J. and Budiansky, B. (1974). Seismic velocities in dry and saturated cracked solids, *J. Geophys. Res.*, 79, 5412–26.

O'Donnell, T. P. and Steif, P. S. (1989). Elastic-plastic compaction of a two-dimensional assemblage of particles, *J. Eng. Mater. Tech.*, 111, 404–8.

Oertel, G. F. (1996). *Stress and Deformation: A Handbook on Tensors in Geology*, Oxford University Press, New York.

Ojala, I. O., Ngwenya, B. T., Main, I. G., and Elphick, S. C. (2003). Correlation of microseismic and chemical properties of brittle deformation in Locharbriggs sandstone, *J. Geophys. Res.*, 79, 2268.

Okubo, S. and Nishimatsu, Y. (1985). Uniaxial compression testing using a linear combination of stress and strain as the controlled variable, *Int. J. Rock. Mech.*, 22, 323–30.

Olsson, R. and Barton, N. (2001). An improved model for hydromechanical coupling during shearing of rock joints, *Int. J. Rock Mech.*, 38, 317–29.

Olsson, W. A. and Brown, S. R. (1993). Hydromechanical response of a fracture undergoing compression and shear, *Int. J. Rock Mech.*, 30, 845–51.

Oron, A. P. and Berkowitz, B. (1998). Flow in rock fractures: The local cubic law assumption re-examined, *Water Resour. Res.*, 34, 2811–24.

Ott, H. (1942). Reflexion und Brechung von Kugelwellen: Effekte 2: Ordnung [Reflection and refraction of spherical waves], *Ann. Physik.*, 41, 443–66.

Paillet, F. L. and Cheng, C. H. (1991). *Acoustic Waves in Boreholes*, CRC Press, Boca Raton, FL.

Palciauskas, V. V. and Domenico, P. A. (1989). Fluid pressures in deforming porous rocks, *Water Resour. Res.*, 25, 203–13.

Pande, G. N., Beer, G., and Williams, J. R. (1990). *Numerical Methods in Rock Mechanics*, John Wiley & Sons, New York.

Panek, L. A. and Stock, J. A. (1964). Development of a rock stress monitoring station based on the flat slot method of measuring existing rock stress, *U. S. Bureau Mines Rep. Invest. 6537.*

Parate, N. S. (1973). The influence of water on the strength of limestones, *Trans. AIME*, 254, 127–31.

Parry, R. H. G. (1995). *Mohr Circles, Stress Paths and Geotechnics*, E&FN Spon, London.

Parton, V. Z. and Morozov, E. M. (1978). *Elastic-Plastic Fracture Mechanics*, Mir Publishers, Moscow.

Paterson, M. S. (1964). Triaxial testing of materials at pressures up to 10,000 kg/sq. cm., *J. Inst. Engrs. Austral.*, Jan-Feb, 23–9.

Paterson, M. S. (1978). *Experimental Rock Deformation – The Brittle Field*, Springer-Verlag, Berlin and New York.

Paterson, M. S. and Olgaard, D. L. (2000). Rock deformation tests to large shear strains in torsion, *J. Struct. Geol.*, 22, 1341–58.

Paterson, M. S. and Weiss, L. E. (1966). Experimental deformation and folding in phyllite, *Bull. Geol. Soc. Am.*, 77, 343–74.

Paterson, M. S. and Wong, T.-F. (2005). *Experimental Rock Deformation – The Brittle Field*, 2nd ed., Springer-Verlag, Berlin and New York.

Patir, N. and Cheng, H. S. (1978). An average flow model for determining effects of three-dimensional roughness on partial hydrodynamic lubrication, *J. Lubr. Tech.*, 100, 12–17.

Patton, F. D. (1966). Multiple modes of shear failure in rock, in *Proc. 1st Int. Cong. Rock Mech.*, M. Rocha, ed., Int. Soc. Rock Mech., Lisbon, pp. 509–13.

Paul, B. (1961). Modification of the Mohr-Coulomb theory of fracture, *J. Appl. Mech.*, 28, 259–68.

Pearson, C. E. (1959). *Theoretical Elasticity*, Harvard University Press, Cambridge, MA.

Peck, L. (1983). Stress corrosion and crack propagation in Sioux quartzite, *J. Geophys. Res.*, 88, 5037–46.

Peltier, B. and Atkinson, C. (1987). Dynamic pore pressure ahead of the bit, *SPE Drill. Eng.*, 2, 351–8.

Penman, A. D. M. (1953). Shear characteristics of a saturated silt measured in triaxial compression, *Géotechnique*, 312–28.

Peselnick, L. and Zeitz, I. (1959). Internal friction of fine-grained limestones at ultrasonic frequencies, *Geophysics*, 24, 285–96.

Pettijohn, F. J., Potter, P. E., and Siever, R. (1987). *Sand and Sandstone*, 2nd ed., Springer-Verlag, Berlin and New York.

Phillips, D. W. (1931). The nature and physical properties of some coal-measure strata, *Trans. Inst. Min. Eng.*, 80, 212–42.

Phillips, O. M. (1991). *Flow and Reactions in Permeable Rocks*, Cambridge University Press, Cambridge and New York.

Pickering, D. J. (1970). Anisotropic elastic parameters for soils, *Géotechnique*, 20, 271–6.

Pickett, G. (1944). Application of the Fourier method to the solution of certain boundary problems in the theory of elasticity, *J. Appl. Mech.*, 66, A176–82.

Piggott, A. R. and Elsworth, D. (1992). Analytical models for flow through obstructed domains, *J. Geophys. Res.*, 97, 2085–93.

Pincus, H. J. (2000). Closed-form / least-squares failure envelopes for rock strength, *Int. J. Rock Mech.*, 37, 763–85.

Pollard, D. D. (1973a). Equations for stress and displacement fields around pressurized elliptical holes in elastic solids, *Math. Geol.*, 5, 11–25.

Pollard, D. D. (1973b). Derivation and evaluation of a mechanical model for sheet intrusions, *Tectonophysics*, 19, 233–69.

Pollard, D. D. and Aydin, A. (1988). Progress in understanding jointing over the past century, *Geol. Soc. Am. Bull.*, 100, 1181–1204.

Pollard, D. D. and Fletcher, R. C. (2005). *Fundamentals of Structural Geology*, Cambridge University Press, Cambridge and New York.

Pollard, D. D. and Holzhausen, G. (1979). On the mechanical interaction between a fluid-filled fracture and the Earth's surface, *Tectonophysics*, 53, 27–57.

Pomeroy, C. D. (1956). Creep in coal at room temperature, *Nature*, 178, 279–80.

Pomeroy, C. D. and Morgans, W. T. A. (1956). The tensile strength of coal, *Brit. J. Appl. Phys.*, 7, 243–6.

Poulos, H. G. and Davis, E. H. (1974). *Elasticity Solutions for Soil and Rock Mechanics*, John Wiley & Sons, New York.

Power, W. L. and Tullis, T. E. (1991). Euclidean and fractal models for the description of rock surface roughness, *J. Geophys. Res.*, 96, 415–24.

Prager, W. (1959). *An Introduction to Plasticity*, Addison-Wesley, Reading, MA and London.

Prandtl, L. (1925). Spannungsverteilung in plastischen Koerpern [Stress distribution in plastic bodies], *Proc. 1st Int. Cong. Appl. Math.*, C. B. Biezano and J. M. Burgers, eds., J. Waltman Jr, Delft, pp. 43–54.

Prasad, R. K. and Kejriwal, B. K. (1984). Effect of specimen size on the long-term strength of coal pillars, *Int. J. Mining Eng.*, 2, 355–8.

Price, N. J. (1960). The compressive strength of coal measure rocks, *Colliery Eng.*, 37, 283–92.

Price, N. J. (1962). The tectonics of Aberystwyth grits, *Geol. Mag.*, 99, 542–57.

Price, N. J. (1964). A study of the time-strain behaviour of coal-measure rocks, *Int. J. Rock Mech.*, 1, 277–303.

Price, N. J. (1966). *Fault and Joint Development in Brittle and Semi-brittle Rock*, Oxford University Press, Oxford.

Price, N. J. and Cosgrove, J. W. (1990). *Analysis of Geological Structures*, Cambridge University Press, Cambridge.

Prigogine, I. (1961). *Introduction to Thermodynamics of Irreversible Processes*, 2nd ed., Interscience, London.

Pruess, K. and Bodvarsson, G. S. (1984). Thermal effects of reinjection in geothermal reservoirs with major vertical fractures, *J. Petrol. Tech.*, 36, 1567–78.

Pyrak-Nolte, L. J. and Cook, N. G. W. (1987). Elastic interface waves across a fracture, *Geophys. Res. Letts.*, 14, 1107–10.

Pyrak-Nolte, L. J. and Morris, J. P. (2000). Single fractures under normal stress: The relation between fracture specific stiffness and fluid flow, *Int. J. Rock Mech.*, 37, 245–62.

Pyrak-Nolte, L. J., Myer, L. R., and Cook, N. G. W. (1990). Transmission of seismic waves across single natural fractures, *J. Geophys. Res.*, 95, 8617–38.

Pyrak-Nolte, L. J., Myer, L. R., Cook, N. G. W., and Witherspoon, P. A. (1987). Hydraulic and mechanical properties of natural fractures in low permeability rock, in *Proc. 6th Int. Cong. Rock Mech.*, G. Herget and S. Vongpaisal, eds., Balkema, Rotterdam, pp. 225–31.

Pyrak-Nolte, L. J., Xu, J. P., and Haley, G. M. (1992). Elastic interface waves propagating in a fracture, *Phys. Rev. Letts.*, 68, 3650–3.

Qu, J. M. (1993). The effect of slightly weakened interfaces on the overall elastic properties of composite materials, *Mech. Mater.*, 14, 269–81.

Rae, D. (1963). The measurement of the coefficient of friction of some rocks during continuous rubbing, *J. Sci. Instr.*, 40, 438–40.

Raghavan, R. and Miller, F. G. (1975). Mathematical analysis of sand compaction, in *Compaction of Coarse-Grained Sediments*, Vol. 1, G. V. Chilingarian and K. H. Wolf, eds., Elsevier, Amsterdam, pp. 403–524.

Rahman, M. and Barber, J. R. (1995). Exact expressions for the roots of the secular equation for Rayleigh waves, *J. Appl. Mech.*, 62, 250–2.

Rahn, W. (1984). Stress concentration factors for the interpretation of doorstopper stress measurements in anisotropic rocks, *Int. J. Rock Mech.*, 21, 313–26.

Raleigh, C. B. and Griggs, D. T. (1963). Effect of toe in the mechanics of overthrust faulting, *Bull. Geol. Soc. Am.*, 74, 819–30.

Raleigh, C. B. and Paterson, M. S. (1965). Experimental deformation of serpentinite and its tectonic implications, *J. Geophys. Res.*, 70, 3965–85.

Ramberg, H. (1964). Selective buckling of composite layers with contrasted rheological properties: a theory for simultaneous formation of several orders of folds, *Tectonophysics*, 1, 307–41.

Ramberg, H. (1967). *Gravity, Deformation and the Earth's Crust*, Academic Press, London and New York.

Ranalli, G. (1995). *Rheology of the Earth*, 2nd ed., Chapman & Hall, London.

Rayleigh, J. W. S. (1885). On waves propagated along the plane surface of an elastic solid, *Proc. London Math. Soc.*, 17, 4–11.

Rehbinder, G. (1985). Stresses and strains around a heated spherical cavity in an elastic medium, *Rock Mech.*, 18, 213–8.

Reiner, M. (1971). *Advanced Rheology*, H. K. Lewis, London.

Rekach, V. G. (1979). *Manual of the Theory of Elasticity*, Mir Publishers, Moscow.

Renshaw, C. E. (1995). On the relationship between mechanical and hydraulic apertures in rough-walled fractures, *J. Geophys. Res.*, 100, 24, 629–36.

Reuschle, T. (1998). A network approach to fracture: The effect of heterogeneity and loading conditions, *Pure Appl. Geophys.*, 152, 641–65.

Reuss, A. (1929). Berechnung der Fliessgrenze von Mischkrystallen auf Grund der Plastizitätsbedingung fur Einkristalle [Estimation of the yield surface of polycrystals based on the plastic behaviour of single crystals], *Zeit. Ang. Math. Mech.*, 9, 44–58.

Reuss, A. (1930). Beruecksichtigung der elastischen Formaenderungen in der Plastizitaet [Consideration of elastic deformation in plasticity theory], *Zeit. Ang. Math. Mech.*, 10, 266–74.

Reynolds, O. (1886). On the theory of lubrication, *Phil. Trans. Roy. Soc. Lond.*, 177, 157–234.

Rice, J. R. and Cleary, M. P. (1976). Some basic stress diffusion solutions for fluid saturated elastic porous media with compressible constituents, *Rev. Geophys. Space Phys.*, 14, 227–41.

Rice, J. R. and Ruina, A. L. (1983). Stability of steady frictional slipping, *J. Appl. Mech.*, 50, 343–9.

Rinehart, J. S. (1975). *Stress Transients in Solids*, Hyperdynamics, Santa Fe, NM.

Ripperger, E. A. and Davids, N. (1947). Critical stresses in a circular ring, *Trans. Am. Soc. Civ. Eng.*, 112, 619–35.

Robertson, E. C. (1955). Experimental study of the strength of rocks, *Bull. Geol. Soc. Am.*, 66, 1275–1314.

Robertson, E. C. (1964). Viscoelasticity of rocks, in *State of Stress in the Earth's Crust*, W. R. Judd, ed., Elsevier, New York, pp. 181–233.

Robin, P.-Y. F. (1973). Note on effective pressure, *J. Geophys. Res.*, 78, 2434–7.

Robinson, K. (1951). Elastic energy of an ellipsoidal inclusion in an infinite solid, *J. Appl. Phys.*, 22, 1045–54.

Roeloffs, E. A. and Rudnicki, J. W. (1984). Coupled deformation-diffusion effects on water-level changes due to propagating creep events, *Pure Appl. Geophys.*, 122, 560–82.

Rudnicki, J. W. and Hsu, T.-C. (1988). Pore pressure changes induced by slip on permeable and impermeable faults, *J. Geophys. Res.*, 93, 3275–85.

Ruina, A. L. (1983). Slip instability and state-variable friction laws, *J. Geophys. Res.*, 88, 10, 359–70.

Rummel, F. and Fairhurst, C. (1970). Determination of post-failure behaviour of brittle rock using a servo-controlled testing machine, *Rock Mech.*, 2, 189–204.

Rutqvist, J. and Stephansson, O. (2003). The role of hydromechanical coupling in fractured rock engineering, *Hydrogeol. J.*, 11, 7–40.

Rutter, E. H. (1972a). The effects of strain-rate changes on the strength and ductility of Solenhofen limestone at low temperatures and confining pressures, *Int. J. Rock. Mech.*, 9, 183–9.

Rutter, E. H. (1972b). The influence of interstitial water on the rheological behaviour of calcite rocks, *Tectonophysics*, 14, 13–33.

Ryder, J. A. and Officer, N. C. (1964). An elastic analysis of strata movement in the vicinity of inclined excavations, *J.South Afr. Inst. Min. Metall.*, 84, 219–44.

Sack, R. A. (1946). Extension of Griffith's theory of rupture to three dimensions, *Proc. Phys. Soc. London*, 58, 729–36.

Sadowsky, M. A. and Sternberg, E. (1949). Stress concentration around a triaxial ellipsoidal cavity, *J. Appl. Mech.*, 16, 149–57.

Saint Venant, B. (1870). Mémoire sur l'établissment des équations différentielles des mouvements internes opérés dans les corps solides ductiles au delà des limites d'élasticité [Memoir on the development of the differential equations of the internal motion in a ductile solid body beyond the elastic limit], *Compt. Rend.*, 70, 473–80.

Salamon, M. D. G. (1964). Elastic analysis of displacements and stresses induced by the mining of seam or reef deposits, Parts 1–3, *J. South Afr. Inst. Min. Metall.*, 64, 128–49, 197–218, 468–500.

Salamon, M. D. G. (1970). Stability, instability, and design of pillar workings, *Int. J. Rock Mech.*, 7, 613–31.

Salamon, M. D. G. (1984). Energy considerations in rock mechanics: fundamental results, *J. South Afr. Inst. Min. Metall.*, 84, 233–46.

Salganik, R. L. (1973). Mechanics of bodies with many cracks, *Mech. Solids*, 8, 135–43.

Sams, M. S., Neep, J. P., Worthington, M. H., and King, M. S., (1997). The measurement of velocity dispersion and frequency-dependent intrinsic attenuation in sedimentary rocks, *Geophysics*, 62, 1456–64.

Santarelli, F. J. and Brown, E. T. (1989). Failure of three sedimentary rocks in triaxial and hollow cylinder compression tests, *Int. J. Rock Mech.*, 26, 401–13.

Sato, H. and Fehler, M. C. (1998). *Seismic Wave Propagation and Scattering in the Heterogeneous Earth*, Springer-Verlag, Berlin and New York.

Savage, J. C. (1966). Thermoelastic attenuation of elastic waves by cracks, *J. Geophys. Res.*, 71, 3929–38.

Savage, J. C. (1969). Comment on Velocity and attenuation of seismic waves in imperfectly elastic rock by R. B. Gordon and L. A. Davis, *J. Geophys. Res.*, 74, 726–8.

Savage, J. C., Byerlee, J. D., and Lockner, D. A. (1996). Is internal friction friction? *Geophys. Res. Letts.*, 23, 487–90.

Savin, G. N. (1961). *Stress Concentration around Holes*, Pergamon, New York.

Sax, H. G. J. (1946). De tektoniek van het Carboon in het Zuid-Limburgsche mijngebied [Tectonics of the south-Limburg coalfields], *Meded. Geol. Stricht. Ser. C-I-I No. 3*.

Sayers, C. M. (1991). Fluid flow in a medium containing partially closed fractures, *Transp. Porous Media*, 6, 331–6.

Sayers, C. M., van Munster, J. G., and King, M. S. (1990). Stress-induced ultrasonic anisotropy in Berea sandstone, *Int. J. Rock Mech.*, 27, 429–36.

Scheidegger, A. E. (1960). On the connection between tectonic stresses and well fracturing data, *Geofis. Pur. Appl.*, 46, 66–76.

Schmitt, D. R. and Zoback, M. D. (1989). Poroelastic effects in the determination of the maximum horizontal principal stress in hydraulic fracturing tests – a proposed breakdown equation employing a modified effective stress relation for tensile failure, *Int. J. Rock Mech.*, 26, 499–506.

Schock, R. N., Heard, H. C., and Stephens, D. R. (1973). Stress-strain behaviour of a granodiorite and two graywackes on compression to 20 kilobars, *J. Geophys. Res.*, 78, 5922–41.

Schoenberg, M. (1980). Elastic wave behavior across linear slip interfaces, *J. Acoust. Soc. Am.*, 68, 1516–21.

Schoenberg, M. and Sayers, C. M. (1995). Seismic anisotropy of fractured rock, *Geophysics*, 60, 204–11.

Scholz, C. H. (1968). Microfracturing and the inelastic deformation of rock in compression, *J. Geophys. Res.*, 73, 1417–32.

Scholz, C. H. (1990). *The Mechanics of Earthquake Faulting*, Cambridge University Press, Cambridge and New York.

Scholz, C. H. and Engelder, J. T. (1976). The role of asperity indentation and ploughing in rock friction – I. Asperity creep and stick-slip, *Int. J. Rock Mech.*, 13, 149–54.

Scholz, C. H., Molnar, P., and Johnson, T. (1972). Detailed studies of frictional sliding in granite and implications for earthquake mechanism, *J. Geophys. Res.*, 77, 6392–406.

Schon, J. H. (1996). *Physical Properties of Rocks: Fundamentals and Principles of Petrophysics*, Pergamon, Oxford.

Secor, D. T. (1965). Role of fluid pressure in jointing, *Am. J. Sci.*, 263, 633–46.

Seeburger, D. A. and Nur, A. (1984). A pore space model for rock permeability and bulk modulus, *J. Geophys. Res.*, 89, 527–36.

Segall, P. (1989). Earthquakes triggered by fluid extraction, *Geology*, 17, 942–6.

Segedin, C. M. (1950). Note on a penny-shaped crack under shear, *Proc. Camb Phil. Soc.*, 47, 396–400.

Selberg, H. L. (1952). Transient compression waves from spherical and cylindrical cavities, *Arkiv Fysik*, 5, 97–108.

Selvadurai, A. P. S. (1982). The additional settlement of a rigid circular foundation on an isotropic elastic halfspace due to multiple distributed external loads, *Géotechnique*, 32, 1–7.

Sezawa, K. and Nishimura, G. (1931). Stresses under tension in a plate with a heterogeneous insertion, *Rep. Aero. Res. Inst. Tokyo*, 6, 25–43.

Shang, J. L., Shen, L. T., and Zhao, J. (2000). Hugoniot equation of state of the Bukit Timah granite, *Int. J. Rock Mech.*, 37, 705–13.

Sharan, S. K. (2003). Elastic-brittle-plastic analysis of circular openings in Hoek-Brown media, *Int. J. Rock Mech.*, 40, 817–24.

Sharpe, J. A. (1942). The production of elastic waves by explosive sources, *Geophysics*, 7, 144–54.

Sheorey, P. R. (1994). A theory for in-situ stresses in isotropic and transversely isotropic rock, *Int. J. Rock Mech.*, 31, 23–34.

Sheorey, P. R. (1997). *Empirical Rock Failure Criteria*, Balkema, Rotterdam.

Sheriff, R. E. and Geldart, L. P. (1995). *Exploration Seismology*, 2nd ed., Cambridge University Press, Cambridge and New York.

Silliman, S. E. (1989). Interpretation of the difference between aperture estimates derived from hydraulic and tracer tests in a single fracture, *Water Resour. Res.*, 25, 2275–83.

Silverman, R. A. (1967). *Introductory Complex Analysis*, Prentice-Hall, Englewood Cliffs, NJ.

Simmons, G. and Wang, H. F. (1971). *Single Crystal Elastic Constants and Calculated Aggregate Properties: A Handbook*, MIT Press, Cambridge, MA.

Sisavath, S., Al-Yaarubi, A., Pain, C. C., and Zimmerman, R. W. (2003). A simple model for deviations from the cubic law for a fracture undergoing dilation or closure, *Pure Appl. Geophys.*, 160, 1009–22.

Sisavath, S., Jing, X. D., and Zimmerman, R. W. (2000). Effect of stress on the hydraulic conductivity of rock pores, *Phys. Chem. Earth A*, 25, 163–8.

Sjöberg, J., Christiansson, R., and Hudson, J. A. (2003). ISRM suggested methods for rock stress estimation – Part 2: Overcoring methods, *Int. J. Rock Mech.*, 40, 999–1010.

Skempton, A. W. (1954). The pore pressure coefficients *A* and *B*, *Géotechnique*, 4, 143–47.

Skjetne, E., Hansen, A., and Gudmundsson, J. S. (1999). High-velocity flow in a rough fracture, *J. Fluid Mech.*, 383, 1–28.

Sneddon, I. N. (1946). The distribution of stress in the neighbourhood of a crack in an elastic solid, *Proc. Roy. Soc. Lond. Ser. A*, 187, 229–60.

Sneddon, I. N. and Lowengrub, M. (1969). *Crack Problems in the Classical Theory of Elasticity*, John Wiley & Sons, New York.

Snow, D. T. (1969). Anisotropic permeability of fractured media, *Water Resour. Res.*, 5, 1273–89.

Sokolnikoff, I. S. (1956). *Mathematical Theory of Elasticity*, McGraw-Hill, New York.

Somerton, W. H. (1992). *Thermal Properties and Temperature-Related Behavior of Rock/Fluid Systems*, Elsevier, Amsterdam.

Soutas-Little, R. W. (1999). *Elasticity*, Dover, New York.

Spencer, A. J. M. (1971). Theory of invariants, in *Continuum Physics*, Vol. 1, A. C. Eringen, ed., Academic Press, New York, pp. 239–353.

Spencer, A. J. M. (1980). *Continuum Mechanics*, Longman, London.

Spetzler, H. and Martin, R. J. (1974). Correlation of strain and velocity during dilatatancy, *Nature*, 252, 30–1.

Spry, A. (1961). The origin of columnar jointing, particularly in basalt flows, *J. Geol. Soc. Austral.*, 8, 191–216.

Starr, A. T. (1928). Slip in a crystal and rupture in a solid due to shear, *Proc. Camb. Phil. Soc.*, 24, 489–500.

Steketee, J. A. (1958). Some geophysical implications of the elasticity theory of dislocations, *Canad. J. Phys.*, 36, 1168–98.

Stephansson, O. (1995). Introduction to special issue on thermo-hydro-mechanical coupling in rock mechanics, *Int. J. Rock. Mech.*, 32, 387.

Stephens, G. and Voight, B. (1982). Hydraulic fracturing theory for conditions of thermal stress, *Int. J. Rock. Mech.*, 19, 279–84.

Sternberg, E. (1960). On the integration of the equations of motion in the classical theory of elasticity, *Arch. Rat. Mech. Anal.*, 6, 34–50.

Sternberg, E. and Sadowsky, M. A. (1952). On the axisymmetric problem of the theory of elasticity for an infinite region containing two spherical cavities, *J. Appl. Mech.*, 19, 19–27.

Stevenson, A. C. (1945). Complex potentials in two-dimensional elasticity, *Proc. Roy. Soc. Lond. Ser. A*, 184, 129–79.

Stoll, R. D. (1989). *Sediment Acoustics*, Springer-Verlag, Berlin.

Stoneley, R. (1924). Elastic waves at the surface of separation of two solids, *Proc. Roy. Soc. Lond.*, 106, 416–28.

Sun, R. J. (1969). Theoretical size of hydraulically induced horizontal fractures and corresponding surface uplift in an idealized medium, *J. Geophys. Res.*, 74, 5995–6011.

Swan, G. (1983). Determination of stiffness and other joint properties from roughness measurements, *Rock Mech. Rock Eng.*, 16, 19–38.

Tada, H., Paris, P. C., and Irwin, G. R. (2000). *The Stress Analysis of Cracks Handbook*, 3rd ed., Am. Soc. Mech. Eng., New York.

Talesnick, M. L. and Ringel, M. (1999). Completing the hollow cylinder methodology for testing of transversely isotropic rocks: Torsion testing, *Int. J. Rock Mech.*, 36, 627–39.

Tang, C., Liu, H., Lee, P. K. K., Tsui, Y., and Tham, L. G. (2000). Numerical studies of the influence of microstructure on rock failure in uniaxial compression–part I: Effect of heterogeneity, *Int. J. Rock Mech.*, 37, 555–69.

Terzaghi, K. (1936). The shearing resistance of saturated soils and the angle between planes of shear, in *Proc. Int. Conf. Soil Mech. Found. Eng.*, Vol. 1, Harvard University Press, Cambridge, MA, pp. 54–6.

Terzaghi, K., Peck, R. B., and Mesri, G. (1996). *Soil Mechanics in Engineering Practice*, 3rd ed., John Wiley & Sons, New York.

Thompson, B. W. (1968). Secondary flow in a Hele-Shaw cell, *J. Fluid Mech.*, 31, 379–95.

Thompson, M. and Willis, J. R. (1991). A reformulation of the equations of anisotropic poroelasticity, *J. Appl. Mech.*, 58, 612–6.

Timoshenko, S. P. and Goodier, J. N. (1970). *Theory of Elasticity*, 3rd ed., McGraw-Hill, New York.

Tincelin, M. E. (1951). Les études de pressions de terrain entreprises dans les mines de fer de Lorraine (France) [Studies of ground stresses undertaken in the iron mines of Lorraine (France)], in Conférence Internationale sur les Pressions de Terrain et le Soutènement dans les Chantiers d'Exploitation, *Annales des Mines de Belgique, Institut National de l'Industrie Charbonnière*, Liege, pp. 164–82.

Todhunter, I. and Pearson, K. (1886). *A History of the Theory of Elasticity and the Strength of Materials*, Vol. 1, Cambridge University Press, Cambridge.

Toksöz, M. N. and Johnston, D. H., eds. (1981). *Sesimic Wave Attenuation*, Soc. Explor. Geophys., Tulsa.

Toksöz, M. N., Cheng, C. H., and Timur, A. (1976). Velocities of seismic waves in porous rocks, *Geophysics*, 41, 621–45.

Toksöz, M. N., Johnston, D. H., and Timur, A. (1979). Attenuation of seismic waves in dry and saturated rocks: I. Laboratory measurements, *Geophysics*, 44, 681–90.

Tolstov, G. P. (1976). *Fourier Series*, Dover, New York.

Tomkeieff, S. I. (1940). The basalt lavas of the Giant's Causeway district of Northern Ireland, *Bull. Volcan.*, 6, 89–143.

Tresca, H. (1864). Sur l'écoulement des corps solides soumis à de fortes pressions [On the flow of solid bodies subjected to large pressures], *Compt. Rend. Acad. Sci. Paris*, 59, 754–63.

Tsukrov, I. and Kachanov, M. (1997). Stress concentrations and microfracturing patterns in a brittle-elastic solid with interacting pores of diverse shapes, *Int. J. Solids Struct.*, 34, 2887–904.

Tsukrov, I. and Novak, J. (2002). Effective elastic properties of solids with defects of irregular shapes, *Int. J. Solids Struct.*, 39, 1539–55.

Tullis, T. E. (1988). Rock friction constitutive behavior from laboratory experiments and its implications for an earthquake field monitoring program, *Pure Appl. Geophys.*, 126, 555–88.

Tullis, T. E. and Tullis, J. (1986). Experimental rock deformation techniques, in *Mineral and Rock Deformation: Laboratory Studies (The Paterson Volume)*, Geophysical Monograph 26, American Geophysical Union, Washington, D.C., pp. 297–304.

Tuma, J. J. (1983). *Handbook of Physical Calculations*, 2nd ed., McGraw-Hill, New York.

Tunbridge, L. W. (1989). Interpretation of the shut-in pressure from the rate of pressure decay, *Int. J. Rock Mech.*, 26, 457–60.

Turner, F. J. (1949). Preferred orientation of calcite in Yule marble, *Am. J. Sci.*, 247, 593–621.

Turner, F. J. and Weiss, L. E. (1963). *Structural Analysis of Metamorphic Tectonites*, McGraw-Hill, New York.

Turner, F. J., Griggs, D. T., and Heard, H. (1954). Experimental deformation of calcite crystals, *Bull. Geol. Soc. Am.*, 65, 883–934.

Turner, F. J., Griggs, D. T., Clark, R. H., and Dixon, R. H. (1956). Deformation of Yule marble, Pt. VII. Development of oriented fabrics at 300°C–500°C, *Bull. Geol. Soc. Am.*, 67, 1259–94.

van Heerden, W. L. (1969). Stress concentration factors for the flat borehole end for use in rock stress measurements, *Eng. Geol.*, 3, 307–23.

Vernik, L. (1997). Predicting porosity from acoustic velocities in siliciclastics: A new look, *Geophysics*, 62, 118–28.

Verruijt, A. (1969). Elastic storage in aquifers, in *Flow Through Porous Media*, R. J. M. De Wiest, ed., Academic Press, New York, pp. 331–76.

Verruijt, A. (1998). Deformations of an elastic half plane with a circular cavity, *Int. J. Solids Struct.*, 35, 2795–804.

Viktorov, I. A. (1967). *Rayleigh and Lamb Waves: Physical Theory and Applications*, Plenum Press, New York.

Voigt, W. (1889). Uber die Beziehung zwischen den beiden Elastizitätskonstanten isotroper Körper [On the relationship between the two elastic constants of an isotropic body], *Wied. Ann.*, 38, 573–87.

Voigt, W. (1928). *Lehrbuch der Krystallphysik* [Textbook of Crystal Physics], Teubner, Leipzig.

von Kármán, T. (1911). Festigkeitsversuche unter allseitigem Druck [Strength measurements under uniform pressure], *Zeit. Ver. dt. Ing.*, 55, 1749–57.

von Kármán, T. and Duwez, P. (1950). The propagation of plastic deformation in solids, *J. Appl. Phys.*, 21, 987–94.

von Mises, R. (1913). Mechanik der festen Körper in Plastisch deformablem Zustand [Mechanics of solid bodies undergoing plastic deformation], *Goett. Nachr. Math. Phys. Kl.*, 1913, 582–92.

Vukuturi, V. S. (1974). The effect of liquids on the tensile strength of limestone, *Int. J. Rock Mech.*, 11, 27–9.

Walsh, J. B. (1965a). The effect of cracks on the compressibility of rocks, *J. Geophys. Res.*, 70, 381–9.

Walsh, J. B. (1965b). The effect of cracks on the uniaxial elastic compression of rocks, *J. Geophys. Res.*, 70, 399–411.

Walsh, J. B. (1965c). The effect of cracks in rocks on Poisson's ratio, *J. Geophys. Res.*, 70, 5249–57.

Walsh, J. B. (1966). Seismic attenuation in rock due to friction, *J. Geophys. Res.*, 71, 2591–9.

Walsh, J. B. (1981). The effect of pore pressure and confining pressure on fracture permeability, *Int. J. Rock Mech.*, 18, 429–35.

Walsh, J. B. and Decker, E. R. (1966). Effect of pressure and saturating fluid on thermal conductivity of compact rock, *J. Geophys. Res.*, 71, 3053–61.

Wang, H. F. (1993). Quasi-static poroelastic parameters in rock and their geophysical applications, *Pure Appl. Geophys.*, 141, 269–86.

Wang, H. F. (2000). *Theory of Linear Poroelasticity*, Princeton University Press, Princeton and Oxford.

Warren, T. M. and Smith, M. B. (1985). Bottomhole stress factors affecting drilling rate at depth, *J. Petrol. Technol.*, 37, 1523–33.

Wawersik, W. and Fairhurst, C. (1970). A study of brittle rock fracture in laboratory compression experiments, *Int. J. Rock. Mech.*, 7, 561–75.

Wei, Z. Q., Egger, P., and Descoeudres, F. (1995). Permeability predictions for jointed rock masses, *Int. J. Rock Mech.*, 32, 251–61.

Weibull, W. (1951). A statistical distribution function of wide applicability, *J. Appl. Mech.*, 18, 293–7.

Werfel, A. (1965). Mohr's circle as an aid to the transformation of symmetrical second-order tensors and to the solution of some other problems, in *Topics in Applied Mechanics*, D. Abir, F. Ollendorff, and M. Reiner, eds., Elsevier, Amsterdam, pp. 299–310.

Westergaard, H. M. (1952). *Theory of Elasticity and Plasticity*, John Wiley & Sons, New York.

Whitney, C. S. (1943). Discussion on a paper by V. P. Jensson, *J. Am. Concr. Inst.*, 39, 584, 2–6.

Whittaker, B. N., Singh, R. N., and Sun, G. (1992). *Rock Fracture Mechanics: Principles, Design, and Applications*, Elsevier, Amsterdam.

Wiebols, G. A. and Cook, N. G. W. (1968). An energy criterion for the strength of rocks in polyaxial compression, *Int. J. Rock Mech.*, 5, 529–49.

Wiederhorn, S. M. and Boltz, L. H. (1970). Stress corrosion and static fatigue of glass, *J. Am. Ceram. Soc.*, 53, 543–8.

Wilkens, R. H., Simmons, G., Wissler, T. M., and Caruso, L. (1986). The physical properties of a set of sandstones, III: The effect of fine-grained pore-filling material on compressional velocity, *Int. J. Rock. Mech.*, 23, 313–25.

Williams, A. (1959). A structural history of the Girvan district, S.W. Ayrshire, *Trans. Roy. Soc. Edin.*, 63, 629–67.

Willis, J. R. and Bullough, R. (1969). The interaction of finite gas bubbles in a solid, *J. Nucl. Mater.*, 32, 76–87.

Winkler, K. W. and Nur, A. (1979). Pore fluids and seismic attenuation in rocks, *Geophys. Res. Letts.*, 6, 1–4.

Winkler, K. W., Nur, A., and Gladwin, M. (1979). Friction and seismic attenuation in rocks, *Nature*, 277, 528–31.

Witherspoon, P. A., Wang, J. S. Y., Iwai, K., and Gale, J. E. (1980). Validity of cubic law for fluid flow in a deformable rock fracture, *Water Resour. Res.*, 16, 1016–24.

Worotnicki, G. (1993). CSIRO Triaxial stress measurement cell, in *Comprehensive Rock Engineering*, Vol. 3, J. A. Hudson, ed., Pergamon, Oxford and New York, pp. 329–94.

Wu, Y. K., Hao, H., Zhou, Y. X., and Chong, K. (1998). Propagation characteristics of blast-induced shock waves in a jointed rock mass, *Soil Dynam. Earthquake Eng.*, 17, 407–12.

Wu, Y. S., Haukwa, C., and Bodvarsson, G. S. (1999). A site-scale model for fluid and heat flow in the unsaturated zone of Yucca Mountain, Nevada, *J. Contam. Hydrol.*, 38, 185–215.

Wyble, D. O. (1958). Effect of applied pressure on the conductivity, porosity, and permeability of sandstones, *Petrol. Trans. AIME*, 213, 430–2.

Wyllie, M. R. J., Gardner, G. H. F., and Gregory, A. R. (1962). Studies of elastic wave attenuation in porous media, *Geophysics*, 27, 569–89.

Wyllie, M. R. J., Gregory, A. R., and Gardner, G. H. F. (1956). Elastic wave velocities in heterogeneous and porous media, *Geophysics*, 21, 41–70.

Yamakawa, N. (1962). Scattering and attenuation of elastic waves, *Geophysical Magazine (Tokyo)*, 31, 63–103.

Yeo, I. W., DeFreitas, M. H., and Zimmerman, R. W. (1998). Effect of shear displacement on the aperture and permeability of a rock fracture, *Int. J. Rock. Mech.*, 35, 1051–70.

Yew, C. H. and Jogi, P. N. (1978). The determination of Biot's parameters for sandstones, Part 1: Static tests, *Exp. Mech.*, 18, 167–72.

Yew, C. H., Jogi, P. N., and Grey, K. E. (1979). Estimating the mechanical properties of fluid-saturated rocks using the measured wave motions, *J. Energy Resour. Tech.*, 101, 112–6.

Yu, H.-S. (2000). *Cavity Expansion Methods in Geomechanics*, Kluwer, Dordrecht.

Yuan, S. C. and Harrison, J. P. (2005). Development of a hydro-mechanical local degradation approach and its application to modelling fluid flow during progressive fracturing of heterogeneous rocks, *Int. J. Rock Mech.*, 42, 961–84.

Yuan, S. C. and Harrison, J. P. (2006). A review of the state of the art in modelling progressive mechanical breakdown and associated fluid flow in intact heterogeneous rocks, *Int. J. Rock Mech.*, 43, 1001–22.

Zemanek, J., Glenn, E. E., Norton, L. J., and Caldwell, R. L. (1970). Formation evaluation by inspection with the borehole televiewer, *Geophysics*, 35, 254–69.

Zenzri, H. and Keer, L. M. (2001). Mechanical analysis of the emplacement of laccoliths and lopoliths, *J. Geophys. Res.*, 106, 13,781–92.

Zhang, X. and Sanderson, D. J. (1996). Effects of stress on the two-dimensional permeability tensor of natural fracture networks, *Geophys. J. Int.*, 125, 912–24.

Zhang, X. and Sanderson, D. J. (1998). Numerical study of critical behaviour of deformation and permeability of fractured rock masses, *Marine Petrol. Geol.*, 15, 535–48.

Zhou, S. (1994). A program to model the initial shape and extent of bore hole breakout, *Comp. Geosci.*, **20**, 1143–60.

Zimmerman, R. W. (1985a). Discussion of the constitutive theory for fluid-filled porous materials, by N. Katsube, *J. Appl. Mech.*, 52, 983.

Zimmerman, R. W. (1985b). Compressibility of an isolated spheroidal cavity in an isotropic elastic medium, *J. Appl. Mech.*, 52, 606–8.

Zimmerman, R. W. (1985c). The effect of microcracks on the elastic moduli of brittle solids, *J. Mater. Sci. Letts.*, 4, 1457–60.

Zimmerman, R. W. (1986). Compressibility of two-dimensional cavities of various shapes, *J. Appl. Mech.*, 53, 500–4.

Zimmerman, R. W. (1988). Stress singularity around two nearby holes, *Mech. Res. Commun.*, 15, 87–90.

Zimmerman, R. W. (1991). *Compressibility of Sandstones*, Elsevier, Amsterdam.

Zimmerman, R. W. (1992). Hashin-Shtrikman bounds on the Poisson ratio of a composite material, *Mech. Res. Commun.*, 19, 563–9.

Zimmerman, R. W. (2000). Coupling in poroelasticity and thermoelasticity, *Int. J. Rock Mech.*, 37, 79–87.

Zimmerman, R. W. and Bodvarsson, G. S. (1996). Hydraulic conductivity of rock fractures, *Transp. Porous Media*, 23, 1–30.

Zimmerman, R. W. and King, M. S. (1986). The effect of the extent of freezing on seismic velocities in unconsolidated permafrost, *Geophysics*, 51, 1285–90.

Zimmerman, R. W. and Yeo, I. W. (2000). Fluid flow in rock fractures: From the Navier-Stokes equations to the cubic law, in *Dynamics of Fluids in Fractured Rocks*, B. Faybishenko, P. A. Witherspoon, and S. M. Benson, eds., American Geophysical Union, Washington, pp. 213–24.

Zimmerman, R. W., Chen, D. W., and Cook, N. G. W. (1992). The effect of contact area on the permeability of fractures, *J. Hydrol.*, 139, 79–96.

Zimmerman, R. W., Kumar, S., and Bodvarsson, G. S. (1991). Lubrication theory analysis of the permeability of rough-walled fractures, *Int. J. Rock Mech.*, 28, 325–31.

Zimmerman, R. W., Myer, L. R., and Cook, N. G. W. (1994). Grain and void compression in fractured and porous rock, *Int. J. Rock Mech.*, 31, 179–84.

Zimmerman, R. W., Somerton, W. H., and King, M. S. (1986). Compressibility of porous rocks, *J. Geophys. Res.*, 91, 12,765–77.

Zizicas, G. A. (1955). Representation of three-dimensional stress distributions by Mohr's circles, *J. Appl. Mech.*, 22, 273–4.

Zoback, M. L. and Zoback, M. D. (1980). State of stress in the coterminous United States, *J. Geophys. Res.*, 85, 6113–56.

Zoeppritz, K. (1919). Über Reflexion und Durchgang seismischer Wellen durch Unstetigkeitsflächen [On the reflection and transmission of seismic waves through interfaces], *Goett. Nachr. Math. Phys. Kl.*, 1919, 66–84.

Index

Printed and bound by CPI Group (UK) Ltd, Croydon, CR0 4YY
27/10/2024
14580391-0003